Springer Advanced Texts in Chemistry

Springer
New York
Berlin
Heidelberg
Barcelona
Budapest
Hong Kong
London
Milan
Paris
Santa Clara
Singapore
Tokyo

Springer Advanced Texts in Chemistry

Series Editor: Charles R. Cantor

Principles of Protein Structure
G.E. Schulz and R.H. Schirmer

Bioorganic Chemistry: A Chemical Approach to Enzyme Action (Third Edition)
H. Dugas

Protein Purification: Principles and Practice (Third Edition)
R.K. Scopes

Principles of Nucleic Acid Structure
W. Saenger

Basic Principles and Techniques of Molecular Quantum Mechanics
R.E. Christofferson

Energy Transduction in Biological Membranes: A Textbook of Bioenergetics
W.A. Cramer and D.B. Knaff

Principles of Protein X-ray Crystallography
J. Drenth

Biomembranes: Molecular Structure and Function
R.G. Gennis

Hermann Dugas

Bioorganic Chemistry

A Chemical Approach to Enzyme Action

Third Edition

With 112 Figures, 6 in Full Color

Springer

Hermann Dugas
Département de Chimie
Université de Montréal
Montréal, Québec
Canada H3C 3J7

Series Editor:
Charles R. Cantor
Boston University
Center for Advanced Biotechnology
Boston, MA 02215, USA

Cover illustrations: Foreground: The figures shows thermolysin with an inhibitor that is believed to be an analog of the transition state (in white) bound to the near active-site cleft. The inhibitor is the tripeptide Cbz-*Phe*^{P}-Leu-Ala. The active site zinc is shown in blue and the four calcium ions are in mauve. β-sheet regions are in green, helices are orange, and irregular regions of the enzyme are shown in yellow. Section 7.3.1 of the book can be consulted for more detail. (Courtesy of Dr. Ingrid Vetter and Prof. Brian Matthews, University of Oregon). **Background:** X-ray structure of the extended thermolysin active-site cleft. Reproduced with permission from *Biochemistry* **26**, 8542–8353 (1987).

Library of Congress Cataloging in Publication Data
Dugas, Hermann.
Bioorganic chemistry : a chemical approach to enzyme action / Hermann Dugas. – 3rd ed.
p. cm. – (Springer advanced texts in chemistry)
Includes bibliographical references and index.
ISBN 0-387-94494-X (alk. paper)
1. Enzymes. 2. Bioorganic chemistry. I. Title. II. Series.
QP601.D78 1996
574.19′25–dc20 95-8361

Printed on acid-free paper.

Acquiring editor: Robert C. Garber
Production coordinated by Chernow Editorial Services, Inc., and managed by Terry Kornak; manufacturing supervised by Jeffrey Taub.
Typeset by Best-set Typesetter Ltd., Hong Kong.
Printed and bound by R.R. Donnelley & Sons, Inc., Harrisonburg, VA.
Printed in the United States of America.

9 8 7 6 5 4 3 2 1

ISBN 0-387-94494-X Springer-Verlag New York Berlin Heidelberg

This book is dedicated to
all those professors, teachers, students, and friends
who have realized and appreciated
the pleasure of transmitting human knowledge.
What else could be more noble?

Series Preface

New textbooks at all levels of chemistry appear with great regularity. Some fields such as basic biochemistry, organic reaction mechanisms, and chemical thermodynamics are well represented by many excellent texts, and new or revised editions are published sufficiently often to keep up with progress in research. However, some areas of chemistry, especially many of those taught at the graduate level, suffer from a real lack of up-to-date textbooks. The most serious needs occur in fields that are rapidly changing. Textbooks in these subjects usually have to be written by scientists actually involved in the research that is advancing the field. It is not often easy to persuade such individuals to set time aside to help spread the knowledge they have accumulated. Our goal, in this series, is to pinpoint areas of chemistry where recent progress has outpaced what is covered in any available textbooks, and then seek out and persuade experts in these fields to produce relatively concise but instructive introductions to their fields. These should serve the needs of one semester or one quarter graduate courses in chemistry and biochemistry. In some cases the availability of texts in active research areas should help stimulate the creation of new courses.

Charles R. Cantor

Preface to the Third Edition

It was over 100 years ago that Emil Fischer postulated his ingenious "lock-and-key" principle, which was subsequently applied to the development of a modern theory of enzyme catalysis. Later, the molecular recognition concept was used as the basis for the elaboration of the different fields of bioorganic chemistry. I am tempted to say that if Emil Fischer had lived in our time, he would undoubtedly be a leader in what is now called supramolecular chemistry, an important discipline of bioorganic chemistry. As we know, enzymes, by their complexity, set chemists a high standard for developing synthetic catalysts to imitate nature's highly selective enzymatic reactions. With the end of the twentieth century and the approach of a new millennium, bioorganic chemistry and enzyme mimetics are becoming more and more fashionable disciplines, particularly with the recent development and application of novel "molecular devices" that are expanding the frontiers of molecular science. With this point of view in mind, the third edition of the book was undertaken.

The publication of this third edition also coincides with the 75th anniversary of the Department of Chemistry of the Université de Montréal and my 25 years of teaching in this department.

To keep the book to a reasonable size, some sections from the second edition have been removed. Indeed, each chapter has been updated with new biomimetic examples, references, and/or review articles. At the same time, this exercise provided an opportunity to correct various graphic and typographic errors present in the previous edition. However, the major change in the third edition is a new chapter on molecular devices, in which the protocols of self-organization and self-assembly at the supramolecular level are exploited. Topologically complex molecules have until recently been regarded as mere academic curiosities, but their potential as components of molecular-scale devices is now being realized. Expressions such as *molecular meccano* and *molecular tectonics* or even *molecular cybernetics* are now used on a daily basis by a growing number of bioorganic chemists, and it is in these new domains, with an obvious analogy to molecular construction kits, that the fastest progress in bioorganic chemistry is taking place.

The reference section is composed of a list, although not exhaustive, of no less than one hundred review articles on bioorganic chemistry that appeared in the last 15 years. This impressive number is a strong barometer of the vivacity and growing popularity of the field.

Again, as in the previous editions, the emphasis is more on the concepts of bioorganic chemistry than on the details of synthetic and mechanistic difficulties of preparing the molecules within each individual project or field of application.

One of the goals of bioorganic chemistry is to mimic and understand via models the living processes in nature. This book is a tribute to this goal. However, the biomimetic approach is not limited to natural processes. Many biomodels go beyond the natural pathways and allow us to learn more about chemistry and nature, and often open new avenues in science. In this book I hope that I have succeeded, in a modest way, in transmitting this taste for knowledge.

Montréal, Canada
January 1995

Hermann Dugas

Preface to the Second Edition

The design and synthesis of molecules to mimic biological events can no longer be considered a new field. This is particularly true since the last decade where we have witnessed a real explosion of new biomodels of enzymatic transformations. The growing importance of bioorganic chemistry is easily measured by the impressive number of review articles (over seventy) written on various bioorganic subjects. The list is given in the reference section, at the end of the book.

Indeed, increasing numbers of chemists and biochemists are studying simple synthetic molecules as models of enzymes. The aim is to fashion molecules that will catalyze useful reactions without the need of the bulky polypeptide backbone of proteins. The major concept behind this approach is *molecular recognition*. The importance of this discipline was recognized worldwide when the 1987 Nobel Prize in chemistry was awarded to D.J. Cram, J.M. Lehn, and C.J. Pedersen, three pioneers and leaders in this field.

In this second edition, Chapters 2 and 3 have been completely restructured to emphasize new developments in amino acid and nucleotide chemistry in the past ten years. Therefore, the aspect of chemical and biological synthesis of proteins and polynucleotides has been replaced by new subjects such as developments in asymmetric synthesis of amino acids, chemical mutations and protein engineering, and other exciting new fields such as antibodies as enzyme and RNA as catalyst and a section on DNA intercalating molecules. Among other modifications, a general presentation of the biomimetic approach to model design has been added in the first chapter, and molecular recognition in Chapter 2 focuses on structure–activity relationship in neuropeptides and drug design. In my experience, this section was particularly appreciated by students. It became also important in this new edition to develop a section on transition state analogs.

Chapters 4 to 7 retain the same basic structure but each section has been updated with new and enlightening examples. All those examples stress the talent of those chemists for imaginative chemical architecture in bioorganic models of biological processes. Chapter 4 has been expanded to include a section on enzyme analogs using polymers, another section on the use of enzymes in organic synthesis, and one on the design of molecular clefts. A new section in Chapter 6 covers recent advances in photosynthetic models.

An honest and judicious selection of pedagogically relevant and appealing biomimetic models among the astonishing number of exciting new synthetic receptors that have emerged in the last decade was not in any way an easy task. Consequently, choices had to be made and not all aspects of bioreceptors could be adequately covered. Detailed synthetic and mechanistic implications related to other applications can be found by looking up the appropriate references where particularly useful aspects can be prepared and developed further. This way, this edition could be kept to a reasonable size.

I tried to incorporate the most recent and key observations, but I also deliberately decided to put the emphasis more on the beauty of the molecular architectures than on their detailed descriptive synthetic difficulties. As such, I believe that there is little need to go into too much elaborated and sophisticated development of mechanistic considerations. Although the course is addressed to the final undergraduate and/or first-year graduate level, the presentation should remain simple and concise, yet lively. This is why more general background materials have been added in the first two chapters. This should create a better link with the rest of the material in the book.

I am once more indebted to my very devoted secretary, Miss C. Potvin, for her cheerful and persistent competence throughout the preparation of the second edition.

Montréal, Canada
February 1988

Hermann Dugas

Preface to the First Edition

Bioorganic chemistry is the application of the principles and the tools of organic chemistry to the understanding of biological processes. The remarkable expansion of this new discipline in organic chemistry during the last ten years has created a new challenge for the teacher, particularly with respect to undergraduate courses. Indeed, the introduction of many new and valuable bioorganic chemical principles is not a simple task. This book will expound the fundamental principles for the construction of bioorganic molecular models of biochemical processes using the tools of organic and physical chemistry.

This textbook is meant to serve as a *teaching* book. It is not the authors' intention to cover all aspects of bioorganic chemistry. Rather, a blend of general and selected topics are presented to stress important aspects underlying the concepts of organic molecular model building. Most of the presentation is accessible to advanced undergraduate students without the need to go back to an elementary textbook of biochemistry; of course, a working knowledge of organic chemistry is mandatory. Consequently, this textbook is addressed first to final-year undergraduate students in chemistry, biochemistry, biology, and pharmacology. In addition, the text has much to offer in modern material that graduate students are expected to, but seldom actually, know.

Often the material presented in elementary biochemistry courses is overwhelming and seen by many students as mainly a matter of memorization. We hope to overcome this situation. Therefore, the chemical organic presentation throughout the book should help to stimulate students to make the "quantum jump" necessary to go from a level of pure memorization of biochemical transformations to a level of adequate comprehension of biochemical principles based on a firm chemical understanding of bioorganic concepts. For this, most chapters start by asking some of the pertinent questions developed within the chapter. In brief, we hope that this approach will stimulate curiosity.

Professor B. Belleau from McGill University acted as a "catalyst" in promoting the idea to write this book. Most of the material was originally inspired from his notes. The authors would like to express their most sincere appreciation for giving us the opportunity of teaching, transforming, and expanding *his* course into a book. It is Dr. Belleau's influence and remarkable dynamism that gave us consant inspiration and strength throughout the writing.

The references are by no means exhaustive, but are, like the topics chosen, selective. The reader can easily find additional references since many of the citations are of books and review articles. The instructor should have a good knowledge of individual references and be able to offer to the students the possibility of discussing a particular subject in more detail. Often we give the name of the main author concerning the subject presented and the year the work was done. This way the students have the opportunity to know the leader in that particular field and can more readily find appropriate references. However, we apologize to all those who have not been mentioned because of space limitation.

The book includes more material than can be handled in a single course of three hours a week in one semester. However, in every chapter, sections of material may be omitted without loss of continuity. This flexibility allows the instructor to emphasize certain aspects of the book, depending if the course is presented to an audience of chemists or biochemists.

We are indebted to the following friends and colleagues for providing us with expert suggestions and comments regarding the presentation of certain parts of the book: P. Brownbridge, P. Deslongchamps, P. Guthrie, J.B. Jones, R. Kluger, and C. Lipsey. And many thanks to Miss C. Potvin, from the Université de Montréal, for her excellent typing assistance throughout the preparation of this manuscript.

Finally, criticisms and suggestions toward improvement of the content of the text are welcome.

Montréal, Canada
January 1981

Hermann Dugas
Christopher Penney

Contents

Series Preface vii
Preface to the Third Edition ix
Preface to the Second Edition xi
Preface to the First Edition xiii

Chapter 1
Introduction to Bioorganic Chemistry 1

1.1 Basic Considerations 1
1.2 Proximity Effects in Organic Chemistry 4
1.3 Molecular Adaptation 8
1.4 Molecular Recognition and the Supramolecular Level 14

Chapter 2
Bioorganic Chemistry of Amino Acids and Polypeptides 21

2.1 Chemistry of the Living Cells 22
2.2 Analogy Between Organic Reactions and Biochemical Transformations 25
2.3 Chemistry of the Peptide Bond 25
2.4 Nonribosomal Peptide Bond Formation 42
2.5 Asymmetric Synthesis of α-Amino Acids 51
2.6 Asymmetric Synthesis with Chiral Organometallic Catalysts 77
2.7 Transition State Analogs 79
2.8 Antibodies as Enzymes 85
2.9 Chemical Mutations 90
2.10 Molecular Recognition and Drug Design 95

Chapter 3
Bioorganic Chemistry of the Phosphate Groups and Polynucleotides 111

3.1 Basic Considerations 111
3.2 Energy Storage 114
3.3 Hydrolytic Pathways and Pseudorotation 125
3.4 DNA Intercalants 142

Chapter 4
Enzyme Chemistry 159

4.1 Introduction to Catalysis 159
4.2 Introduction to Enzymes 171
4.3 Multifunctional Catalysis and Simple Models 181
4.4 α-Chymotrypsin 184
4.5 Other Hydrolytic Enzymes 204
4.6 Stereoelectronic Control in Hydrolytic Reactions 210
4.7 Immobilized Enzymes and Enzyme Technology 230
4.8 Enzymes in Synthetic Organic Chemistry 236
4.9 Enzyme-Analog-Built Polymers 240
4.10 Design of Molecular Clefts 244

Chapter 5
Enzyme Models 252

5.1 Host–Guest Complexation Chemistry 255
5.2 New Developments in Crown Ether Chemistry 277
5.3 Membrane Chemistry and Micelles 317
5.4 Polymers 337
5.5 Cyclodextrins 345
5.6 Enzyme Design Using Steroid Template 361
5.7 Remote Functionalization Reactions 366
5.8 Biomimetic Polyene Cyclizations 373

Chapter 6
Metal Ions 388

6.1 Metal Ions in Proteins and Biological Molecules 388
6.2 Carboxypeptidase A and the Role of Zinc 390
6.3 Hydrolysis of Amino Acid Esters and Amides and Peptides 398
6.4 Iron and Oxygen Transport 407
6.5 Copper Ion 437
6.6 Biomodels of Photosynthesis and Energy Transfer 447
6.7 Cobalt and Vitamin B_{12} Action 460

Chapter 7
Coenzyme Chemistry 482

7.1 Oxidoreduction 483
7.2 Pyridoxal Phosphate 520
7.3 Suicide Enzyme Inactivators and Affinity Labels 542
7.4 Thiamine Pyrophosphate 560
7.5 Biotin 574

Chapter 8
Molecular Devices 593

8.1 Introduction to Self-Organization and Self-Assembly 593
8.2 General Overview of the Approach 595
8.3 Specific Examples 606

References 636

Index 687

Chapter 1

Introduction to Bioorganic Chemistry

"It might be helpful to remind ourselves regularly of the sizeable incompleteness of our understanding, not only of ourselves as individuals and as a group, but also of Nature and the world around us."

N. Hackerman
Science **183**, 907 (1974)

1.1 Basic Considerations

Among the first persons to develop biooriented organic projects was F.H. Westheimer, in the 1950s. He was probably the first physical organic chemist to do serious studies of biochemical reactions. However, it was only twenty years later that the field blossomed to what is now accepted as bioorganic chemistry.

Bioorganic chemistry is a discipline that is essentially concerned with the application of the tools of chemistry to the understanding of biochemical processes. Such an understanding is often achieved with the aid of *molecular models* chemically synthesized in the laboratory. This allows a "sorting out" of the many variable parameters simultaneously operative within the biological system.

For example, how does a biological membrane work? One builds a simple model of known compositions and studies a single behavior, such as an ion transport property. How does the brain work? This is by far a more complicated system than the previous example. Again one studies single synapses and single synaptic constituents and then uses the observations to construct a model.

Organic chemists develop synthetic methodology to better understand organic mechanisms and create new compounds. On the other hand, biochemists study life processes by means of biochemical methodology: enzyme purification and assay, radioisotopic tracer studies in in vivo systems. The former possess the methodology to synthesize biological analogs but often fail to appreciate which synthesis would be relevant. The latter possess an appreciation of what would be useful to synthesize in the laboratory but not the expertise to pursue the problem. The need for the multidisciplinary approach becomes obvious, and the bioorganic chemist will often have two laboratories: one for synthesis and another for biological study. A new dimension results from this combination of chemical and biological sciences, that is, the concept of *model building* to study and sort out the various parameters of a complex biological process. By means of simple organic models, many biological reactions as well as the specificity and efficiency of the enzymes involved have been reproduced in the test tube. The success of many of these models indicates the progress that has been made in understanding the chemistry operative in biological systems. Extrapolation of this multidisciplinary science to the pathological state is a major theme of the pharmaceutical industry—organic chemists and pharmacologists working "side by side," so that bioorganic chemistry is to biochemistry and medicinal chemistry is to pharmacology.

What are the tools needed for bioorganic model studies? Organic and physical organic chemical principles will provide, by their very nature, the best opportunities for model building—modeling molecular events that form the basis of life. A large portion of organic chemistry has been classically devoted to natural products. Many of those results have turned out to be wonderful tools for the discovery and characterization of specific molecular events in living systems. Think, for instance, of the development of antibiotics, certain alkaloids, and the design of new drugs for the medicine of today and tomorrow.

All living processes require energy, which is obtained by performing chemical reactions inside cells. These biochemical processes are based on chemical dynamics and involve reductions and oxidations. Biological oxidations are thus the main source of energy to drive a number of endergonic biological transformations.

Many of the reactions involve combustion of foods such as sugars and lipids to produce energy that is used for a variety of essential functions such as growth, replication, maintenance, muscular work, and heat production. These transformations are also related to oxygen uptake; breathing is a biochemical process by which molecular oxygen is reduced to water. Throughout these pathways, energy is stored in the form of adenosine triphosphate (ATP), an energy-rich compound known as the universal product of energetic transactions.

Part of the energy from the combustion engine in the cell is used to perpetuate the machine. The machine is composed of structural components that must be replicated. Ordinary combustion gives only heat plus some visible light and waste. Biological combustions, however, give some heat but a large portion of the energy is used to drive a "molecular engine" that synthesizes copies of itself and that does mechanical work as well. Since these transformations occur at low temperature (body temperature, 37°C) and in aqueous media, catalysts are essential for smooth or rapid energy release and transfer. Hence, apart from structural components, *molecular catalysts* are required.

These catalysts have to be highly efficient (a minimum of waste) and highly specific if precise patterns are to be produced. Structural components have a static role; we are interested here in the dynamics. If bondbreaking and bond-forming reactions are to be performed on a specific starting material, then a suitable specific catalyst capable of recognizing the substrate must be "constructed" around that substrate.

In other words, and this is the fundamental question posed by all biochemical phenomena, a substrate molecule and the specific reaction it must undergo must be translated into another structure of much higher order whose information content perfectly matches the specifically planned chemical transformation. Only large macromolecules can carry enough *molecular information* both from the point of view of substrate recognition and thermodynamic efficiency of the transformation. *These macromolecules are proteins.** They must be extremely versatile in the physicochemical sense since innumerable substrates of widely divergent chemical and physical properties must all be handled by proteins.

Hence, protein composition must of necessity be amenable to wide variations in order that different substrates may be recognized and handled. Some proteins will even need adjuncts (nonprotein parts) to assist in recognition and transformation. These cofactors are called coenzymes. One can therefore predict that protein catalysts or *enzymes* must have a high degree of order and organization. Further, a minimum size will be essential for all the information to be contained.

These ordered biopolymers, which allow the combustion engine to work and to replicate itself, must also be replicated exactly once a perfect translation of substrate structure into a specific function has been established. Therefore, the molecular information in the proteins (enzymes) must be safely stored into stable, relatively static language. This is where the nucleic acids enter into the picture. Consequently, another translation phenomenon involves protein information content

---

* This too dogmatic opinion is no longer valid since the discovery of RNA molecules with definite catalytic activities (15a,15b).

written into a linear molecular language that can be copied and distributed to other cells.

The best way to vary at will the information content of a macromolecule is to use some sort of backbone and to peg on it various arrays of side chains. Each side chain may carry well-defined information regarding interactions between themselves or with a specific substrate in order to perform specific bond-making or -breaking functions. Nucleic acid–protein interactions should also be mentioned because of their fundamental importance in the evolution of the genetic code.

The backbone just mentioned is a polyamide and the pegs are the amino acid side chains. Why polyamide? Because it has the capacity of "freezing" the biopolymer backbone into precise three-dimensional patterns. Flexibility is also achieved and is of considerable importance for conformational "breathing" effects to occur. A substrate can therefore be transformed in terms of protein conformation imprints and, finally, mechanical energy can also be translocated.

The large variety of organic structures known offer an infinite number of structural and functional properties to a protein. Using water as the translating medium, one can go from nonpolar (structured or nonstructured) to polar (hydrogen-bonded) to ionic (solvated) amino acids; from aromatic to aliphatics; from reducible to oxidizable groups. Thus, almost the entire encyclopedia of chemical organic reactions can be coded on a polypeptide backbone and tertiary structure. Finally, since all amino acids present are of L (or *S*) configuration, we realize that *chirality* is essential for order to exist.

1.2 Proximity Effects in Organic Chemistry

Proximity of reactive groups in a chemical transformation allows bond polarization, resulting generally in an acceleration in the rate of the reaction. In nature this is normally achieved by a well-defined alignment of specific amino acid side chains at the active site of an enzyme.

Study of organic reactions helps to construct proper biomodels of enzymatic reactions and open a field of intensive research: medicinal chemistry through rational drug design. Since a meaningful presentation of all applications of organic reactions would be a prodigious task, we limit the present discussion in this chapter to a few representative examples. These illustrate some of the advantages and problems encountered in conceptualizing bioorganic models for the study of enzyme mechanism. Chapter 4 will give a more complete presentation of the proximity effect in relation to intramolecular catalysis.

The first example is the hydrolysis of a glucoside bond. *o*-Carboxyphenyl β-D-glucoside (**1–1**) is hydrolyzed at a rate 10^4 faster than the

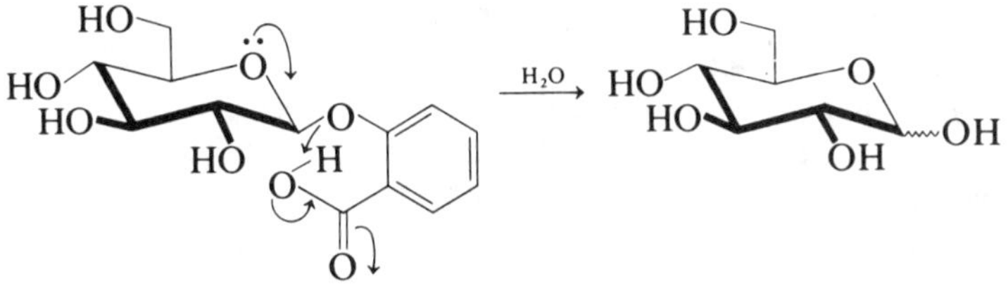

1-1 $\xrightarrow{H_2O}$ 1-2 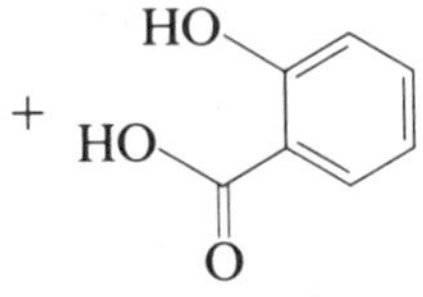 + 1-3

corresponding *p*-carboxyphenyl analog. Therefore, the carboxylate group in the *ortho* position must "participate" or be involved in the hydrolysis.

This illustrates the fact that the proper positioning of a group (electrophilic or nucleophilic) may accelerate the rate of a reaction. There is thus an analogy to be made with the active site of an enzyme such as lysozyme. Of course, the nature of the leaving group is also important in describing the properties. Furthermore, solvation effects can be of paramount importance for the course of the transformation, especially in the transition state. Reactions of this type are called *assisted hydrolysis* and occur by an intramolecular displacement mechanism; steric factors may retard the reactions.

Let us look at another example: 2,2′-tolancarboxylic acid (**1–4**) in ethanol is converted to 3-(2-carboxybenzilidene) phthalide (**1–5**). The rate of the reaction is 10^4 faster than with the corresponding 2-tolancarboxylic or 2,4′-tolancarboxylic acid. Consequently, one carboxyl group acts as a general acid catalyst (see Chapter 4) by a mechanism known as *complementary bifunctional catalysis*.

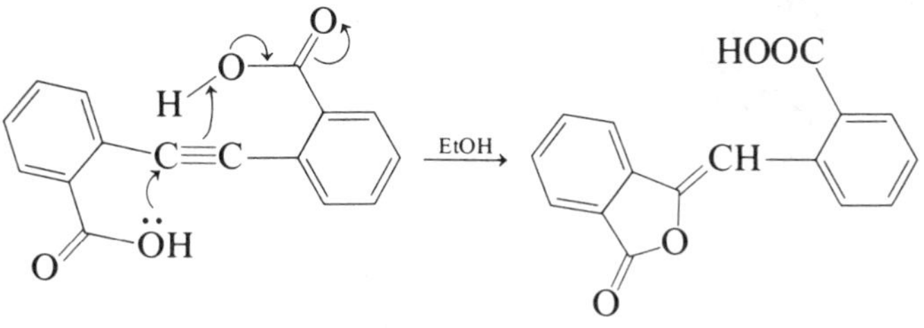

1-4 $\xrightarrow{EtOH}$ 1-5

The ester function of 4-(4′-imidazolyl) butanoic phenyl ester (**1–6**) is hydrolyzed much faster than the corresponding *n*-butanoic phenyl ester. If a *p*-nitro group is present on the aryl residue, the rate of hydrolysis is even faster at neutral pH. As expected, the presence of a better leaving group further accelerates the rate of reaction. This hydrolysis involves the formulation of a tetrahedral intermediate (**1–7**). A detailed discussion of

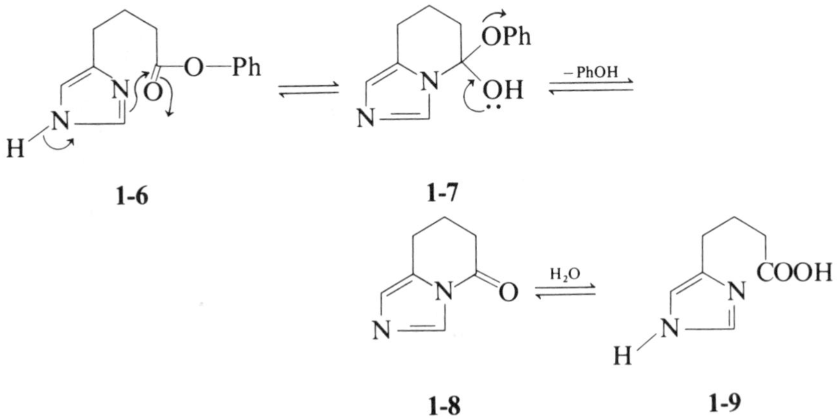

O—Ph   OPh   OH   – PhOH

1-6 **1-7**

H_2O COOH

1-8 **1-9**

such intermediates will be subject of Chapter 4. The imidazole group acts as a nucleophilic catalyst in this two-step conversion, and its proximity to the ester function and the formation of a cyclic intermediate are the factors responsible for the rate enhancement observed. The participation

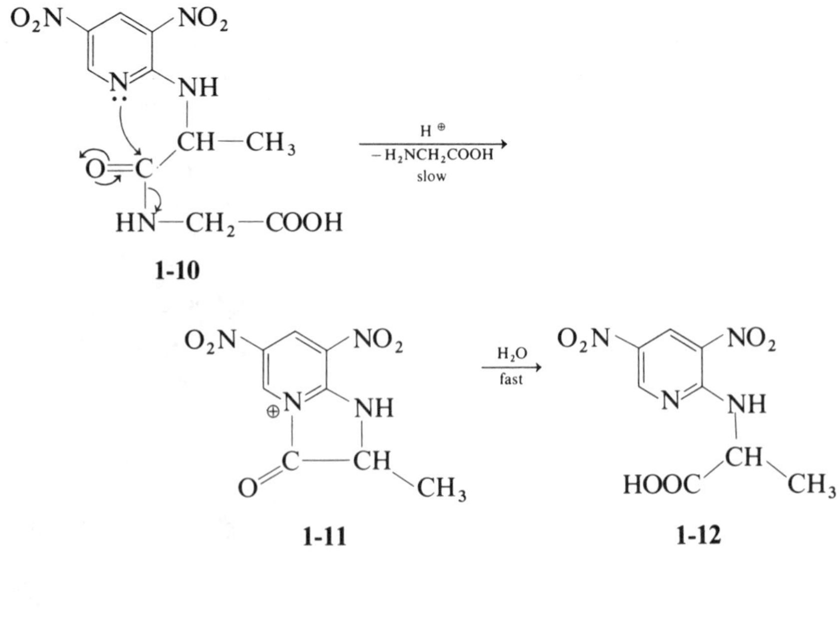

O_2N NO_2 NH $CH—CH_3$ $HN—CH_2—COOH$

$H^{\oplus}$ / $-H_2NCH_2COOH$ / slow

1-10

O_2N NO_2 NH   C—CH CH_3 H_2O / fast O_2N NO_2 NH   CH   HOOC CH_3

1-11 **1-12**

of an imidazole group in the hydrolysis of an ester may represent the simplest model of hydrolytic enzymes.

In a different domain, amide bond hydrolyses can also be accelerated. An example is the following where the reaction is catalyzed by a pyridine ring.

The first step is the rate-limiting step of the reaction (slow reaction) leading to an acyl pyridinium intermediate (**1–11**), reminiscent of a covalent acyl-enzyme intermediate found in many enzymatic mechanisms. This intermediate is then rapidly trapped by water.

The last example is taken from the steroid field and illustrates the importance of a rigid framework. The solvolysis of acetates (**1–13**) and (**1–14**) in CH_3OH/Et_3N showed a marked preference for the molecule having a β-OH group at carbon 5 where the rate of hydrolysis is 300 times faster.

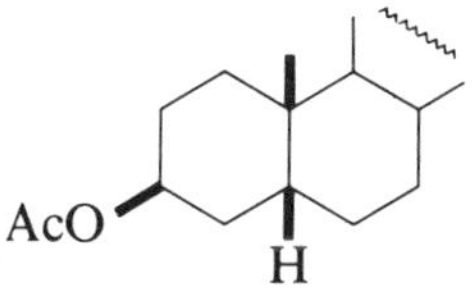

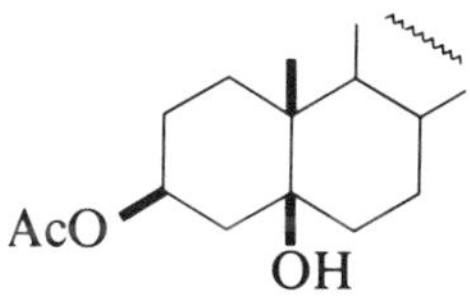

AcO H AcO OH

cis junction

1-13 **1-14**

The reason for such a behavior becomes apparent when the molecule is drawn in three dimensions (**1–15**). The rigidity of the steroid skeleton thus helps in bringing the two functions into proper orientation where catalysis combining one intramolecular and one intermolecular catalyst takes place. The proximal hydroxyl group can cooperate in the hydrolysis by hydrogen bonding, and the carbonyl function of the ester becomes a better electrophilic center for the solvent molecules. In this mechanism, one can perceive a general acid–base catalysis of ester solvolysis (Chapter 4).

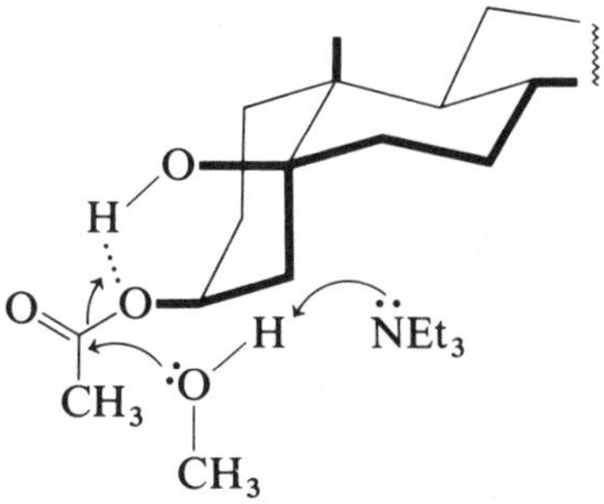

O H O O H NEt_3 CH_3 O CH_3

1-15

These simple examples illustrate that many of the basic active site chemistry of enzymes can be reproduced with simple organic models in the absence of proteins. The role of the latter is of substrate recognition and orientation and the chemistry is often carried out by cofactors (coenzymes) that also have to be specifically recognized by the protein or enzyme. The last chapter of this book is devoted to the chemistry of coenzyme function and design.

1.3 Molecular Adaptation

Other factors besides proximity effects are important and should be considered in the design of biomodels. For instance, in 1950, at the *First Symposium on Chemical-Biological Correlation*, H.L. Friedman introduced the concept of *bioisosteric groups* (1). In its broadest sense, the term refers to chemical groups that bear some resemblance in molecular size and shape and as a consequence can compete for the same biological target. This concept has important application in molecular pharmacology, especially in the design of new drugs through the method of variation, or molecular modification (2).

Some pharmacological examples will illustrate the principle. The two neurotransmitters, acetylcholine (**1–16**) and carbachol (**1–17**), have similar muscarinic action.

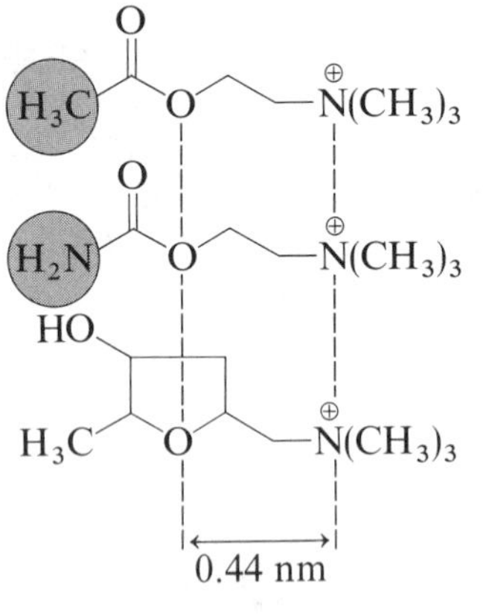

1-16 acetylcholine

1-17 carbachol

1-18 muscarine

0.44 nm

The shaded area represents the bioisosteric equivalence. Muscarine (**1–18**) is an alkaloid that inhibits the action of acetylcholine. It is found, for instance, in *Amanita muscaria* (fly agaric) and other poisonous mushrooms. Its structure implies that, in order to block the action of acetylcholine on receptors of smooth muscles and glandular cells, it must bind in a similar fashion.

5-Fluorocytosine (**1–19**) is an analog of cytosine (**1–20**) that is commonly used as an antibiotic against bacterial infections. One serious

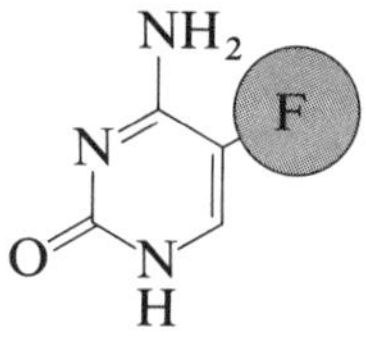

NH$_2$ F N O N H **1-19** NH$_2$ H N O N H **1-20**

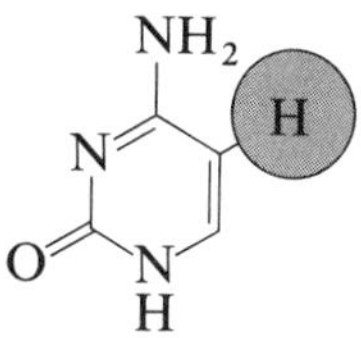

problem in drug design is to develop a therapy that will not harm the patient's tissues but will destroy the infecting cells or bacteria. A novel approach is to "disguise" the drug so that it is chemically modified to gain entry and kill invading microorganisms without affecting normal tissues. The approach involves exploiting a feature that is common to many microorganisms: peptide transport. Hence, the amino function of compound (**1–19**) is chemically joined to a small peptide. This peptide contains D-amino acids and therefore avoids hydrolysis by common human enzymes and entry into human tissues. However, the drug-bearing peptide can sneak into the bacterial cell. It is then metabolized to liberate the active antifungal drug which kills only the invading cell. This is the type of research that the group of W.D. Kingsbury is undertaking at Smith Kline & French Laboratories. This principle of using peptides to carry drugs is applicable to many different disease-causing organisms.

Similarly, 1-β-D-2′-deoxyribofuranosyl-5-iodo-uracil (**1–21**) is an *antagonist* of 1-β-D-2′-deoxyribofuranosyl thymine, or thymidine (**1–22**). That is, it is able to antagonize or prevent the action of the latter in biological systems, though it may not carry out the same function. Such an altered metabolite is also called an *antimetabolite*.

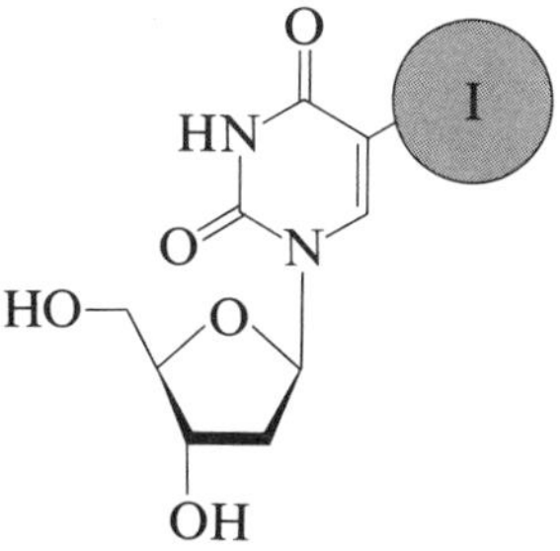

O HN I O N HO O OH **1-21** O HN CH$_3$ O N HO O OH **1-22**

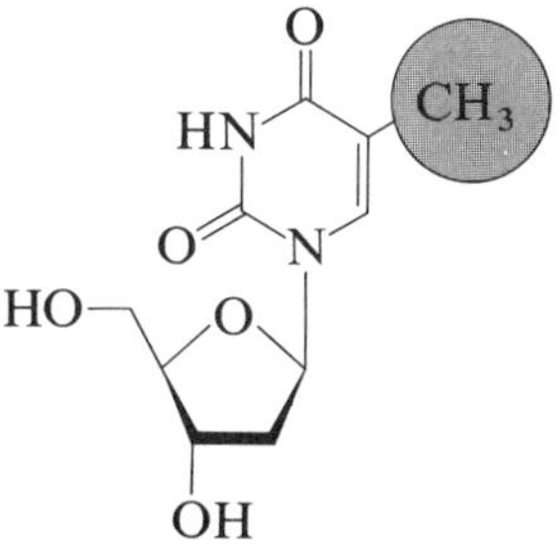

Another example of molecular modification is the synthetic nucleoside adenosine arabinoside (**1–23**). This compound has a pronounced antiviral activity against herpes virus and is therefore widely used in modern chemotherapy.

The analogy with deoxyadenosine (**1–24**), a normal component of DNA, is striking. Except for the presence of a hydroxyl group at the 2′-

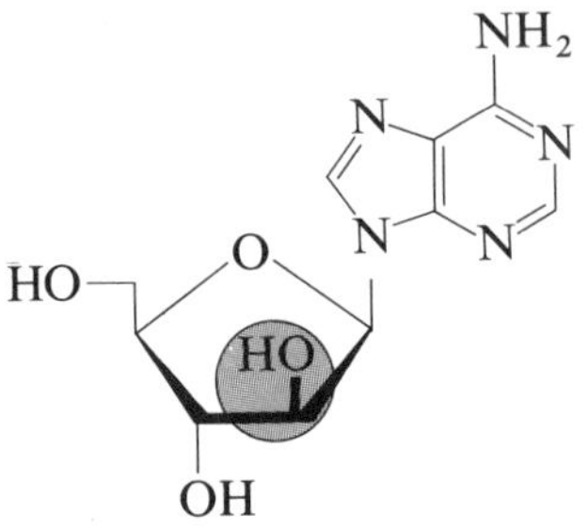

NH$_2$ N N N N HO O HO OH

1-23

NH$_2$ N N N N HO O OH

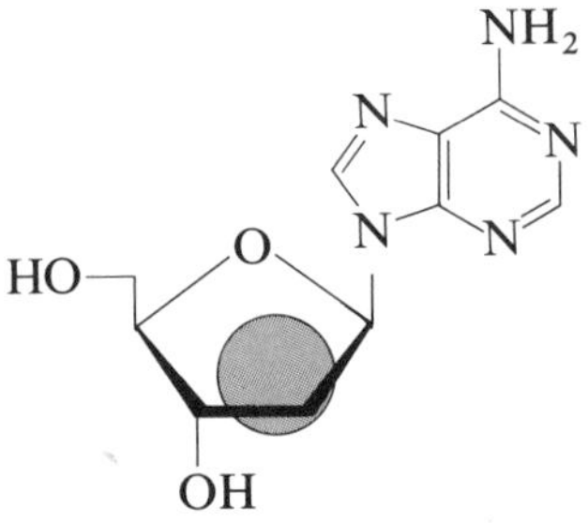

1-24

sugar position, the two molecules are identical. Compared to the ribose ring found in the RNA molecule, it has the inverse or epimeric configuration and hence belongs to the arabinose series. Interestingly, a simple inversion of configuration at C-2′ confers antiviral properties. Its mechanism of action has been well studied and it does act, after phosphorylation, as a potent inhibitor of DNA synthesis. Similarly, cytosine arabinoside is the most effective drug for acute myeloblastic leukemia.

Most interesting was the finding that this antiviral antibiotic (**1–23**) is in fact produced by a bacterium called *Streptomyces antibioticus*. This allows the production by fermentation of large quantities of this active principle.

Even more spectacular is the fact that the acyclonucleotide analog Acyclovir (**1–25**) has now been licensed for clinical use for the treatment of herpes simplex. This potent antiherpes drug inhibits nucleoside phosphorylating enzymes and this affects growth and replication of cells. The knowledge of the mode of action of such a "target enzyme" inhibitor is fundamental for the design of new drugs for cancer chemotherapy.

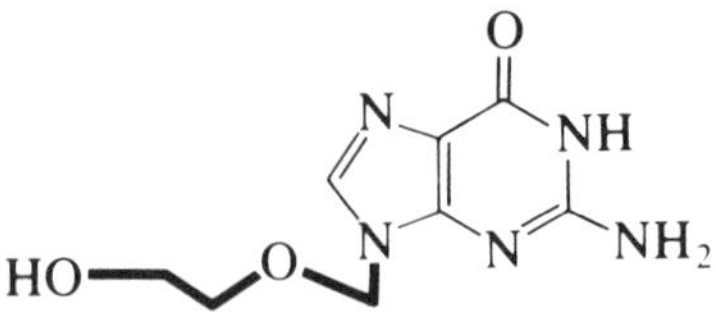

O N NH N N NH$_2$ HO O

Acyclovir
1–25

It is likely that for maximal activity, the open chain molecule adopts a conformation close enough to mimic the furanose ring of the natural nucleosides. An X-ray work has recently confirmed such a molecular arrangement (3).

A number of organophosphonates have been synthesized as bioisosteric analog for biochemically important nonnucleoside and nucleoside phosphates (4). For example, the *S*-enantiomer of 3,4-dihydroxybutyl-1-phosphonic acid (DHBP) has been synthesized as the isosteric analog of

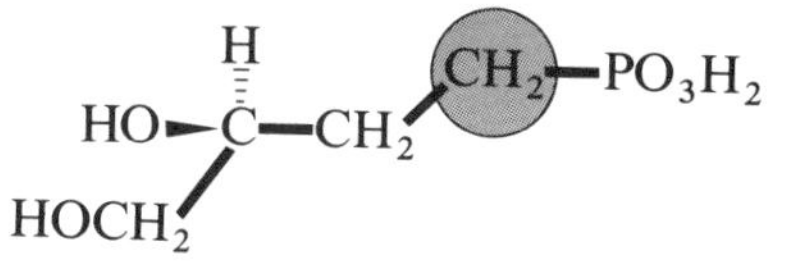

S-DHBP

1–26

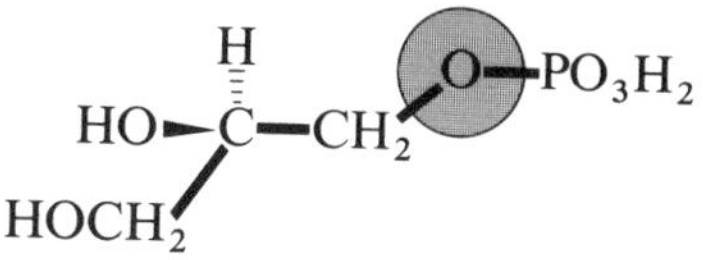

sn-glycerol 3-phosphate

1–27

sn-glycerol 3-phosphate (5). The former material is bacteriostatic at low concentrations to certain strains of *E. coli* and *B. subtilus*. As *sn*-glycerol 3-phosphate is the backbone of phospholipids (an important cell membrane constituent) and is able to enter into lipid metabolism and the glycolytic pathway, it is sensitive to a number of enzyme-mediated processes. The phosphonic acid can participate, but only up to a point, in these cellular reactions. For example, it cannot be hydrolyzed to release glycerol and inorganic phosphate. Of course, the *R*-enantiomer is devoid of biological activity.

The presence of a halogen atom on a molecule sometimes results in interesting properties. For example, substitution of the 9α position by a halogen in cortisone (**1–28**) increases the activity of the hormone by prolonging the half-life of the drug. The activity increases in the following order: X = I > Br > Cl > F > H. These cortisone analogs are employed in the diagnosis and treatment of a variety of disorders of adrenal function and as antiinflammatory agents (2).

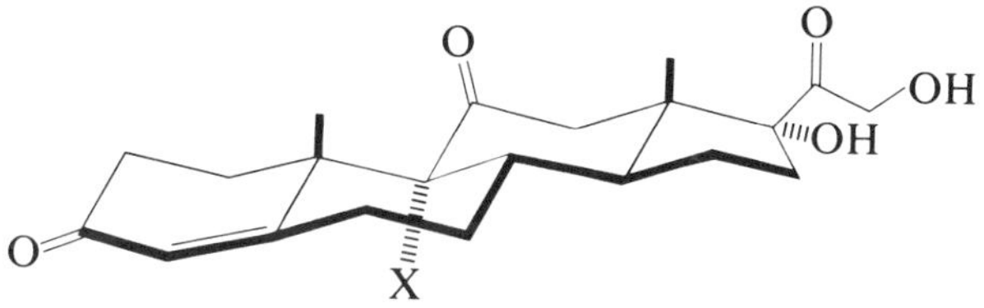

1–28

As another example, the normal thyroid gland is responsible for the synthesis and release of an unusual amino acid called thyroxine (**1–29**). This hormone regulates the rate of cellular oxidative processes (2).

The presence of the bulky atoms of iodine prevents free rotation around the ether bond and forces the planes of aromatic rings to remain perpendicular to each other. However, several halogen-free derivatives still have significant thyromimetic activity. Since there is no absolute requirement for iodine or any halogen for thyroid hormone activity in amphibians and mammals, it is likely that the primary role of iodine is rather to provide appropriate spacial constraints on semirigid biosynthetic amino acid precursors and in a way facilitate the biosynthesis of the

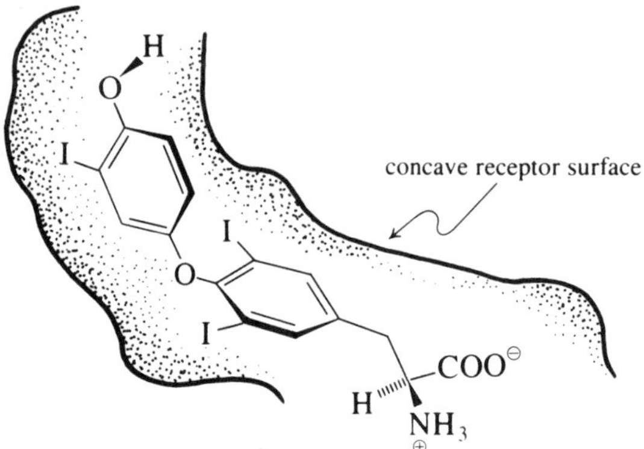

concave receptor surface

thyroxine

1–29

hormone itself in the thyroid gland (6). It is thus an example of a conformational control at the biosynthetic level.

The presence of alkyl groups or chains can also influence the biological activity of a substrate or a drug. An interesting case is the antimalarial compounds derived from 6-methoxy-8-aminoquinoline (**1–30**) (primaquine analog). The activity is greater in compounds in which n is an even number is the range of $n = 2$ to 7. So the proper fit of the side chain on a receptor site* or protein is somehow governed by the size and shape of the side chain.

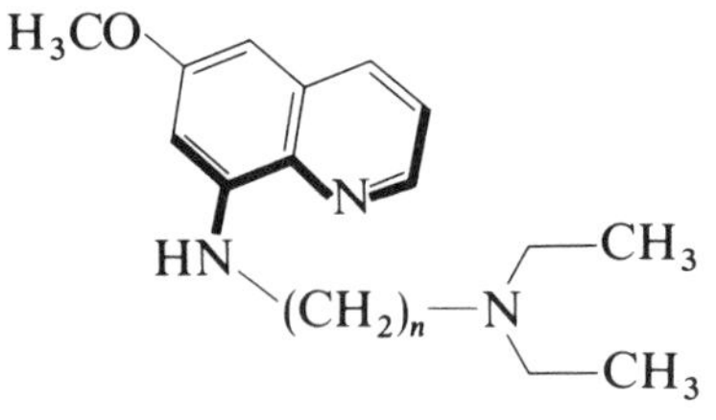

primaquine drug

1–30

Finally, mention should be made of molecular adaptation at the conformational level. The development of a certain "designer drug" is a case history worthy of mention here. In the early 1980s, the sudden

* A discussion of receptor theory is a topic more appropriate for a text in medicinal chemistry. A general definition is that a receptor molecule is a complex of proteins and lipids that upon binding of a specific organic molecule (effector, neurotransmitter) undergoes a physical or conformational change that usually triggers a series of events that results in a physiological response. In a way, an analogy could be made between receptors and enzymes (see Section 2.10).

appearance of parkinsonian symptoms in young drug abusers was very unusual since this disease affects mainly persons over fifty years old. How did this happen? The origin of this "frozen addict" behavior was finally traced back to a "new heroin" type agent. The compound had been manufactured in a clandestine laboratory in San Jose, California, and was called a designer drug.

Meperidine, or Demerol, is a well-known analgesic, less potent than morphine but its action is shorter and it is very addictive. Knowing this, the clandestine chemist decided to synthesize a similar compound, but easier to make, that turned out to be also very active. It was MPPP.

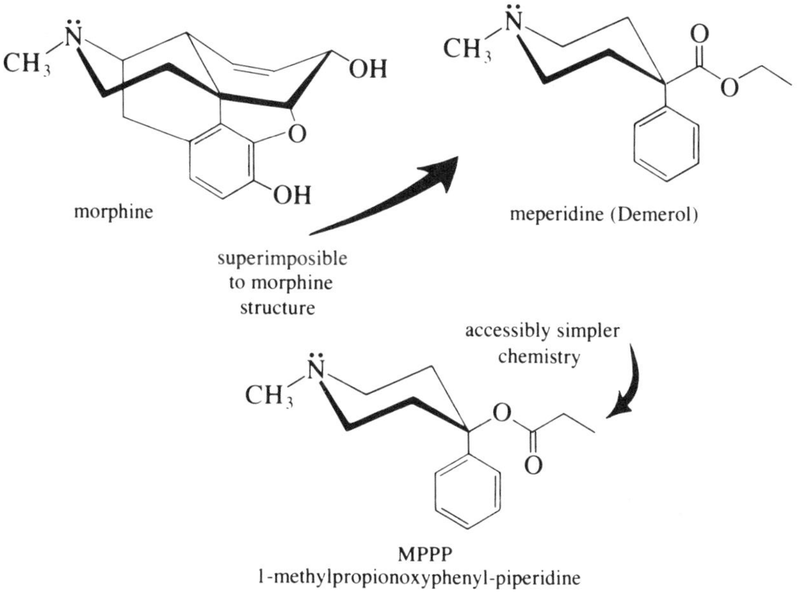

morphine

meperidine (Demerol)

superimposible
to morphine
structure

accessibly simpler
chemistry

MPPP
1-methylpropionoxyphenyl-piperidine

However, on heating to evaporate the solvent and if the pH is not controlled and too low, the molecule has a tendency to hydrolyze and dehydrate to give a by-product, MPTP. This chemistry is reasonable but might not have been anticipated by the underground chemist. Unfortunately, this unnoticed impurity in the faulty batch was the precursor to the key compound responsible for the unexpected side effects of the designer drug, originally conceived to be an easily accessible and simpler analgesic.

Indeed, the neutral MPTP substance is readily oxidized by a monoamine oxidase enzyme (MAO) in the brain to an ionic MPP^+ molecule, a neurotoxic pyridinium metabolite. Similar pyridinium salts are currently used as herbicides. Admittedly, the incidence of Parkinson's disease is distinctly higher in rural regions as compared to urban ones because similar chemicals are used in agriculture. This example illustrates well

MPPP $\xrightarrow[\Delta T]{H^{\oplus}}$ MPTP — becomes toxic by passing through the blood-brain barrier → MPP^+

MPTP
1-methyl-4-phenyl-1,2,3,6,-tetrahydropyridine

MPP^+
1-methyl-4-phenyl-pyridinium
(a MAO metabolite)

how, by accident, a faulty batch of an assumably good drug led to unexpected parkinsonism in young drug abusers (7). This stresses the point that proper conformation can give (sometimes unexpectedly) a compound very unusual therapeutic properties where analogs can be exploited.

In addition to the steric and external shell factors just mentioned, inductive and resonance contributions can also be important. All these factors must be taken into consideration in the planning of any molecular biomodel system that will hopefully possess the anticipated property. Hence, small but subtle changes on a biomolecule can confer to the new product large and important new properties.

It is in this context, that many of the fundamental principles of bioorganic chemistry are presented in the following chapters.

1.4 Molecular Recognition and the Supramolecular Level

By the end of the twentieth century almost half of the chemists will use, during their active life in research, some aspects of molecular recognition, not only in catalysis but also in fields such as affinity chromatography and controlled drug release. In the present context, *molecular recognition* implies not only more binding but also selection and possibly specific function. As such, it represents a "binding mode with a purpose," as does a real biological receptor.

Although Chapter 5 will develop in much detail the biomimetic approach, it could be instructive at this point to draw a global picture of the methodology used to construct bioorganic models of biological systems. The general strategy is summarized in Fig. 1.1.

Basically, two distinct but interconnected chemical levels are involved. On one level, a bioorganic specific receptor is constructed from classical organic synthesis. Such a receptor will be used to recognize and interact specifically with a given substrate and reach the second level of molecular

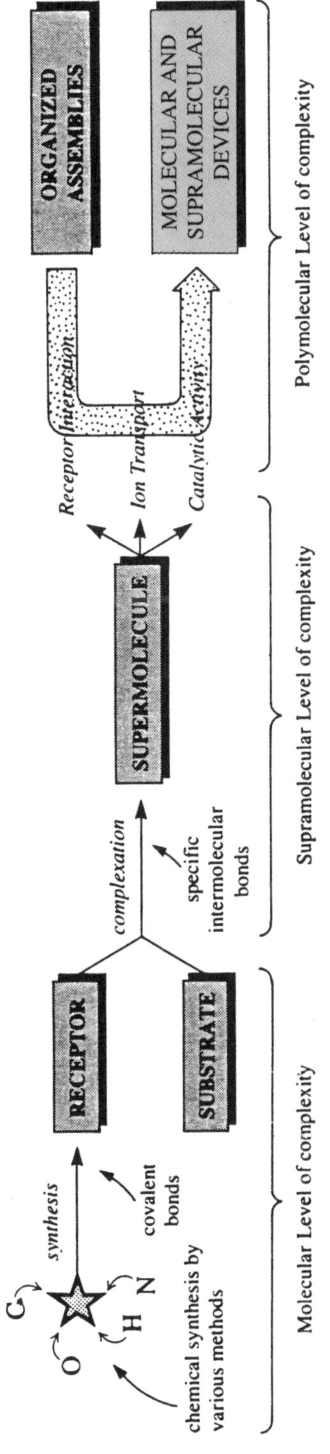

Fig. 1.1. From atoms to molecules to supermolecules. Adapted from (8).

interaction, the supramolecular level. In the development, all resources of molecular chemistry are required. As a result, the efficiency of interaction and activity of the biomodel compound will be limited only by the degree of imagination of the bioorganic chemist.

The supermolecules can then be divided into three distinct areas, each of which has a specific function and must answer particular needs. This is summarized in Fig. 1.2. More specifically, molecular recognition in the design of a receptor involves the understanding of many interactions at the molecular level, such as:

- structural information,
- functional information,
- organic and inorganic molecular adaptation and recognition of intermolecular chemical bonds,
- the energetic of the process,
- the three-dimensional complementarity of molecular shapes, and,
- the concept of molecular cavity as an architectural principle.

Even more complex functions may result from the interplay and association of organized polymolecular assemblies and phases (membranes, vesicules, liquid crystals, etc.) and lead to the development of *molecular devices* (see Section 5.3.2 and particularly Chapter 8).

Furthermore, chemists are now realizing that molecules interact with each other by means of "weak" forces that are stronger than anyone thought. Indeed, a growing number of chemists are seeking to understand the nature of the so-called weak interactions between molecules themselves. This *chemistry beyond the molecule* is of fundamental importance in the biomimetic approach. These small interactions are the forces that, for example, make water a liquid and give cellulose fiber extramolecular

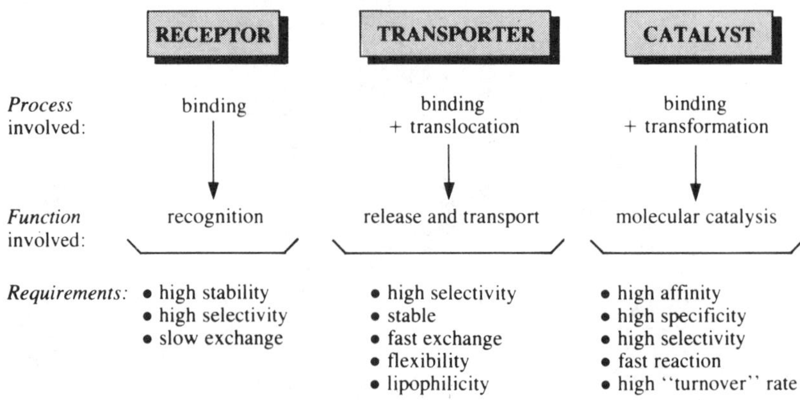

Fig. 1.2. Distinctions between bioorganic models.

strength to stiffen the walls of a plant's cells. F. Stoddart (now at the University of Birmingham) and J.M. Lehn (Université Louis Pasteur, Strasbourg) believe that such dynamic molecular operations as enzyme catalysis and nucleic acid replication can best be understood in terms of what has now come to be known as "supramolecular chemistry" (9). These "weak interactions" dictate in large part many biological processes associated with cooperation, allostery, and regulation and, in a sense, allow molecular signals, or chemical communication, between macromolecules in living cells. It is these weak forces that cause complexes to form and dissociate rapidly and to give rise to molecular recognition phenomena.

Back in 1937, the German chemist K.L. Wolf first used the expression "Übermoleküle," meaning supramolecule (10). It is much later, with the development of new molecules such as crown ethers, polyethers, and cryptands, that Pedersen, Cram, and Lehn revived the expression and developed bioorganic chemistry and supramolecular chemistry as the fields are known today.

Supramolecular chemistry is divided into two broad and overlapping areas. The first one deals with *supramolecules*, that is, well-defined oligomolecular species resulting from the specific intermolecular association of a few components. The second one is concerned with *molecular assemblies* or polymolecular systems, which are formed by the spontaneous association of a large number of components into a specific phase. This phase may have more or less well-characterized microscopic organization and macroscopic domains and properties, depending on its nature (films, layers, membranes, vesicules, micelles, mesophases, surfaces, solids, etc.) (11). This second aspect of supramolecular chemistry is the subject of Chapter 8.

As compared to real biological systems, biomodels have the advantage of being structurally simpler and amenable to chemical modifications at will to make even better molecular adjustments if necessary to optimize the properties. It should also be pointed out that biomimetic or abiotic chemistry is not limited only to the mimicry of living systems for its better understanding but can also lead to new structures, new types of catalysts, and to synthetic transporters capable of functions other than those seen in biological systems with comparable efficiency and selectivity. This challenge is limited only by the imagination and ingenuity of the scientist at designing new and efficient biomodels. Moreover, one should keep in mind that the ultimate goal and satisfaction in bioorganic chemistry is the making of biomodels that will bring about new molecular information and an understanding of the biological systems under study *not* obtainable otherwise.

Important thoughts have been formulated into the conceptualization of chemistry through models (12). In a way, the design and the execution of a molecular biomodel by a chemist are comparable to the approach used by an artist, like a sculptor, working on a masterpiece. They both formally have the same objective in mind, that is:

"Expressing themselves in pictorial forms."

1.4.1 Other Basic Concepts

During the conceptualization of a bioorganic receptor, many association phenomena of chemical origin have to be taken into consideration carefully. Basically, these molecular associations can be grouped into five categories.

1. Electrostatic interactions: They involve dipolar and charged molecules. These interactions are among the strongest.
2. Hydrogen bonds: They are weak but sensitive to angles.
3. Charge-transfer complexes: They use the donor-acceptor concept. They are frequent in flavin chemistry.
4. Van der Waals forces: They are influenced by the polarizibility and structure of the solvent, namely water molecules.
5. Hydrophobic bonds themselves: They are directly concerned with the organic contact between molecules and the lost of ordered water molecules at molecule surface. A favorable entropic factor is involved since water molecules goes back to the "pool" of solvent.

The last two associations are hydrophobic in nature.

To summarize, the design of an artificial enzyme or receptor involves the construction of a molecular architecture or a distinct cavity complementary to a substrate that could be even more efficient if it could also resemble as closely as possible the transition state of an intermediate of the reaction to be mimicked. In other words, the catalytic process to be mimicked would be facilitated if its transition state could be stabilized by specific receptor-substrate interactions. For this, the right geometry and all the factors mentioned above have to be optimized and respected for maximum specificity and efficiency in catalysis. Only if a good intermolecular complementarity between the cavity and the substrate is obtained will selectivity and rate of acceleration in enzymatic reaction be mimicked adequately.

In this context, two examples will illustrate these points in general terms. The first one is taken from the work of Koga (13) and uses functionalized crown ethers as an enzyme model for the synthesis of small peptide chains. The analogy with solid phase peptide synthesis or ribosomal peptide synthesis is apparent. The concepts of proximity of reactive functions and molecular recognition by the crown ether cavity are both operative.

The second example, from Lehn's work, has also been inspired from crown ether chemistry. It uses a macrocyclic framework capable of binding two different metal ions (14).

peptide chain

crown ether ring (receptor)

recognition and binding

thioacylation

new peptide bond formation

elongated peptide chain

This generates a dipodal macrocyclic system that combines a subunit containing a "soft" binding site and one bearing a "hard" site. Such a dinuclear complex can ultimately display novel properties based on the proximity of two different metal ions with different redox potentials.

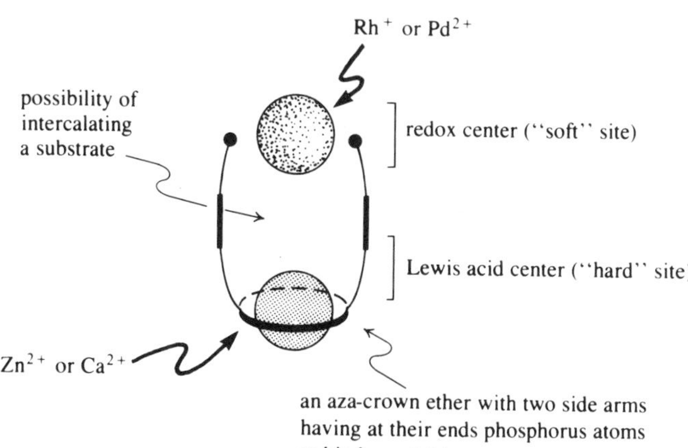

In conclusion, building ligands where the architecture of the molecules has enough thermodynamic stability through secondary interactions with host molecules is crucial for the making of very stable complexes. This principle was recognized by the Swedish Academy who awarded the 1987 Nobel Prize in Chemistry to J.M. Lehn, D.J. Cram, and C.J. Pedersen for their contributions to the study of molecular recognition through the construction of molecular architectures of defined geometry.

Chapter 2

Bioorganic Chemistry of Amino Acids and Polypeptides

"L'imagination est plus importante que le savoir."

A. Einstein

Bioorganic chemistry provides a link between the work of the organic chemist and biochemist, and this chapter is intended to serve as a link between organic chemistry, biochemistry, and protein and medicinal chemistry or pharmacology. The emphasis is chemical and one is continually reminded to compare and contrast biochemical reactions with mechanistic and synthetic counterparts. The organic chemistry of the peptide bond and the phosphate ester linkage (see Chapter 3) are presented "side by side"; this way, a surprising number of similarities are readily seen.

The last decade has witnessed an important breakthrough in the field of asymmetric synthesis of amino acids. We can now tailor make all kinds of α-amino acids with a high control of chirality. Now we have ways to minimize a problem always present before: the danger of racemization. We also have access to asymmetric synthesis in the presence of polyfunctional groups. In pharmaceutical industries this is particularly useful for the synthesis of enantiomerically pure peptides for direct therapeutic use.

Although this chapter is devoted to the chemistry of amino acids and molecules apparented to amino acids, a small introduction to the chemistry of the living cells seems appropriate at the beginning of a book oriented toward the understanding of biological processes at the chemical level. A better understanding of enzyme action and biological transformations in general often relies on a profound knowledge of simpler processes in the cells such as sugar metabolism and energy storage. Therefore, this exercise will serve to bridge comprehension of activation processes in biochemistry with reactivity in organic chemistry.

2.1 Chemistry of the Living Cells

Historically, the metabolism of glucose goes back to the seminal work of A. Szent-Györgyi and H. Krebs, between 1935 to 1937. For the sake of simplicity the presentation will be limited to the combustion of sugar molecules in foods. In the context of bioorganic chemistry, where models of enzymes are developed, it is important to have a general understanding of the chemistry within the biological cell. An appreciation of these processes at a molecular and a supramolecular level is of prime importance since the network connecting these metabolic cycles is an integral part of the "unifying theme" of bioorganic chemistry (16).

When food such as carbohydrate is eaten and digested, a large number of enzymes are called upon to cleave the sugar molecules to smaller fragments that will eventually be further oxidized by mitochondrial enzymes. Because of the exothermic nature of these processes, a good fraction (up to 46%) of the energy liberated will be stored as energy-rich phosphodiester bonds in the form of ATP molecules (see Chapter 3). At the far end of the chain, molecular oxygen is ultimately reduced to water molecules. This is the essence of breathing. Respiration is basically the result of a series of oxidation of carbon molecules to produce water.

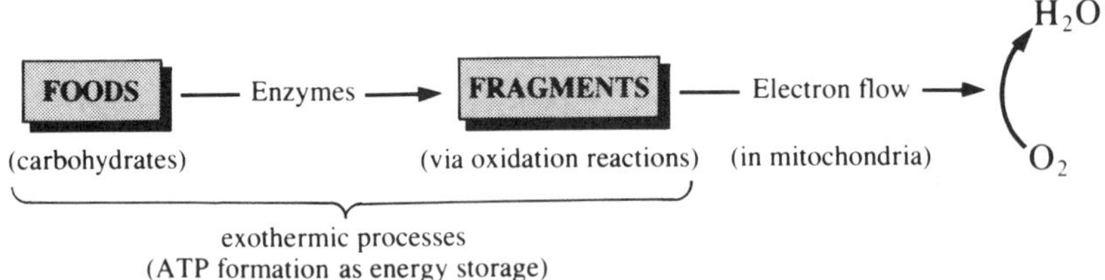

The metabolism of a sugar molecule such as glucose is much more complex than this general presentation seems to indicate. In fact, two distinct processes are involved. First, the six carbon molecules of glucose are cleaved and converted into two molecules of pyruvic acid (a three-carbon molecule) by a series of up to ten different enzymatic reactions. This linear sequence of events is known as *glycolysis*. On the other hand, acetyl-CoA (acetylated coenzyme A) is completely burned to CO_2 molecules via the universally accepted *tricarboxylic acid* (TCA) or Krebs cycle. Interestingly, the bridge between pyruvate ion, the end product of the glycolysis, and the starting material of the TCA cycle is not a direct process but involves a pathway where the coenzyme thiamine pyrophosphate (TPP) and lipoic acid (LA) are both implicated (see

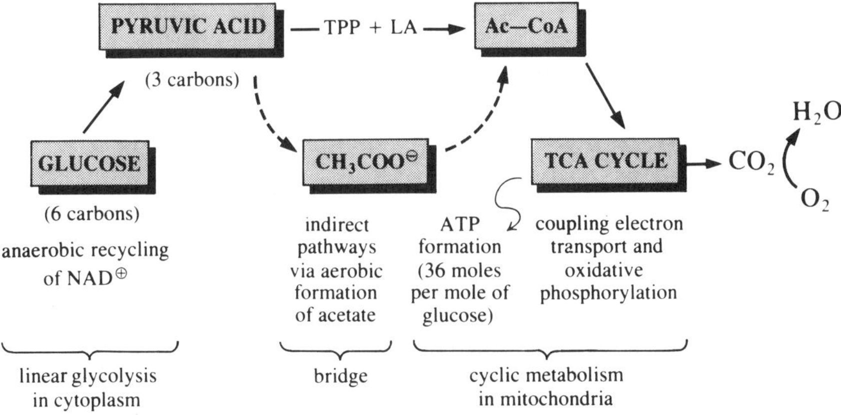

Chapter 7). The whole process can be generalized in the following sequence of transformations (17).

Since these transformations are all occurring in aqueous medium, special types of chemical activators had to be developed by each enzyme to make these processes efficient. This is where *coenzymes* come into play. The mode of action of the coenzymes will be the subject of Chapter 7, but for the sake of discussion, a few words will be said here on coenzyme A and ATP.

Indeed, coenzyme A is of particular interest here. This thiol coenzyme (see p. 520) reacts with acetate ion to form acetyl-CoA, a thioester. As illustrated, this intermediate has a double role in biology, depending on its needs. It is the chemistry of the thioester function that is responsible for it since the electrons of the sulfur atom are poorly delocalized toward the carbonyl.

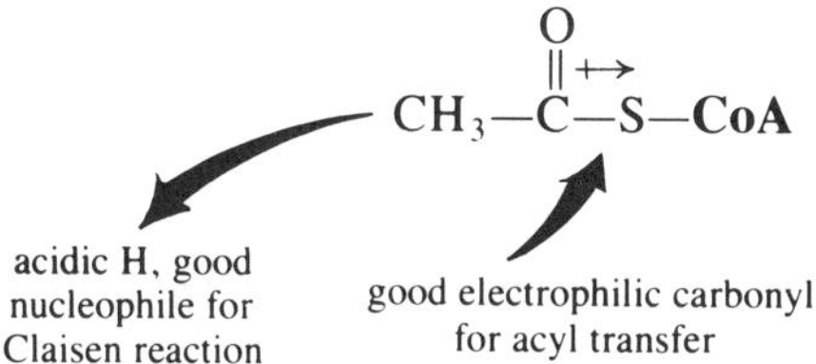

In organic chemistry, acylation of alcohols and amines are routinely done by acetic anhydride treatment in pyridine. However, in the living cell, in aqueous medium, things have to be done differently. Another example is the benzoin condensation which in organic chemistry is catalyzed by cyanide ion. In nature, the coenzyme TPP (mentioned above) is used to do a similar transformation (see p. 562).

In fact, coenzymes are very versatile reagents and modification of their structure can lead to very different properties. For example, it is interesting to realize that the activities of biologically alike adenine molecules can

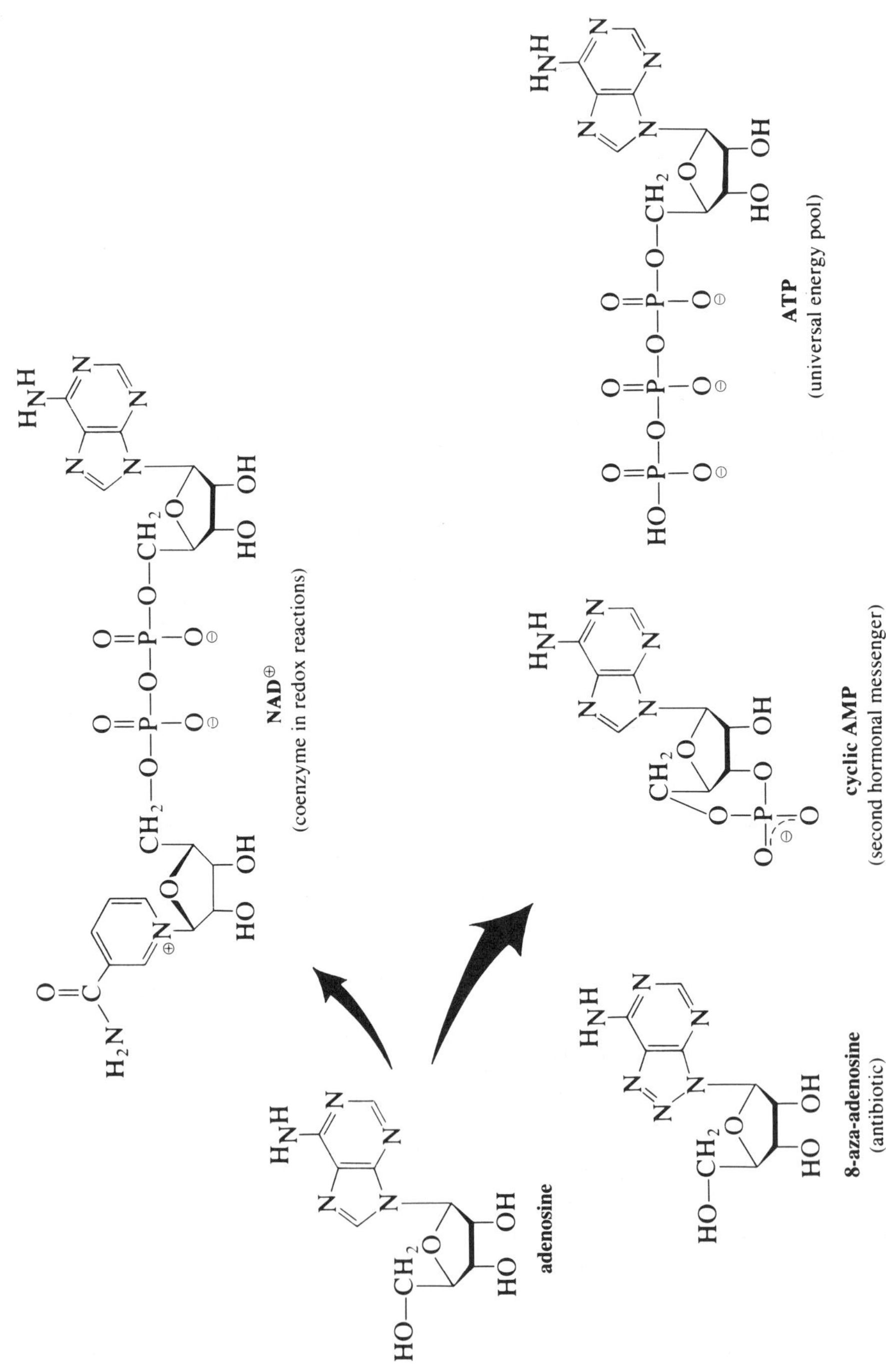
adenosine
NAD⊕
(coenzyme in redox reactions)
8-aza-adenosine
(antibiotic)
cyclic AMP
(second hormonal messenger)
ATP
(universal energy pool)

change dramatically with only minor chemical modifications. The four structures on p. 24 illustrate this point. All these derivatives exhibit very different biochemical behavior.

2.2 Analogy Between Organic Reactions and Biochemical Transformations

There is an instructive comparison to be made between biological processes in nature, using enzymes and coenzymes, and similar reactions in organic chemistry. A list of representative examples are shown in Table 2.1. This table is presented here to emphasize the similarity between what a chemist can do in the organic laboratory as compared to what nature is performing inside the living cell. It also stresses the point that some biological transformations are not so easily carried out in the test tube.

In most cases, the efficiency of each process is the result of the chemistry of the corresponding coenzyme implicated. For clarity these coenzymes are listed in Table 2.2 (page 31) with a few remarks on their role and mode of action. Chapter 7 will develop their individual chemistry in more detail.

Again, to put things into perspective, another parallel between chemical and enzymatic synthesis can be made to stress the gradual degree of complexity involved with bioprocesses as compared to simpler organic analogous transformations. The comparison is shown in Table 2.3. Such comparison is sometimes useful for a better understanding and appreciation of the organic reactions taking place in cellular systems.

As compared to the synthesis of homopolymers on a Ziegler–Natta catalyst, protein synthesis on a ribosome-tRNA complex is more sophisticated by far since a higher degree of stereoregularity and stereospecificity is needed for the synthesis of a large number of different proteins and enzymes. Finally, one should mention that small peptides like gramicidin are synthesized by the bacteria without ribosomes, but again a complex association of different enzymes (see p. 43) are needed to optimize the synthesis.

2.3 Chemistry of the Peptide Bond

The amide bond resulting from the acylation of amino acids is a planar hybrid structure, with an approximately equal distribution between two resonance forms. Because of the greater predominance of the ($C{=}N^{\oplus}{<}$) form, relative to esters, and the ($C{=}O^{\ominus}$) form, the amide bond is a stronger bond. As has been indicated, the amino function of one amino acid may be acylated by the acid function of a second amino acid. The

Table 2.1. Comparison Between Certain Enzymatic Transformations and Their Analogs in Organic Chemistry[a]

Transformations, enzyme cofactors		Comparable reactions in organic chemistry
Peptide synthesis on ribosomes, peptidyl transferase, activation via ATP and tRNA (p. 37)	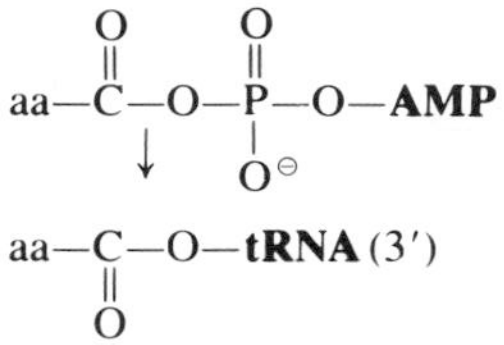 aa—C(=O)—O—P(=O)(O$^\ominus$)—O—**AMP** ↓ aa—C(=O)—O—**tRNA** (3′)	Synthesis of an amide bond, formation of a C—N bond, solid phase synthesis, stereoregulating (template) surface, activation via DCC
Protease, hydrolysis of a peptide bond, α-chymotrypsin (charge relay via Asp-His-Ser), carboxypeptidase (Zn^{2+}, Glu) (pp. 187 and 392)	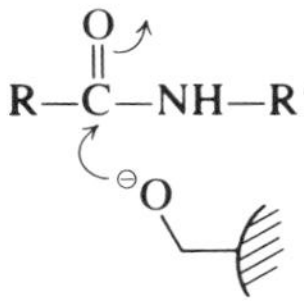 R—C(=O)—NH—R′ ; $^\ominus$O	Cleavage of a C—N bond by a base, a nucleophile
Hydrolysis of polysaccharides, lyzozyme (Asp, Glu) (p. 206)	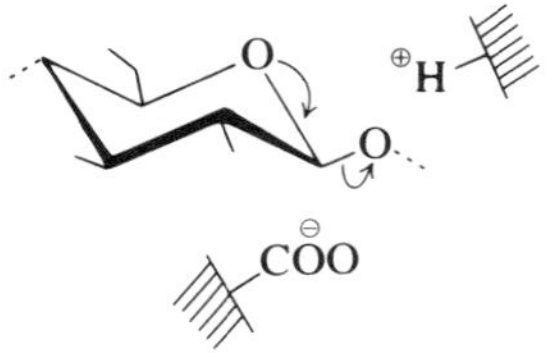 O, O ; $^\oplus$H ; $\overset{\ominus}{C}OO$	Hydrolysis of an acetal in acidic medium, cleavage of a C—O bond
Hydrolysis of phosphodiester bond, RNase (2 His and Lys) (p. 134)	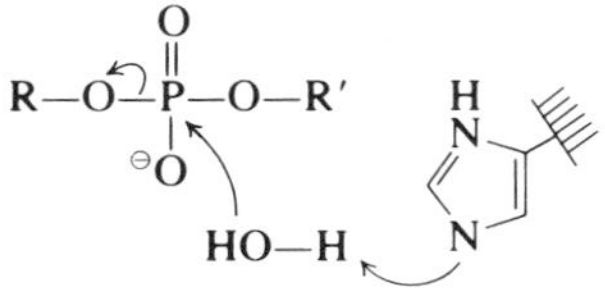 R—O—P(=O)(O$^\ominus$)—O—R′ ; HO—H ; H N N	Cleavage of a P—O bond, pseudorotation

Formation of a phosphodiester bond, RNA and DNA polymerases, nucleotide triphosphates (p. 138)	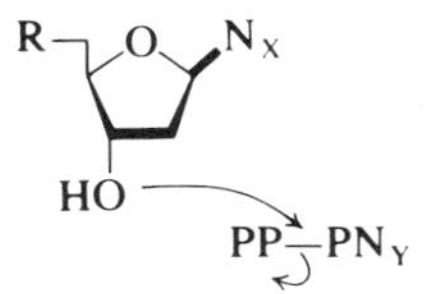 R, O, N_X, HO, $PP{-}PN_Y$	Synthesis of a P—O bond, activation via DCC
Redox reactions, alcohol dehydrogenase, NADH $\leftrightharpoons$ NAD$^+$ (p. 484)	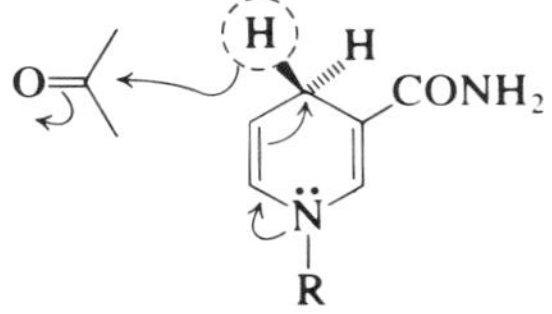 O, H, H, $CONH_2$, N, R	Hydride transfer, $BH_4^{\ominus}$
Sugar (ketoaldose) metabolism, glyoxolase, glutathione (GSH) mediated, inactivated by vitamin C	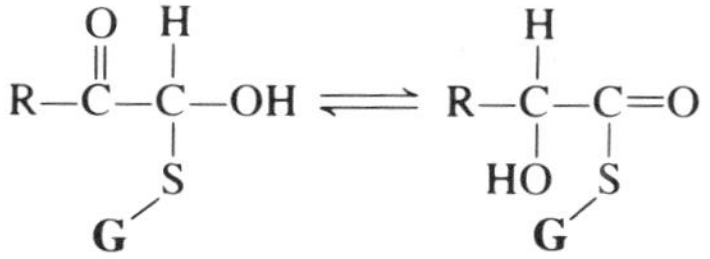 $R{-}C(=O){-}CH(OH){-}S{-}G \rightleftharpoons R{-}CH(OH){-}C(=O){-}S{-}G$	Intramolecular Cannizzaro, transition state analog (p. 79)
Pyridoxal-phosphate–dependent transformations (p. 521)	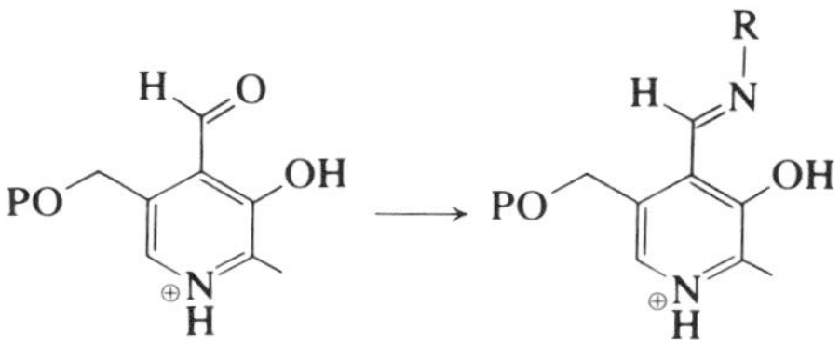 H, O, PO, OH, $\oplus$N, H $\longrightarrow$ R, H, N, PO, OH, $\oplus$N, H	Formation of Schiff base, decarboxylation, racemization
Oxidation reactions dependent on FAD and O_2, via oxene (p. 511)	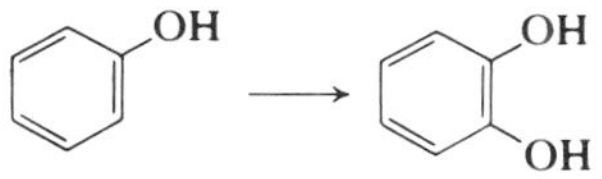 OH $\longrightarrow$ OH, OH	Hydroxylation, insertion of oxygen, ozonelike intermediate

(continued)

Table 2.1. *(continued)*

Transformations, enzyme cofactors		Comparable reactions in organic chemistry
FAD-dependent ketone, monooxygenase in certain bacteria	$R-C(=O)-R' \longrightarrow R-C(=O)-O-R'$	Bayer–Villiger reaction, oxygen insertion
Biotin carboxylation (or dicarboxylation) dependent enzyme, "activated CO_2" (p. 574)		Addition of CO_2, formation of a C—C bond, DCC-like intermediate
Decarboxylation of β-activated acids, Krebs cycle (p. 23)		Decarboxylation of β-keto and β-hydroxy acids, cleavage of a C—C bond
Thiamine-pyrophosphate–dependent decarboxylase, "biological cyanide ion" (p. 567)		Reductive decarboxylation of α-keto acids, benzoin condensation
Aldolases in saccharides, nature's major way to make C—C bonds (p. 174)		Aldol condensation, retro aldol, formation and cleavage of C—C bonds

Synthesis of fatty acids, Krebs cycle, acyl-CoA ligase, "activated acetate" (p. 519)	$CH_3-C(=O)-S-CoA$; $R-C(=O)-X$	Claisen condensation, formation of a C—C bond
Dehydrases, "suicide" substrates (p. 542)	$R-CH=C=CH-C(=O)-R'$; imidazole (N̈, NH)	Michael addition, irreversible formation of a covalent bond
Terpenes synthesis (p. 374)	R, PPO, H, OPP	Formation of a C—C bond, S_N1 (or S_N2) addition of an olefin
Steriod biosynthesis, migration of $H^{\ominus}$ and $CH_3^{\ominus}$, ring rearrangement (p. 377)	⊕ R, H, H, H	Wagner-Meerwein rearrangement in acidic medium
N.I.H. shift, dependent on FAD and O_2 (p. 513)	HO, $COO^{\ominus}$, O, phenylpyruvate → HO, $COO^{\ominus}$, OH, homogentisate	Migration of a C—C bond, formation of an arene oxide

(continued)

Table 2.1. *(continued)*

Transformations, enzyme cofactors		Comparable reactions in organic chemistry
Biosynthesis of aromatic amino acids, chain migration, primary metabolism	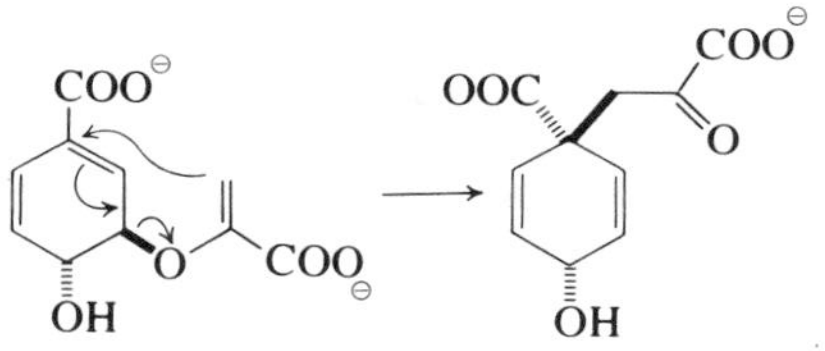 chorismate ⟶ prephenate	Claisen rearrangement, sigmatropic [3.3], Woodward-Hoffman rules, transition state analog (p. 82)
Coenzyme B_{12}–dependent transformations (p. 464)	$CH_3{-}CH(OH){-}CH_2OH \longrightarrow CH_3CH_2CHO$	1,2-migration of covalent bonds

[a] Pages in the text where more material can be found are indicated.

Table 2.2. Coenzyme Activation

Type		Chemistry involved
Acetyl-CoA, thioester (p. 520)	$CH_3-C(=O)-S-$**CoA**	double role (mentioned earlier, p. 23)
acyl phosphate (p. 35)	$CH_3-C(=O)-O-P(=O)(O^{\ominus})-O^{\ominus}$	anhydridelike intermediate
ATP, amino acid activation to load tRNA (p. 37)	$R-C(=O)-O-P(=O)(O^{\ominus})-O-$**AMP**	anhydridelike intermediate for activation of an amino acid
Biotin phosphate intermediate (p. 574)		anhydridelike intermediate or carbodiimide-type chemistry
Thiamine pyrophosphate (p. 560)		"electron sink"

(continued)

Table 2.2. *(continued)*

Type		Chemistry involved
Pyridoxal phosphate (p. 521)	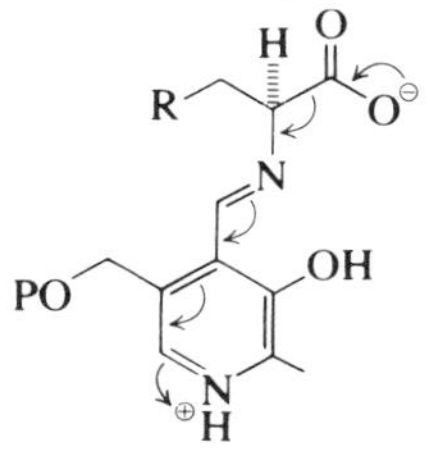	"electron sink"
S-adenosyl methionine (SAM) (p. 40)	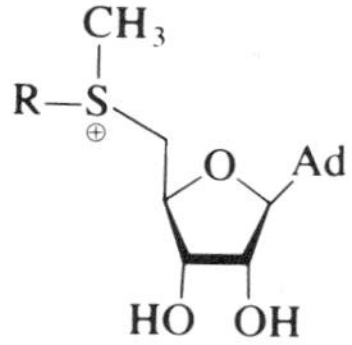	Sulfonium ion intermediate, $CH_3^{\oplus}$ transfer
N^5-methyl-THF, N^5-formyl-THF, N^5-hydroxymethyl-THF, THF = tetrahydrofolic acid (pp. 40, 480)	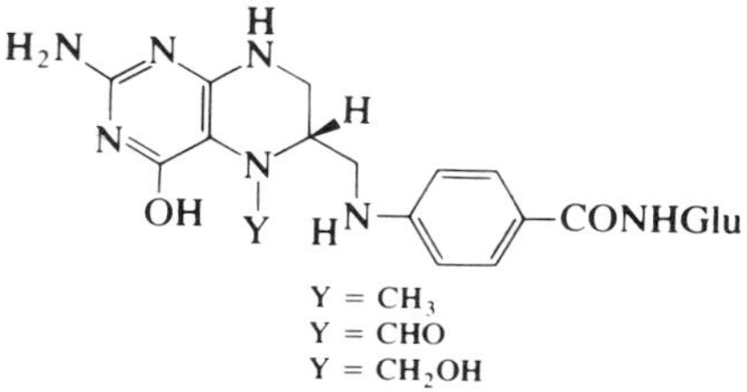 Y = CH_3 Y = CHO Y = CH_2OH	Methyl, formyl, hydroxymethyl, and/or methylene transfer agent

Table 2.3. Parallel Between Chemical Synthesis and Enzymatic Reactions in the Cell

Zeigler—Natta catalyst (p. 169)	Merrifield solid phase peptide synthesis	Cellular protein synthesis (p. 39)
• heterogenous stereoregular surface	• also heterogenous chemistry	• ribosome
• $TiC\ell_4$ + $A\ell Et_3$	• polystyrene and divinylstyrene matrix	• tRNA
• *Catalyst* *Initiator* with a good electronic structure	• the amino acid is first attached to the solid support	• 2 sites on the ribosomal surface
	• the process is automatized by sequential addition of other protected amino acid	• activation of each amino acid by ATP (mixanhydride chemistry again)
• Ti atom has an empty orbital that behaves like a cavity to make a σ bond with C_2H_5 of $A\ell Et_3$. Then this is followed by a π bond with an olefin substrate and finally a σ bond with the substrate occurs and the polymer chain starts to grow	• DCC is used as activating agent (mixanhydride chemistry)	• loading of aminoacyl NA on the ribosome and peptide bond formation
		• Watson–Crick base pair recognition between tRNA and mRNA for coding the right amino acid sequence

Degree of molecular complexity increases from left to right.

amino bond so formed is referred to as a peptide linkage and the product termed a *dipeptide*. Proteins are polymers that have monomeric units connected by amide (peptide) linkages. The amide bond formed between two amino acids is a secondary amide in a *trans* geometry. Free rotation does not readily occur about the C—N bond, since this would destroy the π resonance overlap, with the *trans* geometry being preferable for steric reasons.

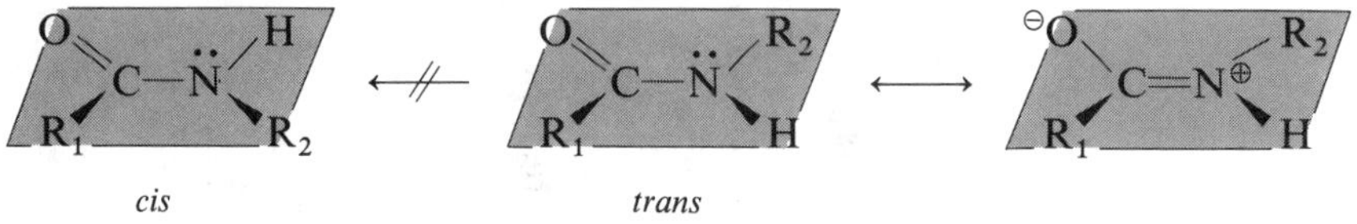

cis *trans*

Structure of the planar sp^2 hydridized amide (peptide) bond; free rotation from a *trans* to a *cis* geometry, with an energy barrier of ~41.8 kJ/mol (10 kcal/mol), does not readily occur.

The peptide bond is a strong bond, and energy is required for its formation. Mixing an aqueous solution of two amino acids, one with an unprotonated amino function (potentially nucleophilic) and the other with a protonated carboxyl function, at room temperature would only result in salt formation. Chemically, the carboxyl function must be converted to a good leaving group. Energetically, the carboxyl function must be activated to compensate for the work done during peptide bond formation. This is reflected in the *free energy of hydrolysis* (ΔG_{hydro}) of the amide bond which is in the range of -12 kJ/mol (-3 to -4 kcal/mol). On the other hand, the ΔG_{hydro} for an acyl chloride is approximately -29.3 kJ/mol (-7 kcal/mol), and the chlorine atom is a good leaving group. It is therefore possible to convert the carboxyl of an amino acid to an acyl chloride (thionyl chloride, phosphorous pentachloride) and react this with the amino function of a second amino acid to form a peptide bond. This is an oversimplification of the problem of peptide bond (protein) synthesis and, as will be seen shortly, a rather elaborate methodology has been developed. However, it may be of interest to first examine how the peptide bond is synthesized within the biological system, with particular attention to the problem of carboxyl function activation. Remember, the same energetics prevail in both the test tube and the organism.

Within the biological context, an anhydride structure, as indicated below, is a high-energy structure or potential energy store. The defini-

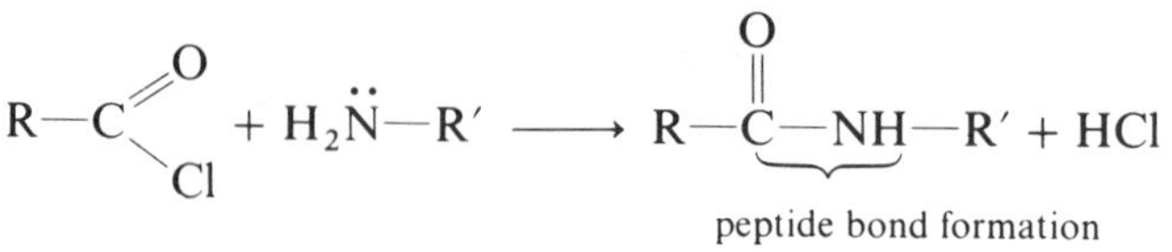

$$R-C(=O)Cl + H_2\ddot{N}-R' \longrightarrow R-C(=O)-NH-R' + HCl$$

peptide bond formation

tion of *high energy* for biological systems and more background into the chemistry of anhydrides is presented in Chapter 3 on bioorganic phosphorus. For the moment, it will suffice to say that any compound with an anhydride structure will exhibit $\Delta G_{\text{hydro}} > 29.3$ kJ/mol (7.0 kcal/mol). Examples include acetic anhydride, acetyl phosphate, and acetyl

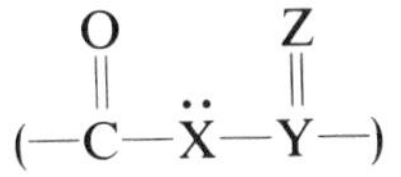

$$(-\overset{\overset{\text{O}}{\|}}{\text{C}}-\ddot{\text{X}}-\overset{\overset{\text{Z}}{\|}}{\text{Y}}-)$$

imidazole. The biological energy store ATP also possesses the anhydride structure in its triphosphate side chain. The fact that ATP can transfer this energy to activate a carboxyl function is shown in a simple biochemical experiment. Incubation of a liver homogenate (liver of an animal homogenized so as to release the enzymes from inside the cell) with glycine, benzoic acid, and ATP leads to the formation of *N*-benzoyl glycine (hippuric acid). Actually, this represents a mechanism of *detoxication* in the mammal or making a harmful substance (benzoic acid) harmless by conjugation with a freely available substance in the body (glycine) and then eliminating the product in the urine. The first step would be nucleophilic attack of the benzoate on ATP, to give an activated benzoate (carboxylate) and inorganic pyrophosphate. The amino function of glycine may now attack this anhydride intermediate to give the *N*-benzoylated product.

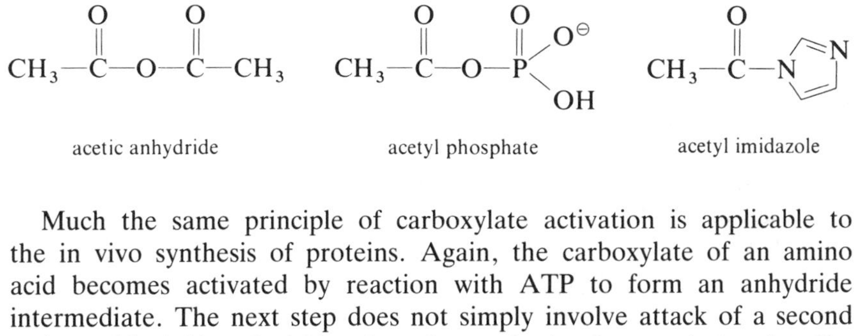

$$CH_3-\overset{\overset{\text{O}}{\|}}{\text{C}}-O-\overset{\overset{\text{O}}{\|}}{\text{C}}-CH_3$$

acetic anhydride

$$CH_3-\overset{\overset{\text{O}}{\|}}{\text{C}}-O-\overset{\overset{\text{O}}{\|}}{\text{P}}(O^{\ominus})(OH)$$

acetyl phosphate

$$CH_3-\overset{\overset{\text{O}}{\|}}{\text{C}}-N\text{(imidazolyl)}$$

acetyl imidazole

Much the same principle of carboxylate activation is applicable to the in vivo synthesis of proteins. Again, the carboxylate of an amino acid becomes activated by reaction with ATP to form an anhydride intermediate. The next step does not simply involve attack of a second amino acid of this anhydride, since the synthesis of a protein involves the precise sequential coupling of many (up to a few hundred) amino acids. A *template* or "ordered surface" must be available to ensure correct sequencing of the protein molecule. The macromolecule that serves such a function is a polynucleotide, transfer ribonucleic acid (tRNA); the constitution of polynucleotides will be described in the next chapter.

Activation of the amino acid by ATP is only an intermediate step catalyzed by the enzyme aminoacyl-tRNA synthetase. The 3′- or 2′-hydroxyl of the terminal adenylic acid of the tRNA molecule then attacks the anhydride intermediate to give an aminoacyl-tRNA molecule.

Ad = adenine

ATP

$-PP_i$

+ AMP (adenosine monophosphate)

hippuric acid

The ester linkage between the hydroxyl of the tRNA is a high-energy bond (because of the adjacent 2′-hydroxyl and cationic amino functions), so that the overall enzyme-catalyzed reaction has a free-energy change close to zero. Each amino acid has a specific tRNA molecule and one specific aminoacyl-tRNA synthetase enzyme. In turn, each aminoacyl-tRNA synthetase enzyme will accept only its particular amino acid as a substrate. However, it has been possible to slightly modify the naturally occurring amino acids and have them serve as substrates for the enzyme. For example, *p*-fluorophenylalanine can substitute to some extent for phenylalanine.

Once synthesis of the aminoacyl-tRNA is complete, the amino acid no longer serves a recognition function. Specificity is dictated by the tRNA portion of the molecule by its interaction with the genetic message (mRNA) and another large surface upon which protein synthesis takes place; a cellular organelle referred to as the *ribosome*.

$$\text{R—CH}\begin{matrix}\text{COO}^{\ominus}\\ \text{NH}_3^{\oplus}\end{matrix} + \text{ATP} + \text{tRNA(3'—OH)}$$

amino acid

$$\Big\downarrow\ -\text{PP}_i \quad \text{Mg}^{2+},\ \text{aminoacyl-tRNA synthetase}$$

intermediate anhydride

$$\Big\downarrow\ -\text{AMP}$$

$$\text{R(H}_3\overset{\oplus}{\text{N}})\text{CH—}\overset{\text{O}}{\overset{\|}{\text{C}}}\text{—O—tRNA}$$ aminoacyl-tRNA

This was demonstrated by taking the aminoacyl-tRNA complex specific for cysteine (abbreviation: cysteinyl-tRNA^{Cys}, showing that the cysteine has combined with its specific tRNA) and modifying the amino acid side chain to form alanine by catalytic reduction.

$$\text{tRNA}^{\text{Cys}}\text{—O—}\overset{\text{O}}{\overset{\|}{\text{C}}}\text{—CH(CH}_2\text{SH)—}\overset{\oplus}{\text{N}}\text{H}_3 \xrightarrow[\text{Raney Ni}]{\text{H}_2} \text{tRNA}^{\text{Cys}}\text{—O—}\overset{\text{O}}{\overset{\|}{\text{C}}}\text{—CH(CH}_2\text{—H)—}\overset{\oplus}{\text{N}}\text{H}_3$$

cysteinyl-tRNA^{Cys} alanyl-tRNA^{Cys}

The modified aminoacyl-tRNA was then incubated in vivo and incorporated alanine into the protein at those positions that normally accepted cysteine.

Protein synthesis begins with the N-terminal amino acid and proceeds from this point. In some bacteria, yeast, and higher organisms, this first aminoacyl-tRNA is known to be *N*-formylmethionyl-$\text{tRNA}^{\text{fMet}}$. Formylation of the amino function can be considered as a protecting group to prevent participation of the amino function in peptide bond formation. The fMet-$\text{tRNA}^{\text{fMet}}$ is then the first aminoacyl-tRNA to bind to the ribosome and mRNA. After the protein is synthesized, the formyl group is removed by enzymatic cleavage (formylase).

The ribosome is a large cellular organelle composed of RNA and a number of different proteins and built of two dissociable subunits. This provides the ultimate ordered surface for protein synthesis, being able to interact precisely with the large tRNA portions of the various aminoacyl-tRNAs. While some of the proteins of the ribosome presumably have a catalytic function, the rest as well as the ribosomal RNA (rRNA) participate in specific conformational interactions that occur during protein synthesis. Protein synthesis is a dynamic process, but again this process occurs in an ordered fashion, which the sequential coupling of amino acids demands. The site to which the fMet-tRNA^{fMet} binds on the ribosome is referred to as the *peptidyl site*. The stage is now set for the synthesis of the peptide linkage.

The second amino acid (aminoacyl-tRNA) also binds on the ribosome (at the so-called *aminoacyl site*) in close proximity to the fMet-tRNA^{fMet}. While no chemical reaction has yet occurred, this binding process requires work, the energy of which comes from a molecule of GTP (guanosine triphosphate; like ATP except adenine is replaced with guanine). The amino function of the aminoacyl-tRNA now attacks the f-methionine, at which point the tRNA^{fMet} becomes a leaving group, and the peptide bond is formed. This reaction is enzyme catalyzed, but neither ATP or GTP is required for the peptide bond formation, since the energy for this process comes from the cleavage of the high energy tRNA^{fMet} ester. While this completes peptide bond formation, obviously some physical changes must occur in order that another incoming aminoacyl-tRNA may attach to the dipeptide. The dipeptidyl-tRNA at the aminoacyl site is physically shifted to the peptidyl site, simultaneously displacing the tRNA^{fMet}. This, very likely, results from a conformational change on the ribosome which again requires energy at the expense of a molecule of GTP. The aminoacyl site is now empty, and the mRNA has also shifted (translocation), so that it can now dictate the entrance of a new aminoacyl-tRNA to the aminoacyl site. Once this has occurred, peptide bond formation (to give a tripeptide) can take place, thus repeating the sequence of events as described (Fig. 2.1).

This provides a basic description of in vivo protein synthesis, neglecting details such as the involvement of protein elongation factors. Obviously the process is very complex, but underlying this complexity is the basic theme: activation of the carboxyl function followed by sequential coupling of amino acids on an ordered matrix that makes an incorrect sequence and other side reactions a virtual impossibility.

In biological systems, the universal methyl group donor is the sulfonium compound *S*-adenosyl methionine (SAM). In turn, this is synthesized from the amino acid methionine and another biologically important compound (a high energy compound or biological energy store) adenosine

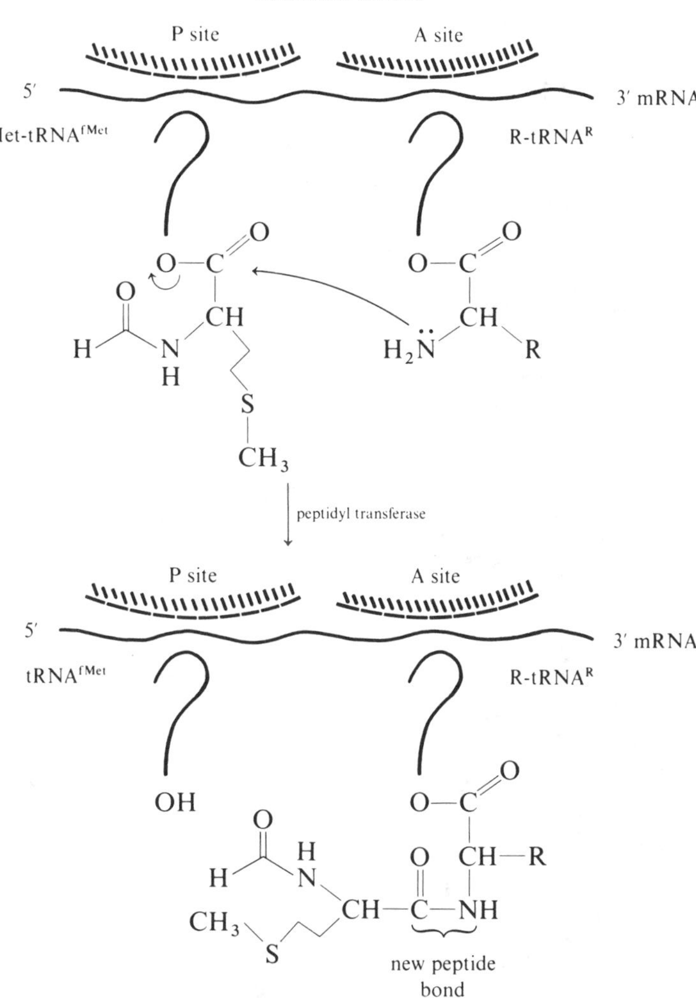

Fig. 2.1. Peptide bond formation in the biological system. The entire process takes place on the ribosome and involves two binding sites, the P (peptidyl) and A (aminoacyl) sites. The reaction is catalyzed by the enzyme peptidyl transferase. It is the message carried by mRNA (which in turn is dictated by the genetic material DNA) that determines by specific interactions as to which aminoacyl-tRNA will bind at the P and A sites.

triphosphate (ATP). As is the case for all chemical reactions that occur in the biological organism, this reaction is catalyzed by a specific enzyme. The reaction is thermodynamically favorable without the presence of the protein catalyst, but the enzyme gives specificity to the reaction. Without the catalyst, other reaction pathways become possible, such as breakdown of the triphosphate chain, but the catalyst serves to bind and orientate the sulfur nucleophile in such a way that only attack on the methylene carbon becomes possible. Much more will be said later about the importance of such binding and proximity effects, but it should be noted that while the adenosine portion of the ATP molecule does not participate in the chemistry of the reaction, it does serve a recognition function. It is recognized by the enzyme catalyst and subsequently binds to the surface of the ATP molecule. In mammals, the methionine substrate is an essential amino acid in the diet.

Met

ATP

enzyme, Mg^{2+}

PPP_i +

SAM

Homocysteine is the amino acid that results after *S*-adenosyl methionine donates its methyl group, and then the product *S*-adenosyl homocysteine is hydrolyzed by a molecule of water. In mammals, homocysteine can be converted to cysteine so that the latter is not an essential amino acid, or it can be converted back to methionine by methylation with a compound that serves as a source of one carbon fragment (methyl, formyl, etc.) in biological systems: *tetrahydrofolic acid*. In some bacteria, homocysteine can be converted back to methionine by methylation with methylcobalamin (the methyl derivative of vitamin B_{12}) in the presence of other required compounds or cofactors. This latter methylation reaction is of interest because it can occur, at a reduced rate, in the absence of any enzymes, and a simpler model system has been developed to mimic this reaction (see details in Chapter 6).

In peptide bond formation, two general approaches are possible. The first is to convert the amino-protected amino acid to an activated form and then react this with the amino function of a second amino acid. As will be recalled, activation is necessary since work must be done during peptide bond formation. Alternatively, it is possible to react two amino acids together (one amino protected, the other carboxyl protected) in the presence of a *coupling agent* that activates the carboxyl function in situ.

In the synthesis of a protein macromolecule, peptide bond formation must be a high-yield reaction. Therefore, another factor peculiar to biological systems must be considered: *optical purity*. Proteins are made of L-amino acids. As such, a chemical synthesis must start with L-amino acids and racemization must be minimized during the synthesis. This is especially true in the synthesis of an enzyme, because catalytic activity is dependent upon optical integrity. Amino acids are particularly susceptible to racemization once they have been acylated (i.e., addition of an acyl protecting group to the amino function) via intermediate azlactone formation. This can occur during protection or coupling procedures:

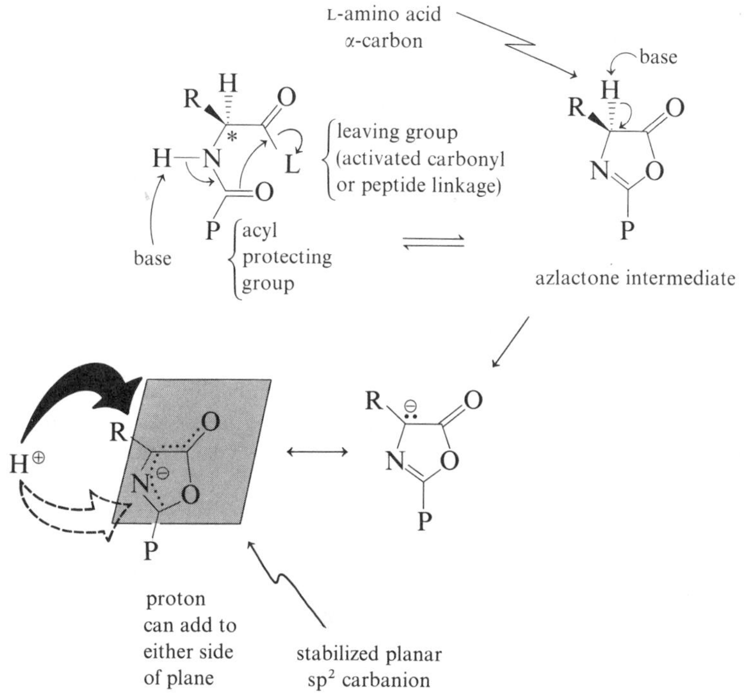

$$+ \quad H_2\ddot{N}{-}R' \longrightarrow P{-}\overset{\overset{O}{\|}}{C}{-}NH{-}\underset{\underset{R}{\wr}}{CH}{-}\overset{\overset{O}{\|}}{C}{-}NH{-}R'$$

optically inactive product

The α-proton of the azlactone is quite base labile (stabilized carbanion), so that a proton may be lost but can later add to either side of the azlactone plane with subsequent loss of optical purity. Further, the azlactone may participate in peptide-bond–forming reactions, so that it disappears during the synthesis, but racemic amino acids are introduced into the product protein.

2.4 Nonribosomal Peptide Bond Formation

Some microorganisms are able to use a simpler, more primitive method of peptide bond synthesis. While perfectly effective for the peptide formation, the system lacks the highly ordered apparatus provided by the ribosome and tRNA structures. As such, only small proteins (polypeptides) are synthesized by this means; the antibiotic gramicidin S providing an important example (18). Gramicidin S is an interesting antibiotic for a number of reasons. First, it contains phenylalanine in the D-configuration. The occurrence of D-amino acids in nature is very rare and only L-amino acids are found in proteins. Second, gramicidin S contains the amino acid ornithine, which is not a normal constituent of proteins.

Gramicidin S is a cyclic decapeptide isolated from *Bacillius brevis* whose antibiotic activity is derived from its ability to complex alkali metal ions and transport them across membranes. It acts as a channel for ion leakage, which is not difficult to imagine when it is noted that it has a doughnut shape capable of encapsulating the metal ion (see Section 5.1.4). Unfortunately this action is not restricted to bacterial membranes and so gramicidin S is used only topically.

We may now examine how a dissymmetric macromolecular complex of proteins, found in nature, is responsible for the synthesis of this cyclic

D-Phe ⟶ L-Pro ⟶ L-Val ⟶ L-Orn ⟶ L-Leu

L-Leu ⟵ L-Orn ⟵ L-Val ⟵ L-Pro ⟵ D-Phe

Orn = ornithine; $H_2N\text{-}(CH_2)_3\text{—}CH(NH_3^{\oplus})COO^{\ominus}$

Structure of the antibiotic gramicidin S; decapeptide consists of two complementary pentapeptide strands; molecule has a twofold axis of symmetry

polypeptide without the need of ribosomes. This process, which occurs only in simple organisms like bacteria, was deciphered by the biochemist F. Lipmann, from Rockefeller University.

Lipmann found that the mere addition of ATP to crude cell homogenate and extracts prepared from *B. brevis* promotes the synthesis of gramicidin S. This activity was found in particle-free supernatant fractions and was resistant to known inhibitors of protein synthesis and to treatment with ribonuclease, excluding the participation of an RNA molecule. In this biological transformation, ATP fulfills the function of a condensing reagent in a manner analogous, for example, to a carbodiimide group.

Two enzymes are involved in the synthesis of gramicidin S: a light (MW = 100,000) and a heavy (MW = 280,000). Synthesis begins on the light enzyme, which also functions as a "racemase," converting L-phenylalanine to the D-enantiomer. A thiol nucleophile on the light enzyme attacks an activated phenylalanine (ATP and the amino acid reacted to form an anhydride) to give a high-energy thiol ester. The ΔG_{hydro} of a thiol ester (base catalyzed) is approximately $-38\,\text{kJ/mol}$ ($-8\,\text{kcal/mol}$). The difference between thiol and normal esters is accounted for by the fact that the oxygen atom will more readily delocalize its nonbonding

light enz.—SH + O=C(CH(NH$_3^{\oplus}$)CH$_2$Ph)–O–P(=O)(O$^{\ominus}$)–O–CH$_2$-ribose(Ad, HO, OH) $\xrightarrow{-\text{AMP}}$ light enz.—S—C(=O)—CH(NH$_3^{\oplus}$)CH$_2$Ph

electrons into the carbonyl function than will the sulfur atom. Such delocalization will reduce the electrophilicity of the carbonyl function. Furthermore, the thiol is a much better leaving group than the corresponding alcohol. Remember, the pK_a of a mercaptan is approximately 10, while that of an alcohol is approximately 15.

In the next step, proline attached to the heavy enzyme by a thiol ester linkage attacks the light enzyme, thus transferring the dipeptide to the heavy enzyme. The function of the light enzyme is terminated.

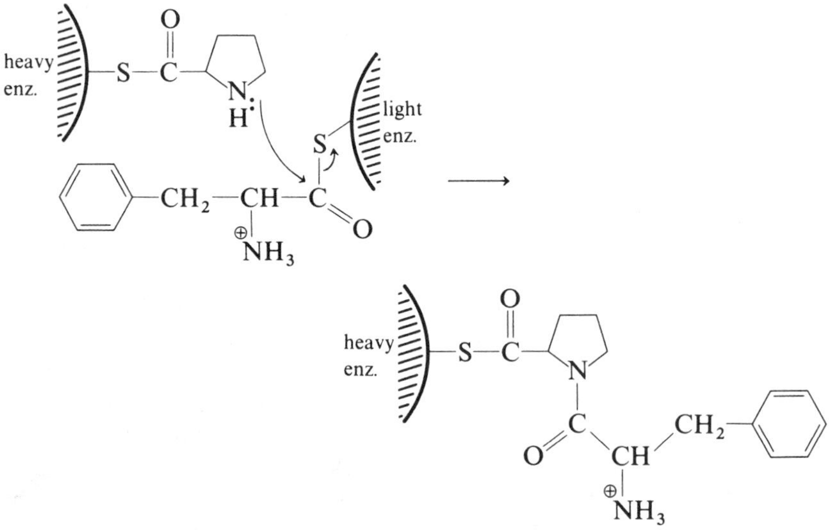

Similarly, on the heavy enzyme, the amino function of valine (attached as the thiol ester) attacks the dipeptide, forming a tripeptide product. Remember, the energy for peptide bond formation is being derived from the aminolysis of the thiol ester.

The next step involves an "arm" on the heavy enzyme. This arm "picks up" or attacks the tripeptide via its thiol function. It is then attacked by ornithine to form a tetrapeptide, but picks up the tetrapeptide (via the same thiol) and is attacked by leucine to form a pentapeptide.

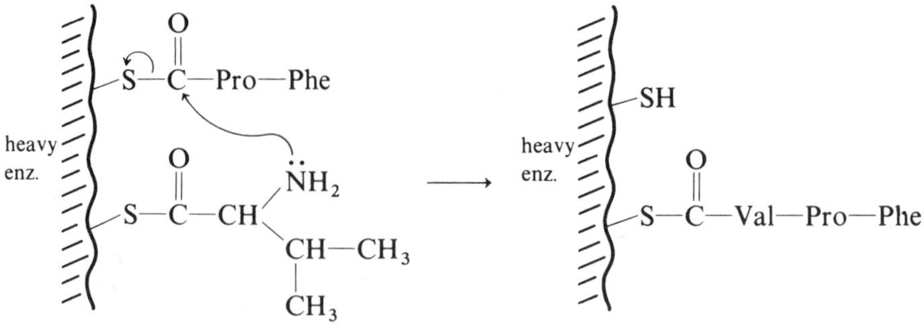

The arm again retrieves the pentapeptide, but this time the pentapeptide is attacked by a complementary pentapeptide (attached to another arm) to give rise to the cyclic decapeptide.

In brief, quaternary interactions between the two enzymes allow the two amino acids to get close enough for peptide bond formation. This process is repeated five times to form the two halves of the molecule. The proteinic complex collapses the two identical chains to give finally the cyclic peptide, schematized in Fig. 2.2.

Figure 2.3 gives a different representation to amplify the fact that only the light chain is responsible for the presence of D-Phe in the sequence. The remaining part of the antibiotic is synthesized by the heavy chain.

In conclusion, these two enzymes have enough tridimensional orientational effect to recognize the right amino acids and to bring them together for the synthesis of a well-defined asymmetric and cyclic peptide molecule. It is the presence of specific protein-protein interactions that enable such an efficient polypeptide synthesis to take place. In other words, all the molecular information needed for the synthesis of gramicidin S is present within this macromolecular enzymatic complex to allow amino acid recognition and proper peptide bond formation. Again, this nonribosomal polypeptide synthesis is a remarkable example of the importance of chirality (see Section 4.2.1). We are still far from being able to produce such an efficient peptide synthesis in the laboratory, even with Merrifield's solid phase method. However, the analogy is apparent.

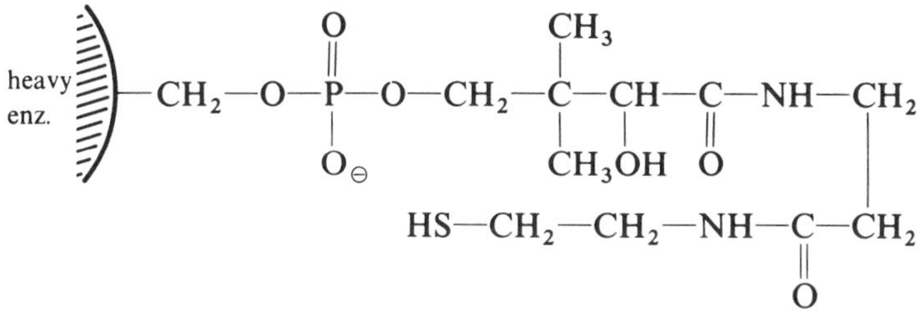

heavy enz.
$$\text{heavy enz.}-CH_2-O-\underset{O^{\ominus}}{\overset{O}{\overset{\|}{\underset{|}{P}}}}-O-CH_2-\underset{CH_3}{\overset{CH_3}{\overset{|}{\underset{|}{C}}}}-\underset{OH}{\underset{|}{CH}}-\underset{O}{\underset{\|}{C}}-NH-CH_2-CH_2-\underset{O}{\underset{\|}{C}}-NH-CH_2-CH_2-SH$$

Structure of the thiol containing "swinging arm" present in "heavy enzyme"; arm consists of β-mercaptoethylamine and pantothenic acid (an important component of coenzyme A) esterified to a serine phosphate of the enzyme

The team of R.L. Letsinger and I.M. Klotz of Northwestern University developed a method for peptide synthesis based on a template-directed scheme that parallels that of the natural mechanism using ribosomes, presented earlier in Fig. 2.1. The strategy employed makes use of a resin support and a polynucleotide template but does not require (like the natural system) temporary amino acid protecting groups in order to achieve fidelity of coupling. The approach has been called the *complementary carrier method for peptide synthesis* and is summarized in Fig. 2.4.

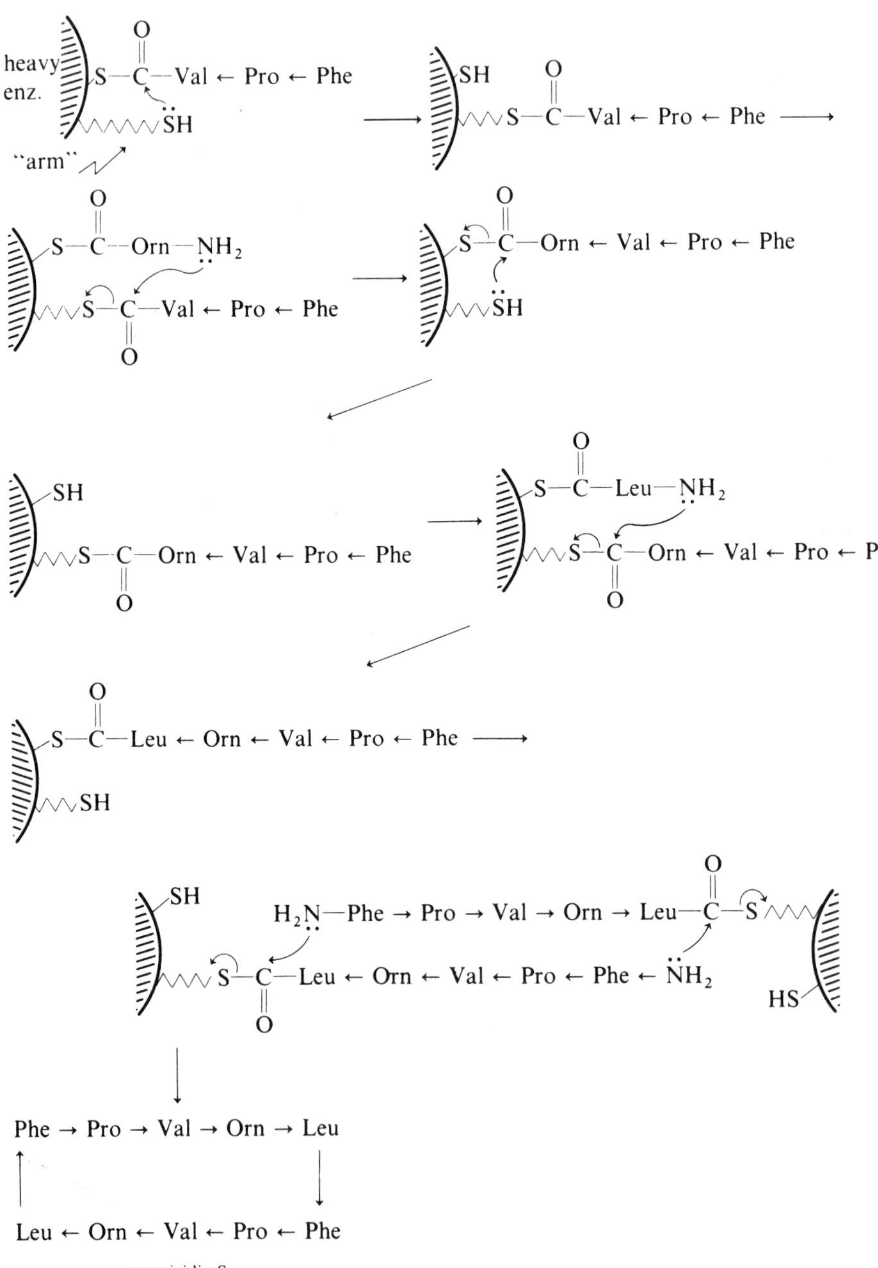

Fig. 2.2. Lipmann's enzyme-controlled synthesis of the antibiotic peptide gramicidin S.

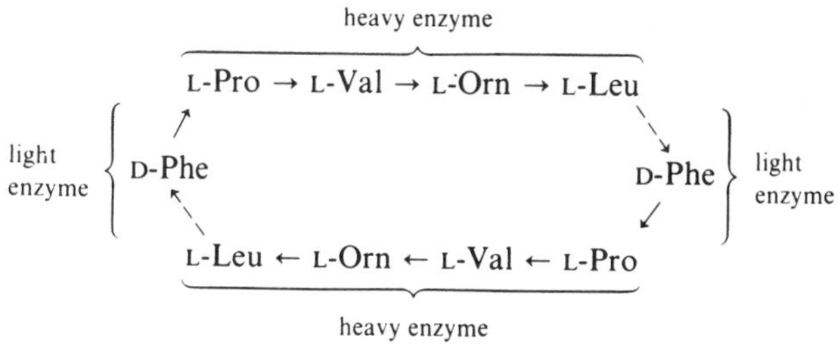

heavy enzyme

L-Pro → L-Val → L-Orn → L-Leu

light enzyme { D-Phe D-Phe } light enzyme

L-Leu ← L-Orn ← L-Val ← L-Pro

heavy enzyme

Fig. 2.3. Sequence of amino acids in gramicidin S. The brackets indicate the enzymes activating the enclosed amino acids.

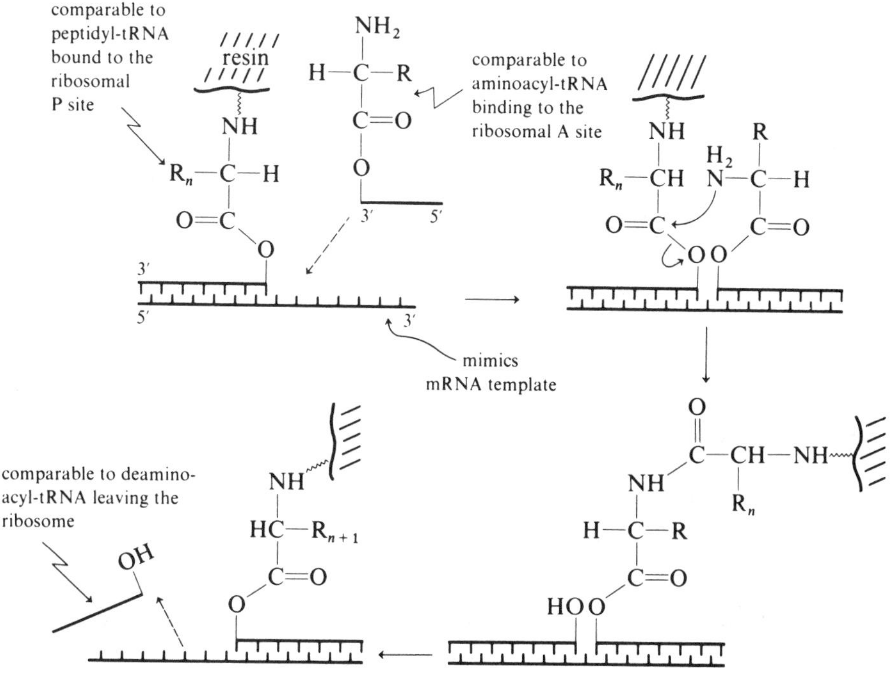

Fig. 2.4. Letsinger and Klotz's complementary carrier scheme for peptide synthesis which has a great similarity with the natural ribosomal mechanism (19).

In practice, the growing polypeptide chain is attached to an oligonucleotide at the ribose 5′-OH by an ester linkage to the terminal carboxyl group. The incoming new amino acid is also linked as an ester but to the 3′-OH group of a second oligonucleotide. From a variety of studies on helix stability, it appears that carriers of eight nucleotides in length are sufficient to bind properly to a longer oligonucleotide template by virtue of Watson–Crick complementary base pairings. Current chemical methods of peptide bond formation are used to extend the polypeptide chain by one amino acid. To proceed to the next coupling step of the synthesis, the unbound oligonucleotides are simply washed from the insoluble solid support under conditions that disrupt base pairing, such as temperature or decreased ionic strength. This circular sequence of reactions can be repeated until the desired polypeptide molecule is constructed. Notice that at each step, the growing polypeptide chain is transferred to the complementary carrier oligonucleotide and at the end of the synthesis the final carrier can then be cleaved from the solid support. The remaining attached oligonucleotide can serve as a chromatographic handle during the purification of the polypeptide molecule and eventually be released under mild basic conditions.

The important feature of this methodology is that fidelity of the amide bond formation is achieved by specific and template-directed coupling of the desired acyl transfer. Therefore, proper juxtaposition of peptidyl and aminoacyl oligonucleotides on a polynucleotide template controls the direction of polypeptide synthesis. This is essentially the way that fidelity of translation in protein biosynthesis is taking place in natural systems.

This scheme (Fig. 2.4) thus shows how oligonucleotides can direct the synthesis of polypeptides in the absence of protein or ribosomal machinery and, as such, is an appealing bioorganic model for the origin of prebiotic protein synthesis. Indeed, it seems most probable that primitive biosystems used a similar concept to carry out primitive protein synthesis where Watson–Crick base pairing provided the intrinsic mechanism for achieving fidelity of replication and direction of protein synthesis. In time, the carrier oligonucleotides could have evolved into more efficient species such as the present-day tRNA molecules.

Solid phase peptide synthesis developed by Merrifield and the classical peptide and protein synthesis on ribosomal surface by the living cells are well covered in biochemistry textbooks. The parallel that can be made between them has been summarized in Table 2.3. The discussion will be focused here to other chemical means to construct polypeptides using molecular templates.

Indeed, by using a matrix or solid support, an innovative method has been developed to carry polypeptide synthesis in the laboratory in a way similar to what is found in nature. This approach has been referred to as *amine capture strategy for peptide synthesis* (20a) by D.S. Kemp of the Massachusetts Institute of Technology.

The conceptual design is based on the following general transformations:

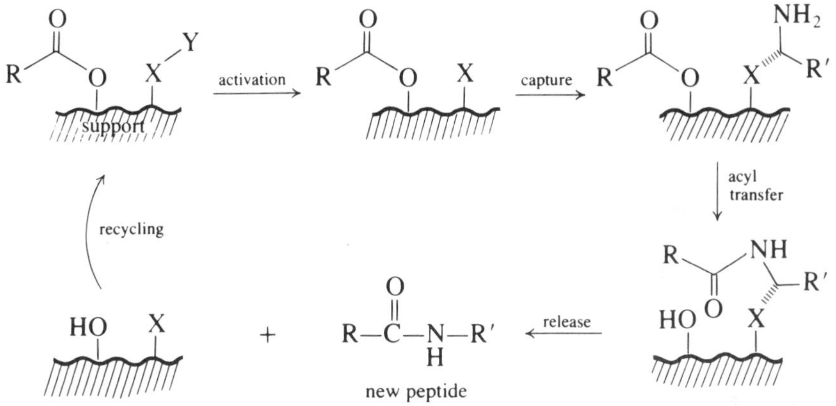

A potential capture site exploited by Kemp's group is the thiol function of a N-terminal cysteine residue. The O,N-acyl transfer occurs via a medium-size ring intermediate. Proximity effect, solvent effect, steric and stereoelectronic effects, all contribute to the success of the designed plan.

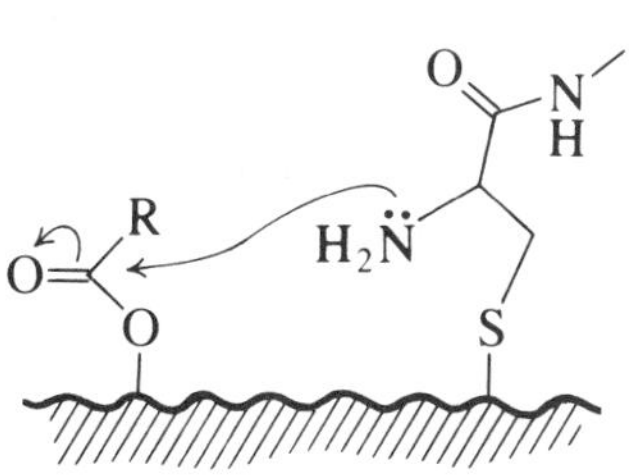

Recently, a variant of this methodology was developed and involves a *thiol capture* for the linkage of a pair of peptide fragments. This is illustrated below. Typically the reaction is carried out in DMSO under N_2 atmosphere in the dark and in the presence of $AgNO_3$ to limit to a minimum the formation of by-products via disulfide interchange during the amide bond formation (20b).

In a different strategy, Koga has also developed an efficient synthesis of small peptide fragments using a functionalized crown ether as the structural support (see Chapter 1).

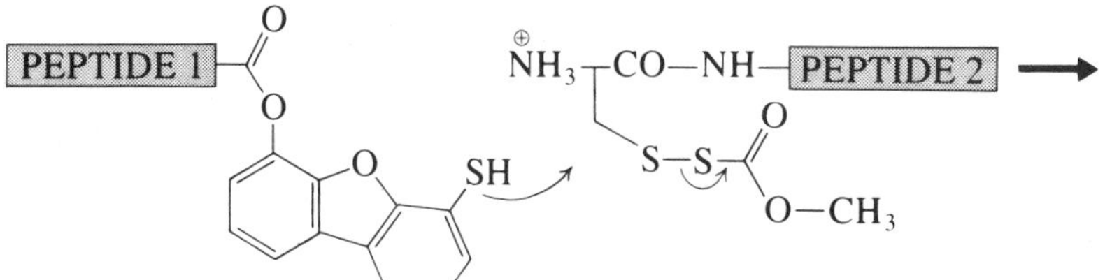

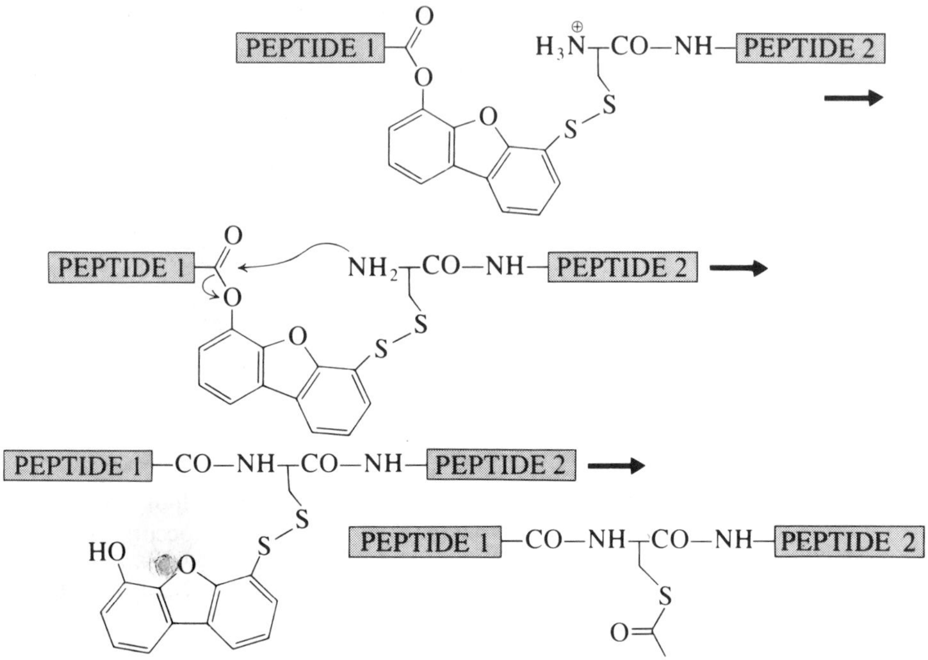

PEPTIDE 1 ... $H_3\overset{\oplus}{N}$—CO—NH—PEPTIDE 2

PEPTIDE 1 ... NH_2—CO—NH—PEPTIDE 2

PEPTIDE 1—CO—NH—CO—NH—PEPTIDE 2

HO

PEPTIDE 1—CO—NH—CO—NH—PEPTIDE 2

Basically, new biological macromolecules with catalytic activity can be created artificially using two approaches. The first exploits a system that selects a few catalytically active biomolecules from a large pool of randomly generated molecules. Catalytic antibodies are obtained that way and are discussed in Section 2.8. The second approach involves the

H_2N-Leu-Ala-Lys-Leu-Leu-Lys-Ala-Leu-Ala-Lys-Leu-Leu-Lys-Lys-$CONH_2$

Because of the properly designed distribution of hydrophobic (Leu) and hydrophilic (Lys) residues, such peptide can fold into an α-helix form.

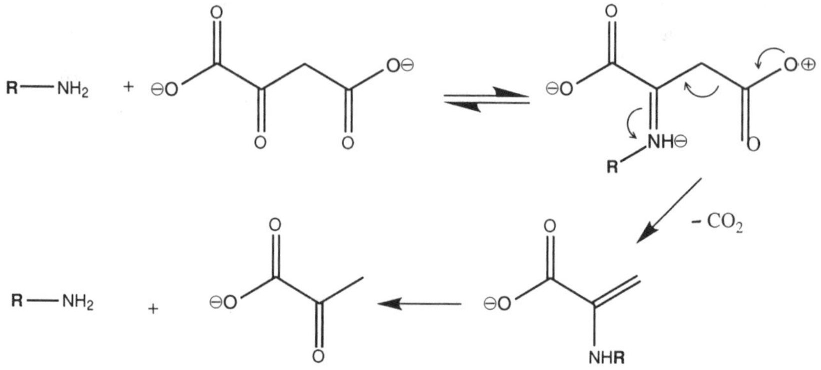

R—NH_2 + ⊖O ... O⊖ ⇌ ⊖O ... NH⊖ R ... O⊕

$- CO_2$

R—NH_2 + ⊖O ... ← ⊖O ... NHR

Fig. 2.5. Rational design of an artificial catalytic polypeptide for the decarboxylation of oxaloacetic acid (21). **R** represents the polypeptide chain.

rational design of a biomolecule that can fold in solution and present to the substrate an array of catalytic functional groups. It is with this objective in mind that S.A. Benner and colleagues recently reported the synthesis of a rationally designed 14-amino acid polypeptide that can catalyze the decarboxylation of oxaloacetate, a β-ketoacid, via an imine intermediate with the catalyst (21). The sequence of the polypeptide and the reaction involved are illustrated in Fig. 2.5.

This well-designed polypeptide, made of lysine and leucine residues, curls into a helix shape, exposing a lysine residue to accomodate the substrate for catalysis. The formation of the imine intermediate by this artificial oxaloacetate decarboxylase is 3 to 4 orders of magnitude faster than can be achieved with simple amines as catalysts. The additional catalytic power of this synthetic peptide is probably due to coulombic binding between its catalytic sites and the anionic transition state following the carbinolamine formation. This approach may have an application in the industrial synthesis of phenylalanine.

2.5 Asymmetric Synthesis of α-Amino Acids

In the course of chemical evolution, nature must have developed selective methods of amino acid synthesis and specific recognition. In this respect, what kind of chemical methods do we have presently to prepare amino acids in optically pure form and to selectively distinguish among enantiomers? We will therefore examine in this section two approaches to asymmetric synthesis of amino acids using the concept of asymmetric induction and specific metal ion complexation.

It is generally accepted that one of the remaining challenges in organic chemistry is to induce efficient asymmetric synthesis on a prochiral precursor, much as the enzyme does. One way to get around this problem is to use a chiral reagent that would produce diastereotopic interactions with a reactant molecule and lead to an asymmetric product.

The fact that α-amino acids are the constituents of proteins bestows on them a great importance. Eight amino acids are classified as "essential" because they cannot be synthesized by mammals and thus must come from food. They are isoleucine, leucine, lysine, methionine, valine, threonine, phenylalanine, and tryptophan. They are all of L-configuration and it is important to dispose of chemical methods giving access to these amino acids. Twenty years ago, mainly biochemical methods were accessible, based on resolution of racemic mixtures.

2.5.1 Corey's Method

In 1968, a French team of chemists under the direction of H.B. Kagan reported the asymmetric synthesis of L-aspartic acid, starting with an optically active amino alcohol (22). The synthesis is outlined in Fig. 2.6.

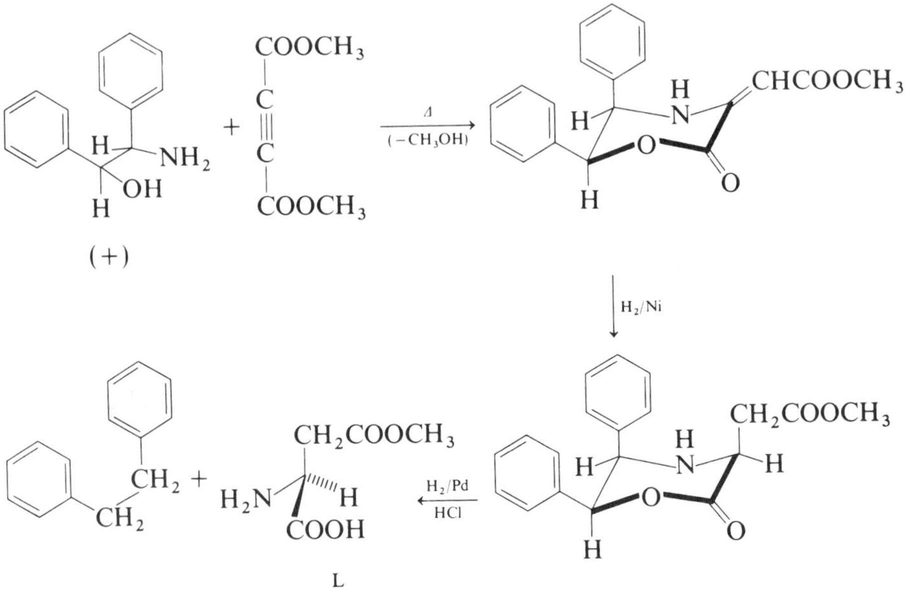

Fig. 2.6. Kagan's synthesis of L-aspartate monomethyl ester (22).

The optically active unsaturated cyclic precursor is hydrogenated from the least hindered side of the double bond in very good yield creating a new asymmetric carbon. The stereochemistry of this reduction is insured by the presence of the bulky pseudoaxial phenyl ring. The chlorohydrate of monomethyl L-aspartate is obtained in 98% optically pure form.* One inconvenience to this approach resides in the loss of the starting asymmetric amino alcohol which is transformed to diphenyl ethane. In addition, the extension to the other α-amino acids appears uncertain.

There was thus a need for a more general method where the asymmetric reagent will not be sacrificed and that insures the recovery of the starting material. To overcome these problems, E.J. Corey and collaborators (23) from Harvard University in 1970 proposed the methodology depicted in Fig. 2.7. The precursor of the α-amino acid is the corresponding α-keto acid. The α-keto acid is combined with a chiral

* The percent optical purity of a compound is defined as

$$\% \text{ optical purity} = \frac{\text{specific rotation of the enantiomeric mixture}}{\text{specific rotation of one pure enantiomer}} \times 100$$

Expressed differently, it means that a racemic mixture that is 50% *R*-isomer and 50% *S*-isomer has 0% optical purity, and a 90% optical yield corresponds to a mixture of 95% of one isomer contaminated with 5% of its antipode.

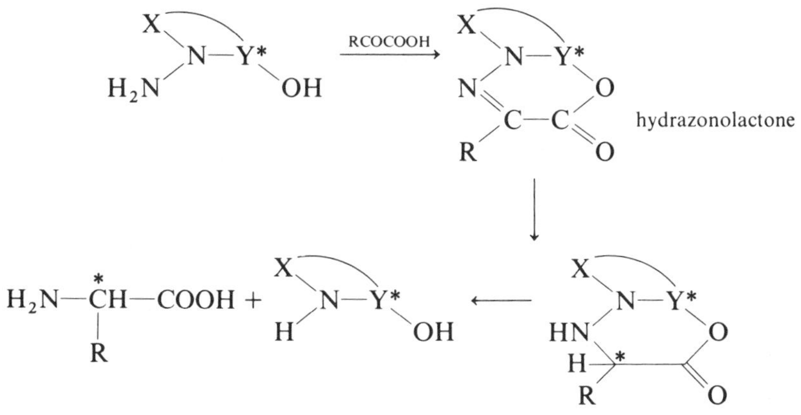

Fig. 2.7. Outline of Corey's methodology (23).*, chiral center.

reagent to form a ring of minimal size possessing a hydrazone function. Specific reduction of the double bond will produce the chiral carbon of the corresponding α-amino acid. Hydrogenolysis of this intermediate will generate the chiral α-amino acid and a chiral secondary amino alcohol, convertible to the original chiral reagent.

To apply this methodology, the chiral reagent chosen was a bicyclic indoline structure and its synthesis is given below:

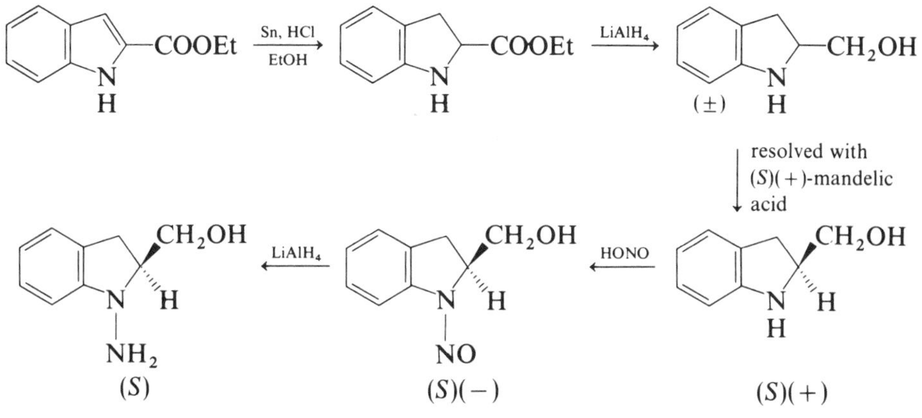

Its absolute stereochemistry was proved by correlation with L(−)-phenylalanine. This simple rigid structure is necessary to insure a good steric control in the reduction of the C=N double bond by asymmetric induction.

Based on this model, the asymmetric synthesis of D-alanine was carried out. All the steps proceed in good yield and the amino acid obtained has an optical purity of 80%. Figure 2.8 outlines this synthesis.

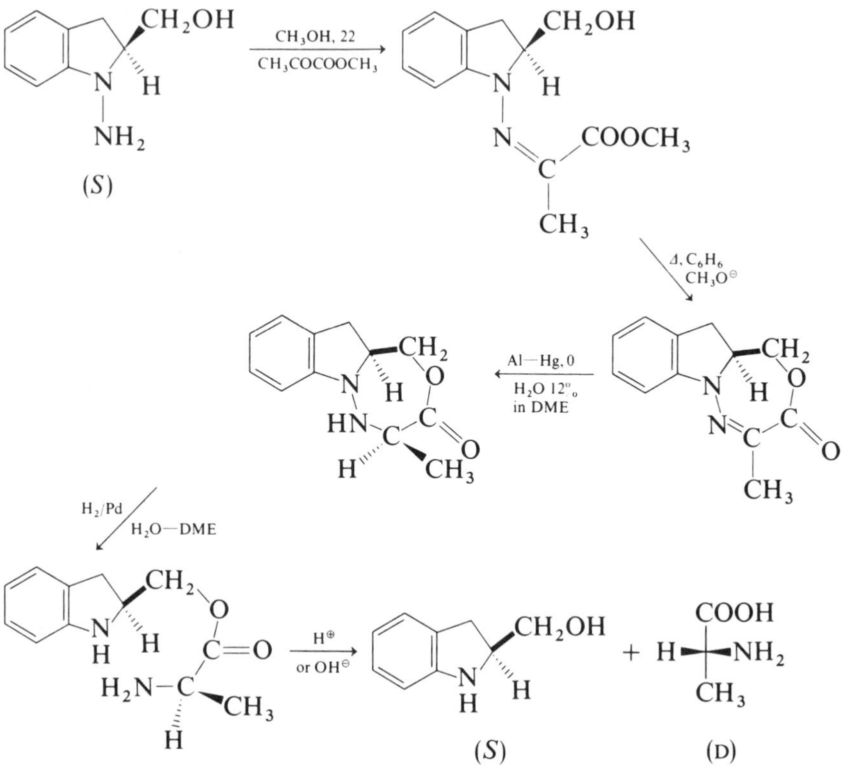
CH₂OH
N H
NH₂
(S)
CH₃OH, 22
CH₃COCOOCH₃
CH₂OH
N H
N
COOCH₃
C
CH₃
Δ, C₆H₆
CH₃O⊖
CH₂
O
N H
N C C
O
CH₃
Al—Hg, 0
H₂O 12%
in DME
CH₂
O
N H
HN C C
O
H CH₃
H₂/Pd
H₂O—DME
CH₂
O
N
H H C=O
H₂N—C CH₃
H
H⊕
or OH⊖
CH₂OH
N H
H
(S)
COOH
+ H NH₂
CH₃
(D)

Fig. 2.8. Asymmetric synthesis of D-alanine (23).

The efficiency of the synthesis could in fact be improved by a modification of the starting material, utilizing a more crowded chiral reagent with two centers of asymmetry:

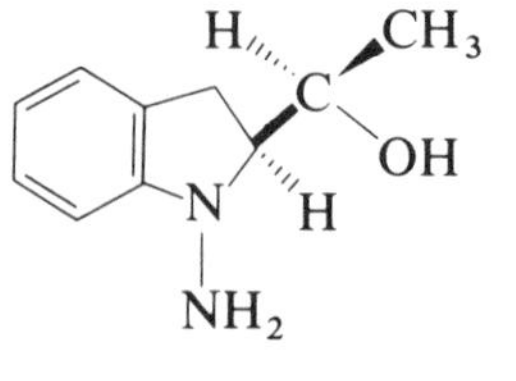
H
CH₃
C
OH
N H
NH₂

The optical yield of the conversion increased to 96% for D-alanine and to 97% for D-valine. Thus, the extra methyl in the front side of the corresponding hydrazonolactone favors a more selective reduction of the double bond from the back side of the tricyclic intermediate.

Another advantage of this approach is the possibility of preparing stereospecifically a series (by varying the R group of the keto acid precursor at will) of α-deuterated amino acids by simply undertaking the chemical reduction in heavy water.

2.5.2 Rhodium(I) Catalyst

Another elegant and useful recent method of production of optically active amino acids is by homogeneous catalytic hydrogenation using rhodium(I) complexes as the catalyst. Indeed, the discovery that the $[Rh(Ph_3P)_3Cl]$ complex (Wilkinson's catalyst) and related derivatives were efficient homogeneous hydrogenation catalysts for many olefins provided a system that could potentially be modified into asymmetric catalysts.

The challenge was to find a catalyst that would stereospecifically reduce an alkene such as

$$\begin{matrix} R_1 \\ R_2 \end{matrix} \!\!> C{=}C <\!\! \begin{matrix} COOR_3 \\ NH{-}R_4 \end{matrix}$$

Early attempts using unidentate phosphines gave only low optical yield. Bidentate systems are more promising but there is still a problem of flexibility and rapid interconversion:

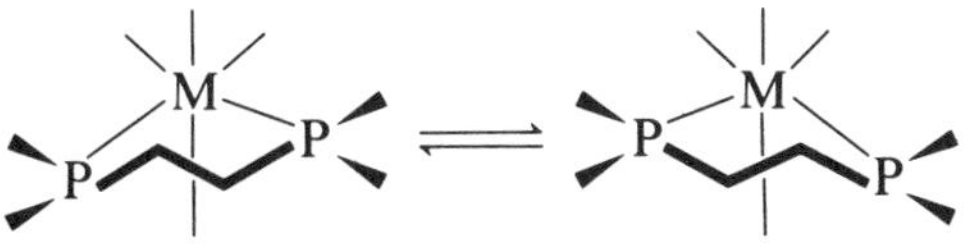

However, the incorporation of chiral bidentate phosphine and phosphite ligands should prevent this equilibrium. For example, if the aliphatic link is substituted and thereby an asymmetric carbon center is produced, the chelate ring may be fixed into a single, static chiral conformation, with the requirement that the substituent be equatorially disposed. Also, a chiral substituent should transmit its chirality to the ring in such a way that the whole molecular framework is twisted into a single chiral conformation. A prochiral olefin would then coordinate this chiral metal complex in a preferential orientation, leading to an asymmetric reduction of the double bond. It was these principles that led B. Bosnich (24) from the University of Toronto in 1977 (now at the University of Chicago), to prepare the chiral ligand (2*S*,3*S*)-bis(diphenylphosphino) butane,

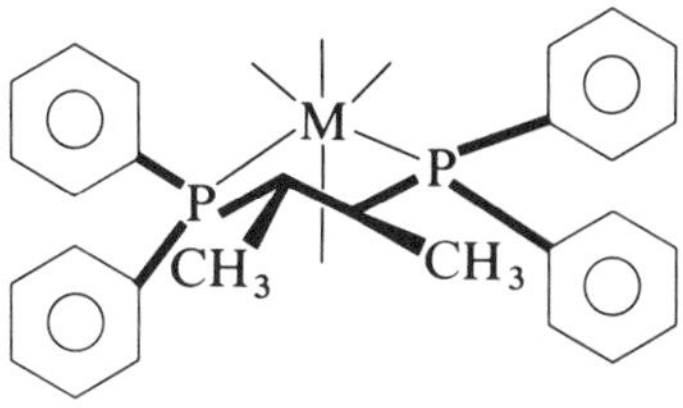

(*S*,*S*)-chiraphos chelated catalyst

abbreviated (*S*,*S*)-chiraphos. It was used to make the corresponding (*S*,*S*)-chiraphos chelate ring, M being a rhodium metal ion.

The two methyl groups are equatorially oriented. By fixing the ring in one conformation only, a chirality is induced in the ring in the disposition of the phenyl residues to allow diastereotopic complex formation with olefins. The synthesis of the catalyst is summarized in Fig. 2.9.

Lithium diphenylphosphide displaces the tosylate groups by an S_N2 mechanism, Ni(II) ions are used to separate the desired product from side products via an insoluble Ni(II) complex, and CN^- ions are finally used to liberate the metal. The (*S*,*S*)-chiraphos is a solid obtained in 30% yield, but that is oxidized slowly by air in solution. It is therefore

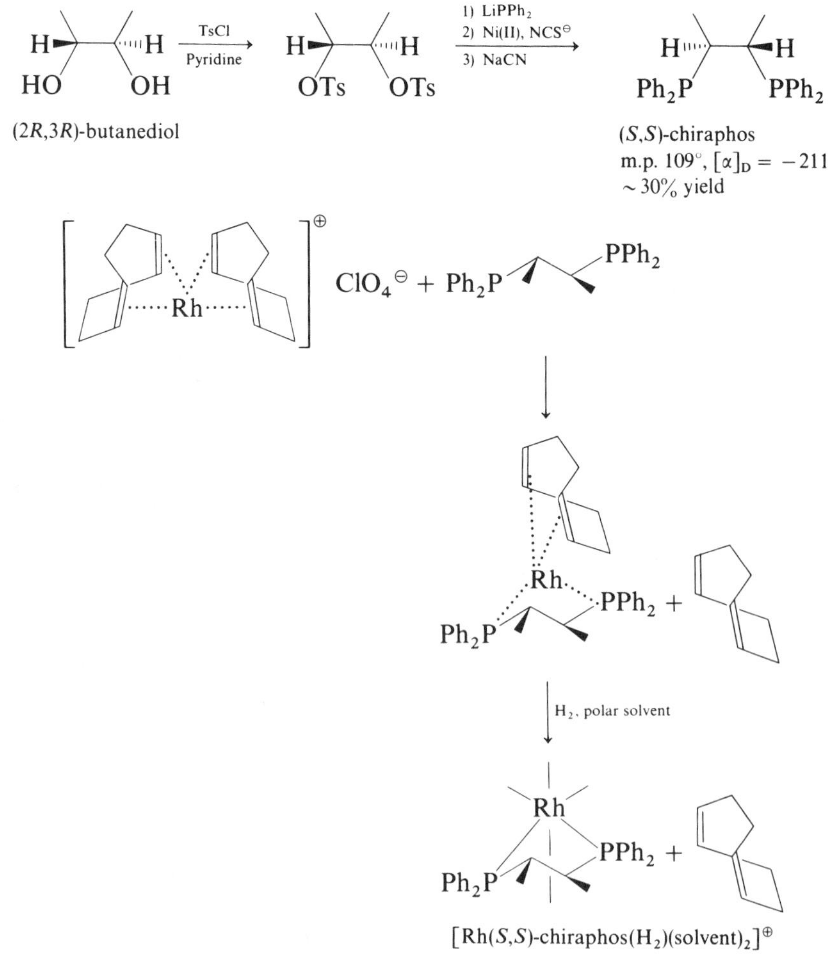

(2*R*,3*R*)-butanediol →(TsCl, Pyridine)→ ditosylate →(1) $LiPPh_2$; 2) Ni(II), $NCS^\ominus$; 3) NaCN)→ (*S*,*S*)-chiraphos, m.p. 109°, $[\alpha]_D = -211$, ~30% yield

$[Rh(\text{cyclooctadiene})_2]^\oplus ClO_4^\ominus$ + Ph_2P–CH(CH$_3$)–CH(CH$_3$)–PPh_2 → Rh(chiraphos)(diene) + diene →(H_2, polar solvent)→ $[Rh(S,S\text{-chiraphos})(H_2)(\text{solvent})_2]^\oplus$ + diene

Fig. 2.9. Preparation of Rh(I) catalyst (24).

immediately converted to the rhodium(I) complex by a displacement reaction involving a di-1,5-cyclooctadiene rhodium(I) precursor. The final product is obtained as an orange-red solid that is stable if kept under nitrogen at 0° to 4°C. It is this species that is capable of hydrogenating a variety of olefins under catalytic conditions. The ratio of catalyst to substrate used is generally 1:100 and the reaction is carried out under nitrogen, at 25°C for 1 to 24 hr. The precursors for the α-amino acids are α-*N*-acylaminoacrylic acids and esters. Table 2.4 shows the optical yield of amino acids obtained. Only *R*-amino acids are obtained with the (*S*,*S*)-chiraphos catalyst. In principle, the corresponding antipode, the (*R*,*R*)-chiraphos catalyst, should give natural *S*-amino acids, and indeed it does.

Table 2.4 shows that the yield of the reaction is sensitive to the nature of the *N*-acyl substituent, the β-vinyl substituent, and the polarity of the solvent. However, the reaction is always stereospecific in that only *cis* hydrogen addition to the olefins is observed. This was verified by

Table 2.4. Optical Yield (%) of *R*-Amino Acid Obtained (24)

Starting olefin	Amino acid	Solvent	
		THF	EtOH
COOH; NHCO—	Phe	99	95
COOEt; NHCO—	Phe	83	—
COOH; $NHCOCH_3$	Ala	88	91
COOH; $NHCOCH_3$	Leu	100	93
COOH; NHCO—	Leu	87	72

undertaking the reaction with $D_2(^2H_2)$ instead of H_2. The resulting product is the [2*R*,3*R*-$^2H_2$]-*N*-benzoylphenylalanine from the corresponding acrylic acid precursor:

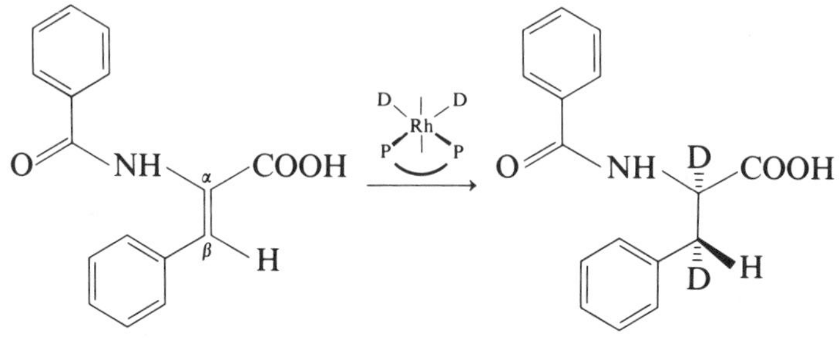

[2*R*,3*R*-$^2H_2$]

This asymmetric deuteration is believed to proceed in two steps involving first the addition of a hydride to the olefin to produce a rhodium alkyl bond. This is rapidly cleaved by the insertion of a proton into the metal alkyl bond:

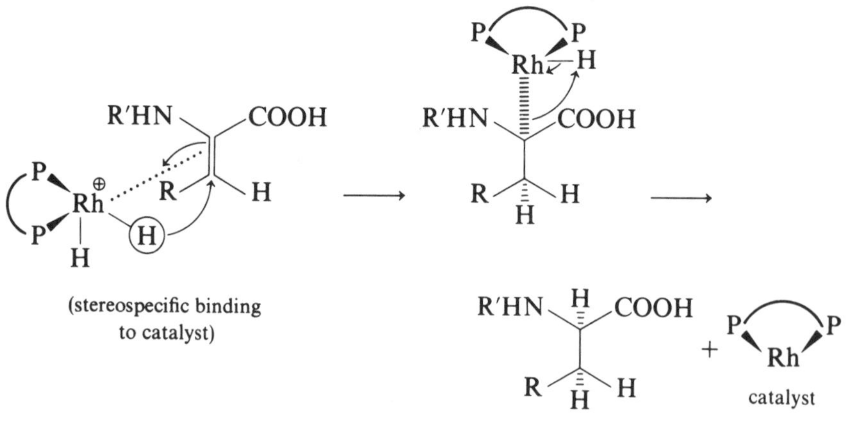

(stereospecific binding to catalyst)

catalyst

Despite the two-step mechanism, the process appears to be completely stereospecific. Furthermore, if a leucine precursor is deuterated in an analogous fashion, the deuterium appears only at the α and β centers. The γ-position carries no deuterium, implying, as expected, that no double bond migration occurs during the reduction. An important gain that is intrinsic to this mechanism is that the optical purity of the β-center is linked to that of the α-center.

The reasons for the effectiveness of the present catalyst in asymmetric hydrogenation of amino acid precursors can be summarized as follows: (1) the conformational rigidity of the chelated diphosphine ligand; (2) the dissymmetric orientation of the phenyl groups, held so by the 5-membered puckered chelate ring. These factors are the major source of the diastereotopic interaction with prochiral substrates to give: (1) a *cis*-endo stereospecific hydrogenation, producing both α- and β-carbons chiral with $^2H_2$*; (b) only *R*-amino acid, irrespective of the solvent.

The success of this approach is helpful for the preparation of specific β-deuterated amino acids. The following transformation is an outline of the method for racemizing the α-carbon center of *N*-benzoyl-phenylalanine via an oxazoline intermediate. The two diastereoisomers can be separated by conventional means.

[2*R*,3*R*-$^2H_2$]

$\xrightarrow{Ac_2O}$

1) Pyr/D_2O
2) $OD^{\ominus}/CH_3OD/D_2O$
3) HCl/H_2O

[2*R*,*S*,3*R*-$^2H_2$]

Thus, as a bonus, this approach allows the rational design of selectively mono- or dideuterated amino acids and analogs. These chirally labeled molecules can be used to study the mechanistic details of biosynthetic

* One should realize that the catalyst approaches the prochiral double bond from the α-*si*, β-*re* face. (For an introduction to the nomenclature used to differentiate the two carbons of a prochiral double bond, see Section 4.2.1.) Deuterium (^2H) and tritium (^3H) atoms are heavy isotopes of hydrogen (^1H), but for the clarity of the structural representations only, the symbols D and T are used throughout this text.

pathways involving amino acid precursors. This approach has been recently extended (25) to allow the synthesis of chiral methyl lactic acid in which the three methyl hydrogens are replaced by a proton, deuteron, and triton.

In 1978, J.K. Stille and his group proposed an interesting extension of the concept of asymmetric synthesis via rhodium complexation by attaching the metallic site to an insoluble polymer (26). The main advantage of this modification is the possibility of recovering the optically active phosphine-rhodium complex catalyst.

One important and challenging problem in polymer-supported catalysis is the proper choice of the polymer matrix and the synthesis of the catalyst site in the matrix. The method used here consists of the introduction of a reactive site on a cross-linked polystyrene bead followed by the reaction of an optically active phosphine-containing ligand at the site. The approach is summarized in Fig. 2.10. The reaction of (−)-1,4-ditosylthreitol with 4-vinylbenzaldehyde affords 2-*p*-styryl-4,5-bis(tosyloxymethyl)-1,3-dioxolane, which is copolymerized radically with hydroxyethyl methacrylate to incorporate 8 mole % of the styryl moiety in the cross-linked copolymer. Further treatment with sodium diphenylphosphide to react with all the hydroxyl functions plus the tosylate groups gives after neutralization a hydrophilic polymer bearing the optically active 4,5-bis(diphenylphosphinomethyl)-1,3-dioxalane ligand. Exchange of rhodium(I) onto the polymer with $[(C_2H_4)_2RhCl]_2$ gives the desired polymer-attached catalyst that swells in alcohol and other polar solvents to allow the penetration of the substrate.

Typical hydrogenations are carried out at 25°C with 1 to 2.5 atm of hydrogen and the catalyst and with an olefin to rhodium ratio of about 50. This way, α-*N*-acylaminoacrylic acids in ethanol are converted to the amino acid derivatives. The optical yields are comparable to those obtained with the previously mentioned homogeneous catalyst. The same absolute (*R*) configuration of the products is observed. The main advantage is that the insoluble catalyst can be reused many times. It may be recovered from the reaction mixture by filtration, under an inert atmosphere, with no loss of catalytic activity or optical purity in the hydrogenated product.

The ultimate variant of this approach was obtained recently by G.M. Whitesides, from M.I.T. (27, 28). He constructed an asymmetric hydrogenation catalyst based on embedding an achiral diphosphine-rhodium(I) moiety at a specific site in a protein. In this case the protein tertiary structure provides the chirality required for enantioselective hydrogenation.

The well-characterized protein avidin, composed of four identical subunits, each of which binds biotin and many of its derivatives, was used. For this, a hydroxysuccinimide substituted biotin was converted to a chelating diphosphine and complexed with rhodium(I) by the following sequence:

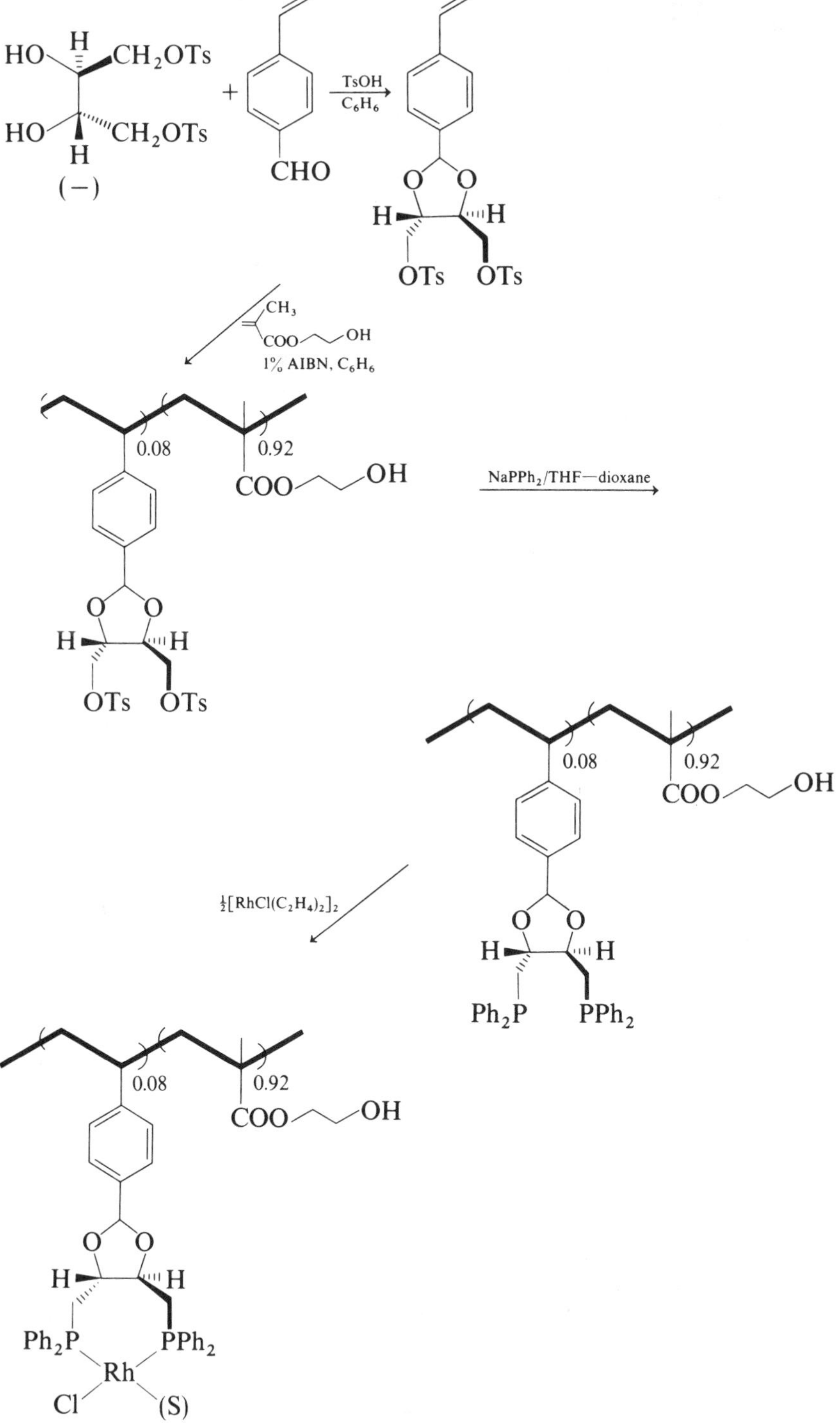

Fig. 2.10. Synthesis of a polymer-supported Rh(I) optically active catalysis (26). AIBN, azadiisobutyronitrile, a free radical initiator for polymerization. S, solvent.

biotin derivative

1) $HN(CH_2CH_2PPh_2)_2 \cdot HCl$
DMF, Et_3N
84%

2) $NBDRh(I)^{\oplus}Tf^{\ominus}$
THF

$Tf^{\ominus}$

$Tf^{\ominus}$ = triflate ion ($CF_3SO_3^{\ominus}$)
NBD = norbornadiene

In this procedure the diphosphine intermediate serves as the basis for the elaboration of a water-soluble rhodium-based homogeneous hydrogenation catalyst. The enantioselectivity of the catalyst was tested by the reduction of α-acetamidoacrylic acid to *N*-acetylalanine. The presence of avidin resulted in a definite increase in activity of the catalyst and in the production of the *S*-enantiomer (natural amino acid) in 40% excess.

COOH, $NHCOCH_3$ → (H_2, avidin·biotin·$RhNBD^{\oplus}TF^{\ominus}$; pH 7.0, phosphate buffer 0.1 *M*, 0°, 48 hr) → CH_3, H, COOH, $NHCOCH_3$

S-isomer

The presence of other enzymes such as lysozyme or carbonic anhydrase had no significant influence on enantioselectivity.

Hence, Whitesides showed that it is possible to carry out in aqueous solution homogeneous hydrogenation using a diphosphinerhodium(I) catalyst associated with a protein. In addition, the chirality of the protein is capable of inducing significant enantioselectivity in the reduction. Finally, as the author pointed out, this technique developed to bind transition metals to specific sites in proteins may find use in biological and clinical chemistry unrelated to asymmetric synthesis (27).

2.5.3 New Developments in Asymmetric Synthesis

Asymmetric syntheses are reactions that can produce optically pure substances from symmetrically constituted compounds with the aid of optically active intermediates (29). Consequently, the asymmetry is induced from an auxiliary material with an asymmetric center. The inductor of asymmetry could also be part of the main molecule or could come from a different compound. Examples of the two aspects will be presented. A general illustration of a set of chemical operations that conform to this definition is shown in the following equations:

achiral

$+H-AX_c$; 1) base, 2) El^+ ; $-H-AX_c$

formation of a metal enolate and trapping by an incoming electrophile (El^+). The chirality of the auxiliary ligand (AX_C) will dictate the outcome of the stereochemistry of the end product.

one chiral acid in large enantiomeric excess

In this example, a carboxylic acid is condensed with an optically pure chiral auxiliary ($H-AX_c$). In the presence of a base, a diastereoselective enolate-mediated C—C bond formation occurs and finally the chiral ligand is removed from the system, affording an enantiomerically enriched α-substituted carboxylic acid. In this instance, the chiral auxiliary (AX_c) is employed in the "intermediate" stage of the set of chemical operations to exert absolute stereochemical control on the new bond formation (30).

Enzymes are systematically using auxiliary or subsites to asymmetrically induce chirality and specificity to substrates. In the laboratory, different strategies have been developed to do asymmetric syntheses. The different approaches will be presented, followed by some more-specific examples, particularly in the field of amino acid synthesis.

One should keep in mind that for practical reasons, a basic requirement in asymmetric synthesis using stoechiometric amounts of a chiral auxiliary is that this reagent be cheap and easily available.

The first two methodologies have already been presented in Section 2.5.1 and will not be commented on here. Furthermore, it will become evident that the other methods are by far more versatile. Since the first-reported oxazoline-mediated asymmetric synthesis, considerable attention has been given to the development and utilization of new versatile

STARTING MATERIAL	CHIRAL AUXILIARY TEMPLATE	END PRODUCT
Kagan's methodology, 1968 (22) $CH_3OOC{-}C{\equiv}C{-}COOCH_3$	(+) via a cyclic intermediate	L
Corey's methodology, 1970 (23)	*S* via a cyclic hydrazonolactone	D
Meyers's methodology 1974 (31)	via metalation of an oxazoline	72-82% ee* α,α′-disubstituted carboxylic acid
Yamada's methodology 1977 (32)	via the hydrazone of *S*-AMP (aminomethoxyproline)	90% ee

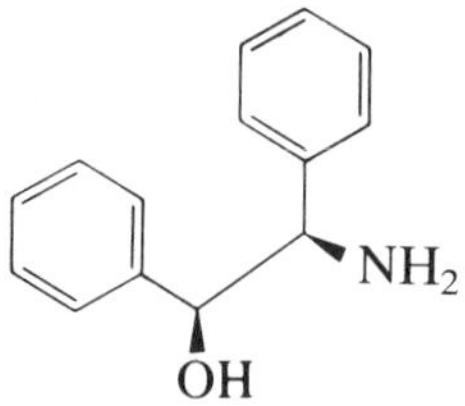

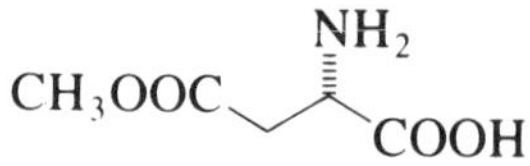

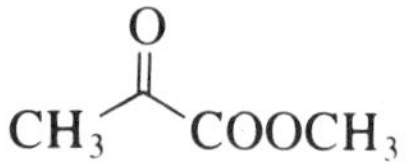

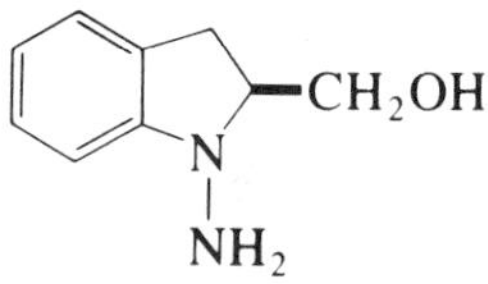

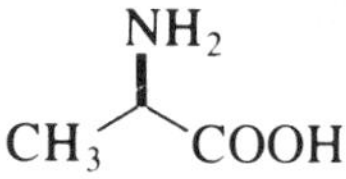

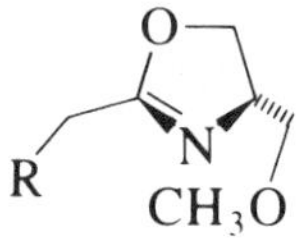

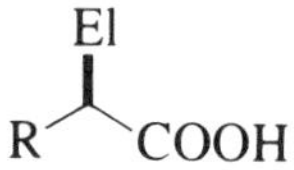

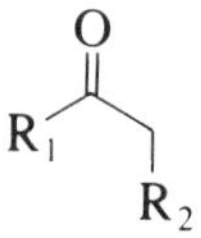

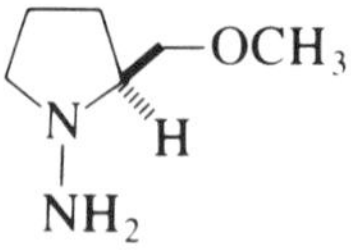

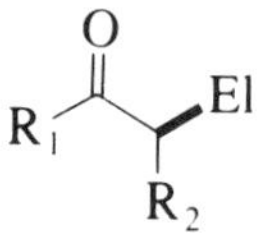

* The percent in enantiomeric excess (ee) is defined as

$$\%\text{ee} = \frac{R\text{-enantiomer} - S\text{-enantiomer}}{R\text{-enantiomer} + S\text{-enantiomer}} \times 100$$

Therefore, 50% ee in *R*-isomer represents a mixture of 75% *R*- plus 25% *S*-isomers.

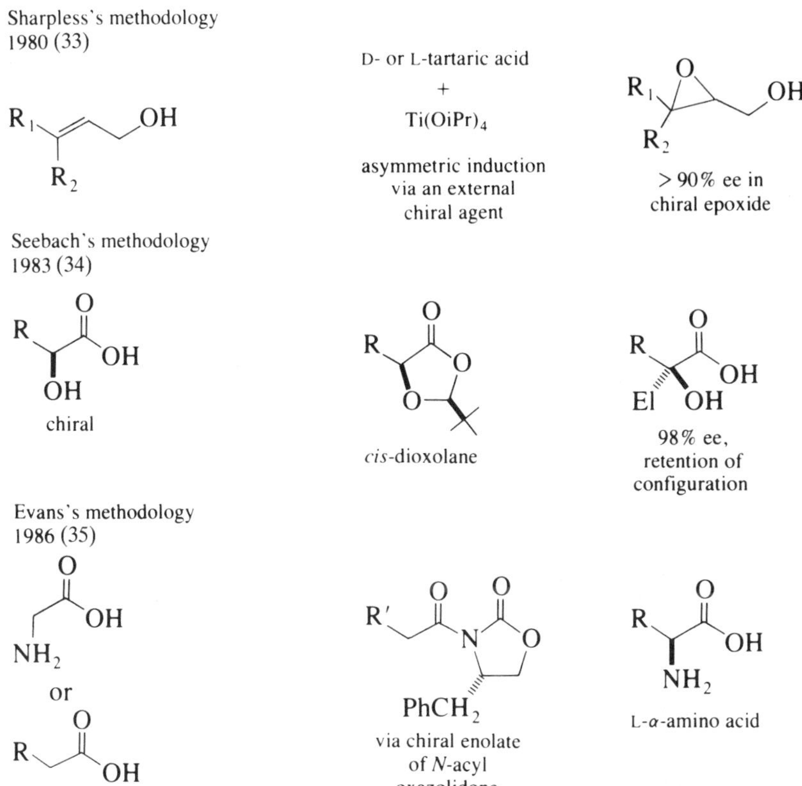

substrates as chiral templates. The Meyers's approach uses (1*S*,2*S*)-1-phenyl-2-amino-1,3-propanol to make the oxazoline. Metalation of the chiral oxazoline with butyllithium gives a stabilized enolate ion where the Z-isomer is largely predominant (95:5). The induction of chirality is caused by the presence of a bulky phenyl ring on one surface of the intermediate.

R1 O H N Li O R_2—X CH_3

the electrophile approaches from the bottom face of the molecule

Z-isomer predominantly

In the work of Yamada, D- or L-proline is used to make the auxiliary template in either optically active form. Again, trapping the anion with various electrophiles leads to regiospecific C—C bond formation. The intermediate would have this structure:

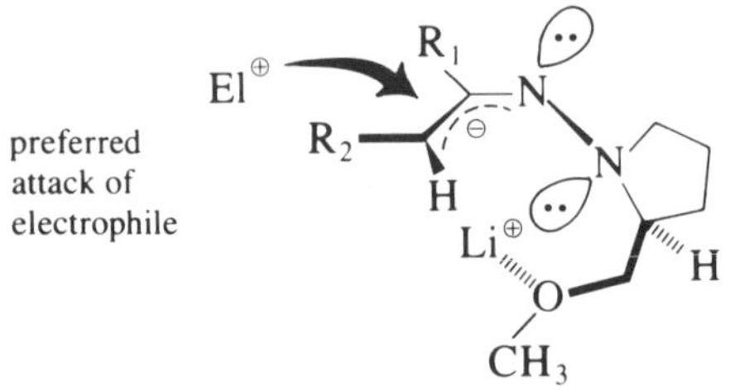

preferred attack of electrophile

At this point it might be instructive to present the work of Schöllkopf (36), one of the most straightforward examples of what is referred to as *intraannular chirality transfer*. In this case, the chiral auxiliary used is the amino acid L-valine. It is condensed with glycine to form a bislactim ether. Enolization and subsequent reaction with the electrophilic acetophenone gives an intermediate with a high level (95%) of diastereoface selectivity. Upon dehydration and hydrolysis, α-vinylamino acid with optical purity greater than 95% is obtained. Easy access to these new amino acids have gained increasing importance as potential enzyme inhibitors (see Chapter 7). The system gives a remarkably high level of 1,4-asymmetric induction. The π-facial selection is dictated by the resident alkyl substituent and bisdeprotonation is strongly disfavored because of the antiaromatic character of the resultant eight π-electron dianion.

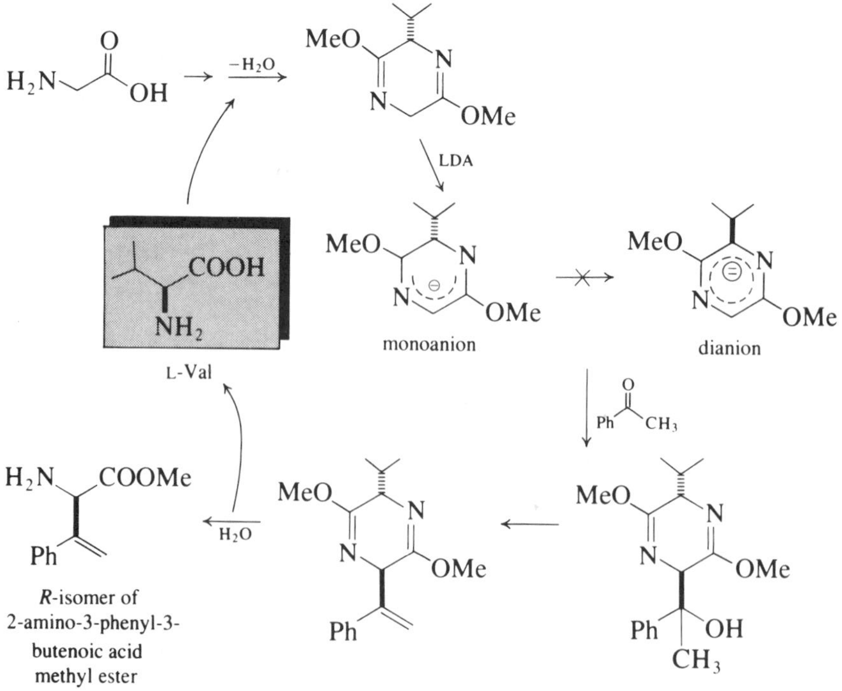

L-Val

monoanion

dianion

R-isomer of 2-amino-3-phenyl-3-butenoic acid methyl ester

The approach developed by Seebach is similar but very cleverly designed to have access to substituted α-hydroxy acids, again by intraannular chirality transfer. The transformations start with either *S*-mandelic or *S*-lactic acid to build a derivatized counterpart with overall net retention of stereochemistry. Notice that the basic concept behind this asymmetric C—C bond formation is distinctly different from that outlined in the preceding example. In the present case, the internal relay of chirality (acid → dioxalane), which precedes the enolization step, conserved the predisposed diastereofacial chirality of the ring system.

R = Me, lactic
R = Ph, mandelic

cis-disubstituted 1,3-dioxolan-4-one

1) LDA
2) $PhCH_2Br$

$H_3O^{\oplus}$

95% optically pure (81% yield)

LDA ≡ lithium diisopropylamide

When addressing the issue of enolate π-facial selectivity in reactions with electrophiles, one is faced with an analysis of both stereoelectronic and steric effects in the diasteriomeric transition states. In recent years, the orientation of reagent approach to π-bonded systems has been reevaluated in the light of new data.

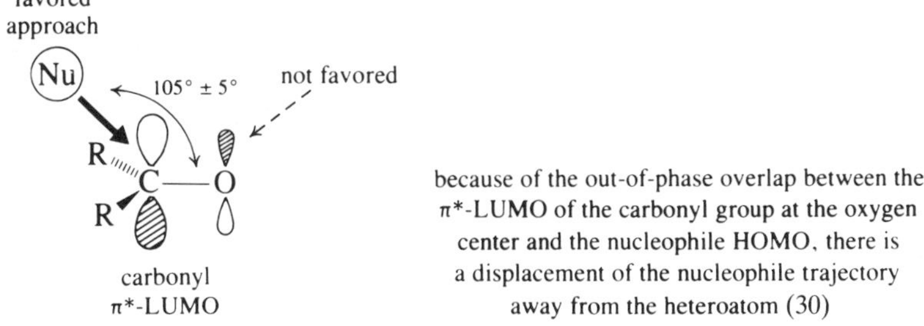

because of the out-of-phase overlap between the π*-LUMO of the carbonyl group at the oxygen center and the nucleophile HOMO, there is a displacement of the nucleophile trajectory away from the heteroatom (30)

Through the elegant crystallographic studies of Dunitz and co-workers, it is now accepted that *nonperpendicular* attach by nucleophiles on carbonyl centers is not only possible but preferable (37, 38).

For the enolate-electrophile reactions, similar secondary orbital considerations may also be applied. In the enolate HOMO, the related repulsive interaction between the electrophile LUMO and the enolate oxygen might provide an analogous perturbation away from perpendicularity.

nonperpendicular electrophilic trajectory

enolate π-HOMO

Baldwin also pointed out, maybe for similar reasons, the importance of trajectory considerations to synthetic organic chemistry. For instance, in the formation of 5-membered rings, the hydroxy ester below cyclized to give exclusively the lactone with loss of methanol. From observations on a vast array of compounds, he derived a set of qualitative rules that are now accepted as useful predictions (39). He concluded that the intramolecular attack prefers an angle of trajectory greater than 90° (EXO) for efficient cyclization at a trigonal carbon.

5-EXO-TRIG 5-ENDO-TRIG

With thiols, the situation is somewhat reversed because the sulfur atom is large, its electrons are polarizable, and as a consequence, the system accommodates better with a curved trajectory (smaller angle).

a thioether is formed

No trajectory theory caused more controversy than Koshland's "orbital steering" theory, particularly in the lactonization of bicyclic hydroxyacids such as:

HO COOH or OH COOH

This concept of orbital steering (40) was the subject of a long discussion in the first edition of the book and will not be covered here. In 1980, Menger revised all the structures and the arguments and presented strong evidence against the original theory (41). On the other hand, enzymes can impose orientations on interacting orbitals at active sites, thereby accounting for a large portion of enzymatic catalysis. The role played by water molecules is also important, and Lipscomb has pointed out the importance of solvent effects of transition states in high dielectric medium (42).

Coming back to amino acid chemistry, it is an attractive goal to do substitutions on an asymmetric center of simple chiral compounds without racemization. This is particularly important since amino acids and α- or β-hydroxyacids are among the abundantly available compounds in enantiomerically pure form. In 1981, Seebach's group (43) developed an economical method that does not require recovery, and recycling, of the mostly valuable chiral auxiliary to achieve the overall process of stereoselective substitution. He labeled the process "self-reproduction of the stereogenic center."

For instance, L-proline is condensed with privalaldehyde under acidic catalysis to give a bicyclic chiral lactone (44). Upon LDA treatment, a nonracemic enolate results that can undergo reactions with a variety of electrophiles *cis* (or *re*-facial) to the *t*-butyl group. The yields are moderately good but the diastereoselectivity is quantitative in most cases.

H N H COOH + CH_3—C(CH_3)(CH_3)—CHO $\xrightarrow{H^{\oplus}}$ H N O O

$\downarrow$ LDA −78°C

N O OLi $\xleftarrow{R-X}$ $\xleftarrow{H_3O^{\oplus}}$ R N H COOH

R—X ≡ PhCHO
$PhCH_2Br$
CH_3I

Subsequent removal of the aldehyde moiety provides the optically pure α-substituted L-proline derivatives possessing the same sense of chirality as the starting amino acid.

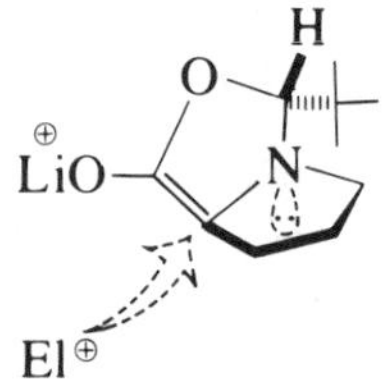

the bicyclic enolate intermediate has a V-shaped form with a pseudoequatorial *t*-butyl anchor group. Such arrangement favors attack of the electrophile almost exclusively from the convex face

LiO
El$^{\oplus}$

In a variant approach, the carboxylic group of serine (a β-hydroxyacid) is electrochemically decarboxylated to form enantiomerically pure products via an oxazolidine intermediate. Either nucleophilic or electrophilic reagents can be used to attach new substituents with *retention* of configuration at the former position of the carboxylic group (44).

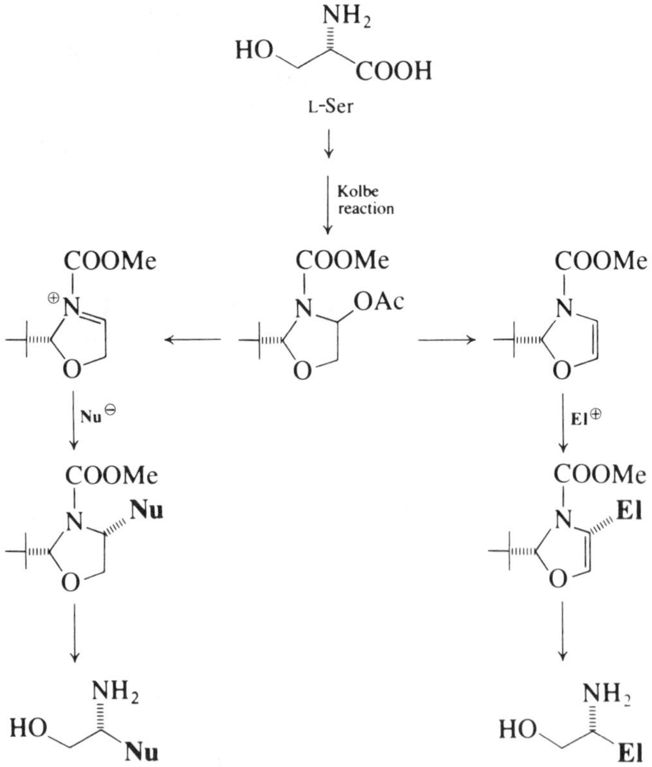

L-Ser

Kolbe reaction

COOMe

OAc

Nu$^{\ominus}$

El$^{\oplus}$

Nu

El

Recently Seebach made an interesting observation for enolate chemistry. He found that if one takes the tripeptide H-Ala-Gly-Gly-OH and adds LiCl in THF, an insoluble complex is formed. If then one adds

LDA, the enolate becomes soluble in THF, and upon further addition of methyl iodide, the amide bonds become *N*-methylated. This methodology has been applied to methylation of seven nitrogen atoms of the well-known immunosuppressive drug cyclosporine, a cyclic undecapeptide of 1202 in molecular weight without loss of optical purity. Limited structural activity studies have demonstrated that modification of the amino acid moiety dramatically affects the immunosuppressive activity to the resultant cyclosporine analog (45).

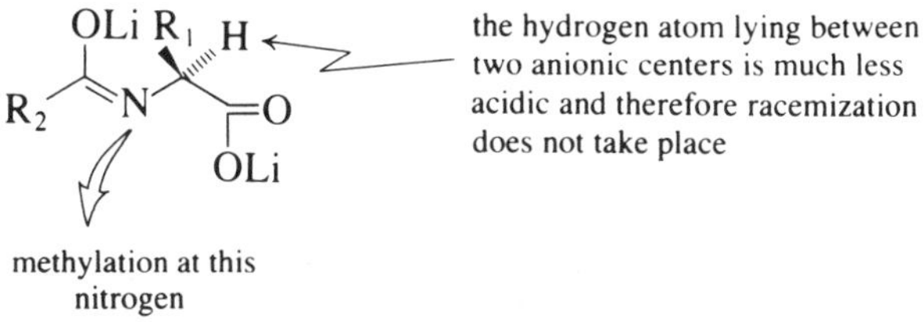

OLi R_1 H
R_2 N =O
OLi

the hydrogen atom lying between two anionic centers is much less acidic and therefore racemization does not take place

methylation at this nitrogen

Let's remind ourselves once more that the danger of racemization with loss of optical and biological activities is a constant preoccupation facing the peptide chemist. Furthermore, when dealing with a polyfunctional group or peptides, asymmetric synthesis at a given center is not always obvious.

We will end this section by a brief presentation of Evans's approach to asymmetric synthesis of amino acids.

One of the useful chiral auxiliaries used by Evans's group is an oxazolidinone. The approach was successfully applied to the synthesis

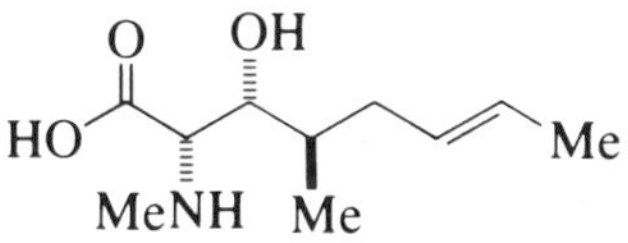

O OH
HO Me
MeNH Me

MeBmt

(4*R*)-4-[(*E*)-2-butenyl]-4,*N*-dimethyl-L-threonine

of the unusual C_9 amino acid MeBmt, found in the cyclic peptide cyclosporine. For this, chiral glycine synthon, as its derived stannous enolate, has been demonstrated to undergo a highly *syn* diastereoselective aldol addition reaction with representative aldehyde in good yield (46).

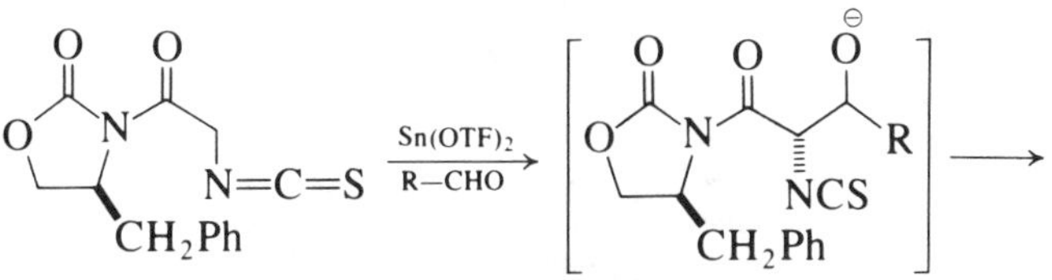

O O
O N
N=C=S
CH_2Ph

chiral glycine synthon

$Sn(OTF)_2$
R–CHO

O O O⊖
O N R
NCS
CH_2Ph

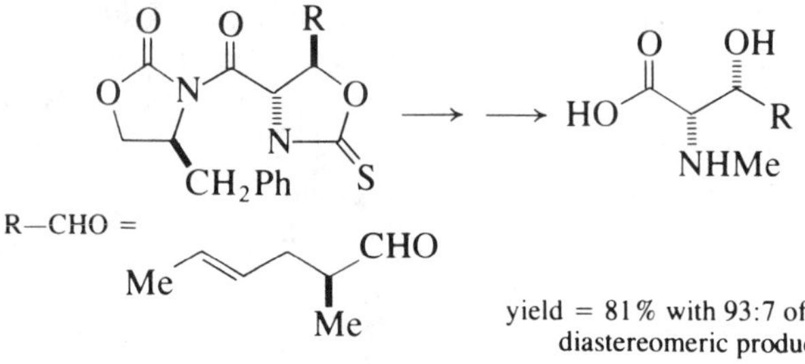

R—CHO =

yield = 81% with 93:7 of desired diastereomeric products

A retrosynthetic analysis of asymmetric amino acid synthesis can be pictured as:

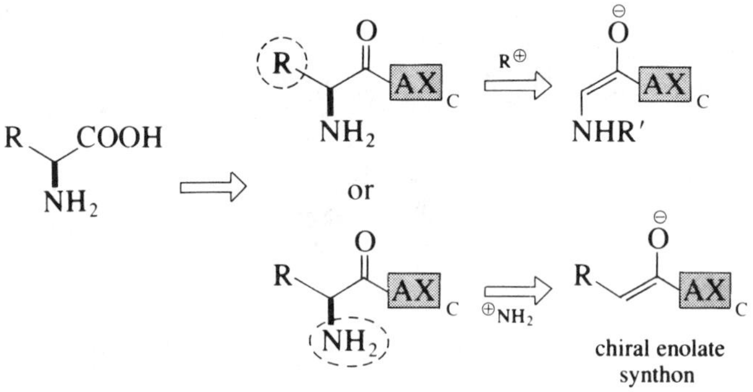

chiral enolate synthon

The amination of an acid analog, via its chiral enolate, has the advantage that the basic bond of the amino acid (the R group) is not perturbed in contrast to the other approach.

Among the various amine transfer group envisaged, Evans found that the trisyl-N_3 turned out to give the best results. The efficient electrophilic kinetic azido transfer to chiral enolates is now accepted as a convenient and general method for the asymmetric synthesis of α-amino acids (47).

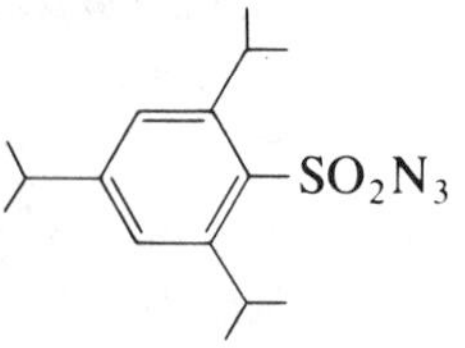

trisyl-N_3

the efficient electrophilic kinetic diazo transfer to chiral enolates is now accepted as a convenient and general method for the asymmetric synthesis of α-amino acids (47)

R COOH → R-CO-N(oxazolidinone)-CH$_2$Ph (chiral auxiliary template)

$KN(SiMe_3)_2$

recycling

1) $R'—N_3$ (R = trisyl)
2) AcONa

$LiOH/H_2O_2$

recovery of chiral auxiliary

reduction

α-D-amino acid

Examination of the enolate amination in more detail revealed that the structure of the triazine intermediate can be proven by NMR. The amide resonance keeps the two amide functions in the same plane giving enough time for triazine decomposition without racemization. This prevents racemization from occurring at the new asymmetric center. One can thus

3:1

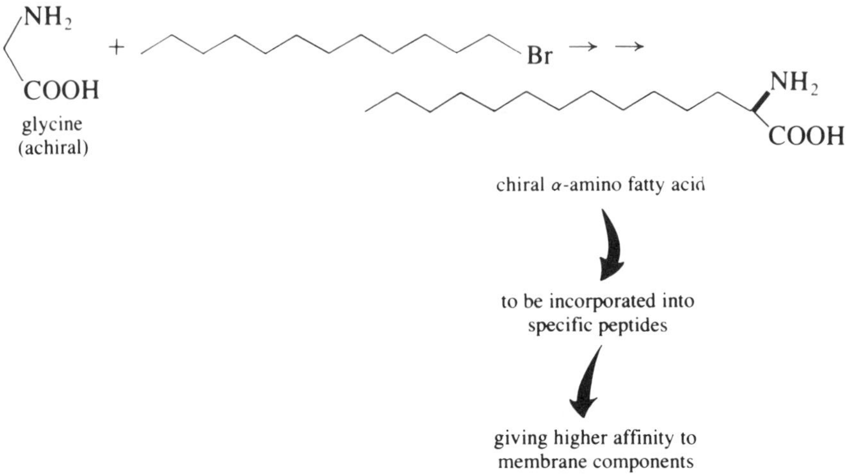

NH$_2$ COOH + Br → →

glycine (achiral)

NH$_2$ COOH

chiral α-amino fatty acid

to be incorporated into specific peptides

giving higher affinity to membrane components

do "molecular surgery" on chiral molecules without losing the stereogenic center. It became evident that such methodology also gives a rapid and direct access to highly hydrophobic or lipophilic chiral amino acids. This could open a new dimension to the overlapping field of membrane chemistry and lipoproteins. Interestingly, the slowreleasing substance leukotriene LTC$_4$ could be viewed as a highly unsaturated fatty acid branched on a tripeptide. Construction of chiral molecules using amino acid synthons is becoming more and more a la mode among synthetic chemists. A book has been recently devoted entirely to this subject (48).

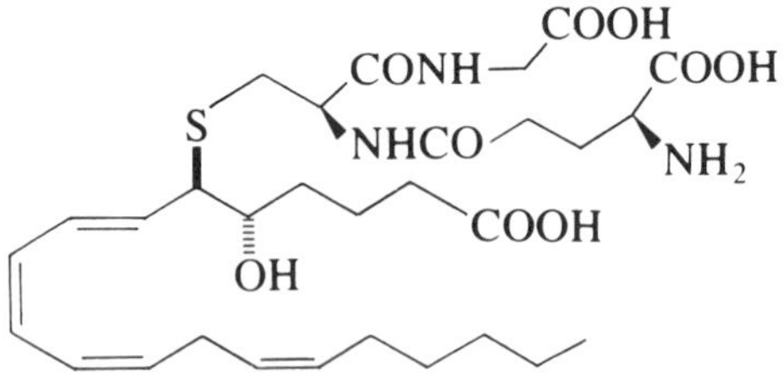

COOH CONH COOH S NHCO NH$_2$ COOH OH

LTC$_4$

It is not possible to treat asymmetric synthesis in organic chemistry without mentioning the extremely valuable contribution of H.C. Brown to the development of *organoboranes*. Indeed, organoboranes have turned out to be among the most versatile intermediates available to the organic chemist. Essentially all structural types of organic compounds can be made by utilizing organoborane intermediates, and enantiomerically pure compounds can be made via chiral organoboranes. Accordingly, hydroboration of *cis*-2-butene in diglyme with diisopinocamphenylborane (made from α-pinene) provided 2-butanol in 87% enantiomeric purity.

α-pinene

BH_3/0°C

1) [alkene] 2) H_2O_2

1) [alkene] 2) $HOSO_2NH_2$

HO H
R-(−)
87% ee

H_2N H
R-(−)
75% ee

The same chiral intermediate can also be readily converted into optically active 2-aminobutane. Other aspects will not be treated here since organoborane chemistry is adequately covered in most organic chemistry textbooks.

Finally, mention should be made of the stereoselective epoxidation of olefins by titanium-tartrate catalysts. This reaction is particularly impressive. How can high levels of asymmetric induction consistently be achieved in a system that tolerates wide variations in the steric features of its components? To date, no exception has been found among prochiral

D(−)-diethyl tartrate (unnatural)

top face attack

R_2 R_1 R_3 OH

+OOH

$Ti(DiPr)_4$
−20°/CH_2Cl_2

O OH
70–90% yield with more than 90% ee

bottom face attack

L(+)-diethyl tartrate (natural)

OH O

substrate to the enantioface selection rule shown below. Thus, this asymmetric epoxidation is a rare and truly successful blend of substrate generality and high enantioselectivity. The mechanistic principles behind this reaction are not well understood but involved specific coordination of a transition metal ion with a π-electron system. The selectivity of recognition is governed by the external chiral auxiliary diacid molecule. Historically, chirality was introduced into chemistry in the last century by Louis Pasteur, working on the resolution of tartaric acid salts. Furthermore, the controlled formation of new σ-bonds from π-bonds is reminiscent of the surface catalysts developed by Ziegler and Natta for the stereospecific synthesis of polymers (see Chapter 4).

It is to be hoped that the comprehension of the mechanism of stereoselective epoxidation will be applied to the rational design of new selective catalysts (49).

Sharpless and collaborators (50) have now developed a new and very promising procedure for chiral dihydroxylation of olefins using a mixture of the alkaloids dihydroquinidine or dihydroquinine as the *p*-chlorobenzoate derivatives, a catalytic amount of osmium tetroxide and

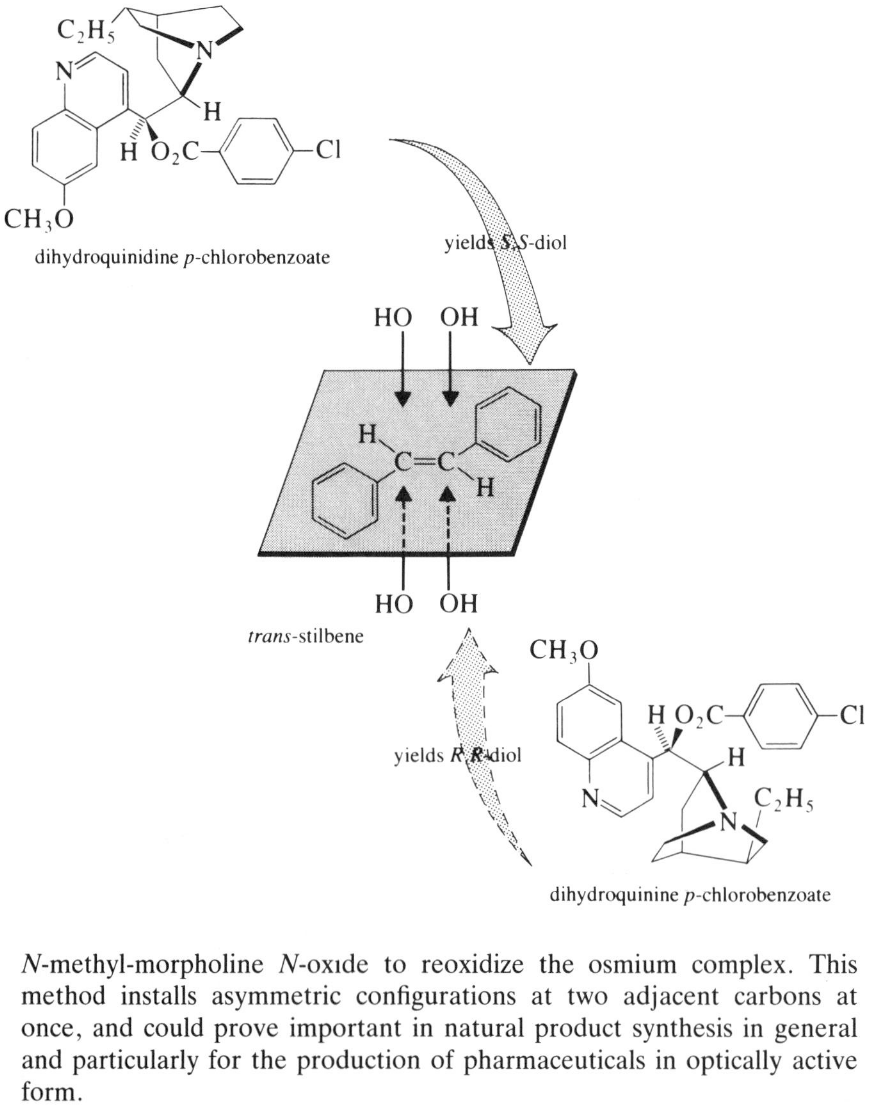

dihydroquinidine *p*-chlorobenzoate

yields *S,S*-diol

trans-stilbene

yields *R,R*-diol

dihydroquinine *p*-chlorobenzoate

N-methyl-morpholine *N*-oxide to reoxidize the osmium complex. This method installs asymmetric configurations at two adjacent carbons at once, and could prove important in natural product synthesis in general and particularly for the production of pharmaceuticals in optically active form.

2.6 Asymmetric Synthesis with Chiral Organometallic Catalysts

A metalloenzyme uses a metal ion in its active site to perform its role as a catalyst. More will be said on this subject in Chapter 6. By analogy, chemists have developed a large variety of organometallic chiral ligands for asymmetric synthesis. This approach has led to a dramatic increase in new methods for the preparation of chiral molecules, particularly the industrial construction of optically pure pharmaceuticals. Perhaps the most impressive achievement has been in the selective hydrogenation of olefins. Some of these reactions on prochiral precursors rival enzyme stereospecificity.

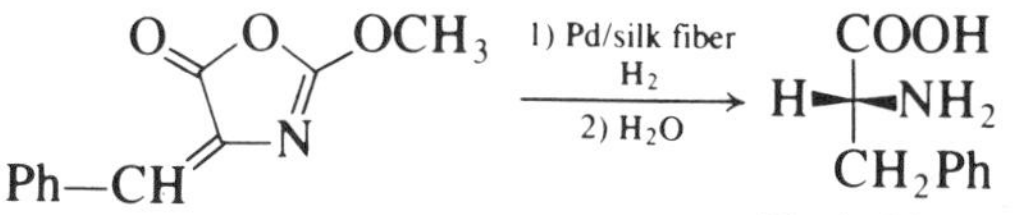

1) Pd/silk fiber H_2 2) H_2O

COOH H NH$_2$ CH$_2$Ph

Phe in 35% ee

One of the first examples of asymmetric hydrogenation was in the synthesis of phenylalanine using an heterogeneous catalyst composed of palladium and silk fiber. It is presumably the surface of the natural peptide that was responsible for the asymmetric induction. For practical reasons, however, a more rational approach to asymmetric synthesis of a large variety of optically pure compounds was the use of homogeneous catalysts. Among the most popular are the biphosphine ligands. The (*R*,*R*)-(−) DIOP was among the first one to be developed in the 1970s by Kagan. A few more are listed below. These bidentate chiral phosphines form stable complexes with rhodium and are all efficient optically active ligands for asymmetric homogeneous hydrogenation (51–53).

Homogeneous transition metal catalysts now provide the selectivity needed for synthesizing chiral pharmaceutical compounds (54). Perhaps

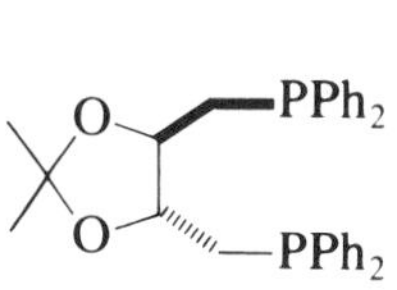

(*R*,*R*)-(−)-DIOP
(Kagan)

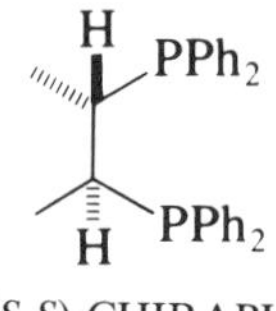

(*S*,*S*)-CHIRAPHOS
(Bosnish)

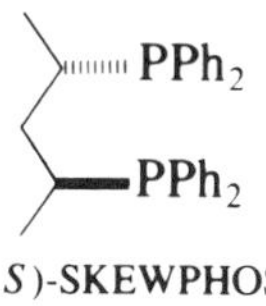

(*S*,*S*)-SKEWPHOS
(Bosnish)

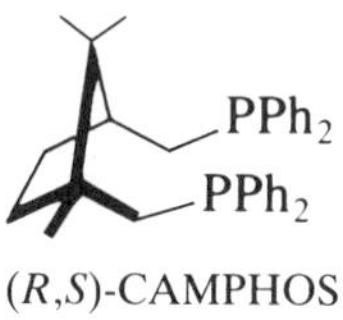

(*R*,*S*)-CAMPHOS

one of the most dramatic examples of the importance of chirality control was the tragedy associated with thalidomide, which was sold as a racemic mixture. Unfortunately, one enantiomer, originally assured to be inert, produced fetal deformities in newborn babies. Classic industrial demonstrations are the manufacture of the artificial sweetener, Aspartame by Searle & Company and the stereoselective production of L-dihydroxyphenylalanine (L-DOPA) by Monsanto Company, a drug currently used to treat patients suffering from Parkinson's disease.

CH_3O COOH AcO NHAc —(+)-DIOP + Rh, $+H_2$→ → HO COOH HO NH_2

L-DOPA

A rigid structure of the chiral ligand is not necessarily a crucial factor since some flexibility plays an important role in enabling an "induced fit" of the chiral metal complex molecules corresponding to the shape of the starting material (55). The extension of this methodology to the preparation of optically active dipeptides is also appealing (56).

To conclude this section, a last example will be given. One of the most exciting new developments in this field is the stereoselective hydrogenation of β-keto carboxylic esters to β-hydroxy esters. They are extremely important precursors in various natural product syntheses. So far, access to such optically pure synthons has to rely mainly on biochemical transformations using bakers' yeast to effect stereoselective reduction of β-keto esters. Noyori's group found, however, a practical application with the use of the excellent chiral recognition ability of BINAP-coordinated Ru(II) complexes (57).

PPh_2 PPh_2 —$Ru(OAc)_2$→ Ph Ph P $Ru(OAc)_2$ P Ph Ph

(*R*)-2,2′-bis(diphenylphosphino)-
1,1′-binaphthyl
(BINAP)

O O R OCH_3 —BINAP—$RuCl_2$, H_2→ OH O R OCH_3

R = CH_3, C_2H_5, C_2H_5, $(CH_3)_2CH$

with *R*-catalyst → 3*R*-product
S-catalyst → 3*S*-product

More impressive is:

(±)-racemic mixture —[R-BINAP]RuCl$_2$→ 3R; 49% yield (97% ee) + 51% yield (90% ee)

From a practical view, this hydrogenation method is operationally simple and economical and hence capable of large-scale production of a variety of optically active β-hydroxy esters. Notably, this transition-metal catalysis, unlike the biological version, allows a high degree of enantio-selective transformation with high yield of substrate conversion.

2.7 Transition State Analogs

Many compounds having a structure very similar to a substrate for a given enzyme will not be transformed by the enzyme but will act as enzyme inhibitors. They are basically of two types. The *reversible inhibitors* are those that compete for the substrate binding site but can readily be removed from the enzyme, thus enabling the enzyme to regain its activity. On the other hand, the *irreversible inhibitors* cannot be removed from an enzyme, and so the activity is not restored when the enzyme is separated from the solution containing the inhibitor. In such a case, a covalent bond is usually formed between the enzyme and the inhibitor. There are also intermediate cases of inhibition that will not be discussed here (58). However, a unique class of reversible inhibitors are the *transition state analogs*. They are stable products, conceived to mimic the structure of an intermediate in the path of the substrate's transformation by the enzyme. They behave as very potent inhibitors, not transformed into products and having an association constant for the enzyme larger than the substrate itself. Since they are structurally mimicking the transition state of a chemical transformation, they are based on Pauling's postulate (see p. 207), whereby an enzyme recognizes and binds more tightly to the transition state than to the ground state of the substrate product interconversion.

$$E + S \underset{k_{-1}}{\overset{k_1}{\rightleftharpoons}} [ES] \rightleftharpoons [ES]^{\ddagger} \rightleftharpoons [EP] \rightleftharpoons E + P$$

[ES]: Michaelis complex; [ES]‡: transition state

E + S $\underset{}{\overset{+I}{\rightleftharpoons}}$ EI — this represents competitive inhibition

the fundamental equation is

$$V = \frac{k_c[E]_0[S]_0}{K_m + [S]_0}$$

For clarity, a few fundamental concepts of enzyme catalysis are summarized here. The following equation of enzymatic events are governed by Michaelis-Menten kinetics. The velocity (v) of an enzymatic reaction is governed by two important constants, characteristic of a given substrate. They are the k_c, the catalytic constant, and K_m, the Michaelis binding constant. A good substrate must have a k_c/K_m ratio as large as possible; that is, the K_m should be small.

On the other hand, a good inhibitor will have a small K_i where

$$K_i = \frac{[E][I]}{[EI]}.$$

A few examples of transition state analogs will illustrate these points.

1. The antibiotic coformycin inhibits the enzyme adenosine deaminase:

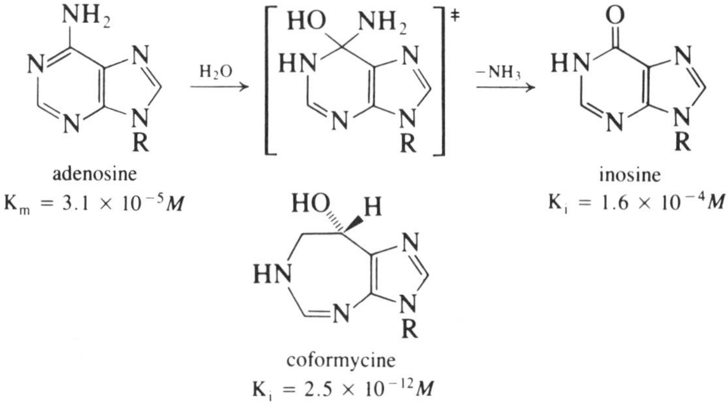

adenosine $K_m = 3.1 \times 10^{-5} M$ $\xrightarrow{H_2O}$ [transition state]‡ $\xrightarrow{-NH_3}$ inosine $K_i = 1.6 \times 10^{-4} M$

coformycine $K_i = 2.5 \times 10^{-12} M$

The product of the reaction, inosine, is a good inhibitor of the enzyme, but coformycin, which resembles more closely the transition state intermediate, is many orders of magnitude superior as an inhibitor.

2. 2-phosphoglycollate blocks the triosephosphate isomerase (see p. 174), an analog of the enediolate intermediate.

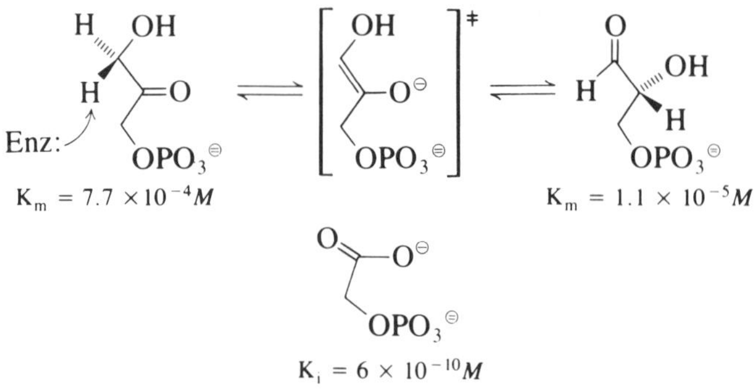

Enz: $K_m = 7.7 \times 10^{-4} M$ $\rightleftharpoons$ [enediolate]‡ $\rightleftharpoons$ $K_m = 1.1 \times 10^{-5} M$

$K_i = 6 \times 10^{-10} M$

3. Benzylsuccinate inhibits the enzyme carboxypeptidase A (see p. 392) by stabilizing the charges generated.

$$\text{Bz—NH—CH}_2\text{—C(=O}\cdots\text{Zn}^{2\oplus})\text{—NH—CH(CH}_2\text{Ph)—COO}^{\ominus}\cdots{}^{\oplus}\text{Arg-145} \xrightarrow{\text{C Pase}}$$

Bz-Gly-Phe
$K_m = 2 \times 10^{-3} M$

$$\text{Bz—NH—CH}_2\text{—C(=O)—O}^{\ominus}\cdots\text{Zn}^{2\oplus} + \text{H}_2\text{N—CH(CH}_2\text{Ph)—COO}^{\ominus}\cdots{}^{\oplus}\text{Arg-145}$$

Phe
$K_i = 5.5 \times 10^{-3} M$

$$^{\ominus}\text{O—C(=O)—CH}_2\text{—CH(CH}_2\text{Ph)—COO}^{\ominus}$$

$K_i = 6.0 \times 10^{-9} M$

In fact, inhibition of enzymatic reactions by transition state analogs has been an extremely important approach for drug design (59). As stated above, the principle underlying this is that nature has developed enzymes for binding efficiently to the transition states of the reactions they catalyze.

For example, phosphonates have been used as analogs of biological phosphates (60).

$$\text{CH}_2\text{=C(COOH)—O—P(=O)(OH)OH} \qquad \text{CH}_2\text{=C(COOH)—CH}_2\text{—P(=O)(OH)OH}$$

phosphoenolpyruvate 2-phosphonomethylacrylate analog

Phosphonamidates has also been used by P.A. Bartlett's group as transition state analog inhibitors of proteases such as thermolysin (61), carboxypeptidase A (62), and, more recently, leucine aminopeptidase

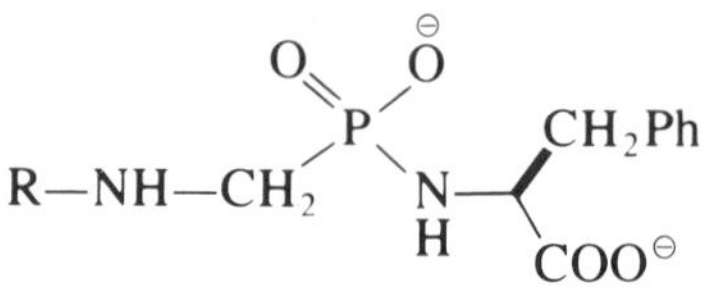

phosphonamidate peptide
(to be compared with example 3 above)

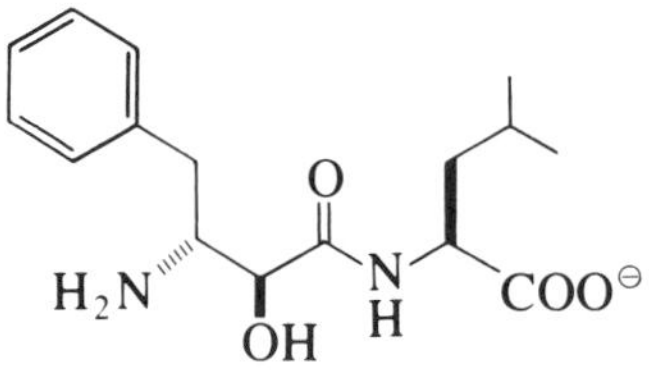

bastatin, a potent aminopeptidase inhibitor

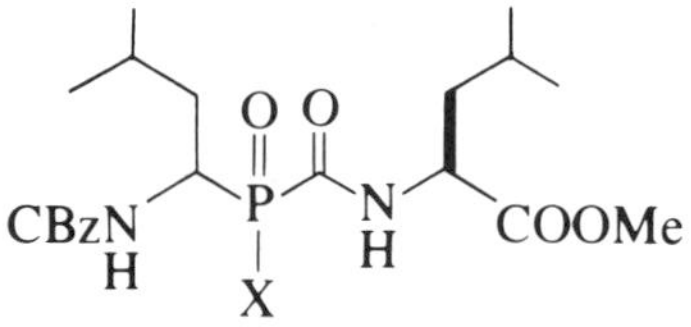

X = OH, OMe, NH_2
phosphorus analogs to mimic unstable tetrahedral intermediates or naturally occurring peptide inhibitors like bastatin

(63). All these examples as well as those in the references could be developed by the instructor at one point or another during the lecture.

An interesting effort in the transition state analog strategy is the design of molecules that could mimic the transition state of the chorismate to prephenate conversion. This specific biological transformation occurs via a Claisen rearrangement in a [3,3]-suprafacial mechanism (64). This intramolecular transformation in biology links a family of sugars with

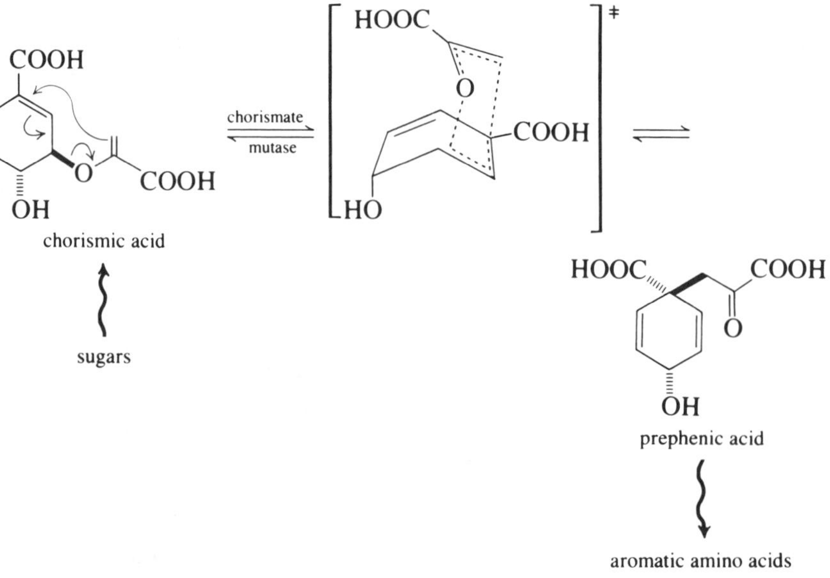

chorismic acid

sugars

chorismate mutase

prephenic acid

aromatic amino acids

aromatic amino acids. Very likely, well-substituted [3,3,1]-bicyclic compounds could imitate the transition state and bind to chorismate mutase. It turned out that the following compound is a good inhibitor of the reaction. However, the axially oriented analog, which would probably prefer a boat conformation, does not bind to the enzyme. This constatation raises the question as to whether the transition state does prefer a chair or a boat conformational intermediate.

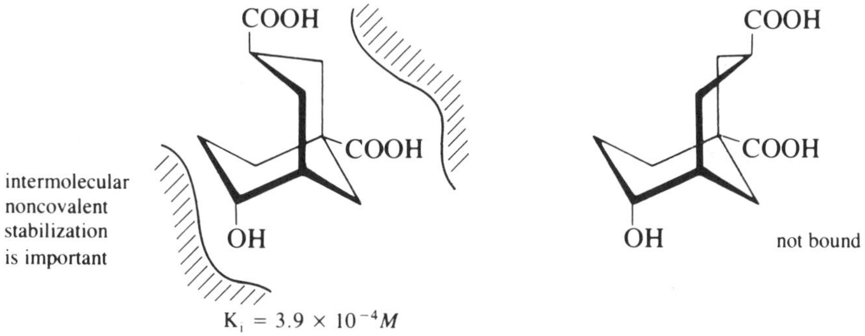

intermolecular noncovalent stabilization is important

COOH COOH OH

$K_i = 3.9 \times 10^{-4} M$

COOH COOH OH

not bound

This intriguing problem was elegantly solved by Copley and Knowles (Harvard University). By an ingenious double-labeling of chorismic acid they were able to prove that a nonenzymatic rearrangement of chorismate to prephenate prefers a transition state of chairlike geometry. This is illustrated below (65). The stereochemical course of the reaction can be

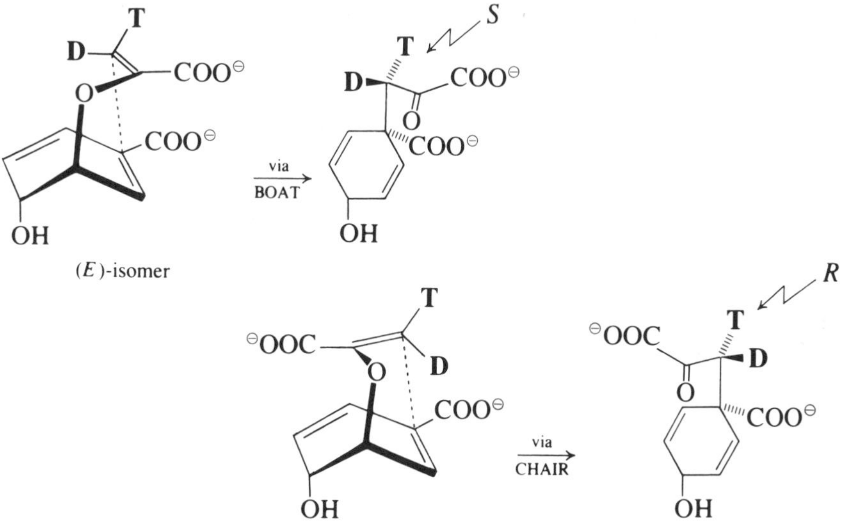

T D COO⊖ O COO⊖ OH

(*E*)-isomer

via BOAT

T D COO⊖ O COO⊖ OH *S*

⊖OOC T D O COO⊖ OH

via CHAIR

⊖OOC T D O COO⊖ OH *R*

determined by analyzing the configuration at the methylene carbon of prephenate that derives from chorismate stereoselectivity labeled with deuterium (D or ^2H) and tritium (T or ^3H) at carbon-9. Indeed, the prephenate that is formed from (Z)-[9-^{2}H,^{3}H] chorismate will have a tritium with pro-S position if the reaction proceeds through a chairlike transition state and in the pro-R position if the transition state is boatlike.

As expected, for (Z)-chorismate, close to 80% of the tritium was found in the pro-S position of prephenate whereas with the (E)-isomer only 36% of tritium was transferred to the pro-S position. These figures agree quite well with predictions and confirm the chairlike nature of the intramolecular Claisen rearrangement.

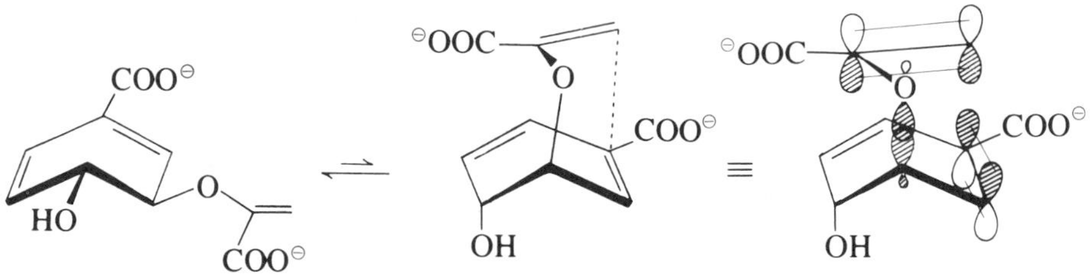

more stable pseudodiequatorial conformer (deduced from NMR coupling constants)

less stable pseudodiaxial conformer, but more prone to Claisen rearrangement

These results and others suggest that this nonenzymatic transformation involves a dipolar transition state having some of the character of a tight ion pair between the enol pyruvate anionic moiety and the cyclohexadienyl cationic portion.

The relatively small difference in the free energies of the two conformers of chorismate in aqueous solution ($\Delta G_0 = 0.9$ to 1.4 kcal/mol) suggest that the enzyme chorismate mutase can directly select the pseudodiaxal conformer from solution (66). Consequently, the conformer of chorismate from which rearrangement occurs has the hydroxyl group and the enolpyruvoyl group axial, yet the more stable conformer that predominates in solution is the pseudodiequatorial one.

A molecule resembling even more closely the transition state conformation was designed by Bartlett and Johnson (67). It turned out to be the most potent inhibitor of chorismate mutase yet reported. The molecule as its *endo*-form adopts a chair conformation of the tetrahydropyran ring.

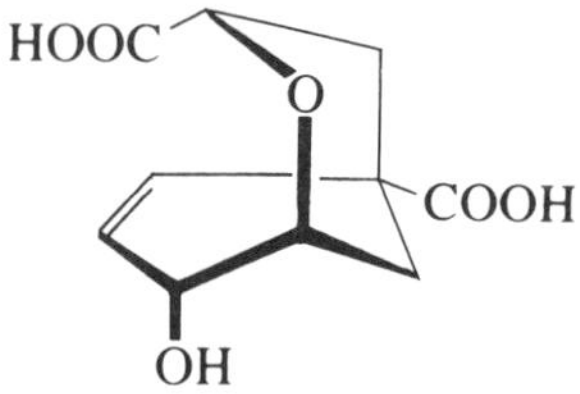

endo-isomer $K_i = 4 \times 10^{-8} M$

This molecule has also been used to raised antibodies to catalyze the reaction.

The *exo*-isomer is a hundredfold less potent as an inhibitor with values comparable to the saturated analog treated above.

2.8 Antibodies as Enzymes

Immunoglobulins are a very important class of soluble proteins in the blood serum since they serve as antibodies to combat the invasion of our body against foreign materials. They are made of four polypeptide chains (two light and two heavy), bound together by disulfide bridges and forming a quasisymmetrical network. It is in the *N*-terminal regions of the molecule that the primary sequence is the most variable (V) among different antibodies. It is also in this region that the antigens bind to the antibodies. Furthermore, these molecules are bivalent, capable of binding to two antigen molecules.

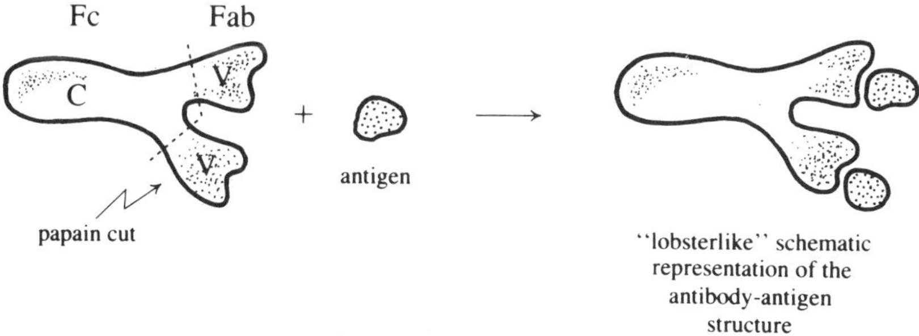

"lobsterlike" schematic representation of the antibody-antigen structure

The action of the proteolytic enzyme papain gives three fragments, 2 Fab (antibody binding) and one Fc (constant region) and can be used to characterize the antibodies. The noncovalent molecular recognition of the antigens probably also causes some structural perturbations in the non-variable (C) region of the molecule.

It is their structural diversity that confers on them an almost unlimited capacity to recognize so many different antigens, the foreign molecules in our system. These antigens could be protein, like those found at the surface of a viral capsule; they could also be small molecules called *haptens*, and specific antibodies can be "raised" against them. Like enzymes, antibodies have the capacity to recognize complementary molecular structures and form very specific complexes with high association constants. However, one basic difference between antibodies and enzymes is that the former bind molecules in their ground states whereas enzymes bind their substrates in higher energy states.

How do you go about trying to associate the prodigious capacity of molecular recognition of antibodies with potential enzymatic activity? It was with this goal in mind that P.G. Schultz (University of California at Berkeley) and R.A. Lerner (Scripps Research Institute) developed in 1986 the first antibodies with catalytic activities. Lerner later called them *abzymes*.

In fact, the strategy is simple. It is based on the principle of transition state analogs seen in the previous section. Pauling had predicted that stable substrate analogs of a transition state would bind tightly to the enzyme and increase their activity. Based on Pauling's hypothesis, Lerner's and Schultz's groups "looked" for antibodies that could stabilize the transition state of a given reaction such as ester hydrolysis. The objective was thus to produce antibodies that can catalyze the hydrolysis of organic esters and carbonates with a certain degree of specificity. The antigenic molecules used to elicit these antibodies were either phosphonate or phosphate analogs of the esters or carbonates to be hydrolyzed. Indeed, they mimic the tetrahedral transition state of the hydrolysis reaction (68,69). The concept of the transition state analog is the focal point in this endeavor, and the idea is thus to design antigens (haptens) based on mechanistic chemical principles (70). (See structures on page 87.)

They thus raised antibodies against phosphonate species to produce a binding site on the antibody structures that would preferentially recognize and bind the tetrahedral structure of the transition state of carbonate hydrolysis. It turned out to be a challenging but successful project to the point that the resulting antibodies were so specific to catalyzes the hydrolysis of the *para*-nitro substituted carbonates but not the corresponding *ortho*-analogs.

In the case of antiphosphorylcholine antibodies, they display saturation kinetics with a catalytic constant k_{cat} of $0.4\,min^{-1}$ and a Michaelis constant K_m of $208\,\mu M$. This represents rates of acceleration greater than 1000 over the uncatalyzed reaction. The hydrolysis was also inhibited by *para*-nitrophenylphosphorylcholine.

Obviously, with the access to *monoclonal antibody* technology, homogeneous ligand-binding sites with enzymelike binding affinities and specificities can now be generated for most biologically active macromolecules as well as smaller synthetic molecules like the previous examples. The expectation of this property makes antibodies one of the most versatile and important classes of receptors in biology and medicine today.

Since antibodies can be produced against virtually any haptens with precise structural specificity, the possibility exists for developing general catalytic methods for the synthesis and functionalization of simple organic molecules by antibodies. In this regard, R.A. Lerner and S.J. Benkovic (Pennsylvania State University) examined the possibility of performing an intramolecular cyclization reaction, having noted that structural constraints imposed by the antibody binding pocket should confer a modest degree of rate acceleration depending on the extent of reduction of rotational entropy (71).

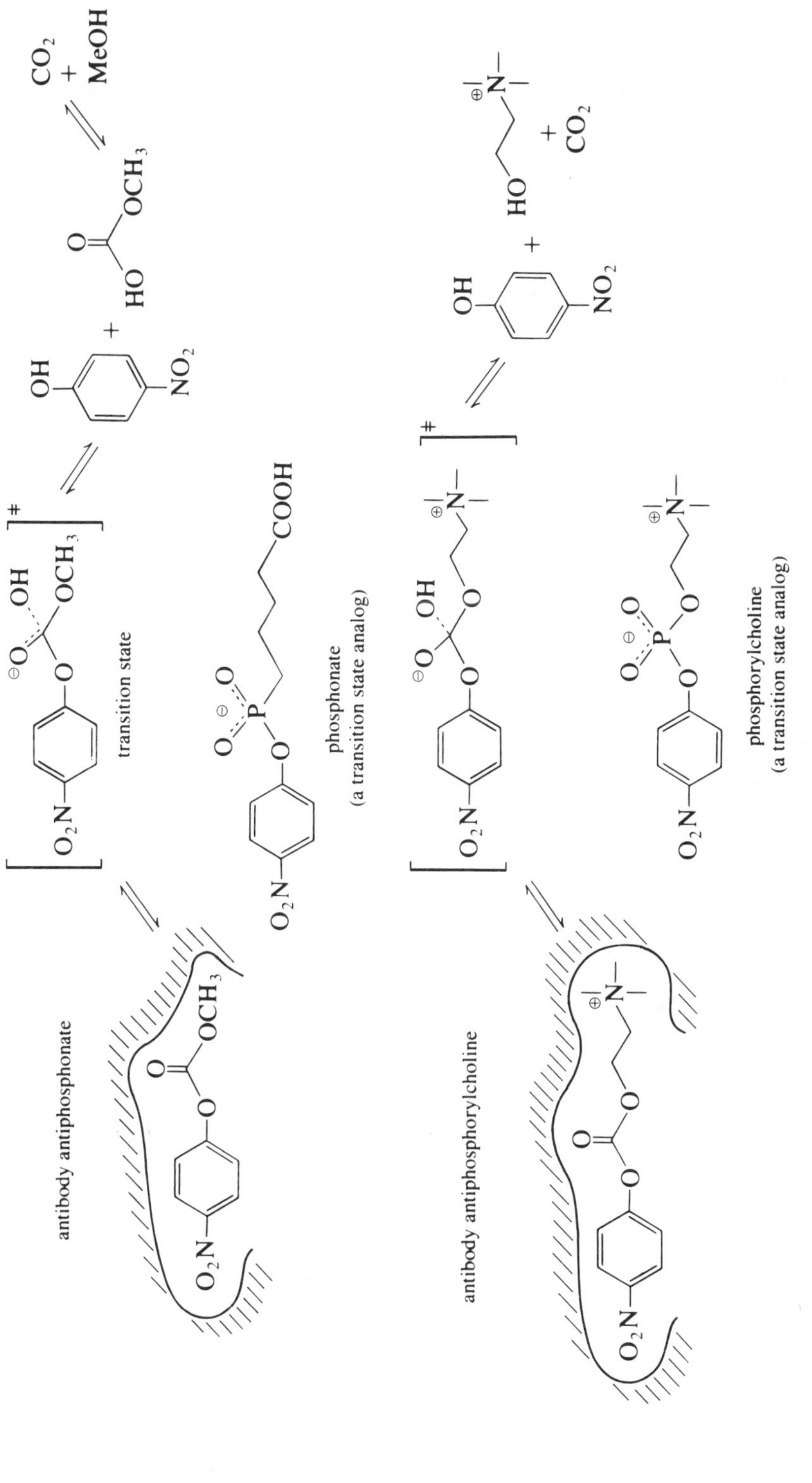

transition state

phosphonate
(a transition state analog)

antibody antiphosphonate

phosphorylcholine
(a transition state analog)

antibody antiphosphorylcholine

They chose the formation of a lactone from an hydroxy ester.

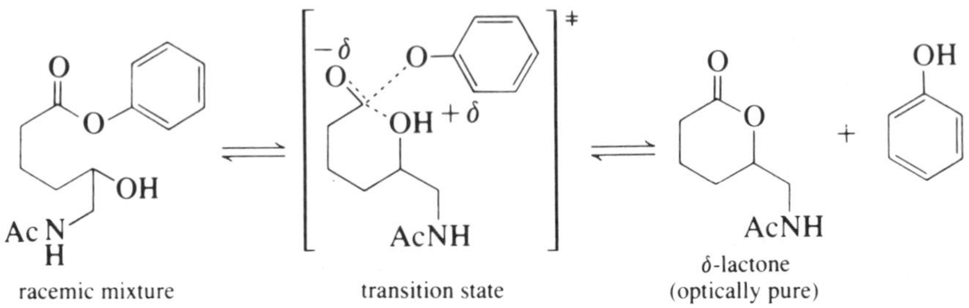

racemic mixture

transition state

δ-lactone (optically pure)

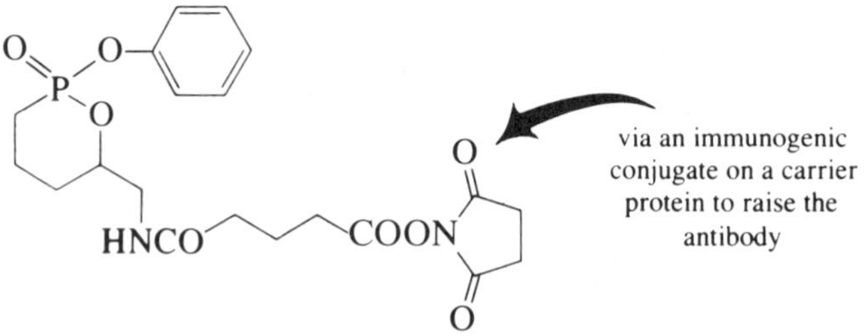

via an immunogenic conjugate on a carrier protein to raise the antibody

cyclic phosphonate ester analog

(resembling a "productlike" intermediate along the cyclization route)

Interestingly, only one single enantiomer of the δ-lactone in 94% ee (by NMR) was formed from the corresponding δ-hydroxy ester. Furthermore, the stereospecific ring closure reaction was accelerated by the antibody by about a factor of 170. This simple transformation demonstrates the feasibility of a tailor-made generation of catalytic antibodies for chemical reaction that requires stereochemical control.

Recently, this field has been extended to the development of antibodies that can bind specifically to the "illicit" drug cocaine and break it down into two inert by-products (72). Researchers have also created antibodies that can catalyze disfavored cyclization reactions and others that can carry stereo- and regioselective reduction of diketones (73). Even multigram-scale, enantioselective hydrolysis of an enol ether has been obtained with a catalytic antibody (74). The transformation involved is the following:

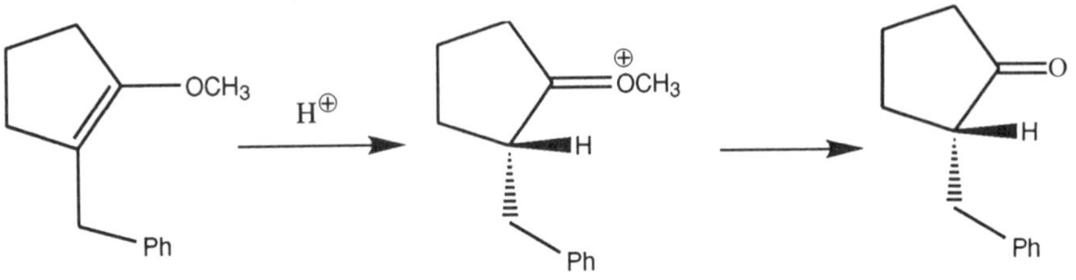

The antibody was raised against the following pyridinium compound that mimics the charged transition state of the reaction.

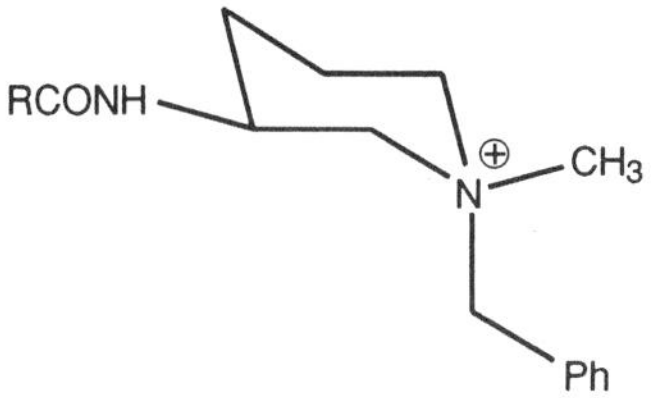

Stable transition-state analog of the enol ether hydrolysis
(R = carrier protein)

Although catalytic antibodies may never find large-scale industrial application, they may be expected to give to chemists access to many chemical transformations that are otherwise unattainable.

Finally, the recent work of P.G. Schultz's group towards the development of an antibody-catalyzed prodrug activation should be mentioned (75). The target is the anticancer drug 5-fluoro-uracil (5-FdU) (already mentioned in Section 1.3). This product is then converted *in vivo* into 5-fluoro-deoxyuridine 5′-phosphate and acts as a mechanism-based inhibitor of thymidylate synthetase, an enzyme necessary to construct the building blocks of DNA molecules. Since the amino acid analogue, 5′-D-Val ester of 5-FdU is less toxic than 5-FdU, Schultz reasoned that he could generate an antibody capable of catalyzing the conversion of the 5′-amino acid analogue to 5-FdU.

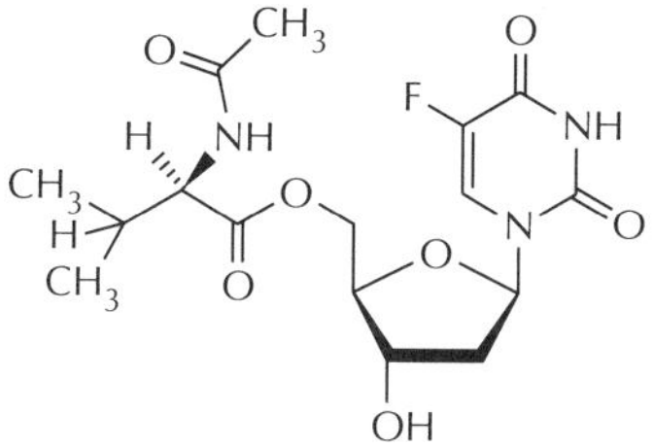

5' -amino acid analog, prodrug

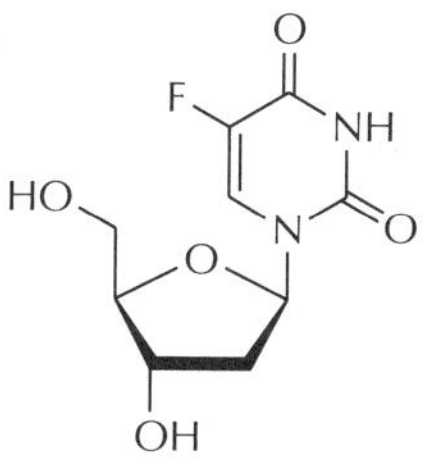

5–FdU

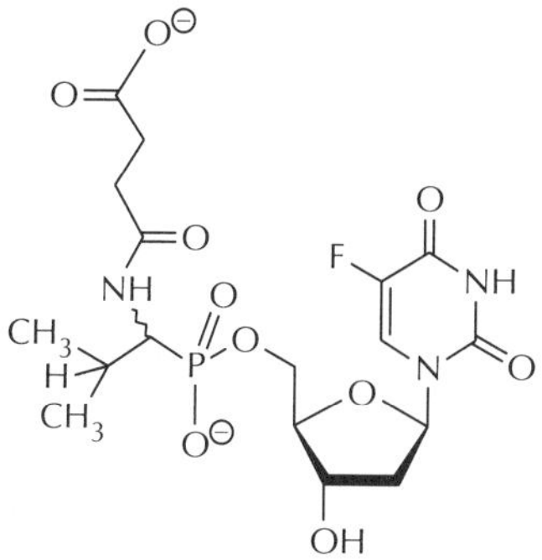

transition state analog

The antibodies were raised against a phosphonate hapten, a transition state analog for the hydrolysis reaction. In the presence of the prodrug and the antibody, the growth of the bacteria was completely inhibited, whereas the antibody alone did not affect the growth of the bacteria. This promising area may find applications in therapeutic drugs and could potentially be used to treat diseases ranging from allergies to viral infections.

For the future, there is a strong belief that the field of enzymes will not be limited to antibodies selectively mimicking transition states of hydrolytic reactions but might include, for instance, introducing a metal ion or another cofactor into the antibody recognition site to carry other catalytic functions. This direction will open the way to the production of "semisynthetic" catalytic antibodies. A large number of important and new problems will then be addressed by putting chemistry and immunology together. Such a multidisciplinary approach will allow new molecular tools to emerge in bioorganic chemistry (70). For instance, one can imagine production of antibodies that can not only hydrolyze peptide bonds with a high degree of specificity but can also be sequence specific. This would give to catalytic antibodies the capacity to cut and paste proteins. Such a tool would open up new strategies for protein engineering. With the right combination of technological advances and clever chemistry, catalytic antibodies would have the potential to make major breakthroughs in key areas of science and medicine.

2.9 Chemical Mutations

For a long time medicinal chemists, through quantitative structural-activity relationships (QSARS), have tried to optimize the properties of enzyme inhibitors in terms of steric, lipophilic, and electronic factors for a more efficient binding to enzyme binding sites. On the other hand, there has been a constant effort, in recent years, to mimic the catalytic action of enzymes. However, is it possible to design or modify an enzyme active site? The answer is yes. It is now possible by changing one or more amino acids on an enzyme to generate a new catalyst. This approach is called *site-directed chemical mutation* of proteins and enzymes.

The first controlled modification of an enzyme was reported independently in 1966 by Koshland's group and Bender's associates (76,77). They used a series of chemical reactions to change the hydroxyl group of the active site serine residue of subtilisin, a bacterial serine protease, to a sulphydryl group. This corresponds to the replacement of serine residue by a cysteine. However, the resultant thiosubtilisin was inactive against normal substrates. Since then, the reverse operation was also carried out. The replacement of the active cysteine residue of papain by a serine residue also resulted in an inactive enzyme (78).

OH $\xrightarrow{PhCH_2SO_2F}$ $O{-}SO_2CH_2Ph$ $\xrightarrow{CH_3{-}C(=O)S^{\ominus}}$

active site serine residue
of subtilisin
(active enzyme)

SH $\xleftarrow{H_2O}$ $S{-}C(=O){-}CH_3$

thiosubtilisin
(inactive)

Similarly, aspartic acid–102 of α-chymotrypsin (see p. 184) has been converted to asparagine and the enzyme retains some activity. On the other hand, conversion of the carboxylic group of aspartic acid–52 of lysozyme (see p. 205) to a primary alcohol completely inactivates the enzyme.

All these site-directed modifications have been useful for the elucidation of enzyme mechanisms. However, the problem with such chemical modifications of amino acid residues is that rather drastic conditions have to be used that one would generally try to avoid. Also, they are often nonspecific. Alternative approaches have thus been sought. Two of them deserve special attention and representative examples in each case will now be presented.

2.9.1 Site-Directed Mutagenesis

With the advance in genetic engineering technology, it is now possible to make amino acid substitutions on proteins on an almost routine basis. Such site-directed mutagenesis involves the cloning of the gene that evolves for the given protein and incorporates it into a suitable vector such as a plasmid. Since these highly sophisticated *recombinant DNA* techniques are beyond the scope of this book and are more appropriate in a textbook on molecular biology, only one example will be used to illustrate this point. Two useful review articles on molecular biology written for chemists by Parish and McPherson could be consulted (79,80).

D.A. Estell's group at Genencor and chemists and biochemists at Genentech have made many mutants of subtilisin, the endopeptidase already mentioned and produced by a variety of *Bacillus* species in order to attempt to change its substrate specificity (81). The enzyme cleaves bonds adjacent to neutral amino acids, and consequently, glutamic acid is one of subtilisin's worst substrates with a k_c/K_m of about $16\,M^{-1}\,\text{sec}^{-1}$. By contrast, the k_c/K_m for tyrosine, subtilisin preferred substrate is close to $10^6\,M^{-1}\,\text{sec}^{-1}$.

From X-ray work and model studies, there seems to be an unfavorable electrostatic interaction between the negatively charged side chain of a glutamic acid substrate and the glutamic residue (Glu-156) in the enzyme active site. Glu-156 normally stabilizes the tetrahedral intermediate of neutral substrates by ion pair stabilization. Modification of Glu-156 for glutamine (Gln-156) produced a mutant enzyme with a k_c/K_m of 260 $M^{-1}\,sec^{-1}$ for glutamic substrates. Another possible change is to introduce a positive charge on the enzyme to neutralize the incoming electrostatic interaction caused by the presence of glutamic acid substrates. For this, the conserved glycine residue (Gly-166) in the bottom of the active site clift was changed to a lysine residue. Alone, that change produces an enzyme with a k_c/K_m for glutamic acid substrates greater than 12,000 $M^{-1}\,sec^{-1}$. The double mutant, containing both Gln-156 and Lys-166, has a k_c/K_m of 50,000 $M^{-1}\,sec^{-1}$, indicating that the effects of the two changes are independent and at least to some degree, additive. It turned out, however, that the mutant does not bind tyrosine substrates any more and cleaves their amide bonds only slowly. These results are summarized in Table 2.5.

Both Gly-166 and Glu-156 have been the target of studies using site-directed mutagenesis. In fact, Gly-166 has been replaced by many other amino acids such as Ala, Ser, Thr, Val, Leu, Ile, Phe, Tyr, Trp, Cys, Met, and Pro. This technique developed by Gentech scientists is called *cassette mutagenesis*. It allows the relatively straightforward insertion of a series of synthetic oligonucleotides into a gene. Using this method, all nineteen possible amino acid substitutions at a given site in an enzyme can be made (81).

Finally, other effects have been examined with modified subtilisin, namely the pH profile on the catalytic activity. Sometimes the changes made on the enzyme produced results almost the opposite of what was originally anticipated or predicted.

Ultimately, the use of site-directed mutagenesis in combination with chemical mutation of enzyme active sites should increase considerably the range of possibilities in the design of more effective semisynthetic enzymes (82,83).

Table 2.5. Site-Directed Mutagenesis on Subtilisin

Subtilisin	k_c/K_m for Glu ($M^{-1}\,sec^{-1}$)	k_c/K_m for Tyr ($M^{-1}\,sec^{-1}$)
Native (classical Asp-His-Ser triad in the active site)	16	10^6
Mutant Glu-156 → Gln-156	260	
Mutant Gly-166 → Lys-166	12,000	
Double mutant	50,000	low

2.9.2 Chemical Mutations and Semisynthetic Enzymes

As early as 1978, Whitesides elaborated an elegant approach to semisynthetic enzymes by the use of a modified biotin molecule (see p. 62). An alternative approach for the elucidation of enzyme mechanisms is based on modification of the structure of amino acid functional groups that are present in the active site. One way to alter the specificity of enzymes is to attach organic molecules or organometallic catalysts to their surface. These have produced some interesting changes in reactivity. A leader in this field was the late E.T. Kaiser (Rockefeller University, N.Y.). His approach is to design a new catalyst that relays the combination of an existing protein-binding site with a chemically reactive coenzyme analog, thus exploiting the binding specificity of the protein or enzyme but expressing the characteristic chemical reactivity of the new covalently attached cofactor (82). This represents a significant step towards the production of enzyme-like molecules or semisynthetic enzymes by a process that is now called *chemical mutation*.

In this way, flavopapain was prepared by covalent modification of the thiol group of the active site Cys-25 of papain by a flavin derivative. The design was first based on the X-ray diffraction studies of covalent papain-inhibitor complexes. By modeling the binding interactions, few flavin-papain complexes were prepared without destruction of the enzyme as a catalytic species. (See structure on page 94.)

The flavopapain FP-3 was the first effective semisynthetic analog to be prepared (84). In the oxidation of *N*-alkyl NADH derivatives, the k_c/K_m values were almost two orders of magnitude larger than rate constants observed for the corresponding model reactions. Basically, a proteolytic enzyme has been converted to an enzyme now doing oxidoreduction reactions. The new semisynthetic enzyme exhibits staturation kinetics, as would be expected for an enzymatic reaction involving the obligatory participation of a binding site.

Since FP-3 seems to be a better catalyst than FP-2 and FP-1, it is believed to be stabilized at the active site by favorable hydrogen bonds with the neighboring amino acid Gln-19. Biophysical measurements further suggest that a charge–transfer complex between the flavin moiety and the aromatic amino acid Trp-26 is disrupted prior to the redox reaction with NADH analogs. From all these observations, a model of the active site of the semisynthetic enzyme FP-3 was postulated (82).

In a similar approach but with a different strategy, H. Zemel showed that a semisynthetic enzyme can be made from a nonenzymatic protein (85). The material used was the sperm whale Metmyoglobin from which the prosthetic heme group had been removed. The resulting apoprotein possesses a deep hydrophobic cavity, resembling an enzymatic binding site to accommodate hydrophobic substrates. In proteases, ester hydrolysis is catalyzed by imidazole residues. Within this new pocket of

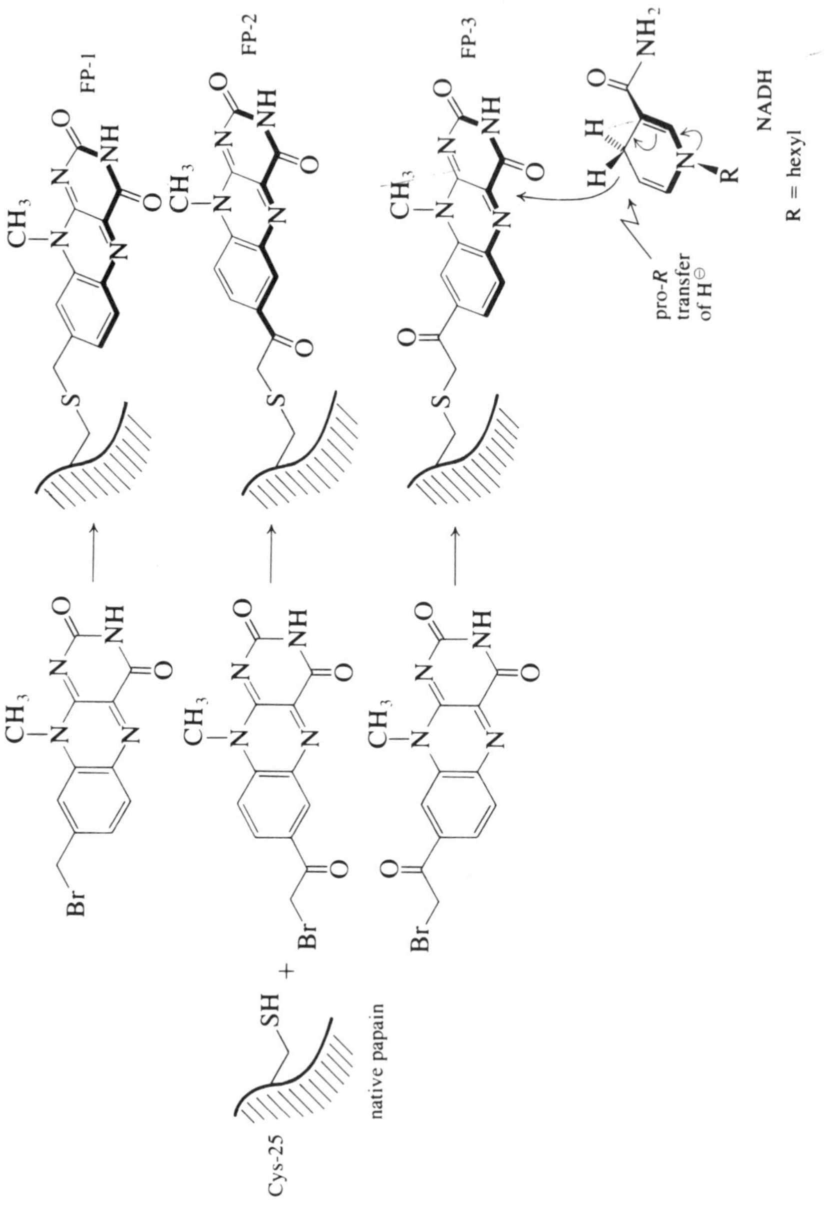

Cys-25

native papain

FP-1

FP-2

FP-3

pro-*R* transfer of $H^{\ominus}$

NADH

R = hexyl

Metmyoglobin, there are indeed two histidine residues capable in principle of performing hydrolytic activities. As expected, the combination of the hydrophobic cavity with the catalytic groups resulted in an excellent hydrolytic capacity for a fatty acid ester of the following structures.

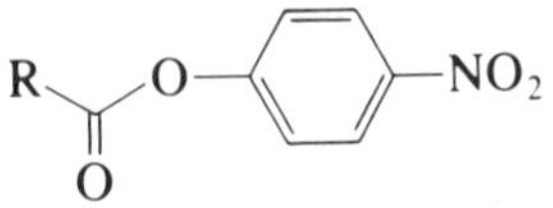

R = C_5H_{11} (caproate)
R = C_9H_{19} (caprate)

The new proteinic catalyst exhibits saturation kinetics with k_c/K_m of 600 M^{-1} sec^{-1}, an acceleration rate of 3000 times greater than imidazole itself. Its activity is even 15 times larger than poly(ethylenimine)imidazole polymer, developed by Klotz (see p. 342).

We will conclude this section by presenting the list than Kaiser has addressed for the design of an efficient semisynthetic enzyme (82).

1. The enzyme to be used should be readily available in highly purified form.
2. The X-ray structure of the enzyme should be know.
3. The enzyme should have a suitably reactive amino acid functional group at or near the active site.
4. The covalent modification of the enzyme should result in a significant change in the activity characteristic of the native enzyme.
5. The attachment of the coenzyme analog should not cause the entry of substrates to the binding site to be blocked.

Furthermore, the choice of a suitable cofactor analog to act as a chemically modifying agent should also fulfill three basic criteria.

1. The coenzyme should have the potential to act as a catalyst when bound to an enzyme active site without a requirement for specific functional groups of amino acid residues in the enzyme to participate in the catalytic act.
2. Model building should indicate that the placement of the coenzyme analog is compatible with the spatial requirements of the enzyme template.
3. Model building should indicate that the covalently bound coenzyme should be capable of interacting with a potential substrate in a productive fashion.

2.10 Molecular Recognition and Drug Design

The nervous system is very complex and involves many different kinds of specific interactions with different neurotransmitters. Because of this

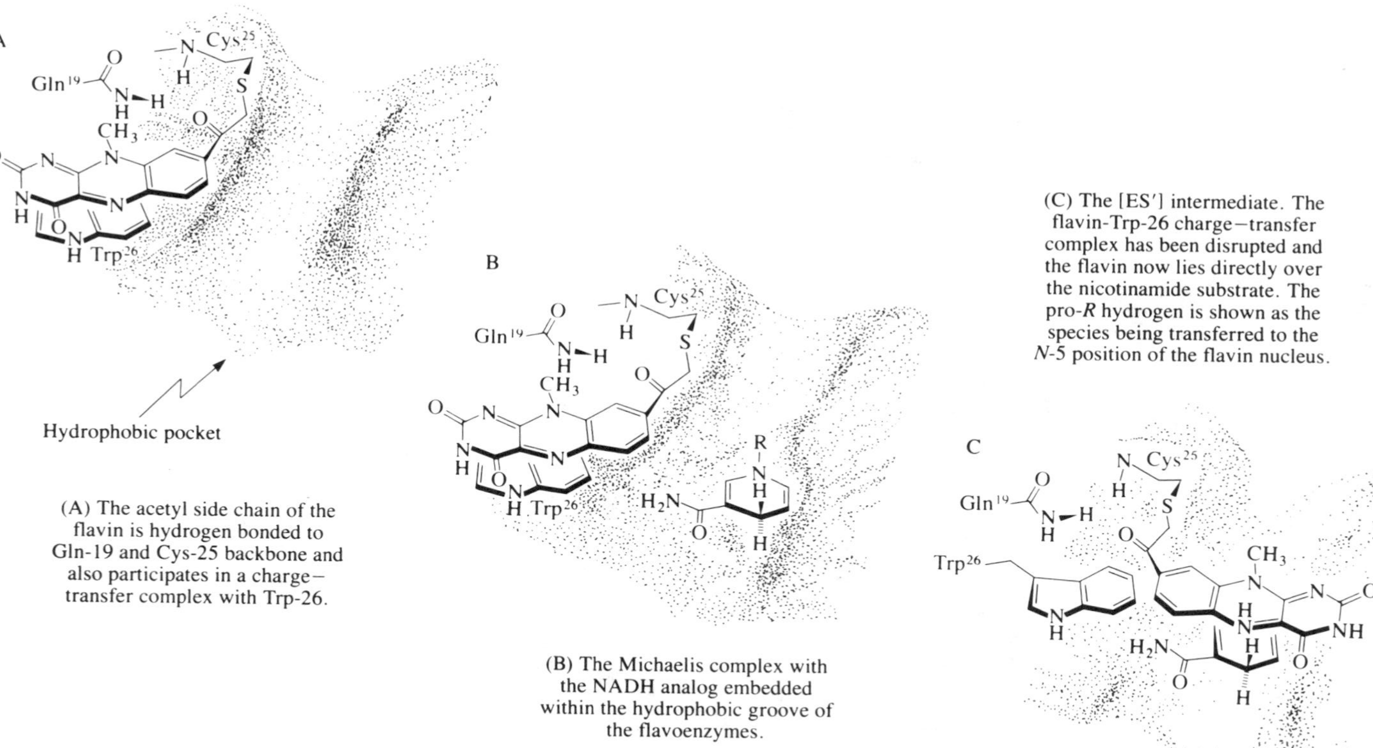

(A) The acetyl side chain of the flavin is hydrogen bonded to Gln-19 and Cys-25 backbone and also participates in a charge–transfer complex with Trp-26.

(B) The Michaelis complex with the NADH analog embedded within the hydrophobic groove of the flavoenzymes.

(C) The [ES′] intermediate. The flavin-Trp-26 charge–transfer complex has been disrupted and the flavin now lies directly over the nicotinamide substrate. The pro-*R* hydrogen is shown as the species being transferred to the *N*-5 position of the flavin nucleus.

Fig. 2.11. Active site of flavopapain FP-3. Adapted from (82).

complexity, we will limit our presentation to only one fascinating aspect of neurochemistry: the development of analgesic drugs or painkilling molecules. Morphine and synthetic analogs are the best examples. They produce their biological effects by mimicking the action of small peptides produced naturally by the body. With the characterization of these neuropeptides in the brain, a new facet in peptide and medicinal chemistry has been enlightened in recent years. A brief discussion on the development of morphine analogs will be presented and, in parallel, their molecular interactions with specific receptors. In medicinal chemistry, a *receptor* is believed to be a complex of proteins and lipids which, on binding of a specific organic molecule (effector, neurotransmitter), undergoes a physical or conformational change that usually triggers a series of events that results in a physiological response. Molecules that stimulate a receptor in a way similar to the corresponding neurotransmitter are called *agonists*. Those that act on a specific receptor but do not produce a biological response are *antagonists*.

The discussion will be completed by a presentation of the endogeneous opiate peptide enkephalin and its derivatives.

2.10.1 Painkiller Chemistry

Since immemorial times, people have been in search of drugs to reduce pain. Alkaloids of the morphine family belong to this group of molecules called *analgesic* drugs. However, they all possess side effects ranging from addiction to difficulty in respiration and constipation. Despite many efforts, a miracle drug with no side effects, nonaddictive, and more potent than morphine is yet to be found.

Heroin, the 3,6-diacetyl derivative of morphine, is a popular drug among drug abusers. Being more hydrophobic than morphine, it does cross the blood-brain barrier more efficiently. As a result, the effect is faster and the euphoria more intense. However, it does not bind directly to the opiate receptors and must first be metabolized to 6-acetyl morphine or to morphine itself to produce its action (86). In a sense it is considered as a "pro-drug." Codeine, a much weaker analgesic, is about ten times less potent than morphine. It is the 3-methoxy analog of morphine and is commonly used in syrup preparation as an anticough agent.

Based on molecular modification and structure–activity relationship considerations (87), we will look at one aspect of how new analgesic drugs have been developed.

Atropine is a well-known anticholinergic drug acting on acetylcholine receptor sites. By a chance discovery, the analgesic activity of a 4-phenyl piperidine molecule, Demerol, was developed by screening compounds that were designed as anticholinergics, based on the structure of atropine. Demerol is about as active as codeine and also causes dependence and

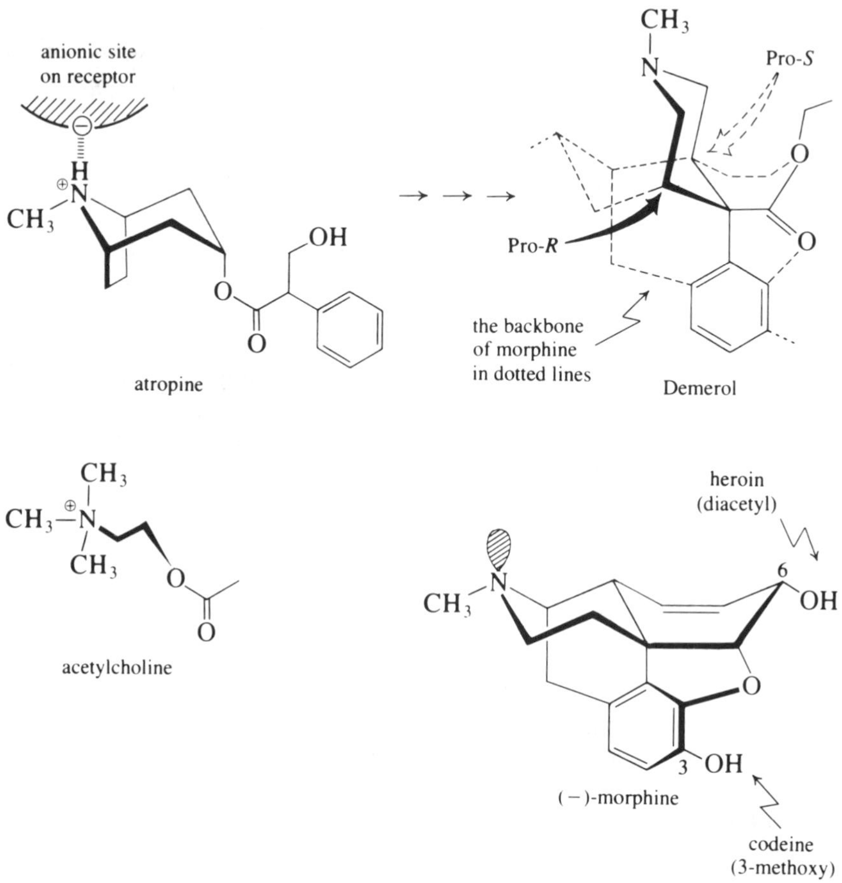

anionic site on receptor

atropine

the backbone of morphine in dotted lines

Demerol

acetylcholine

heroin (diacetyl)

(−)-morphine

codeine (3-methoxy)

respiratory depression. Its structure can be nicely superimposed on the morphine skeleton, and on this structural basis its analgesic action can be explained. Therefore, the structure of atropine turned out to be a lead compound for the development of a new drug, devoided of anticholinergic activity but with a somewhat a priori unpredicted analgesic action.

In a search for more-active compounds and probably also in order to improve the hydrophobic nature of the N$-$$CH_3$ region complementary to the receptor, Benzitramide was designed. It is about seven times more potent than morphine and is used as a long-acting analgesic in man. It has a close structural similarity with Fentanyl, another potent analgesic, and surprisingly with Droperidol, a potent neuroleptic. *Neuroleptics* are antipsychotic agents acting on dopaminergic receptors and community used to treat schizophrenia. This one belongs to the class of neuroleptics having a butyrophenone fragment.

This example alone shows how unpredictable the result can be, even after careful thinking, of designing new drugs. Finally, in another mole-

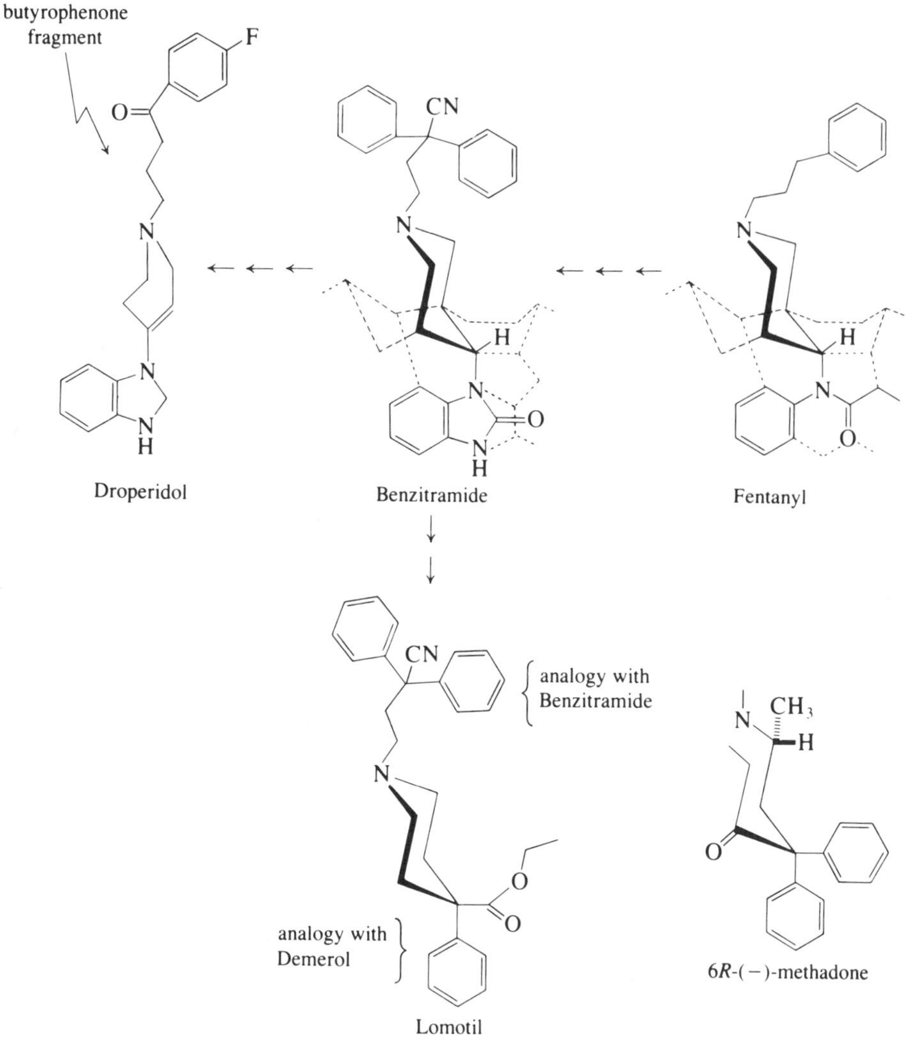
butyrophenone fragment
F
O
N
N
N
H
Droperidol
CN
N
H
N
O
N
H
Benzitramide
N
H
N
O
Fentanyl
CN
analogy with Benzitramide
N
O
O
analogy with Demerol
Lomotil
N
CH3
H
O
6R-(−)-methadone

cular variation of Benzitramide with the added structural elements common to Demerol, Lomotil came to birth. A flexible molecule like methadone also binds to opiate receptors.

Amazingly, Lomotil is devoid of the analgesic activity found in its conceptual precursors but maximizes one of the secondary effects of opiate drugs, namely, inhibiting gastrointestinal elimination. It is thus used as an antidiarrheic agent. All these examples are good illustrations that structural and conformational factors of molecules are very crucial in the understanding of drug-receptor interactions. Small molecular modifications can lead sometimes to rather unexpected results. This short presentation suffices to emphasize the complexity of the opiate receptors.

Absolute configuration and chirality of new compounds are equally important in drug design. For example (−)-levorphanol with a *trans*-ring junction is an active analgesic binding to opiate receptors. The isomer isolevorphanol having instead a *cis*-ring junction is, on the other hand,

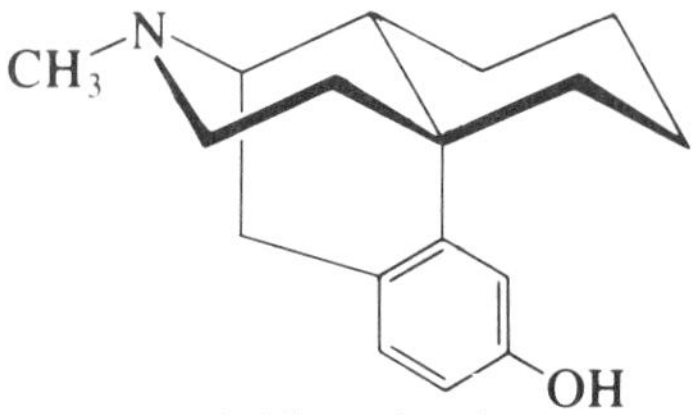

CH_3–N … OH
(−)-levorphanol

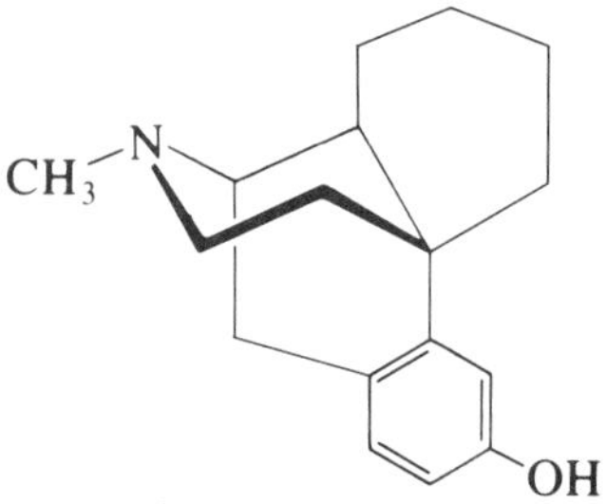

CH_3–N … OH
isolevorphanol

inactive. Even more spectacular is the following comparison between the (+)- and (−)-isomers in the morphan series. The contrasting absolute chiral specificity of the receptor toward morphan is very much like that of many enzymes toward their substrates (see p. 173, Chapter 4 for more details).

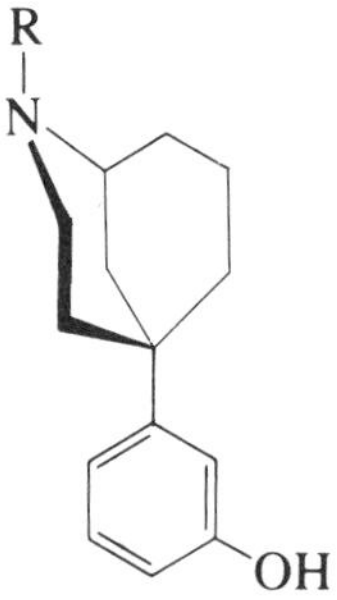

R–N … OH
(+)-isomer
active

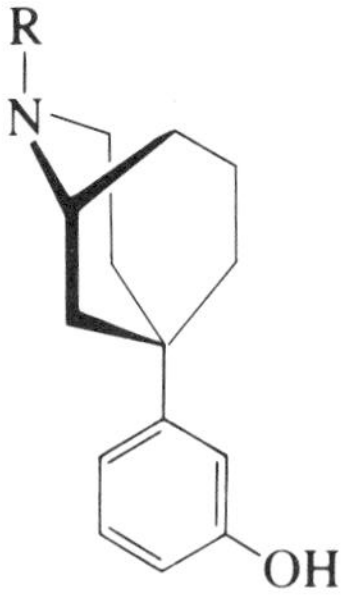

R–N … OH
(−)-isomer
inactive

R = CH_3
R = (allyl)

From such observations and many others with rigid and semirigid analogs, a general picture of the morphine receptor can be schematically represented as shown on page 101.

A protonated anionic site on the receptor serves to anchor the nitrogen atom either in the morphan or morphinan orientation. Almost perpendicular to the plane of the cyclohexane ring, a large aromatic auxiliary site serves to accommodate the phenol ring or the benzimide nucleus of other analogs. This illustrates the structural latitudes of the drug-receptor complex.

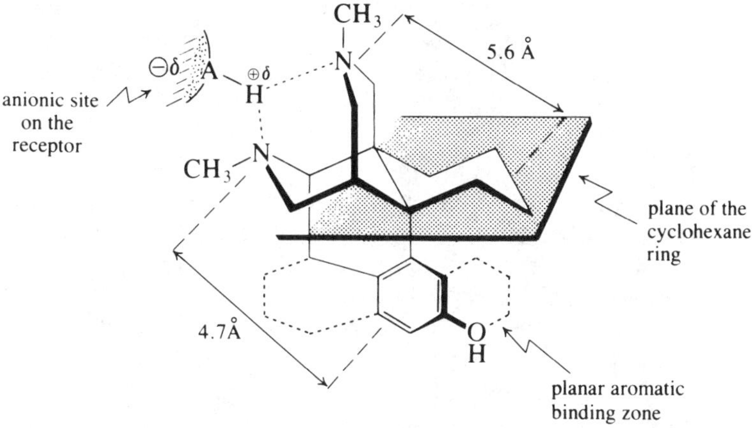

anionic site on the receptor

plane of the cyclohexane ring

planar aromatic binding zone

5.6 Å

4.7Å

Ionic channels are also important in the modulation of the opiate receptors, namely the Na^+/K^+ ratio that directly influences the binding of drugs that mediate analgesia. For instance, a high level of Na^+ ions decreases the binding affinity for agonists while a high K^+ concentration increases the binding of agonists to the receptors. At the same time, an increase in K^+ permeability indirectly results in a decrease in the entry of Ca^{2+} ions into the nerve terminal.

The nuance between agonist and antagonist is rather complex in this context, but among the new developments, there is now concrete evidence that the relative spatial orientation of the nitrogen lone pair of electrons of analgesic drugs is of capital importance for productive interaction with the opiate receptors. In this regard, Belleau (88) has recently pointed out that one aspect of conformation-activity relationships that has escaped attention concerns the importance of stereoelectronic effects (see Chapter 4). He made some interesting observations with a series of morphinan

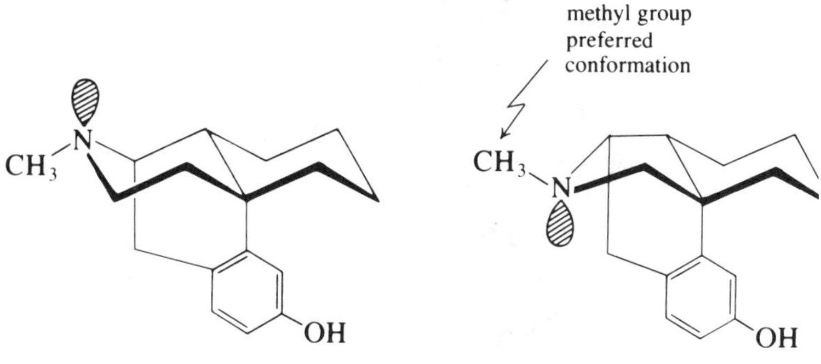

methyl group preferred conformation

morphinan (C_6) active

D-normorphinan (C_5) inactive

molecules and arrived at the hypothesis that if the lone pair of electrons on the nitrogen atom is far from the phenolic ring, the molecule should have a good affinity for the receptor. On the other hand, if the lone pair of electrons is oriented toward the phenolic ring, the molecule should be inactive.

For example, the above morphinan with a six-membered piperidine ring is active while the five-membered analog, D-normorphinan has no receptor affinity. It is concluded that D-normorphinan is inactive because the nitrogen lone pair responsible for proper binding to the anionic site is not properly oriented. There is thus an important stereoelectronic factor taking place at the receptor level. A further strong evidence in this direction is provided from the binding results obtained with the isomeric 16-α-and 16-β-butanomorphinans.

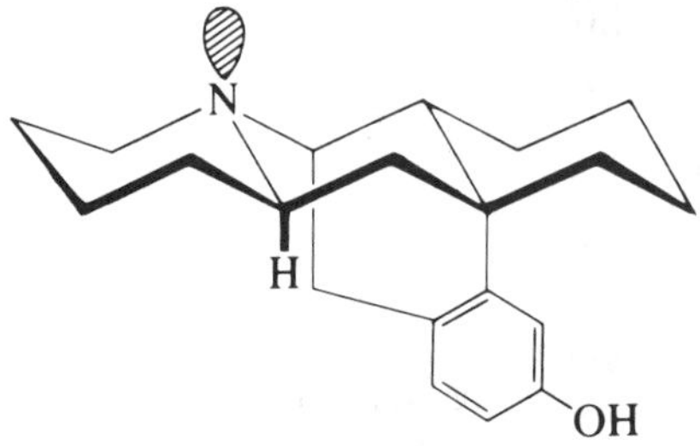

preferred boat form

butanomorphinan (16-β) active

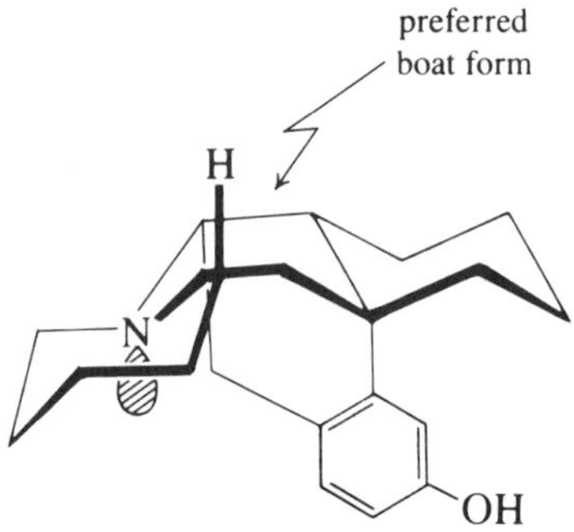

butanomorphinan (16-α) inactive

In the 16-α-isomer it is known that the piperidine ring D is locked in the boat conformation, forcing the nitrogen electron pair to be pointed toward the phenyl ring. The 16-α-isomer is inactive. In contrast, the 16-β-isomer displays analgesic activity and exists in a conformation where the nitrogen lone pair is accessible to the receptor surface.

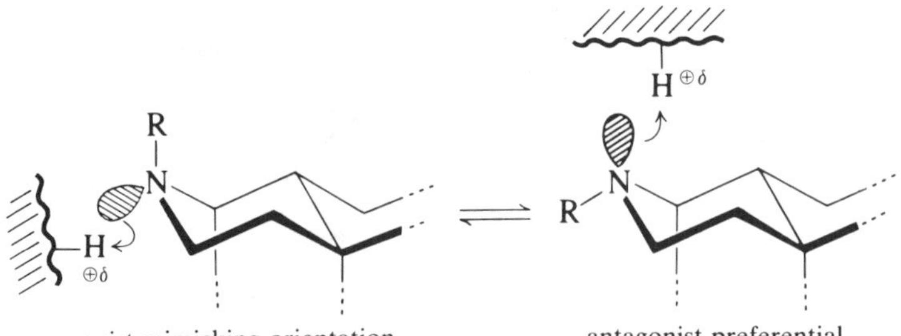

agonist mimicking orientation

antagonist preferential orientation

It was also argued that the nitrogen lone pair must participate in a proton transfer when it interacts with the receptor anionic site. Thus, proton transfer would be required for strong analgesics. For this to occur, the electron pair of the nitrogen atom of opiate molecules should be properly oriented for facile proton transfer and induction of analgesia. The preferred orientation of the lone pair of electrons could even discriminate between agonist and antagonist behavior.

This is particularly true for the molecule of naloxone, one of the most effective opiate receptor antagonists that block the euphoric effects of opiates. Secondary interactions with the axial OH group and the π-

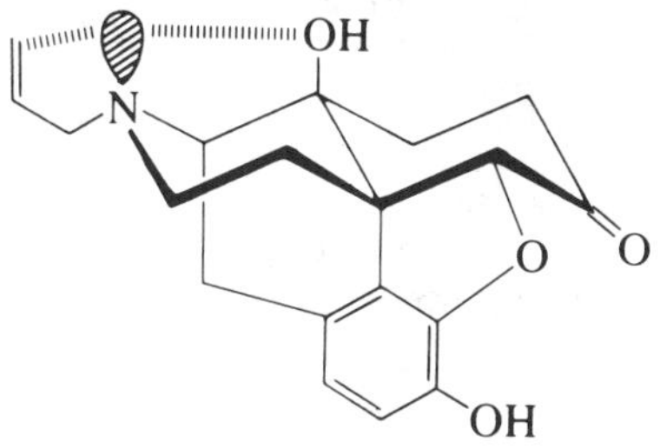

naloxone

electrons of the N-allyl chain would largely favor an orientation of the lone pair of electrons away from the phenyl ring.

2.10.2 Neuropeptide Chemistry

Analgesic drugs suppress pain information selectively over other sensory functions because of the high density of opioid receptors on nerves that respond to painful stimuli than on nerves that subserve other physiological functions. There are at least three distinct types of receptors through which analgesics exert their biological effects. They have been termed μ, δ, and κ.

Alkaloids of the morphine family bind preferentially to μ-receptors. Endogeneous neuropeptides like enkephalins show mainly δ activities while its larger precursor, β-endorphin, shows mostly μ activity. Another peptide, dynorphin, has κ-receptor activity. It has seventeen amino acids and the first five amino acids of its N-terminal contains the Leu-enkephalin sequence.

Other small peptidic substances are produced directly or indirectly by the brain pituitary glands (hypothalamus) and are used to signal the presence of foreign materials to the body and the sensation of pain derived from damaged tissues.

The enkephalin molecules are made from a common large precursor called the *POMC* (proopiomelanocortin). As such, POMC does not do

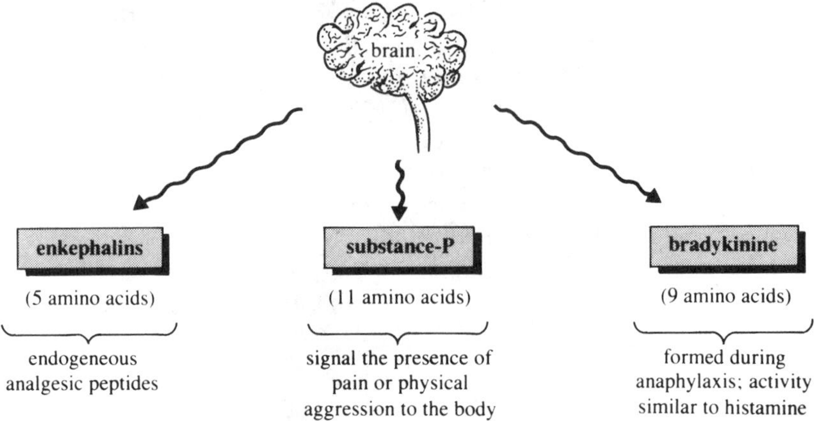

much, but upon maturation by specific enzymatic cleavages many peptidic intermediates are formed. Some will be used for pigmentation of cell tissues, others will act against stress, and finally endorphin and enkephalin molecules will have analgesic properties. Endorphin is the fragment 60 → 91 of β-LPH. The fragment 60 → 65 is Met-enkephalin.

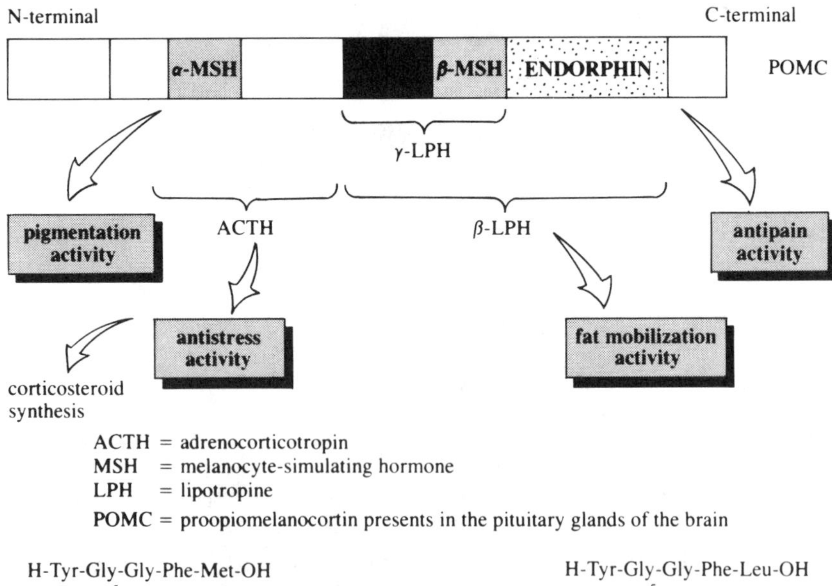

ACTH = adrenocorticotropin
MSH = melanocyte-simulating hormone
LPH = lipotropine
POMC = proopiomelanocortin presents in the pituitary glands of the brain

H-Tyr-Gly-Gly-Phe-Met-OH
Met[5]-enkephalin

H-Tyr-Gly-Gly-Phe-Leu-OH
Leu[5]-enkephalin

We are probably now less afraid of analgesics since the discovery in 1975 by Hughes (89) that our brain is naturally making some for our own protection against pain from the outside world. The calming action of acupuncture treatment is believed to be due to the production of enkephalin, being stimulated by the penetration of needles at specific points of the body.

Does morphine bind to the same receptor as enkephalin molecules? Is the μ-receptor of morphine structurally different from the β-receptor for the neuropeptides? Precise answers to these questions are hard to get since not enough information is available yet on the topological chemical architecture of either receptor site. One cannot eliminate also a possibility of $\mu \rightleftharpoons \delta$ interconversion of receptors via an allosteric perturbation by the drugs.

A solution 3D-structure of Met-enkephalin has been proposed by Roques (90) based on NMR evidence. The molecule has a U-shaped arrangement with the three amino acids Tyr, Phe, and Met pointing away from the main peptide backbone. The β-turn curled form of the pentapeptide is maintained by at least one intramolecular hydrogen bond and the lipophilic side chain of Phe remains accessible for additional interactions with the receptor surface. At first glance, the highly flexible structure of the pentapeptide does not appear to resemble a classical opiate structure like morphine. However, it is chemically rewarding to find that the morphine molecule can be superimposed to the N-terminal moiety of the peptide. Furthermore, oripavine, another analgetically active alkaloid, can also fit relatively well on the peptide backbone.

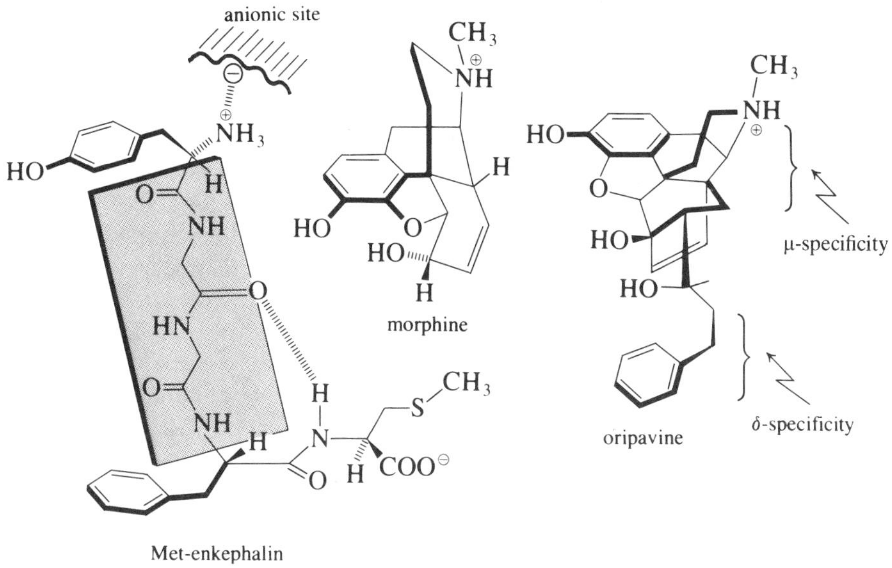

morphine

oripavine

Met-enkephalin

From such molecular adaptation exercises, it is believed that molecules that can fit better to the N-terminal portion of the peptide would have a marked μ-receptor specificity. Those offering a superposition toward the C-terminal would be inclined to have a greater δ-specificity.

However, more recent ^{15}N- and ^{13}C-NMR studies indicated an extended conformation for enkephalin, in contrast to the β-turn bend structure suggested previously (91). These last findings are in better agreement with the X-ray crystal study of Leu-enkephalin (92). In fact, the peptide chain folding is solvent dependent, and enkephalin is a flexible molecule capable of assuming several different conformations of comparatively low energy.

A molecular modeling study was performed recently on Leu-enkephalin bounded to ions such as lithium, sodium, or potassium, since these ions are important to the activity and water permeability of the peptidic hormone on membranes (93). The work showed that in the presence of sodium, but not the other ions, the enkephalin-ion complex adopts a quasicyclic conformation. The hydrophilic side-chain residues are at the surface of a modeled membrane, whereas the hydrophobic residues are pointing toward the interior of the membrane. This conformation could be responsible for the increase in membrane permeability to water by enkephalin in the presence of sodium ion.

Positions 2 and 4 of the peptide chain can accommodate various amino acids with hydrophobic side chains such as isopropyl and *t*-butyl groups. They can even accept amino acids of the D-configuration. The physiological roles of the different receptor classes (μ, δ, κ, and even σ) and their structural characteristics remain to be elucidated. For that reason, various tetra- and pentapeptides with modified amino acid sequences and cyclic enkephalin analogs were made and their conformation–activity relationships analyzed (94,95). Molecular modeling allows the development of cyclic polypeptides with selectivity for enkephalin receptors. Furthermore, modification of certain amino acids in those cyclic peptides were used to obtain new types of polypeptides with efficient and selective action. From such molecular information it becomes possible to conceptualize the topology of receptor sites. A major goal in drug development is the design and synthesis of neuropeptide analogs with high specificity for a particular class of receptors.

However, most bioorganic projects are now designed and developed with the help of computers using molecular modeling, particularly in the field of *peptidomimetic* and drug design. The peptidomimetic approach uses nonnatural synthetic models of amino acids or peptides to better understand the mode of action of natural peptides. The structures on the cover of this book are a good illustration of this trend. Section 7.3 should also be consulted for additional examples.

Some representative examples are illustrated below with a comparison of inhibitory effects of either μ- or δ-specificity from rat brain homogenates.

For comparison, [Leu^5]enkephalin has a K_i^{δ}/K_i^{μ} of 0.27, corresponding to a marked δ-receptor affinity. The conformationally restricted analogs

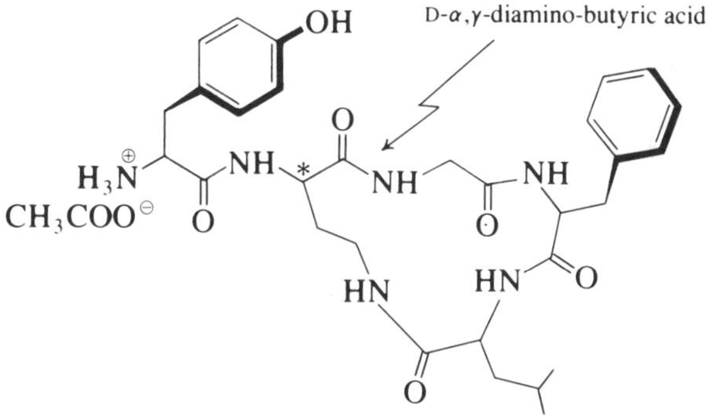

D-α,γ-diamino-butyric acid

Tyr-cyclo[N^{γ}_{u}-D-A_2bu-Gly-Phe-Leu]
$K_i^{\delta}/K_i^{\mu} = 8.3$
(μ-selective)

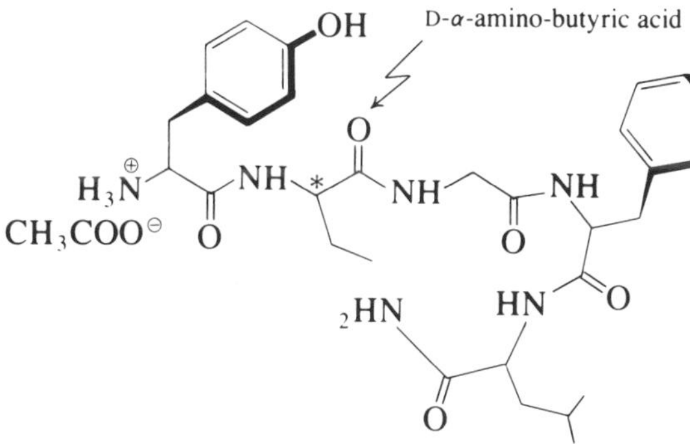

D-α-amino-butyric acid

[D-Abu^2, Leu^5]enkephalinamide
$K_i^{\delta}/K_i^{\mu} = 0.73$

often show a high potency in μ-receptor selectivity whereas the open chain peptides show nearly equal affinity for μ- and δ-receptors. Indeed, bioassays and binding assays reveal that introduction of a conformational constraint by ring closure in enkephalin analogs is directly and uniquely responsible for its decreased δ-character compared with the linear peptides. The lack of structual rigidity may be the reason for the low receptor selectivity of many natural neuropeptides since conformational adaptation to different receptor topographies is always possible.

The results just described represent one of the first direct demonstrations of a difference in conformational requirements between receptor subclasses in a situation of peptide receptor heterogeneity. As P. Schiller (Institut de Recherches Cliniques de Montréal) pointed out, the demonstration of the existence of receptor subclasses with topographically distinct features opens up an avenue for the design of receptor-selective peptide lighands by incorporation of appropriate conformational constraints.

A few more examples will show that subtle variations of the conformational restriction by means of different types of cyclizations can result in cyclic enkephalin analogs displaying various degrees of receptor selectivity.

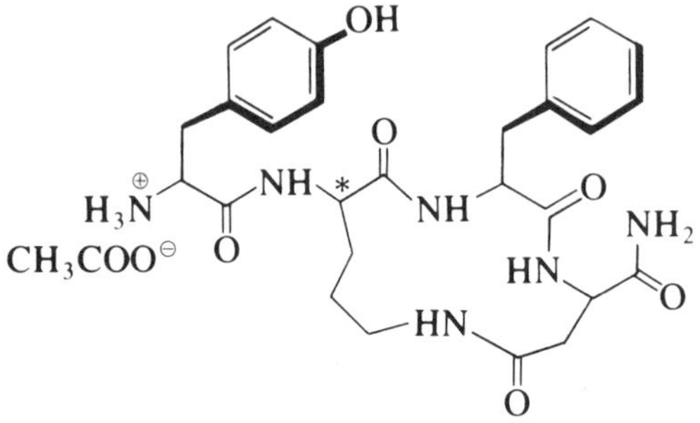

H—Tyr—D-Orn—Phe—Asp—NH$_2$

$K_i^{\delta}/K_i^{\mu} = 213$

(smaller ring and μ-selective)

The cysteine-containing cyclic peptide, (X = NH_2), a cysteine-bridged enkephalin analog, was found to be as potent as [Leu^5]enkephalin in a guinea pig ileum (GPI) assay, but not selective. The corresponding analog with a free C-terminal carbonyl group (X = OH) showed a considerably improved δ-receptor selectivity as compared to the natural enkephalins. Analogous cyclic tetrapeptides (no Ala^3), however, containing a rather rigid 13-membered ring structure, showed a high μ-receptor selectivity and a very weak affinity for the δ-receptor.

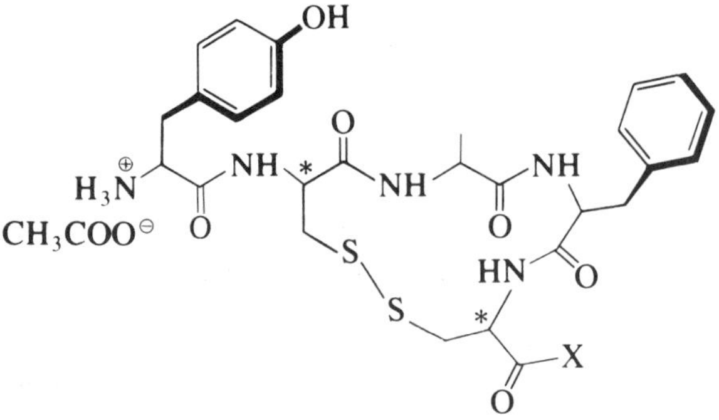

H—Tyr—D-Cys—Ala—Phe—D(or L)-Cys—X

X = NH_2

X = OH

* ≡ D configuration

Based on these results and others obtained so far with a variety of cyclic peptides, it appears that cyclic opioid peptides containing relatively small and rigid ring structures (13- to 14-membered) are more selective than their linear analogs, whereas in the case of cyclic analogs with larger and more flexible ring structures, the opposite may be true. For instance, H-Try-D-Lys-Gly-Phe-Glu-NH$_2$ with a larger ring similar to the cysteine-containing analogs shows high affinity for both the μ- and the δ-receptor and is therefore also nonselective.

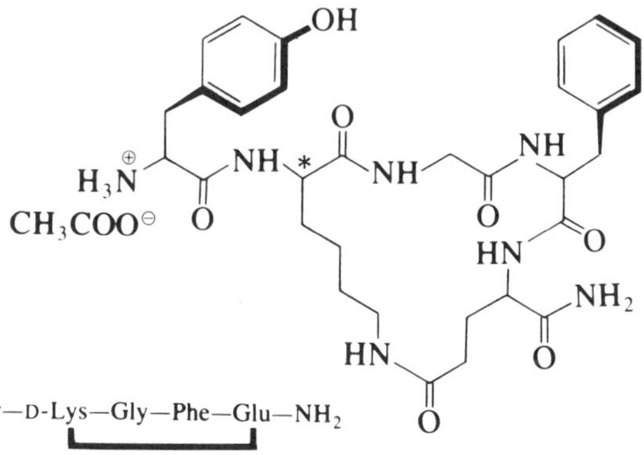

H—Tyr—D-Lys—Gly—Phe—Glu—NH$_2$

$K_i^{\delta}/K_i^{\mu} = 0.53$

Dipolar interactions must also play a crucial role in receptor specificity since Leu-enkephalin with an Arg$^{\oplus}$-Arg$^{\oplus}$ sequence extention at the C-terminal ending develops μ-receptor selectivity, whereas corresponding analogs with C-terminal Asp$^{\ominus}$-Asp$^{\ominus}$ endings have marked δ-receptor affinity.

In conclusion, these studies have led to one hypothesis that conformational restriction may alter the intrinsic activity of neuropeptides. Although loss of "efficacy" occurs with many cyclic analogs, they were shown to be extremely resistant to enzymatic degradation and some of them do produce long-lasting effects. They would therefore have interesting implications for peptide drug design and may represent a first step toward developing neuropeptide mimetics.

A last consideration concerns the suggestion that μ and δ opiate receptors may coexist as distinct recognition sites on an opiate receptor complex in the brain. If this is so, it is conceivable that a hybrid bivalent ligand, containing a μ-selective opiate and a δ-selective enkephalin pharmacophore separated by a spacer of appropriate length could possess analgesic activity that is considerably greater than the sum of the individual activities if both opiate sites are capable of being occupied simultaneously.

It is with these considerations in mind that the team effort of Portoghese and Schiller (96) developed the first bivalent ligand possessing both alkaloidlike and peptidelike pharmacophores. It consists of an oxymorphamine moiety and a [D-Glu^2]enkephalin portion linked through a monoor diglycyl chain.

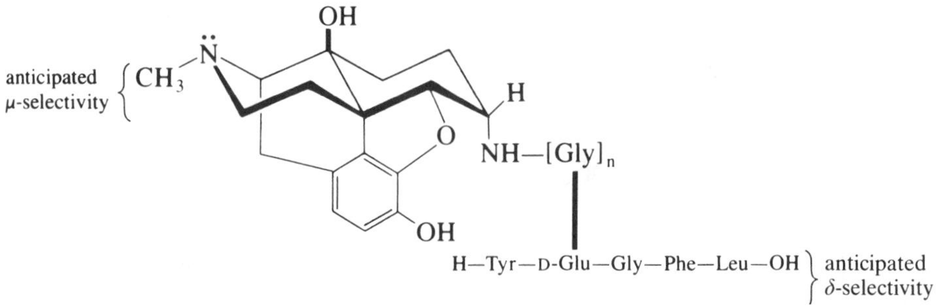

anticipated μ-selectivity { CH_3–N (morphine skeleton with OH, O, OH, H) NH—[Gly]$_n$

H—Tyr—D-Glu—Gly—Phe—Leu—OH } anticipated δ-selectivity

Testing compounds with n = 0, n = 2 as well as those with a glycyl chain to either the morphine skeleton or the enkephalin molecule revealed that the antinociceptive (antipain) potency of the bivalent ligand is substantially higher than either of its monovalent analogs, but the affinity to the δ-receptor is reduced. The presence of the spacer does not seem to contribute significantly to the pharmacologic effects of the ligand. Considering that the opiate (alkaloid) pharmacophore is about twentyfold more potent than the enkephalin (peptide) pharmacophore in monovalent ligands, the results conclude to a potency enhancement by a factor of approximately eightfold for the bivalent ligand but only for n = 0.

Apparently the glycylglycyl spacer for n = 2 separates the pharmacophores in such a way as to prevent interaction of the enkephalin portion with a binding site that could be vicinal to the site that binds the alkaloid pharmacophore. Furthermore, that the binding is greatly decreased on attachment of the enkephalin portion to the opiate component demonstrates that the combined pharmacophores influence the selectivity of binding in a predictable way. However, to end on a careful note, the enhanced opiate potency found for this bivalent ligand (with n = 0) could also be explained by postulating the presence of an accessory site, vicinal to the μ or opioid recognition zone.

Chapter 3
Bioorganic Chemistry of the Phosphate Groups and Polynucleotides

"La chimie crée son objet . . ."

M. Berthelot
1860

Too often a detailed description of the synthesis and properties of peptides is given without consideration of the analogous, but equally important, synthesis and properties of phosphodiesters, and vice versa. Indeed many of the problems and strategies (i.e., use of DCC, common protecting groups, polymeric synthesis, etc.) are similar, if not the same, yet they are never presented "side by side." It is with this purpose in mind that this chapter is written.

3.1 Basic Considerations

In order to maintain the life process, it is necessary to *pass on specific information* from one generation of organisms to the next. Such information is vital, for it allows the continuation of those traits or properties of the organism that insure survival. This information must be stored, to be used at an appropriate time, much as data may be stored and played on a tape recorder. As such, the job of information storage and processing must have a molecular basis. Indeed, it is not difficult to imagine some of the properties that are requisite for a molecular biological "tape."

Considering that a great deal of information must be stored on this molecule (i.e., the type of organism, physical characteristics, chemical pathways, etc.), it is expected that it must be a biopolymer. Perhaps a protein could function as a storage molecule? Probably not, since proteins already play an important structural and functional (enzyme catalyst) role in the cell. Such an important job as data storage must require a unique macromolecular structure that cannot possibly be confused with routine cellular processes. This special biopolymer is expected to be rather homogenous in structure, for it must fulfill one crucial role. It is not expected to be as diverse as enzyme structures for the latter are capable of participating in a wide variety of chemical reactions. On the other hand, it must have a heterogenous component in order to carry the different information necessary to the cell. This biopolymer is expected to be of a rigid or defined shape, because it must be able to interact with the cellular apparatus when it becomes necessary to transmit the stored information.

The molecules that function as the monomeric units for genetic information storage are nitrogen heterocycles: derivatives of *purine* and *pyrimidine*. The polymeric molecule that functions to both store and pass on genetic information is *deoxyribonucleic acid* (DNA). The related polymer *ribonucleic acid* (RNA) helps in the passage of this genetic information. It acts as the messenger that converts a specific genetic message to a specific amino acid sequence. The common purine bases of DNA and RNA are *adenine* and *guanine*. The common pyrimidine bases of DNA are *cytosine* and *thymine* (5-methyluracil), while for RNA they are cytosine and *uracil*. The purines and pyrimidines occur in combination with the sugar *ribose* (in RNA) or *deoxyribose* (in DNA) via the anomeric carbon of the latter. The glycosidic linkage occurs through the ring nitrogen of the base: either nitrogen-9 of the purines or nitrogen-1 of the pyrimidines. This chemical combination of a sugar and base (with elimination of water) is referred to as a *nucleoside*.

Other bases are also present in the DNA and RNA polymers, but to a much lesser extent. Examples include 5-methyl cytosine (present in the DNA of some plants and bacteria), 5-hydroxymethyl cytosine (present in the DNA of T-even phages; bacterial viruses) and 4-thiouracil (present in some bacterial tRNA).

It is possible to esterify a sugar hydroxyl with either phosphoric acid, pyrophoshoric acid, or triphosphoric acid to give the nucleoside mono-, di-, or triphosphates, respectively. The example of adenosine triphosphate (ATP) has already been encountered. The combination of a base, sugar, and phosphate is referred to as a *nucleotide*. Just as the amino acid is the monomeric unit of the protein polymer, the nucleotide (nucleoside monophosphate) is the monomeric unit of the DNA and RNA polymers. Further, just as the monomeric amino acids are linked via an amide (peptide) linkage, the monomeric nucleotides are linked via a phosphoester linkage.

It is important to realize that *it is the purine or pyrimidine base that carries the genetic information, just as the side chain of an amino acid determines its functional chemistry*. The DNA hereditary molecule is organized in the cell into units called *genes*. These are in turn located within structures called *chromosomes*, the latter being found in the nuclei of plant and animal cells. It is the gene that contains the information necessary for the determination of a specific trait: color of eyes, hair, height, sex, and so forth. However, the gene is a rather complex concept to describe at the molecular level, for the number of molecular events required for the exhibition of a particular trait may be many. Given that any genetic trait will involve protein synthesis (either structural proteins or enzymes), it becomes possible to define a simpler unit: the *cistron*. The cistron is defined as that part of DNA that carries the genetic information (codes) for the synthesis of one polypeptide chain. A chromosome contains many hundreds of cistrons. The total DNA content of the cell is referred to as the *genome*.

The genetic information is passed from the parent cell to the daughter cell by *replication* (synthesis) of DNA. The genetic information is stored in DNA until it is needed, at which time it is converted to the appropriate message for the synthesis of a protein of specified sequence via the process of *transcription*. The genetic message is transcribed to the RNA (messenger RNA) biopolymer. This in turn interacts with the appropriate specific aminoacyl-tRNAs resulting in the sequential coupling of amino acids. The transformation of the genetic information from RNA to a specific amino acid sequence is referred to as *translation*. The terms *replication*, *transcription*, and *translation* may be understood by considering the analogy of printing a book. When a book goes to print, many replicas or copies are made; hence replication. When a passage is copied by hand from the book, it is transcribed; hence transcription. When the book is written in another language: it is translated; hence translation (from nucleotides to amino acids). The entire process is described below, and only a few more words will be said shortly to relate this to the molecular level.

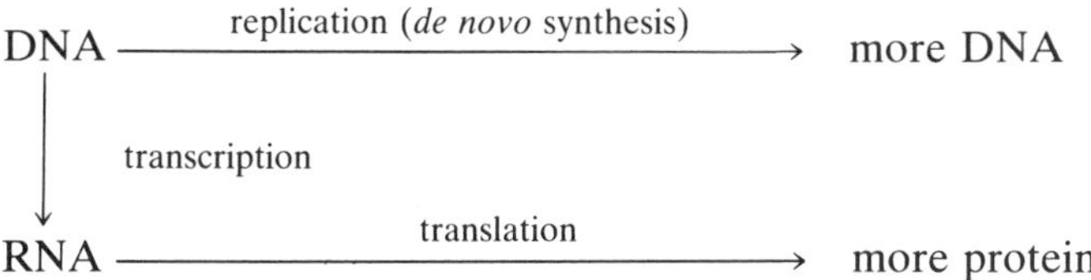

Considerable efforts have been made recently in the structural modification of oligonucleotides, particularly in the replacement of the phosphodiester linkage. Thus, amide linkage has been developed as a new type of backbone modification in oligonucleotides (97), as shown below:

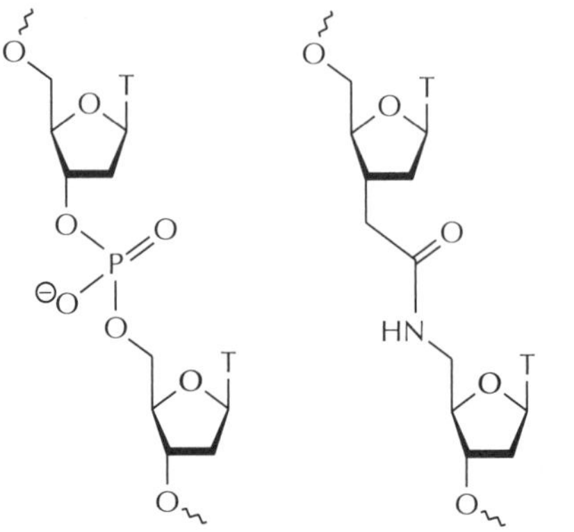

normal DNA amide linkage analog

An advantage of this modification is that the lower overall charge of oligonucleotides that contain neutral amide groups should facilitate penetration into negatively charged cell membranes. Furthermore, the amide moiety is readily accessible by simple synthetic methods and is achiral. This is not the case in phosphodiester chemistry.

Interestingly, these modified DNA oligonucleotides, with a new amide bond, not only display a similar or even slightly higher affinity for an RNA target than the natural analogs, but are substantially more stable toward nucleases. These modified DNA molecules open new prospects for the stability of antisense oligonucleotides and applications in biotechnology and chemotherapy.

3.2 Energy Storage

Nucleoside phosphates (nucleotides) are important not only because of their participation in biological polymers. A number of monomeric nucleotides are important to such diverse functions as energy storage (ATP), regulation (cyclic nucleotides), and cofactor chemistry (NAD^+ and $NADP^+$, see Chapter 7).

3.2.1 Phosphates

The importance of phosphate in DNA and RNA macromolecules has been described. It serves as an integral part of the molecular "backbone" as well as the means by which the monomeric units are connected. It has important binding properties, being able to participate in strong (electrostatic) bonding interactions with metal and amine cations.

Free inorganic phosphate may be represented as a tetrahedral structure with three negative charges distributed over four oxygen atoms. Binding of appropriate ligands localizes these charges. The tetrahedral structure of inorganic phosphate ($PO_4^{3\ominus}$, or P_i)

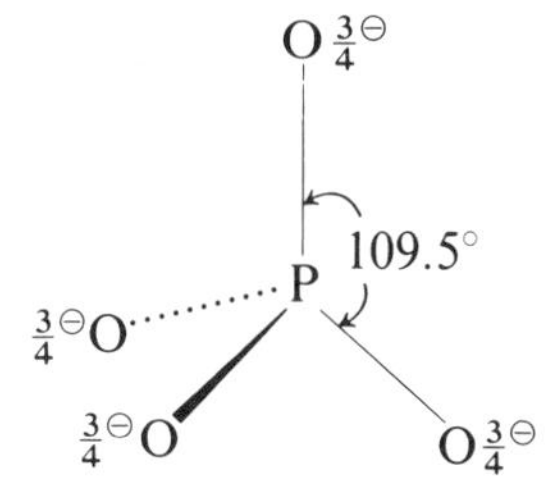

structure, the evenly distributed charge-density (per oxygen atom), and the pK_a values for the proton dissociation of phosphoric acid (H_3PO_4:2.1,7.2, and 12.3) may be considered as a consequence of the electrostatic repulsion between each charge of the trianionic species. As such, the phosphoryl function (P=O) is not a localized entity in phosphate salts or all organophosphates, where it can migrate between two or more oxygen atoms. This is similar to the delocalization of the carbonyl function over two oxygen atoms in carboxylic acids. Here also, esterification is required to localize the carbonyl function. In many respects, carboxylic acid derivatives and phosphoric acid derivatives will be seen to have similar properties.

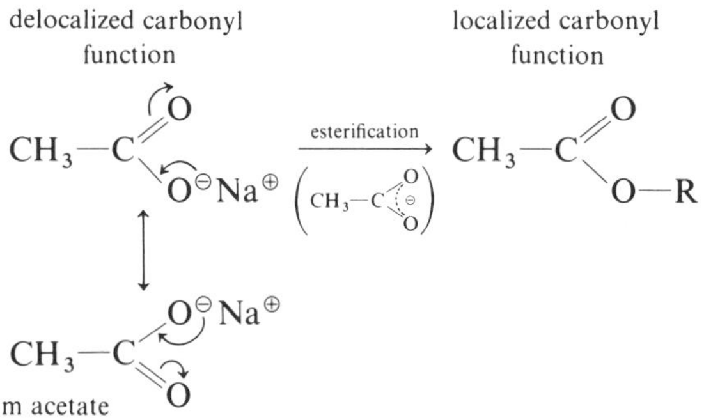

delocalized carbonyl function

esterification

localized carbonyl function

sodium acetate

esterification

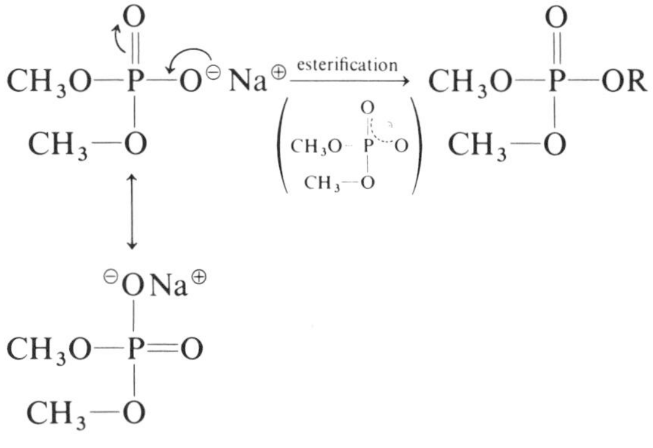

sodium dimethyl phosphate

However, the phosphoryl function does not consist of a π-bond between two p orbitals as does the carbonyl function. The phosphoryl π-bond involves an overlap between the p orbital of the oxygen atom and the d orbital of the phosphorus atom. This bond is a hybrid structure, even when localized, as is illustrated for triphenylphosphine oxide:

$$(Ph)_3P{=}O \leftrightarrow (Ph)_3\overset{\oplus}{P}{-}\overset{\ominus}{O} \text{ (ylid form)}$$

The phosphoryl function has significant double bond character, but the ylid is also important.

Phosphate readily forms a number of covalent compounds, ranging from simple esters (trimethyl or triethyl phosphate) to complex macromolecules (DNA or RNA). Many are of biological importance, and the role of phosphate is by no means confined to DNA and RNA structure. In fact, just as a number of amino acids do not have to be part of a protein structure in order to have a biological function, many monomeric nucleotides are also of biological importance. Examples include ATP and the biological regulatory molecules cyclic adenosine and guanosine monophosphate (cAMP and cGMP) (Section 3.2.3). Furthermore, a number of nonnucleoside phosphates are also of biological importance. Examples include sugar phosphates and creatine phosphate; an alternative

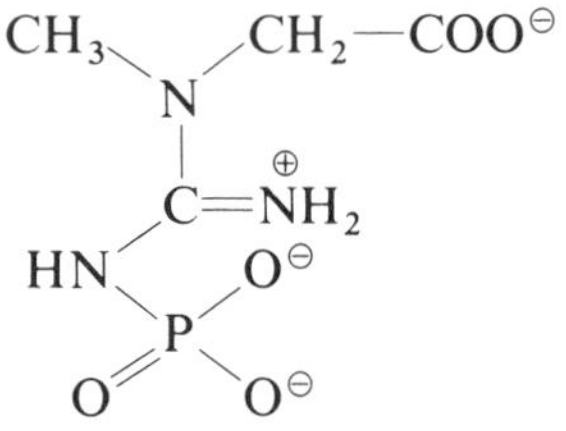

CH_3 $CH_2{-}COO^{\ominus}$
N
$C{=}\overset{\oplus}{N}H_2$
HN $O^{\ominus}$
P
O $O^{\ominus}$

creatine phosphate

energy store containing an acid labile phosphoramidate linkage. However, no further discussion will be given to the role of nonnucleoside phosphates, since these are overshadowed by their nucleotide counterparts and have been adequately described in numerous biochemistry texts. Of course, phosphate readily forms inorganic salts that also can be an important constituent of the biological organism (i.e., blood and urine phosphorus).

3.2.2 ATP

As has been noted, ATP (adenosine triphosphate) functions as a biological energy store, thus enabling the synthesis of a number of biochemically important compounds. For example, the synthesis of the methyl donor *S*-adenosyl methionine from ATP and methionine has

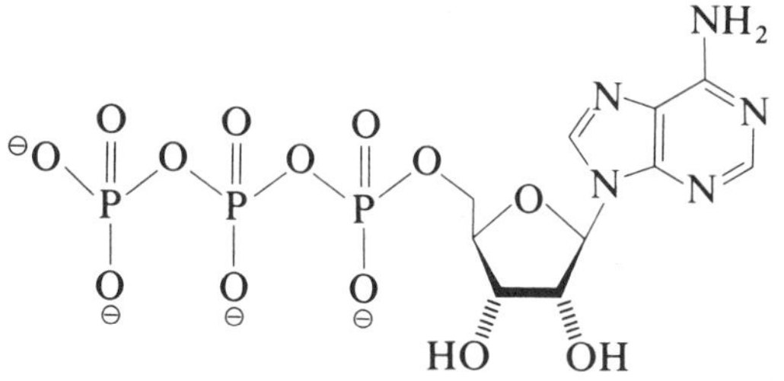

ATP

already been described (Chapter 2). The "high-energy" content arises from the presence of the anhydride structure in the triphosphate chain; the adenosine portion of the molecule functions as a recognition and binding site for the various ATP-utilizing enzymes. As such, simpler phosphate anhydrides (i.e., acetyl phosphate, pyrophosphoric acid) also possess a similar energy content. The reactivity and energy content of anhydrides may be attributed to "competing resonance," or two functions (phosphoryl, carbonyl, etc.) competing for the nonbonding electrons from one oxygen atom. Upon hydrolysis, this competition is relieved and a greater resonance stabilization is observed. Hence, product formation is favorable, relative to the reactant anhydride. The function (carbonyl, phosphoryl, etc.) will be more electrophilic in the anhydride than in the

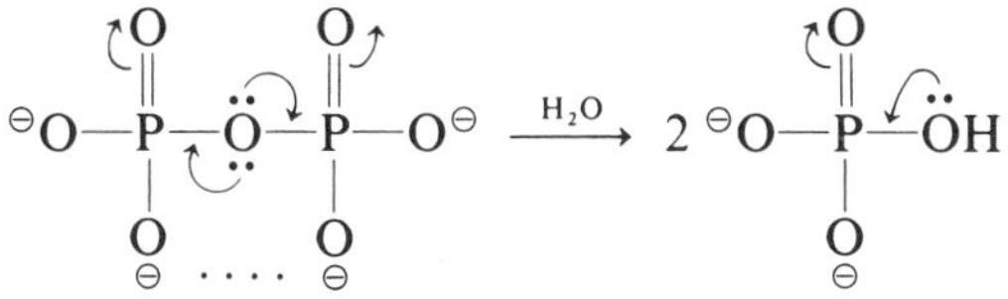

$$\ominus O-P(=O)(O^{\ominus})-O-P(=O)(O^{\ominus})-O^{\ominus} \xrightarrow{H_2O} 2\ \ominus O-P(=O)(O^{\ominus})-OH$$

ester analog, and thus more reactive, since in the ester there will be one oxygen atom per function and not one oxygen atom shared by two functions. In the case of phosphoric anhydrides, product formation will be favorable since it will reduce electrostatic repulsion along the di- or triphosphate chain. Multiple anionic charges will no longer be covalently linked together.

At physiological pH (7.35), three of the four ionizable protons will be lost. The fourth has a pK_a value of 6.5 and so it will be mostly ionized. In the cell, this polyanionic species will tend to bind magnesium and so exist as a 1:1 complex with the latter. In vitro, ATP will also bind other divalent metal ions, such as calcium, manganese, and nickel. In addition to the two phosphate oxyanions, chelation of the metal ion can also involve the adenine base (i.e., N-7 of the purine ring). The metal ion can serve as an electrophilic catalyst (Lewis acid) for the hydrolysis of ATP.

Of course, the presence of the metal ion bound to the phosphate chain would partially neutralize the total negative charge, thus making it easier for the approach of a negative nucleophile, such as hydroxide.

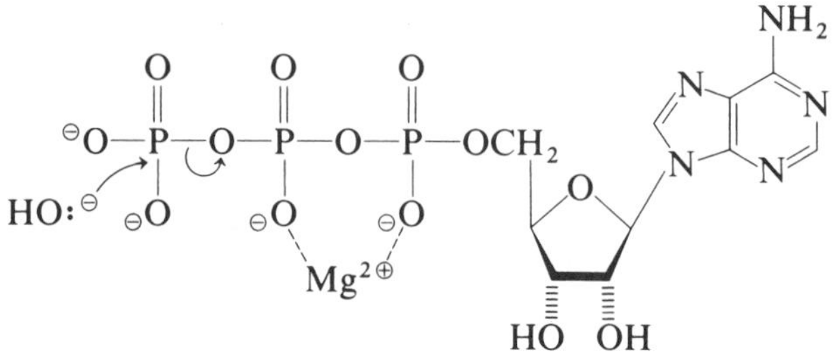

The $\Delta G^0{}_{hydro}$ at pH 7.0 for hydrolysis of the γ-(terminal) and β-phosphate is approximately −31.2 kJ/mol (−7.5 kcal/mol). This value is pH dependent and defined under standard conditions. However, within the cell the value has been estimated at approximately −50 kJ/mol (−12 kcal/mol). At any rate, ATP is, for the purpose of chemical reactions within the biological system, a "high-energy" compound. While the definition is rather arbitrary, a high-energy compound may be defined as one that has, at physiological pH, a ΔG_{hydro} in excess of −29.4 kJ/mol (−7 kcal/mol) or ΔH_{hydro} in excess of −25 kJ/mol (−6 kcal/mol).

ATP can transfer this potential energy to a variety of biochemically important compounds. For example, it has already been demonstrated that the energy for peptide bond formation (via intermediate acid-anhydride formation) may be acquired from the ATP molecule. Similarly, it will be shown shortly (see below) that cyclic nucleotides, which are also high energy, may be synthesized from an ATP (or a GTP) precursor.

Guanosine triphosphate (GTP) is another high-energy compound that has the same structure as ATP, except the adenine base is replaced by guanine. While of less use than ATP in biological systems, it still finds some use in energy-requiring processes, such as peptide bond synthesis on the ribosome.

The biosynthesis of adenosine and guanosine triphosphates proceeds by phosphorylation of the low-energy monophosphate precursors. For example, GTP may be synthesized from guanosine monophosphate (GMP), at the expense of two ATP molecules:

$$\text{GMP} + \text{ATP} \rightleftharpoons \text{GDP} + \text{ADP}$$

$$\text{GDP} + \text{ATP} \rightleftharpoons \text{GTP} + \text{ADP}$$

These reactions are catalyzed by the enzymes nucleoside monophosphokinase and nucleoside diphosphokinase, respectively. Note that these reactions are reversible, so that ATP may be synthesized at the expense of GTP or another nucleoside triphosphate. The precursor ADP (adenosine diphosphate) may also be synthesized from the reaction of AMP with ATP, catalyzed by the enzyme adenylate kinase:

$$\text{AMP} + \text{ATP} \rightleftharpoons 2\ \text{ADP}$$

All the interconversions above are thermodynamically acceptable, since they involve the formation of an anhydride structure at the expense of one that already exists. Perhaps, as this suggests, the total ATP, ADP, and AMP in the cell is in a steady-state concentration. It should be realized that ATP may be synthesized by phosphorylation of ADP with other phosphate donors besides the nucleoside triphosphates and that within the cell pathways exist by which the energy from the breakdown of sugars (glucose) is utilized for this ATP synthesis.

An elegant large-scale synthesis of ATP is one that was recently developed by G.M. Whitesides, M.I.T., that utilizes acetyl phosphate as the phosphate donor and *immobilized enzymes* (see Section 4.7) as a catalyst (98). The reaction occurs under mild conditions (approximately neutral pH, room temperature), the insoluble polymeric catalyst is easily removed (centrifugation) and need be present only in small amounts relative to the substrate, and the reaction is more specific than a non-enzymatic synthesis (i.e., adenosine tetraphosphate is not formed as a by-product). The net reaction is:

$$3\left(CH_3{-}\overset{\overset{\displaystyle O}{\|}}{C}{-}O{-}\overset{\overset{\displaystyle O}{\|}}{P}(O^{\ominus})_2\right) + \text{HO—(ribose: O, Ad; HO, OH)}$$

$$\Big\downarrow\ \text{3 enzymes, pH = 6.7–6.9, 20°C}$$

$$3\left(CH_3{-}\overset{\overset{\displaystyle O}{\|}}{C}{-}O^{\ominus}\right) + \text{ATP}$$

The reaction pathway in detail is as follows:

1) Adenosine + ATP	$\xrightarrow{\text{adenosine kinase}}$	AMP + ADP
2) AMP + ATP	$\xrightarrow{\text{adenylate kinase}}$	2 ADP
Total: Adenosine + 2 ATP	$\longrightarrow$	3 ADP
3) 3 ADP + 3 acetyl phosphate	$\xrightarrow{\text{acetate kinase}}$	3 ATP + 3 acetate

While the equations do not necessarily indicate this, only a small amount of ATP (less than 1/1000 of the amount produced by weight) needs to be added in order to initiate the reaction, since the reaction will be perpetuated by the product ATP. The three enzymes are immobilized to a cross-linked polyacrylamide gel by reaction of their primary amino functions with "active esters" on the polymer surface.

Finally, mention must be made of the recent synthesis of ATP chiral at the γ-phosphorus by the group of J.R. Knowles at Harvard University (99). This was accomplished by reaction of ADP with a phosphorylating agent isotopically labeled to form a [γ-^{16}O,^{17}O,^{18}O]-ATP product, as outlined below:

$^{17}O=PCl_3$ + (−)-ephedrine hydrochloride $\xrightarrow{NEt_3/Et_2O}$ cyclic phosphoramidate chloride $\xrightarrow{Li^{18}OH}$ $Li^{\oplus}\,{}^{18}O^{\ominus}$ cyclic phosphoramidate $\xrightarrow{\text{ADP (free acid)}}$ ephedrine-ATP adduct $\xrightarrow{Pd/C}$

(−)-ephedrine

Reaction of ^{17}O-labeled phosphorus oxychloride with (−)-ephedrine produces a reactive five-membered ring whose chlorine atom may be displaced in excellent yield by reaction with ^{18}O-labeled lithium hydroxide. The product is an excellent phosphorylating agent by virtue of the strain of the five-membered ring, and the acid lability of the phosphoramidate linkage. The benzylic linkage of the acyclic phosphomonoester is now susceptible to catalytic hydrogenolysis, to give rise to the chiral ATP product.

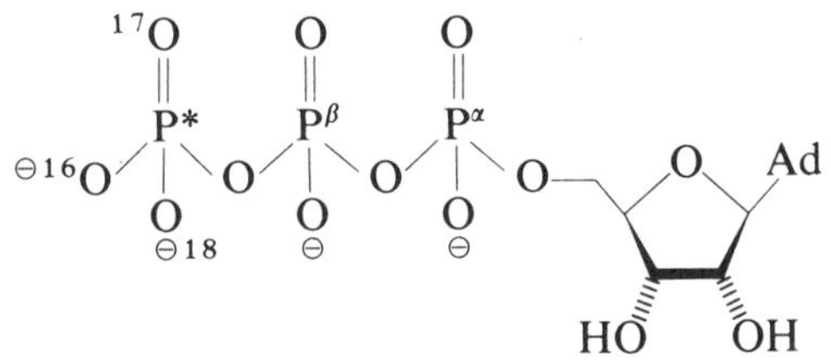

More recently, G. Lowe's group (University of Oxford) has reported the synthesis of adenosine 5′[(*R*)-α-^{17}O] triphosphate by a series of enzymatic manipulations of 5′[(*S*)-α-thio] triphosphate (100).

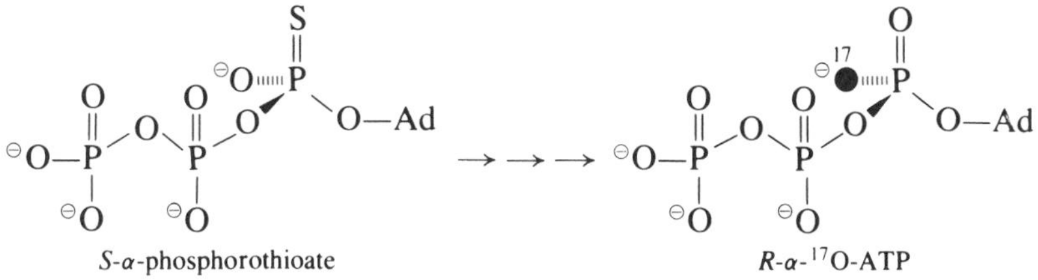

S-α-phosphorothioate → → → *R*-α-^{17}O-ATP

The synthetic ATP is an excellent probe for the stereochemistry and hence reaction mechanism of phosphorylating (kinase) enzymes (see page 125). At least two pathways are possible for the enzyme-catalyzed donation of the γ-phosphate of ATP to a substrate. This may simply proceed via direct displacement on the enzyme surface, with an overall inversion of configuration in the case of a chiral γ-phosphate:

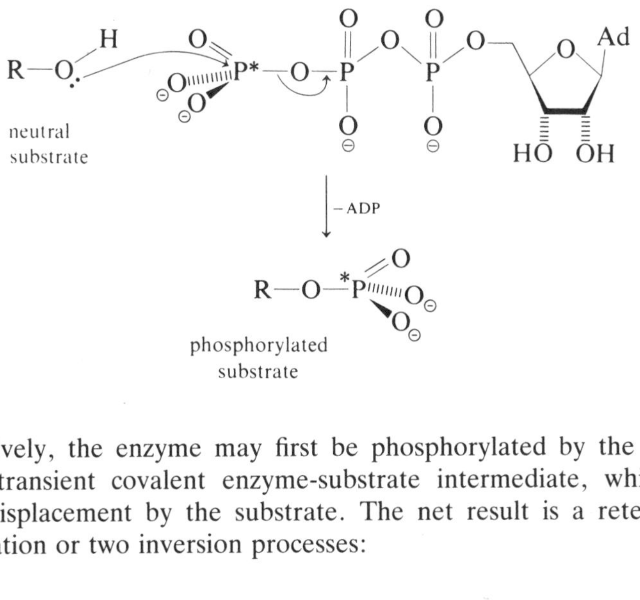

neutral substrate

− ADP

phosphorylated substrate

Alternatively, the enzyme may first be phosphorylated by the ATP to form a transient covalent enzyme-substrate intermediate, which then suffers displacement by the substrate. The net result is a retention of configuration or two inversion processes:

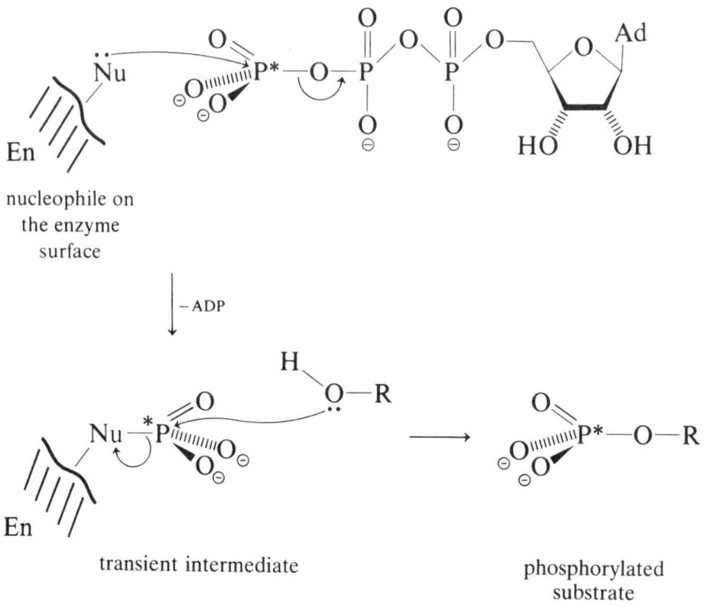

nucleophile on the enzyme surface

−ADP

transient intermediate

phosphorylated substrate

In the case of the enzymes glycerol kinase, pyruvate kinase, and hexokinase, inversion of configuration was observed at the chiral γ-phosphorus (101).

3.2.3 cAMP and cGMP

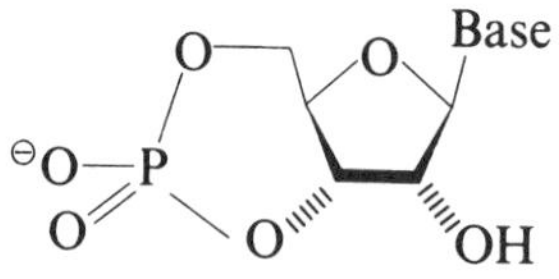

Base = adenine or guanine

Cyclic nucleotides, 3′,5′-cyclic adenosine monophosphate (cAMP) and 3′,5′-cyclic guanosine monophosphate (cGMP), function to regulate cell-to-cell communication processes (102). Cellular communication follows primarily three pathways. The first involves the transmission of electrical impulses via the nervous system. The second involves chemical messengers or hormonal secretions. The third involves *de novo* protein synthesis. All three processes are usually in response to some demand or stimulus and involve, at least to some extent, regulation by cyclic nucleotides.

The enzymes that synthesize cAMP and cGMP are adenyl cyclase and guanyl cyclase, respectively.

There are numerous other examples of cyclic nucleotide-mediated processes that fall into the three categories noted above: hormonal secretion, neuronal transmission, and protein synthesis. For example, cAMP, again via a protein kinase, activates the enzyme tyrosine hydroxylase.

Tyrosine hydroxylase, which catalyzes the conversion of the amino acid tyrosine to DOPA (dioxyphenylalanine), is the rate-limiting step in the biosynthesis of the neurotransmitters dopamine and norepinephrine. A *neurotransmitter* is a substance that is released at the junction of two nerve cells or a nerve and muscle cell (a so-called *synapse*) when it is required that an electrical (nerve) impulse be transmitted from one cell to the next. This is a chemical alternative to an electrical discharge traveling between the two cells, for the neurotransmitter is released only upon arrival of the electrical impulse at the synapse and once the neurotransmitter binds to the second cell, a potential difference is created that results in the generation of an electrical impulse in the second cell. Interestingly, creation of this *action potential* in certain nerve cells appears to require cyclic nucleotide-mediated protein kinase.

HO–C$_6$H$_4$–$CH_2CH(COO^{\ominus})(\overset{\oplus}{N}H_3)$ (tyrosine) —(cAMP) ⇓ tyrosine hydroxylase, O_2→ (HO)$_2$C$_6$H$_3$–$CH_2CH(COO^{\ominus})(\overset{\oplus}{N}H_3)$ (DOPA)

DOPA —DOPA decarboxylase, $-CO_2$→ (HO)$_2$C$_6$H$_3$–$CH_2CH_2\overset{\oplus}{N}H_3$ (dopamine)

dopamine —dopamine-β-hydroxylase→ (HO)$_2$C$_6$H$_3$–$CH(OH)CH_2\overset{\oplus}{N}H_3$ (norepinephrine)

cAMP also activates the enzyme RNA polymerase I (mammalian), as always by protein kinase-mediated phosphorylation. This enzyme catalyzes the synthesis of the mRNA polymer, which is a process essential to the synthesis of proteins of correct amino acid sequence. Other enzymes that are known to be mediated by cyclic nucleotides include glycogen synthetase (synthesis of glycogen), phosphofructokinase (phosphorylation of fructose 6-phosphate), ornithine decarboxylase (decarboxylation of the amino acid ornithine), and ATPase (hydrolysis of ATP).

It is noteworthy that the cyclic phosphate ring of cAMP and cGMP is a high-energy structure (103). It has a ΔG_{hydro} at pH 7.3 of approximately -12.5 kJ/mol (-3.0 kcal/mol) smaller than the biological energy store ATP. It is the enthalpy term that accounts mostly for this free energy value; the ΔH_{hydro} of cAMP and cGMP are -59.2 (-14.1) and

−44.1 kJ/mol (−10.5 kcal/mol), respectively. Of course, hydrolysis represents the conversion of cyclic nucleotide to a 5′-acyclic structure:

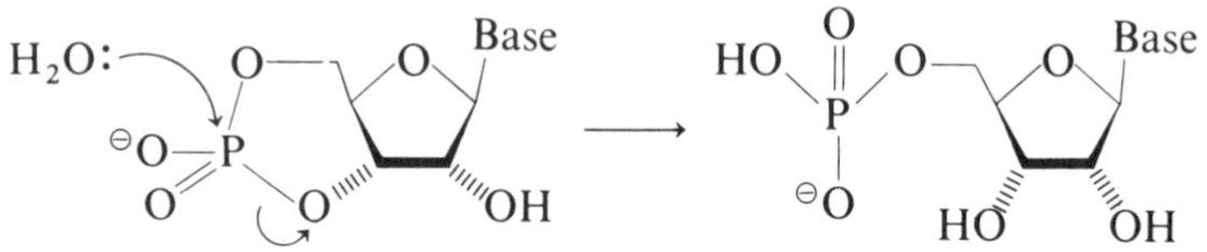

Base = adenine or guanine

The importance of these high-energy cyclic structures to the biological function of cyclic nucleotides is not yet certain. However, it has been suggested that such is necessary for the binding of cyclic nucleotides to the regulatory subunit of protein kinase via a covalent intermediate. For example, a carboxylate residue of the regulatory subunit might open the cyclic phosphate to form a high-energy intermediate. This in turn could break an ionic attraction between the regulatory and catalytic subunits, thus causing dissociation of the two structures:

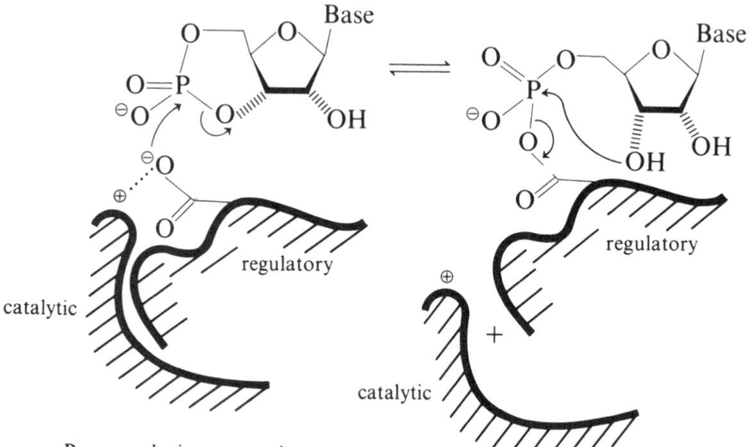

Base = adenine or guanine

It is known that six-membered cyclic phosphates are not typically high-energy structures. That is, the six-membered cyclic phosphate, unlike the five-membered ring, is not strained (angular distortions). This is indicated by the enthalpy data for ethylene, trimethylene, and tetramethylene phosphoric acid:

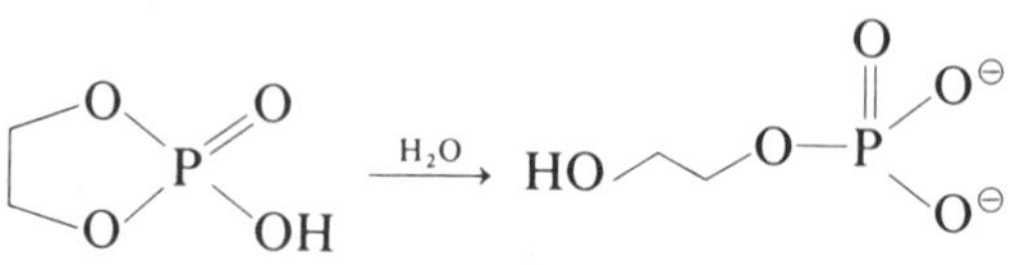

$\Delta H_{\text{hydro}} = -26.9$ kJ/mol (-6.4 kcal/mol)

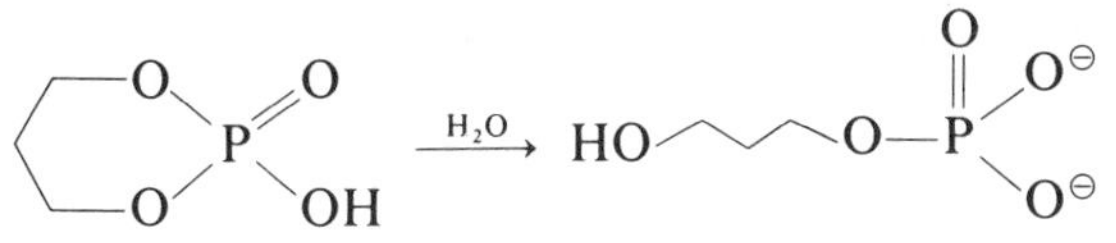

ΔH_{hydro} = −12.6 kJ/mol (−3.0 kcal/mol)

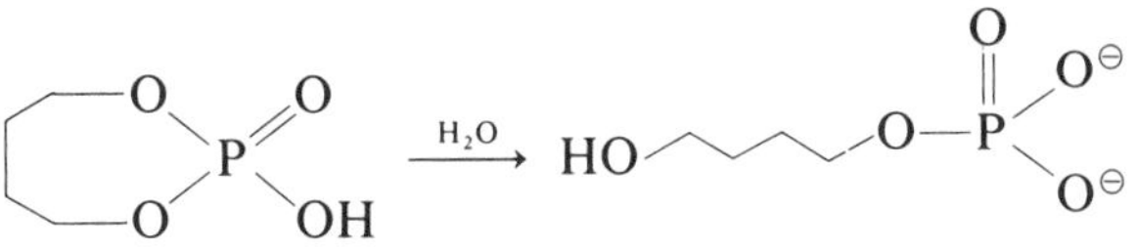

ΔH_{hydro} = −9.2 kJ/mol (−2.2 kcal/mol)

Further, the base is not responsible for this high-energy content. It is the presence of the ether oxygen of the ribofuranoside ring *trans* fused to the six-membered cyclic phosphate that is believed to be responsible for the high-energy content of the cyclic phosphate. Hence, replacement of the furanose ring with a pyranose (glucose) or a cyclopentane structure (i.e., the ether oxygen with a methylene group) results in a cyclic phosphate of significantly lower energy.

3.3 Hydrolytic Pathways and Pseudorotation

Most chemical reactions involving phosphates in the biological system are either the addition (phosphorylation) or removal (hydrolysis) of phosphate. In the case of biological phosphorylations, the phosphate source or donor is the energy store ATP, and the reaction is catalyzed by an enzyme often referred to as a kinase. A simple example is the phosphorylation of the sugar glucose:

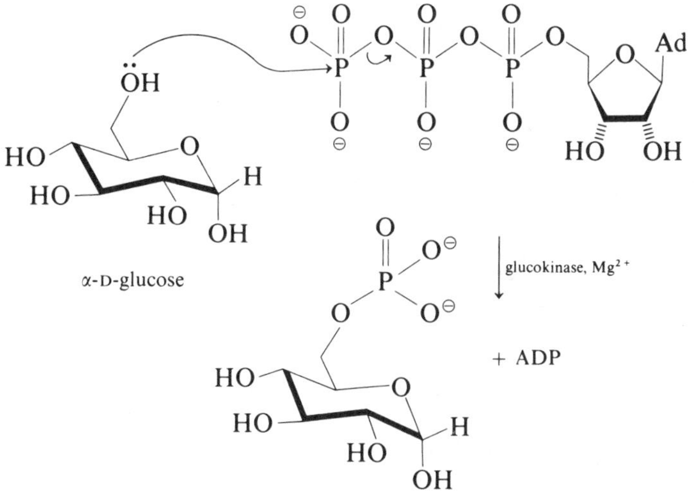

α-D-glucose

glucokinase, Mg^{2+}

+ ADP

α-D-glucose 6-phosphate

Acid and base-catalyzed hydrolysis of carboxylate esters usually proceeds by *acyl fission*. However, a few cases of *alkyl fission* are known.

$$R{-}C(=O){-}O{-}CH_3 + {}^{\ominus}OH \longrightarrow R{-}C(=O){-}O^{\ominus} + HOCH_3$$

acyl-oxygen fission

For example, dimethyl ether will form upon reaction of methoxide with methyl benzoate. Acyl fission merely regenerates starting materials. The

$$Ph{-}C(=O){-}O{-}CH_3 + {}^{\ominus}OCH_3 \longrightarrow Ph{-}C(=O){-}O^{\ominus} + CH_3{-}O{-}CH_3$$

acid-catalyzed hydrolysis of *tert*-butyl acetate proceeds by an S_N1 process (stable carbonium intermediate) with alkyl fission.

$$CH_3{-}C(=O){-}O{-}C(CH_3)_3 \xrightarrow{H^{\oplus}/H_2O} CH_3{-}C(=O){-}OH + HO{-}C(CH_3)_3$$

Neutral phosphate triesters will readily undergo acid- or base-catalyzed hydrolysis with alkyl and/or *phosphoryl fission*. For example, trimethyl and triethyl phosphate will undergo hydrolysis in neutral water via an S_N2 mechanism with alkyl fission. Acid-catalyzed hydrolysis also proceeds with alkyl fission.

$$EtO{-}P(=O)(OEt){-}O{-}CH_2{-}CH_3 + :OH_2 \xrightarrow[-\,EtOH]{H_2O,\ pH\ 7} (EtO)_2P(=O){-}O^{\ominus}$$

The base-catalyzed hydrolysis of trimethyl phosphate proceeds by phosphoryl fission and allows a preparative synthesis of dimethyl phosphate.

$$CH_3O{-}P(=O)(OCH_3){-}OCH_3 \xrightarrow[1\ N\ NaOH]{24\ hr,\ 20\ C} CH_3O{-}P(=O)(O^{\ominus}){-}OCH_3\ Na^{\oplus}$$

The anionic product causes electrostatic repulsion of any incoming hydroxide, so that monomethyl phosphate will not readily form. The half-life of this latter reaction is 16 days at 100°C, and the mechanism involves exclusively alkyl fission since the phosphorus atom is less accessible:

$$CH_3O-P(=O)(O^{\ominus})-O-CH_3 + {}^{\ominus}OH \longrightarrow CH_3O-P(=O)(O^{\ominus})_2$$

In the case of the base-catalyzed hydrolysis of diphenyl phosphate, only phosphoryl fission can occur. This proceeds very slowly (the half-life for the hydrolysis of diphenyl phosphate in neutral water, at 100°C, has been estimated at 180 years!) enhanced only by the better leaving ability of phenoxide, relative to alkoxide. Indeed, appropriately substituted diphenyl phosphates (i.e., bis-2,4-dinitrophenyl phosphate) with more acidic leaving groups will hydrolyze at a significantly greater rate. This data suggests that DNA, which functions as a "tape" for the storage of genetic information, is resistant to decomposition by hydrolysis. Indeed, after 1 hr at 100°C, in 1 *N* sodium hydroxide, no degradation of DNA is observed.

On the other hand, 2-hydroxyethyl methyl phosphate will easily undergo base hydrolysis: the half-life of the reaction is 25 min, at 25°C in 1 *N* sodium hydroxide. This may be accounted for by the intermediate formation of a strained five-membered cyclic phosphate by nucleophilic

$$HO-CH_2CH_2-O-P(=O)(O^{\ominus})-OCH_3 \xrightarrow{HO^{\ominus}} \left[\text{cyclic } (CH_2O)_2P(=O)O^{\ominus}\right] \xrightarrow{{}^{\ominus}OH} HO-CH_2CH_2-O-P(=O)(O^{\ominus})_2$$

strained five-membered ring

attack of the vicinal hydroxyl (*anchimeric assistance*) function. An indication of the strain present in a five-membered cyclic phosphate lies in the observation that the rate of hydrolysis of ethylene phosphate is 10^6 times that of its acyclic analog, dimethyl phosphate. Further, its hydrolysis is considerably more exothermic reflecting the inherent strain energy present in the cyclic structure:

$$(CH_2O)_2P(=O)O^{\ominus} \xrightarrow{H_2O} HO-CH_2CH_2-O-P(=O)(O^{\ominus})_2$$

ethylene phosphate

$\Delta H = -26.8$ kJ/mol
(-6.4 kcal/mol)

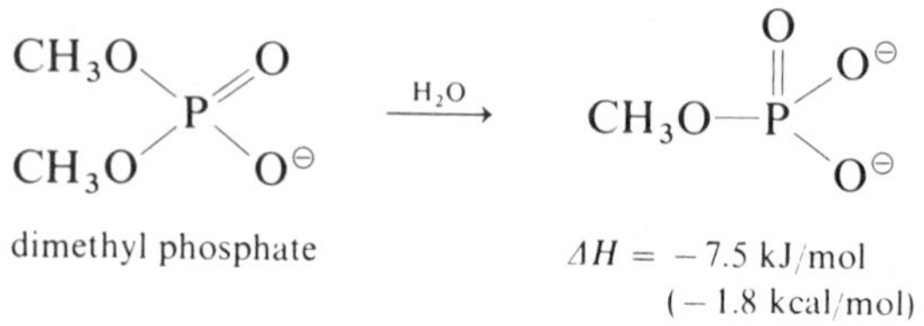

$(CH_3O)_2PO_2^{\ominus} \xrightarrow{H_2O} CH_3O{-}PO_3^{2\ominus}$

dimethyl phosphate

$\Delta H = -7.5$ kJ/mol
(-1.8 kcal/mol)

These data suggest that RNA, which functions as a carrier of genetic information and thus has a high turnover rate, is unstable or readily able to undergo degradation. Indeed, RNA will be degraded in $0.1\,N$ sodium hydroxide, at room temperature, to give a mixture of 2′- and 3′-phosphates:

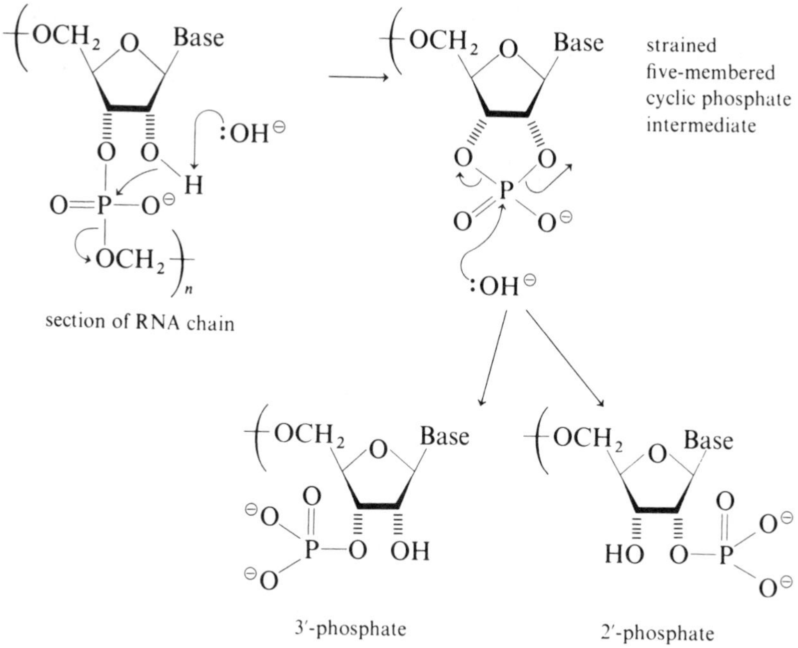

section of RNA chain

strained five-membered cyclic phosphate intermediate

3′-phosphate

2′-phosphate

Thus far, hydrolysis mechanisms have been written as if they proceed by a simple displacement process. However, just as the hydrolysis of carboxylate esters proceeds by a tetrahedral intermediate (as has been already noted), phosphate ester hydrolysis proceeds by a dsp^3 hybridized "pentacoordinate" intermediate. The geometry of this intermediate is that which is representative of five electron pairs about a central (phosphorus) atom: trigonal bipyramidal. In fact, a number of stable pentacoordinate phosphorus compounds adopt such a geometry. For example, gaseous phosphorus pentachloride (PCl_5) exists as a trigonal bipyramid. It is important to note, as will be seen shortly, that while it has five equivalent (chlorine) ligands, the dsp^3 hybridization scheme provides five nonequivalent bonding lobes. As such, two of the chlorine-phosphorus bond lengths (designated as *apical* chlorines; 0.22 nm) are longer than the remaining chlorine-phosphorus bond lengths (designated as *equatorial* chlorines; 0.20 nm). The three equatorial chlorine atoms lie in a plane, while the two apical atoms are perpendicular to that plane.

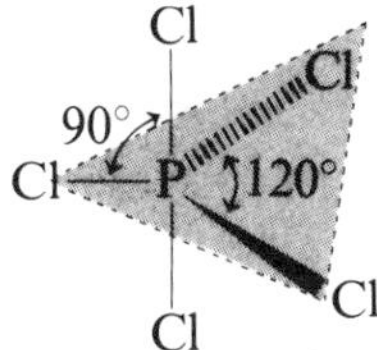

trigonal bipyramidal structure of gaseous phosphorus pentachloride

During a hydrolysis reaction of an initially tetrahedral phosphate ester, the entering group would approach from the distance, perpendicular to three ligands so that it will be perpendicular to the plane of the equatorial atoms in the pentacoordinate intermediate. Such is expected, for the new bond that is forming will, at least at first, be long in nature, which is compatible with the long bond of an apical ligand. This is analogous to the initial long bond formation that occurs in S_N2 displacements. Similarly, the leaving group will depart from an apical position only, going through a long-bond geometry during its departure. This is illustrated for the hydrolysis of a phosphomonoester:

tetrahedral phosphomonoester

"long-bond" formation

departing electron density

$\longrightarrow$ R—OH + HPO_4^{2-}

A consideration of the hydrolysis of methyl ethylene phosphate indicates that the hydrolysis of organophosphates does not necessarily proceed by a simple displacement mechanism. If this were the case, then exclusive opening of the strained ring would be expected to occur:

Notice that the five-membered ring in the pentacoordinate intermediate is represented as spanning an apical and equatorial position. This is the most favorable geometry since the strain in five-membered rings arises from angular distortions. A five-membered ring spanning two equatorial positions or two apical positions would be too unstable.

In fact, acid-catalyzed hydrolysis of methyl ethylene phosphate leads to significant displacement of the methoxy function:

$$\text{(cyclic methyl ethylene phosphate, P(=O)(OCH}_3\text{))} \xrightarrow{H^{\oplus}/H_2O} HO\text{–}CH_2CH_2\text{–O–P(=O)(OH)(OCH}_3) + \text{(cyclic ethylene phosphate, P(=O)(OH))}$$

70% 30%

This observation is accounted for by the phenomenon of *pseudorotation* (104,105). Westheimer and Dennis developed, from 1966, rules for when pseudorotation could occur, which can be used to predict both the speed of ester hydrolysis and whether it can occur without ring opening. Pseudorotation is the process of phosphorus-ligand bond deformation that occurs in pentacoordinate intermediates such that an appropriate leaving group that could not leave prior to the process is made able to do so. Pseudorotation will occur as dictated by the five "preference rules":

1. Hydrolysis of phosphate esters proceeds by way of a pentacoordinate intermediate that has the geometry of a trigonal bipyramid.
2. During the hydrolysis process, groups will enter and leave from apical positions only.
3. Because of strain, a five-membered ring will not be formed between two equatorial or two apical positions but, instead, between one apical and one equatorial position.
4. The more electropositive ligands tend to occupy equatorial positions, while the more electronegative ligands tend to occupy apical positions.
5. Ligands can exchange positions by the process of pseudorotation.

By pseudorotation two apical ligands become equatorial and simultaneously two equatorial ligands become apical. The dirving force for the process is transforming a good leaving group from an equatorial to an apical geometry and thus allowing it to leave. Such bond deformations are not surprising when it is remembered that covalent bonds are not static entities but, instead, are able to undergo various bending and stretching modes, as indicated by infrared spectroscopy. During pseudorotation, the two apical bond lengths may be thought of as being "squeezed together" to form two new equatorial bonds, while two of the three equatorial bonds (the third bond remains "neutral" or of constant bond length during the process; this is referred to as the "pivot") may be thought of as being "pushed out" (lengthened) in response to form two new apical bonds. This is illustrated (arrows indicate increasing or decreasing bond length):

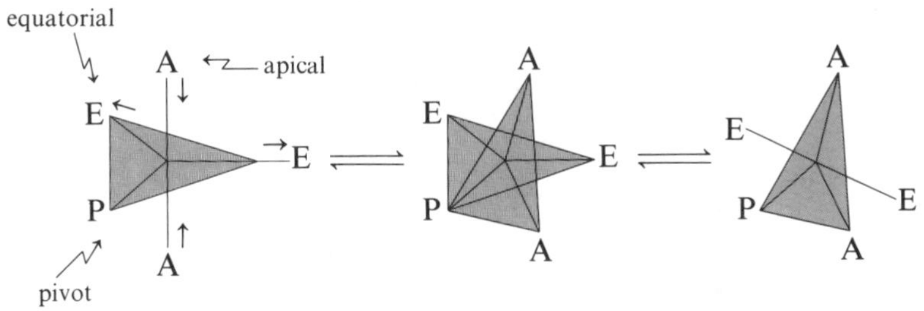

equatorial
A apical
E
E
P
A
pivot

Hence, via pseudorotation, the methoxy function of methyl ethylene phosphate is able to leave.

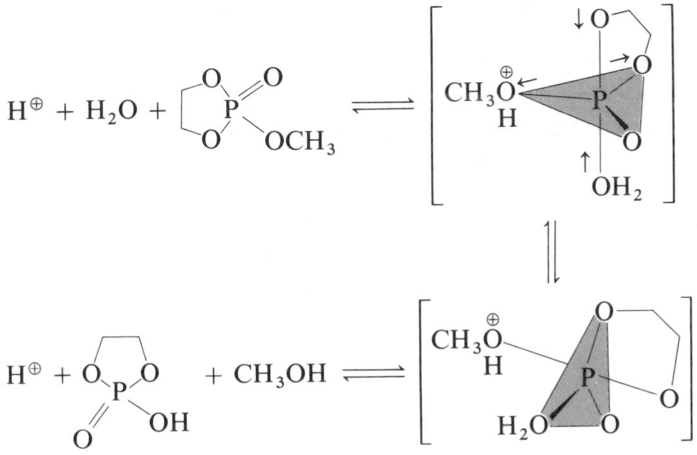

$H^{\oplus} + H_2O +$ [cyclic methyl ethylene phosphate] $\rightleftharpoons$ [pentacoordinate intermediate] $\rightleftharpoons$ [pseudorotated intermediate] $\rightleftharpoons H^{\oplus} +$ [cyclic ethylene phosphate, P–OH] $+ CH_3OH$

Hydrolysis of the phosphonate analog proceeds exclusively with ring opening

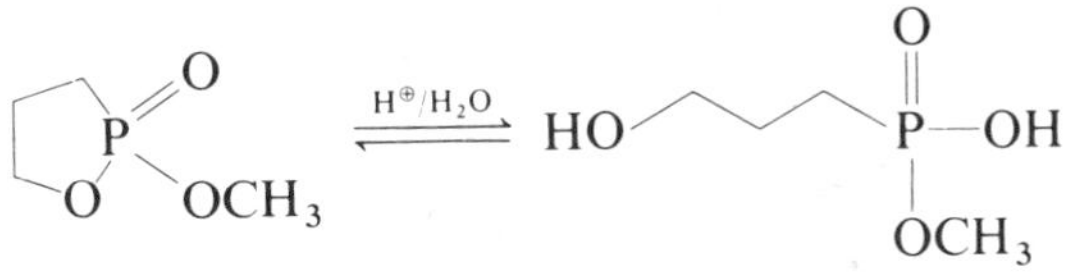

[cyclic phosphonate, P–OCH_3] $\xrightleftharpoons{H^{\oplus}/H_2O}$ HO–$(CH_2)_3$–P(=O)(OH)OCH_3

This is a consequence of the fourth preference rule, which states that electropositive ligands prefer to be equatorial (close to phosphorus) and electronegative ligands prefer to be apical (away from phosphorus). Phosphorus, itself a nonmetal, is fairly electronegative and so prefers to be away from other electronegative atoms. As such, the carbon atom of the cyclic phosphonate will acquire an equatorial geometry and the oxygen atom an apical geometry:

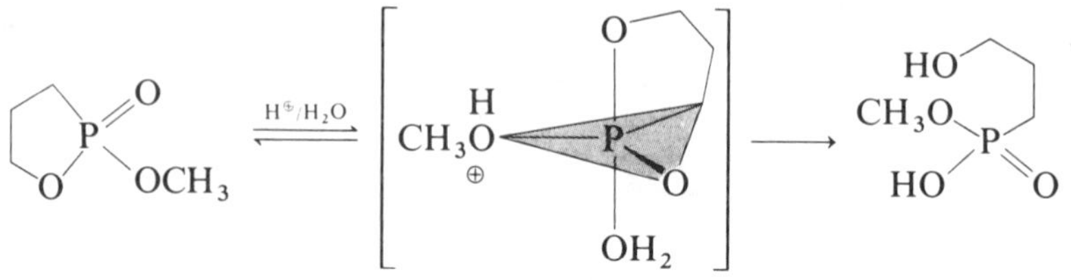

Once the pentacoordinate intermediate is formed, it cannot undergo a pseudorotation since this would require the oxygen and carbon atoms of the ring to exchange positions. Inhibition of pseudorotation is the result and thus direct displacement by water is the only available pathway.

It has already been noted that within the organism, hydrolysis of organophosphates is an important reaction pathway.

Having considered some of the mechanisms of hydrolysis, it is of interest to compare these with the active site mechanism of a few enzymes that catalyze such reactions. The enzyme ribonuclease A (RNase A)* (bovine pancreas, MW = 13,680, one polypeptide chain consisting of 124 amino acid residues) catalyzes the degradation of RNA by a two-step mechanism: transesterification followed by hydrolysis of the five-membered cyclic intermediate. The enzyme will not cleave each phosphodiester linkage of the RNA polymer but will instead attack only along certain points of the chain. The susceptibility of a particular phosphorus atom to attack by the enzyme is dependent on the purine or pyrimidine base present at the sugar esterified to the phosphate by a 3′-hydroxyl. Further, the base type is dependent on the source of the enzyme. For example, bovine pancreatic RNase will attack the phosphoester linkage if the nucleoside at the 3′-linkage contains a pyrimidine. On the other hand, RNase from the bacteria *B. subtilis* will attack the phosphorus linkage if the nucleoside at the 3′-linkage contains a purine.

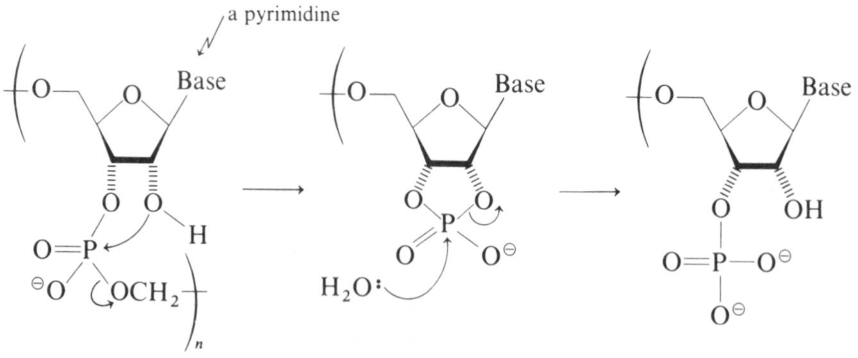

*The fundamental basis of enzyme function will be presented only in Chapter 4. However, Sections 4.1 and 4.2 could be read at this point for a better understanding of RNase A mode of action.

Various studies indicate the participation of at least three amino acid residues in the active site chemistry of ribonuclease: two histidines and one lysine. RNA hydrolysis (Fig. 3.1) proceeds by two steps: transesterification followed by hydrolysis. Note that at physiological pH, one of the two imidazole rings is protonated while the other is not. The imidazole rings function as a general-base-general-acid catalyst while the cationic lysine probably aids in the stabilization of the pentacoordinates intermediate.

This lysine residue apparently is not involved in the initial binding of the phosphate moiety to the enzyme (106). The phosphate substrate is instead bound by means of two hydrogen bonds between the phosphoryl oxygens and an amide backbone hydrogen and a hydrated glutamine residue. These hydrogen bonds will also increase the electrophilicity of the phosphorus atoms. Another important hydrogen bond occurs between the protonated imidazole ring and the oxygen atom of the substrate that will be cleaved from the phosphorus atom. In addition to a binding function, this hydrogen bond will polarize electron density in a direction that will facilitate complete proton transfer from the imidazole ring during hydrolysis of the substrate.

The mechanism of hydrolysis is straightforward and resembles that for the hydroxide-catalyzed hydrolysis of RNA; with exception of specificity since only certain phosphodiester linkages are cleaved to give 3′- (and not 2′-) monophosphates (Fig. 3.2). However, with a knowledge of pseudorotation, two stereochemical mechanisms become possible for each step of the enzyme-catalyzed reaction.

The two mechanisms are referred to as "in-line" or "adjacent": the type depending upon the relative geometry of the incoming and leaving groups during the displacement process. Referring to the plane of the three equatorial ligands, the stereochemical mechanism is in-line if the entering group (nucleophile) enters on one side of the plane while the leaving group departs on the other side. On the other hand, if the entering group (nucleophile) enters on one side of the plane while the leaving group is situated in the plane, the mechanism is adjacent. According to the preference rules, the adjacent mechanism will require a pseudorotation to transform the leaving group from an equatorial to apical geometry.

Transesterification begins with an unprotonated imidazole at the active site abstracting a proton from the 2′-hydroxyl (general-base catalysis) of the RNA substrate. As an alkoxide ion is generated, it becomes a potent nucleophile that attacks the phosphorus atom. If the leaving group is in an apical geometry, it may do so, being protonated simultaneously by a second imidazole (imidazolium cation) situated below the plane of the three equatorial ligands. This in-line mechanism requires the participation of two histidine residues, for it would be physically impossible for the imidazole ring that removes the 2′-hydroxyl proton to extend below the

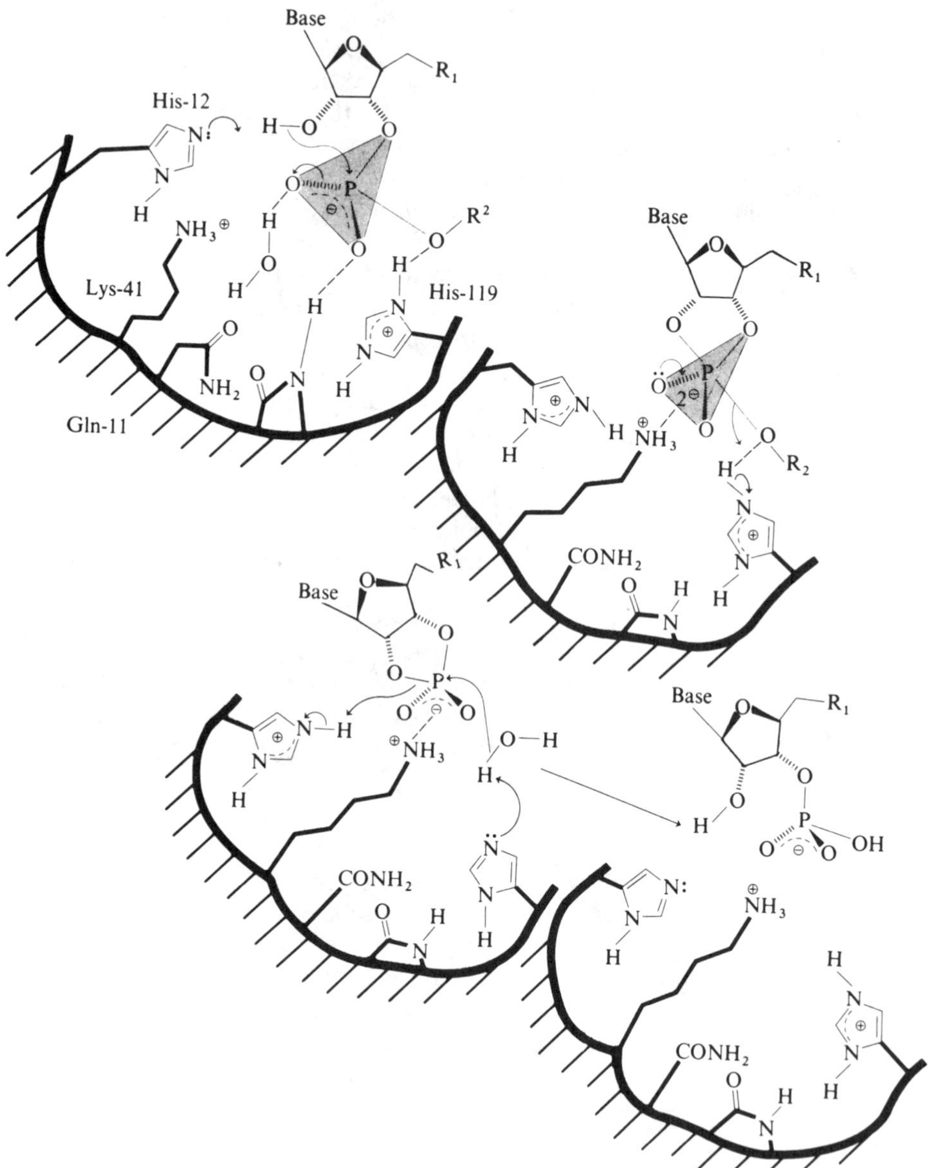

Fig. 3.1. Mechanism for the ribonuclease-catalyzed hydrolysis of RNA (106). Amino acid residue number refers to its position relative to the N-terminal amino acid of the polypeptide (protein) chain.

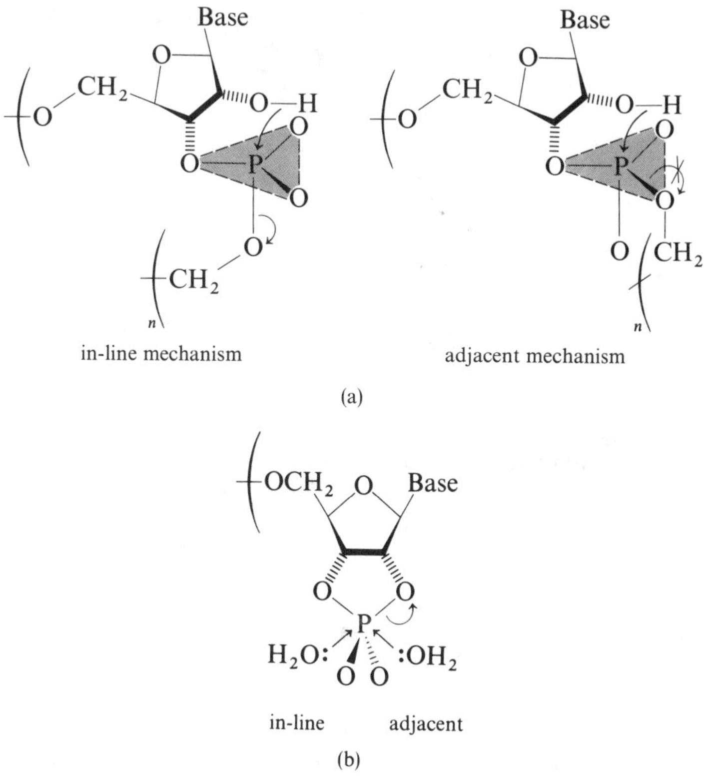

in-line mechanism adjacent mechanism

(a)

in-line adjacent

(b)

Fig. 3.2. Differences between in-line and adjacent mechanism of phosphodiester bond hydrolysis. (a) First step (transesterification) in the RNase-catalyzed hydrolysis of RNA illustrating two possible stereochemical pathways. The in-line mechanism allows a direct displacement while the adjacent mechanism requires a pseudorotation. (b) Second step (hydrolysis) in the RNase-catalyzed hydrolysis of RNA illustrating two possible stereochemical pathways. Again, the in-line mechanism allows a direct displacement to form the 3′-phosphate, while the adjacent requires a pseudorotation(108).

plane and protonate the leaving group (see Fig. 3.1). If the leaving group is in an equatorial geometry, it must first pseudorotate, or change bond length. However, it would still be situated close enough to the incoming 2′-hydroxyl to be protonated, after the pseudorotation, by the same imidazole that abstracted the 2′-hydroxyl proton from the substrate. This adjacent mechanism requires an additional step (a pseudorotation) but the participation of only one histidine residue.

Much the same arguments apply for the hydrolysis of the 2′,3′-cyclic phosphate. Here again, an imidazole ring functions as a general-base catalyst with abstraction of a proton from water, and depending on an in-line or adjacent mechanism, a second or the same imidazole can protonate the leaving group. Note that after hydrolysis is complete, the active site is fully regenerated and the initially unprotonated and protonated imidazole rings are again formed.

Later on, in 1986, Breslow and Labelle (109) studied the cleavage of poly-U by imidazole buffer and proposed that a sequential base-acid mechanism is involved. As shown above, the basic imidazole of RNase (His-12) acts to remove the proton from the attacking C-2′ hydroxyl group, but in a variant mechanism, they suggest that it then transfers that proton to one of the new equatorial phosphate oxygens, so that the intermediate formed is still a phosphate monoanion paired with the lysine cationic side chain. Following this, the deprotonated imidazolium ion of the enzyme (His-119) reaches up and removes the proton from one of the phosphate equatorial oxygens so as to form a product monoanion. Thus, the ribonuclease A imidazole groups serve not simply to remover or add protons but instead to move them from one part of the substrate to another. This proton movement is reminiscent of the function of the imidazole group of His-57 in α-chymotrypsin (see p. 187).

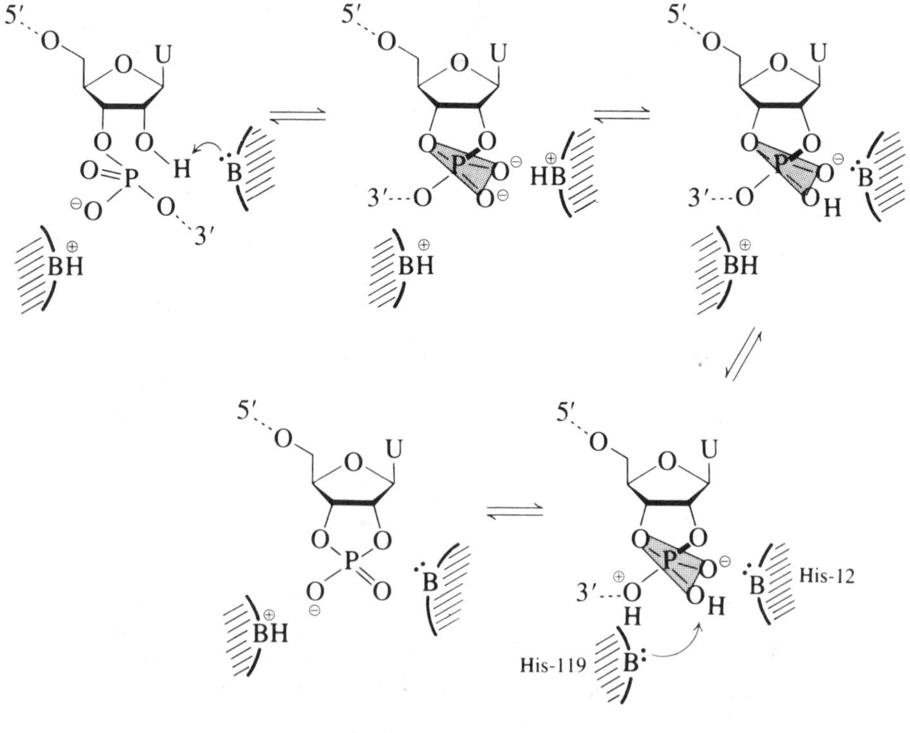

His-12

His-119

Subsequent to this work, further evidence showed that the alternative acid-base mechanism, which could not be excluded on the basis of earlier kinetic data, may in fact be excluded by using stereoelectronic considerations (110) (see also Section 4.6.3).

By a series of elegant experiments it has been determined that both steps of the RNase mechanism proceed by an in-line mechanism. The detailed mechanism was elucidated by D. Usher from Cornell University (107,108).

Key to the solution is the isolation of two diastereoisomers of uridine 2′,3′-cyclic phosphorothioate. Both isomers were used for hydrolysis in

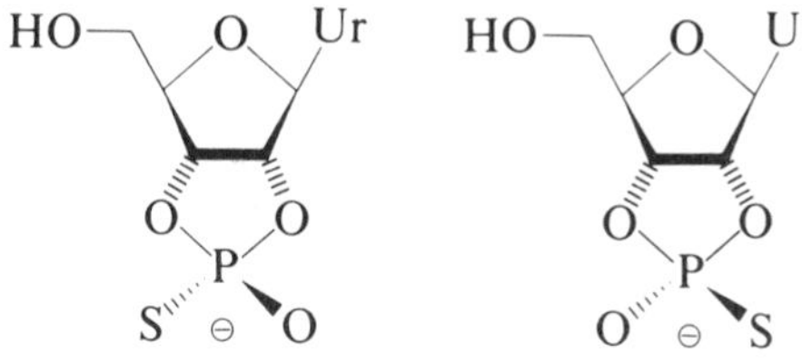

isomer A

isomer B

$H_2{}^{18}O$ by RNase A. The two mechanisms (adjacent or retention and in-line or inversion) are presented below for isomer A. Since the subsequent chemical ring closure with diethyl phosphorochloridate in pyridine is known to proceed with inversion, the stereochemistry of the first step (the enzymatic one) can be determined by isolation of the epimeric cyclic diesters and measuring their ^{18}O content. As expected, ^{18}O was not

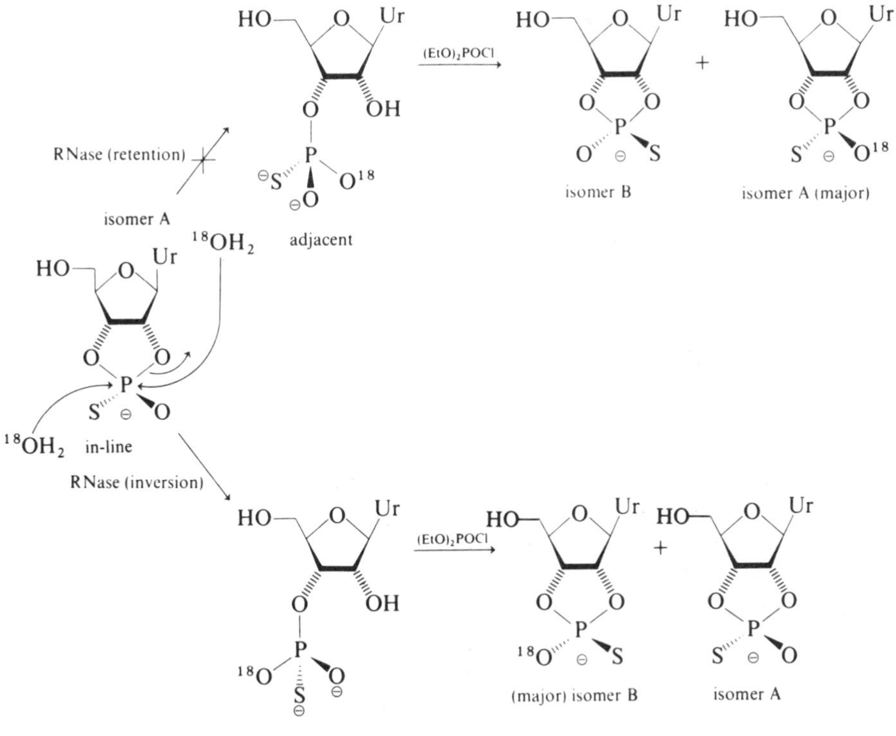

RNase (retention)

isomer A

adjacent

$(EtO)_2POCl$

isomer B

isomer A (major)

in-line

RNase (inversion)

$(EtO)_2POCl$

(major) isomer B

isomer A

present in the isomer A used in the enzymatic step, but in its epimer B. Therefore, the loss of ^{18}O of the nucleotide by a reaction that involves inversion requires that the introduction of ^{18}O be with inversion. These data preclude pseudorotation as a step in bond cleavage and help to identify the position that H_2O must occupy in models derived from crystallographic studies in RNase A.

The increased use of nucleoside phosphorothioates in bioorganic chemistry is based mainly on the general higher stability of phosphorothioates toward hydrolysis cleavage relative to the corresponding phosphates (111). It is a powerful tool for the investigation of many biological processes and enzymatic reactions at a phosphorus center (112).

A similar mechanistic study of DNA-dependent RNA polymerase from *E. coli* using phosphorothioate analogs has been carried out by the group of F. Eckstein (113).

A number of other enzymes that catalyze the hydrolysis of phosphoesters are of biological importance. These include cyclic purine phosphodiesterase (little is known about its active site chemistry at present, but more will be said about its biological role shortly) and the phosphatases. Acid and alkaline phosphatase catalyze the hydrolysis of phosphomonoesters to the corresponding alcohol and inorganic phosphate. Their pH optimums are 5.0 and 8.0, respectively; hence their names. Both form covalent enzyme-substrate intermediates:

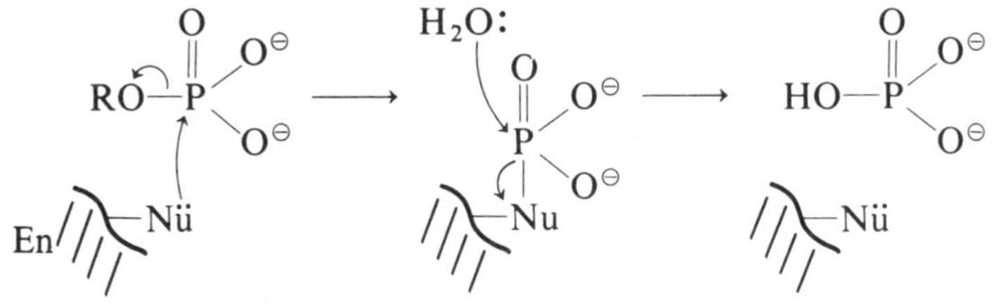

In the case of acid phosphatase, the nucleophile is an imidazole (the intermediate phosphoramidate is susceptible to acid pH) and in the case of alkaline phosphatase, the nucleophile is a serine. Alkaline phosphatase is known to contain a zinc atom which probably functions as a binding site and a Lewis acid for the phosphoryl function. This is understandable when it is remembered how difficult the hydrolysis of phosphomonoesters is at alkaline pH.

The stereochemical course of a hydrolysis reaction catalyzed by staphylococcal nuclease has been determined (114). This enzyme requires calcium ions for both DNase and RNase activities.

The R_p diastereomer of $[^{17}O,^{18}O]$-*p*-nitrophenyl-pT was hydrolyzed in water at pH 8.8 and 42 °C in the presence of 10 m*M* Ca^{2+} by the nuclease as catalyst. ^{31}P-NMR spectroscopy was used to characterize the product by using Knowles's procedure in which the chiral phosphoryl group is transferred to chiral 1,2-propanediol with alkaline phosphatase and the resulting mixture of 1- and 2-phosphopropanediols is chemically cyclized and methylated with diazomethane (115). Measurements of the ^{18}O perturbations of the ^{31}P-NMR resonances of the resulting cyclic triesters allows the configurations of the 1-phosphopropanediol and, therefore, of the [^{16}O,^{17}O,^{18}O]-phospho-*p*-nitrophenol to be assigned.

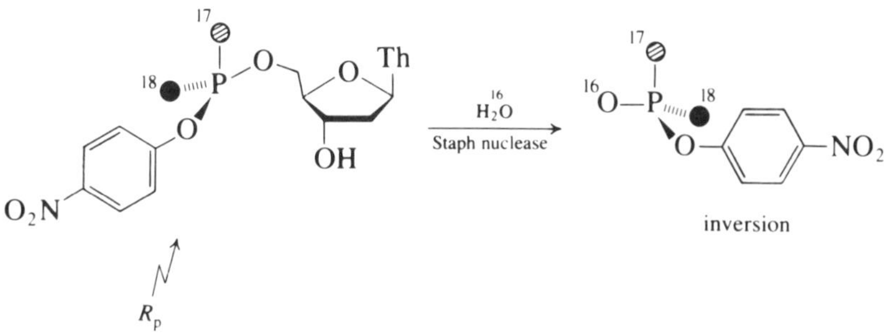

The established *retention* of configuration of the alkaline phosphatasecatalyzed phosphoryl transfer leads to the conclusion that the hydrolysis reaction catalyzed by staphylococcal nuclease proceeds with *inversion* of configuration at phosphorus.

In conclusion, no enzymatic reaction has so far been reported that would call for the involvement of pseudorotation.

3.3.1 Cyclic Phosphates and Other Considerations

As we have just witnessed, pentacovalent phosphorus forms molecules that have trigonal bipyramidal geometry about the phosphorus atom. In pseudorotation, internal forces in the molecule change the bond angles about the phosphorus atom so that the trigonal bipyramidal appears to have been rotated about one of the interatomic bonds.

In a series of papers, W.S. Wadsworth, Jr. has outlined the factors that give rise to either retention or inversion of configuration at the phosphorus atom, particularly with cyclic phosphate molecules. He proposed that back bonding between the attacking nucleophile and the phosphorus atom is of prime importance for the stereochemical course of the reaction (116).

The outcome of the displacement reaction is that efficient back bonding leads to retention while inversion is favored by nucleophiles that are poor back donors, and by leaving groups weakly bonded to phosphorus. Retention is commonly pictured as processing via a trigonal bipyramidal intermediate or transition state, whereas inversion is depicted as a direct S_N2 displacement.

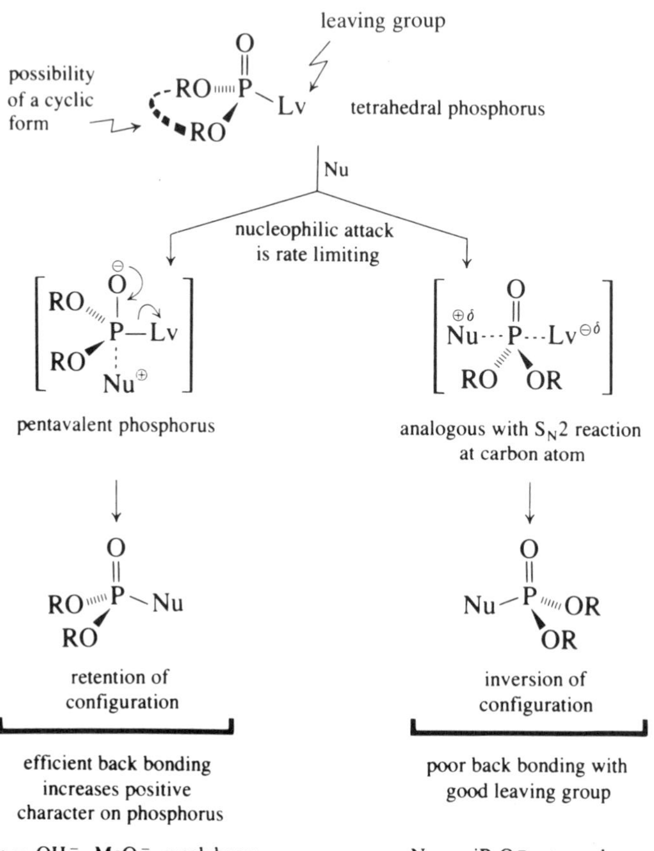

The next example is to illustrate that a neighboring carboxylate group could also be a powerful nucleophilic catalyst for the hydrolysis of phosphate esters. This reaction is thought to involve a pentacovalent intermediate, for which pseudorotation is inhibitied by the presence of two negatively charged oxygen atoms in the equatorial plane. This chemistry is somehow similar to the one encountered in RNase A since an intermediate cyclic phosphate is formed (see top of page 141).

For the same reason, the hydrolysis of the dianion of bis-2-carboxyphenylphosphate has been studied (117). The product is hydrolyzed at a rate about 10^{10} times faster than diphenylphosphate. The reaction

RO Base His-12 O O—H :N NH P=O HN NH O O R' His-119

is believed to involve an intramolecular general acid catalysis by the *ortho*-COOH of the salicyl group and the breakdown of a penta-covalent phosphorus intermediate. However, the driving force of the reaction is more likely due to the breakdown of a high-energy intermediate to two

O O O P COOH OH O O

relatively stable fragments, salicylate and cyclic acyl phosphate, since the presence of a second carboxylic group is not sufficient by itself to account for such a large rate of acceleration in hydrolysis.

COOH O O O P O O O O H O P O O O O + salicylate

D.G. Gorenstein (Purdue University) has studied exhaustively the hydrolysis of cyclic phosphate esters (118), particularly the following two epimeric phosphate triesters. Under basic conditions, the isomer with the equatorial phenoxide leaving group hydrolyzes considerably faster than the axial isomer. This observation is in agreement with the stereoelectronic control theory (see Chapter 4). Since the addition of the nucleophile ($OH^{\ominus}$) is rate limiting, the two antiperiplanar lone pairs of electrons on

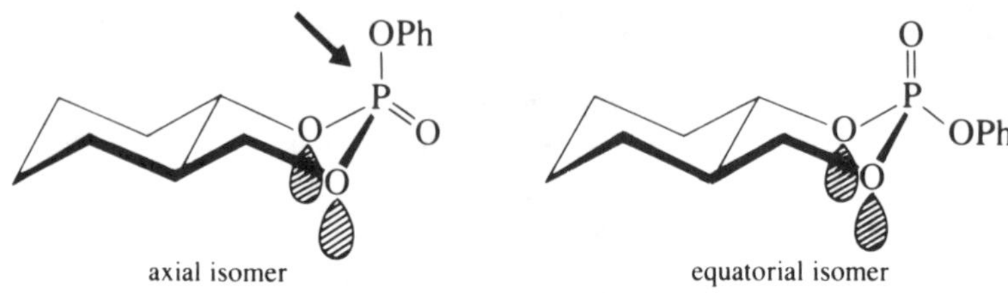

the endocyclic ester oxygens require hydroxide attack from the top side of both isomers. This will force an $O^{\ominus}$ into an apical position of the trigonal bipyramid intermediate in the axial isomer. Since this is unfavorable according to pseudorotation theory, the equatorial isomer is the more reactive. Although for sterical reasons the equatorial isomer should be the more reactive for attack from the bottom side of the molecule in an "in-line" mechanism, this pathway is not favored.

3.4 DNA Intercalants

Both DNA and RNA have been described as polymers whose monomeric units are nucleotides, connected together by a phosphoester linkage. This may present a picture of a long, single-stranded chain. However, both DNA and RNA may be double-stranded, or exhibit a *double-helix* structure. Such is not totally unusual, for the structural protein collagen exists as a triple helix: three polypeptide chains wrapped around one another and held together by interchain hydrogen bonds. Here also, hydrogen bonds serve to hold two strands together: either two strands of DNA, RNA, or a so-called DNA-RNA hybrid. The hydrogen bonds arise between the purine and pyrimidine bases of the two strands in a very specific fashion. That is, the bases of one strand align to the bases of the second strand such that a purine base is always hydrogen bonded to a pyrimidine base, and vice versa. More precisely, adenine and thymine (uracil), which are capable of forming two hydrogen bonds with each other, are paired with each other, just as guanine and cytosine, which are capable of forming three hydrogen bonds with each other, are paired together. Such specific hydrogen bonding (*Watson–Crick pairs*) is important for replication purposes, as the two strands are complementary to each other.*

The DNA double-helix consists of two strands of DNA wrapped one about the other such that there are ten base pairs stacked one upon the

---

* It should also be noted that the extensive hydrogen-bonding potential of the nucleotides provides the possibility for numerous types of hydrogen bonding in addition to the classical Watson–Crick base-pairing scheme. For the most part, these interactions are found with synthetic polynucleotides. However, non-Watson–Crick base pairing and hydrogen bonds involving the sugar-phosphate backbone are important elements in the structure of tRNAs.

other, per helical turn (119,120). This is analogous to protein molecules: the DNA core will be of a hydrophobic nature, while the exterior phosphodiester backbone will face toward the solvent. Of course, during the synthesis of new DNA, it becomes necessary to unravel these strands so as to expose that portion of the molecule containing the genetic information. Furthermore, the two chains of the double helix are "antiparallel": they run in opposite directions. That is, at each end of the double helix there will be a 3′-hydroxyl function on one strand, and 5′-phosphate function on the other strand. These strands are very long. For example, in T_2 phage, the molecular weight of the DNA is 3.2×10^7, while in *E. coli* (a bacterium whose chromosome is a single molecule of DNA) the moldecular weight is 2.0×10^9. The structure of the information storage from that of DNA is most often a double-stranded chain. One exception is the DNA of the phage φX174 (which attacks *E. coli*) which is single stranded.

Polynucleotides, like their monomeric units, absorb UV light. They exhibit the phenomena of *hypochromism:* a solution of polynucleotide will have a lesser optical density of absorbance than will an equimolar solution of the monomeric units. Double-stranded polynucleotides exhibit an even lesser absorbance than single-stranded polynucleotides. This simply reflects the fact that the more ordered the bases are in solution (i.e., the better they are able to undergo cooperative interactions), the lesser is the observed extinction coefficient. This allows the monitoring of the cooperative unwinding (hydrogen bond breaking) that occurs upon heating of a double-stranded polynucleotide by means of "melting curves."

Mention must be made of drugs that are able to inhibit DNA synthesis. These are particularly important to the problem of *cancer chemotherapy*, although they have the undesirable property of not being able to inhibit the synthesis of DNA in the cancer cell without affecting DNA synthesis in the normal cell. Their value does lie in the fact that in many cancers (e.g., leukemia) the rate of cancer cell proliferation greatly exceeds normal cell growth. Such drugs are one of two categories: nucleoside and nonnucleoside analogs.

Perhaps best known of the nucleoside analogs are the arabinose compounds: Ara(C) (cytosine arabinoside, 1-β-D-arabinofuranosylcytosine) and Ara(A) (adenine arabinoside, 9-β-D-arabinofuranosyladenine). Both compounds are triphosphorylated [converted to Ara(CTP) and Ara(ATP) in vivo] upon which they may exert a dual action on DNA polymerase. They may inhibit DNA polymerase (e.g., Ara(CTP) inhibits DNA polymerase II and III) and they may be incorporated into newly synthesized DNA. In fact, Ara(C) but not Ara(A) is also incorporated into RNA. Such uptake into DNA is lethal since a subtle difference will develop in the DNA geometry as a result of the cytosine base adapting a new glycosidic bond angle as a result of steric hindrance from the 2′-hydroxyl function.

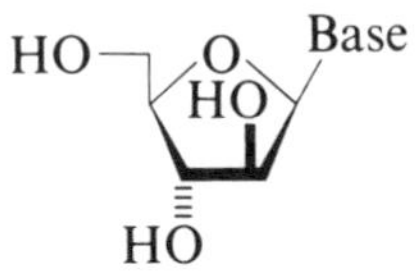

Base = cytosine: Ara(C)
Base = adenine: Ara(A)

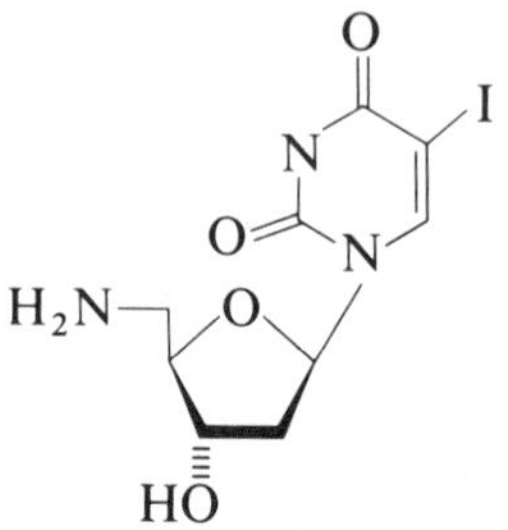

5-iodo-5′-amino-2′,5′-dideoxyuridine (IAdU)

One highly specific drug has recently been developed: 5-iodo-5′-amino-2′,5′-dideoxyuridine (IAdU). It inhibits herpes simplex virus (type-I) DNA replication without affecting the mammalian host DNA replication process. Unfortunately, this specificity has not been extended to oncogenic (cancer-causing) viruses, but work is under way with this compound and related derivatives [i.e., the 5′-amino analog of Ara(C)] that may lead to similar specificity. The mechanism of selectivity appears to originate at the level of the synthesis of the triphosphate substrate. Only viral-infected cells appear able to prepare IAdUTP, which then becomes incorporated into the viral DNA. Such DNA will contain labile phosphoramidate linkages which presumably eventually interferes with the replication process (i.e., perhaps by decreasing the stability of the phosphodiester backbone).

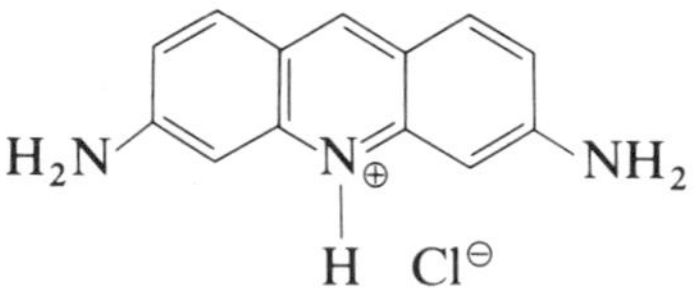

proflavine

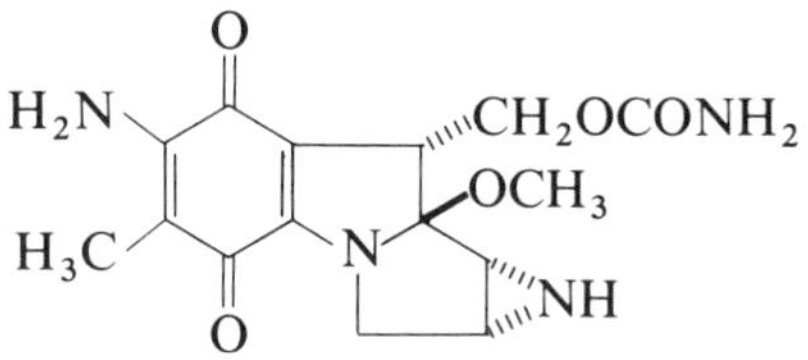

mitomycin C

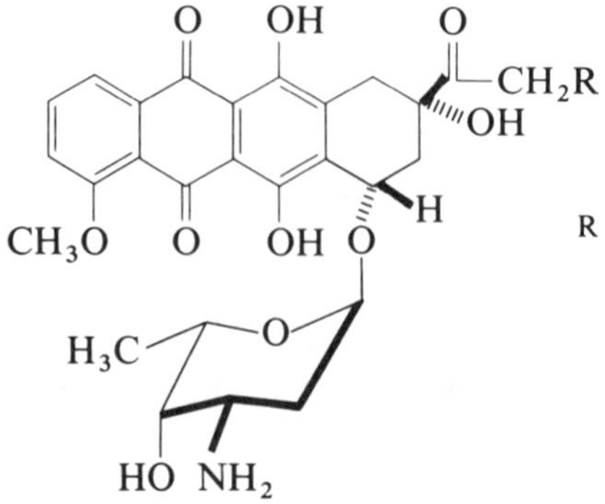

R = OH, adriamycin
= H, daunomycin

Examples of nonnucleoside drugs that inhibit DNA synthesis by binding to the DNA double-helix include the acridines (e.g., proflavine) and various antibiotics (e.g., mitomycin C, adriamycin, daunomycin). The ability of the acridines to bind to DNA and RNA has led to their use as biological stains for these molecules. Binding is achieved by "intercalation"; the planar ring system of proflavine, for example, intercalates (squeezes) between the stacked base pairs of the double-helix. Such a "sandwich" of DNA with bound molecules would be expected to geometrically distort (i.e., lengthen) the double-helical structure; this has been observed (121,122). A similar mechanism of binding is operative for the antibiotics. As such, all three possess antitumor properties.

3.4.1 Mono- and Bis-Intercalant Molecules

In 1969, B. Rosenberg discovered that *cis*-dichlorodiamine platinum(II) was particularly active against many forms of cancer while the *trans*-isomer remains inactive. Platinum ion binds to N-7 of adjacent guanidine

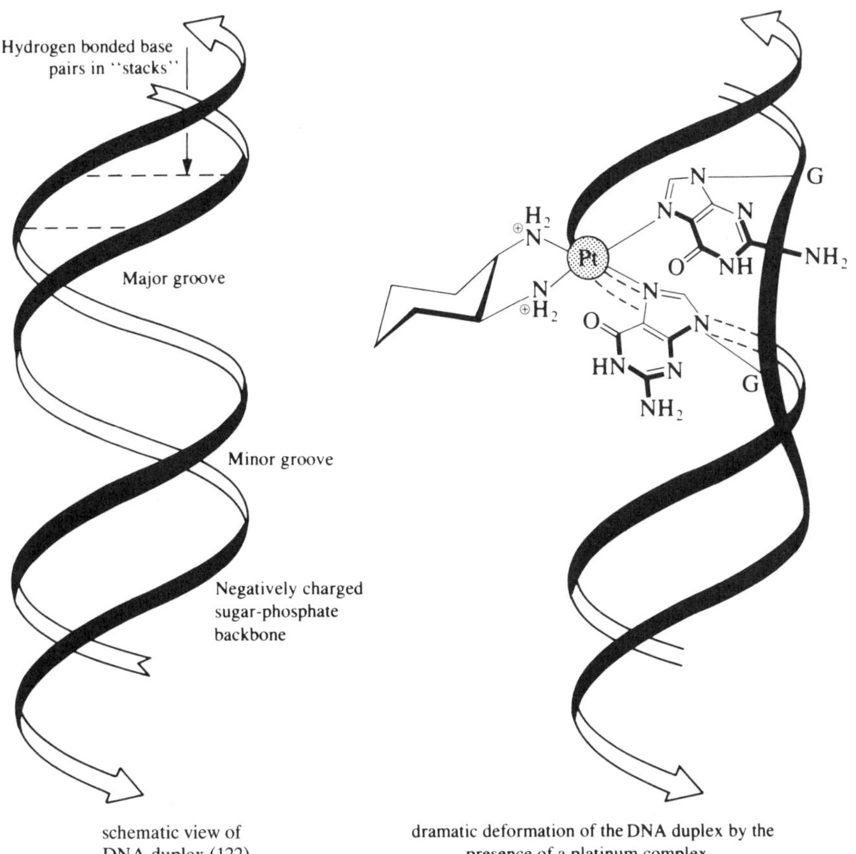

schematic view of DNA duplex (122)

dramatic deformation of the DNA duplex by the presence of a platinum complex

bases, and as a consequence, disturbs considerably the double-helix structure of the DNA. The *cis*-dichloroplatinum-*trans*-1,2-diaminocyclohexane complex is a good example. In some cases, the DNA structure is so distorted by the binding agent or drug, particularly with intercalants, that a right-handed helix is almost converted to a left-handed helix in the region of binding.

S.J. Lippard (Massachusetts Institute of Technology) has developed in 1984 a variant of the antitumor drug *cis*-diaminedichloroplatinum(II). This molecule possesses an intercalating acridine orange group, linked to the platinum center by a polymethylene chain (123). This novel intercalant has now the double capacity of recognizing and binding to DNA at one end of the molecule and at the same time to intercalate at the other end via the aromatic residue. In the presence of light, this aromatic intercalant is able to nick DNA.

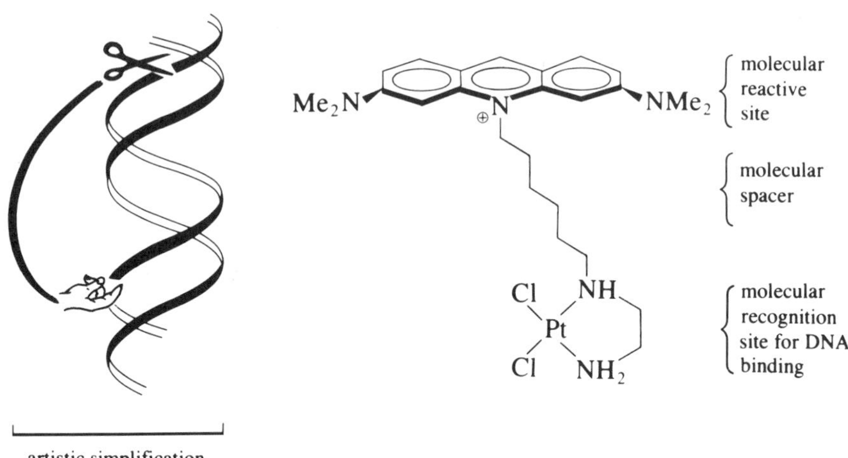

in the presence of O_2 and light, acridine orange acts as a "molecular scissor," snapping the segment of DNA to which it is attached

The preceding molecule could be classified as a *bis-intercalant*. Few natural products are known as bis-intercalant among which the cyclic peptide triostin A and the macrocyclic antibiotic carzinophilin A are good representatives (124,125). Both possess two aromatic residues at 10 Å apart that can bind to the DNA at contiguous sites in a *syn*-fashion.

Synthetic bis- and tris-intercalating DNA ligands with potential antitumor activity have been made in recent years. Particularly promising are carbazole analogs (126) and diacridine and triacridine derivatives (127). Indeed, many drugs can bind specifically with DNA by noncovalent interactions (128,129).

With these considerations, P.B. Dervan's group (California Institute of Technology) launched a program toward the synthesis of a variety of new

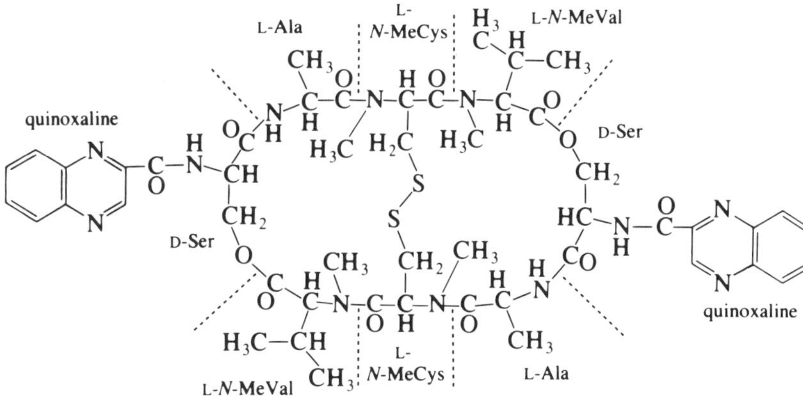

triostin A

bis-intercalant molecules. The first compound made was an analog of the well-used intercalant ethidium bromide. They develop a synthesis to make the corresponding dimer, bis-(methidium) spermine (BMSp) (130). This double intercalant has a binding site of four base pairs and binds at

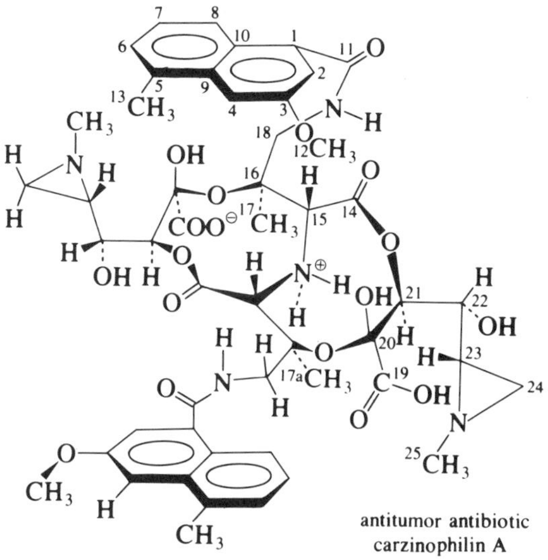

antitumor antibiotic carzinophilin A

least 10^4 times stronger to DNA than the simple monomer. That DNA is a defined macromolecular receptor allows a rational approach to drug design and permits a unique opportunity for studying site-specific drug-binding processes. The molecular recognition of DNA by small molecules is indeed an important field of chemotherapy.

Beside binding to the DNA, can these intercalating molecules be extended to molecular devices that can cut DNA structure to mimic the

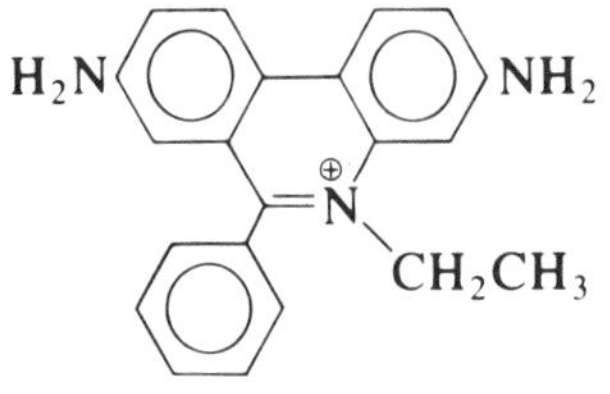

ethidium bromide

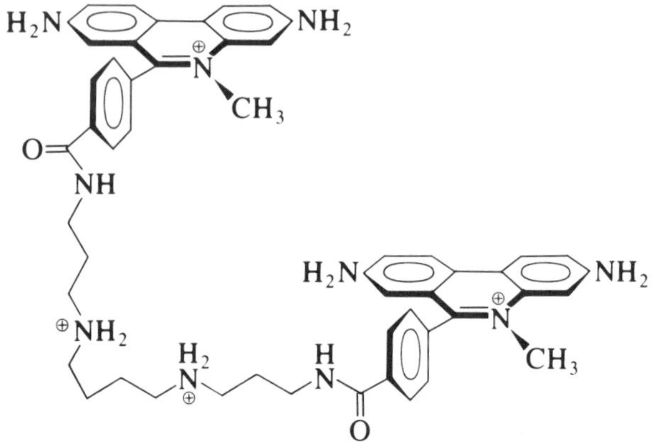

BMSp

bleomycins? The bleomycins are part of a family of clinically useful metalloglycopeptide antitumor antibiotics (129,131). Their therapeutic effects are believed to occur primarily by DNA strand scission (nicking) in an oxidative process involving either Fe(II) or Cu(II) and molecular O_2 bound to the molecules. For this, all the bleomycin molecules are composed of two distinct domains: a DNA-intercalating region involving a bisthiazole group plus a positively charged guanidium or sulfonium moiety and a metal-binding region made of a imidazole and a pyrimidine nucleus

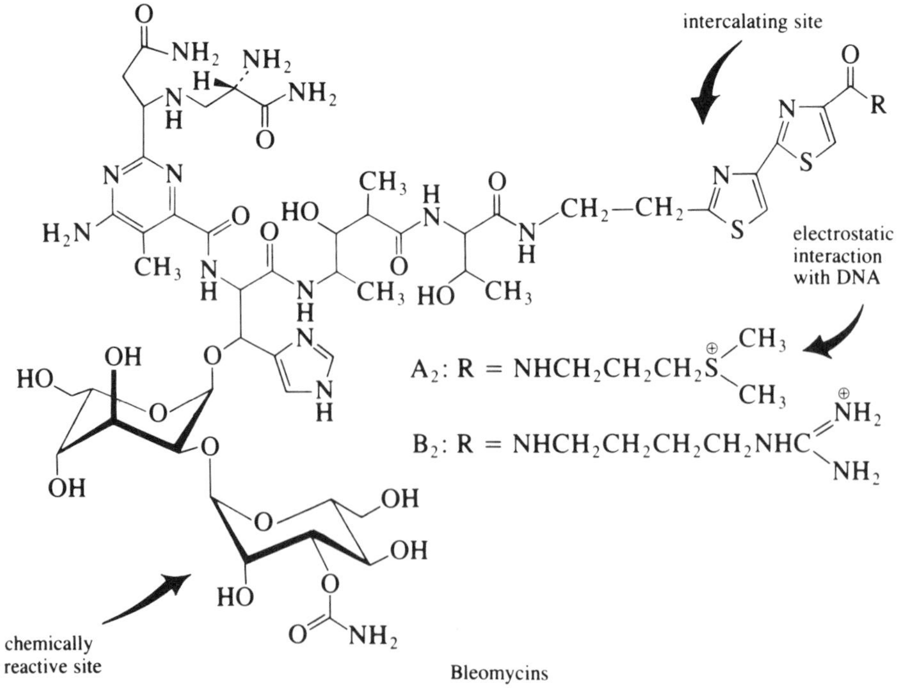

intercalating site

electrostatic interaction with DNA

A_2: R = $NHCH_2CH_2CH_2S^{\oplus}(CH_3)_2$

B_2: R = $NHCH_2CH_2CH_2CH_2NHC(=NH_2^{\oplus})NH_2$

chemically reactive site

Bleomycins

plus amide side chains that act as ion ligands. These two regions are linked together by a linear tripeptide chain.

In this venture, Dervan's group has presented the synthesis of two bioorganic models of the bleomycins by preparing (methidium-propyl-EDTA) iron(II), and (distamycin-EDTA) iron(II) complexes (132,133).

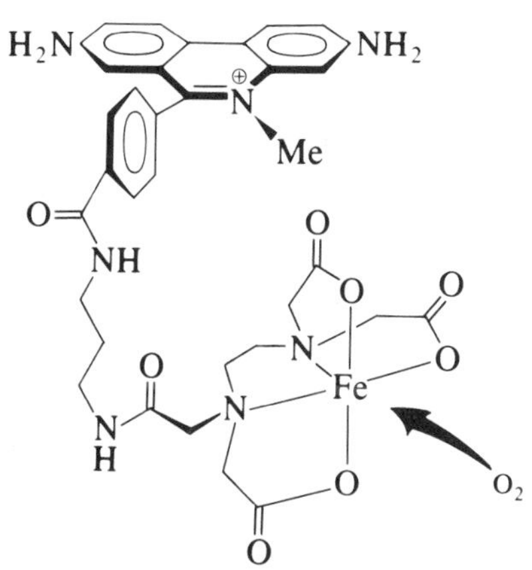

(methidiumpropyl-EDTA) Fe(II)
MPE-Fe(II)

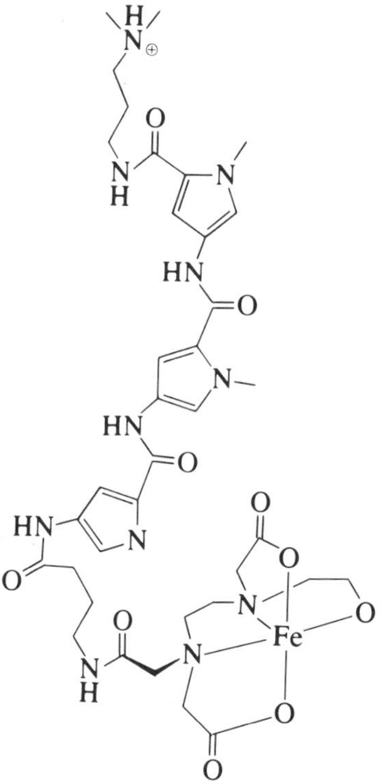

(distamycin-EDTA) Fe(II)
DE-Fe(II)

They both cleave double helical DNA in the presence of dithiothreitol (DTT) with efficiencies comparable to those of bleomycin—Fe(II)/DTT. Unlike bleomycin—Fe(II), the cleavage with MPE-Fe(II) is non-sequence specific. In principle, however, the attachment of EDTA-Fe(II) to a sequence-specific DNA binding molecule could generate a *sequence-specific DNA cleaving molecule*. Such was the case with DE-Fe(II) where the three-methylpyrrole carboxamide portions of the natural antibiotic distamycin are known to bind in the minor groove a double helical DNA with a strong preference for A + T rich regions.

Dervan argued that such an approach of attaching a EDTA-Fe(II) fragment to sequence-specific DNA binding molecules (antibiotics, polypeptides, oligonucleotides, etc.) should provide a new class of "DNA affinity cleaving molecules" and may form the basis for the design and construction of artificial restriction endonucleases with defined target sequences and binding site sizes.

In fact, using a combination of MPE-Fe(II) cleavage of drug-protected DNA fragments and the Maxam-Gilbert gel method of nucleic acid sequence analysis, they were able to determine the preferred binding zones of the natural intercalator actinomycin D as well as for the minor groove binders netropsin and distamycin A (134).

The three approaches can be schematized as follows:

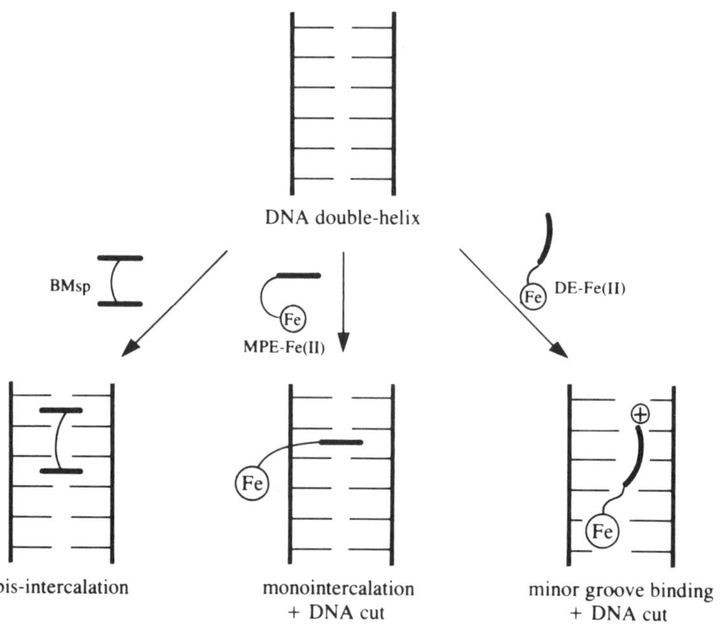

In an even more elaborated model, a bis(distamycin)phenoxazone intercalator was synthesized and found to bind and cut specifically DNA having the sequence $(A,T)_4(G,C)_2(A,T)_4$. This novel hybrid molecule possesses an aromatic intercalating nucleus flanked by two groove binders ended by iron-EDTA "molecular scissors" (135). The DNA cleavage patterns reveal a major cleavage site separated by a ten base pair sequence 5′-TATAGGTTAA-3′. It also represents a mimic of the antibiotic actinomycin D which consists of two peptide loops attached to a planar phenoxazone ring, which intercalates into DNA specifically at GpC sites and not at CpG sites.

Similarly, a bis(netropsin)-polyether-EDTA-Fe(II), also designed by Dervan's group, turned out to bind to DNA in the presence of Ba^{2+} or Sr^{2+} ions to the same ten base pair sequence (136). In this particular case, the crescent-shaped molecule was modulated by the presence of an ion bound to the polyether chain to form a pseudo-crown ether, facilitating its adjustment to the DNA molecule. This work demonstrates "metalloregulation" in the sequence specific binding of a small synthetic molecule to DNA duplex. (See structure on page 153.)

In both cases, EDTA-Fe(II) attached to a DNA-binding molecule generates, under controllable redox conditions, a highly reactive short-lived diffusible hydroxyl radical (HO·) that oxidizes the deoxyribose backbone in DNA structure, resulting in cuts in both strands of the DNA.

Let's hope that DNA recognition by well-designed agents should contribute in the future to the control of molecular diseases at the level of DNA itself. Using the tools of synthetic and mechanistic organic chemistry in combination with nucleic acid techniques, Dervan is actually developing a set of rules for the three-dimensional readout of right-handed double-helical DNA (137).

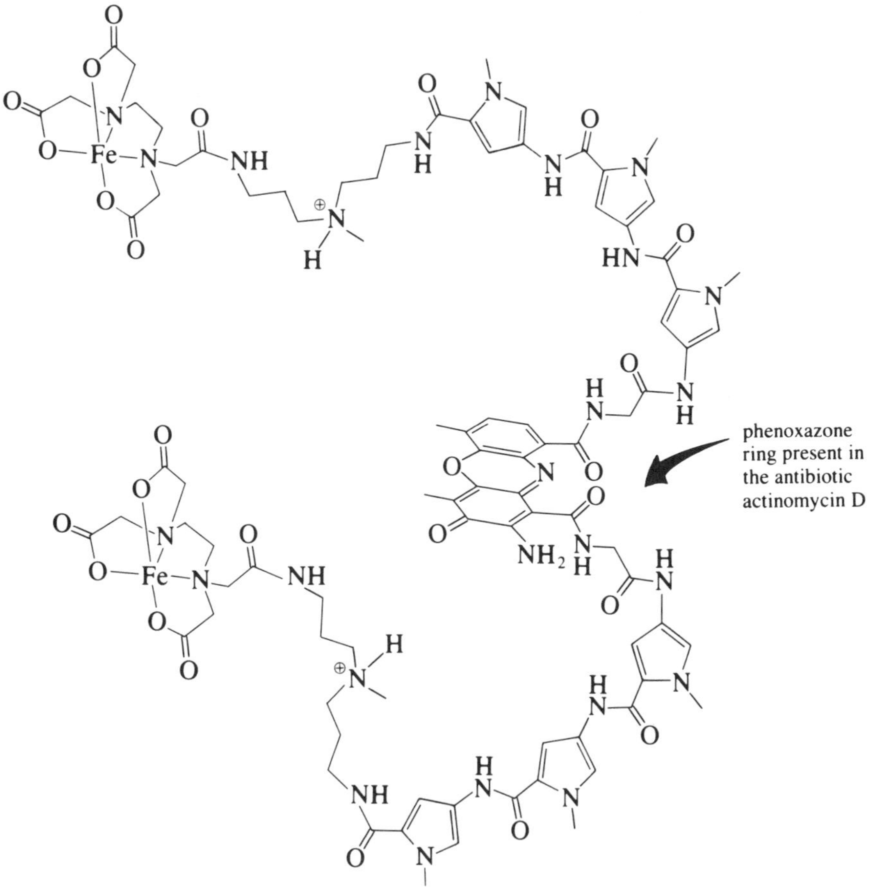

phenoxazone ring present in the antibiotic actinomycin D

Among the recent developments, he has designed a covalent peptide heterodimer for the sequence-specific recognition in the minor groove of

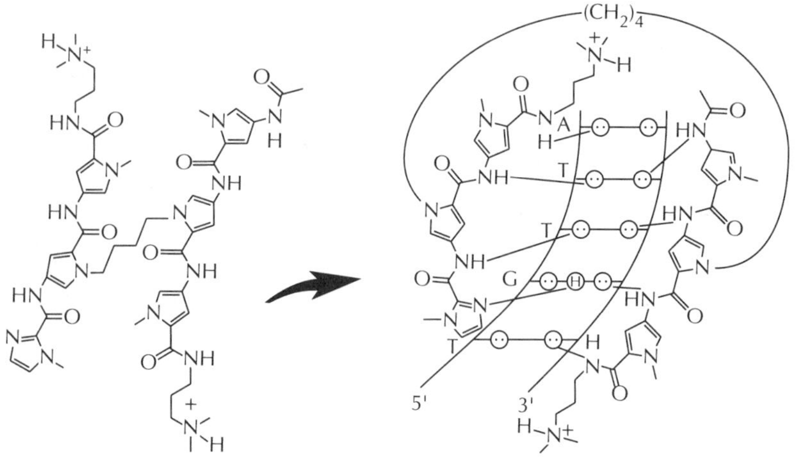

2-ImN-C_4-P3

2-ImN-C_4-P3 • 5'-TGTTA-3'

Fig. 3.3. Structure of the peptide heterodimer (2-ImN-C_4-P_3) and a model for the complex with the 5′-TGTTA-3′ site (138). Reproduced with permission. Copyright © 1994 by the American Chemical Society.

double-helical DNA (138). Fig. 3.3 illustrates this. The nonnatural peptide heterodimer 2-ImN-C_4-P_3 is composed of two synthetic peptides connected through the nitrogens of the central pyrrole rings with a butyl linker. This dimer intercalates and binds specifically to a designated sequence 5′-TGTTA-3′ of a synthetic DNA molecule.

A variant of this heterodimer would consist of replacing the carbon chain linker with a suitable small peptide sequence that can fold back on itself and form a hairpin motif.

Dervan's work has also been extended recently to the synthesis of second-generation, artificial, sequence-specific DNA-cleaving proteins, containing the iron-binding ligand of bleomycin (139).

The design and synthesis of transition-metal complexes, based on ruthenium chemistry, to probe DNA conformation and to try to determine the principles governing site-specific recognition of DNA, is the central effort of J.K. Barton's group (California Institute of Technology) (140).

With a similar approach, the group of C.R. Cantor (Columbia University) has made some interesting mono- and bis-psoralen intercalating molecules that can be photocrosslinked to DNA duplex (141). The photochemical addition products formed between psoralen and the pyrimidine bases (viz., dT and U) of DNA and RNA have been characterized extensively (142). Monoadducts and diadducts are formed where the furan and the coumarin double bonds are implicated as the covalent crosslinking sites with the pyrimidine rings. However, in all cases, the spacer chain is quite flexible.

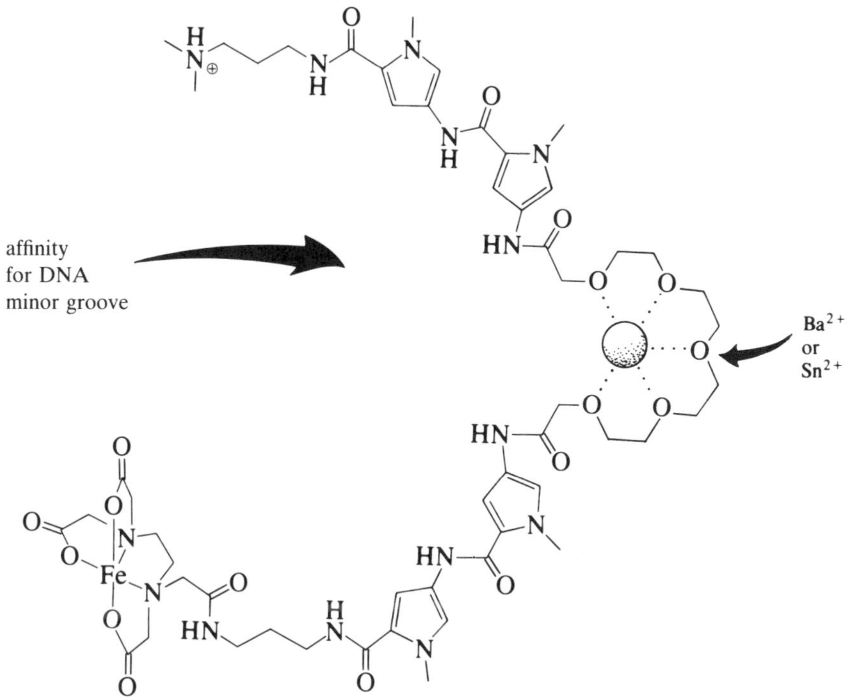

affinity for DNA minor groove

Ba^{2+} or Sn^{2+}

In a slightly different strategy, the synthesis of a series of analogous mono- and bis-intercalating psoralen (PS) molecules that are covalently attached to a crown ether (CrE) was undertaken (143). This approach offers the advantage that being now on a semirigid crown ether backbone with well-disposed orientation, the intercalating functions would be expected to show more specificity and stronger binding properties with DNA receptor molecules Such an assumption is indeed justified because an earlier NMR study with bisacridine derivatives showed that increasing the rigidity of the linking chain between the two intercalating functions

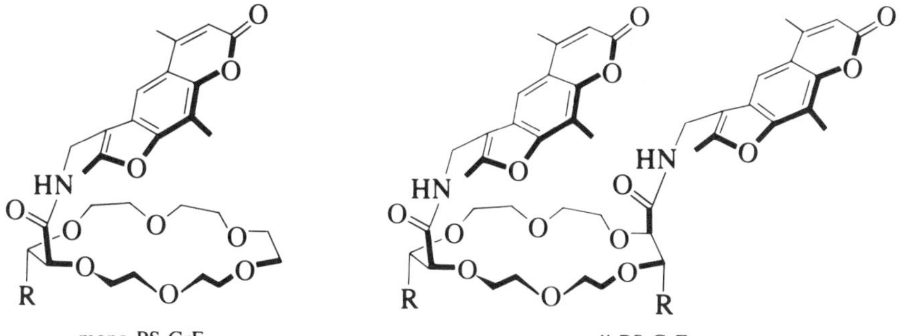

mono-PS-CrE

syn-di-PS-CrE

$R = CO_2H$

$R = CO_2CH_3$

enhances both their stability and their binding efficiency to a model DNA duplex.

Bioassays to evaluate the selectivity to inhibit the restriction endonucleases aHa III and/or Bgl I on the pBR327 plasmid were performed

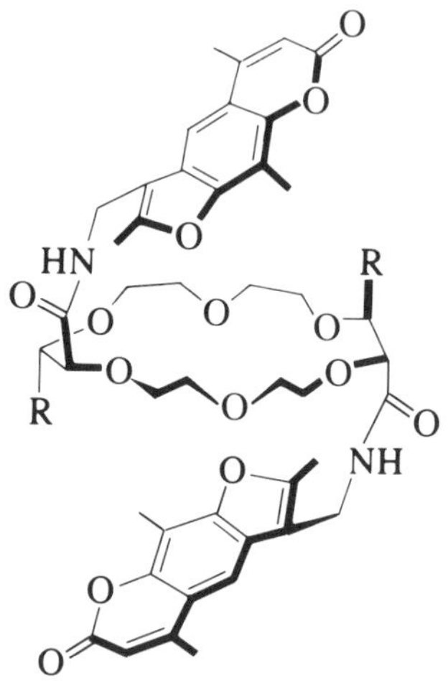

anti-di-PS-CrE

under standard conditions. The first results showed that the monocompound and the *anti*-compound are reactive but, surprisingly, the bis-*syn*-psoralen was not binding to the DNA. The corresponding dimethyl ester (R = CO_2CH_3) did however bind weakly and photocross to the DNA structure. One reason why the *syn*-compound does not intercalate into DNA is probably due to a favorable stacking between the psoralen rings, a phenomenon in fact associated to a strong hypochromism observed in the UV spectrum. A second reason is the electronegativity of the crown ether itself, especially with R = CO_2H. Such a molecule would have less affinity for DNA since the backbone of DNA structure behaves as an anionic polyelectrolyte.

In principle, an intercalator with two binding sites should form a much tighter complex with DNA than one with a single binding site. The following compound, a methidium-dinucleotide (M-TpT), may therefore be viewed as a single representative of an oligonucleotide derivative with two different types of binding sites: the methidium ring and the pyrimidine bases that could act cooperatively. R.L. Letsinger (Northwestern University) reasoned that an oligonucleotide derivative possessing a structural feature that could enhance its binding to complementary sequences in a polynucleotide should have interest as a potential site-specific inhibitor (or promotor) of enzymatic processes involving the polynucleotide (144). Examination of molecular models indicates that the linker arm should be

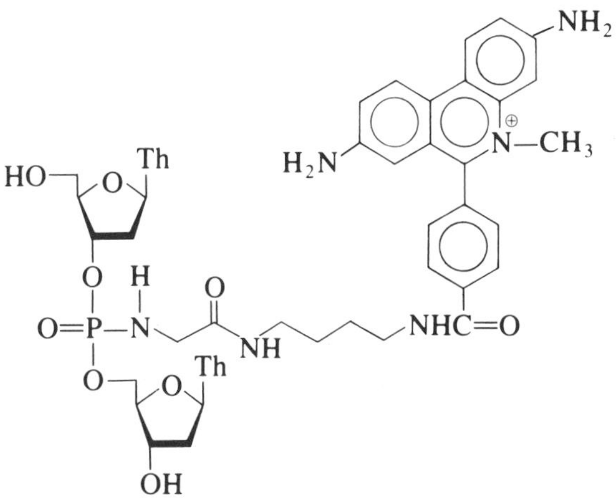

M-TpT

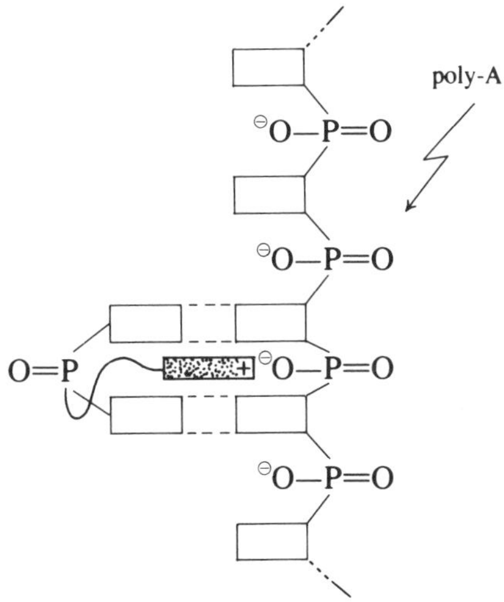

model for binding to the oligonucleotide

long enough to permit the methidium moiety to fold back and insert into the pocket formed by the adjointed nucleoside bases and to complementary bases in a polynucleotide (see schematic representation). Interaction of M-TpT with poly-A in aqueous solution was examined spectrophotometrically in a temperature range of 0° to 85°C. Titration of the solution indicated formation of a complex with a melting transition, Tm, near 47°C. These data are consistent with a model where M-TpT is aligned along the poly-A chain with the methidium moiety interspersed between the Watson-Crick base pairs. By comparison, poly-G, poly-C, and poly-U showed little or no interaction. This work can be extended to more effective and versatile recognition–delivery systems.

Hydrogen bonding and hydrophobic and electrostatic forces are key recognition features common to all nucleotide bases for recognition. Thus, development of new biomimetic receptor molecules that can re-

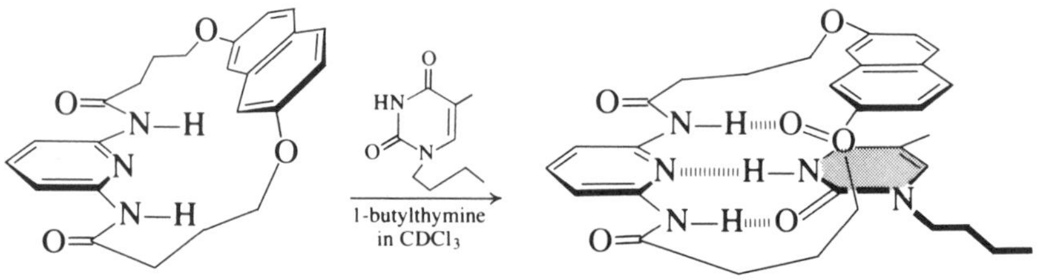

1-butylthymine in $CDCl_3$

cognize and bind to specific nucleotides or nucleotide base pairs is an important goal in contemporary bioorganic chemistry. The following macrocyclic napthalene molecule was synthesized with this objective in mind (145). The X-ray structure showed an open conformation with the naphthalene poised away from the pyridine ring at an interplane angle of 127.5°. Addition of one equivalent of 1-butylthymine results in several characteristic changes in the NMR spectrum, and the X-ray of the complex shows that the naphthalene ring now lies approximately parallel (14°) to the plane of the thymine substrate at a closet interplane contact of 3.37 Å. Thus, on substrate complexation by hydrogen bonding, the synthetic receptor molecule acts like a "molecular hinge" and swings the napthalene unit through a 34.1° arc to within van der Waals distances of the thymine ring. Such induced-fit behavior in a synthetic molecule directly mimics the recognition of nucleotides by ribonuclease T_1 in which a tyrosine residue moves into place above the bound guanine.

A.D. Hamilton (now at the University of Pittsburgh) has extended this approach to the molecular recognition of phosphate esters by an ingenious balance of hydrogen bonding and proton-transfer interactions (146). The structure below is an illustration of this approach.

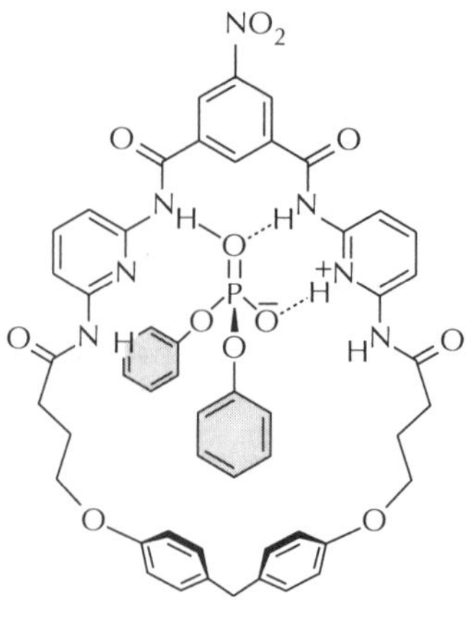

A receptor for phosphate ester (146)

Extremely exciting work has came out from the laboratory of E. Anslyn (University of Texas at Austin) on the design of phosphate receptors and selective hydrolysis of deoxymonophosphate nucleotides. The strategy is based on the construction of a bis-alkyl guanidinium receptor that can bind cooperatively dialkyl-phosphate molecules via four hydrogen bonds. Their concept was then extended to the binding and hydrolysis of phosphonucleotides in the presence of imidazole as a general base catalyst, similar to the enzyme ribonuclease A (see Fig. 3.1). As illustrated below, this emphasizes the fact that the oxygen of the phosphorane intermediate, in the first step of the hydrolysis, must be more basic (p$K \sim 15$ to 17) than the guanidinium ion itself in order to permit a bifunctional proton-transfer catalysis via a tautomeric array.

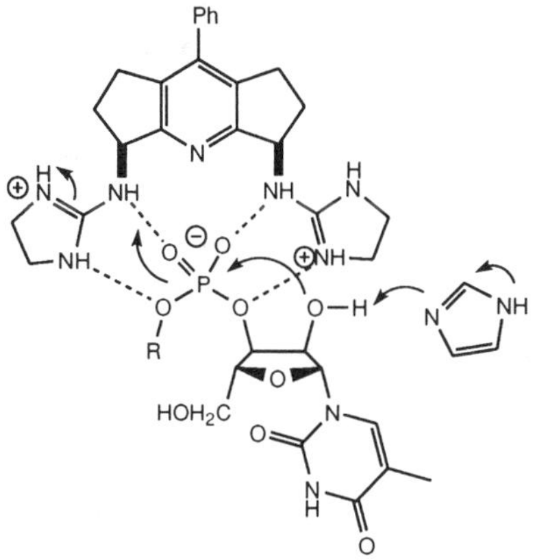

Semirigid receptor for nucleotides and hydrolysis of the phosphate ester in the presence of an external imidazole base. E. Anslyn, personal communication.

With similar objectives in mind, a highly aromatic receptor was recently designed to recognize and bind uric acid (147). Although the molecule looks very lipophilic, it has the right concave geometry for proper hydrogen bonds with the purine base in organic solvents. The control of such alignment would appear central to the development of efficient synthetic receptor surfaces.

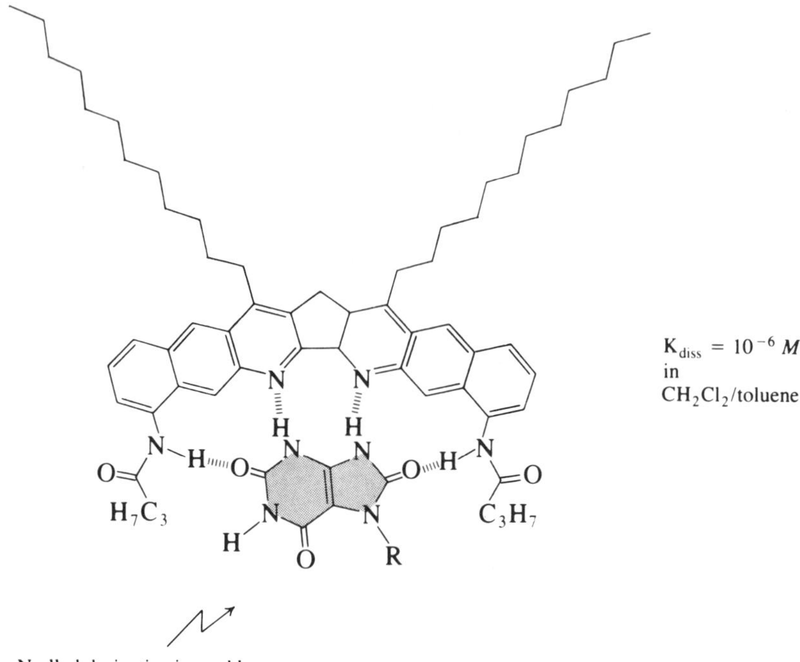
$K_{diss} = 10^{-6}\ M$
in
CH_2Cl_2/toluene
N-alkyl derivative is used because
uric acid itself is too insoluble
in organic solvents

An interesting review article on the computer molecular modeling study of antitumor active drugs as intercalators of DNA has been published (148).

Chapter 4
Enzyme Chemistry

> *"Imagination and shrewd guesswork are powerful instruments for acquiring scientific knowledge quickly and inexpensively."*
>
> van't Hoff

An *enzyme* is characterized by having both a high degree of specificity and a high efficiency of reaction. The factors involved in enzyme-catalyzed reactions are the main subject of this chapter.

For this, hydrolytic enzymes will be used to present the concept of the *active site*. However, an introduction to the general concepts of catalysis, which are based on *transition state* (T.S.) *theory*, is needed first. Proximity and orientation of chemical groups will also be illustrated as factors responsible for the magnitude of enzyme catalysis. This will eventually allow the bridging of nonenzymatic heterogeneous catalysis and enzymatic catalysis.

4.1 Introduction to Catalysis

It was noted earlier that enzymes are proteins that function as catalysts for biological reactions. When it is remembered that body temperature is 37°C and that many organic reactions occur at temperatures well above this, the need for these catalysts becomes apparent. It becomes of interest to understand how these proteins perform their catalysis function. The exact mechanism of enzyme action is a fundamental problem for the bioorganic chemist. Most of the "action" occurs on the surface of the protein catalyst at an area designated as the *active site*, where chemical transformations follow the basic principles of organic and physical chemistry. Several parameters operate simultaneously, and these must be sorted and examined individually by such techniques as model building. However, an appreciation of the catalytic conversion of reactant (substrate) to product first requires a basic understanding of the phenomenon of *catalysis*. The word *substrate* is commonly given to the chemical reactant whose reaction is catalyzed by an enzyme.

By definition, a *catalyst* is any substance that alters the speed of a chemical reaction without itself undergoing change. This is true of the enzymes, for they are of the same form (i.e., conformation and chemical integrity) before and after the catalytic reaction. This is mandatory for catalytic efficiency. If the enzyme were altered after the chemical reaction with a first molecule of substrate, then it would not be able to interact with a second substrate molecule. Of course, as catalysts, enzymes need be present only in small amounts.

A catalyst may either increase or decrease the velocity of a chemical reaction. However, in current usage, a catalyst is a substance that increases the reaction velocity; a substance that decreases the rate of a reaction is called an *inhibitor*. This definition also implies that a catalyst is not consumed during the course of the reaction but serves repeatedly to assist molecules to react. In biochemistry many enzyme-catalyzed reactions require other substances that may also properly be called catalysts but that are consumed (or modified) in the course of the reaction they catalyze. They are the *coenzymes* and are often restored to their original form by a subsequent reaction, so that in the larger context the coenzymes are unchanged. The detailed chemistry of these substances will appear in Chapter 7.

In M.L. Bender's view, the function of a catalyst is to provide a new reaction pathway in which the rate-determining (slowest) step has a lower free energy of activation than the rate-determining step of the uncatalyzed reaction. Furthermore, all transition state energies in the catalyzed pathway are lower than the highest one of the uncatalyzed pathway. Figure 4.1 illustrates this with a free-energy diagram of an exothermic reaction.

According to transition state theory, the processes by which the reagents collide are ignored. The only physical entities considered are the reagents, or ground states, and the most unstable species on the reaction pathway, the *transition state*.

The importance of transition state theory is that it relates the rate of a reaction to the difference in Gibbs free energy ($\Delta G^{\ddagger}$) between the transition state and the ground state. This theory may be used quantitatively in enzymatic reactions to analyze structure reactivity and specificity relationships involving discrete changes in the structure of the substrate.

Catalysts may be classified as *heterogeneous*, or not in solution. A typical example is the hydrogenation of ethylene in the presence of a

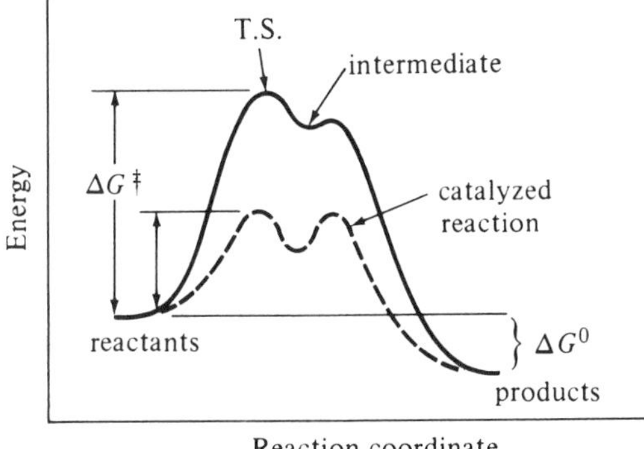

Fig. 4.1. Hypothetical free-energy diagram of a reaction.

metal catalyst such as palladium, platinum, or nickel. Here the catalyst functions to "bring together" the hydrogen and ethylene molecules so that they may react with each other. Whereas the reactants do have an affinity for the metallic surface (i.e., through the π electrons of the ethylene molecule), the product (ethane) does not, and so is "desorbed" to make room for more reactant molecules.

Catalysts may also be *homogeneous*, or limited to one phase, in which a surface or phase boundary is absent. Two examples are the acid-catalyzed hydrolysis of esters, in which the proton is a catalyst:

$$H_2O + CH_3{-}\overset{\displaystyle O}{\overset{\|}{C}}{-}O{-}R \underset{}{\overset{H^{\oplus}}{\rightleftharpoons}} CH_3{-}\overset{\displaystyle O}{\overset{\|}{C}}{-}OH + ROH$$

and the benzoin condensation, in which cyanide is the catalyst:

$$2\ Ph{-}C\genfrac{}{}{0pt}{}{\nearrow O}{\searrow H} \xrightarrow{:CN^{\ominus}} Ph{-}\overset{\displaystyle O}{\overset{\|}{C}}{-}\overset{\displaystyle OH}{\overset{|}{C}H}{-}Ph$$

In the first case, the mechanism is as follows:

$$CH_3{-}\overset{\displaystyle O}{\overset{\|}{C}}{-}OR + H^{\oplus} \rightleftharpoons CH_3{-}\overset{\displaystyle {}^{\oplus}OH}{\overset{\|}{C}}{-}OR \longleftrightarrow CH_3{-}\overset{\displaystyle OH}{\overset{|}{\underset{\oplus}{C}}}{-}OR$$

$$\curvearrowleft :OH_2$$

$$\Updownarrow$$

$$\mathrm{ROH} + \mathrm{CH_3{-}C(={}^{\oplus}OH){-}OH} \rightleftharpoons \mathrm{CH_3{-}C(OH)(OH){-}\overset{\oplus}{O}HR} \rightleftharpoons \mathrm{CH_3{-}C(OH)({}^{\oplus}OH_2){-}OR}$$

$$\mathrm{CH_3{-}C(={}^{\oplus}OH){-}OH} \rightleftharpoons \mathrm{CH_3{-}C(=O){-}OH} + \mathrm{H^{\oplus}}$$

The protonation of the ester makes the carbonyl carbon more electrophilic (note the carbonium ion resonance form) and hence more susceptible to nucleophilic attack by water.

The benzoin condensation proceeds as follows:

$$\mathrm{Ph{-}C(=O){-}H} + \mathrm{:\overset{\ominus}{C}N} \rightleftharpoons \mathrm{Ph{-}C(\overset{\ominus}{O})(H){-}CN} \rightleftharpoons \mathrm{Ph{-}\overset{\ominus}{C}(OH){-}CN} + \mathrm{Ph{-}C(=O){-}H}$$

$$\mathrm{Ph{-}C(OH)(CN){-}C(\overset{\ominus}{O})(H){-}Ph} \rightleftharpoons \mathrm{Ph{-}C(\overset{\ominus}{O})(CN){-}C(OH)(H){-}Ph} \rightleftharpoons \underset{\text{benzoin}}{\mathrm{Ph{-}C(=O){-}C(OH)(H){-}Ph}} + \mathrm{\overset{\ominus}{C}N}$$

The hydrogen of benzaldehyde is not acidic, so it cannot attack a second molecule of the same. However, by addition of cyanide (cyanohydrin formation), the hydrogen atom becomes acidic (the carbanion will be resonance stabilized by the cyano function) and the reaction proceeds.

Enzymes incorporate the features of both classes of catalysts. They may bring reactants together on a protein surface or draw them from an aqueous phase into a hydrophobic environment. However, they may interact with the reactants in a manner such that the rate of chemical reaction is greatly improved. For example, as will be seen shortly, the enzyme-catalyzed hydrolysis of an amide bond does not proceed merely by hydrolysis on the protein surface but, instead, involves chemical interaction of the enzyme to form a more susceptible ester intermediate which then undergoes hydrolysis.

Homogeneous catalysts may be further subdivided into *specific-acid*, *specific-base*, *general-acid*, *general-base*, *nucleophilic*, and *electrophilic catalysts*. A specific acid is merely a proton, a specific base is a hydroxyl ion. General-acids and -bases are any other acidic or basic species, respectively. The bromination of acetone illustrates these different catalysts. It may proceed by specific-acid-catalyzed enolization (149):

$$CH_3{-}\overset{\overset{\displaystyle O}{\|}}{C}{-}CH_3 + H^{\oplus} \rightleftharpoons CH_3{-}\overset{\overset{\displaystyle {}^{\oplus}OH}{\|}}{C}{-}\underset{\underset{\displaystyle H}{|}}{C}H_2 \xrightleftharpoons{-H^{\oplus}} CH_3{-}\overset{\overset{\displaystyle OH}{|}}{C}{=}CH_2 \;\; Br{-}Br$$

$$\xrightleftharpoons{-HBr} CH_3{-}\overset{\overset{\displaystyle O}{\|}}{C}{-}CH_2{-}Br$$

It may proceed by specific-base enolization:

$$CH_3{-}\overset{\overset{\displaystyle O}{\|}}{C}{-}CH_3 + O\overset{\ominus}{H} \rightleftharpoons CH_3{-}\overset{\overset{\displaystyle O}{\|}}{C}{-}\overset{\ominus}{C}H_2 \;\; Br{-}Br \rightleftharpoons CH_3{-}\overset{\overset{\displaystyle O}{\|}}{C}{-}\underset{\underset{\displaystyle Br}{|}}{C}H_2 + Br^{\ominus}$$

It may also proceed by general-acid and/or general-base catalysis. For example, in a sodium acetate buffer, both general-acid and general-base catalysis may occur.

$$\begin{array}{c} AcO{-}H \\ CH_3{-}\overset{\overset{\displaystyle O}{\|}}{C}{-}\underset{\underset{\displaystyle H}{|}}{C}H_2 \\ \overset{\ominus}{O}Ac \end{array} \longrightarrow \begin{array}{c} AcO^{\ominus} \\ CH_3{-}\overset{\overset{\displaystyle HO}{|}}{C}{=}CH_2 + AcOH \\ \downarrow Br_2 \\ \text{products} \end{array}$$

Nucleophilic catalysis may be defined as the catalysis of a chemical reaction by a (rate-determining) nucleophilic substitution. Thus, if a particular nucleophilic reaction is slow, then the reactant may undergo attack by a nucleophilic catalyst to give rise to an intermediate that is more susceptible to the desired nucleophilic displacement than the reactant. A classic example of nucleophilic catalysis is the acetylation of alcohols with acetic anhydride in pyridine:

$$R{-}OH + CH_3{-}\overset{O}{\overset{\|}{C}}{-}O{-}\overset{O}{\overset{\|}{C}}{-}CH_3 \xrightarrow{\text{pyridine}} CH_3{-}\overset{O}{\overset{\|}{C}}{-}O{-}R + CH_3COOH$$

At first glance, it might be thought that pyridine merely functions to scavenge any acetic acid. However, this is not correct:

$$CH_3{-}\overset{O}{\overset{\|}{C}}{-}O{-}\overset{O}{\overset{\|}{C}}{-}CH_3 \longrightarrow CH_3{-}\overset{O}{\overset{\|}{C}}{-}\overset{\oplus}{N}C_5H_5 + CH_3COO^{\ominus}$$

(pyridine $\ddot{N}$ attacks the carbonyl carbon; $R{-}\ddot{O}H$ then attacks the acylpyridinium → products)

Without the pyridine, the acetylation of the alcohol would be much slower.

Another well-known example of nucleophilic catalysis is hydrolysis of *p*-nitrophenyl acetate, as catalyzed by imidazole:

$$CH_3{-}\overset{O}{\overset{\|}{C}}{-}O{-}C_6H_4{-}NO_2 \xrightarrow{\text{slow}} CH_3{-}\overset{O}{\overset{\|}{C}}{-}N(\text{imidazole}) + HO{-}C_6H_4{-}NO_2$$

(imidazole $\ddot{N}$ attacks the carbonyl carbon)

$$\xrightarrow[\text{fast}]{H_2O} CH_3{-}COOH + HN(\text{imidazole})N$$

The imidazole is a nucleophilic catalyst for the reaction, but under basic conditions, an OH^- ion can act as a proton acceptor to assist the nucleophilic catalyst (the imidazole) in attacking the substrate. Thus, the OH^- ion is acting as a general-base. A transition state for the tetrahedral intermediate can be postulated:

$$\left[CH_3{-}\overset{O}{\overset{\|}{C}}{-}O{-}C_6H_4{-}NO_2 \;\; (C\cdots N \text{ imidazole } N\cdots H\cdots OH) \right]^{\ominus}$$

tetrahedral T.S. intermediate

This reaction is a typical example of nucleophilic catalysis. It is also called *covalent catalysis* when the substrate is transiently modified by forming a covalent bond with the catalyst to give a reaction intermediate.

Just as nucleophilic catalysis involves the addition of electrons from the catalyst to the substrate, *electrophilic catalysis* involves the abstraction of electrons or electron density from the substrate to the catalyst. Metal ions are excellent electrophilic catalysts. Especially in the chemistry of phosphates, where anionic charges tend to repel nucleophiles from the phosphorus atom, electrophilic catalysts can be important. For example, the synthesis of 3′,5′-cyclic guanosine monophosphate (cGMP, see Section 3.2.3) from guanosine triphosphate is markedly accelerated by divalent metal cations [e.g., Mg(II), Mn(II), Ba(II), Zn(II), Ca(II)].

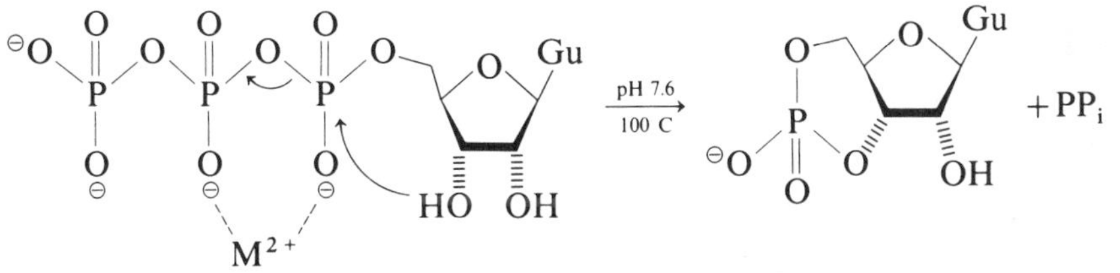

The hydrolysis of ATP is also subject to metal ion catalysis. In both of the above cases, the metal chelates to the oxygen atoms of the triphosphate moiety and withdraws electron density from the phosphorus atom, making the latter more electrophilic.

Enzymes may use any of the modes of catalysis mentioned above in order to catalyze a particular chemical reaction. For example, the imidazole ring of a histidine residue of the enzyme α-chymotrypsin (Section 4.4) can function as a general-base catalyst, while in the enzyme alkaline phosphatase, the same residue can function as a nucleophilic catalyst. Indeed, enzymes are complex catalysts that employ more than one catalytic parameter during the course of their action. It is by this successful integration of a combination of individual catalytic processes that a rate enhancement as high as 10^{14} may be achieved. Furthermore, it is this combination of factors that results in a specific catalyst.

To further illustrate the importance of imidazole in catalysis, let us examine its role as a basic catalyst which is extremely common in chemical reactions. Hydrolysis of carboxylic acid derivatives, for instance, is assisted by both general-acid and -base catalysis.

The example on the following page shows that hydroxyl ions and imidazole are proton acceptors and can participate in the rate-determining step of ester hydrolysis by general-base catalysis.

Why is the reaction accelerated by participation of a base? Many reasons can be given. Principally, the base (imidazole) accepts in the transition state a proton from the attacking water molecule, so the oxygen of the latter acquires a greater share of electron density. Thus, the oxygen of water becomes more negatively charged and the ability to denote an electron pair to the carbonyl group is enhanced. The net result is that the presence of a base lowers the free energy of activation of the reaction. In the uncatalyzed reaction, another molecule of water, which is less basic and thus a less efficient catalyst, would accept the proton.

An important experimental question from the standpoint of catalysis is: How can general-base catalysis be distinguished from nucleophilic catalysis?

$$R{-}CH_2{-}C(=O){-}OEt + H_2O + \text{imidazole} \rightleftharpoons \left[R{-}CH_2{-}C(=O)(\cdots O(H)\cdots H \cdots N\text{-imidazole}){-}OEt\right]$$

T.S.

$$\rightleftharpoons$$

$$R{-}CH_2{-}COOH + EtOH + \text{HN-imidazole} \rightleftharpoons R{-}CH_2{-}C(O^{\ominus})(OH){-}OEt + \text{HN-imidazole-NH}^{\oplus} \leftrightarrow \text{HN}^{\oplus}\text{-imidazole-NH}$$

T.I.
(tetrahedral
intermediate)

As seen in the reaction sequence above, water is involved in the transition state. Thus, a possible procedure is to perform the reaction in ^2H_2O. A $^2H{-}O$ bond is stronger than a H—O bond, so the energy of transfer of a deuterium ion is higher. Experimentally, base catalysis in ^2H_2O shows a difference of the order of 2 to 3 times smaller for the rate of reaction as compared to H_2O.

On the other hand, catalysis by a nucleophile will give the same rate in both solvents. An example is the hydrolysis of anhydrides by formate ions. Formate is less basic than acetate but is a better nucleophile. Furthermore, in a nucleophilic catalysis an unstable acyl-intermediate is formed and can be accumulated by having an excess of nucleophile.

In a less polar solvent (water plus alcohol), the individual groups are less solvated and they interact together more often. This situation involves *multiple catalysis*, where the presence of both acid and base might have a cooperative effect greater than the sum of the individual effects. However, four molecules, substrate, nucleophile general-acid, and general-base, have to line up in the right way before such cooperativeness of catalysis can take place. This situation is thus more unlikely to occur than in general-acid or -base catalysis, where collision among three molecules has a higher probability of occurrence.

This situation is, however, more probable in an enzyme active site that combines several types of catalytic groups that operate simultaneously. It explains in part why enzymes are such efficient catalysts.

In cases where the nucleophile is on the same molecule, *intramolecular catalysis* takes place. We can illustrate this by the amino group participation in the hydrolysis of an ester. Four types of mechanisms are possible (150).

(a) intramolecular nucleophilic catalysis:

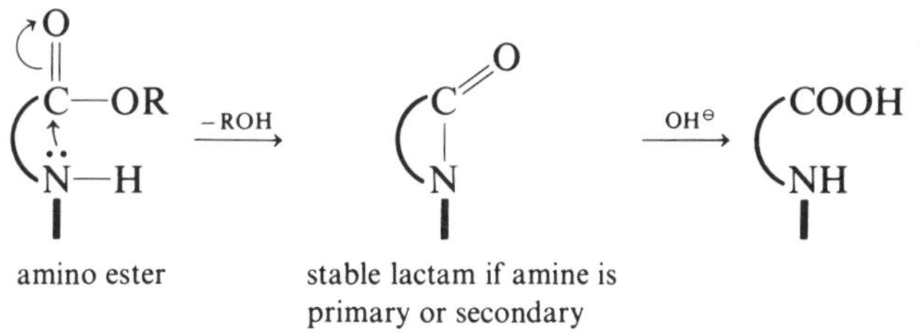

(b) intramolecular general-base catalysis:

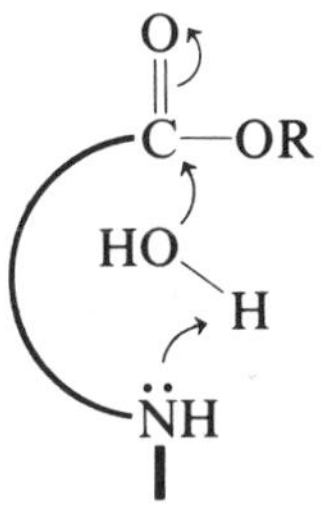

(c) intramolecular general-acid specific-base catalysis:

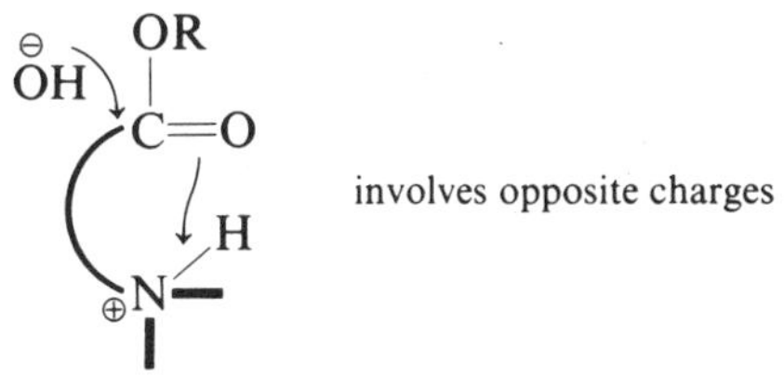

(d) electrostatic facilitation:

the only mechanism with a quaternary amine

Mechanisms (a), (b), and (c) are kinetically indistinguishable but, in principle, (b) and (c) can be differentiated from (a) by deuterium solvent isotope effects. Mechanism (c) should be sensitive to ionic strength effects, but not (b) because it is electrically neutral. Mechanisms (c) and (d) are alike but the most favored one is (c). However, with a good leaving group, process (a) is the most likely to occur (151).

Thus far we have considered only *homogeneous catalysis* in which all reactants and the catalyst are in the same phase (in solution). In *heterogeneous catalysis*, however, at least two phases are present in the reaction mixture. As mentioned earlier, the most common types are systems with a solid catalyst in contact with substrates in the gaseous or liquid phase.

Many analogies can be made between surface heterogeneous catalytic polymerization of olefins and the way that nature catalyzes chemical reactions. It is in this context that nonenzymatic and enzymatic catalysis can be bridged. Indeed, heterogeneous catalysis is similar to enzyme catalysis in many respects. A substrate molecule collides with an *active site* on the surface of the solid catalyst to form an adsorptive complex. The adsorbed substrate reacts in one or more steps under the influence of catalytic groups at the active site. Finally, the product molecules desorb (or escape) from the active site. Thus, the concepts of an active site and of complex formation between substrate and active site are common both to enzyme catalysis and heterogeneous catalysis. A comprehension of these concepts serves to bridge nonenzymatic and enzymatic catalysis. They are nonetheless fundamentally different in the sense that most enzymes have only one active site per molecule whereas heterogeneous catalysts have many active sites per particle. Furthermore, in heterogeneous catalysts, the sites are not necessarily identical.

The concept of constrained geometry is also important, and this is why π-complexes of metal and olefins can lead to structured molecules. For example, a model of cyclododecatriene-nickel shows that the nickel atom is situated exactly in the center of the ring. This picture represents a "lock

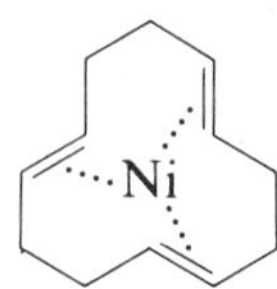

and key" fit very precisely, another analogy with an enzyme–substrate complex.

Heterogeneous catalysts are exceedingly useful and important in the petroleum industry and in the synthesis of synthetic polymers and plastics. In this respect, a well-known catalyst for the polymerization of olefins is the Ziegler–Natta catalyst, developed in the 1950s. It is formed by the couple $TiCl_4 \cdot AlEt_3$. Ti(IV) is the coordinating transition metal and the organometallic portion acts as an activator (a reductor) by alkylating the active site titanium ion and removing one chlorine ligand:

$$TiCl_4 + Al(Et)_3 \longrightarrow [Cl_3Ti(\mu\text{-}Cl)(\mu\text{-}C_2H_5)Al(C_2H_5)_2] \longrightarrow (C_2H_5)TiCl_3 + Al(Et)_2Cl$$

Initiation

σ-bond

active center

π-complex

regeneration of former active center

chain migration

active center (σ-bond)

transition state

Propagation

Titanium is the real catalyst; aluminum serves only at the beginning to initiate the reaction by alkylation. The alkylated titanium surface-complex produced in this way has a ligand vacancy (empty orbital) oriented in a cavity for an olefin to make a π-complex (Fig. 4.2).

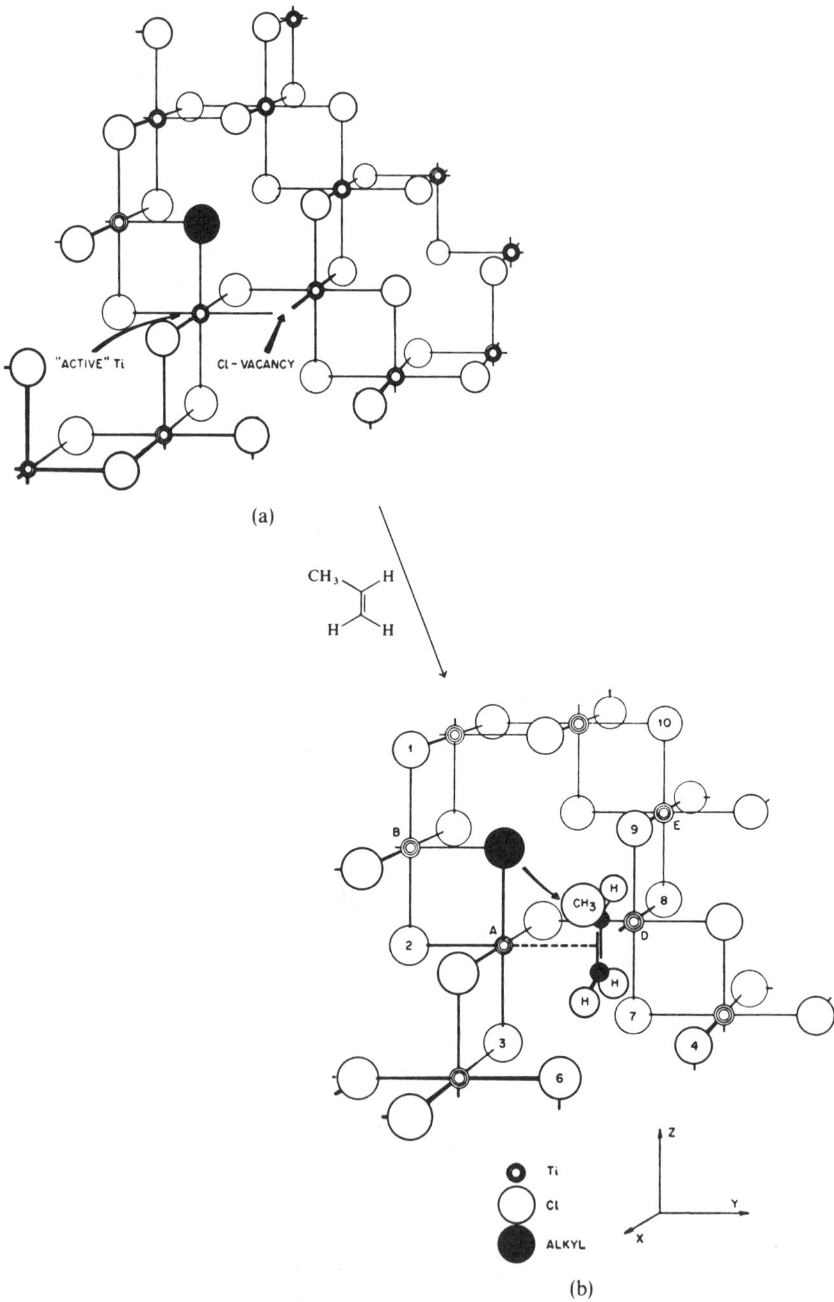

Fig. 4.2. Surface lattice structure of titanium trichloride showing (a) active site titanium ion with a ligand vacancy and (b) a propylene molecule complexed to the titanium ion of the active site. The black spheres represent the growing polymeric chain (151). Copyright © 1962 by The Chemical Society.

Therefore, the titanium atom has just the right electronic structure to form a strong covalent bond with an alkyl chain which becomes labile only in the presence of an olefin to form a π-bond which allows an easy migration of the ligand. This complex evolves to a new Ti—Cσ bond with the olefin, and the extension of the chain occurs by ligand migration to regenerate the free ligand in the original orientation.

The reaction is stereospecific because each new olefin monomer always enters and reacts from the same ligand position on the titanium active site. A *stereoregular* polyolefin results with all substituents pointing in the same direction. It is an *isotactic* polymer. If the groups alternate on the chain, it is a *syndiotactic* polymer. If the orientation is random, the molecule is said to be *atactic*.

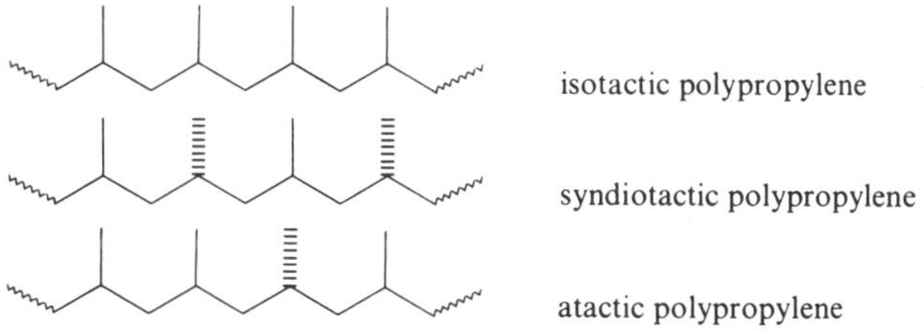

isotactic polypropylene

syndiotactic polypropylene

atactic polypropylene

Thus, the polymerization of olefins under the influence of a metal catalyst can proceed with a high degree of specificity to give crystalline polymers in which the monomer units are all arranged in the same configuration. This demands that there be regio- and stereospecific interactions between the growing polymer, the catalyst, and the reacting monomer as the polymer is synthesized.

This unique example of solid surface catalysis has many analogies with an enzymatic reaction. In particular, there is an active site located in a specific region of the metal surface to accommodate the "substrate," and the transformation is always stereospecific.

4.2 Introduction to Enzymes

In its broadest sense, an *enzyme* is a protein capable of catalytic activity. More specifically, it can be defined as a polypeptide chain or an ensemble of polypeptide chains possessing a catalytic activity in its native form. It is a sophisticated copolymer made of amino acid monomers of the same configuration. The catalysis takes place in a specific region of the enzyme referred to as the *active center* or catalytic cavity. The active center or *active site* consists of all the amino acid residues implicated in the mode of binding and the specificity of the substrate as well as the side chains of the amino acids that are directly implicated in the catalytic process. An integral part of the concept of the active site is that there are a small number of groups on the enzymes that produce its high catalytic activity and very often in a complementary way. The simple template "lock and key" theory, which assumes immobility, is only a first approximation for the binding of substrate to enzyme (152,153).

The outstanding characteristic of enzyme catalysis is that the enzyme specifically binds its substrate and the reactions take place in the confines of the enzyme–substrate complex. Thus, to understand how an enzyme works, we need to know not only the structure of the native enzyme but also the structure of the complexes of the enzyme with its substrates, intermediates, and products.

The amino acid sequence or primary structure of an enzyme dictates the secondary and tertiary (three-dimensional) structure that is the folding of the peptide chain to form a globular macromolecule with a well-defined cavity to accommodate the substrate, and the coenzyme if needed. Enzymes have complex and compact structures where the polar amino acids are at the surface of the molecule directed toward the solvent, while the nonpolar ones are generally oriented toward the interior of the molecule away from the solvent. The three-dimensional structure is maintained by a large number of intramolecular noncovalent apolar or hydrophobic interactions as well as by ionic interactions, disulfide bridges, hydrogen bond interactions, and sometimes salt bridges (154). *Hydrophobic forces* are probably the most important single factor to explain the large binding free energies that are observed in enzyme–substrate interactions.

The review article of R.J.D. Miller (155) should be consulted for a better understanding of the energetics and dynamics factors governing protein motion.

Why are enzymes macromolecules? The reason is that the active center must be of a defined or highly ordered geometry if it is to contain all the binding and catalytic amino acid residues in a correct alignment for optimal catalysis. This imposes a heavy entropy demand upon the system which can be compensated for at the expense of another already ordered region of the biopolymer (156).

As suggested by Srere, enzymes are big, but their size may well be related to their need to have sufficient surface area to provide specific binding sites for their mobilization and their localization in the living cell, as well as for their integration into specific metabolic pathways (157).

The most obvious means by which an enzyme may increase the rate of a bimolecular reaction is simply to bring the reacting molecules together at the active site of the enzyme. Therefore, the two most important considerations here are, first, the magnitude of the rate enhancement that might be expected from such an approximation of the reactants and, second, the mechanism by which this rate enhancement is brought about. This chapter will try to clarify these points. Change in solvation may also provide a significant rate-acceleration effect in both intramolecular and enzymatic reactions. The nonpolar interior of an enzyme is similar to the low dielectric constant of an organic solvent. Electrostatic interactions are no longer shielded by solvent molecules and thus become stronger with the substrate. Before evaluating the factors responsible for enzyme specificity, we should emphasize again the importance of chirality.

4.2.1 Molecular Asymmetry and Prochirality

Enzymes catalyze biochemical reactions stereospecifically. For this reason asymmetric syntheses are very common in nature and are often essentially unidirectional. Consequently, most natural products are optically active as a result of having been constructed by the catalytic action of three-dimensional enzymes. In its simplest terms, the substrate "fits" into the active site of the enzyme in a precise geometric alignment. For example, the enzyme triosephosphate isomerase catalyzes the conversion of achiral dihydroxyacetone monophosphate to D-glyceraldehyde 3-phosphate (158).

The substrate here has a *prochiral* center* and one hydrogen in transferred specifically from only one face (the *re*-face) of the double bond to the carbonyl function. It is primarily the chirality of the enzyme that determines the correct course of the reaction. Another example of this is when these same two substrates are in the presence of the enzyme aldolase, to give fructose diphosphate, the H_S rather than the H_R-hydrogen is exchanged with water.

* The concept of prochirality and a proposed system of pro-*S*, pro-*R* configurational nomenclature was definitively developed by K.R. Hanson in 1966. Hanson's general definition of prochirality is as follows: "If a chiral assembly is obtained when a point ligand in a finite achiral assembly of point ligands is replaced by a new ligand, the original assembly is prochiral." Point ligands are the chemical groups attached to the center of prochirality. From a practical point of view, a group designated as the pro-*R*(H^R) substituent is the one that upon replacement or modification leads to a chiral molecule with the *R* configuration (159). Hanson also proposed a systematic way of labeling the faces of trigonal atoms such as the carbon of the C=O group. A trigonal prochiral center is planar and is viewed from one side as in the drawing below. The three groups surrounding the carbon atom are given priorities, a, b, and c, according to *R,S* nomenclature of Cahn–Ingold–Prelog.

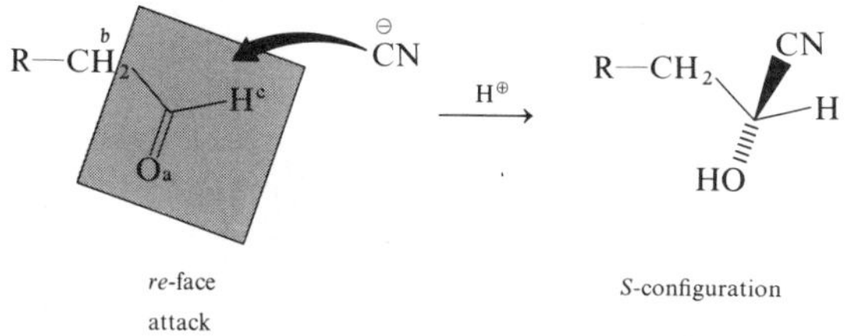

re-face attack

S-configuration

If the sequence a, b, c of priorities of the groups is clockwise, the face toward the reader is *re* (rectus); if counterclockwise, *si* (sinister). If necessary, "plantom atoms" attached to C and O must be added to assign priorities on the basis of the atomic numbers of atoms surrounding the carbon of the carbonyl function.

$CH_2OPO_3^{2-}$, =O, HO—, —OH, —OH, $CH_2OPO_3^{2-}$ — fructose 1,6-diphosphate

$\xrightleftharpoons{\text{aldolase}}$

H_S, OH, C, H_R, =O, H, C, H, $OPO_3^{2\ominus}$ — dihydroxyacetone monophosphate

$\xrightleftharpoons{\text{isomerase}}$

H, O, C, H, OH, H, C, H, $OPO_3^{2\ominus}$ — D-glyceraldehyde 3-phosphate

The concept underlying this specific behavior is called the *Ogston effect* (160). A.G. Ogston suggested in 1948 that in order for an enzyme to be capable of asymmetric attack upon a symmetric substrate, it must bind the substrate to at least three points. Only then is stereoselective recognition or synthesis possible.

Fundamentally, for a compound of the type $yzCx_2$ the enzyme can discriminate between the like groups x because the two transition states for attack of any chiral reagent on either x groups are diastereoisomers and consequently the rates of reaction proceeding through them will differ. This hypothesis laid the foundation for the concept of prochirality and provided the impetus for investigation of the stereochemistry of enzymatic reactions occurring at prochiral centers.

For example, yeast alcohol dehydrogenase (YADH) was the first alcohol dehydrogenase to be crystallized and the first one to be shown that direct hydrogen transfer between substrate and coenzyme can occur. Deuterium-labeled ethanol was used by F.H. Westheimer (161) in 1953 to prove that the process is stereospecific and only one of the enantiotopic protons attached to C-1 of the primary alcohol is transferred to C-4 of the 3-acetamidopyridinium ring of NAD^+ coenzyme (RPPRA = the rest of the coenzyme). If *R*-deutero-1-ethanol is used, all the deuterium is found on the top face of the aromatic portion of the coenzyme (more details on this coenzyme will be given in Chapter 7). The new asymmetric carbon of NAD^2H has the *R* stereochemistry. If *S*-deutero-1-ethanol is used instead, all the label is found on the acetaldehyde and none on the reduced coenzyme. This is a particularly good example of the Ogston effect where coenzyme and substrate are held at the active site of the enzyme in a specific manner. The enzyme has the capacity to distinguish without ambiguity between the two identical hydrogens at the prochiral carbon of ethanol. Of course, the reverse reaction is also stereospecific.

A more dramatic example of asymmetric synthesis is the photosynthetic process of plants which can convert solar energy into chemical energy with the assistance of chlorophyll molecules. In this complex process, achiral carbon dioxide ends up as D-glucose.

In summary, a large number of natural products (hormones, vitamins, biopolymers, etc.) become chiral substances because of the dissymmetric influences exerted by enzymes and other organic materials during biochemical pathways.

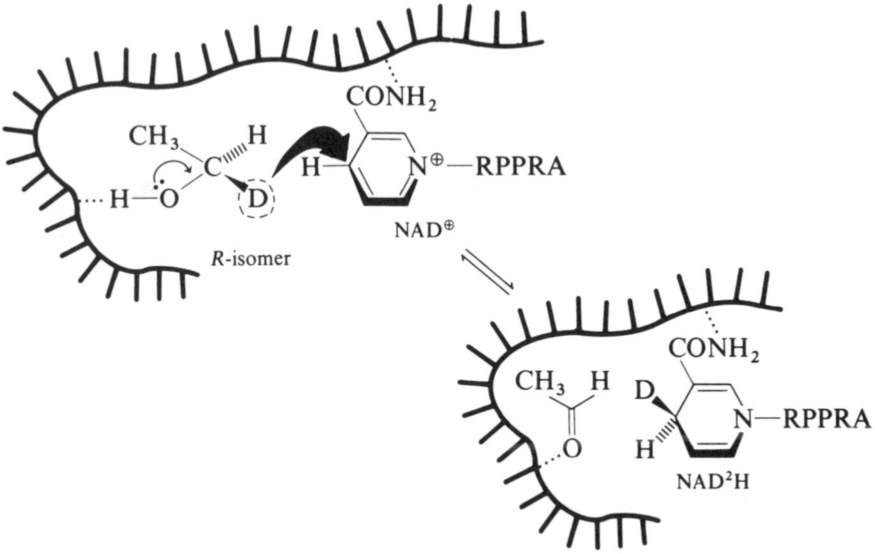

CONH₂
CH₃ H
H—O C D H N⊕—RPPRA
NAD⊕
R-isomer
CH₃ H D CONH₂
O H N—RPPRA
NAD²H

4.2.2 Factors Responsible for Enzyme Specificity

It has been recognized that an enzyme has three levels of specificity: *structural specificity*, *regiospecificity*, and *stereospecificity*. An enzyme must first recognize some common structural features on a substrate (and a coenzyme) to produce a specific catalysis. Second, catalysis must occur at a specific region on the substrate (or the coenzyme) and the stereochemical outcome must be controlled by the enzyme (152,153).

Reduction of ketones by the reduced coenzyme nicotinamide adenine dinucleotide (NADH) and alcohol dehydrogenase provides again a good example (see Section 7.1.2 for pertinent applications in organic synthesis).

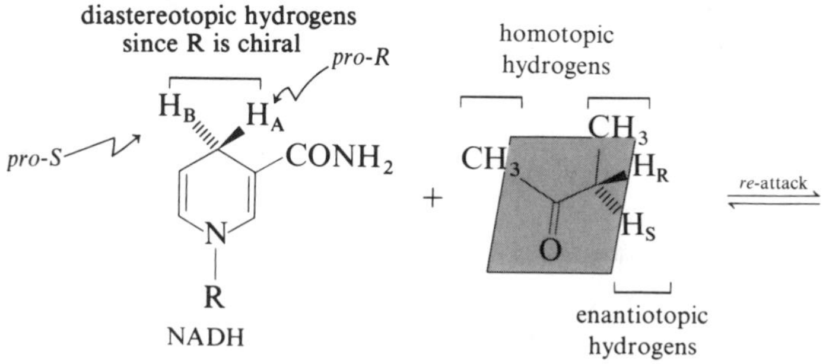

diastereotopic hydrogens since R is chiral
pro-R
pro-S
H_B H_A CONH₂
N
R
NADH
+
homotopic hydrogens
CH_3 CH_3 H_R H_S O
enantiotopic hydrogens
re-attack

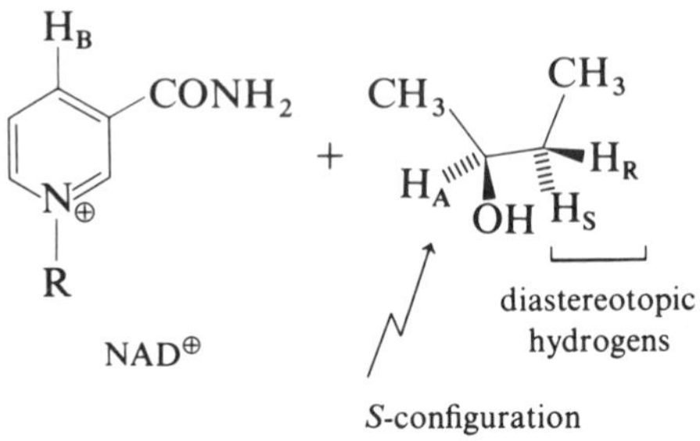

H_B CONH$_2$ + CH_3 CH_3 H_R H_A OH H_S

N$^{\oplus}$ R

NAD$^{\oplus}$

diastereotopic hydrogens

S-configuration

The carbonyl function of the ketone substrate has two faces and is thus said to be *enantiotopic*. In the presence of the enzyme, the coenzyme NADH delivers selectively the hydrogen H_A and only from the rear side of the carbonyl function, producing a chiral alcohol where the methylene hydrogens then become *diastereotopic*. The methyl hydrogens, however, cannot be distinguished by the enzyme and are called *homotopic*. In this example, the H_A (*pro-R*) hydrogen of the coenzyme migrated on the substrate, but other alcohol dehydrogenases have the opposite regiospecificity where the H_B (*pro-S*) hydrogen is used.

Enzyme-catalyzed reactions obey the principle of *microscopic reversibility*. This states that the mechanism of any reaction in the backward direction is just the reverse of the mechanism in the forward reaction. This thermodynamic concept is very useful for elucidating the nature of a transition state from knowledge of that for the reverse reaction. Certainly enzymes are nature's most specific and powerful catalysts. No man-made catalysts have been capable of duplicating the great catalytic efficiency exhibited by enzymes under mild physiological conditions. Rate enhancements by factors of the order of 10^{10} to 10^{14} compared with similar nonenzymatic reactions have been observed.

One property that most profoundly distinguishes enzymes from other catalysts is their ability to bind their substrates in close proximity to each other and to the catalytic groups of the enzyme. Thus, enzymes accelerate reactions by juxtaposition of reacting atoms, the *proximity effect*, and by proper orientation of the relevant chemical groups. Certainly proximity and orientation effects are very important in catalysis. Random collision between molecules is too hazardous a process to bring about specific and efficient catalysis. Direction must be given. This direction can be brought about by a stereospecific complexing of the substrate. Correct stereochemistry between catalyst and substrate would then give a favorable entropy of activation, leading to efficiency of catalysis.

Generally speaking, the factors involved in the catalytic activity of an enzyme can be of four types. First, a *chemical apparatus* is needed at the active center that can deform or polarize bonds of the substrate to make the latter more reactive. Second, a *binding site* that can immobilize the substrate in the correct geometry relative to other reactive groups that are participating in the chemical transformation. Third, a correct and precise *orientation* of the substrate which permits each step of the reaction to proceed with minimal translational or rotational movement about the bonds of the substrate. Finally, the mode of fixation of the substrate must *lower the energy of activation* of the enzyme–substrate complex in the transition state. The proper distribution of charges at the active site and the geometry of the active center are among the factors responsible for a decrease in the overall entropy in the transition state. All these factors act, to a different degree, on the structure of the active site of an enzyme and cannot be separated from one another. Together they accelerate the rate of the enzymatic reactions and allow the enzyme to function as a powerful catalyst (162).

These factors can be evaluated experimentally with the aid of model systems. Rate enhancements have been obtained in several biomodel chemical reactions that are of similar magnitude to those observed in analogous enzyme-catalyzed reactions.

4.2.3 Intramolecular Catalysis

Intramolecular acid–base catalysis is an effective way of catalyzing reactions in organic systems. However, it would be useful to know the contribution of this to enzyme catalysis. There is a fundamental difference between enzyme and solution chemistry. The rate constants for solution catalysis are second order: the rate increasing with increasing concentration of catalyst. The reactions in an enzyme–substrate complex are first order, the acids and bases being an integral part of the molecule. The importance of acid and base catalysts in the latter system may be evaluated by synthesis of model compounds with the catalytic group as part of the substrate. This may be compared with the reaction rates for the corresponding intermolecular reactions.

Let us examine the following two examples:

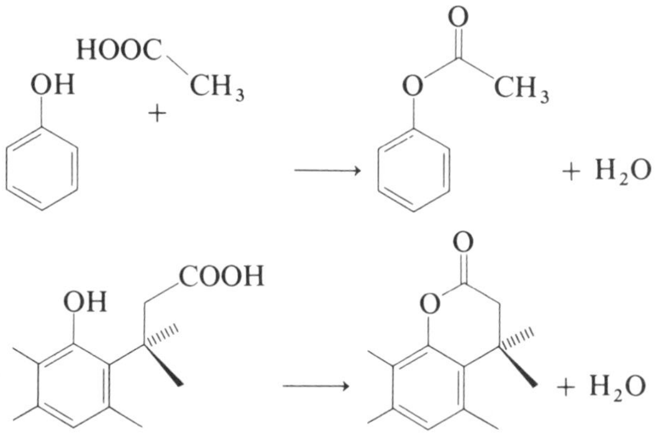

The intramolecular reaction shows an enhancement in rate relative to the intermolecular reaction by a factor of 10^{15} and the molecule lactonizes even in alkali! The aromatic methyl group fits between the geminal methyl groups of the side chain because of steric compression and "locks" the system. Cyclization also relieves the large ground state strain built into the molecule. Furthermore, the "effective concentration" of the carboxyl group in the proximity of the attacking hydroxyl group is very high.

As mentioned already, the high *effective concentration** of intramolecular groups is one of the most important reasons for the efficiency of enzyme catalysis. One function of an enzyme is thus to bring the substrate, by binding at the active site, into proximity with functional groups of the enzyme. This involves a change in *entropy* of the system. Hence, the catalytic consequence of an intramolecular reaction over its intermolecular counterpart is due to an entropic effect. The intermolecular reaction involves two or more molecules associating to form one, leading to an increase in "order" and a consequent loss of entropy.

Other factors can also be invoked, in particular, solvation factors. An intramolecular nucleophile is less heavily solvated in the ground state than an intermolecular nucleophile in dilute solution. Furthermore, an intramolecular nucleophile is more rigidly held with respect to the reaction center. The relief of strain in an intramolecular reaction can also account in a large part for the rate enhancement. Finally, nonproductive binding is sometimes suggested to account for the low reactivity of "poor" substrates (163).

Another interesting example of intramolecular catalysis is the following (163):

HO O NH $\overset{\oplus}{N}H_3$ ⇌

intramolecular nucleophilic attack

general acid catalysis

HO$^{\oplus}$ $O^{\ominus}$ N H H–$\overset{\oplus}{N}H_2$ $\xrightarrow{-H^{\oplus}}$ O O + H_2N NH_2

*The effective concentration of the group concerned is obtained by dividing the observed first-order rate constant for the intramolecular reaction by the second-order rate constant for the corresponding intermolecular process (164). Values in the range of 10^7 to $10^9\,M$ are anticipated in enzymatic reactions.

The reaction showed a rate enhancement of about $10^8\,M/L$, attributable almost entirely to an entropic factor. The intramolecular participation of the ammonium group, which acts as a general-acid catalyst, accounts for less than $10\,M/L$ in rate enhancement.

Since the entropy of a molecule is composed of the sum of its translational, rotational, and internal entropies, T.C. Bruice in 1960 argued that the rate increases found in intramolecular reactions could be due to the restriction of unfavorable rotomer distribution (165). To prove this point, he studied the intramolecular displacement of *p*-bromophenol (to form an anhydride intermediate) in the following series of compounds as a function of increasing rigidity in the molecule:

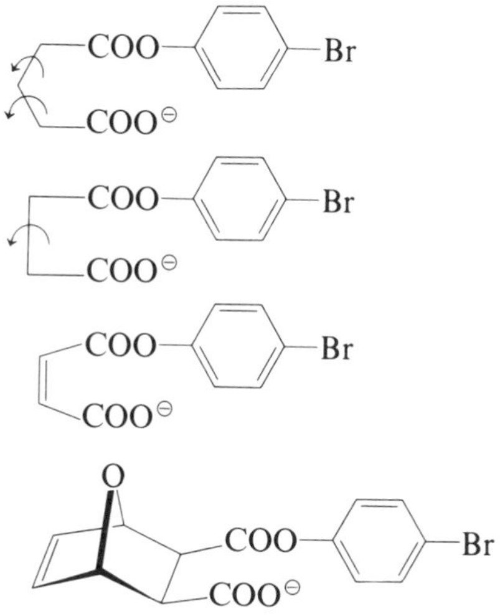

	relative rate of diacid formation
COO–C₆H₄–Br / COO⊖ (two free rotations)	1
COO–C₆H₄–Br / COO⊖ (one free rotation)	200
COO–C₆H₄–Br / COO⊖ (cis double bond)	10,300
O-bridged bicyclic COO–C₆H₄–Br / COO⊖	53,000

Note that removal of one degree of freedom increases the rate by about 200. With the rigid molecules, the reactive groups are properly aligned for reaction, and the reaction rates are much larger. The resulting rate enhancement in these reactions is a direct consequence of the *proximity effect* or closeness of the reactive groups. This results in favorable changes in translational and rotational entropy of activation. Bruice believes that "freezing" internal rotation of the substrate accounts for the principal factor in catalytic efficiency of enzymes, plus the entropic effect.

In contrast, W.P. Jencks of Brandeis University argues that loss of *translational entropy* (less motion) is among the most important factors to explain the large rate increase in intramolecular catalysis. In enzymatic reactions this entropy loss is offset by a favorable binding energy of the substrate to the enzyme which provides the driving force for catalysis. In other words, besides a lowering of the degree of rotation and of the translational entropy, Jencks suggests that the concept of *intrinsic binding energy*, which results from favorable noncovalent interactions with the substrate at the site of catalysis, is to a large degree responsible for the remarkable specificity and the high rates of enzymatic reactions (166).

By studying intramolecular lactonization reaction, where rate enhancements of 10^3 to 10^6 over the bimolecular counterpart were found, D.E. Koshland, Jr. (40) in 1970 proposed a new concept: *orbital steering*. His experimental evidence suggests that "steering" reacting atoms can lead to (or explain) large rate factors. Not only must the reacting groups be close, but they must also be properly oriented. In this concept, both favorable proximity and orientation are important. The more rigid molecules show the largest rates. The loss of translational entropy is not as important as Jencks originally suggested, according to Koshland's views. However, Menger and Glass have now presented new and convincing experimental facts that invalidate the orbital steering theory (41).

Nevertheless, the Bruice–Koshland controversy forms part of the large question of why intramolecular reactions are so favorable. The truth is probably a blend of both effects, and other factors. In Section 4.6 we will see that proper orientation of lone-pair orbitals is an important stereoelectronic requirement to explain the rate enhancement in hydrolytic reactions.

Besides *proximity* effect and *orientation* effect, *steric compression* effect could be a third factor to consider. Nonetheless, there can still be an enhancement of 10^2 to 10^4 in rate constants not accounted for by these factors. Among the likely candidates, electrostatic stabilization of the transition state and release of ground state strain should be mentioned. The notion of "freezing of substrate specificity" by Bender is also a factor for consideration. An example is the "aromatic hole" in α-chymotrypsin (see below) which allows a favorable steric situation for the amino acid side chain of the substrate.

In summary, entropy is one of the most important factors in enzyme catalysis. For chemical reactions catalyzed in solution, the catalyst and the substrate molecules have to come together in close contact and this involves a large loss of entropy. Enzymatic reactions, however, can be looked at as being restricted to an understanding of the chemistry going on at the enzyme–substrate complex where the catalytic groups are acting cooperatively on the same molecule. As a consequence, the catalytic groups on an enzyme have very high effective concentrations compared with bimolecular reactions in solution. Thus, there is no loss of translational and/or rotational entropy in the transition state. In other words, the rotational and translational entropies of the substrate are lost on formation of the enzyme–substrate complex and not during the subsequent catalytic steps (156).

A comprehensive review on enzyme reactivity from an organic perspective, published by F.M. Menger (Emory University), should be consulted for a more critical view of modern enzymology (167).

4.3 Multifunctional Catalysis and Simple Models

Another fundamental idea that has been invoked to explain enzymatic catalysis is that such reactions utilize bifunctional or multifunctional catalysis. That is, several functional groups in the active center are properly aligned with the substrate so that *concerted catalysis* may occur.

One of the earliest examples (168) in solution chemistry is the study of the rate of mutarotation of *O*-tetramethyl-D-glucose in the presence of α-pyridone. A 10^{-3} *M* solution of α-pyridone was 7,000 times more effective than an equivalent concentration of phenol and pyridine.

A concerted mechanism was proposed where the catalyst acts as a base and an acid simultaneously:

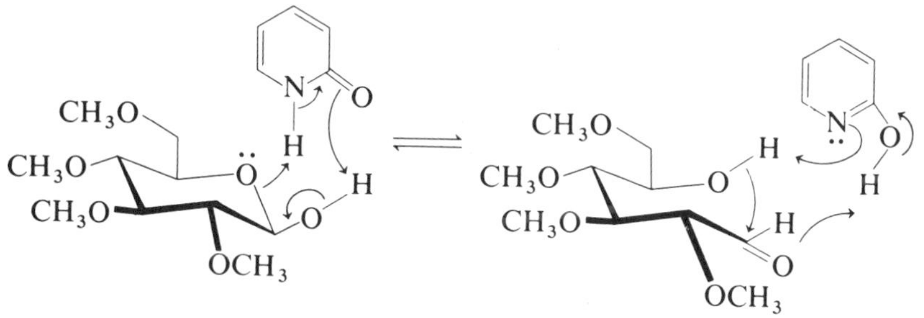

A key feature of enzyme catalysis is observed: regeneration of the catalyst (via the reversible pathway). 2-Aminophenol, which is both a stronger acid and base than α-pyridone, is not as potent a catalyst. In the presence of 2-aminophenol a concerted mechanism of catalysis is not possible.

Measurement of primary and secondary deuterium isotope effects were performed on models to obtain a deeper understanding of the nature of this two-proton transfer reaction in bifunctional catalysis (169). The study agreed that the two protons are in transit in the transition state of the rate-limiting step.

Another classic example (170) of multifunctional catalysis is the hydrolysis of the monosuccinate ester of hexachlorophene:

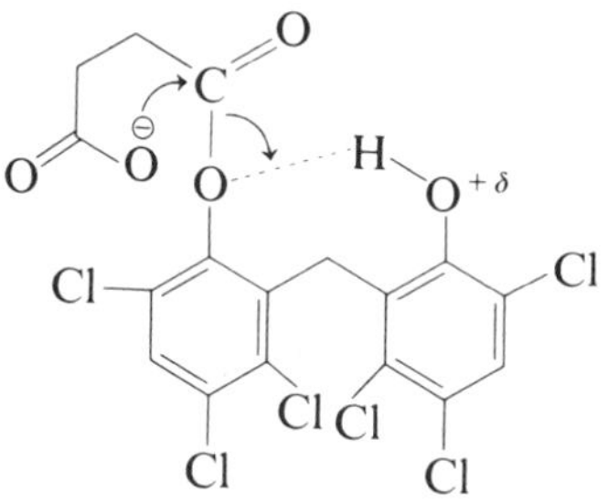

The proximity and participation of a hydroxyl group greatly accelerates the reaction rate. Similarly, comparison of the corresponding monoacetate and diacetate shows that the monoacetate is hydrolyzed 500 times faster than the diacetate below pH 5. A plot of the rate of reaction versus pH has a bell-shaped* profile and indicates intramolecular nucleophilic and general-acid hydrolysis. It is also indicative of ionic species involved with opposite pK_a values for maximal activity.

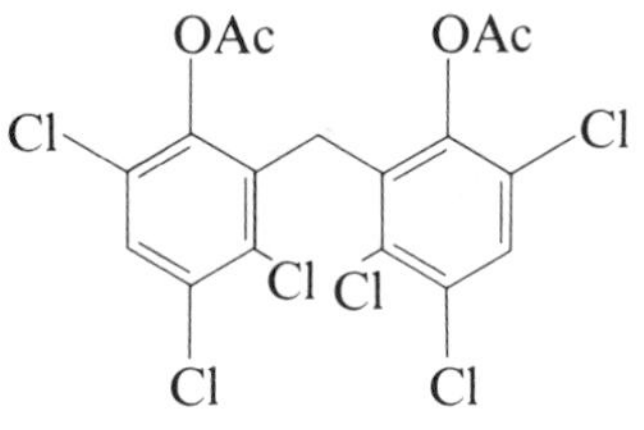

diacetate

monoacetate

Perhaps of more relevance to the understanding of enzyme function is a model for bifunctional catalysis in an aqueous milieu. One of the first reactions to show a significant rate enhancement in water was the hydrolysis of an iminolactone. This work was undertaken by Cunningham and Schmir (171) who observed that phosphate buffer was at least 200 times as effective as imidazole buffer (even though both buffers have approximately the same pK_a in catalyzing the expulsion of aniline.

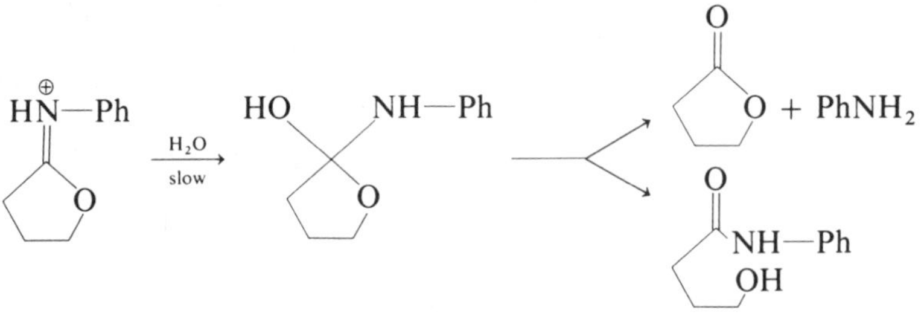

At basic pH, amide formation would be expected to be the preferred pathway, since the amide nitrogen would be a poor leaving group. This is observed in imidazole buffer. However, with phosphate, bifunctional catalysis allows ejection of aniline to become significantly more favorable:

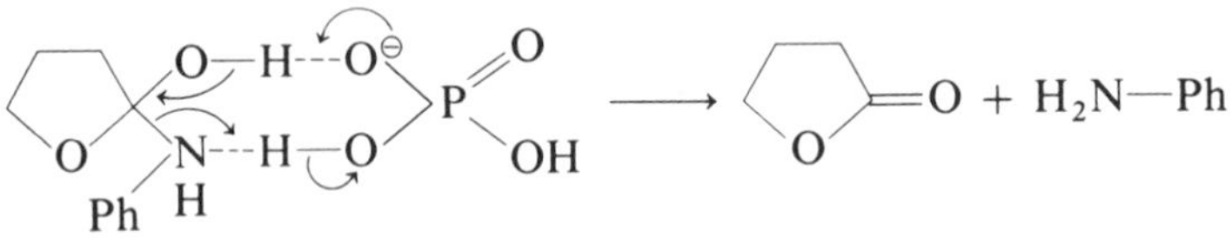

* A bell-shaped rate profile as a function of pH is characteristic of acid–base concerted catalysis.

Similar catalysis was observed with bicarbonate and acetate buffers.

More recently, Lee and Schmir (172,173) have studied similar bifunctional catalytic mechanisms in more detail with acyclic imidate esters. Again, hydrolysis proceeds via the rate-determining formation of a tetrahedral intermediate followed by breakdown to an amide or ester product:

$$CH_3{-}C(OEt){=}\overset{\oplus}{N}(Ph)CH_3 \xrightarrow[\text{slow}]{H_2O} CH_3{-}C(OEt)(\ddot{O}H){-}N(Ph)(CH_3)$$

$$CH_3{-}\overset{O}{\overset{\|}{C}}{-}N(Ph)(CH_3) + EtOH \qquad CH_3{-}\overset{O}{\overset{\|}{C}}{-}OEt + HN(Ph)(CH_3)$$

In the absence of buffer, the yield of amide will increase with increasing pH. At pH 7.5, 50% amide and 50% ester is formed. However, in the presence of various bifunctional catalysts amide formation will become the favored pathway. This is illustrated in Table 4.1.

Table 4.1. Hydrolysis of Acyclic Imidate Esters in Various Buffers (172,173)

Buffer	pH range studied	Rate ratio[a]
$HCO_3^{\ominus}$	7.49–9.73	290
$CH_3C(=N{-}OH){-}CH_3$	7.73–8.85	8700
2-pyridone (NH, C=O ring)	7.92–8.55	1460
$HPO_4^{2\ominus}$	6.96–9.11	52
$HAsO_4^{2\ominus}$	7.18–8.82	61
$H_2AsO_3^{\ominus}$	7.42–8.99	150
$CF_3C(O^{\ominus})(OH)CF_3$	7.13–8.90	47

[a] Refers to the ratio of amine formation for the bifunctional catalyst, and its monofunctional equivalent of the same pK_a.

With the exception of acetone oxime, all of the bifunctional catalysts contain an acidic and basic group in a 1,3-relationship. The cyclic transition state required for a concerted proton transfer would consist of an eight-membered ring. This is illustrated for phosphate:

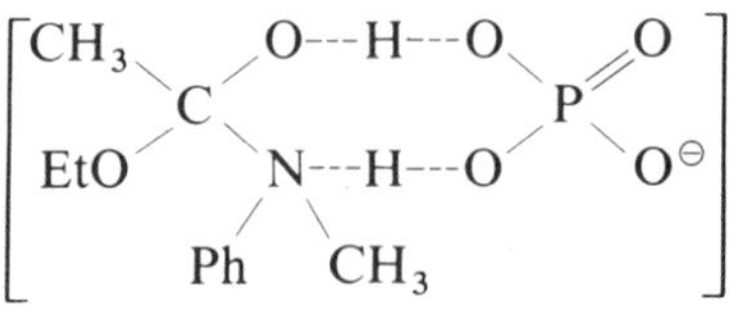

$$\left[\begin{array}{c} CH_3\diagdown \quad O\text{---}H\text{---}O \diagdown \quad \diagup\!\!\!\!= O \\ C \qquad\qquad\qquad P \\ EtO\diagup \quad N\text{---}H\text{---}O\diagup \quad \diagdown O^{\ominus} \\ Ph \quad CH_3 \end{array}\right]$$

Once the cyclic transition state has been achieved, breakdown will occur in a fashion analogous to that described for the iminolactone, to give ejection of aniline and ester formation. Notice that while imidazole does possess an acidic and basic function in the 1,3-geometry, these groups are sterically unsuited for a cyclic proton transfer.

In 1972, Jencks (174) concluded that concerted bifunctional acid–base catalysis is rare or nonexistent in H_2O because of the improbability of the reactant and catalyst meeting simultaneously. Furthermore, bifunctional catalysis does not necessarily represent a favorable process in aqueous solution even when a second functional group is held sterically in proper position to participate in the reaction. Caution should then be used in assuming that most enzymes are utilizing bifunctional or multifunctional catalysis. Nevertheless, this idea has played a leading role in concepts of enzyme catalysis.

4.4 α-Chymotrypsin

The catalytic efficiency of enzymes is a subject of great fascination, particularly when crystallographic structures are available and a great deal is known about the physical organic chemistry of the enzymatic mechanism. In this regard, the most studied enzyme of a group of enzymes called serine proteases is α-chymotrypsin. The term *serine protease* derives from the fact that this class of enzymes contain at their active site a serine hydroxyl group that exhibits unusual reactivity toward the irreversible inhibitor diisopropylphosphorofluoridate (DFP).

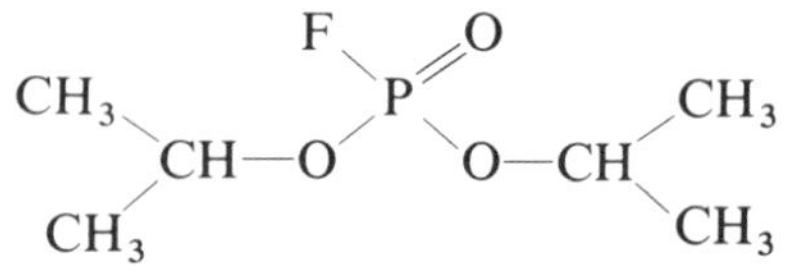

$$(CH_3)_2CH\text{—}O\text{—}P(=O)(F)\text{—}O\text{—}CH(CH_3)_2$$

DFP

The highly reactive (P—F) linkage readily suffers displacement by nucleophiles such as the hydroxyl of the active serine of proteolytic enzymes. Other enzymes in this group include trypsin, thrombin, and subtilisin.

The mechanism of action of α-chymotrypsin is probably understood in more detail than any other enzyme at the present time. Its physiological function is to catalyze the hydrolysis of peptide bonds of protein foods in the mammalian gut. It is secreted in the pancreas as an inactive *zymogen* precursor, chymotrypsinogen, having a single polypeptide chain of 245 amino acids. Such an inactive form is necessary to prevent "self-digestion." The precursor is activated by trypsin hydrolysis (scission of the peptide backbone) at two specific regions to form active α-chymotrypsin having three peptide chains held by disulfide bridges. The three-dimensional structure was deduced by the X-ray diffraction study of D.M. Blow (162) and the mechanism of action was elucidated in collaboration with B.S. Hartley and J.J. Birktoft (175,176), all at the MRC Laboratory of Molecular Biology, Cambridge, England.

In addition to the active serine-195 residue, the other active site residues are histidine-57 and aspartic acid-102, deduced from the X-ray work. The other histidine residue, His-40, is not implicated in the catalysis. The enzyme has a specificity for aromatic amino acids. Esters of aromatic amino acids are also good substrates for the enzyme and most of the kinetic data were obtained with ester substrates. The enzyme cuts up proteins on the carboxyl side of aromatic amino acids. After the formation of the Michaelis complex, the uniquely reactive Ser-195 is first acylated to form an acyl-enzyme intermediate with the substrate.

The transformation of the Michaelis complex to the acyl-enzyme involves first the formation of a tetrahedral intermediate (T.I.) (see following section). Finally, attack by a water molecule hydrolyzes the acyl-enzyme, which normally does not accumulate.

$$\text{En—OH} + \text{R—}\overset{\overset{\text{O}}{\|}}{\text{C}}\text{—X} \longrightarrow [\text{En—OH}\cdot\text{R—COX}] \longrightarrow \text{En—O—}\underset{\underset{\text{R}}{|}}{\overset{\overset{\text{O}^{\ominus}}{|}}{\text{C}}}\text{—X} \xrightarrow{\text{HX}}$$

X = OR′, NHR′
R = Phe, Trp, Tyr

Michaelis complex

T.I.

$$\text{En—O—}\overset{\overset{\text{O}}{\|}}{\text{C}}\text{—R} \xrightarrow{\text{H}_2\text{O}} \text{En—O—}\underset{\underset{\text{OH}}{|}}{\overset{\overset{\text{O}^{\ominus}}{|}}{\text{C}}}\text{—R} \longrightarrow \text{En—OH} + \text{HO—}\overset{\overset{\text{O}}{\|}}{\text{C}}\text{—R}$$

acyl-enzyme

T.I.

It has long been thought that the basicity of the imidazole residue His-57 increases the nucleophilicity of the hydroxyl of Ser-195 by acting as a general-base catalyst.

H—N N:⌒H—O Ser-195

His-57

However, on the basis of the crystallographic structure for α-chymotrypsin, a new form of general catalysis was proposed. It is called the *charge-relay* system and originates with the alignment of Asp-102, His-57, and Ser-195, linked by hydrogen bonds (Fig. 4.3).

The acid group (Asp-102) is believed to be buried and its unique juxtaposition with the unusually polarizable system of the imidazole ring (His-57), which in its uncharged form can carry a proton on either of the two ring nitrogens, seems to be the key to the activity of this enzyme, and other serine proteases (177). The function of the buried Asp-102 group is to polarize the imidazole ring, since the buried negative charge induces a positive charge adjacent to it. This gives a possibility of proton transfer along hydrogen bonds, to allow the hydroxyl proton of Ser-195 to be transferred to His-57. The active serine residue becomes then a reactive nucleophile capable of attacking the scissile peptide bond.

Asp-102—C(=O)—O⊖---H—N N:---H—O Ser-195

His-57

R

X O

$pK_a \sim 6.5$ to 7.0 X = NHR′, OR′

The catalytic efficiency of α-chymotrypsin cannot be attributed solely to the presence of the charge-relay system. X-ray work (178) has indicated the many parameters operative in the catalytic process. Nine specific enzyme-substrate interactions have been identified in making the process more efficient. For example, stabilization of the tetrahedral intermediate, and thus lowering of the transition state energy barrier, is accomplished by hydrogen bond formation of the substrate carbonyl function with the amide hydrogen of Ser-195 and Gly-193. In chymotrypsinogen, the latter hydrogen bond is absent. Indeed, refinement of the X-ray structures of chymotrypsinogen and α-chymotrypsin indicates a difference between the structure of the catalytic triad in the zymogen and the enzyme. This conformational change in the overall three-dimensional structure of the enzyme may produce significant changes in the chemical properties of the

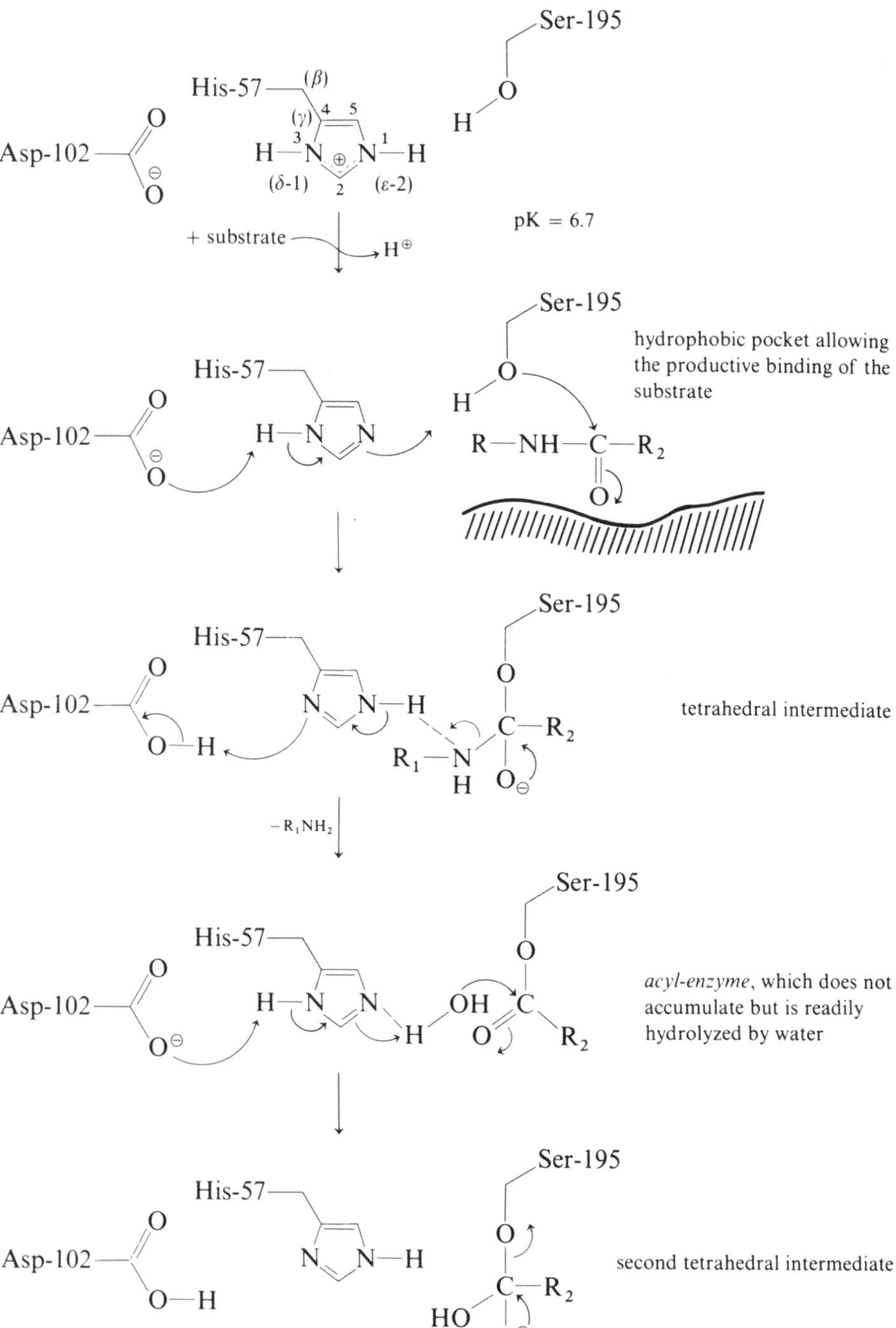

Fig. 4.3. The charge-relay system of α-chymotrypsin (position 3 of the imidazole group is often referred to as Nδl or π and position 1 as Nε2 or τ).

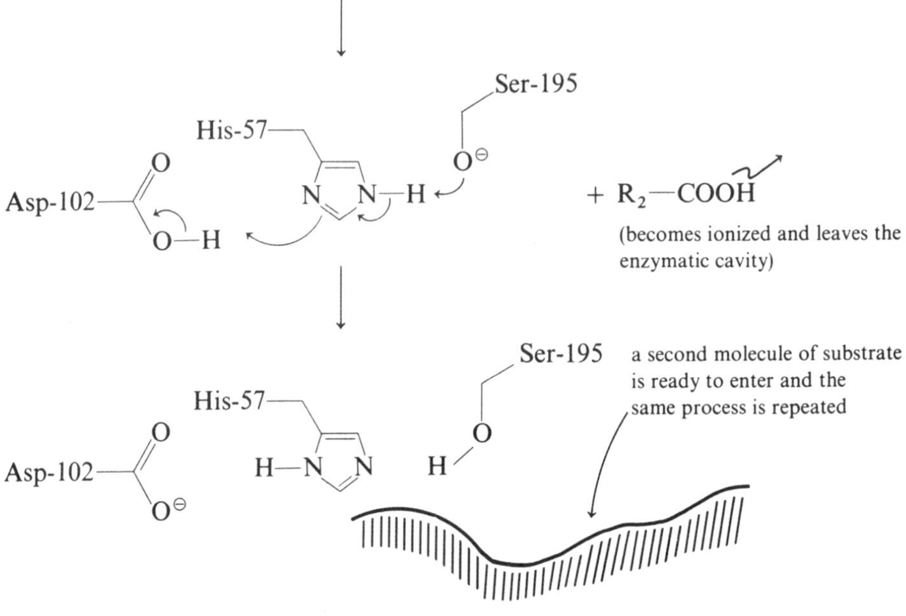

Fig. 4.3. (continued)

catalytic residues, which may play an important role in the amplification of enzyme activity upon zymogen activation.

Evidence has been accumulated for the existence of a tetrahedral intermediate and a concerted proton transfer mechanism from Ser-195 to His-57 and from His-57 to Asp-102 during the hydrolysis of *p*-nitroanilide peptide substrates catalyzed by a serine protease from *Myxobacter* 495 named α-lytic protease. It possesses only one histidine residue (179,180). The rate-limiting step of the hydrolysis is the decomposition of the tetrahedral intermediate to acyl-enzyme and *p*-nitroaniline. The experimental results support the view that the two proton transfers that occur during this step take place in a concerted, not stepwise, manner.

Surprisingly, ^{13}C-NMR work (181) suggested that the group ionizing with a pK_a of about 7 is the buried aspartate and not the expected His-57, suggesting that the imidazole ring remains unprotonated during catalysis in the pH range 7–8. However, careful high resolution ^{1}H-NMR work by J.L. Markley's group (182) at Purdue University on chymotrypsinogen, α-chymotrypsin, and trypsin supports the previous conclusion of Robillard and Shulman (183) that the pK_a of His-57 is higher than that of Asp-102 in both the zymogen and enzyme. Based on this argument and others, Markley favors the following mechanism, a variant of the one in Fig. 4.3, for formation of the tetrahedral intermediate:

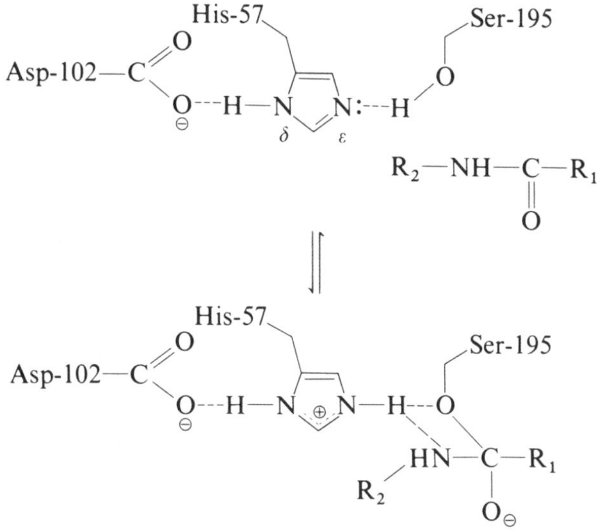

Therefore, both of the ^{1}H-NMR evidence clearly support the existence of a His-57–Asp-102 ion pair (183), and a concerted transfer of two protons becomes unlikely.

A recent X-ray work (184) on α-lytic protease showed that Asp-102 is in a strongly polar environment with a pK_a of 4.5. Furthermore, histidine, enriched in ^{15}N in the imidazole ring, has been incorporated into α-lytic protease, using a histidine auxotroph of *Myxobacter* 495 (185). The behavior of the ^{15}N-NMR resonances of this labeled α-lytic protease in the "catalytic triad" as a function of pH indicated clearly the presence of a hydrogen-bonded interaction between NH at the position-3 of histidine (Nδ1) and the adjacent buried carboxylate group of aspartic acid.

Furthermore, Komiyama and Bender (186) proposed that the proton abstracted from the hydroxyl group of the serine by the imidazole group of the histidine is donated to the nitrogen atom of the leaving group of the amide before the bond between the carbonyl carbon atom of the amide and the attacking serine oxygen atom is completed.

It would be most interesting to know the pK_a of the active site His-57 in the transition state of an α-chymotrypsin-catalyzed reaction. Efforts in this direction come again from the group of Markley (187). They have looked at the pH dependence of the ^{31}P-NMR chemical shifts of a series of diisopropylphosphyryl-serine (DFP-serine) proteases. The pK_a values obtained from these titration studies agree with earlier pK_a values derived from ^{1}H-NMR data for peaks assigned to the hydrogen at position Cε1-H (C-2) of His-57 of each of these DFP derivatives. Interestingly, the pK_a of His-57 of three enzymes studies (α-chymotrypsin, trypsin, and α-lytic protease) increases (by as much as 2.0 pK_a units for α-lytic protease) when Ser-195 is derivatized. Since inhibitors such as DFP can be considered as transition state analogs (188) of serine proteases, Markley argued that the increased pK_a values of His-57 observed in his derivatives may represent an approximation of the transition state pK_a values. The fact that His-57 has a lower pK_a in free enzymes ensures that the imidazole residue is in the unprotonated form at physiological pH (187). After formation of the substrate–enzyme complex and as the transition state is approached, the pK_a of His-57 is raised, making the imidazole a more efficient base for accepting the hydroxyl proton of Ser-195. Obviously, biophysical techniques such as nuclear (^{1}H, ^{15}N, ^{13}C, and ^{31}P) magnetic resonance spectroscopy have played a key role in the evaluation of the correct mechanism of action of serine proteases.

A recent study using ^{15}N-NMR to follow the hydrogen-bonding interactions in the active site of α-lytic protease resulted in further evidence for a catalytic mechanism involving directed movement of the imidazole ring of the active site histidyl residue (189). The use of ^{15}N-NMR spectroscopy in biological systems has been developed in a review article (190).

Trypsin, a mammalian protease, and subtilisin, a bacterial protease, have both been shown to have a mechanism of action similar to α-chymotrypsin. While α-chymotrypsin and subtilisin have totally different foldings of their polypeptide backbones, the residues involved in catalysis (serine, histidine, aspartic acid) have the same spatial relationships. This similarity of active centers is a prime example of *convergent evolution* of active center geometries in enzymes (162).

4.4.1 Tetrahedral Intermediates

The existence of a tetrahedral intermediate in the enzyme-catalyzed reaction of an ester or an amide has been a long-standing question of major concern in mechanistic enzymology (191,192). For this reason much attention has been given to the catalytic mechanism operative in serine proteases, based on the availability of much data from chemical and X-ray studies. In 1979 (193), tetrahedral intermediates were observed with the mammalian enzyme elastase. With specific di- and tripeptide *p*-nitroanilide substrates, tetrahedral intermediates have been detected, accumulated, and stabilized at high pH (pH of 10.1) by using subzero temperatures (−39°C) and fluid aqueous/organic cryosolvents. When corrected for the effect of temperature and cosolvent, the rate of intermediate formation was in good agreement with results obtained at 25°C in aqueous solution by stopped-flow techniques. These recent findings are necessary prerequisites for eventual crystallographic cryoenzymological studies. The implication of an imidazole function in the mechanism of action of the serine proteases has led to extensive investigations of imidazole catalysis of ester hydrolysis. For instance, many model compounds indicated that histidine can act as a nucleophile in intramolecular reaction. An early example is the hydrolysis of p-nitrophenyl γ-(4-imidazolyl) butyrate (194):

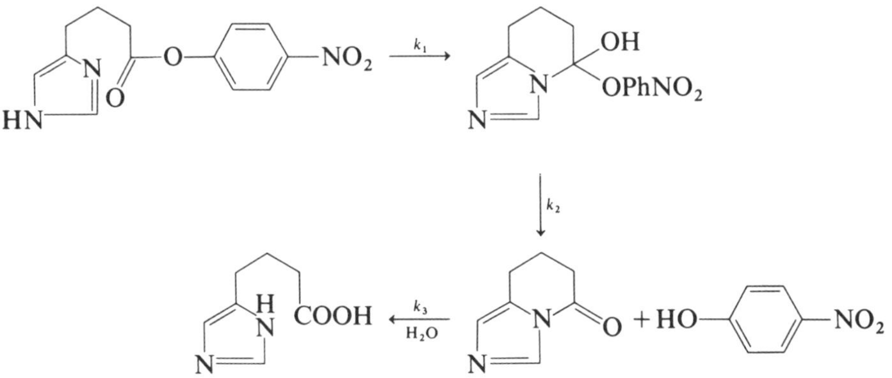

$\xrightarrow{k_1}$ OH OPhNO$_2$ $\downarrow k_2$ COOH $\xleftarrow[H_2O]{k_3}$ + HO—⟨⟩—NO$_2$

acyl-enzyme analog

At pH 7, 25°C, the rate is comparable ($200\,\text{min}^{-1}$) to the release of p-nitrophenolate by α-chymotrypsin ($180\,\text{min}^{-1}$). The rate-limiting step here is the acylation step k_2. In the first approximation it is a satisfactory model to mimic the formation of an acyl-enzyme intermediate, although His-57 in α-chymotrypsin acts as a general-acid–general-base catalyst and not as a nucleophilic catalyst.

As shown in the previous section, the geometrical arrangement of the triad Asp-His-Ser in the active site of serine proteases has been referred to as the charge-relay system. The ensemble is situated in a hydrophobic environment inside the enzyme and attack by the hydroxyl of the serine residue is assisted by general-base removal of the proton by the imidazole ring of the histidine residue. This results in the (partial) generation of a potent (unsolvated) alkoxy nucleophile.

G.A. Rogers and T.C. Bruice have made an extensive study with model compounds to evaluate the validity of the charge-relay system (195–198). The following model compounds show an enhancement in acetate hydrolysis of up to 10^4-fold relative to a mixture of phenylacetate and imidazole.

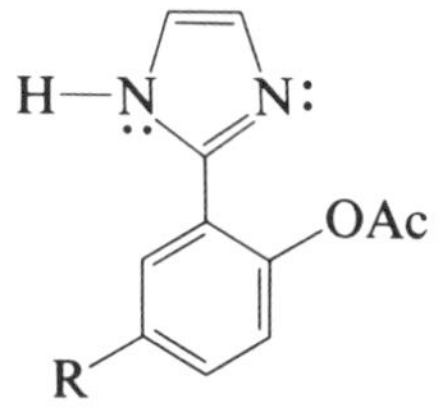

H—N N: OAc R

R = H, $SO_3^{\ominus}$

Three schemes of catalysis have been proposed, depending on experimental conditions.

(a) at low pH

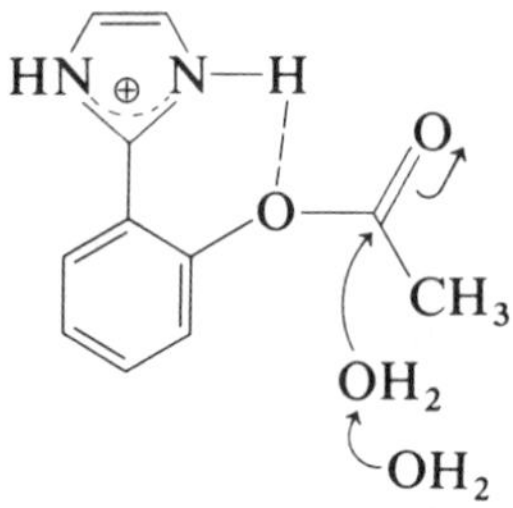

general-acid–assisted
H_2O attack

(b) at neutral pH

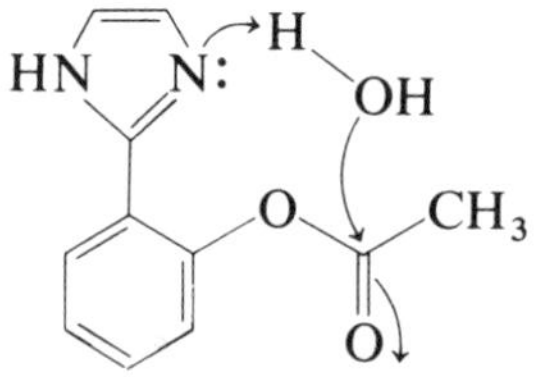

general-base–assisted
H_2O attack

(c) at high pH

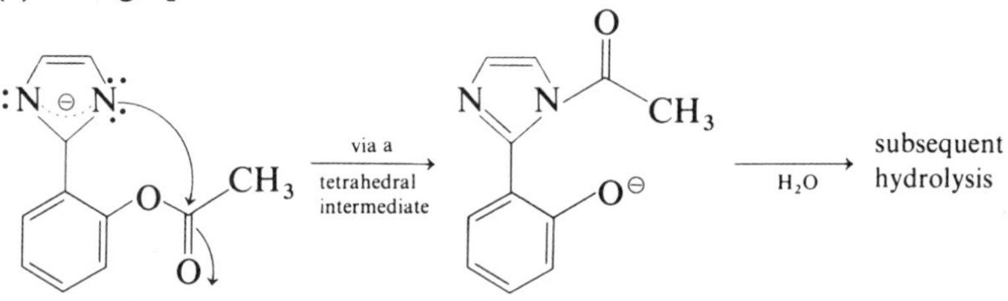

nucleophilic mechanism

At high pH the possibility of O → N acyl-transfer to imidazole anion occurs via a nucleophilic mechanism. Subsequent hydrolysis of the acyl-imidazole intermediate takes place because the rate of hydrolysis exceeds the thermodynamically unfavorable N → O transfer. This model does have some analogy with serine proteases mode of action although only the neutral pH mechanism is relevant. The introduction of a carboxylate group, however, provides a more accurate assessment of the charge-relay system:

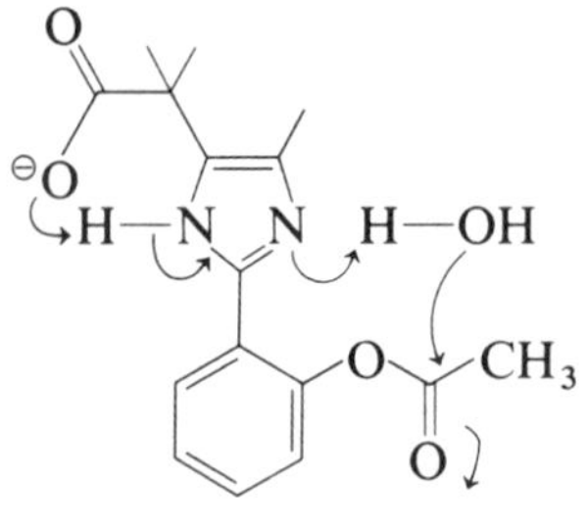

The introduction of the hydrogen-bonded carboxylate does enhance catalysis by the neighboring imidazole and assesses the role of tandem general-base hydrolysis. However, the enhancement in rate of ester hydrolysis is still poor (only by a factor of three). So it is negligible from the standpoint of enzymatic catalysis.

What is the effect of solvent? Is water an integral part of the mechanism? Experiments with a less polar solvent did not alter the role of the carboxylate group. A hydrogen bond should be expected between the carboxylate group and an imidazolium ion. In fact, hydrogen bonding has been definitely established for the zwitterion:

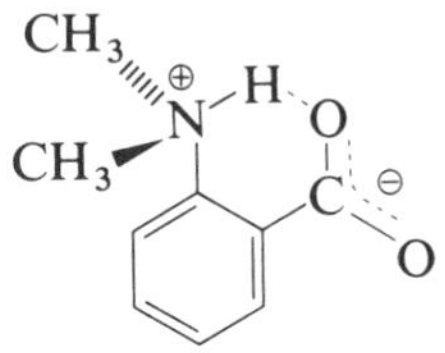

It was found that the reaction with OH^- at the bridged-proton is 10^3 slower than diffusion-controlled values for simple anilinium ions (197). So if the concentration of water decreases in the medium (dielectric constant decreases), both dipole–dipole interactions and hydrogen bonding are expected to increase, a situation not encountered in Bruice's model compound.

Nevertheless, Rogers and Bruice's model does support the charge-relay hypothesis because larger rate enhancements have been reported when nearly anhydrous acetonitrile or toluene is employed. Under these conditions, the hydrogen bond system is "locked" inside and does not exchange with the medium. In this situation scheme (b), general-base–assisted water attack, is possible where a dipolar transition state is formed from a neutral ground state. This would happen in acetonitrile only if a nearby carboxylate (anion) is present.

The model also had the merit of allowing the first isolation of a tetrahedral intermediate in an acetyl-transfer reaction (195). The scheme is as follows on page 194.

The conversion **4-1** → **4-3** involves an *N*-acyl imidazole intermediate. The evidence follows: Competition experiments in the presence of an amine acting as a base indicate that the carboxyl group of **4-1** is attacked to give the corresponding amide of **4-3**. NMR spectroscopy shows the nonequivalence of the two geminal methyl groups in **4-1** but not in **4-2** or **4-3**. This is caused by the rigidity and the asymmetric nature of the tetrahedral intermediate. Furthermore, borohydride reduction of **4-1** to **4-4** shows the presence of a lactone function. No reduction occurred with compounds **4-2** or **4-3**.

The formation of **4-1** from **4-2** probably proceeds through a carboxyl attack on the *N*-acetyl pyridinium salt to give an anhydride intermediate

both 1.57δ

4-2
$pK_a = 3.1, 6.2, 9.8$
IR: 1710, 1640 cm^{-1}

CH_3COCl, pyridine
5 days

1.58 and 1.70δ

4-1
IR: 1780, 1645 cm^{-1}

CH_3COCl, pyridine
1 day

H_2O

$NaBH_4$

both 1.65δ

both 1.33δ

4-3
IR: 1780, 1730, 1645 cm^{-1}
δ = chemical shift (in ppm)

4-4
$pK_a = 5.8, 8.9$

before imidazole displacement. This represents the first isolation and unequivocal characterization of a labile acyl tetrahedral intermediate **4-1** capable of a migration to give an acyl-transfer product **4-3.***

In 1977, Komiyama and Bender (200) also studied the validity of the charge-relay system in serine proteases. They examined the general base-catalyzed hydrolysis of ethyl chloroacetate by 2-benzimidazoleacetic acid as a model system.

---

* Other tetrahedral adducts between a strongly electrophilic ketone and tertiary amines have been isolated and characterized (199).

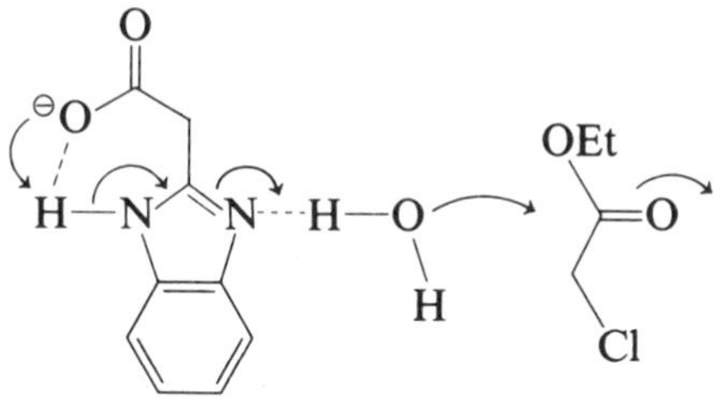

Compared to benzimidazole itself, an 8-fold increase in rate of hydrolysis is observed, implying general-base catalysis assisted by the carboxyl group.

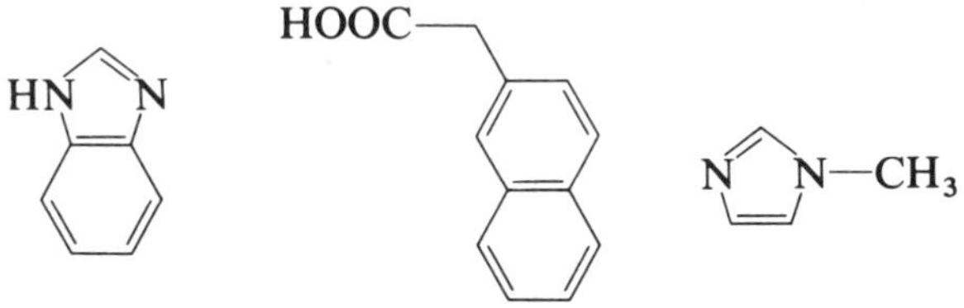

benzimidazole β-naphthyl acetic acid N-methylimidazole

But the carboxyl group alone, as in β-naphthyl acetic acid, cannot catalyze the reaction. Therefore, the mechanism involves the cooperative function of the carboxylate, the imidazole and water.

N-Methylimidazole was also tested as a model comparable to methylated α-chymotrypsin. It was completely inactive toward the hydrolysis of ethyl chloroacetate and indicates that in the charge-relay system both nitrogens of the imidazole ring must be free to participate in the relay.

The properties of α-chymotrypsin methylated at His-57 were also examined to explain the mechanism of this enzyme which is about 10^5 times less active than α-chymotrypsin (201). The results suggest that general-base catalysis remains an integral feature of the hydrolytic mechanism of the modified enzyme. While only subtle alterations occur in the active site upon methylation of His-57, the transition state and the tetrahedral intermediate are destabilized relative to the native enzyme.

Hydroxyl group participation in hydrolysis of esters and amides also proceeds through the formation of a tetrahedral intermediate (T.I.).

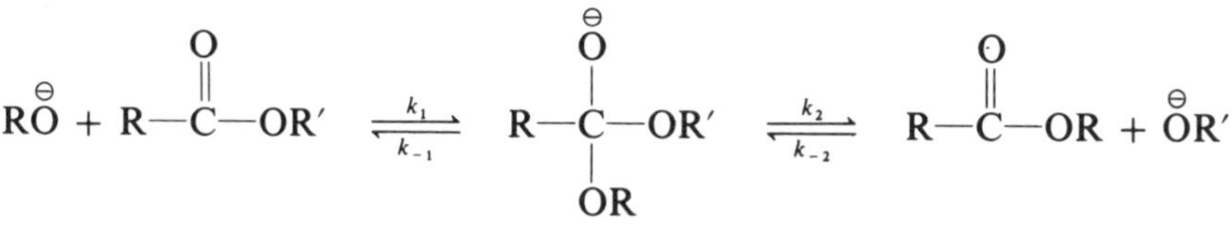

$$RO^{\ominus} + R-\overset{\overset{O}{\|}}{C}-OR' \underset{k_{-1}}{\overset{k_1}{\rightleftharpoons}} R-\underset{\underset{OR}{|}}{\overset{\overset{O^{\ominus}}{|}}{C}}-OR' \underset{k_{-2}}{\overset{k_2}{\rightleftharpoons}} R-\overset{\overset{O}{\|}}{C}-OR + {}^{\ominus}OR'$$

formation T.I. breakdown

Depending on the pK_a value of the leaving group ($R'O^-$), either of these transition states can be reached:

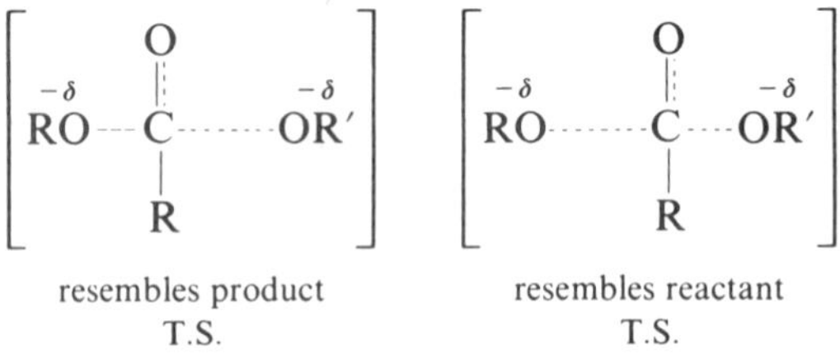

$$\left[\overset{-\delta}{\mathrm{RO}} \cdots \underset{\mathrm{R}}{\overset{\mathrm{O}}{\mathrm{C}}} \cdots\cdots \overset{-\delta}{\mathrm{OR'}} \right] \qquad \left[\overset{-\delta}{\mathrm{RO}} \cdots\cdots \underset{\mathrm{R}}{\overset{\mathrm{O}}{\mathrm{C}}} \cdots \overset{-\delta}{\mathrm{OR'}} \right]$$

resembles product T.S.

resembles reactant T.S.

If the transition state of the tetrahedral intermediate resembles the product, the rate-limiting step is the breakdown of the tetrahedral intermediate. In other words, $k_2 < k_{-1}$; that is, k_2 is the uphill process (slow) and k_1 is fast. This situation arises when RO^- is a good nucleophile, or its pK_a is less than the pK_a of $R'O^-$.

In the other situation, where the transition state resembles the reactant, the rate-limiting step is the formation of the tetrahedral intermediate where $k_2 > k_{-1}$ and k_1 is slow. In this case $R'O^-$ is a good leaving group and its pK_a is smaller than the pK_a of RO^-, the nucleophile. These principles are readily applicable to the hydrolysis of *p*-nitrophenyl esters by a methoxy group.

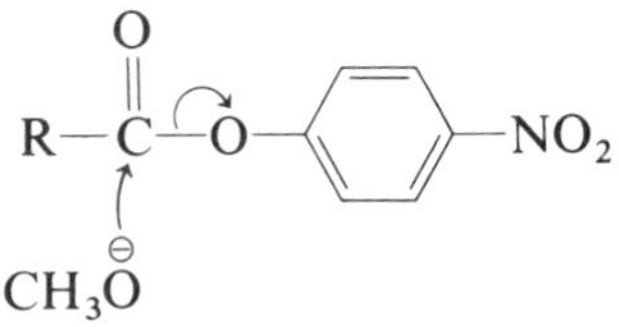

$$\mathrm{R{-}C(=O){-}O{-}C_6H_4{-}NO_2} \quad \leftarrow \; \mathrm{CH_3O^{\ominus}}$$

It is a downhill process when the nucleophile is the conjugate base of a weak acid (methanol) and the leaving group is the conjugate base of a strong acid (pK_a is small). Consequently, the reverse reaction is much more difficult.

In summary, with a good leaving group (weak conjugate base) the rate-limiting step (slow) in ester hydrolysis is the formation of the tetrahedral intermediate. With a poor leaving group the rate-limiting step is the breakdown of the tetrahedral intermediate.

Of course, hydroxyl group participation in catalysis has some analogy with the participation of the serine residue in serine proteases. For this reason model compounds have been prepared and studied.

An example is the following intramolecular transformation:

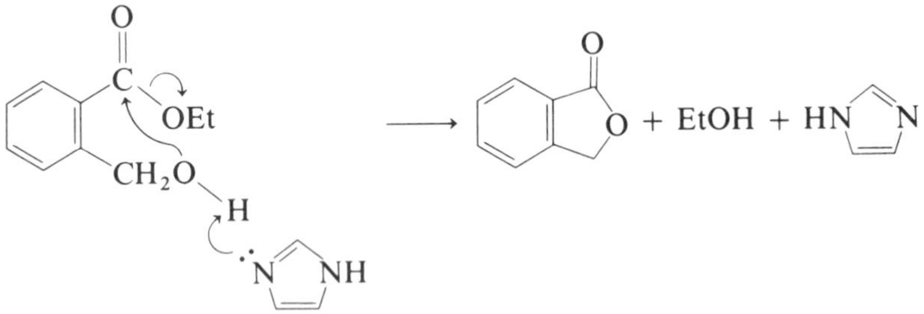

$$\text{(o-}CH_2OH\text{ benzoate ethyl ester, } C(=O)OEt,\ CH_2O{-}H \cdots :N\text{ imidazole NH)} \longrightarrow \text{phthalide} + \mathrm{EtOH} + \mathrm{HN}\text{ imidazole N}$$

The rate of this reaction is 10^5 larger than for ethyl benzoate. The reaction is base catalyzed, and as expected in the presence of ^2H_2O instead of H_2O solvent, the rate is reduced by a factor of at least two ($k_H/k_D = 3.5$). Thus, proton-transfer occurs in the transition state and the analogy with α-chymotrypsin acylation is apparent.

Another interesting simple model is the amide analog (202):

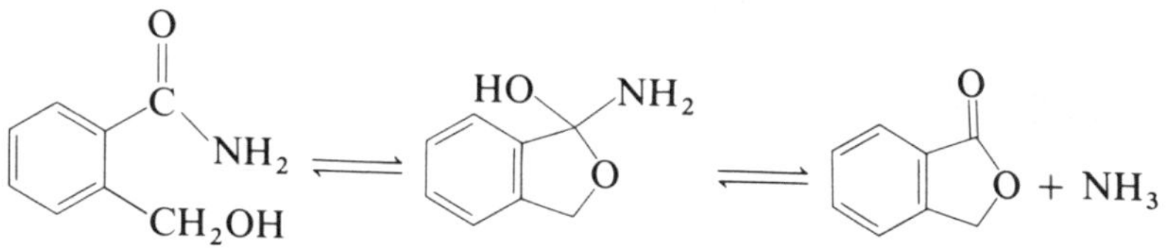

O; C; NH_2; CH_2OH; HO; NH_2; O; O; O + NH_3

At low pH, protonation of the nitrogen occurs and $—\overset{+}{N}H_3$ becomes a good leaving group. At high pH, however, the hydroxyl group of the tetrahedral intermediate becomes a better leaving group. The reaction was studied in the stable imido form:

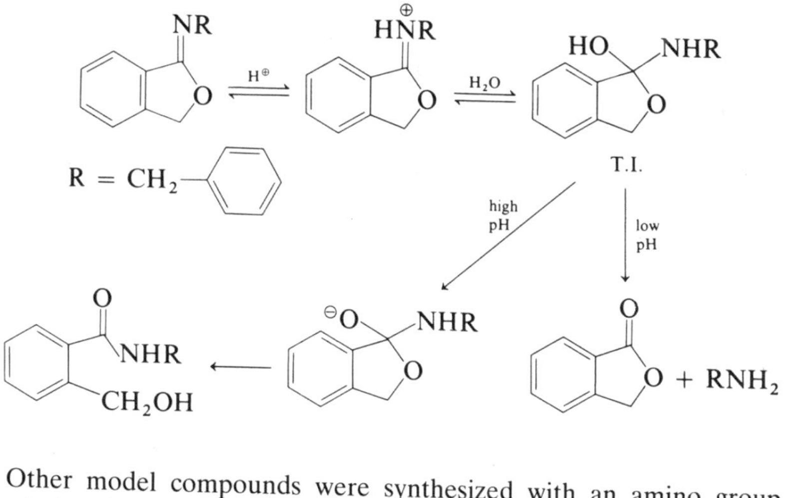

NR; O; $H^{\oplus}$; $H\overset{\oplus}{N}R$; O; H_2O; HO; NHR; O

$R = CH_2$—(phenyl)

T.I.

high pH

low pH

O; NHR; CH_2OH; $^{\ominus}O$; NHR; O; O; O + RNH_2

Other model compounds were synthesized with an amino group in proximity. At position 6 it was hoped that the amino group would help to decompose the tetrahedral intermediate and hence accelerate the rate of the reaction. However, the presence of the amino group resulted in a reduction in rate by a factor of ten.

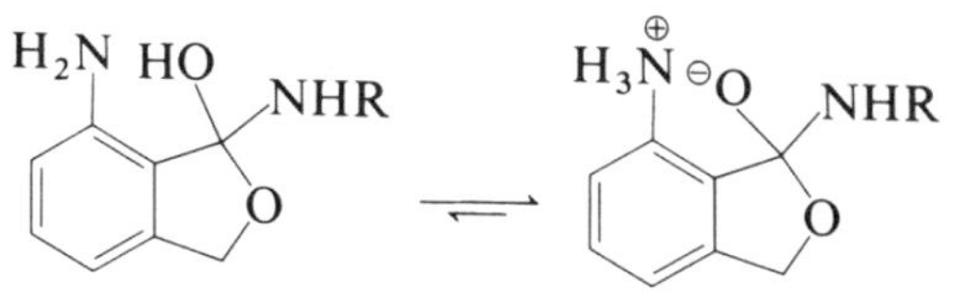

H_2N HO; NHR; O; $H_3\overset{\oplus}{N}$ $^{\ominus}O$; NHR; O

stable ion pair

Possibly the formation of an ion-pair-attractive interaction stabilizes the tetrahedral intermediate from being broken down.

What would happen if the amino group were at position 3 instead?

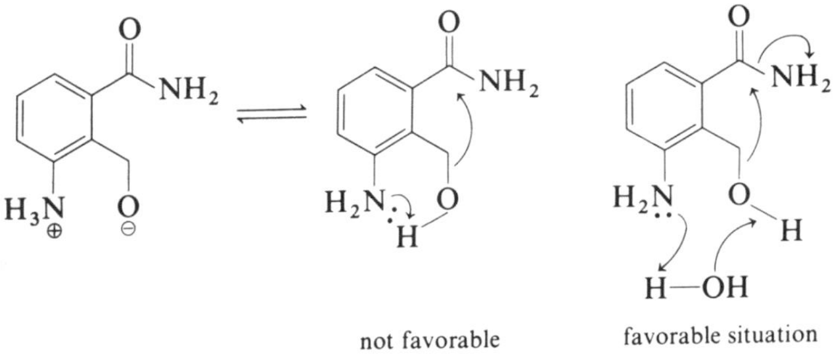

not favorable favorable situation

The rate of the reaction becomes 10^3 faster than with the 6-amino analog. In ^2H_2O a k_H/k_D value of 2.82 is obtained, suggesting an intramolecular base catalysis through solvent participation.

For both papain and trypsin, A.I. Scott's group (Texas A&M University) reported a novel direct application of NMR spectroscopy to the study of the structure and the stereochemistry of well-designed tetrahedral inhibitor complexes (203). In the case of papain, for example, a ^{13}C-NMR study of the complex between the enzyme and *N*-acetyl-L-phenylalaninylglycinal revealed the formation in equal quantities of *two* tetrahedral intermediates where acetal carbonyl resonances appeared respectively at 75.02 ppm and 74.68 ppm.

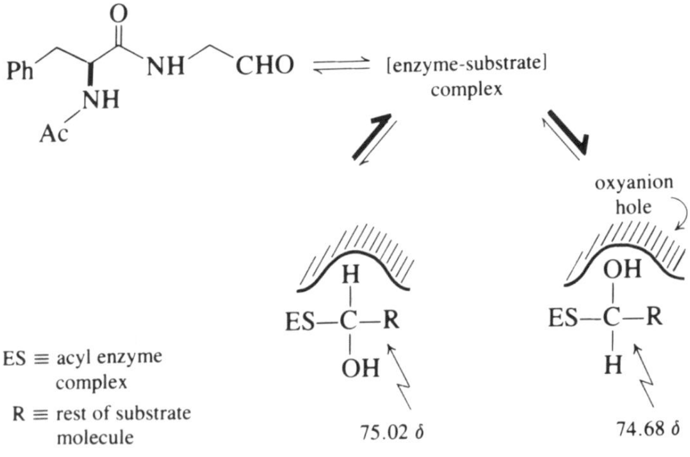

[enzyme-substrate] complex

oxyanion hole

ES ≡ acyl enzyme complex

R ≡ rest of substrate molecule

75.02 δ 74.68 δ

The intermediate that gives the high field resonance was predominant and, as a result, was designated the intermediate resulting from a "productive" binding mode with the OH group sitting in the oxyanion hole. The other diastereoisomeric intermediate is subject to general acid–base catalyzed formation/breakdown by nearby His-159 on the enzyme. This work was the first spectroscopic proof of the existence of a tetrahedral intermediate in proteases.

Similarly for trypsin, characteristic resonances were observed for substrate and tetrahedral intermediates (204).

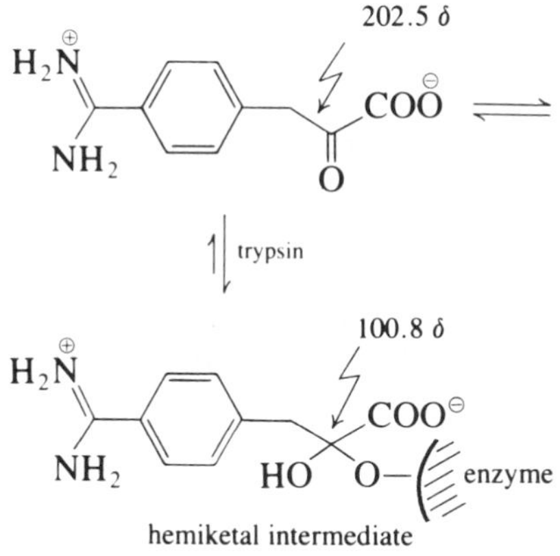

202.5 δ

$H_2\overset{\oplus}{N}$ ... $COO^{\ominus}$ ⇌ hydrated intermediate (95.4 δ)

trypsin

100.8 δ

hemiketal intermediate

4.4.2 Absolute Conformation of Bound Substrate

What is the exact orientation of a substrate in the active center? What is the position of the scissile bond relative to the catalytic groups? Those are some of the questions asked by C. Niemann of the California Institute of Technology and B. Belleau of McGill University.

One way to approach this problem would be through the use of simple model substrates for the enzyme α-chymotrypsin. In this regard, Hein and Niemann (205) first attempted the elucidation of the conformation of well-selected chymotrypsin-bound substrates by using a conformationally constrained molecule as a model for the active conformation of a typical open-chain substrate, *N*-acetyl L-phenylalanine methyl ester (L-APME). For this, Niemann studied the kinetic behavior of D- and L-1-keto-3-carbomethoxy-1,2,3,4-tetrahydroisoquinoline (KCTI). It is a rigid analog of L-APME as well as *N*-formyl L-phenylalanine methyl ester (L-FPME) and *N*-benzoyl L-alanine methyl ester (L-BAME), all of which are good substrates of α-chymotrypsin.

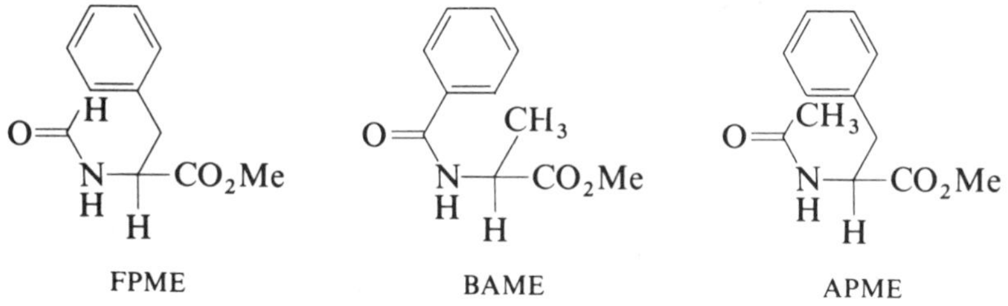

FPME BAME APME

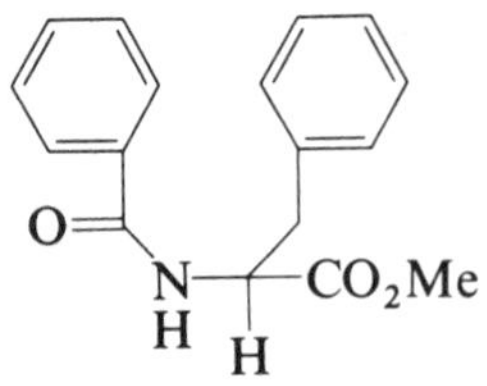

BPME

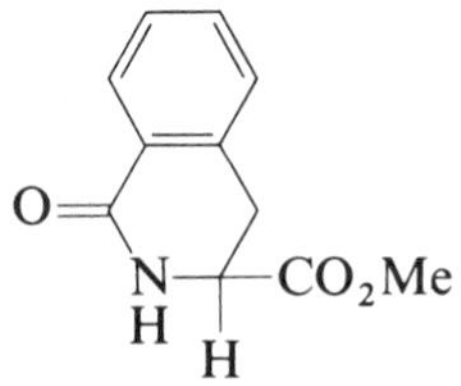

KCTI
(Niemann)

Although α-chymotrypsin is stereospecific toward the L-isomer of most amino acid substrates, Niemann showed that the stereospecificity is reversed in the case of KCTI. The D-isomer of this conformationally restricted ester is hydrolyzed at a rate comparable to that of N-acetylated L-phenylalanine methyl esters while the L-isomer is hydrolyzed very slowly. Hein and Niemann (205,206) pointed out that this anomaly is consistent with a requirement for the carboxylate group of D-KCTI to be in an axial conformation, a conformation that matches a probable conformation of open-chain L-amino acid ester substrates.

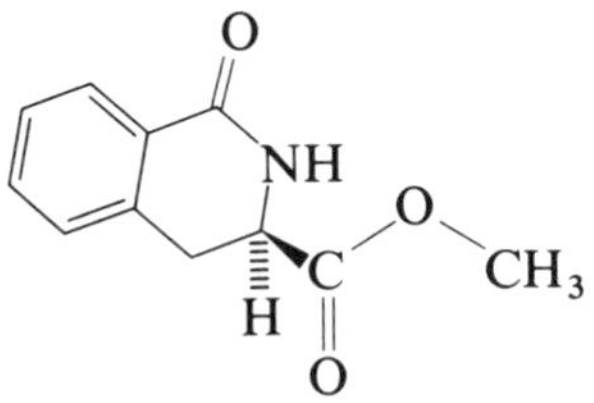

D-KCTI

Subtilisin, a serine protease of bacterial origin, also shows the same inversion of stereospecificity toward D-KCTI as α-chymotrypsin (207). This suggests that both enzymes have similar specificity with respect to the configuration of their substrates. This common specificity neatly confirms and reflects the close structural analogy between the primary binding sites of the two enzymes. This close structural analogy shows how α-chymotrypsin and subtilisin, which have completely different phylogenetic origins, actually "freeze" their substrates in the same active conformation.

The work of Silver and Sone (208) favors, however, the equatorial orientation of the ester group. Thus, the results of these studies gave rise to very intense controversies regarding the orientation (axial or equatorial) of the carbomethoxy function of D-KCTI, and hence of enzyme-bound L-APME (208,209). Whether the orientation of this ester function is axial or equatorial has not been unambiguously resolved with Niemann's compound and its derivatives. The reason for this probably lies in the fact that in most of the compounds used, the structural features of the normal substrates of α-chymotrypsin are so grossly altered as to cast strong doubts on the validity of the models.

However, one uniquely constrained substrate, 3-methoxycarbonyl-2-dibenzazocine-1-one, was synthesized in 1968 by Belleau and Chevalier from diphenic anhydride (210). This 2,2′-bridged biphenyl compound is a constrained analog of *N*-benzoyl L-phenylalanine methyl ester (BPME).

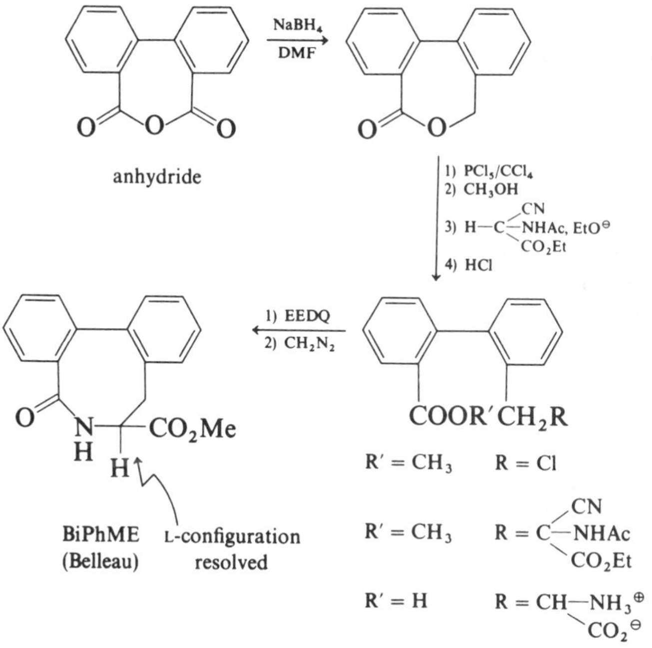

$NaBH_4$ / DMF

anhydride

1) PCl_5/CCl_4
2) CH_3OH
3) $H{-}C(CN)(NHAc)(CO_2Et)$, $EtO^{\ominus}$
4) HCl

1) EEDQ
2) CH_2N_2

CO_2Me

BiPhME (Belleau)

L-configuration resolved

$COOR'CH_2R$

$R' = CH_3$ $R = Cl$

$R' = CH_3$ $R = C(CN)(NHAc)(CO_2Et)$

$R' = H$ $R = CH(NH_3^{\oplus})(CO_2^{\ominus})$

This biphenyl model compound was shown to possess the so-called *primary optical specificity* of α-chymotrypsin. That is, only the enantiomer related to L-phenylalanine was hydrolyzed by the enzyme. Generally speaking, there are three more levels of specific substrate recognition in enzyme catalysis. Let us consider a peptide bond in a polypeptide chain.

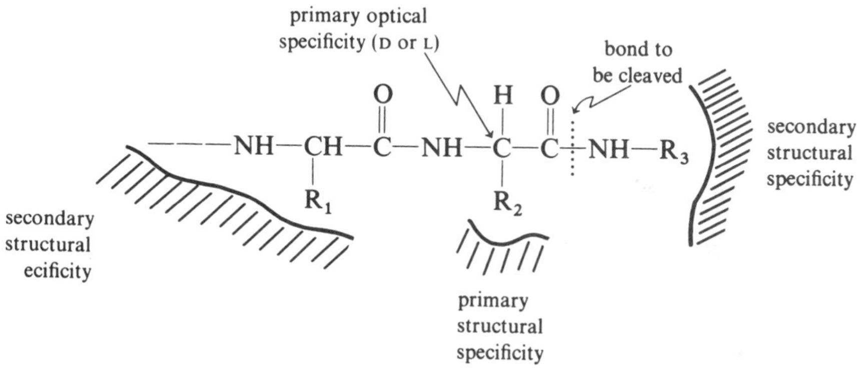

primary optical specificity (D or L)

bond to be cleaved

$-NH-CH(R_1)-CO-NH-CH(R_2)-CO \vdots NH-R_3$

secondary structural specificity

secondary structural ecificity

primary structural specificity

The lateral side chain R_2 is responsible for the normal specificity of the enzyme. For α-chymotrypsin, R_2 is an aromatic side chain and the hydrophobic cavity, an "aromatic hole" in the active center, is there to accommodate the amino acid to be recognized by the enzyme. This is referred to as the *primary structural specificity*.

The influence of the adjacent amino acid residues R_1 and R_3, in the active center subsites, is also important to maintain favorable interaction and proper orientation of the substrate. This is responsible for the *secondary structural specificity* of the enzyme. Finally, there is a *tertiary* level of structural specificity, which will be discussed shortly.

One fascinating aspect of the biphenyl model is that in solution it exists as two sluggishly interconvertible forms in which the ester function occupies either an outside (equatorial) or an inside (axial) orientation relative to the biphenyl system. α-Chymotrypsin shows marked specificity for the S,S_{eq} conformer, the other forms being essentially inert toward the enzyme. The rate of hydrolysis and Michaelis constant for the active conformer are virtually identical to those of the corresponding normal substrate *N*-benzoyl phenylalanine methyl ester.

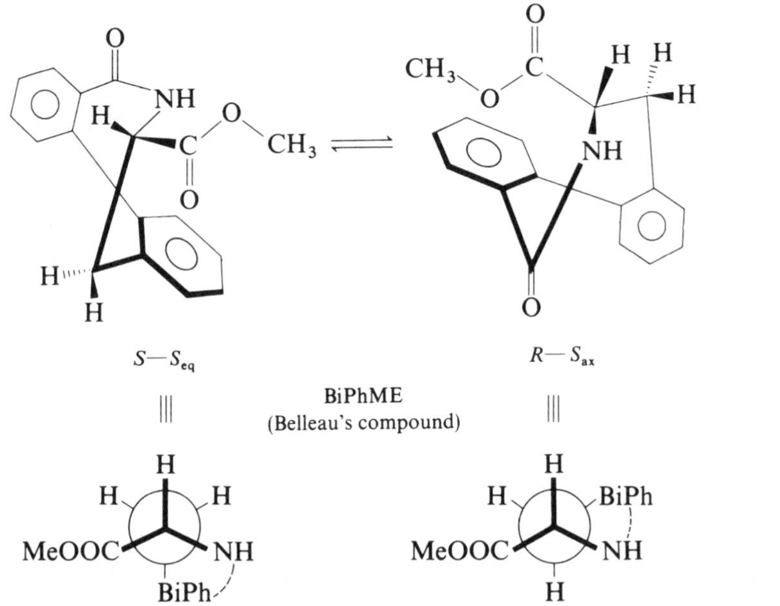

S—S_{eq} ⇌ R—S_{ax}

BiPhME
(Belleau's compound)

When a fresh solution of the R,S_{ax} conformer (in 95:5 water–dioxane) is incubated with α-chymotrypsin, no hydrolysis occurs. However, with an aged stock solution in dioxane, α-chymotrypsin-catalyzed hydrolysis takes place readily. Hence, isomerization to an enzymatically active conformer is gradually taking place upon destruction by dissolution in an organic solvent of the crystal structure of the R,S_{ax} conformer.

A careful comparison of molecular models leads to the conclusion that Niemann's compound (D-KCTI) and Belleau's compound (S, S_{eq} conformer) are related as a key is to its lock only when the ester function

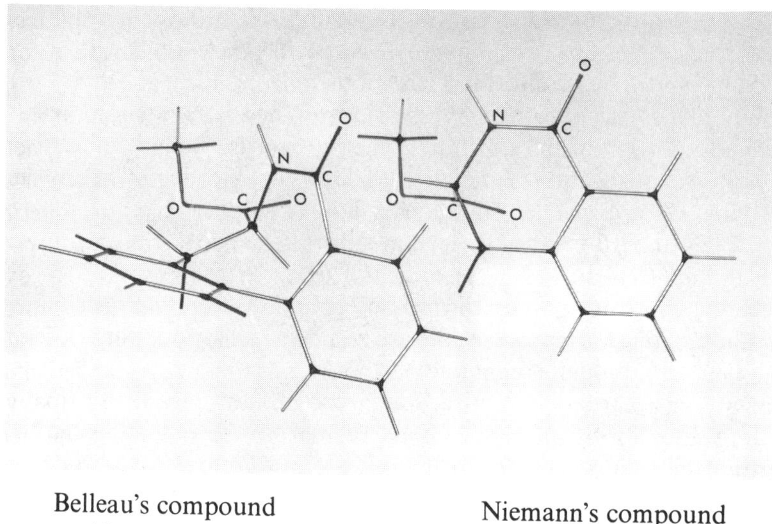

Belleau's compound ($S-S_{eq}$ form) Niemann's compound (D-isomer)

Fig. 4.4. Dreiding stereomodels showing that the D-configuration of Niemann's compound can be superimposed on Belleau's biphenyl model compound.

of D-KCTI is *axially* oriented in the α-chymotrypsin-bound state (Fig. 4.4). Therefore, Belleau's and Niemann's compounds are identical at the molecular level if the ester function is axially oriented in the Niemann's product. The versatility of Belleau's compound makes it a better model for analyzing the conformation of α-chymotrypsin-bound substrates.

Also important is the finding that the R,S_{ax} conformer is not only inert to hydrolysis by α-chymotrypsin, but it also failed to inhibit enzymatic hydrolysis of the active S,S_{eq} conformer. In marked contrast, L-KCTI has been shown to strongly inhibit chymotryptic hydrolysis of D-KCTI. This pattern of competitive inhibition has also been demonstrated for other enantiomeric pairs of chymotrypsin substrates. To understand this behavior, it should be realized that the two conformers of Belleau's compound differ in two important aspects: orientation of the carbomethoxyl group and the chirality of the biphenyl system. Consequently, it must be concluded that in its reaction with this constrained substrate, α-chymotrypsin displays specific recognition of molecular asymmetry. This is referred to as *tertiary structural specificity*. The specificity of the biphenyl compound thus serves to extend the concept that appropriately constrained substrates can serve as very useful tools.

Is this "molecular chiral specificity" the same for enzymes having such different phylogenic origins as papain (thiol protease of plants) and subtilisin (serine protease from bacteria)? The constrained biphenyl compound is naturally a good choice for answering this question because it is a molecular asymmetric analog of BPME, which is a good substrate for all three proteases. It turns out that subtilisin behaves exactly like α-chymotrypsin toward the biphenyl isomers. However, with papain, the R,S_{ax} and S,S_{eq} conformers are both inactive. The inactivity of the compound toward papain might be due to the *cis* configuration of the acylamido portion of the substrate. In other words, papain would require a *transoid* configuration about the amide bond in order to hydrolyze its substrates. However, the situation is not as simple because *cisoid* and *transoid* 9- and 10-membered lactam substrate analogs were synthesized by Elie and Belleau (211). They found that the *trans* isomers have no detectable substrate activity toward any of the three proteases tested.

Although these negative results are somewhat disappointing (it is often the case with model compounds), they nevertheless suggest that the reactivity of a particular substrate seems to be determined principally by the conformation of the scissile bond relative to the overall shape of the substrate molecule. The crucial parameter is whether or not the scissile bonds are held in the correct orientation for catalysis.

As already mentioned, there exists a considerable cross-specificity among α-chymotrypsin, papain, and subtilisin. The results of such studies on chiral specificity will eventually help bring to light new aspects of the evolutionary divergences between mammalian, bacterial, and plant proteases. In addition, zymogen activation is often involved in the biosynthesis of proteases themselves as well as in a great variety of biological processes such as blood coagulation, complement reaction, hormone production, fibrinolysis, and the like. Such precise and limited proteolysis by enzymes, exhibiting a generally broad primary specificity, illustrates also the overwhelming importance of the tertiary structural specificity of proteases in their interaction with their natural substrates (211).

In parallel fashion, one of the key problems in *molecular pharmacology* is the cross-specificity exhibited by different classes of receptors. As a result, drugs having the structural features indispensable for activity toward a given receptor often cause undesirable side effects through interactions with other related receptor sites. Knowledge of the *chiral specificity* of a particular receptor (where applicable) could allow the design of drugs with a much improved specificity for this receptor. It is hoped that a systematic examination of the chiral specificities of proteases could provide, at a simpler level, a more rational basis for the development of receptor-specific effectors because enzyme–substrate interactions are fundamentally of the same nature. Proteases offer the advantage of being less complex and more accessible than the often elusive receptors.

4.5 Other Hydrolytic Enzymes

Among the best-known hydrolytic enzymes of which enough structural and mechanistic information is available are the exopeptidase carboxypeptidase A (see Chapter 6), ribonuclease A (see Chapter 3), and lysozyme. In this chapter we shall examine the chemistry of this last enzyme.

Lysozyme is an important enzyme that catalyzes the hydrolysis of a polysaccharide that is the major constituent of the cell wall of certain bacteria. The polymer is formed from β(1 → 4) linked alternating units of *N*-acetylglucosamine (NAG) and *N*-acetylmuramic acid (NAM) (Fig. 4.5).

The enzyme is small, having a polypeptide chain of 129 amino acids. It was the first one, in 1967, to have its tertiary structure elucidated by X-ray crystallography (212). Unlike α-chymotrypsin, lysozyme has a well-defined deep cleft running down one side of the ellipsoidal molecule for binding the substrate.

The cleft is divided into six subsites: ABCDEF. NAM residues can bind only in sites B, D, and F, while NAG residues of synthetic substrates may bind in all sites. The bond that is cleaved lies between sites D and E.

The carboxyl group of Glu-35 in the unionized form and the carboxyl group of Asp-52 in its ionized form are the two functions implicated at the active site.

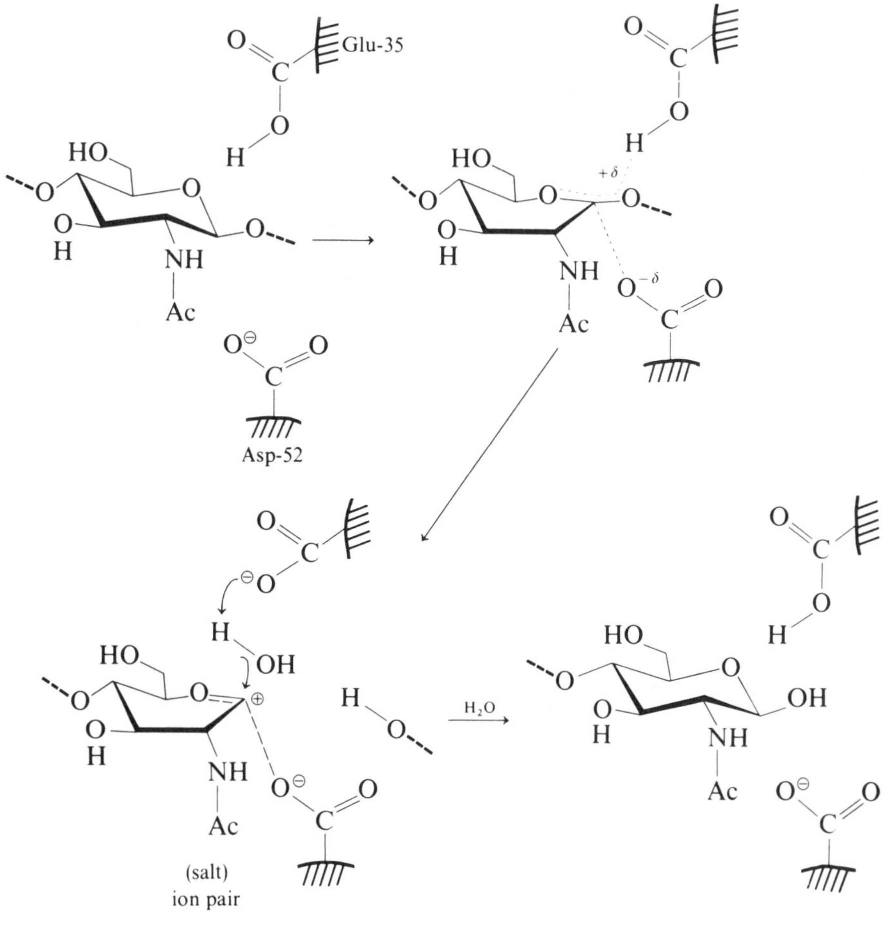

Glu-35

Asp-52

H_2O

(salt)
ion pair

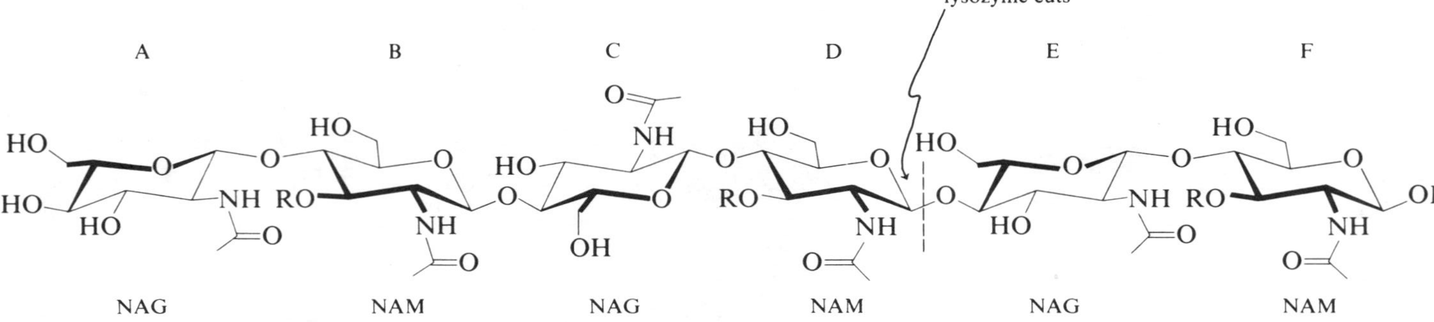

Fig. 4.5. Structure of an hexasaccharide that can bind at the active center of lysozyme. Upon binding to the enzyme, the sugar ring D of the substrate becomes distorted and catalysis proceeds through the promotion of an oxocarbonium ion (see p. 205). This results in a polar transition state. However, an important feature of enzyme is the capacity to stabilize (neutralize) the enzyme–substrate complex by *electrostatic interactions* with amino acid residues at the active site.

Glu-35 serves as a general-acid catalyst and Asp-52 stabilizes, by electrostatic interaction, the oxocarbonium ion intermediate. It helps the development of a positive charge at the anomeric carbon. Interestingly, the intermediate develops a half-chair conformation to relieve the strain created by the introduction of a transition state with sp^2 character (double bond) in the ring. Such structural change of the substrate allows departure of the leaving group (ring E) via stereoelectronic control (Section 4.6). The mechanism of action of lysozyme is also a good illustration of a concept stated in 1948 by L. Pauling: *The active site of an enzyme is complementary to the transition state of the reaction it catalyzes.* In other words, the enzyme has evolved to bind more efficiently a transition state intermediate between substrate and product than the substrate itself in its ground state (164).

A great deal of theoretical work has been carried out to evaluate the binding energy involved between substrates in enzymes. In this respect, the efforts of H.A. Scheraga's group from Cornell University have given promising results (214). They have investigated, by conformational energy calculations, the most favored binding modes of oligomers of NAG to the active site of lysozyme. In order to do this, both the substrate and the side chains of the enzyme were allowed to undergo conformational changes and relative motions during energy minimization. It was found that $(NAG)_6$ had a clear preference for binding to the active site cleft of lysozyme with its last two residues on the "left" side of the cleft region, thus in agreement with earlier X-ray work (213). The calculations also showed that the presence of the *N*-acetyl groups on NAG oligomers is important to provide sufficient interactions with the enzyme and accounts for the fact that glucose oligomers bind with much lower affinity.

Study of substrates' distortion could, of course, in model compounds have important mechanistic consequences as well as an effect on reaction rate.

For example, the hydrolysis of the following acetals has been studied (215):

O—CH_2—O—CH_3 COOH ; O-glucoside (OH, OH, OH, OH) COOH $\xrightarrow{H_2O}$ OH COOH

The presence of an *o*-carboxyphenyl facilitates the rate of hydrolysis by a factor of 10^4 in the glucose series and by 300 in the other simpler molecule relative to the *para* compound. But the participation of the carboxyl group in the mechanism of the latter example remains obscure. Again, reaction in ^2H_2O versus H_2O indicates that proton-transfer occurs in the critical transition state.

Three possible intramolecular mechanisms are possible: two involve a nucleophilic attack by the carboxylate and one involves an intramolecular general-acid catalysis.

(a) [Scheme: o-(CH₂–O⁺(H)–CH₃)-oxy benzoate (COO⁻) → cyclic intermediate + CH₃OH; cyclic intermediate —H₂O (blocked)→ salicylic acid (OH, COOH) + HCHO]

However, this first mechanism does not occur because synthesis of the cyclic intermediate by a different route showed that it is stable to the reaction conditions.

(b) [Scheme: phenyl O⁺(H)–CH₂–O–CH₃ with o-COO⁻ → o-hydroxybenzoate ester C(=O)–O–CH₂–OCH₃ → products]

This corresponds to a $\overset{+}{C}H_2OCH_3$ migration. However, no isosbestic points* in the UV are seen between the spectrum of the reactant and product, as would be expected if this mechanism were correct.

(c) [Scheme: $\overset{+\delta}{C}H_2$–$\ddot{O}CH_3$ on phenolic O···H–O$^{-\delta}$–C(=O) → o-hydroxybenzoate (OH, COO⁻) + $CH_2 \overset{\oplus}{\cdots} O$–$CH_3$ oxocarbonium ion]

* Isosbestic points are those points of equal amplitude at a given wavelength that occur in superpositions of three or more spectra of a system under different concentrations. This can be observed only for a system containing an equilibrium between *two* spectrally active states in which only the relative populations of the states have been changed.

This last mechanism is the favored one. In a way, it mimics lysozyme because the carboxylate behaves as a general acid (Glu-35) and the CH_3O- group as the glucose ring oxygen.

Mention should be made that the participation of neighboring carboxylic acids in the hydrolysis of amides is also a phenomenon pertinent to the understanding of enzyme-catalyzed amide hydrolysis. One such enzyme is the acid protease pepsin, found in the gastric juice, and it obeys general-acid catalysis. R. Kluger and C.H. Lam, of the University of Toronto, prepared rigid model compounds to study the carboxylic acid participation in amide hydrolysis (216). They found that anilic acid derivatives of endo-*cis*-5-norbornene meet the criteria of rigid geometric proximity of the interacting functional groups.

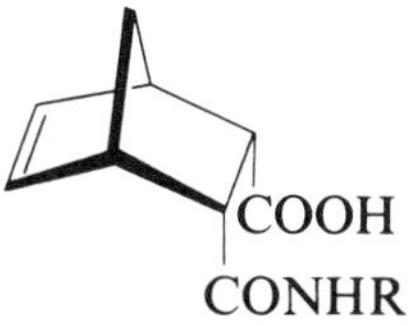

$R = C_6H_5$
$= C_6H_4{-}4'{-}OCH_3$
$= C_6H_4{-}4'{-}Cl$
$= C_6H_4{-}3'{-}NO_2$
$= C_6H_4{-}4'{-}NO_2$

These substrates are rigid, so that the entropy barrier to formation of an intermediate is minimized. Both general-acid and general-base catalysis are readily observable over a wide pH range for certain of these compounds.

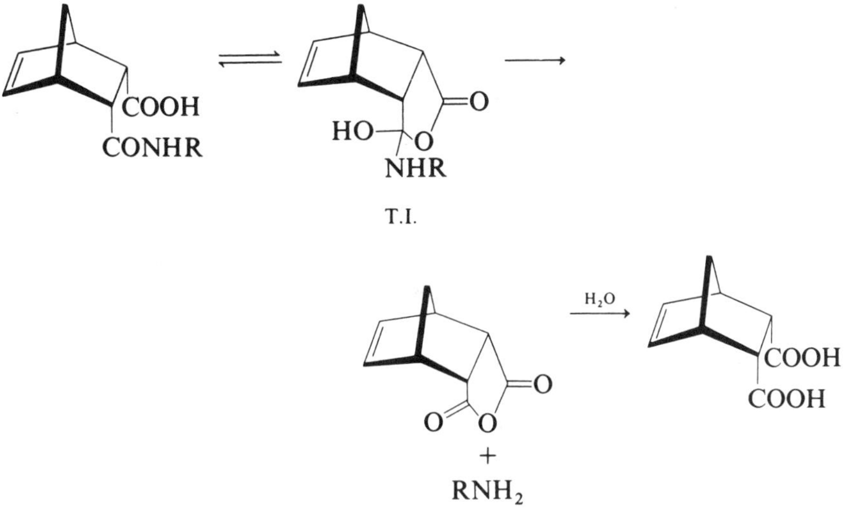

T.I.

$+ RNH_2$

The possibility that the acid protease, pepsin, may form a covalent amino enzyme derivative between its active site aspartic acid group and the substrate peptide suggests that a reactive anhydride may form. Since a second carboxylic acid is available at the active site (217), a mechanism analogous to the one above may be relevant.

A biomimetic model for the deacylation step of papain, a thiol protease, has been developed by following the intramolecular solvolysis of thio esters by a proximal imidazole group (218). The pH profile of the reaction has a plateau between 7.5 and 10. This finding can be explained in terms of two kinetically indistinguishable processes where imidazole could act as a general-base or via an imidazolium ion acting as a general-acid, promoting the attack of the base on the carbonyl. No *S*- to *N*-acyl transfer seems to occur during the course of the solvolysis.

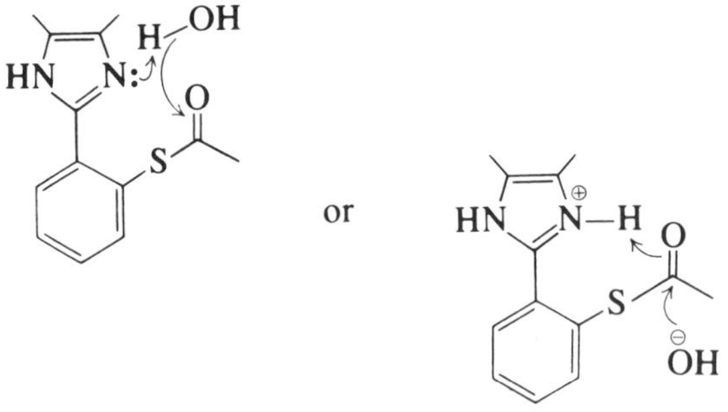

HN N: H OH O S

or

HN N⊕—H O S ⊖OH

4.6 Stereoelectronic Control in Hydrolytic Reactions

Now that we have seen examples of hydrolytic reactions and acetal hydrolysis by enzymes, we may wonder how important the stereochemistry of the products and reactants is in these transformations.

This section is devoted to this question through the presentation of a relatively new concept in organic chemistry, *stereoelectronic control,* exploited by P. Deslongchamps from the University of Sherbrooke (219,220). It uses the properties of proper orbital orientation in the breakdown of tetrahedral intermediates in hydrolytic reactions. This concept is quite different from Koshland's "orbital steering" hypothesis where proper orbital alignment is invoked for the formation of a tetrahedral intermediate. Here we are interested by the process that follows: the cleavage of the tetrahedral intermediate in the hydrolysis of esters and amides.

It is generally accepted that the most common mechanism for the hydrolysis of esters and amides proceeds through the formation of a tetrahedral intermediate. Deslongchamps argued that the conformation of this tetrahedral intermediate (hemiorthoester from ester and hemiorthoamide from amide) is an important parameter in order to obtain a better understanding of the hydrolysis reaction.

$$\mathrm{R{-}\overset{\displaystyle O}{\overset{\|}{C}}{-}X} + \mathrm{H{-}OR'} \longrightarrow \left[\mathrm{R{-}\overset{\displaystyle OH}{\overset{|}{\underset{\underset{\displaystyle OR'}{|}}{C}}}{-}X}\right] \longrightarrow \mathrm{R{-}\overset{\displaystyle O}{\overset{\|}{C}}{-}OR'} + \mathrm{H{-}X}$$

T.I.

X = OR″ or NR_2'' R′ = H or alkyl

Since 1971, Deslongchamps has developed a new stereoelectronic theory in which the precise conformation of the tetrahedral intermediate plays a major role. In other words, the stereochemistry and the ionic state of the tetrahedral intermediate, the orientation of nonbonded electron pairs, and the relative energy barriers for cleavage and for molecular rotation are the key parameters in the stereoelectronically controlled cleavage of the tetrahedral intermediate formed in the hydrolysis of an ester or an amide. He postulated that the precise conformation of the tetrahedral intermediate is transmitted into the product of the reaction and that the specific decomposition of such an intermediate is controlled by the orientation of the nonbonded electron pairs of the heteroatoms.

trans *gauche*

X = OR″, NR_2''

The *trans* conformer (the two R groups are away from each other) gives a lower energy pathway than the *gauche* which cannot effectively compete in the cleavage process. *Consequently, the cleavage is stereoelectronically controlled when and only when two heteroatoms of the intermediate, each having one nonbonded electron pair, orient antiperiplanar to the departing O-alkyl or N-alkyl group.*

Experimental evidence to prove this postulate comes from four sources: (1) a study of the oxidation of acetals by ozone, (2) acid hydrolysis of cyclic orthoesters, (3) concurrent carbonyl-oxygen exchange and hydrolysis of esters by using oxygen-18 labeling, and (4) basic hydrolysis of N,N-dialkylated imidate salts. We will briefly examine results from the last three approaches. Then we will try to apply this concept to the hydrolysis of ester and amide substrates by serine proteases.

4.6.1 Hydrolysis of Orthoesters

The acid hydrolysis of an orthoester gives a hemiorthoester intermediate that cleaves to an ester and an alcohol.

$$\mathrm{R'O{-}\underset{\displaystyle R}{\overset{\displaystyle OR'}{C}}{-}OR'} \longrightarrow \mathrm{R'O{-}\underset{\displaystyle R}{\overset{\displaystyle OH}{C}}{-}OR'} + \mathrm{R'OH} \longrightarrow \mathrm{R{-}COOR'} + \mathrm{R'OH}$$

For a cyclic orthoester, a similar equation can be written:

O OEt, OEt ⟶ O OH, OEt + EtOH ⟶ OH, COOEt (hydroxyester); ⟶̸ lactone (=O) + EtOH

The hemiorthoester intermediate decomposes in a completely specific manner yielding exclusively a hydroxyester, and no trace of a lactone is detected. To explain this result, we have to examine all the possible conformers of the intermediate. Theoretically, there are nine different *gauche* conformers for a hemiorthoester tetrahedral intermediate. Figure 4.6 represents the nine possibilities for the case of the cyclic orthoester.

The task is then to define which conformer should be taken into consideration to explain the formation of a hydroxyester as the only product of hydrolysis. The stereoelectronic theory predicts that the precise conformation of the tetrahedral intermediate is transposed into the product of the reaction and a cleavage of C—O alkyl bond is allowed only if the other two oxygens of this intermediate each have an orbital oriented antiperiplanar to the C—O alkyl bond to be broken.

Consequently, a detailed examination of each conformer shows that five of them (B, D, G, H, and I) are readily eliminated because of severe 1,3-*syn* periplanar interactions between the two ethyl groups or with the methylenes of the ring. Since the population of these compounds will be small at equilibrium, they can be neglected. Conformer C must also be eliminated as a reactive conformer simply because it does not have proper orbital orientation on two oxygen atoms to permit the cleavage of the C—O bond of the third oxygen atom. The remaining three conformers,

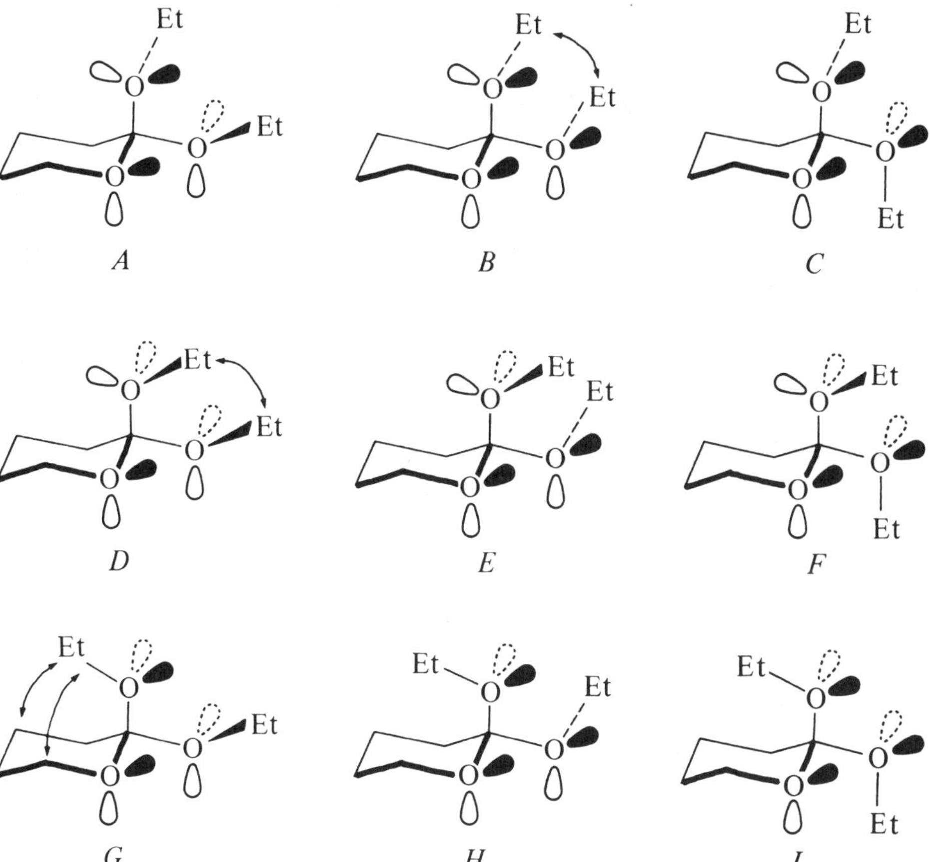

Fig. 4.6. The possible conformers of a cyclic orthoester (219). Reprinted with permission. Copyright © 1975 by Pergamon Press Ltd.

A, E, and F, do not suffer any strong steric interactions and each have two oxygens with proper orbital orientation to cleave the C—O bond of the third atom.

The cleavage of each of these three conformers will give a corresponding dioxolenium ion according to the proper orbital orientation rule just mentioned. Now if we look at the relative stability of each of these dioxolenium ions for conformers A, E, and F, we see that the ion from E is *cis* and the ones from A and F are *trans*. It is known that *trans*-dioxolenium ions are more stable than *cis*-dioxolenium ions, just as *trans* esters are more stable than *cis* esters. Consequently, the cleavage of conformer E should be a higher energy process and can thus be eliminated on that basis.

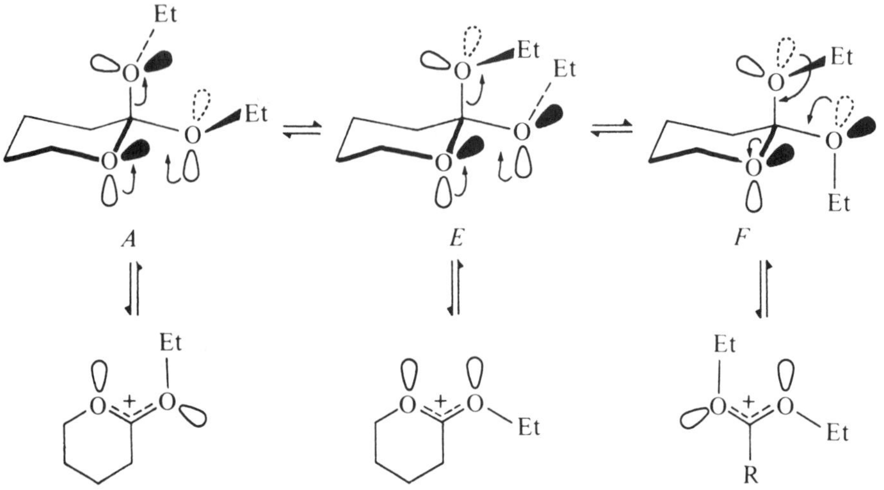

It is more difficult to find arguments to differentiate between conformers A and F. In principle, the formation of a cyclic dioxolenium ion should be favored over an acyclic one when the starting orthoester is cyclic. Furthermore, cleavage of conformer A gives two molecules, while conformer F gives only one molecule. This entropy factor alone should favor conformer A over conformer F as the reactive species. In conclusion, the cyclic orthoester is hydrolyzed preferentially via conformer A only even if it exists in rapid equilibrium with conformers F and E (214).

A large number of geometrically different constrained acetals, esters, and imidate salts have been examined, with uniform consistency with this theory. For instance, a methyl lactonium salt can be prepared by reaction of a conformationally rigid bicyclic lactone with trimethyloxonium tetrafluoroborate:

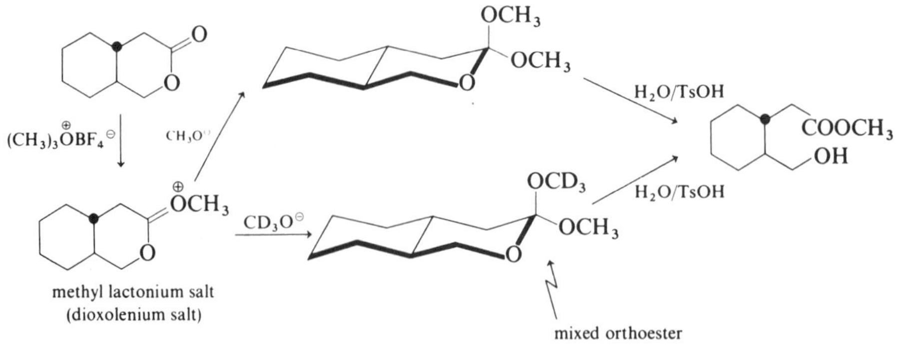

Reaction with sodium methoxide in methanol afforded the dimethoxyester. In the presence of deuterated sodium methoxide the mixed cyclic orthoester was obtained in over 90% yield with the deuterated methoxy group exclusively in the axial orientation. The hydrolysis of the orthoester in water containing *p*-toluenesulfonic acid resulted in the formation of only the nondeuterated hydroxyester. These results confirm that the hydrolysis of cyclic orthoesters proceeds by loss of the axial alkoxy group. It is a further proof that conformer F can be eliminated as a reactive conformer and consequently the hydrolysis of the cyclic dialkoxyorthoester takes place through the reaction of conformer A.

A further experimental proof of the stereoelectronic control in ester (and amide) hydrolysis comes from concurrent carbonyl–oxygen exchange and hydrolysis using oxygen-18 labeled esters (221).

$$\mathrm{R{-}C({=}O^*){-}OR + H_2O} \underset{k_2}{\overset{k_1}{\rightleftharpoons}} \mathrm{R{-}C(^*OH)(OH){-}OR} \xrightarrow{k_3} \mathrm{R{-}C({=}O){-}\overset{*}{O}H + ROH}$$

$$\mathrm{R{-}C({=}O){-}OR + H_2O^*} \underset{k_2}{\overset{k_1}{\rightleftharpoons}} \mathrm{R{-}C(^*OH)(OH){-}OR}$$

$$^{18}\mathrm{O} \equiv \mathrm{O^*}$$

If it is indeed true that there is no conformational change in the tetrahedral intermediate when stereoelectronically controlled cleavage is allowed, then concurrent carbonyl–oxygen exchange during hydrolysis can be used to demonstrate both the lack of conformational change and the stereoelectronic theory. Application of these postulates to the hydrolysis of esters led to the following predictions: (*Z*)-esters can undergo carbonyl–oxygen exchange but (*E*)-esters cannot.* An ^{18}O-labeled δ-lactone is an example of an (*E*)-ester. Reaction with hydroxide ion gives a tetrahedral intermediate that has the required orientation of electron pairs to breakdown in two directions only to give either the starting labeled ester or the product of the reaction.

---

* The (*Z*) for *Zusammen* and (*E*) for *Entgegen* nomenclature for double bonds and carbonyl functions in molecules is another application of the Cahn–Ingold–Prelog priority rules. If the two groups of highest priority at the ends of the double bond are on the same side of the double bonds (*cis*), the double bond configuration is designated (*Z*). When the two groups of highest priority are on opposite sides on the double bond (*trans*), the double configuration is (*E*). These rules can also be applied to ester conformations by looking at the C—O bond. Since the lone pair of electrons has the lowest priority, the conformation is said to be (*Z*) when the R group of the alkoxy is on the same side as the oxygen of the carbonyl. If it is on the other side (such as with lactones), the conformation is the (*E*).

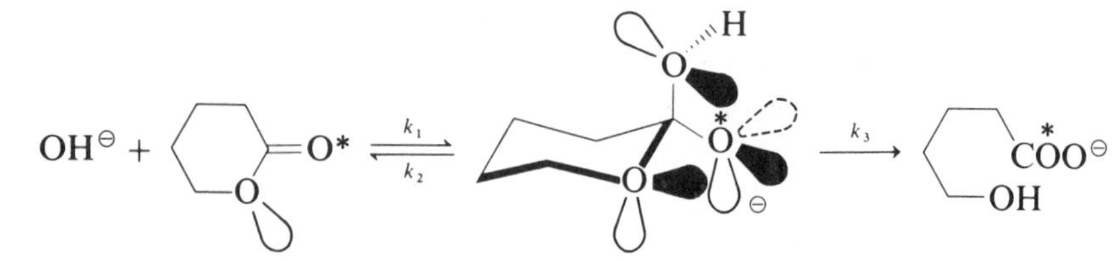

Since the tetrahedral intermediate cannot yield the unlabeled ester, carbonyl–oxygen exchange has not occurred with the (*E*)-ester. The experimental result is in accord with results previously described in the literature which show that lactones do not undergo carbonyl–oxygen exchange. On the other hand, hydrolysis of (*Z*)-esters always occurs with carbonyl–oxygen exchange with the solvent. The results with lactones can also be explained by using a kinetic argument. Indeed, the extent of carbonyl exchange can vary depending on the relative value of k_3 and k_2. If k_3/k_2 is greater than 100, there will be a very low exchange which will be difficult to detect. This argument agrees with experimental results obtained with (*E*)-esters.

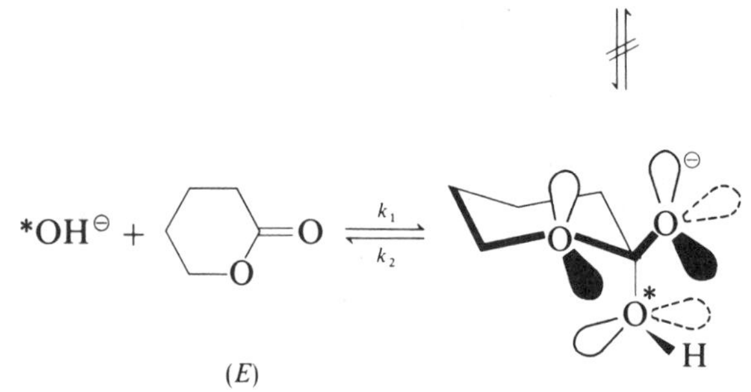

More information on the theory of stereoelectronic control can be found in a book on the subject written by Deslongchamps (222).

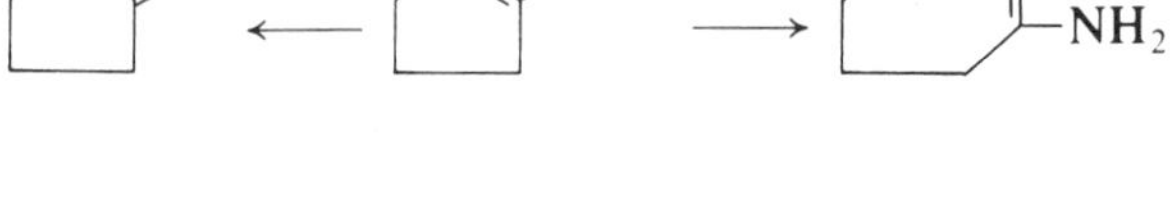

Recently C.L. Perrin and co-workers (University of California at San Diego, La Jolla) questioned the importance of stereoelectronic control in the hydrolysis of cyclic amidines (223). In the case of the basic hydrolysis of 5-membered and 7-membered cyclic amidines, they found that besides the open-chained aminoamide, which is expected by the theory, up to 50% of the lactam is also produced. The first tetrahedral intermediate formed will give the expected aminoamide after the appropriate proton transfer. However, Deslongchamps argued that in order to form the lactam, the tetrahedral intermediate must undergo a conformational change followed by a proton transfer. The lactam can also be slowly formed from the aminoamide via the second tetrahedral hemiorthoamide intermediate. It is also the thermodynamically more stable product.

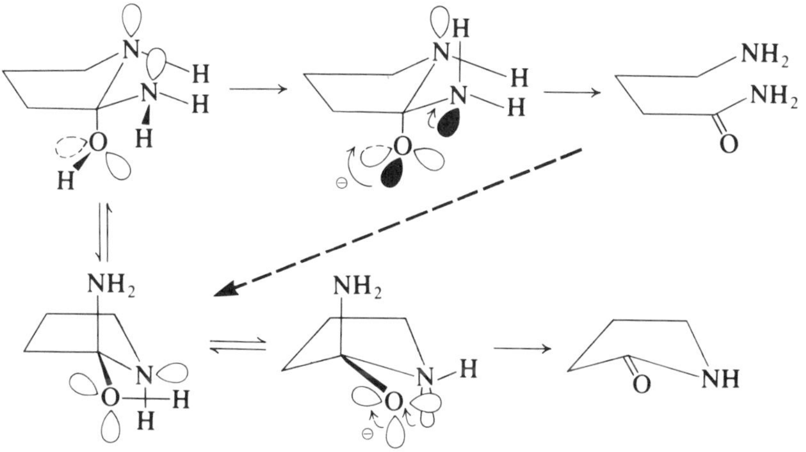

A last but important point is that the energy of stabilization of amidines by resonance structures is in the order of 12 kcal/mol, compared to 17 kcal/mol and 26 kcal/mol for esters and amide respectively. The outcome of the stereoelectronic effect could consequently be less noticeable with amidine than with amide and ester functions.

4.6.2 Application to Serine Proteases

It is reasonable to inquire whether recognition of such a concept might apply to mechanistic ideas about enzyme reactions. It now appears that the stereoelectronic theory that Deslongchamps and co-workers have developed is specifically applicable to peptide hydrolyses by serine proteases. Hydrolysis of simple esters will be used first to illustrate the approach.

A hemiorthoester intermediate can be cleaved to yield two different esters, each of which has a (*E*) or (*Z*) conformation.

According to the principle of microscopic reversibility, the generation of a hemiorthoester via alkoxide ion attack on an ester should also occur with stereoelectronic control.

Since a *transoid* (*Z*)-ester is generally more stable than the corresponding *cisoid* (*E*)-ester, we will use it to analyze the attack by an active site serine-OH. It is expected that the serine residue will approach the ester with an orientation away from the more bulky R group of the ester substrate.

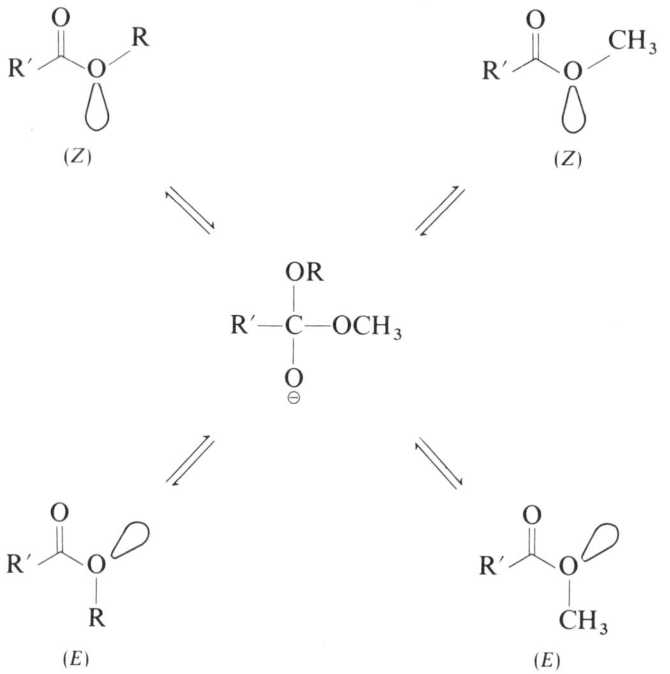

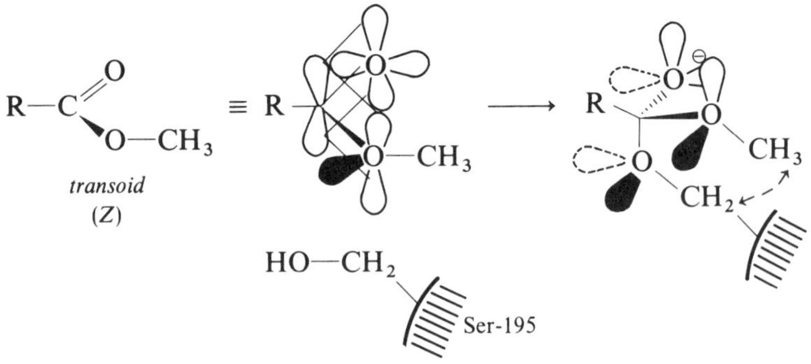

However, the introduction of a serine residue in this tetrahedral intermediate form will cause a large steric hindrance (~33.4 kJ or ~8 kcal) with the alkoxy chain. To reduce this strain, a conformational change *must* occur at the active site by a rotation of 120° around the C—O bond connecting the tetrahedral carbon to the serine-OH group.

This new tetrahedral intermediate will eventually generate a *cisoid* acyl-enzyme intermediate and it should be the preponderant product.

Another possibility is a rotation of 120° of the $—OCH_3$ group. Here the ester group has changed its conformation from *trans* to *gauche* relative

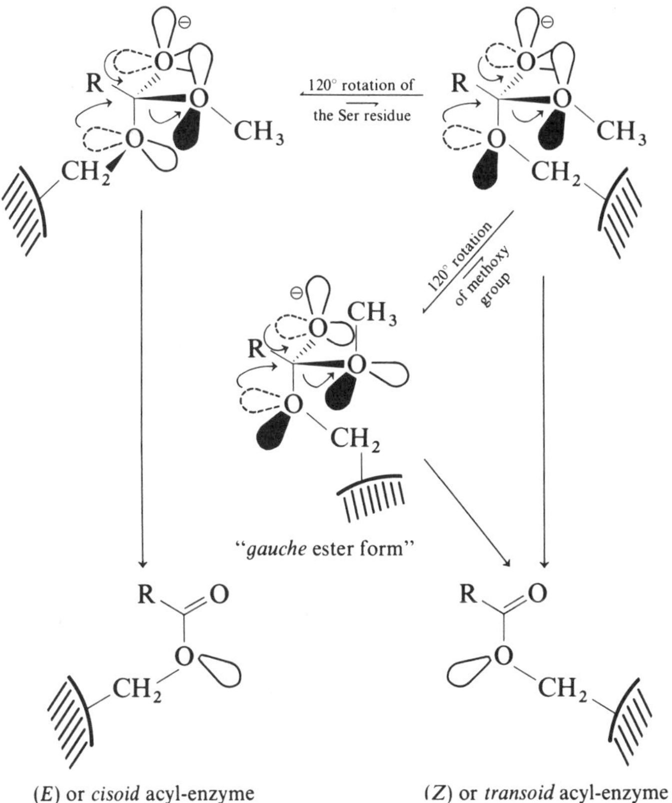

(*E*) or *cisoid* acyl-enzyme (*Z*) or *transoid* acyl-enzyme

to the R group. This new tetrahedral intermediate and the former one will both yield a *transoid* acyl-enzyme intermediate after orbital-assisted cleavage of the —OCH_3 group.

The third possible conformer can be excluded since it cannot break down, even after rotation of the —OCH_3 group. It is a nonproductive binding intermediate.

If, however, a *transoid* → *cisoid* enzyme-mediated isomerization of the substrate ester takes place prior to the attack by the serine residue, a different tetrahedral intermediate will be formed:

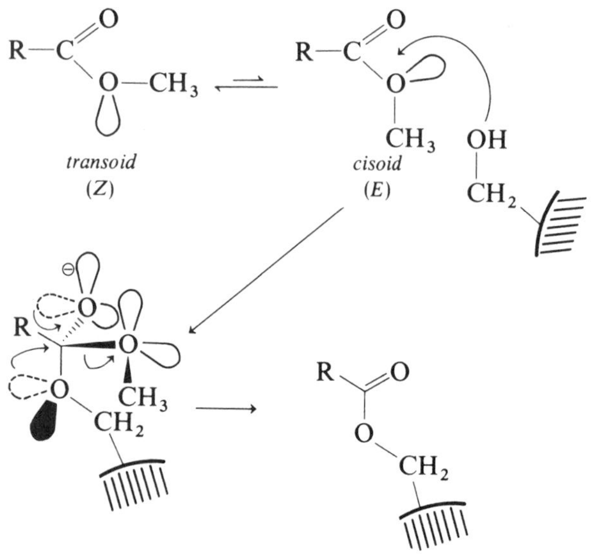

transoid (Z) cisoid (E)

(Z) or *transoid* acyl-enzyme

In this case the steric strain is not as serious as before and this intermediate will break down to give mainly a *transoid* acyl-enzyme.

The hydrolysis of N,N-alkylated imidate salts is also under stereoelectronic control and has been used to understand the mechanism of amide hydrolysis. Two paths are possible: ejection of ROH or R_2NH.

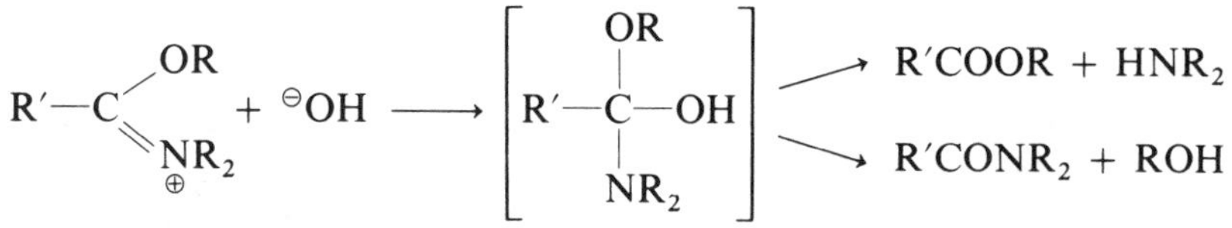

$$R'-C(OR)=\overset{\oplus}{N}R_2 + {}^{\ominus}OH \longrightarrow \left[R'-C(OR)(OH)-NR_2 \right] \begin{array}{l} \nearrow R'COOR + HNR_2 \\ \searrow R'CONR_2 + ROH \end{array}$$

Here again, the ejection of a residue critically requires the assistance of two antiperiplanar lone-pair orbitals on the remaining heteroatoms of the incipient amide. In the event that only one pair is potentially available, facile decomposition will be observed only after bond reorientation (224).

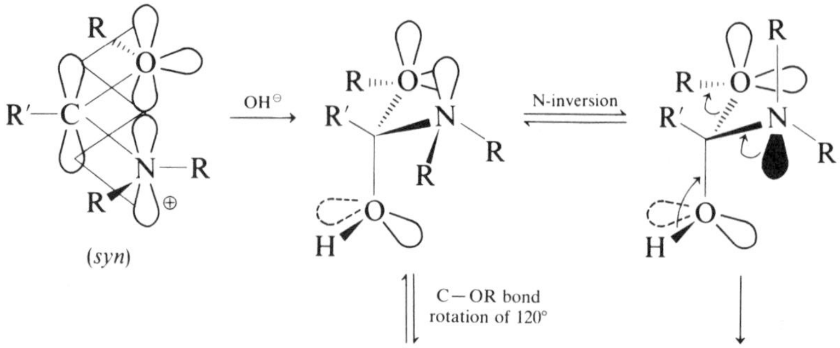

(*syn*)

$OH^{\ominus}$

N-inversion

C—OR bond rotation of 120°

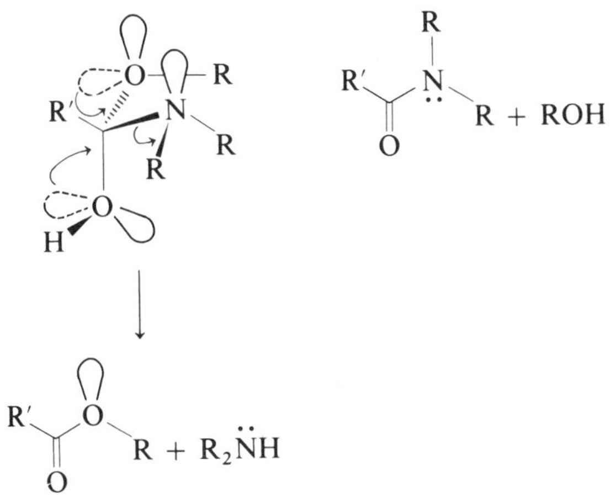

This model is applicable to amide bond hydrolysis by proteases and it would be surprising if enzyme hydrolysis does not obey this principle.

This simple presentation leads to interesting speculation on the pathways preferred by the enzyme. Does the conformational change occur at the level of the substrate tetrahedral intermediate or at the level of the enzyme's active site residue? Obviously, rigid bioorganic models are necessary to solve this dilemma. Such models are yet to come but will offer the capacity of defining the enzyme mechanism of proteases from relatively limited conformations of the tetrahedral intermediate.

Furthermore, we suspect that the concept of stereoelectronic control hydrolysis of esters and amides is broadly applicable to other fields of enzymology, and consequently some well-accepted catalytic principles will require reconceptualization.

One of the closest approaches so far developed is by Bizzozero and Zweifel (225), who tried to explain in 1975 why a proline residue involved in a peptide bond is resistant to α-chymotrypsin cleavage. The objective was to find if the unreactivity of the peptide bond results from an unfavorable interaction of the methylene groups of the proline ring with the enzyme active site or whether the steric hindrance occurs upon formation of the enzyme–substrate complex or during the subsequent bond-change steps, and whether this steric hindrance is related to the ring structure of proline or simply to substitution of the amido nitrogen. In order to answer these questions, the dipeptides *N*-acetyl-L-phenylalanyl-L-proline amide and *N*-acetyl-L-phenylalanyl-sarcosine amide were synthesized and their behavior as model substrates of α-chymotrypsin studied.

Both products were found to be unreactive but proved to be good competitive inhibitors of a specific substrate, *N*-Ac-L-Phe-OMe. This indicates that they form enzyme–substrate complexes of normal stability and that the reason for their unreactivity has to be sought in the nature of

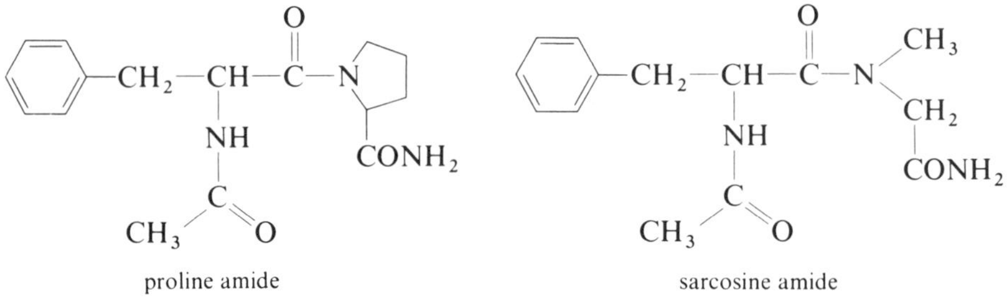

proline amide

sarcosine amide

the enzyme–substrate interactions occurring during the subsequent bond-change steps. In other words, their unrectivity can be understood by considering the stereoelectronic course of the transformation leading to the acyl-enzyme intermediate.

The attack of a *trans*-peptide bond by the hydroxyl of Ser-195 of α-chymotrypsin will give the steric situation depicted below:

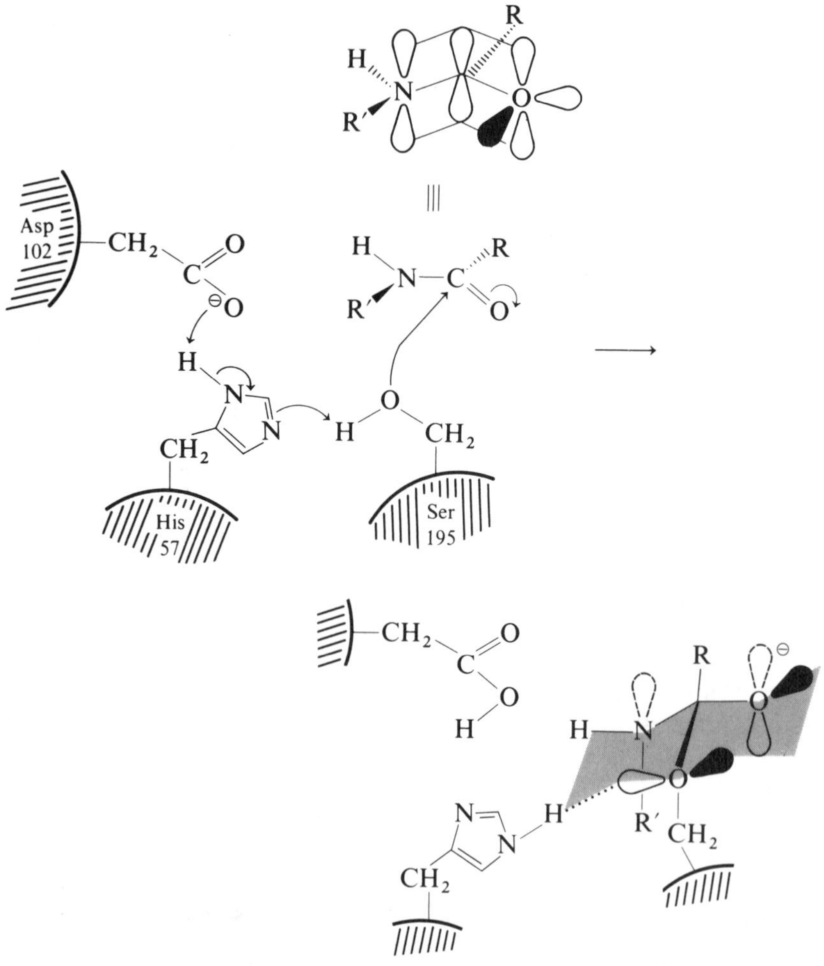

From the principle of microscopic reversibility, the stereoelectronic control theory predicts that the developing lone-pair orbitals on the heteroatoms must be antiperiplanar to the new C—O (from Ser-OH and amide carbonyl) or C—N bond. The important point to be considered here is that the nonbonded pair of electrons on the nitrogen atom points toward the solvent and the N—H bond toward the inside of the enzyme active site. To facilitate the spacial view, the relevant atoms are superimposed on a *trans*-decalin frame (shaded area).

When the N—H hydrogen is replaced by an alkyl group, as in the case of a proline residue, this substituent would come too close to the imidazole ring of His-57. Hence, a dipeptide containing a proline residue is inactive toward α-chymotrypsin hydrolysis because the steric hindrance prevents formation of the tetrahedral intermediate.

With regard to the mechanism of the acylation step, the stereoelectronic requirements for cleavage of the C—N bond to form the acyl-enzyme are fulfilled, but the cleavage step still requires protonation of the nitrogen atom. Since it is generally accepted that the imidazole of His-57 is the protonating agent, the situation shown above could not occur since the nonbonded orbital of the leaving nitrogen points in the wrong direction.

Assuming that a protonation–deprotonation mechanism mediated by the solvent does not occur, an inversion mechanism at the leaving nitrogen is needed to produce a new intermediate in which the orientation of the N—H bond and the nonbonded orbital are interchanged.

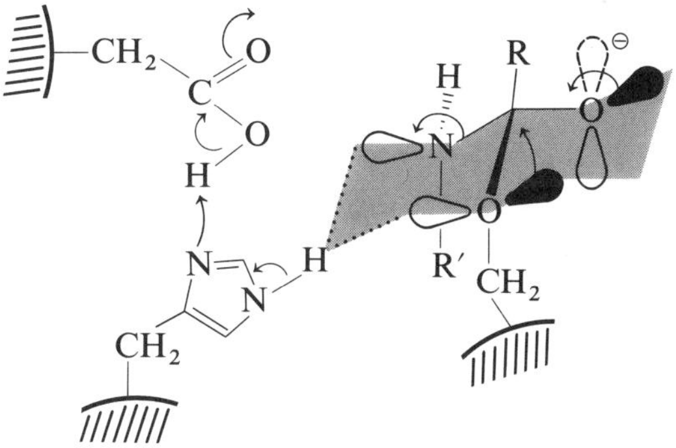

Such an inversion would not only make direct protonation of the leaving nitrogen by His-57 possible, it would also stabilize the C—O (Ser-195) bond since this would now be antiperiplanar to the N—H bond rather than to the nonbonded electron pair. This conformational change would probably not take place in the presence of a proline residue.

Deslongchamps kindly provided us with his own view of the mechanistic path by which α-chymotrypsin and other serine proteases can hydrolyze secondary amides by stereoelectronic control (Fig. 4.7). Petkov *et al.* (226) also arrived at a similar proposal by studying the influence

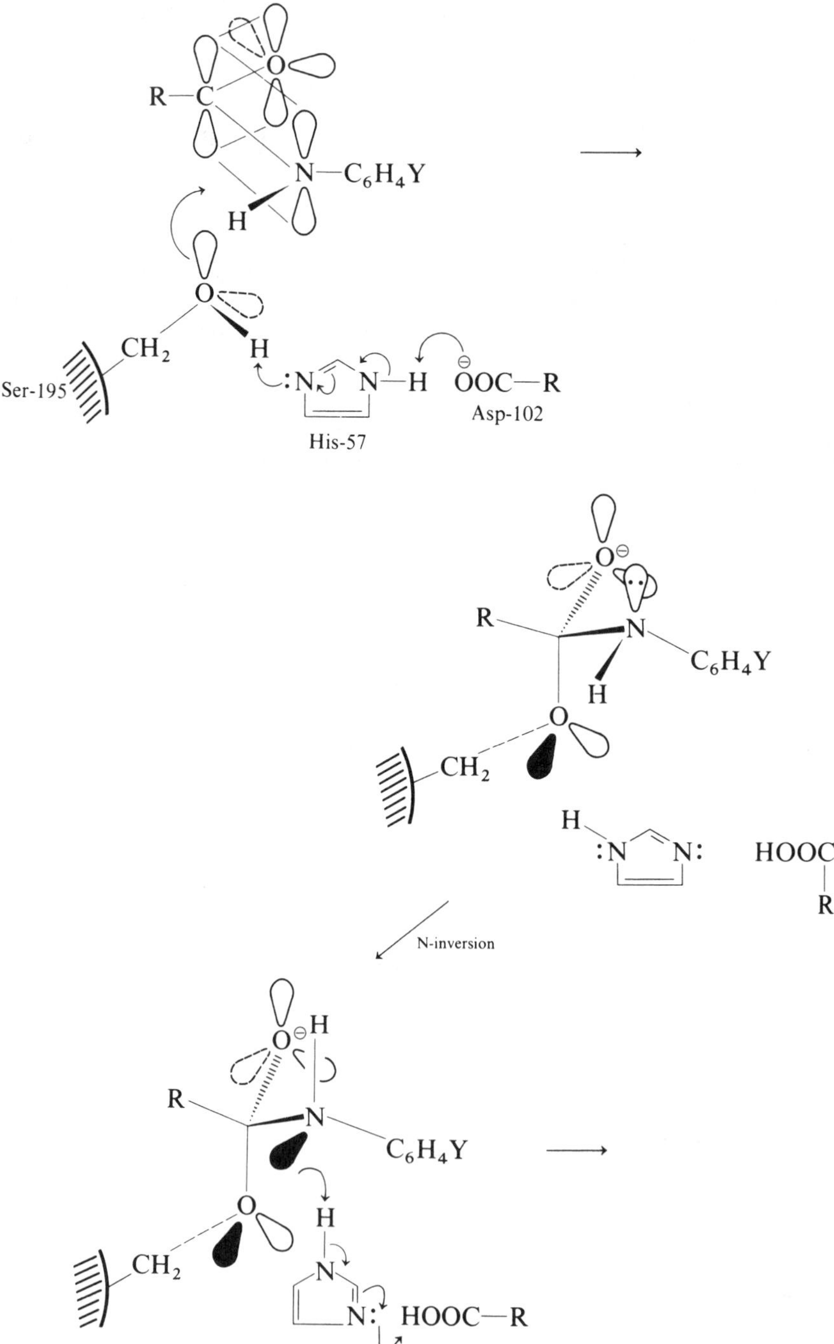

Fig. 4.7. Deslongchamps's proposal of stereoelectronic control in secondary amide hydrolysis by α-chymotrypsin.

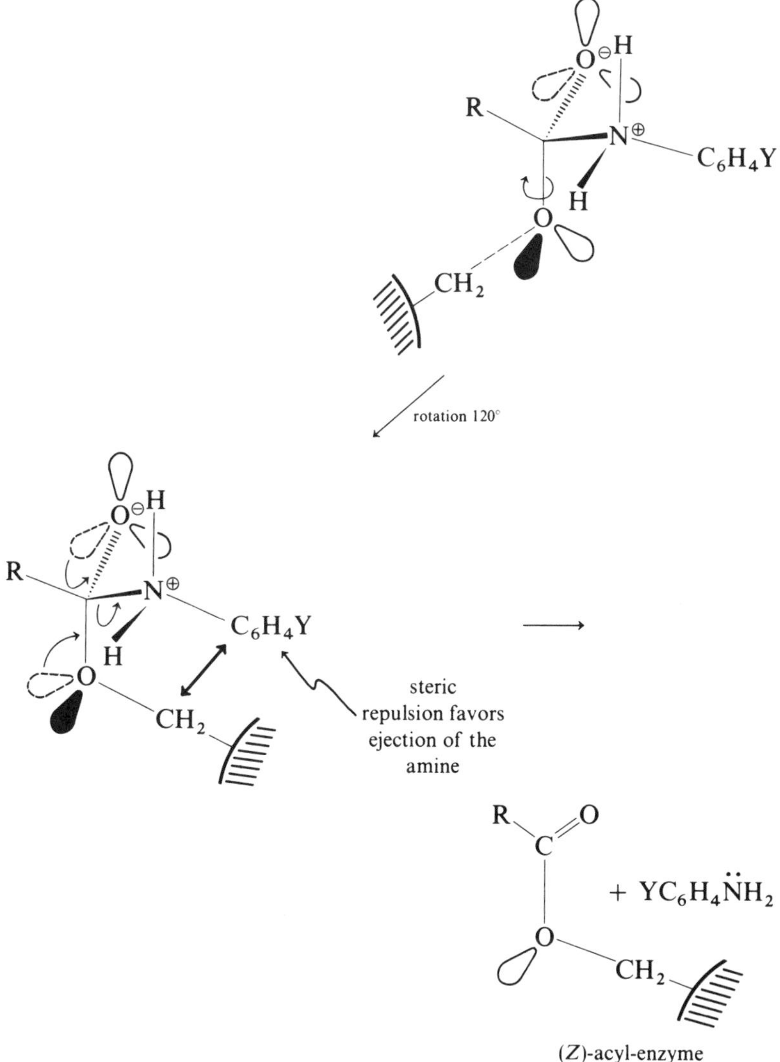

Fig. 4.7. (continued)

of the leaving group on the reactivity of specific anilides in α-chymotrypsin-catalyzed hydrolysis. Furthermore, the stereoelectronic control theory has been applied to the mode of action of ribonuclease A, staphylococcal nuclease, and lysozyme (227).

Finally, it should be mentioned that a concept of torsional strain of amide bonds was also developed by W.L. Mock in 1976 and applied to the mode of action of carboxypeptidase A (228). It involves a torsion of the amide bond to allow either *cis* or *trans* addition of the nucleophile and the proton on the developing orbitals in the transition state. This principle complements Deslongchamps's theory of optimum geometry of the transition state tetrahedral intermediate.

4.6.3 Stereoelectronic Effects in Other Systems

This theory is extremely powerful as an explanation of the outcome of hydrolytic reactions. Rate of acceleration can also be accounted for by the stereoelectronic control (229). In the following intramolecular model of carboxypeptidase A, the amide hydrolysis occurs with a rate enhancement factor of over 10^6. The perpendicular attack of OH^- and opening of the lactam complex are favored stereoelectronically by the antiperiplanar orientation of lone pairs of both oxygens of the tetrahedral intermediate.

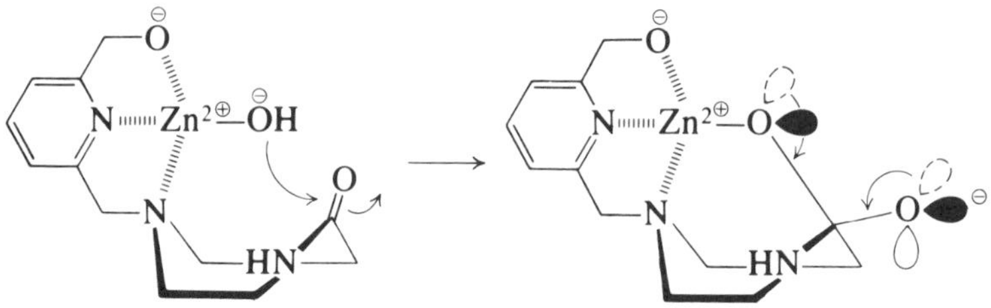

The importance of stereoelectronic effects on acetal hydrolysis is well illustrated in the mode of action of the enzyme lyzozyme (230). It has been pointed out in this chapter (p. 205) that there is good evidence that in the hydrolysis of β-glucosides by lyzozyme, one ring of the substrate must take a boat conformation in order to produce a half-chair oxocarbonium ion. The stereoelectronic theory plays a key role in the pathway of this reaction.

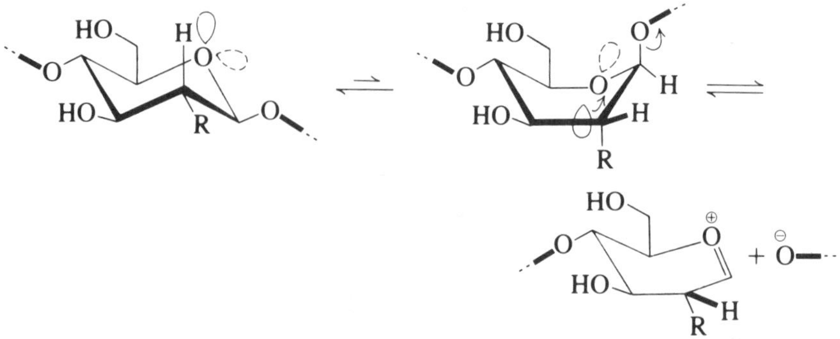

Regarding the mechanism of action of lyzozyme, Post and Karplus recently proposed a very interesting novel mechanistic suggestion that takes into consideration both the stereoelectronic arguments and the accepted molecular dynamic situations known for the enzyme–substrate complex (231). Using molecular dynamic calculations and energy minimization results, they proposed a mechanical pathway that does not involve the substrate distortion mechanism (see p. 207). The alternative mechanism they proposed is shown in Fig. 4.8.

Basically, the alternative mechanism (top part) is an attempt to try to reconcile stereoelectronic effects and molecular dynamic simulations. The endocyclic $C_1—O_5$ bond is first cleaved with stereoelectronic assistance by the participation of an antiperiplanar lone pair of electrons on the O_4' oxygen. As a result, the stereoelectronic effect provides an entropic advantage over the distortional enthalpic strain energy implicated in the previous mechanism (bottom part). This way, if the proper *gauche* conformation of the $C_1—O_5$ bond already exists, a distortion of the chair conformation of the sugar ring does not have to be assumed to occur. Another important difference between the two mechanisms is that the formal one produces an intermediate oxocarbonium ion (exocyclic cleavage) that remains cyclic, while in the new proposal the system evolves through an open chain intermediate (endocyclic cleavage).

The specificity of action of alcohol dehydrogenases, which are NADH dependent (see Chapter 7), is also under the control of the stereoelectronic theory. The hydride ion transfer from NADH to a carbonyl compound is stereospecific and can occur via two different stereochemical pathways, depending on the class of enzyme. One class favors the transfer of the *pro-R* hydrogen, while the other catalyzes the transfer of the *pro-S* hydrogen of NADH.

S.A. Benner (E.T.H., Zurich) has proposed an elegant explanation of this observation (232). He observed that the more reactive carbonyl compounds are reduced by the *pro-R* hydrogen and the less reactive by the *pro-S* hydrogen of the coenzyme. For this, he proposed that the enzymatic transfers of the *pro-R* and *pro-S* hydrogens occur only when the nicotinamide ring is in the *anti* and the *syn* conformation respectively. It was further concluded that the difference in reactivity can be explained by proper orbital overlap between the nitrogen lone pair of electrons and the π-electron in the ring. For this to be optimized, the nicotinamide ring has to adopt a boat conformation and the *anti*-bonding orbitals of adjacent C—O bond of the ribose ring will overlap the nitrogen lone pair of electrons. As a result, the *syn*-conformation puts *pro-S* hydrogen (H_S) in axial position and the *anti*-conformation puts the *pro-R* hydrogen (H_R) in the axial orientation for proper expulsion to reduce carbonyl compounds. Because of orbital overlap, the axial hydrogen is the most easily transferred.

Since in the *pro-R* and *pro-S* hydrogens of dehydrogenases examined, the difference between these two redox potentials is about 1.3 kcal/mol, Benner concluded that evolutionary selective pressures are capable of extraordinarily fine tuning of the free energies of enzyme reactions.

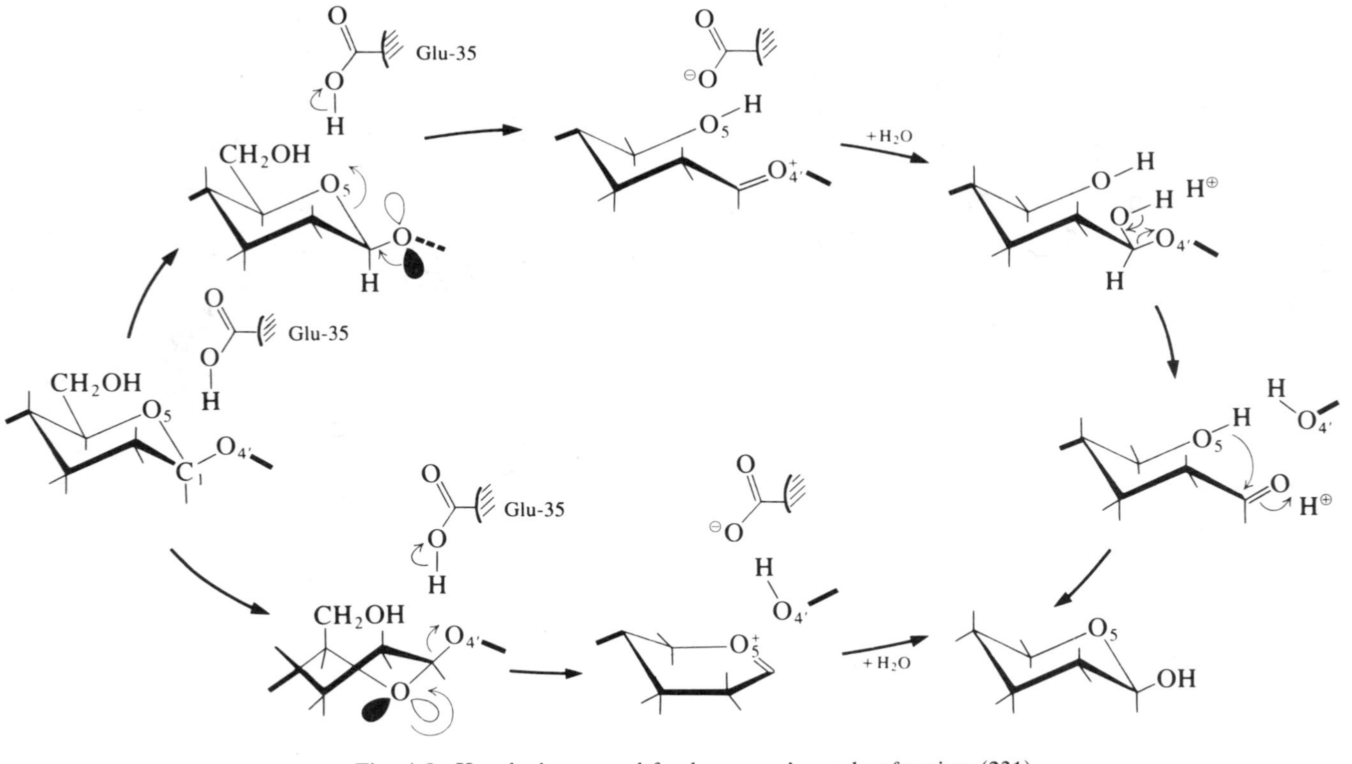

Fig. 4.8. Karplus' proposal for lyzozyme's mode of action (231).

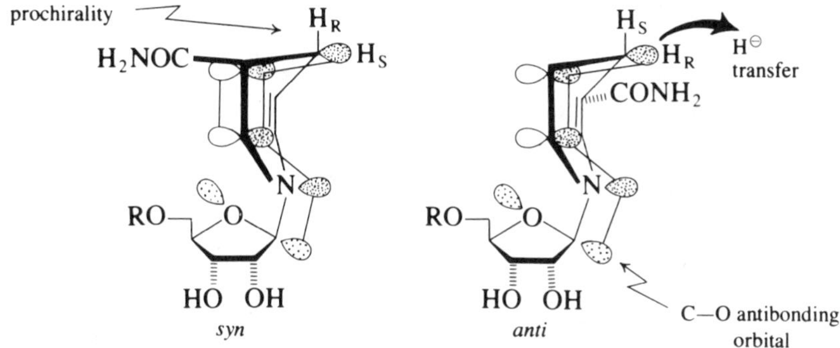

to stabilize each conformer, the electron density of the nitrogen lone pairs is shifted to the antibonding orbital of the neighboring C—O bond

experimentally it is observed that this conformation of NADH is more reactive toward more stable carbonyls and the H_s hydrogen is transferred

in this conformation, the reduction power of NADH is weaker and more reactive carbonyls are reduced by a H_R transfer

Therefore, the stereoelectronic effect is considered to arise from the mixing of a lone-pair orbital with the antibonding σ^* of an adjacent polar bond. The *anomeric effect*, so familiar in sugar chemistry, has a similar stereoelectronic basis. The term *anomeric effect* was introduced by R.U. Lemieux (University of Alberta) in 1959 and refers to the tendency of an alkoxy group (—OR) at the C_1 carbon of a pyranose ring to assume the axial rather than the equatorial orientation, despite unfavorable steric interactions. The *exo*-anomeric effect, also introduced by Lemieux, concerns the preferred orientation of the O—R bond of the alkoxy group at the anomeric center (233).

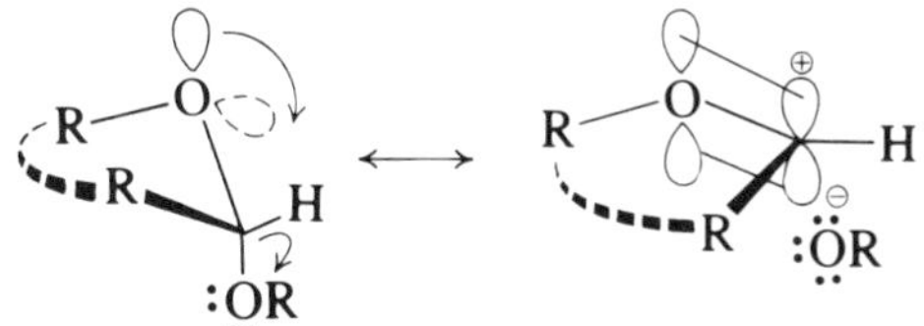

overlap with σ^* of C—OR, only if axially oriented

stabilization *via* a double bond–no bond resonance concept

Orbital considerations are also important in phosphorus species (234). For example, the rate of hydrolysis of the following bicyclic phosphate is almost 10^4 times that of the acyclic triethyl phosphate. On the other hand, in triethyl phosphate, freezing of one conformation is required to put two lone pairs antiperiplanar to the P—O bond to be broken. This conformational restriction is entropically disfavored.

stereoelectronic effects in the transition state of the intermediate

Remember that in the case of ribonuclease A, mentioned in Chapter 3 (p. 132), only the P—O 5′-bond is cleaved. It is then possible to postulate a possible pertagonal bipyramidal transition state for the transesterification step that will maximize the advantage of stereoelectronic participation.

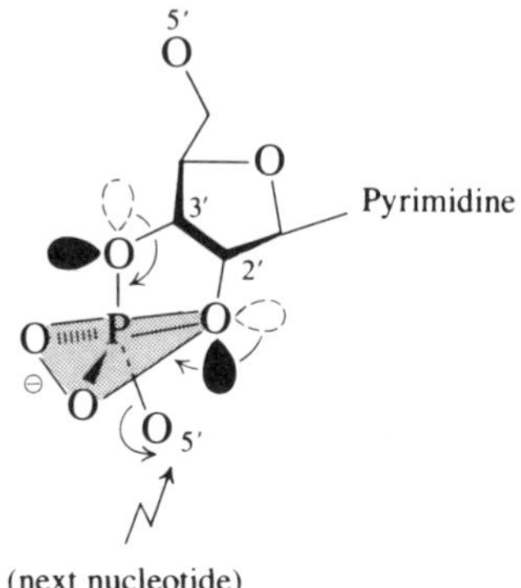

4.7 Immobilized Enzymes and Enzyme Technology

There is a growing technology where enzymes are anchored to a solid matrix and used to perform specific transformation of biomaterials. The enzyme of interest is covalently attached to a carrier polymer by a “spacer” molecule.

The inert carrier could be a polyurethane or other type of resin or a natural polymer such as collogan, easily available from animal skin. A flexible "spacer-arm" is attached to a functional group on the polymer gel. The length of the arm is an important parameter because the free end must be accessible to a functional group of the enzyme to make a second covalent bond without affecting the enzymatic activity. Such a matrix-bound enzyme is usually referred to as an *immobilized enzyme* (235–238). Contrary to the well-exploited method of *affinity chromatography*, here the enzyme rather than the substrate analog is covalently fixed on a solid support. But the concept of biospecific recognition is similar.

In conventional affinity chromatography, agarose and cross-linked sepharose are used as support to immobilize substrates. Usually BrCN is the activating agent and the spacer-arm is an α,ω-diamine. These polysaccharide supports are biodegradable and, consequently, an organic polymer gel is a more useful matrix and is amenable to a wider range of chemical modifications. It is these reasons that prompted the group of G.M. Whitesides from M.I.T. to develop in 1978 a new procedure for the immobilization of enzymes in cross-linked organic polymer gels (239). The procedure surpasses in its operational simplicity and generality the earlier methods. It is also especially valuable in immobilization of relatively delicate enzymes for enzyme-catalyzed organic synthesis for application in large-volume enzymatic reactors.

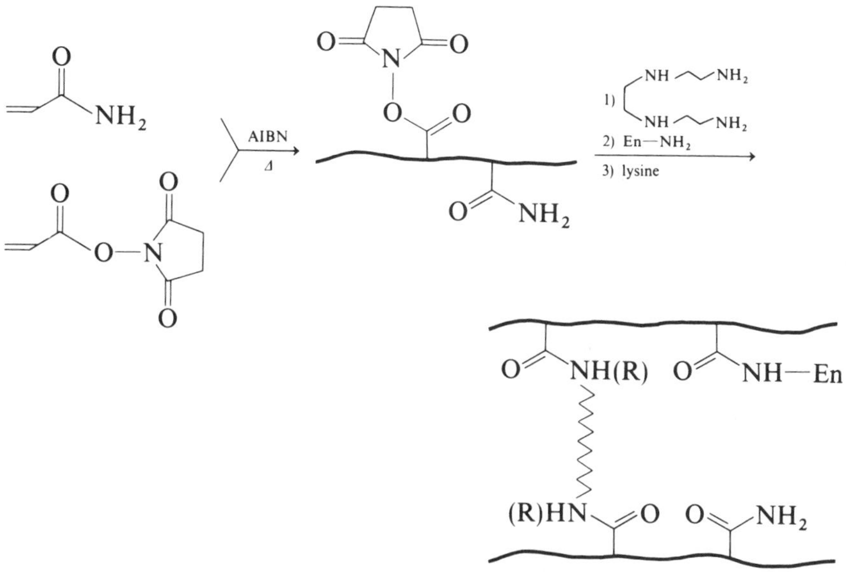

First, a non-cross-linked water-soluble copolymer bearing active ester groups is prepared by heating acrylamide and *N*-acrylyloxysuccinimide with azadiisobutyronitrile (AIBN), an initiator of radical polymerization.

Reaction of this polymer with an α,ω-diamine as a cross-liking agent and with the enzyme of interest results in enzyme immobilization and gel formation. Operations involving oxygen-sensitive enzymes are carried out under argon. A solution of lysine is added at the end to destroy residual active ester groups.

An important feature of this procedure is that addition of the enzyme to the reaction mixture during the formation of the gel minimizes enzyme deactivation. Furthermore, covalent incorporation of the enzyme into the gel provides some protection against proteases. Second, the procedure is simple and of general use and should be directly applicable to a variety of enzyme systems as well as immobilization of whole cells and organelles. Finally, the gel can be rendered susceptible to magnetic filtration by including a ferrofluid in the gel formation step.

It is important to realize that developing methods for stabilizing the enzyme in its native form is also a major goal in this field (236). Successful progress along these lines has been obtained recently by the use of bifunctional reagents to cross-link the peptide chains within the enzyme tertiary structure.

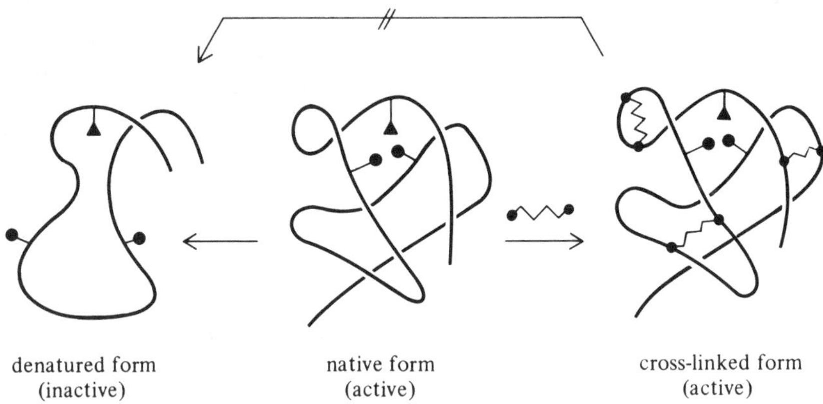

Cross-linking prevents denaturation from occurring under the severe conditions of manipulation and keeps the active site region intact. The most used cross-linking agents are imino esters. By reacting with available lysine side chains, amide bonds are eventually formed between two sections of the polypeptide chain.

However, the number of carbons separating the reactive imino ester functions must be kept small. Otherwise, intermolecular links will also occur.

This technique of immobilized enzymes combines the unique features of both forms of catalysts: the specificity of the enzymes with the stability and ease of handling and storing of supported heterogeneous catalysis. Furthermore, immobilized enzymes can be reused and are applicable to flow systems. Consequently, they find increasing applications in several analytical areas and in medicine.

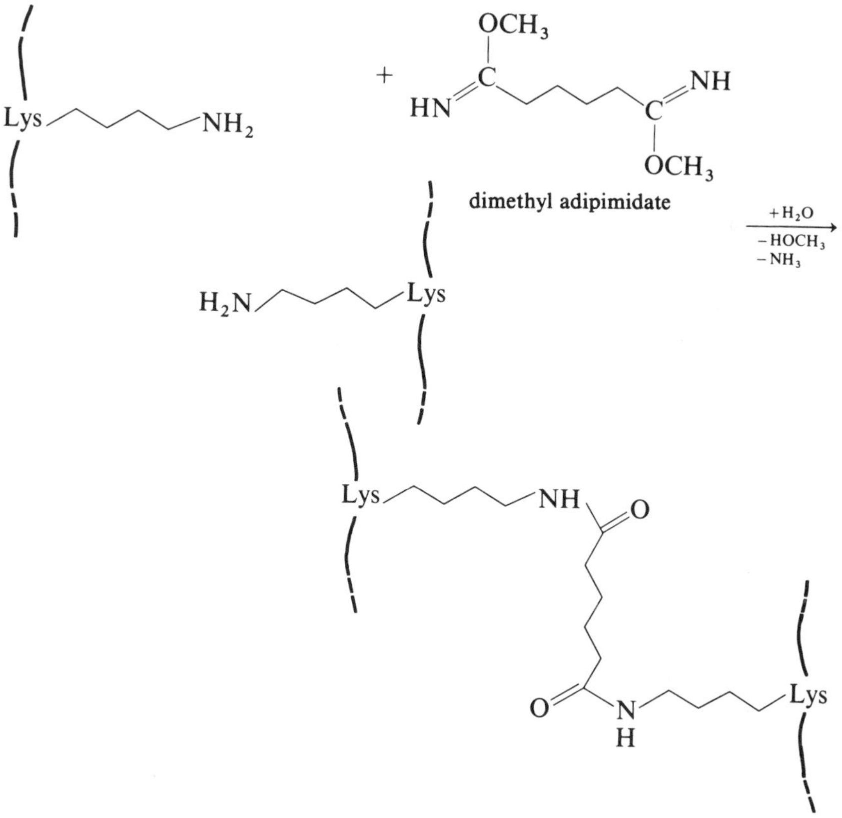
OCH3
+
NH
HN
Lys
NH2
OCH3
dimethyl adipimidate
+H2O
-HOCH3
-NH3
H2N
Lys
Lys
NH
O
O
N
H
Lys

A useful application of the system is in commercial processes and the term *enzyme engineering* or *enzyme technology* has been attached to it. Enzyme technology has been called a "solution in search of problems" (235). Actually, the technology has yet to fulfill much of the promises that many of its advocates are convinced it holds. Nevertheless, we should be aware of its potential and sensitized to this future technology.

Let us examine a few applications, especially in the field of food processing. One effort that has been made to provide better quality foodstuff is the following. Immobilized β-galactosidase is bound to a polyisocyanate polymer molded to a magnetic stirring bar. This fiber-entrapped enzyme is used to reduce the lactose content of milk to overcome the problem of lactose intolerance. Furthermore, by this process milk can be stored frozen for a longer period of time without thickening and coagulating, caused before by lactose crystallization in the milk.

Other efforts in the food industry have been made to utilize immobilized enzymes. One is to use cellulose from waste paper, wood chips, or sugar cane, degrade it to glucose, and convert this sugar unit back to starch, as edible material. All these processes are enzyme catalyzed and should be applicable to enzyme technology.

A more spectacular example is the possible conversion of fossil fuel (oil) derivatives to edible carbohydrates. This transformation needs an industrial breakdown of petroleum products to glyceraldehyde. Then glyceraldehyde can be enzymatically converted to fructose, glucose, and starch.

The ultimate application of enzyme technology to carbohydrate synthesis will be the mimicking of nature's method: the fixation of CO_2. This will require, in addition to immobilized enzymes, *immobilized coenzymes*. Many efforts have been made in this direction (237).

For instance, an NAD^+ analog bound to a water-soluble dextran polymer has been applied in the preparation of an *enzyme electrode* (240) and a *model enzyme reactor*. Medicinal applications in this field are obvious. One example will illustrate the principle. The oxidoreduction reaction, $NAD^+ + \text{substrate} \rightarrow NADH + H^+ + \text{product}$, can be coupled to immobilized enzymes. In this model system, the substrate is pumped into a chamber containing the dextran-bound NAD^+ and two NAD^+-linked dehydrogenases. At the other end the product of the reaction is removed at the same rate by ultrafiltration. Hence, the process can be recycled.

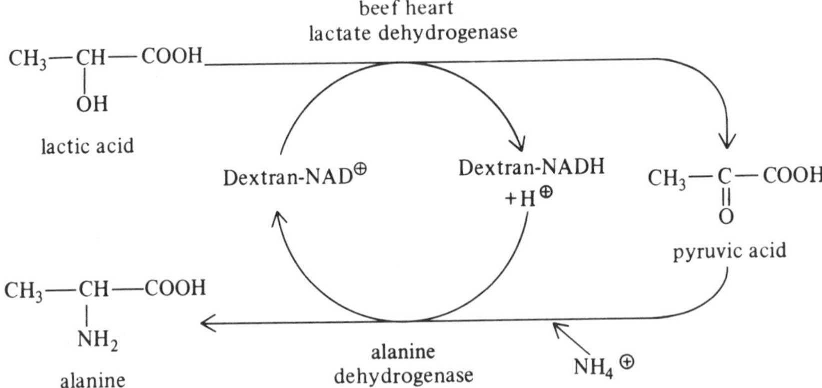

The feasibility of such a reactor has been demonstrated by the production of alanine from lactic acid. The unfavorable equilibrium in this lactate dehydrogenase-alanine dehydrogenase reactor is offset by using a high concentration of substrate and by the rapid consumption of pyruvic acid by the second enzyme. It is worthy of mention that such a system also serves as a model for potential therapeutic applications whereby enzymes and coenzymes immobilized together could function as a self-contained unit to correct a metabolic imbalance.

Some of the examples presented may be seen by skeptics as less than reality. However, it is the author's belief that these projects or a variant of them will be developed on a practical scale in the near future. As K. Mosbach pointed out, the examples discussed above may be looked upon as first steps in the field of *synthetic biochemistry* (237).

In this context, two processes now have useful and practical application in the pharmaceutical industry. The first one is the synthesis of a cortisone analog, prednisolone, that is used as a drug against rheumatoid arthritis. A steroid precursor, Reichstein's compound, is passed through a series of two different columns, each containing a specific enzyme attached to a polyacrylamide polymer support (241). The hydroxylation reaction at the prochiral C-11 position occurs with retention of configuration.

Reichstein's compound

11β-hydroxylase

$\Delta^{1,2}$-dehydrogenase

cortisol

prednisolone

This way, the synthesis is rapid, regio- and stereoselective, and economical. These transformations were in part responsible for the lowering in the price of cortisone from about $30.00 per gram in the 1950s to a few cents per gram in the 1970s.

The second example, which is also in daily use, is the specific hydrolysis of natural penicillin G by penicillin amidase in preparative scale, again through a chromatographic process (242). It produces a clean hydrolytic product (6-aminopenicillanic acid) free of possible contaminant and it can then be used to prepare chemically all kinds of semisynthetic penicillin derivatives. Unlike the rather unstable soluble enzyme, the insolubilized preparation shows no loss of activity after up to eleven weeks of continuous operation at 37°C. Furthermore, the procedure is free from potentially allergy-inducing contaminants.

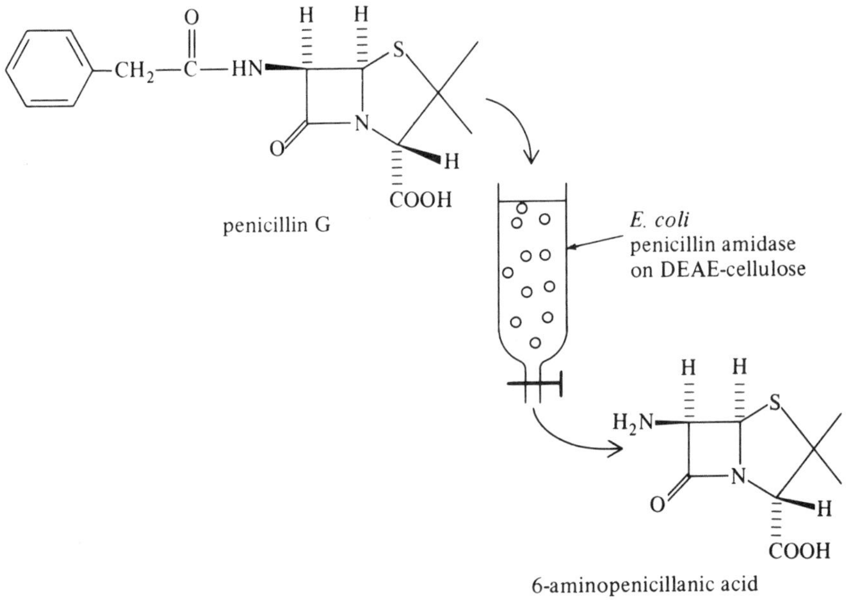

penicillin G

E. coli penicillin amidase on DEAE-cellulose

6-aminopenicillanic acid

Since enzymes are valuable not so much for their purely catalytic abilities as for their selectivity during reactions, enzyme technology offers many capabilities in synthetic chemistry applications and will attract more organic chemists to use enzymes to make biological molecules.

4.8 Enzymes in Synthetic Organic Chemistry

Although more than two thousand different enzymes have been described, only a few hundred are commercially available in small quantities and less than two dozen in industrial amounts. Nonetheless, biochemical procedures are slowly becoming accepted as routine in organic synthesis. The main reasons why enzymes have not been used more often in organic chemistry are a lack of motivation and a lack of familiarity on the part of synthetic chemists (243,244). On the other hand, chemists are appreciating more and more the potential of enzymes as selective catalysts to construct chiral synthons of complicated compounds, along with the possibility of recycling the biocatalyst if mounted on an immobilized support (see preceding section).

The immense potential of enzymes as catalysts for synthesis in organic solvents is well documented in a recent review article by A.M. Klibanov from M.I.T. (245).

The following two examples illustrate that chiral molecules can be obtained directly from achiral precursors.

$COO^{\ominus}$—CH=CH—$COO^{\ominus}$ (achiral precursor) + $NH_4^{\oplus}$ —aspartase→ $^{\ominus}OOC$—CH_2—CH($^{\oplus}NH_3$)—$COO^{\ominus}$ (L-aspartic acid)

achiral precursor

L-aspartic acid

prochiral center

MeOOC—CH_2—C(X)(Y)—CH_2—COOMe —esterase→ MeOOC—CH_2—C(X)(Y)—CH_2—COOH

chiral molecule from mono-hydrolysis

Of course, one major problem associated with the use of enzymes in organic chemistry is that many of the substrates of interest are "unnaturals." However, enzymes can operate on unnatural substrates by changing the reaction conditions or solvent to change the activity of the enzymes. Enzymes have been divided into six groups and, in each of them, specific enzymes have been used in organic syntheses.

1. *Oxidoreductase* for oxidation and reduction reactions (see Section 7.1.2)
2. *Transferase* for transfer of functional groups
3. *Hydrolase* in hydrolysis reactions of esters, amides, phosphates, et cetera
4. *Lyase* for addition to double bonds
5. *Isomerase* particularly for racemization
6. *Ligase* for the formation of chemical bonds with ATP cleavage

Particularly useful for the synthetic chemists is the formation of new C—C bonds with stereospecific introduction of chiral center. Along these lines, Whitesides has exploited the potential of aldolase-catalyzed aldol reactions as a route to isotopically labeled sugar molecules on a molar scale (246). For this, dihydroxyacetone is first phosphorylated by a kinase. In the presence of the aldolase (from rabbit muscle), the ketosubstrate forms a Schiff base with a lysine residue of the enzyme to give eventually an enamine intermediate that acts as a nucleophile and reacts with an incoming chiral aldehyde. The stereochemistry of the aldol condensation catalyzed by the enzyme indicates that *two new chiral centers* are formed enantiospecifically. This approach has been used for the synthesis of ^{13}C-labeled glucose and fructose.

The use of enzymes as organic tools in the fields of medicinal chemistry, endocrinology, immunology, and biotechnology will become more and more frequent in the coming years. All these disciplines require chemical

aldolase

kinase

* = ^{13}C-enriched molecule
R = widely variable
R′ = must be H or OH

manipulation of complicated compounds as well as biologically derived or biologically related substances. The utility of high selectivity and the fast rate of enzyme reactions could contribute greatly to the development of rapidly accessible chiral precursors.

It is well known that enzymes have the ability to discriminate between enantiomers of racemic substrates. The following presentation will explore this capacity by giving few examples of enantiomeric specificity and prochiral specificity.

1. Resolution of an epoxy ester (247).

porcine pancreatic lipase

(2*S*, 3*R*)

(2*R*, 3*S*) 90% ee

Despite the success of the Sharpless approach for stereospecific *trans* epoxide preparations (see Section 2.5.3), the above epoxy alcohol has not been readily available so far.

2. Hydrolysis of a *meso*-diester of prochiral alcohol (248).

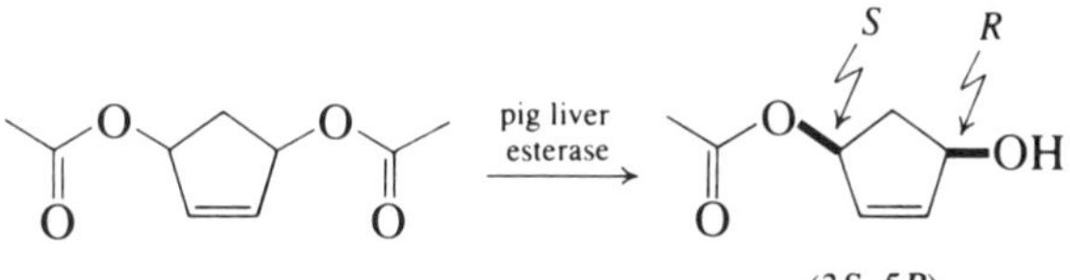

pig liver esterase

(3*S*, 5*R*)
82% ee in
83% yield

In fact, pig liver esterase (PLE) exhibits a broad tolerance of structural variation in its *meso*-diester substrates. Its enantiotopic specificity has been exploited recently for the preparation of chiral cyclic acid-esters and bicyclic lactones (249).

For example:

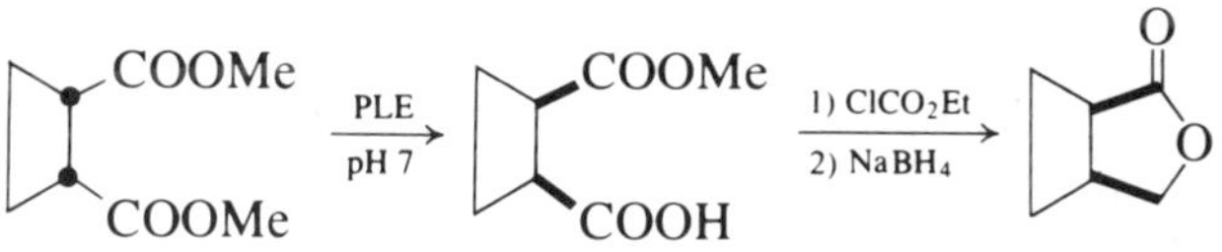

PLE, pH 7; 1) $ClCO_2Et$ 2) $NaBH_4$

(−)-(1*R*, 2*S*)
87% yield

37% yield
97% ee

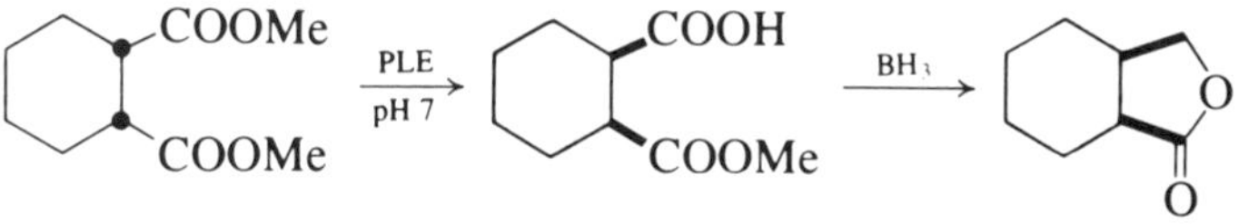

PLE, pH 7; BH_3

(+)-(1*S*, 2*R*)
88% yield

90% yield
97% ee

Sabbioni and Jones observed a dramatic change of stereospecificity on going from cyclobutane (or cyclopropane) diesters to the cyclohexane substrate, with analogous cyclopentane diester hydrolysis representing the changeover point within the series. They argued that this reversal of enzyme stereospecificity is explicable in terms of a two binding-pocket active site model.

In both cases, the ester group to be hydrolyzed is located adjacent to the nucleophilic hydroxyl group of the catalytically vital serine residue of the enzyme in the active site. For the cyclohexane diester substrate, the steric requirements of the cyclohexyl ring cannot be accommodated in the small hydrophobic region of the active site, as are those of the cyclobutyl ring, so it must therefore bind entirely in the second auxiliary larger site. This results in exclusively *R*-center ester hydrolysis. The equatorial

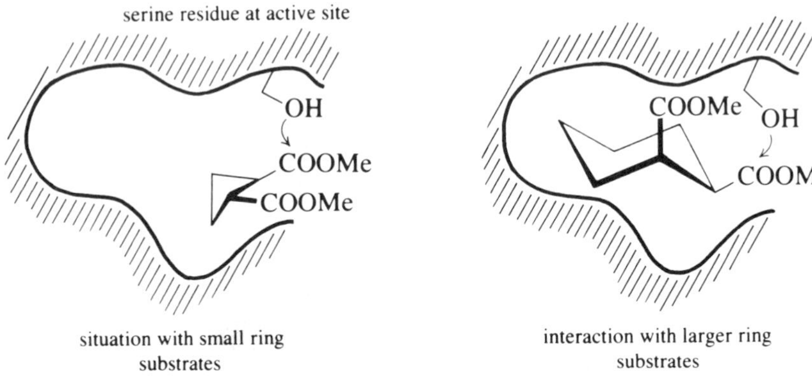

orientation of the ester group attacked by the serine hydroxyl had already been established by Jones's co-workers.

In conclusion, these biochemical methods of enzymatic asymmetric catalysis have the great advantage of being able to generate in high yields bifunctional chiral synthons with optical purities not obtainable otherwise. As pointed out by Whitesides (243): "Those unwilling to use these and other biologically derived synthetic techniques may find themselves excluded from some of the most exciting problems in molecular science."

4.9 Enzyme-Analog-Built Polymers

Because of the selectivity and efficiency of enzymes, there has been considerable interest in the synthesis of organic macromolecules that in some way could simulate enzyme action. For this, chemists have been making progress in building specific binding sites or cavities into polymer networks as models of biological receptors and enzymes. One of the challenging problems in the design of such enzymelike systems is the selective introduction of suitable functional groups in a stereorational manner with the maintenance of their stereochemical integrity over a period of time. Such enzyme-analog-built polymers might also serve as catalysts and as tailored supports for chemoselective affinity chromatography by selectively recognizing small molecules complementary in shape to the built-in cavities.

Pioneer efforts in this direction came from the group of G. Wulff (University of Düsseldorf). This technique, which was termed *template synthesis method* by K.J. Shea (University of California, Irvine), consists of developing a general methodology for the controlled introduction of multiple organic functionality in organic macroporous polymers (250,251).

One of Wulff's early achievements in the field was the partial resolution of D- and L-mannose by incubating a solution of the racemate with a resin made from a D-mannose template. The approach consisted in preparing a template by esterification of nitrophenyl-α-D-mannose with 4-vinyl boronic acid. The template molecule was then copolymerized with ethylene dimethylacrylate to give a cross-linked polymer. Basic hydrolysis of the mannose derivative from the resin left cavities containing boronic acid groups fixed in chiral environments.

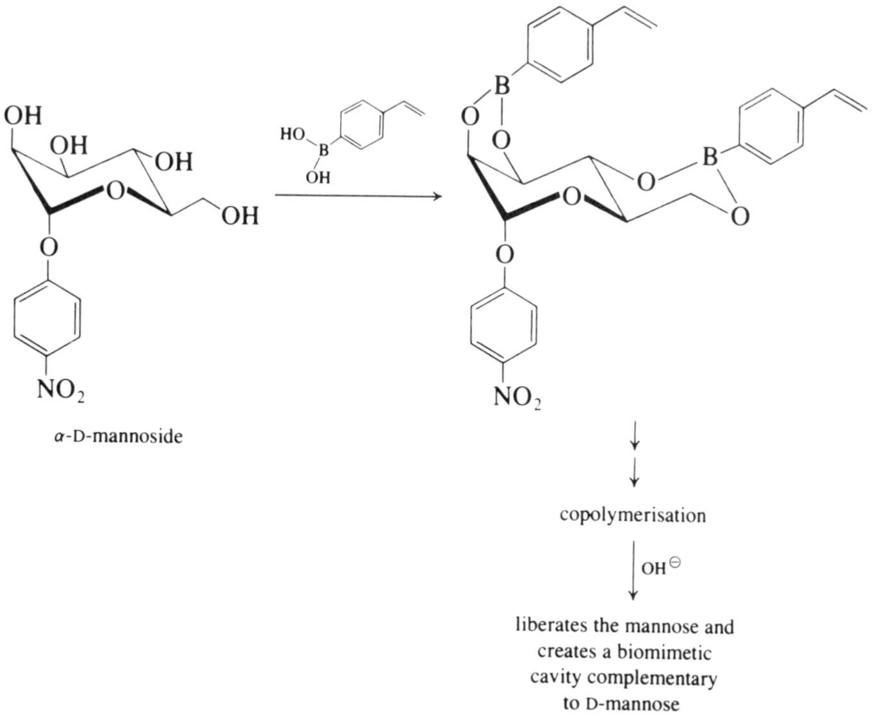

α-D-mannoside

copolymerisation

$OH^{\ominus}$

liberates the mannose and creates a biomimetic cavity complementary to D-mannose

Since the hydrolyzed polymer can now exhibit a "molecular memory" for the original template molecule, incubation of this resin with a racemic mixture of mannose yielded a solution 13% enriched in L-mannose; the sugar bound to the resin was 40% enriched in the D-isomer.

In this methodology, template molecules create chiral sites on polymer matrices for specific molecular recognition. This is somehow analogous to antibody–antigen recognition (see Section 2.8).

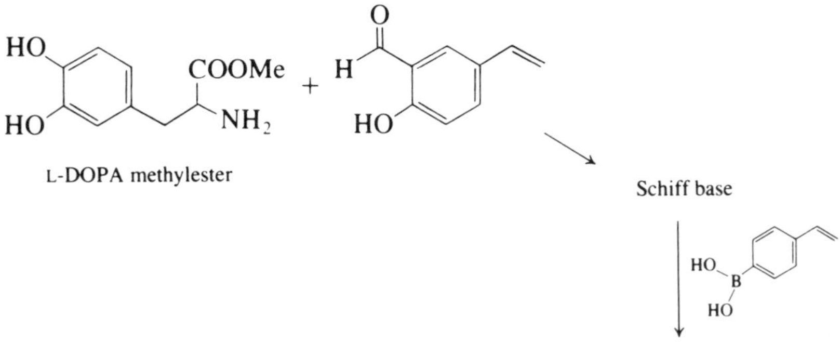

L-DOPA methylester

Schiff base

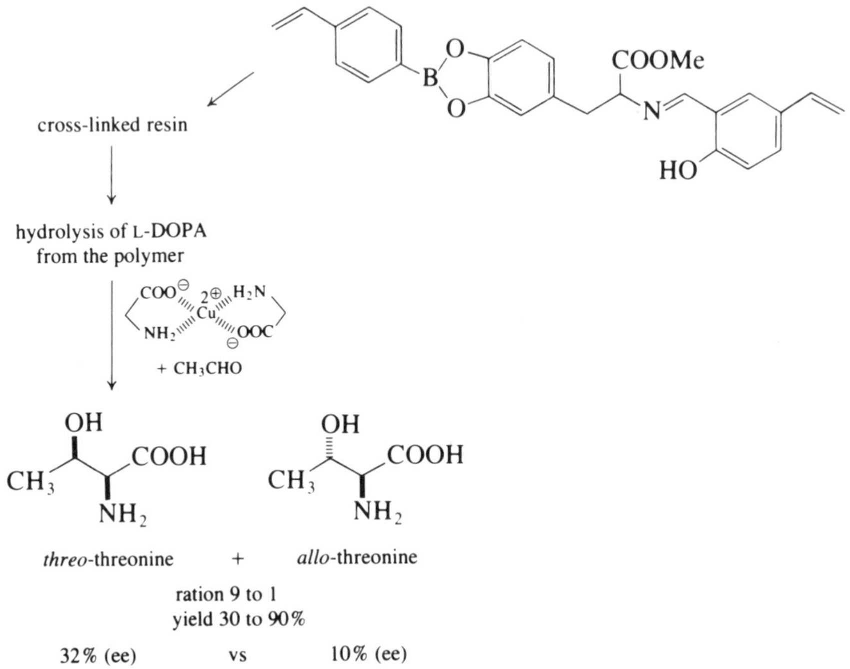
COOMe
HO
cross-linked resin
hydrolysis of L-DOPA
from the polymer
+ CH3CHO
threo-threonine
+
allo-threonine
ration 9 to 1
yield 30 to 90%
32% (ee)
vs
10% (ee)

Wulff also demonstrated the potential of these cross-linked polymers by performing a stereospecific synthesis of the amino acid threonine by condensation of glycine with acetaldehyde.

Similarly, Shea used a three-dimensional polymeric matrix made of styrene and divinylbenzene (251). Electron microscopy of the microspheres of the macroporous resin shows the presence of a network of channels similar to those found in bone char or alumina. Treatment of the

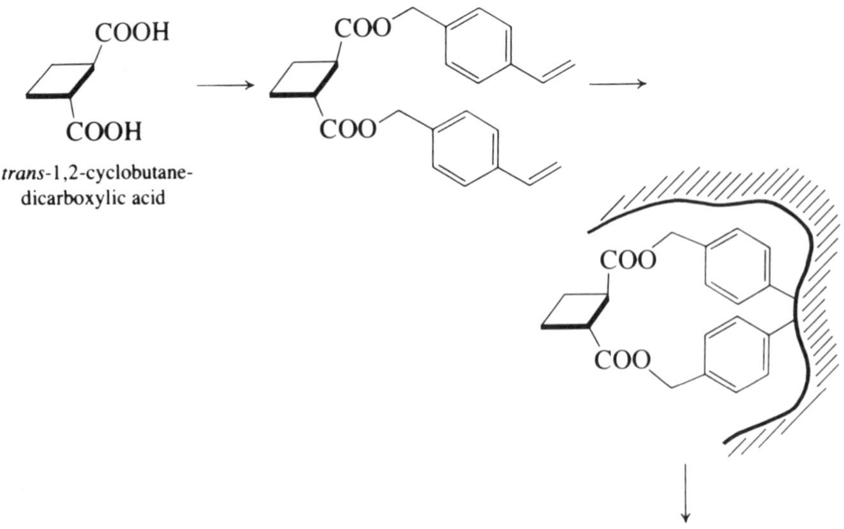
COOH
COOH
trans-1,2-cyclobutane-
dicarboxylic acid
COO
COO
COO
COO

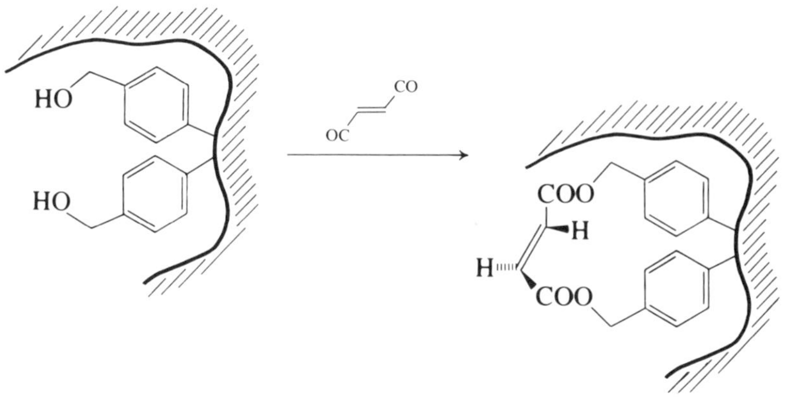

resin with fumaryl chloride results in covalent attachment of the fumarate group to the polymer. This demonstrates that the original geometry of the cavities has been conserved. That fumaric acid is indeed covalently fixed to the polymer was established by the finding that the acid can be liberated only by a second hydrolysis. Furthermore, the polymer-bound fumaric ester can be converted to a *trans*-1,2-cyclopropyldicarboxylic ester, with a methylene-transfer reagent, in a slight enantiomeric excess.

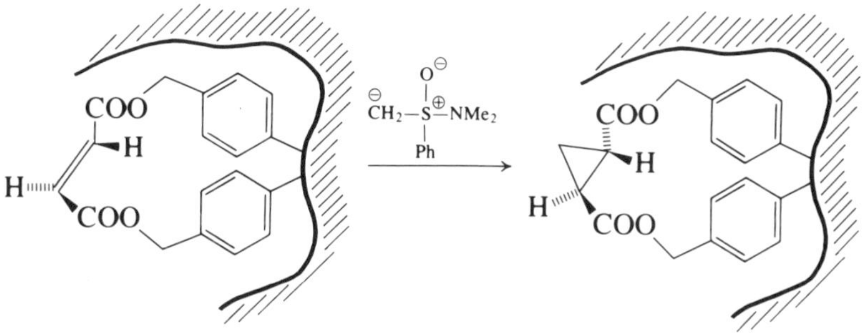

Recently, a template polymer matrix that can incorporate *N*-benzyl-D-valine with almost 100% stereospecificity was synthesized by using cobalt coordination chemistry (252). The protocol is shown below.

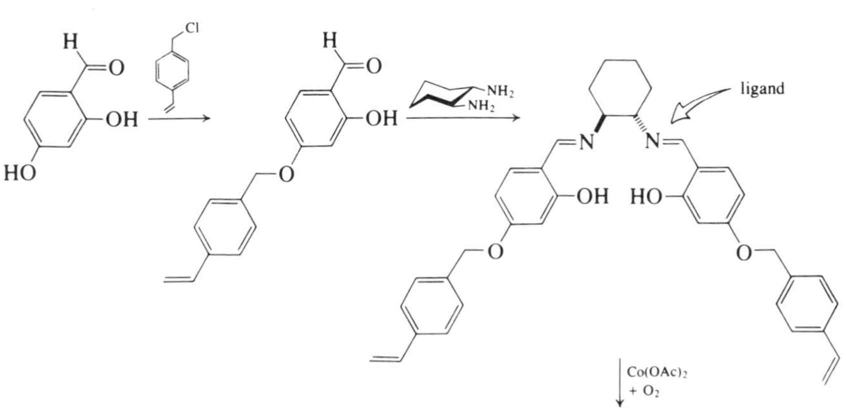

S = styrene
DVB = divinylbenzene

copolymerization S + DVB

TEMPLATE

HCl 3*M* (-amino acid)

Bz-DL-Val (-amino acid)

TEMPLATE + Bz-L-Val

Bz-D-Val

The cavity for the chiral amino acid was made by template synthesis of a polymer Schiff base Co(III) complex. Dissociation of the coordinated amino acid yields a cavity that possesses the capacity of chiral recognition. The chiral discrimination was significantly higher than that of *trans*-[Co(ligand$H_2O)_2$]Cl.

Matching of chemical structure to a macromolecular target of known architecture has been used in designing drugs to fit a macromolecular receptor (253). The methodology has also been applied for the construction of chiral cavities as specific receptor sites in chemoselective and bioselective affinity chromatography (254). Other aspects of *molecular imprinting* of polymers with the chirality on the main chain have also been developed (255). Therefore, main-chain chirality results from optical activity in the configurational or conformational arrangement of the main chain within the polymer.

4.10 Design of Molecular Clefts

An important contribution to bioorganic chemistry in recent years has been the conceptualization by J. Rebek, Jr. (now at M.I.T.) of molecular clefts using convergent functional groups on receptors (256). This was

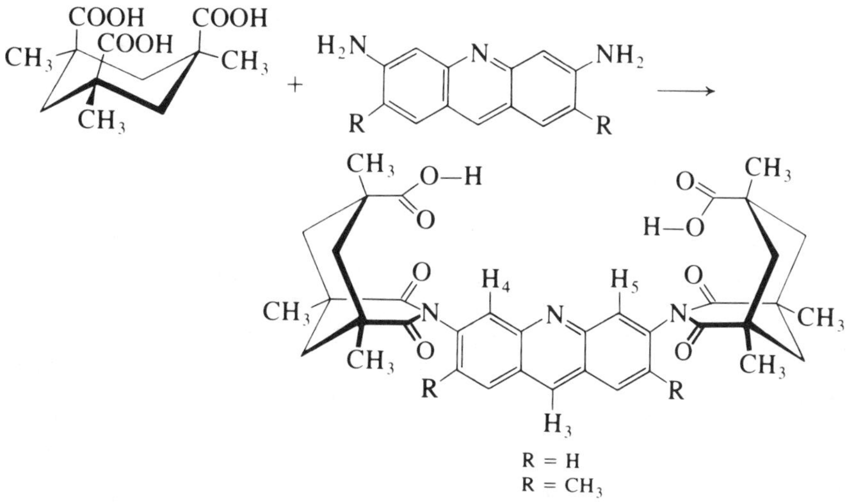

COOH COOH
CH_3 COOH CH_3 + H_2N N NH_2 →
CH_3 R R

CH_3 O—H O CH_3
O H—O
O H_4 H_5 O
CH_3 N N N CH_3
CH_3 O O CH_3
R R
H_3

R = H
R = CH_3

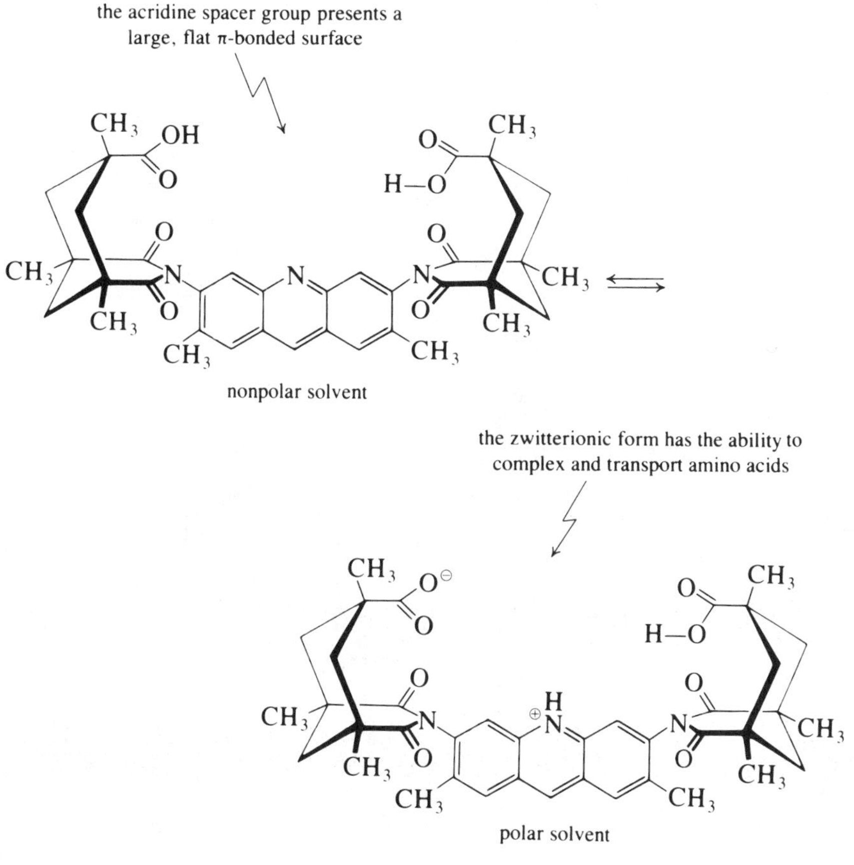

the acridine spacer group presents a large, flat π-bonded surface

CH_3 OH O CH_3
O H—O
O O
CH_3 N N N CH_3 ⇌
CH_3 O O CH_3
CH_3 CH_3

nonpolar solvent

the zwitterionic form has the ability to complex and transport amino acids

CH_3 $O^{\ominus}$ O CH_3
O H—O
O H O
CH_3 N ⊕N N CH_3
CH_3 O O CH_3
CH_3 CH_3

polar solvent

elegantly illustrated by constructing an acridine derivative possessing two carboxylic functions facing each other. The receptor is rapidly and efficiently assembled from the appropriate aromatic diamine and merely heating it with a cyclohexyl triacid (Kemp's triacid) in the absence of solvent (257).

The structure incorporates two acidic groups that *converge* on a molecular cleft in a manner that resembles the convergence of functional groups at the active sites of enzymes. Such an arrangement of groups in a U-shaped molecular cleft permits the molecule to recognize substrate of complementary size, shape, and functionality. A smaller cleft is obtained by using a naphthalene spacer instead of acridine orange.

The convergent nature of the carboxyl groups, the polar microenvironment of the cleft, and the lipophilic character of the skeleton are features that prove useful in the transport of polar molecules and amino acids across liquid membranes. The receptor can also preferentially complex diamines (pyrazine and imidazole) and purines, even in the presence of a stronger base like pyridine (258). These examples emphasize the contribution of ionic and aromatic stacking interactions of complementary groups in molecular recognition. The association constants were measured by using the NMR techniques and are in the range of 10^3 L/mol.

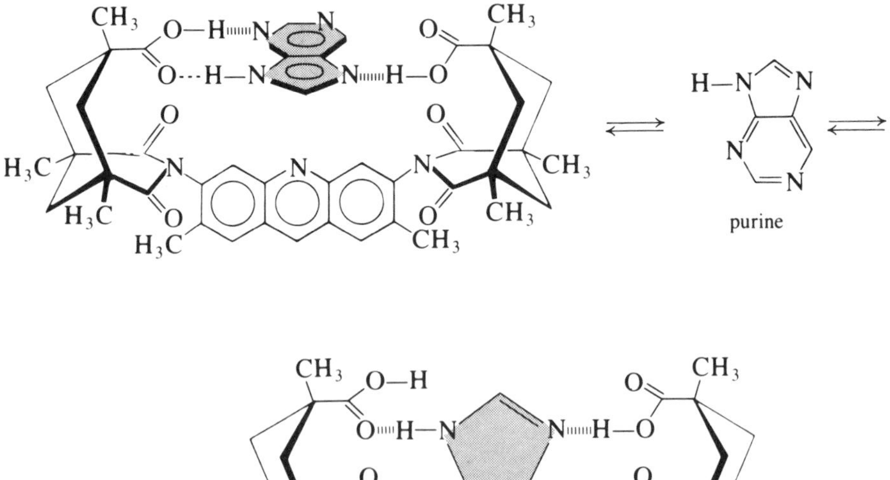

purine

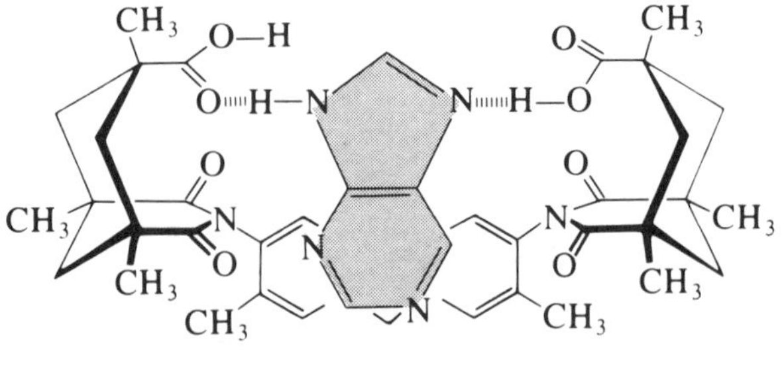

The combination of ionic and stacking interactions resulted also in the highly selective binding of the acridine receptor to physiologically important β-arylethylamines, like dopamine and tryptamine. In the following ternary complex, a K_a of $10^7\,M^{-2}\,L^{-2}$ was measured.

$X = H, OCH_3, NO_2$

Oxalic acid, a diacid, is also recognized by this receptor where a mimic of cyclic hydrogen-bond dimer is observed spectroscopically. A nice example of this donor–acceptor approach is the formation of a complex with benzylmalonic acid. Notice the presence and the role of two extra methyl groups on the receptor molecule. This prevents rotation of the imide bond and increases the stability of the complex (259).

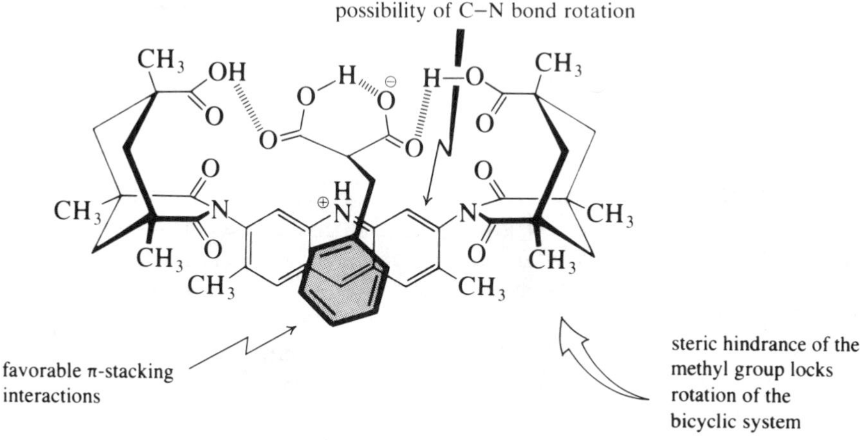

The convergent nature of the two carboxyl groups in such geometrically restrained receptors provides a good opportunity to measure the contribution of stereoelectronic factors at the carboxyl oxygens, since it is

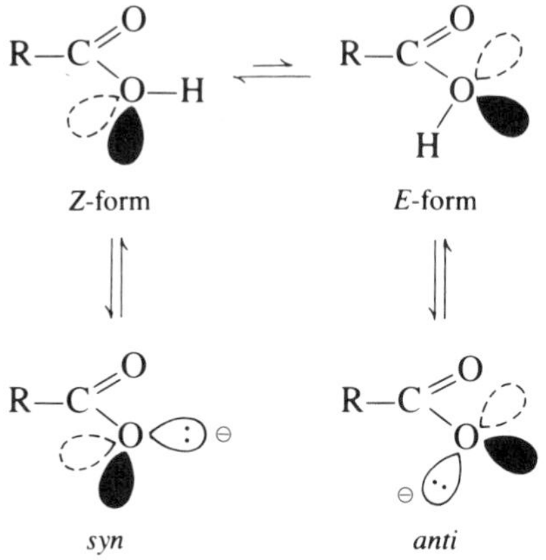

Z-form

E-form

syn

anti

accepted that for carboxylic acids, the *Z*-form is regarded as more stable than the *E*-form by about 5 kcal.

R.D. Gandour noticed that this difference could indeed have biochemical implication (260), since he observed that carboxylates found at the active sites of enzymes (lyzozyme, α-chymotrypsin, carboxypeptidase A) generally employ the more basic *syn* lone pairs rather than the less basic *anti* lone pairs.

Rebek's molecules were measured for pK_a values in $EtOH/H_2O$ (1:1, w/w) and the results are shown below (261).

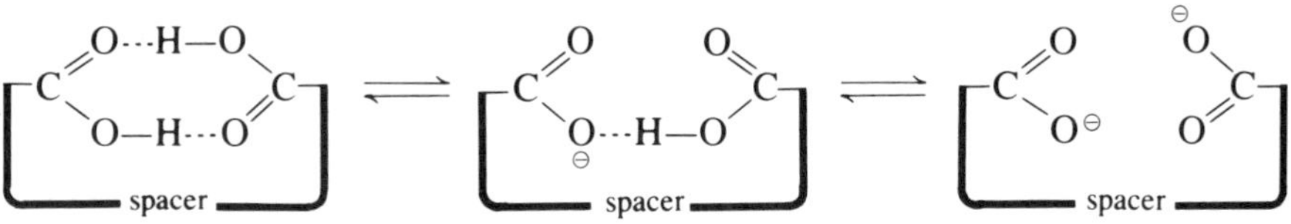

spacer

	pKa_1	pKa_2	
spacer = dimethyl-3,6-naphthalene	5.5	7.5	$\Delta pKa = 2.0$
spacer = dimethyl-3,6-acridine	6.5	7.8	$\Delta pKa = 1.3$
spacer = dimethyl-2,4-benzene	4.8	11.1	$\Delta pKa = 6.3$

For the receptor where the spacer is a benzene ring, the distance separating the two carboxyl groups is about 3 Å, and the conformation forces the convergence of the *syn* lone pairs in the dianion. The results of such high electron density in a limited volume is seen in the unusually high value for pKa_2 ($\Delta pK_a = 6.3$). The situation is more normal with the other two derivatives. The effects of intramolecular hydrogen bonding are difficult to quantify in these systems and alternate interpretations are possible. However, the fact that the benzene derivative binds strongly to Ca^{2+} or Mg^{2+} ions seems to indicate that in this geometrically favored microenvironment, *syn* lone pairs on carboxyl oxygens must be involved in the stabilization of the organometallic complex. Large ΔpK_a values have also been observed in crown ether derivatives having two carboxyl

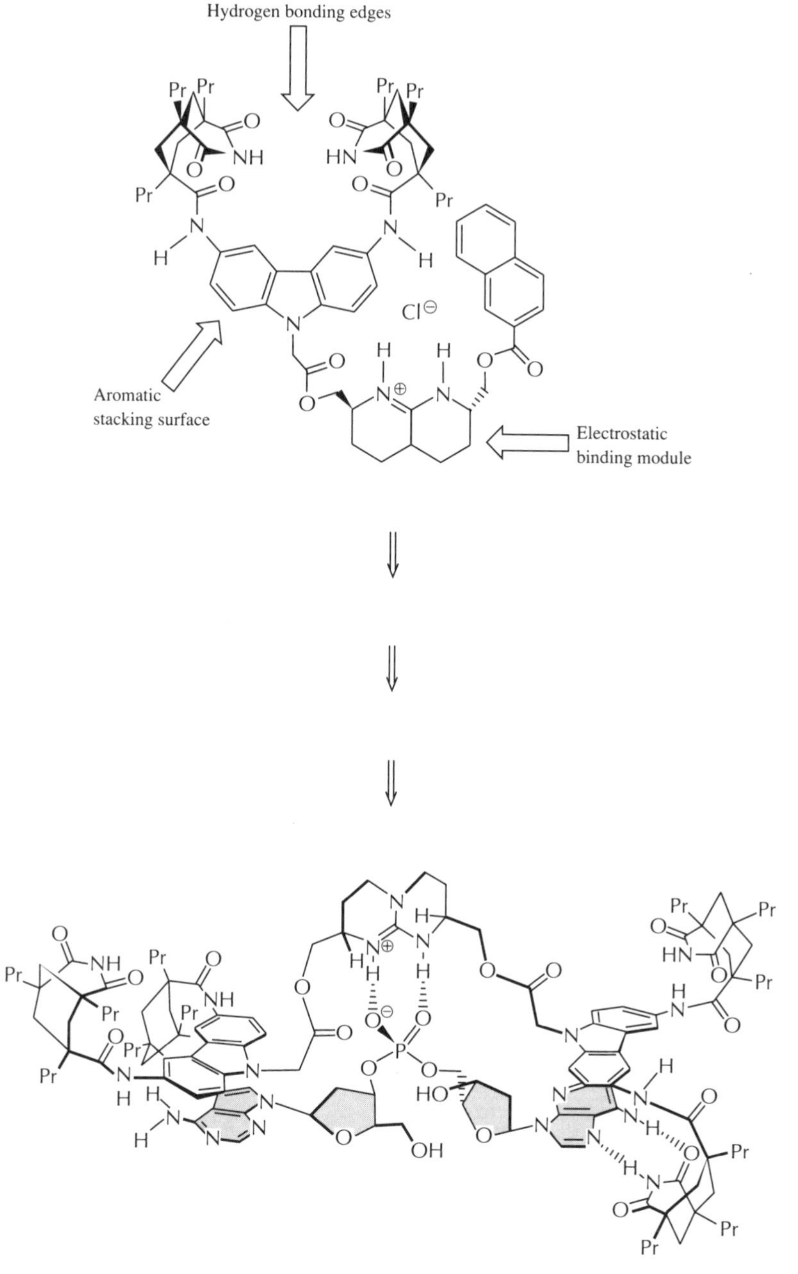

Hydrogen bonding edges

Aromatic stacking surface

Electrostatic binding module

Fig. 4.9. Receptor for the dinucleotide *d*(AA) (in gray), adapted from (265).

groups facing each other in a *syn*-fashion (318). Such molecules also bind strongly to Ca^{2+} ions (see Section 5.2.2).

Other aspects related to acid-base catalysis by such artificial enzymes and receptors are covered in a recent review article (262).

The concept behind the construction of molecular clefts using this convergent functional group approach can be extended to other types of molecular recognitions. The structure of these receptors can eventually be tailored for other specific applications such as mimicking Watson–Crick base pair stacking interactions or modeling antibody–antigen interactions (263). These molecular clefts also offer the possibility of building an asymmetric microenvironment complementary to asymmetric hosts. Furthermore, such conformationally constrained molecular structure with two acidic convergent functions could possess catalytically advantageous properties to be exploited in the future.

In the past few years, more sophisticated receptors have appeared for the recognition of amino acids and nucleic acids (264,265). The strategy consists of constructing a receptor cavity with a series of modules, one for hydrogen binding, one for aromatic stacking, and one for electrostatic interactions. The general structure in Fig. 4.9 (see previous page) is an example of a complex assembly of molecular recognition sites that was used for the design of a receptor for adenine dinucleotide [*d*(AA)].

In a different line of thought, S.C. Zimmerman (University of Illinois) has also contributed greatly in designing a new class of synthetic receptors for aromatic substrates (266). Known as *molecular tweezers*, these receptors form very stable sandwich-type complexes with π-deficient aromatic substrates in organic solvents. This is illustrated by the complex below:

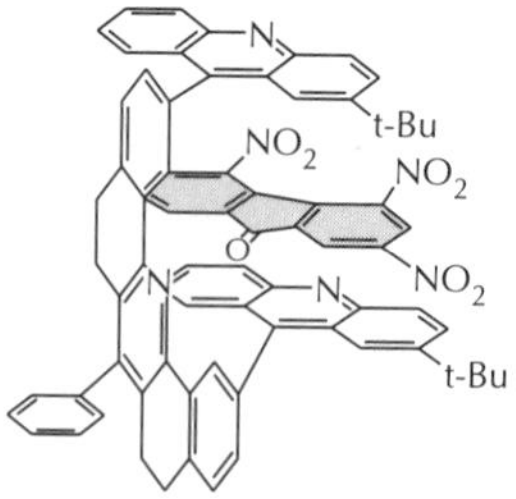

N
t-Bu
NO_2
NO_2
O
NO_2
N
N
t-Bu

Molecular tweezers containing a carboxylic acid group in their binding clefts have also been synthesized and have been shown to form very stable inclusion complexes with 9-propyladenine, in chloroform solution, and with other nucleoside bases.

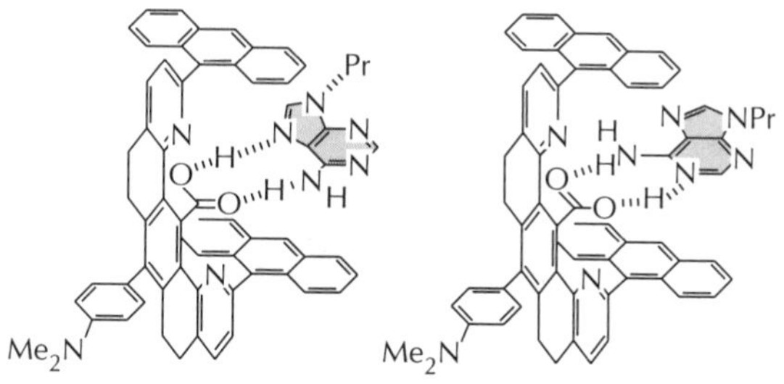

In this complex, II-overlap is maximized and the left structure is the most probable geometry although the contribution from the right one cannot be ruled out.

Chapter 5
Enzyme Models

"We dance round in a ring and suppose
But the Secret sits in the middle and knows!"

R. Frost
from "The Secret Sits"

Enzyme models are generally organic synthetic molecules that contain one or more features present in enzymatic systems. They are smaller and structurally simpler than enzymes. Consequently, an enzyme model attempts to mimic some key parameter of enzyme function on a much simpler level. To dissect out a particular factor responsible for the catalytic efficiency of the enzyme within the biological system would be a tremendous task requiring a knowledge of each of the components that would contribute to the overall catalysis. Instead, with appropriate models, it is possible to estimate the relative importance of each catalytic parameter in the absence of those not under consideration. One noticeable advantage of the use of "artificial structures" for modeling enzymatic reactions is that the compounds can be manipulated precisely for the study of a specific property. The state of the art may be further refined by combining those features that contribute most and designing models that actually approach the efficiency of the enzyme. That is, with the tools of synthetic chemistry, it becomes possible to construct a "miniature enzyme" which lacks a macromolecular peptide backbone but contains reactive chemical groups correctly oriented in the geometry dictated by an enzyme active site. It is often referred to as the *biomimetic chemical approach* to biological systems.* Therefore, biomimetic chemistry represents the field that

* The word "biomimetic," introduced by R. Breslow in 1972, generally refers to any aspect in which a chemical process imitates a biochemical reaction (267).

attempts to imitate the acceleration and selectivities characteristic of enzyme-catalyzed reactions (405). It is hoped that such an approach will eventually bridge or at least reduce the gap between the known complex structures of organic biomolecules and their exact functions in life. In order to do this, many factors related to the mechanism of action of a particular enzyme must be known. These include (a) the structure of the active site and the enzyme–substrate complex, (b) the specificity of the enzyme and its ability to bind to the substrate, and (c) the kinetics for the various steps and a knowledge of possible intermediates in the reaction coordinate.

Enzymes are complicated molecules and only a few mechanisms have been definitively established. This is one of the reasons why model systems are necessary. Among the functional groups found on polypeptide chains, those generally involved in catalytic processes are the imidazole ring, aliphatic and aromatic hydroxyl groups, carboxyl groups, sulphydryl groups, and amino groups.

How can such a limited number of functional groups participate in the large variety of known enzymatic reactions, and how can the rate of the enzymatic reactions be accounted for in mechanistic terms? These are the fundamental questions that should be asked during the planning of a bioorganic model of an enzyme (268).

In general, an enzyme model should fulfill a twofold purpose: (a) it should provide a reasonable simulation of the enzyme mechanism, and (b) it should lead to an explanation of the observed rate enhancement in terms of structure and mechanism. Of course, all the information obtained with enzyme models must ultimately be compared and extended to the in vivo enzymatic system under study in order to correlate the bioorganic models to the real natural system.

A model can represent general features for more than one enzyme! Viewed from a different angle, the requirements necessary for the design of a good enzyme model can be summarized in these five criteria:

1. Because noncovalent interactions are the key to biological flexibility and specificity, the model should provide a good (hydrophobic) binding site for the substrate.
2. The model should provide the possibility of forming electrostatic and hydrogen bonds to help the substrate bind in the proper way.
3. Carefully selected catalytic groups have to be properly attached to the model to effect the reaction.
4. The structure of the model should be rigid and well defined, particularly with respect to substrate orientation and stereochemistry.
5. Of course, the model should preferably be water soluble and catalytically active under physiological conditions of pH and temperature.

These criteria are of only limited application but implicit in the summary is the understanding that an efficient catalyst will be constructed by a proper choice of a matrix that can bring catalytic groups and substrates together. This assumption implies that the matrix does not take an active part in the catalysis other than holding and orienting the substrate and the catalytic group (or groups) rigidly in proximity and correctly with respect to one another. Hopefully, the matrix, like an enzyme, can in the binding process raise the ground state energy of the substrate by rigidification and bond distortion. In addition, proper stereochemistry between the model catalyst and the substrate will result in better specificity and efficiency of reaction. These principles are of fundamental importance in this chapter.

In summary, for an enzyme model to be operative, a certain number of criteria, characteristic of enzyme catalysis, must be fulfilled, among which is substrate specificity—that is, selective differential binding. The enzymelike catalyst must also obey Michaelis–Menten kinetics (saturation behavior), lead to a rate enhancement, and show bi- and/or multifunctional catalysis (269).

Enzyme processes imitated to date have been mostly hydrolytic (270), but stepwise assembly of macromolecules such as proteins and nucleic acids may be possible soon. For instance, structures resembling drug receptors may be incorporated into synthetic membranes, facilitating studies of these receptors without immunological and toxic complications. Furthermore, the ability of membranes to segregate charged species may find commercial use in systems for energy storage or hydrogen generation.

This chapter on enzyme models will start by describing the chemistry and properties of some man-made organic host molecules that possess the capacity of enantiomeric discrimination. Six additional aspects of enzyme models will be developed, and other interesting and fascinating enzyme and coenzyme models will be presented in the forthcoming chapters, especially Chapter 7 on coenzyme design and function.

Among the enzymes that have been mimicked in one way or the other, six have been particularly well studied. All are hydrolytic, four have proteolytic activities, one is glycolytic, and the other is nucleolytic. Their mode of action is summarized below.

The amino acids implicated in the catalysis are indicated as well as the pages in the text where these enzymes are discussed. Biomodels of each are presented in this chapter and Chapter 4. For comparison with organic mechanisms, Table 2.1 can be consulted.

Of particular importance for the modeling of an efficient tailormade catalyst with high specificity is the positioning and adequate distribution of charges. Also important is the proper distribution of hydrogen bonds and the presence of a well-oriented hydrophobic auxiliary region for good binding. It is hoped all this will allow good complementary intermolecular interactions as well as a chiral discrimination if needed. Furthermore, it is of prime importance in catalysis that the transition state of the reaction be stabilized. To meet this criterion, one needs a system where any geometric fluctuations of the substrate to reach the transition state occur at the same time as the transition state evolves toward the products of the reaction.

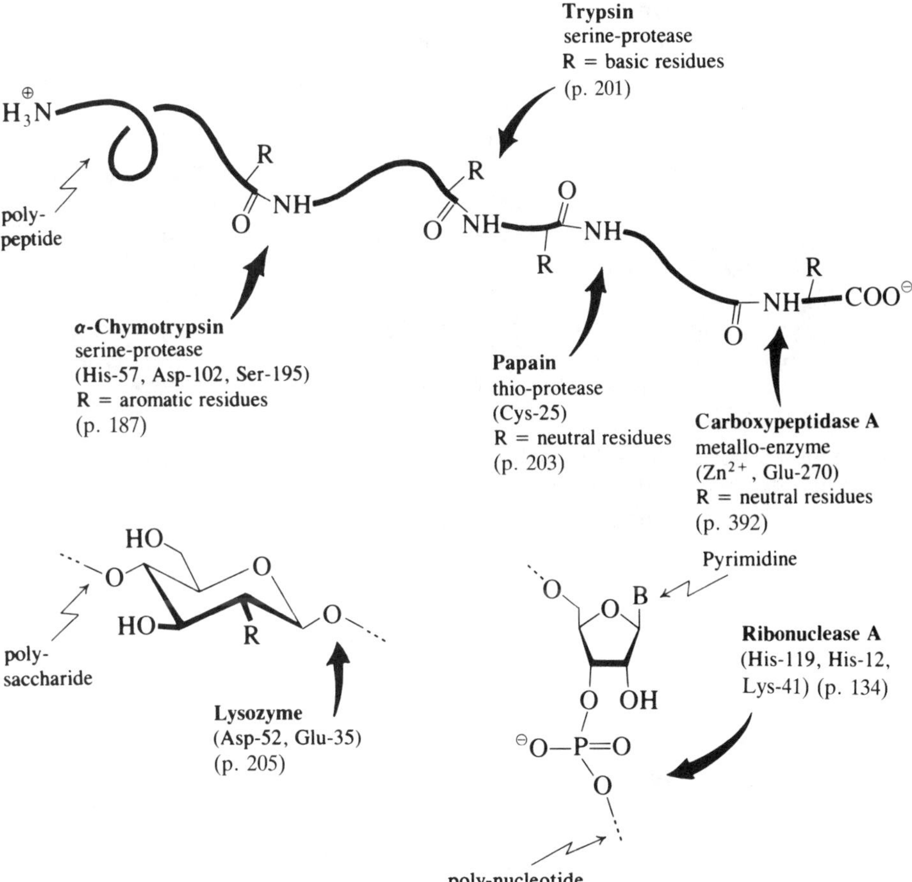

5.1 Host–Guest Complexation Chemistry

The discovery in 1967 by C.J. Pedersen (271,272) that crown ethers have the unique ability to form stable complexes with metal ions and primary alkyl ammonium cations opened new horizons in organic chemistry (273–275).

The abbreviated nomenclature used is simple; the first number represents the total number of atoms in the ring and the second the total

12-crown-4 ether

number of heteroatoms. It is easy to see an analogy between such complexes having a "cavity" to bind the ligand (L) and the active site of an enzyme that recognizes its specific substrate. The size of the macroring can be varied to allow the binding of ligands of different shapes. The cyclic polyethers of the "crown" types are relatively easy to make and are subject to wide structural modification. To this field of chemistry, D.J. Cram from the University of California at Los Angeles coined the name *host–guest complexation chemistry** (276–278). We recall Emil Fischer's lock and key metaphor back in 1894 to describe the match of an enzyme with its substrate in an enzyme–substrate complex. Besides enzyme catalysis and inhibition, complexation plays a central role in biological processes, such as replication, genetic information storage and retrieval, immunological response, and ion transfer. Enough structural information is now known about the complexes involved to inspire organic chemists to design highly structured molecular complexes and to study the chemistry unique to complexation phenomena.

A highly structured molecular complex is composed of at least one host and one guest component. The host component is defined as an organic molecule or ion whose binding sites converge in the complex. The guest component, on the other hand, is defined as any molecule or ion whose binding sites diverge in the complex (278). Guests can be organic compounds or ions, or metal ions, or metal–ligand assemblies. In general, simple guests are abundant, whereas hosts usually have to be designed and synthesized.†

A host–guest relationship involves a complementary stereoelectronic arrangement of binding site in host and guest. Therefore, any man-made synthetic host–guest complex must have binding sites (polar and dipolar) and steric barriers located to complement each other's structures. Micelles and cyclodextrins are naturally occurring hosts and their properties will be the subject of forthcoming sections in this chapter. The prosthetic groups of hemoglobin, chlorophyll, and vitamin B_{12} are also in this category because they bind selectively iron, magnesium, and cobalt ions.

Figure 5.1 represents some available synthetic host compounds. Can chemists use such organic crown ethers for enantiomeric discrimination (or racemic resolution) as mimic of an enzyme? Cram and others have reported that chiral crown ether complexes have this remarkable property of binding selectively one antipode of amino acid derivatives (276–278). In the design of host molecules, Corey–Pauling–Koltun (CPK) molecular models are invaluable (279,280). Space-filling scale models provide a

---

* Although this expression is very useful and directly understandable conceptually, it is not easily translated into other languages. We thus suggest that the expression *receptor–substrate complexation* could be of more universal use.

† Association constants between simple organic ammonium salts and 18-crown-6 ether in $CDCl_3$ at 25°C are on the order of $10^6\,M^{-1}$, corresponding to a $\Delta G°$ of complexation of about −33.5 kJ/mol (−8 kcal/mol).

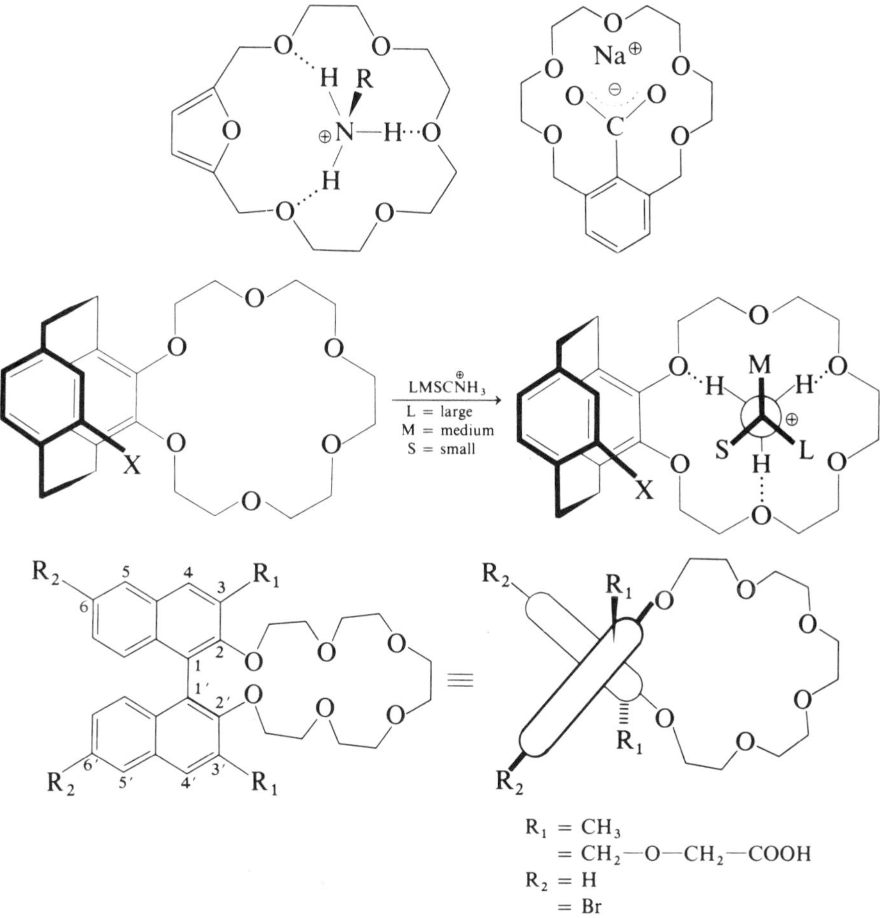

Fig. 5.1. Examples of host molecules.

possible guide to searching for host structures to bind given amino acid guest compounds. For instance, a fundamental question in host design is that of the effect of preorganization of binding sites on binding ability. Another problem is that of the placement of substituents in positions that converge on the functional or binding sites of guest compounds (279).

After many trials in the molecular design of chiral crown ethers, a 1,1′-binaphthyl unit incorporated in a macroring by substitution in the 2,2′-positions proves to possess the desirable properties. The naphthalene-containing system, chosen for practical and strategic reasons, imparts rigidity and lipophilicity to conventional cyclic polyethers. The synthesis of such a host is presented in Fig. 5.2.

This host is chiral and possesses a C_2 axis of symmetry, and the dihedral angle between the planes of the two naphthalene rings attached

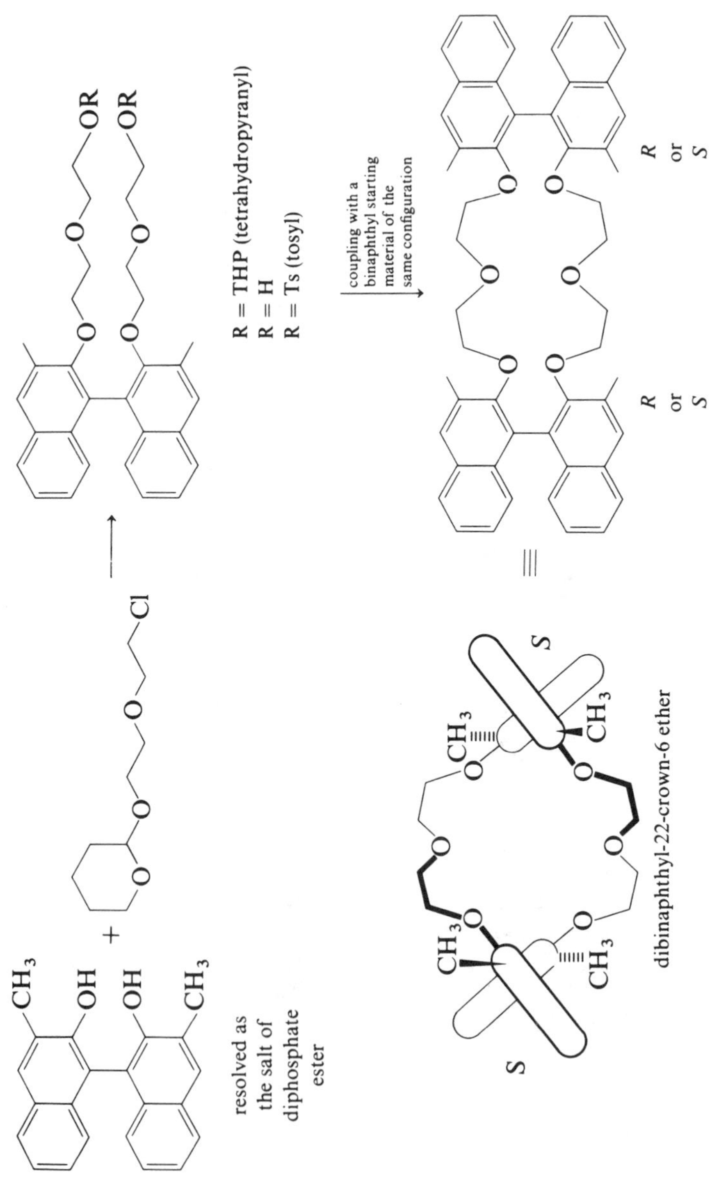

CH_3
OH
OH
CH_3
resolved as the salt of diphosphate ester
+
Cl
OR
OR
R = THP (tetrahydropyranyl)
R = H
R = Ts (tosyl)
coupling with a binaphthyl starting material of the same configuration
R or S
R or S
S
S
CH_3
CH_3
CH_3
CH_3
dibinaphthyl-22-crown-6 ether

Fig. 5.2. Synthesis of a chiral host molecule.

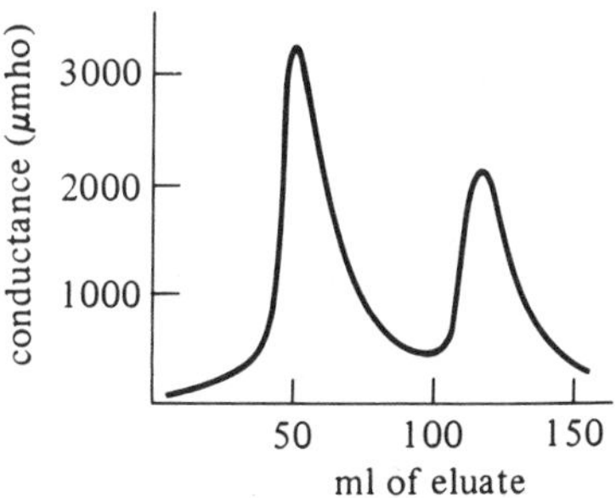

Fig. 5.3. Chromatographic optical resolution by (*R*,*R*)-host of methyl phenylglycinate hexafluorophosphate salt (278).

to one another can vary between 60° to 120°. Including a binaphthyl system in a crown ether causes the macrocyclic ring to twist like a helix rather than being planar. Both the (*S*,*S*)- and (*R*,*R*)-configurations are known, and being optically active, they can be used to resolve racemic primary amine salts and amino esters.

The corresponding diastereomeric "activated complexes" formed are highly structured, and their differences are the basis of what Cram calls *chiral recognition*. A liquid–liquid chromatography system is employed for the optical resolution of racemic amine and amino ester salts by the chiral hosts, mimicking enzyme stereospecificities. The ability to resolve optically active compounds by passage of racemic mixtures through columns of such complexing agents could have commercial importance.

Experimentally, a silica gel or celite support is saturated with an aqueous solution of $NaPF_6$ or $LiPF_6$. The mobile phase is a chloroform solution of optically pure host. The racemic mixture is then added to the column, and after equilibrium a "selective elution" of diastereomeric "activated complexes" is obtained. The appearance of salt in the eluate is monitored by its relative conductance. With racemic methyl phenylglycinate, the elution pattern in Fig. 5.3 is obtained.

Pure material is recovered from each peak after neutralization and optical rotation is taken to determine the conformation by comparison with authentic samples. Finally, NMR investigation was used to deduce the structure of the complexes and two diastereomeric complexes were proposed:

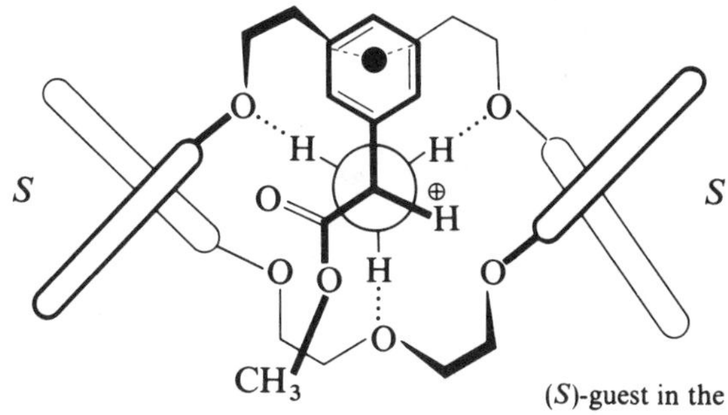

(*S*)-guest in the three-point binding model

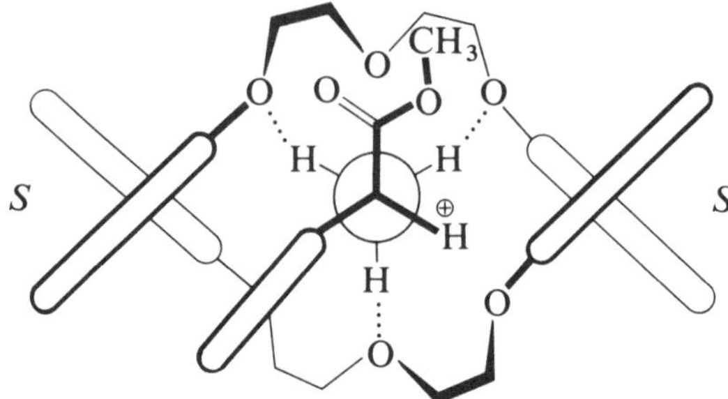

(R)-guest in the
four-point binding model

The terms *three-point binding* and *four-point binding* are names given to represent configurational relationship between host and guest. The names, while imperfect and incomplete, are convenient and useful (278). At the same time, they identify the more stable diastereomer and point to what is probably the main structural feature responsible for the difference in stabilities of the two diastereomers. These Newman projections show that for both models, the guest molecule binds the (S,S)-host via three N—H hydrogen bindings to the ether oxygens of the macroring. The three substituents of the asymmetric carbon (small, medium, and large) are distributed in space to minimize steric effects. The four-point binding model has an extra dipole–dipole interaction with the ester function because of the stacking imposed by the aromatic rings of the guest and the host.

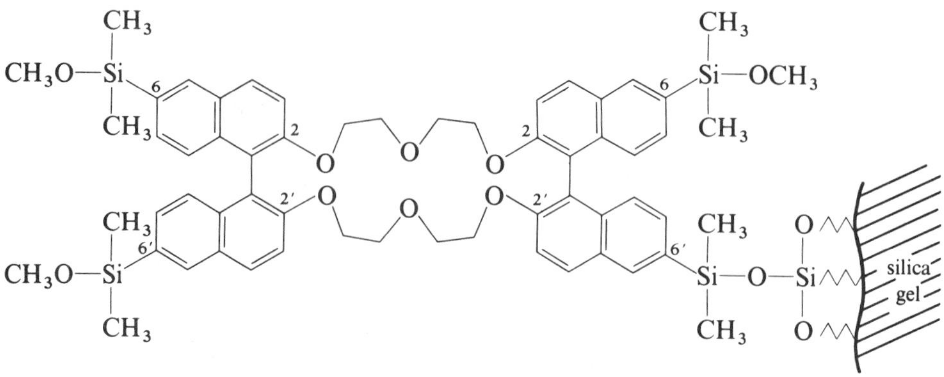

Nevertheless, the three-point binding model is sterically more stable. The reason is that introduction of substituents at 3- and 3′-positions increases the bulkiness of the complex, causing the system to become more selective and favoring the three-point binding model. In other words, as the complex becomes more crowded by increasing the steric requirements of either the host or the guest, the complexation becomes more stereoselective. As a consequence the (S,S)-host has a preferential selection for the *S*-isomer of the guest molecule. Diastereomeric association constant ratios as high as 18 have been reported.

The advantages of a 1,1′-bisbinaphthyl-22-crown-6 ether system can be summarized in three points: (1) The binaphthyl unit is rigid and offers a good chiral barrier. (2) The structure is such that substituents in 2,2′-positions converge, where in 3,3′-positions they diverge. This permits a higher degree of stereoselectivity. (3) A functional group at the 6- or 6′-position will not interfere in the complex formation between guest and host, and allows the possibility for the host to be grafted on a solid support.

The naphthalene ring acts as a "spacer" between the solid-phase surface and the host. In this way, optically pure (*R*,*R*)-host was covalently bound to a silica gel solid support, and the resulting material used to resolve primary amine and amino ester salts by solid–liquid chromatography. This technique could be termed *affinity chromatography* specific for enantiomers.

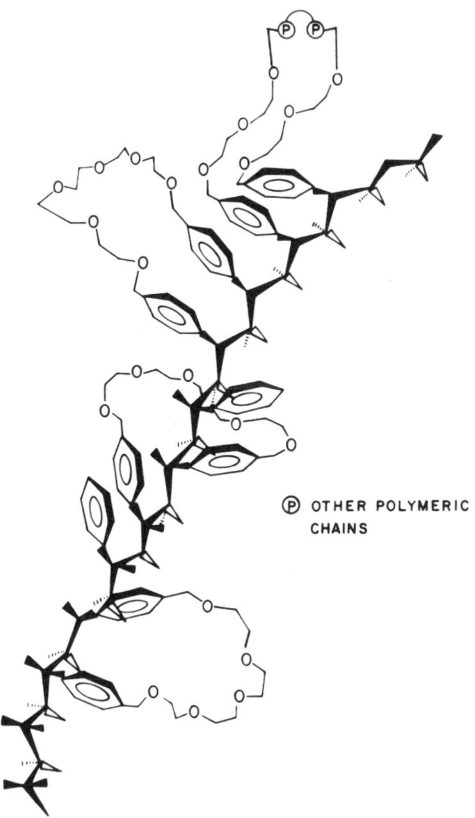

Polymeric pseudo crown ether (281). **Reprinted with permission. Copyright © 1979 by the American Chemical Society.**

This approach has been extended to the synthesis of macrocyclic ethers (pseudocrown ethers) incorporated as part of a macromolecular network (styrene–divinylbenzene copolymer) and results in polymers of high coordinating power for various ions (281). The combination of macrocyclic structures with polymer will allow, in the near future, the development of new catalysts containing specific binding properties together with effective catalytic behavior (269).

Moreover, some 1,1′-binaphthyl macrocyclic ethers solubilize calcium in hydrocarbon solvents by complexation. Because this behavior resembles transport of calcium across hydrocarbonlike environments of cell membranes, such compounds could be important in studying mechanisms of nerve and muscle function.

To learn more about the structural parameters involved in complexation, other studies used a pyridinyl unit to form the following host compounds. The tripyridine host strongly complexes the *tert*-butyl ammonium thiocyanate salt. The mixed-host, *S*,*S*-dipyridinyl-dinaphthyl ether, also complexes *S*-phenylglycinate methyl ester preferentially like the

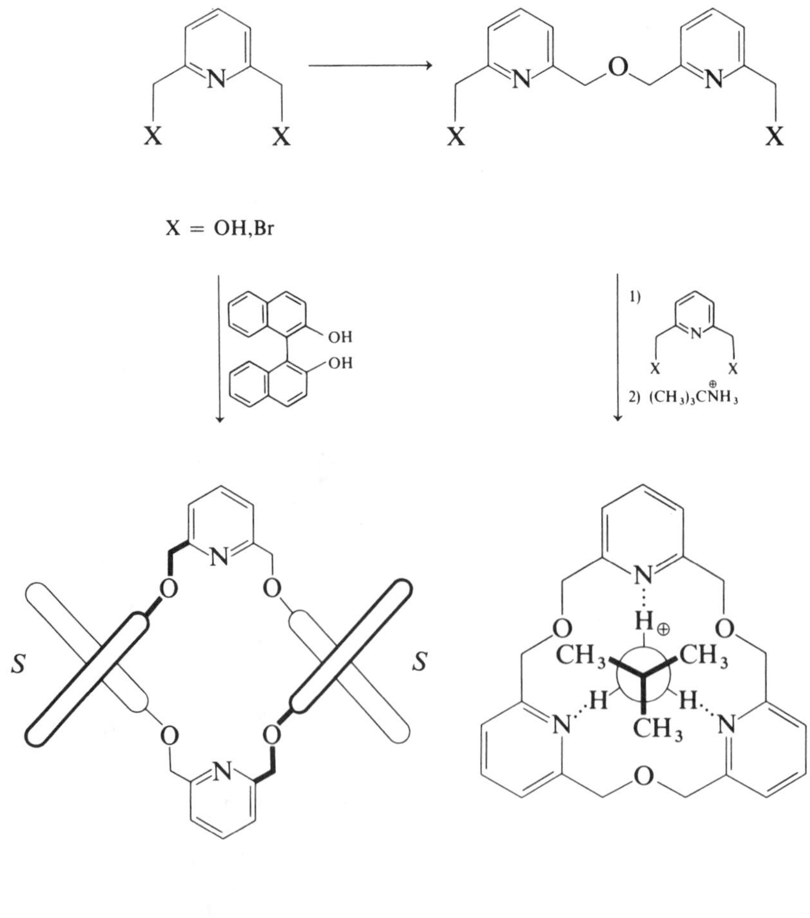

X = OH,Br

1)

2) $(CH_3)_3C\overset{\oplus}{N}H_3$

S S

polyether analog. Again, the organization of three hydrogen bonds and three $N^{\oplus} \cdots O$ pole–dipole interactions appear critical for strong and efficient binding (280).

In 1977, a British group under the direction of J.F. Stoddart started to use sugar derivatives for the synthesis of new host compounds (282,283). Carbohydrates and their derivatives are rich in substituted bismethylendioxy units for incorporation into an 18-crown-6 constitution. They also provide a relatively inexpensive source of chirality and are usually well endowed with functionality.

Starting from L-tartaric acid or D-mannitol, Stoddart's group prepared a series of 18-crown-6 ethers. The D-mannitol precursor permits the

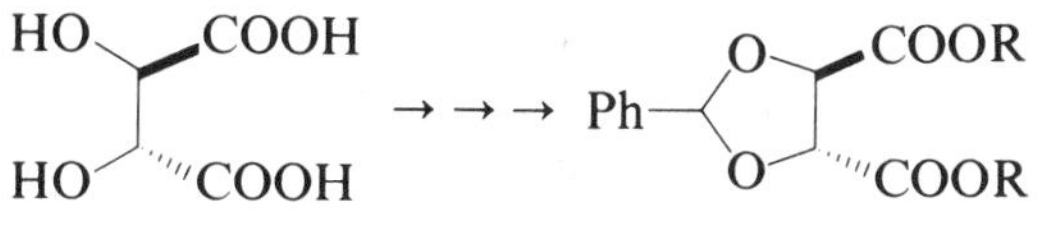

L-tartaric acid

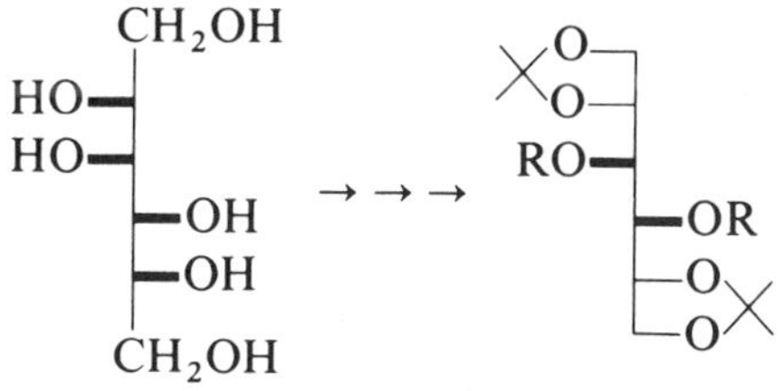

D-mannitol

host precursors

association of bulky substituents more intimately with the crown ether, and at the same time doubles the number of chiral centers from four to eight.

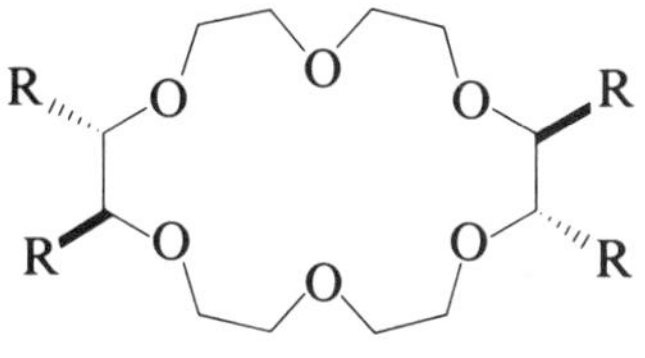

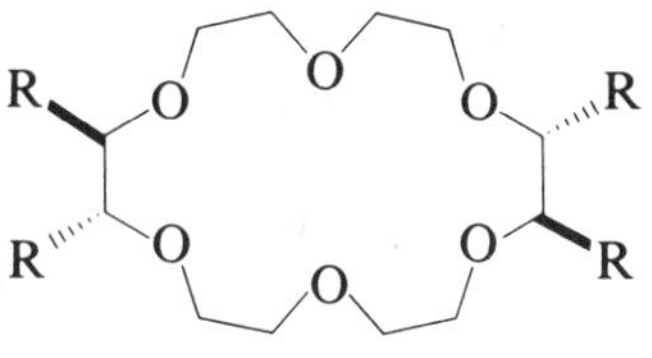

L,L-host from L-tartaric acid
R = CH_2OH
R = CH_2—$OCPh_3$
R = $CONHR_1$

D,D-host from D-mannitol
R =
R =

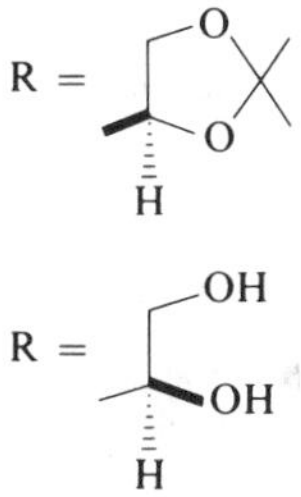

These hosts exhibit enantiomeric differentiation in complexation toward primary alkylammonium salts. Newman projections led to the prediction that the (D,D)-host-(R)-guest complex is more stable. Furthermore, the nature of the anion is important in promoting complexation. *Soft anions** such as SCN^-, ClO_4^-, and PF_6^- favor complex formation, whereas *hard anions* such as OH^-, Cl^-, and Br^- form very stable salts and mitigate against the formation of the host–guest complex.

Such an extension in host–guest complexation chemistry is a good example of a long-term objective to build enzyme analogs by lock and key chemistry with crown compounds. Of course, matching of sizes, shapes, and electronic properties of binding portions of hosts and guests is a necessary requisite to strong binding. In this respect, carbohydrates and carbohydrate derivatives are "nature's gifts to chiral synthesis" because they can be transformed into the structural framework of noncarbohydrate targets (284). This concept will be more and more exploited in the coming years.

A macrocyclic enzyme model system has also been developed in recent years by Murakami's group in Japan (287,288). They found that 11-amino-[20] paracyclophane-10-ol catalyzes the deacylation of *p*-nitrophenylhexadecanoate with a rate 1000-fold greater than 2-aminocyclodecanol.

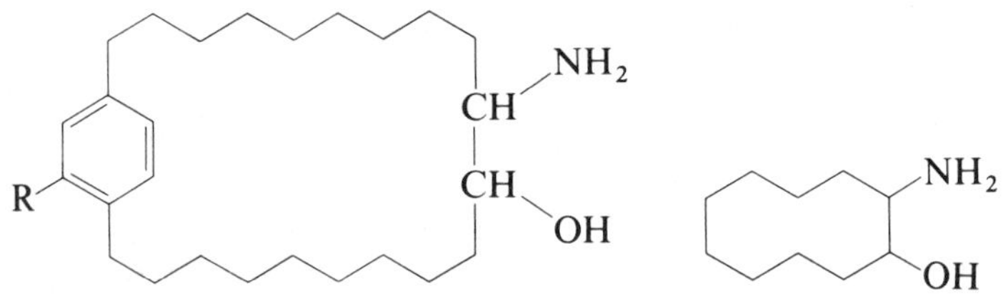

[20] paracyclophane ring

2-aminocyclodecanol

The greater effectiveness of the paracyclophane ring as compared to the cyclodecanol ring suggests that a sufficient hydrophobicity must be provided by the macrocycle where the cyclodecanol molecule has no ability to incorporate the substrate ester in its cavity. Furthermore, the functional group of the macrocycle must be oriented geometrically in favor of a pseudointramolecular reaction with the ester bond of the bound substrate. Substituted paracyclophanes ($R \neq H$) are readily amenable to design, and because of their hydrophobic nature, paracyclophanes remain among the simplest enzyme models. More will be said about these structures in Section 5.2.7.

* According to Pearson's nomenclature hardness is asociated with high electronegativity and small size of the ion. On the other hand, a soft base is one in which the donor atom is of high polarizability and of low electronegativity and is easily oxidized. The general principle of hard and soft acids and bases is that hard acids prefer to associate with hard bases and soft acids prefer to associate with soft bases (285,285).

5.1.1 Macrocyclic and Cryptate Effects

Pedersen's pioneer research on crown ethers proved to chemists that they did not have to rely on compounds from natural sources to achieve selective recognition and complexation of small molecules and ions by larger ones. He was able to show that these macrocyclic crown ethers form complexes with alkali and alkaline earth metals. More important, he observed that by varying the ring sizes, he created a series of crown ethers that can complex preferentially certain of these metals according to their respective ionic radii. From this, chemists have long used these compounds as phase-transfer catalysts for transferring ionic materials from an aqueous to an organic medium and to solubilize ionic reagents in organic solvents. Furthermore, crown ether chemistry has also inspired biologists in their own efforts to understand selective ion transport across phospholipidic cell membranes.

Later, Lehn extended Pedersen's concepts to three dimensions with the design of polycyclic compounds having heteroatoms other than oxygen in the rings. These macrobicyclic structures form cavities that he names *cryptands*; the corresponding complexes are called *cryptates*. These terms are now an established part of organic nomenclature.

Polyethers can chelate to ions. However, the asociation constant for the same ion with a crown ether having the same number of oxygen as the polyether or podand is many orders of magnitude higher.

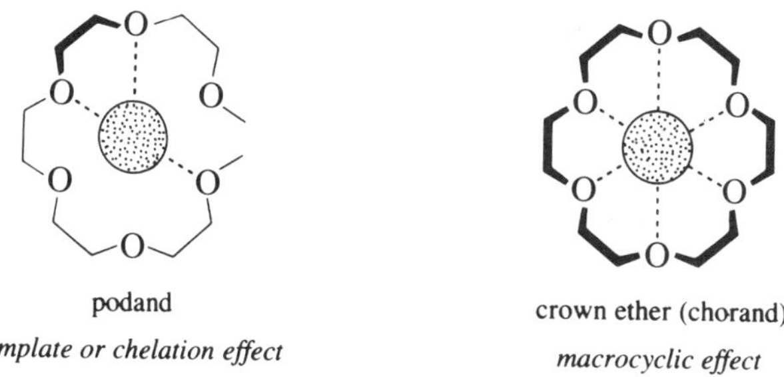

podand
template or chelation effect

crown ether (chorand)
macrocyclic effect

comparison of the association constant
in MeOH: 1:10^4 (for K^+)

The molecular organization found in the crown ether contributes greatly to the stability of the complex. The so-called *macrocyclic effect* is thus directly related to an entropic factor, or more order. A favorable enthalpy change is also associated to the phenomenon of chelation. Now, if we extend this reasoning to an aza-crown ether with a side arm, the following comparison can be made.

Here the effect is even more spectacular. The spheric K^+ ion is symmetrically surrounded by six oxygen atoms, all contributing to the stability of the complex by pole–dipole interactions. This is the *macro-bicyclic effect.* In a way, it is the oxygens of the cryptand (see p. 273) that

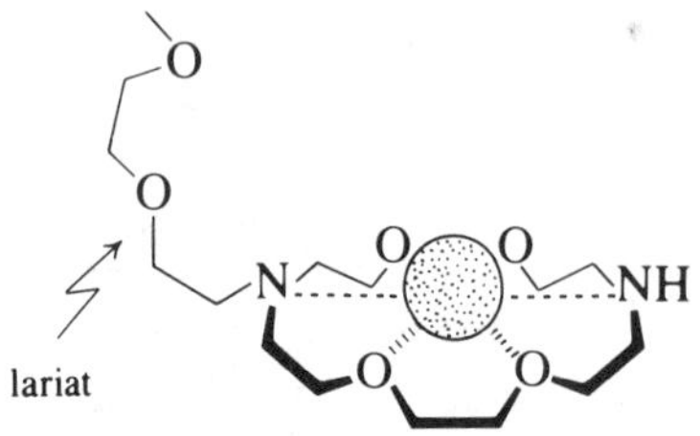

lariat

aza-crown ether

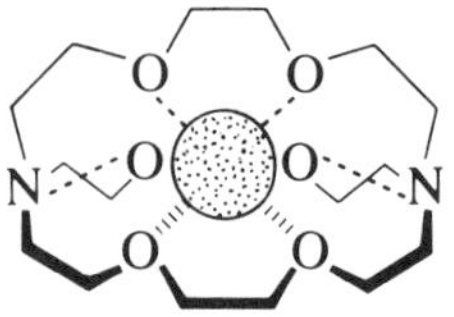

cryptand

comparison of the association constant
in MeOH: 1:10^5 (for K^+)

solvates the ion. This cryptand can equally well bind NH_4^+ ions. The interactions are, however, different than those with K^+ ions, since an ammonium ion is tetrahedral and hydrogen bonds are implicated.

By going one step further, one could compare the affinity of bicyclic and tricyclic compounds of same size for ammonium and potassium ions.

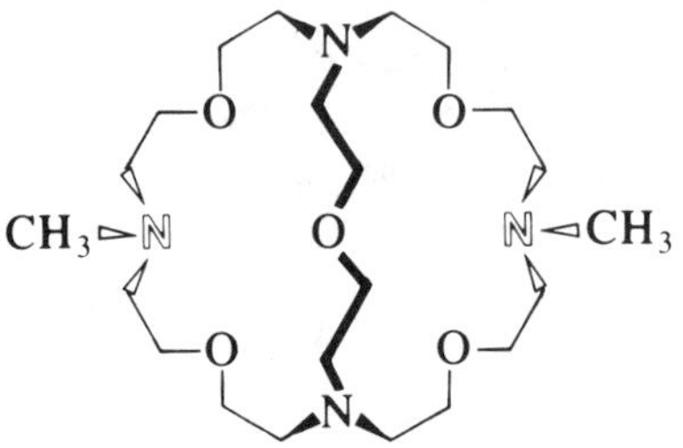

ratio NH_4^+/K^+ affinity in water

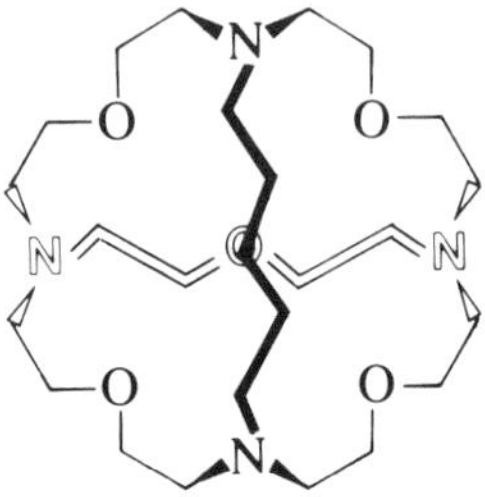

2.5 — cryptand (bicyclic)

more organized →

63 — (tricyclic) hydrophobic chain "carrier"

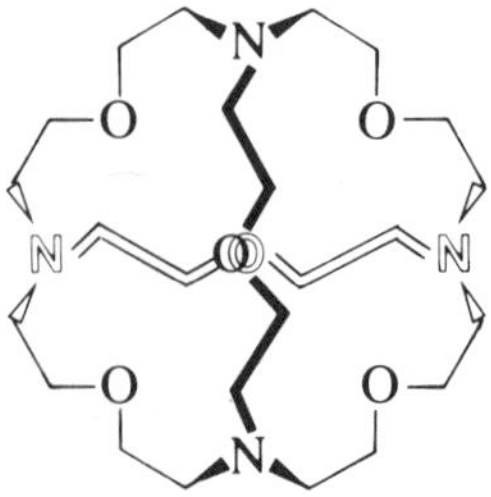

500 — (tricyclic) hydrophilic chain "receptor"

As the molecule becomes more structurally organized and symmetrical, the selectivity of NH_4^+ over K^+ is dramatically higher. More spectacular is that having a tricyclic molecular cage with more hydrophilic contour stabilizes considerably the complex to the point that the molecule can be considered a good receptor for NH_4^+ ions. On the analogous system with a hydrocarbon chain, the ion is not as tightly bound in the interior of the cryptand and therefore can dissociate more easily. This ligand thus has the character of a "carrier" rather than a "receptor." This situation illustrates well the duality in designing specific ion receptors or carriers.

These aza-tricyclic crown ethers, with their nitrogen atoms pointing inward in the cavity, can be protonated in acidic media. As a result, the corresponding cryptate tetraammonium, which is positively charged, can now bind chlorine ions. Interestingly, a receptor for cations has been transformed to a receptor for anions. This has been proven by ^{35}Cl-NMR studies (289). The facility to recognize and bind anions is a general

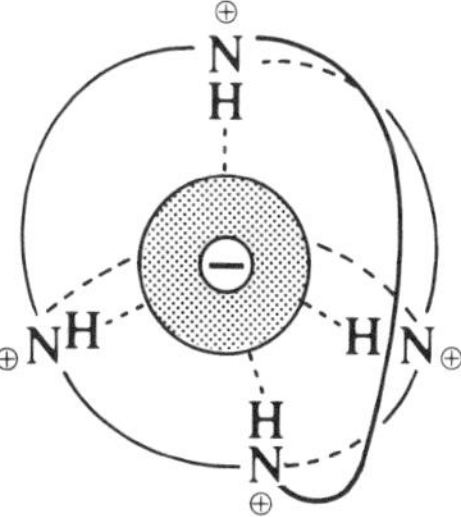

tetrahedral recognition by a macrotricyclic cryptand illustrates the molecular engineering required in abiotic receptor chemistry

property of protonated cryptands. For instance, the following elongated cryptand accommodates the linear azide anion.

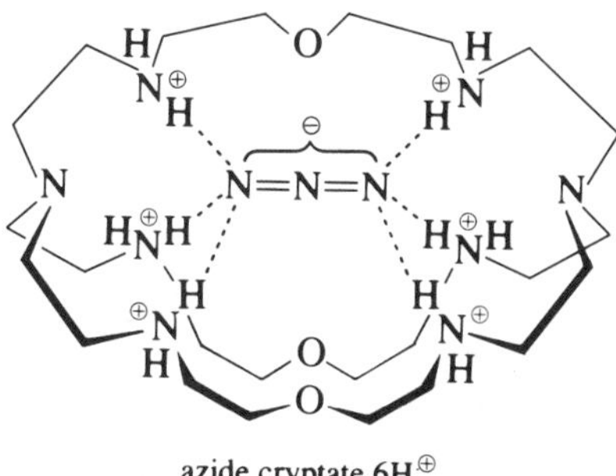

azide cryptate $6H^{\oplus}$

5.1.2 Chiral Recognition and Catalysis

The preceding section demonstrated that chiral macrocyclic polyether hosts discriminate in complexation reactions in chloroform solution between enantiomers of amino ester salt guests. With these results can we go one step further and mimic a catalytic site? We will now describe the design of a host that upon complexation with α-amino ester salts produces a transition state intermediate corresponding to a transacylation (thiolysis) reaction between the chiral host catalytic group (thiol) and the enantiomeric guest salts (290). However, it should immediately be realized that these model systems mimic only the acylation step encountered in serine protease catalysis. So far, no acceleration in rate has been observed for the deacylation step.

The following cyclic chiral host has been prepared, and the properties for complexation have been compared to the corresponding open chain system.

The hosts are used to study the hydrolysis of L- and D-amino acid *p*-nitrophenyl esters. The reaction is carried out in 20% EtOH-CH_2Cl_2 buffered with 0.2 *M* AcOH and 0.17 *M* NaOAc at pH 4.8. A burst of *p*-nitrophenol is followed by ethyl ester formation.

SH SH SH HS SH HS S O O O O O O O O O O O O O O O O O O CH_3 CH_3

R—CH(COO—C_6H_4—NO_2)($\overset{\oplus}{N}H_3$) $\xrightarrow{\text{EtOH}}$ R—CH(COOEt)($\overset{\oplus}{N}H_3$) + HO—C_6H_4—NO_2

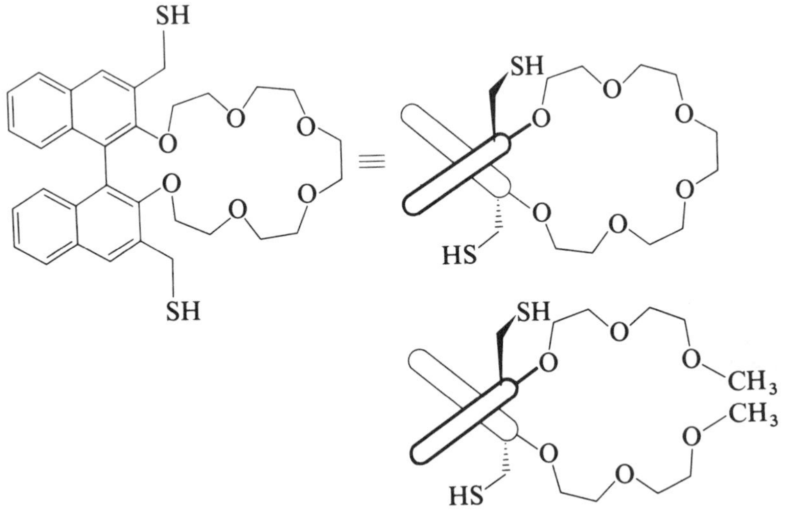

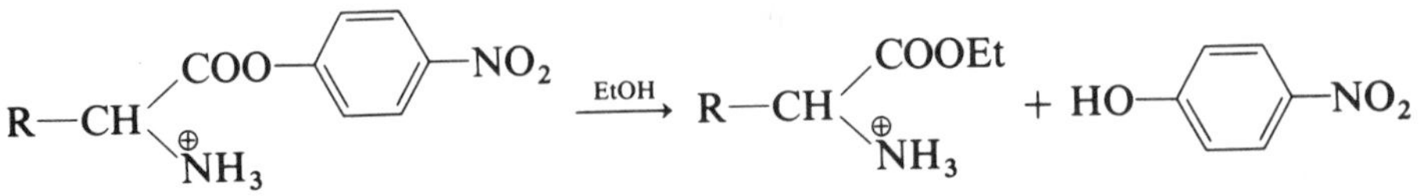

It was observed that the L-amino acid esters react 10^2 to 10^3 times faster with the cyclic host than with the open host analog. Clearly, enforced convergence of binding sites enhances complexation by substantial factors. Second, proline esters react at equal rate with both hosts. Hence one needs, as shown before, three protons on the α-nitrogen atom for efficient complexation. In a more polar solvent (40% H_2O-CH_3CN) the rate decreases by a factor of ten. This indicates that the water molecules hydrogen bond to the α-NH_3^+ group and thus compete with the host. In all cases, the *S*-host reacts faster than the *R*-host for the natural L-amino acid. The ratio of reactivity depends on the size of the substituent (R) on the glycine moiety.

R	*S*-host/*R*-host
$CH_3-CH(CH_3)-$	9.2
$Ph-CH_2-$	8.2
$(CH_3)_2CHCH_2-$	6.0
CH_3-	1.0

These reactions resemble a transacylation where the designed host has some of the properties of trypsin recognition of an NH_3^+ group and papain (a cysteine residue at the active site).

The following structures show the "transition state"-like relationships between *S*-host and either L- or D-amino ester guest (290).

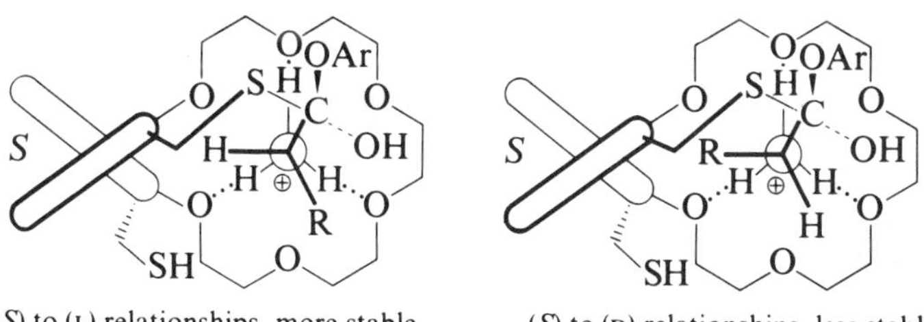

(*S*) to (L) relationships, more stable (*S*) to (D) relationships, less stable

It is rather easy to understand now why the *S*-isomer was preferentially selected and hydrolyzed. The side chain R being away from the chiral barrier minimizes the steric factors and leads to the formation of a more stable intermediate.

In the same line of thought, Lehn and Sirlin (291) have prepared a chiral macrocyclic molecular catalyst bearing cysteinyl residues. The catalyst complexes primary ammonium salts and displays enhanced rates of intramolecular thiolysis of the bound substrates with structural selectivity for dipeptide esters and high chiral recognition for the L-enantiomer (70 times faster) of a racemic mixture of glycylphenylalanine *p*-nitrophenyl esters. The representation below shows the complex between the chiral crown ether and the dipeptide glycylglycine *p*-nitrophenyl ester salt.

The rate accelerations observed (10^3–10^4 times) are due to complexation of the primary ammonium salt in the crown ether cavity and the participation of an –SH group of the cysteinyl residues to give an *S*-acyl intermediate. This "artificial enzyme" model displays molecular complexation, rate acceleration, and structural and chiral discrimination analogous to true biological catalysts.

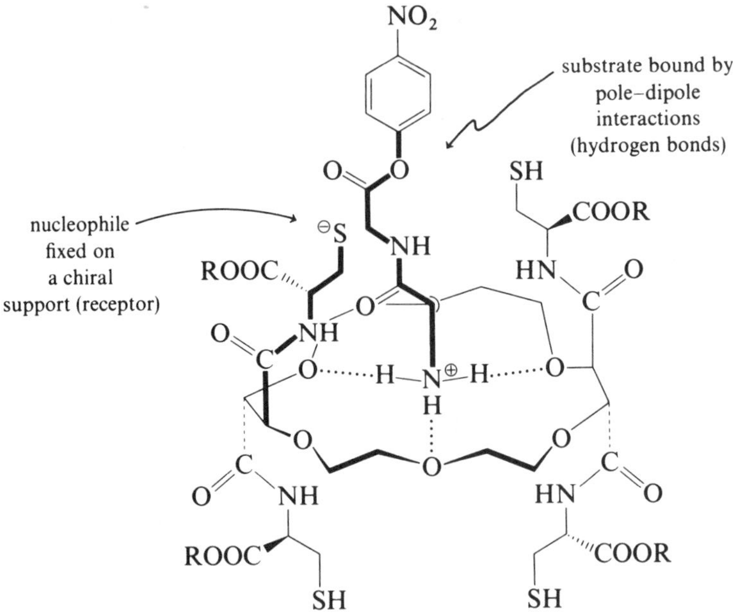

A more detailed description of these studies as well as additional data have recently appeared (292). From this, a global picture has emerged where this supermolecular catalytic process is composed of four distinct steps. They are presented on page 271.

5.1.3 Stereoselective Transport

Since the discovery of crown ethers it was predicted that chiral carriers would provide a possibility for chiral specificity of guest transport through liquid membranes. This prediction was confirmed in 1974 when Cram and his colleagues found that *enantiomeric differentiation* occurs when designed, neutral, lipophilic, and chiral host compounds carry amino ester salts (guest compounds) from an aqueous solution through bulk chloroform to a second aqueous solution (278).

A binaphthyl dissymmetric (*R*,*R*)-host and racemic phenylglycinate methyl ester were used as guest molecules:

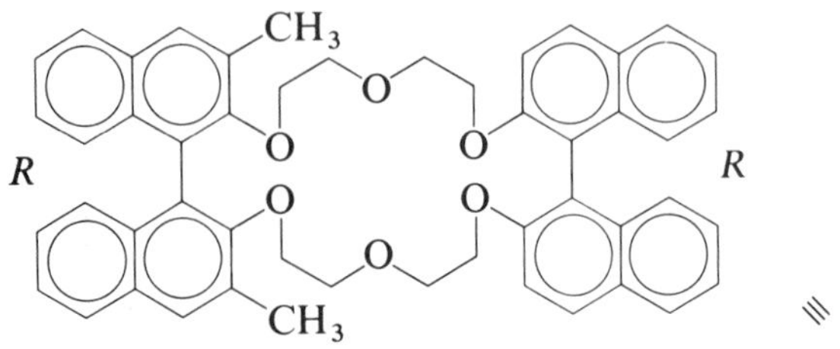

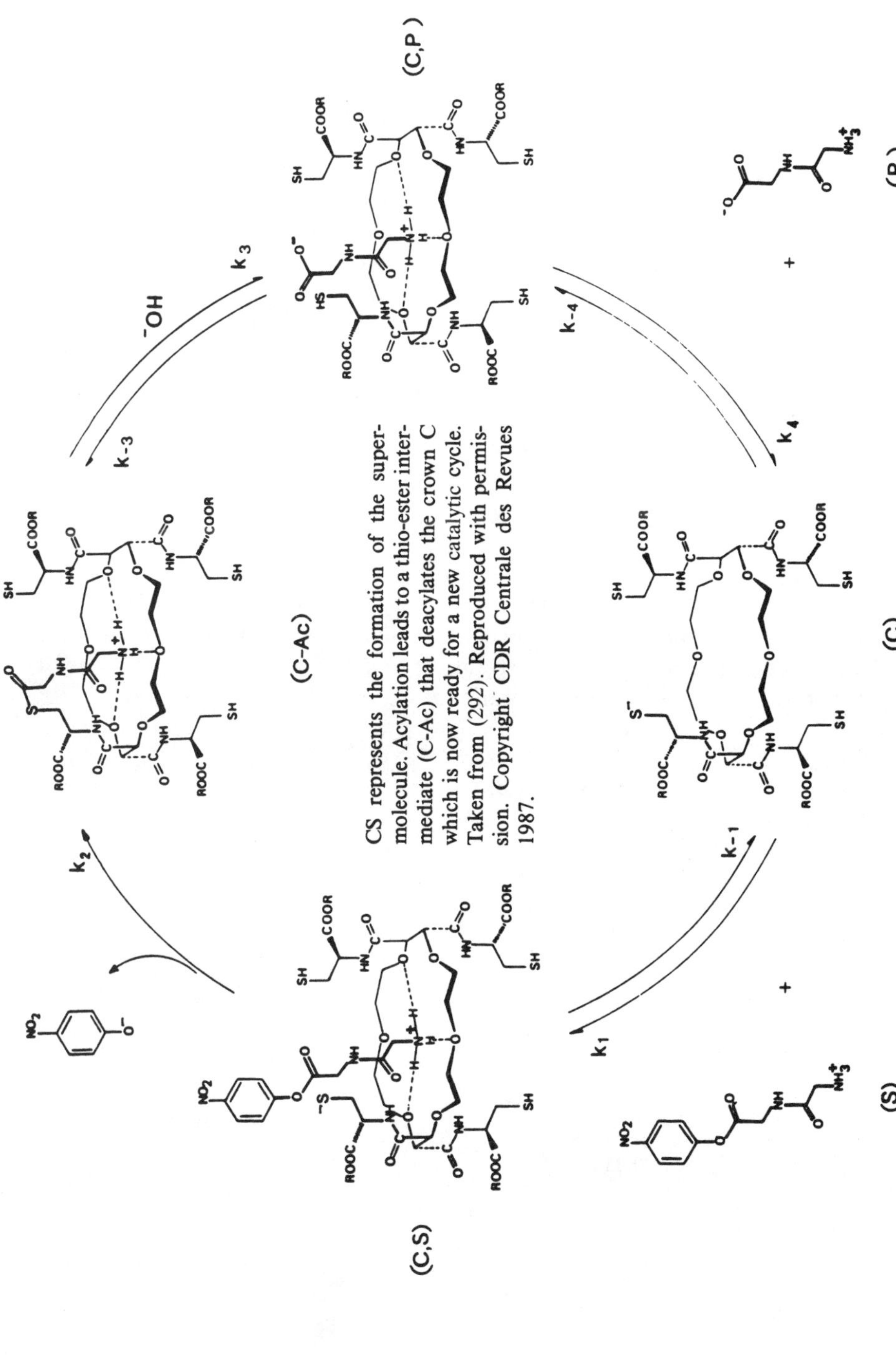

CS represents the formation of the supermolecule. Acylation leads to a thio-ester intermediate (C-Ac) that deacylates the crown C which is now ready for a new catalytic cycle. Taken from (292). Reproduced with permission. Copyright CDR Centrale des Revues 1987.

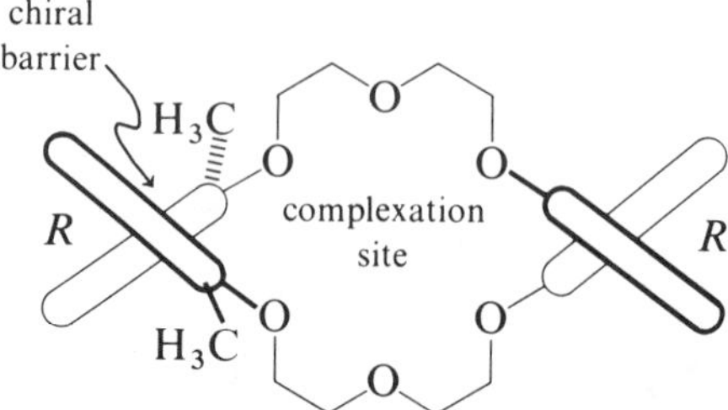

The cell used was a U-shaped tube filled as shown in Fig. 5.4.

A chloroform solution of the host is placed at the bottom of the tube. The guest is present in the α-arm. Ultraviolet absorbance and specific rotation can be measured at different times in the β-arm. In this way, the rate constants for transport were measured for the fast moving enantiomer and the slow moving enantiomer. After 12–19 hr the *R*-isomer was selectively "pulled" in the β-arm. It has an optical purity of about 80%. The entropy of dilution and the changes in solvation energies associated with inorganic salt, "salting out" the organic salt from its original organic solution, provided the thermodynamic driving force for transport.

As mentioned in Section 5.1.1, by adding an additional bridge to a crown host molecule, a third dimension is added to the host, and J.M. Lehn from Strasbourg called these molecules *cryptands* (293). These macropolycycles contain an intramolecular cavity (or crypt) and have recently been used by Lehn's group for selective transport of alkali metal cations through a liquid membrane (294).

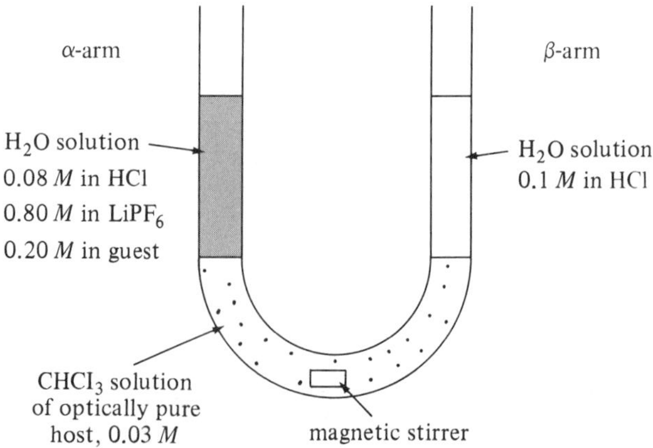

Fig. 5.4. Cell for chiral recognition in transport (278). Reproduced with permission from *Techniques of Chemistry* (1976), A. Weissberger, ed., Vol. X, part II, Wiley-Interscience, New York.

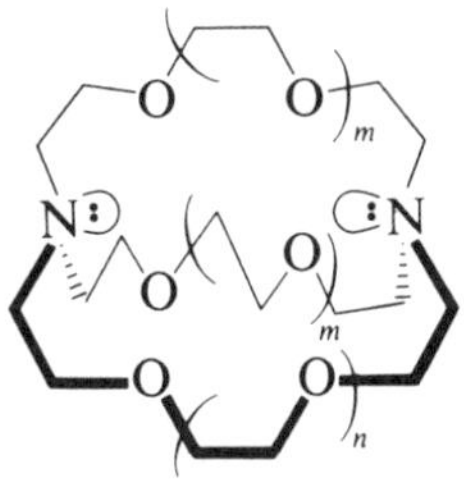

5-1 $m = n = 1$ [2.2.2]
5-2 $m = 1, n = 2$ [3.2.2]
5-3 $m = n = 2$ [3.3.3]
5-4 $m = 1, n = \text{—}(CH_2)_8\text{—}$ [2.2.C$_8$]

These cryptands form *cryptate*-type inclusion complexes with Na^+, K^+, or Cs^+ picrates. The cryptands function as cation carriers by dissolving the alkali metal picrate into a bulk liquid chloroform membrane as a 1:1 cryptate–picrate ion pair and releasing it from an *in* to an *out* aqueous phase interface (293). Comparison of properties shows, for instance, that **5-4** carries Na^+ and K^+ much faster than **5-1**. This means that by removal of two oxygen-binding sites, the cryptand is transformed from a specific K^+ *receptor* (**5-1**) into a specific K^+ *carrier* (**5-4**). The work of Lehn on cryptates allowed the design of ligands that can be either a cation receptor or carrier depending on their structure. For instance, **5-1** has the transport sequence $K^+ < Na^+ \ll Cs^+$ opposite to the stability sequence of the complexes $Cs^+ \ll Na^+ < K^+$. Similarly, **5-2** has opposite complexation $Na^+ < K^+, Cs^+$ and transport $Cs^+, K^+ < Na^+$ selectivities, whereas **5-3** has $Na^+ < K^+, Cs^+$ for both complexation and transport selectivities. The origin of these differences in receptor and carrier behavior lies in the extent of carrier saturation; the most stable cryptates, like **5-1**, have low cation dissociation rates. Thus, the transformation of a cation receptor into a cation carrier may be achieved by simple structural changes and may be accomplished simply by replacing one or two oxygen-binding sites by nonbinding CH_2 groups.

This new coordination chemistry can be extended to anions and other organic ligands. The cryptands have the interesting property of showing a preference for binding one ion over others. At will, the cavities can be constructed either spherical or cylindrical in shape. Such a vast domain of macrocyclic and macropolycyclic metal cation complexes thus has a future in the construction of specific receptors for ions and molecular recognition and the design of molecular catalysts and selective carriers. It allows the design of minimolecules (miniature catalysts) that embody properties of much larger proteins. This chemistry is developed further in Section 5.2.

These examples all represent *passive* but stereoselective transport where asymmetric recognition can be achieved with organic model systems. However, there is an analogy to be made with these results and the process of mediated transport across biological membranes. All lipidic membranes are practically impermeable to intracellular proteins and to highly charged organic and inorganic ions that surround either side of the membrane. The diffusion of Na^+ out of a cellular membrane and K^+ into the cell occurs along a negative gradient of chemical potential and is called *passive transport*. The passive transport of ions through membranes may be prompted by ionophores (defined in the next section). Fortunately, the concentrations of cations on either side of the membrane are different and the situation is maintained by *active transport*, which is dependent on metabolic energy. The mechanism for this process is termed the *sodium pump* which functions to keep intracellular K^+ concentrations high and Na^+ concentrations low. Calcium is also thought to be actively pumped out of the cells. For these cases, the energy for the transport is provided by the hydrolysis of ATP. Diffusion of sugars and amino acids to strategic targets in the cell, however, is an example of simple facilitated passive transport.

5.1.4 The Ionophores

It is logical at this point to mention the role and importance of some of nature's chelating agents, the *ionophores*. Figure 5.5 gives representative examples. They contain polyamide, polyester, and polyether functions and most are cyclic. They have the capacity to selectively bind metal ions and act as carriers across membranes (295).

An ion (cation) is too hydrophilic to efficiently cross a thick (~10 nm) hydrophobic layer of lipids and lipoproteins such as those found in natural or artificial membranes. However, by binding selectively with the polar

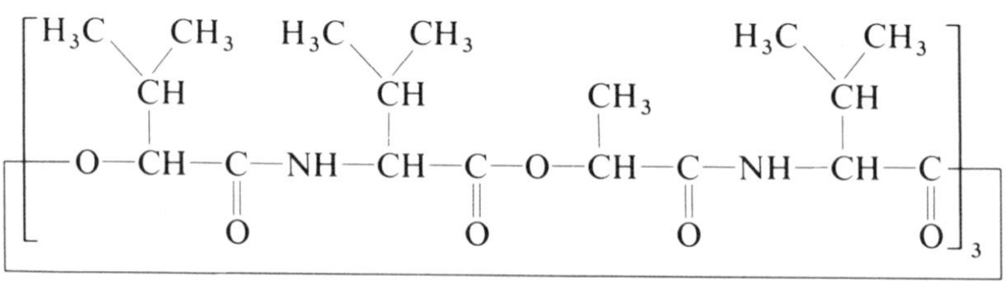

valinomycin

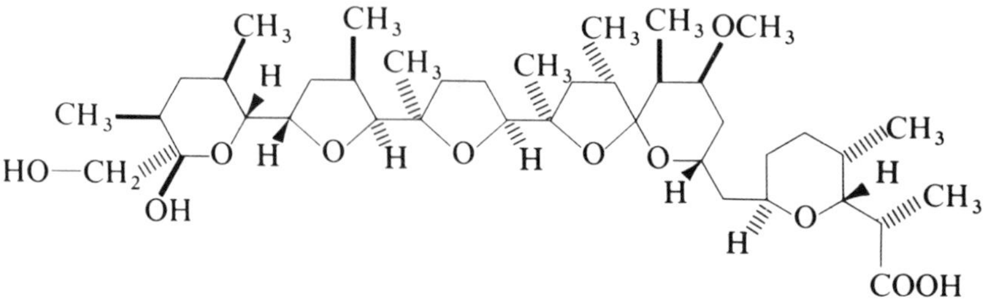

nigericin

Fig. 5.5. Structures of some ionophores.

Pro → Ala → Phe → Phe → Pro
↑ ↓
Pro ← Val ← Phe ← Phe ← Pro

antamanide

a cyclic peptide that neutralizes the effect of phalline B, a toxin (also a cyclic peptide) present in the deadly mushroom *Amanita phalloides*

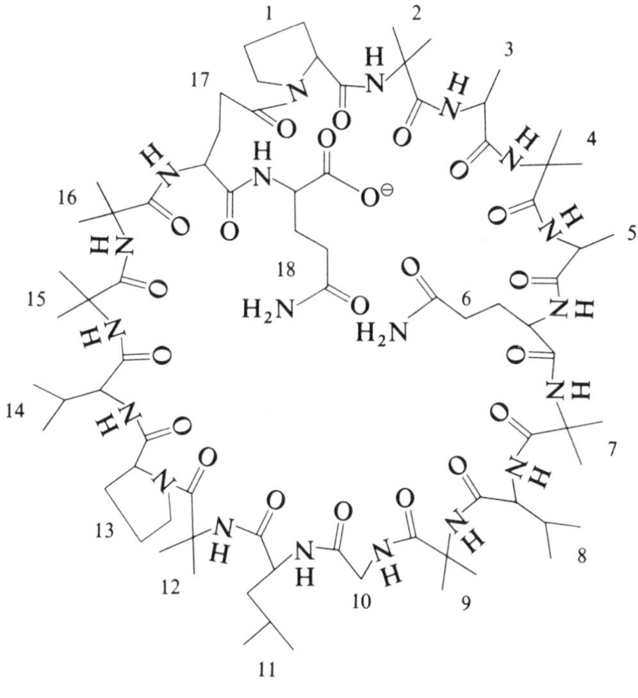

alamethicin

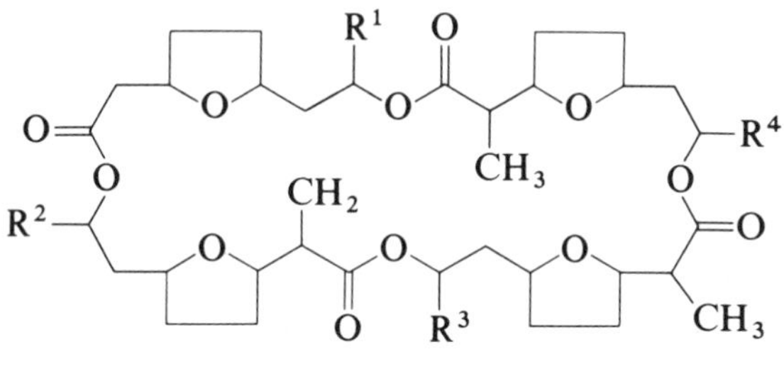

nonactin $R^1=R^2=R^3=R^4=CH_3$
monactin $R^1=R^2=R^3=CH_3$, $R^4=C_2H_5$
dinactin $R^1=R^3=CH_3$, $R^2=R^4=C_2H_5$

Fig. 5.5. (continued)

functions located inside the macrocyclic ring, the cation is now coated with a hydrophobic shell and can then more readily pass across the membrane.

The naturally occurring ionophores are found mainly in microorganisms and many are used as antibiotics. By complexing with metal ions, they apparently disrupt the control of ion permeability in bacterial membranes. An example is nonactin, an antibiotic that functions by transporting sodium ions into the bacteria until the resulting osmotic pressure causes rupture of the cell wall. Valinomycin coordinates selectively K^+ ions where the permeability of the membrane for potassium is offset. In contrast, antamanide, a decapeptide containing only L-amino acids, has a binding cavity of a different geometry than valinomycin and shows a strong preference for Na^+ ions over K^+ ions. Nigericin also invokes the opposite effect induced by valinomycin and its analogs, and reversible ring closure occurs upon binding to metal ions. Besides carrying ions, ionophores are probably associated with other inhibition processes in the cell such as the control of hormone transport across membranes, regulation of certain metabolisms, and the permeation of neurotransmitters in nerve cells.

It is interesting to realize that some ionophores contain amino acids of the D-configuration. Such a situation is not found in higher organisms.

Stimulated by the discovery of natural ionophores in the 1960s, chemists have successfully synthesized a number of compounds, composed of natural building blocks, that can complex inorganic and organic ions. One example will be presented here. In 1974, E.R. Blout (296) from Harvard University, reported the enantiomeric differentiation between D- and L-amino acid salts in complexes with cyclo-(L-Pro-Gly)$_n$ peptides (n = 3,4) (Fig. 5.6). In chloroform, (±)-Pro-OBz · HCl, Phe-OMe · HCl, or

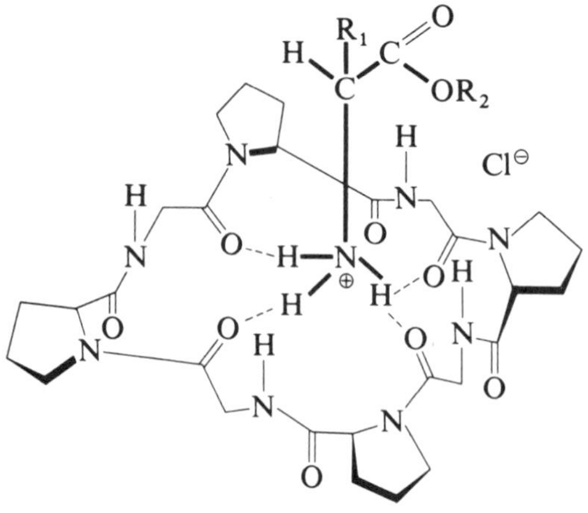

Fig. 5.6. Complex of cyclo-(L-Pro-Gly)$_4$ with an amino acid ester (296). Reprinted with permission. Copyright © 1974 by the American Chemical Society.

Val-OMe · HCl were mixed with the cyclopeptide and the complexes analyzed by ^{13}C-NMR. Diastereomeric pairs of complexes were formed and distinct resonances were observed and attributed to different orientational effects upon complexing D- or L-amino acids.

The binding scheme shows that the four carbonyl functions of the glycine residues are oriented toward the interior of the ring and allow hydrogen bonding with the α-$\overset{+}{N}H_3$ group of the guest molecule in the cavity. Blout's compounds, although structurally resembling ionophores, are closely related to Cram's compounds in function.

In leaving this subject we note that, in the near future, systems like those of Cram and Blout will find application in the synthesis of dissymmetric molecules by an asymmetric induction of prochiral molecules.

5.2 New Developments in Crown Ether Chemistry

It is certainly in the field of crown ether chemistry that the largest number of new receptor models has emerged in recent years. These macrocycles offer the advantage that they are amenable to various chemical and stereochemical modifications.

The most frequently used crown ethers are the following three:

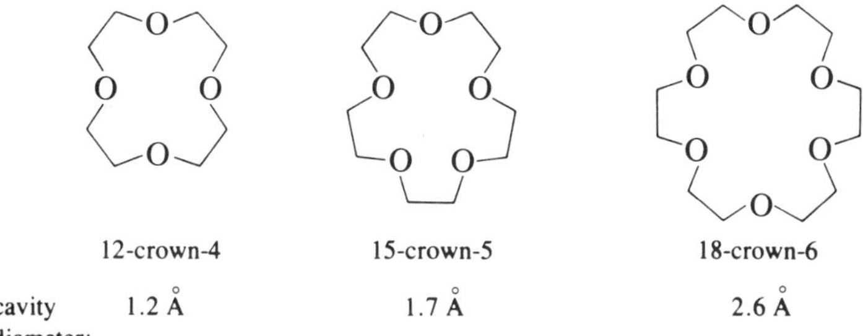

Each crown ether allows the binding of ions having comparable ionic diameter. Most of the ions listed below are found in biological transformations.

	diameter		diameter
Li^+	1.36 Å	Ca^{2+}	1.98 Å
Na^+	1.96 Å	Mg^{2+}	1.38 Å
K^+	2.66 Å	Rb^+	2.98 Å
NH_4^+	2.66 Å	Cs^+	3.30 Å

The "gripping power" of crown ethers for ions of complementary sizes and the development of chiral crown ethers with well-defined functional groups on the macrocyclic ring open the way to the construction of an impressive number of new receptors, based on the molecular recognition concept (297). In this section we will be looking in more detail at some of them.

As mentioned earlier, tartaric acid could be used to introduce chirality in crown ethers. By this approach, Lehn's group was able to prepare either *syn*- or *anti*-disubstituted crown ethers from the corresponding tetra-acid 18-crown-6 (298). The opening of the dianhydride by an amine gives a mixture of *syn*- and *anti*-isomers. However, in the presence of triethylamine, only the *syn*-product is formed. The reason for this selectivity is that in the transition state of the first intermediate, the formation of an amine salt on one face of the macrocyclic ring forces the nucleophile to approach almost exclusively from the other face. The formation of an ionic intermediate also helps the polarization of the carbonyl group on the second anhydride function.

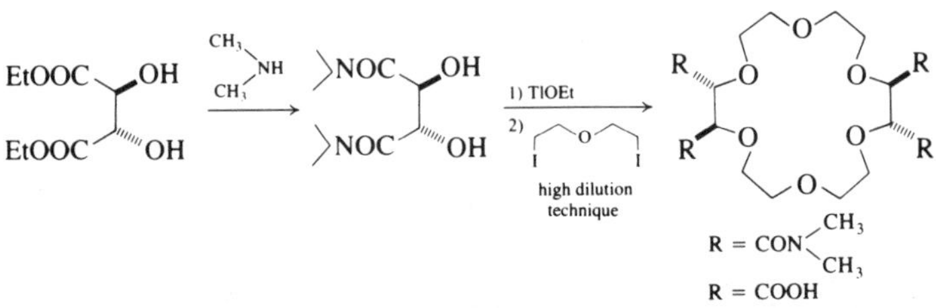

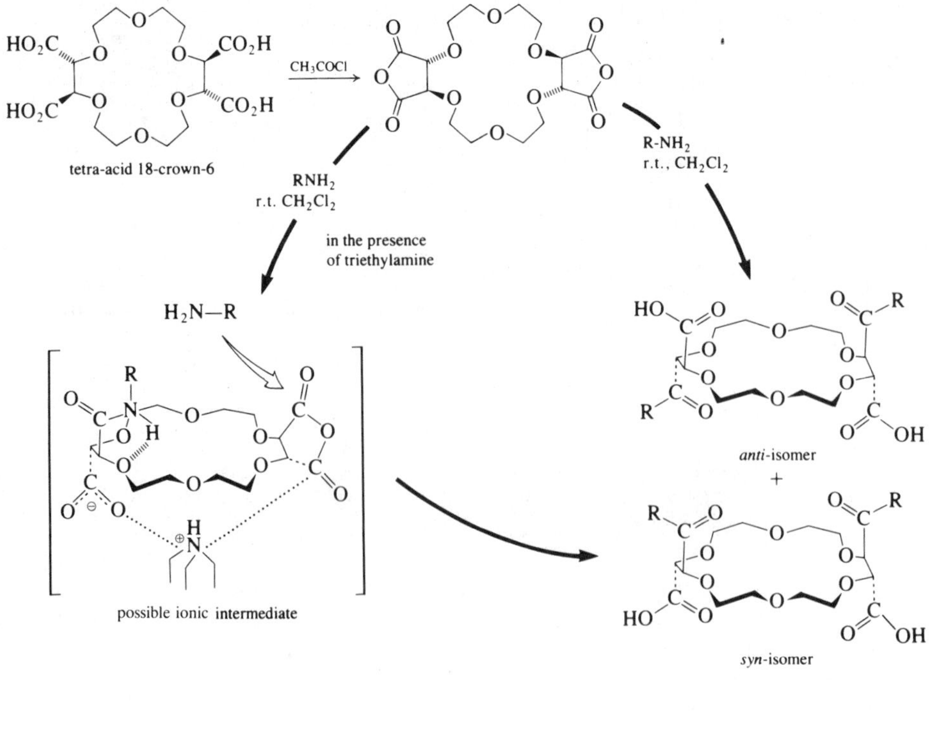

The synthesis of these molecules was a very important step for the elaboration of other face-discriminated and side-discriminated macrocyclic polyethers.

For example, direct bridging of the dianhydride was obtained by treatment with a diamine (299). These novel macrobicyclic polyfunctional cryptands of defined stereochemistry could have diverse applications in the design and synthesis of other molecular receptors, catalysts, and carriers.

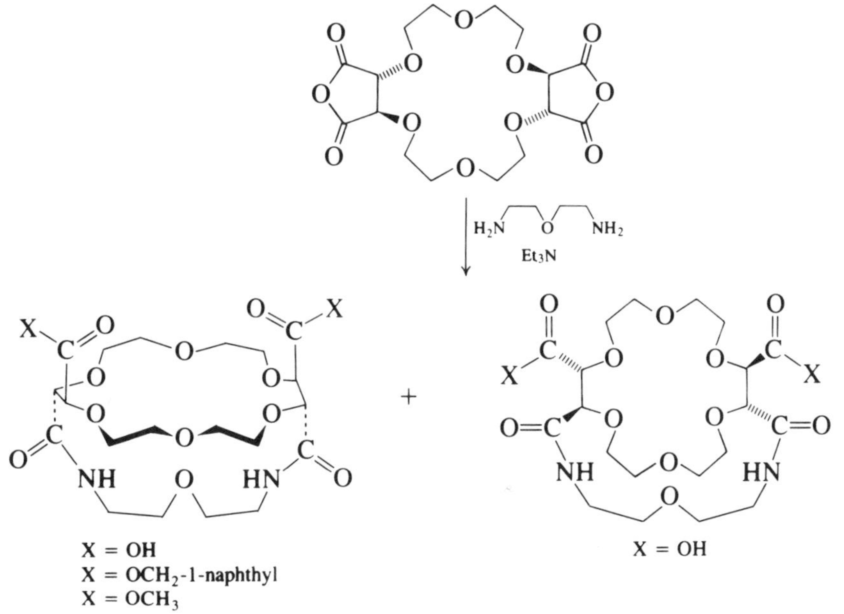

Et_3N

X = OH
X = OCH_2-1-naphthyl
X = OCH_3

X = OH

This is also illustrated by the recent synthesis of a macropentacyclic ditopic receptor molecule (300).

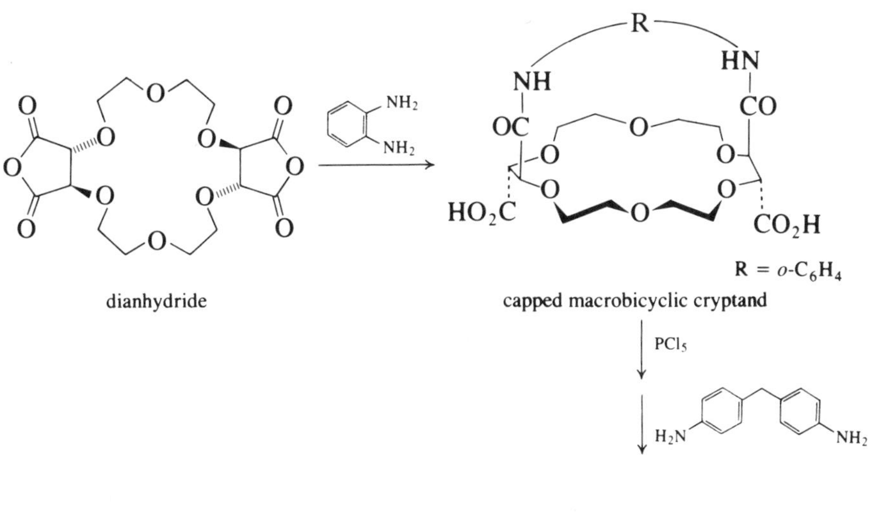

R = o-C_6H_4

dianhydride

capped macrobicyclic cryptand

PCl_5

R = o-C_6H_4
Ar = $CH_2(p\text{-}C_6H_4NHCO)_2$

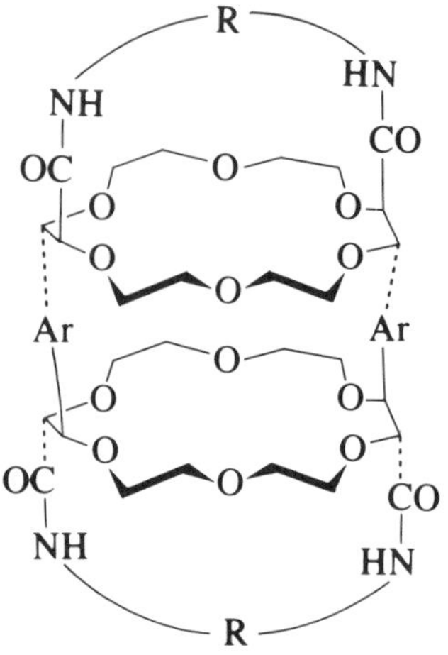

This coreceptor molecule* contains two or more binding subunits and may complex either several singly bound substrates or a multiply bound polyfunctional substrate. Characteristic ^1H-NMR shifts indicate that a strong 1:1 association occurs with terminal diammonium ions of appropriate length such as $H_3\overset{+}{N}-CH_2CH_2-C_6H_4-CH_2CH_2-\overset{+}{N}H_3$, giving slow exchange spectra.

5.2.1 Aza-Crown Ethers and Macrocyclic Polyamines

As shown above, when two binding subunits are located at the poles of a coreceptor molecule, the complexation of a difunctional substrate will depend on the complementarity and the distance between the two binding sites in the receptor and the distance between the two corresponding functional groups in the substrate (9). Such linear recognition by ditopic coreceptors has been achieved for both dicationic and dianionic species. The following cylindrical macrotricycle is a good example of a bis-azacrown ether capable of binding$-\overset{+}{N}H_3$ groups, to yield a molecular cryptate with diammonium ions. With a biphenyl spacer, the length of the molecular cavity can selectively bind to cadaverine over putrecine ($\overset{+}{N}H_3-CH_2CH_2CH_2CH_2-\overset{+}{N}H_3$). Relaxation times ($T_1$) of the CH_2 groups of the diamine substrate were determined by a ^{13}C-NMR study and the values were characteristic of bound substrate (301). Thus, complementarity between components of this supramolecular species expresses itself in both structural and dynamic properties.

---

* Coreceptors are defined as polytopic receptor molecules combining two or more discrete binding subunits within the same macropolycyclic architecture.

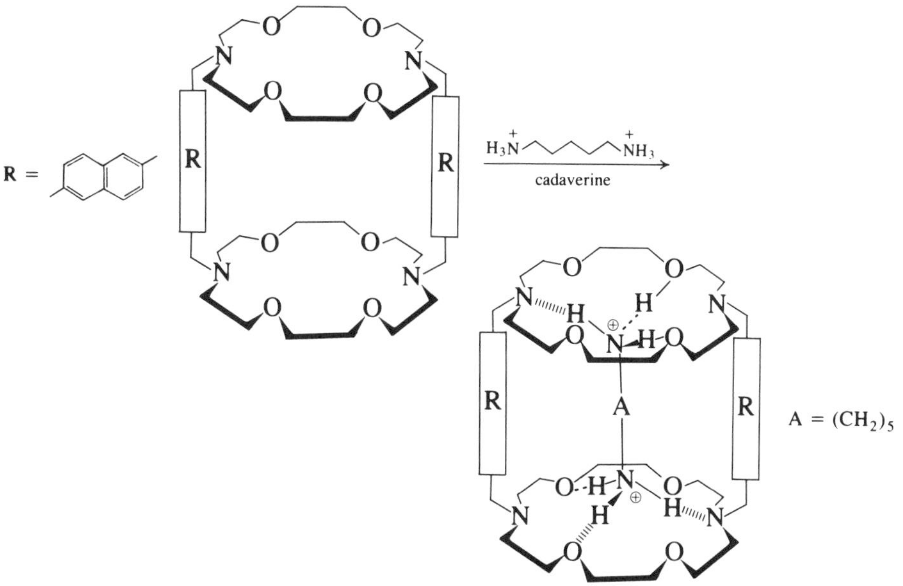

R =

R R

$H_3\overset{+}{N}$ $\overset{+}{N}H_3$

cadaverine

A = $(CH_2)_5$

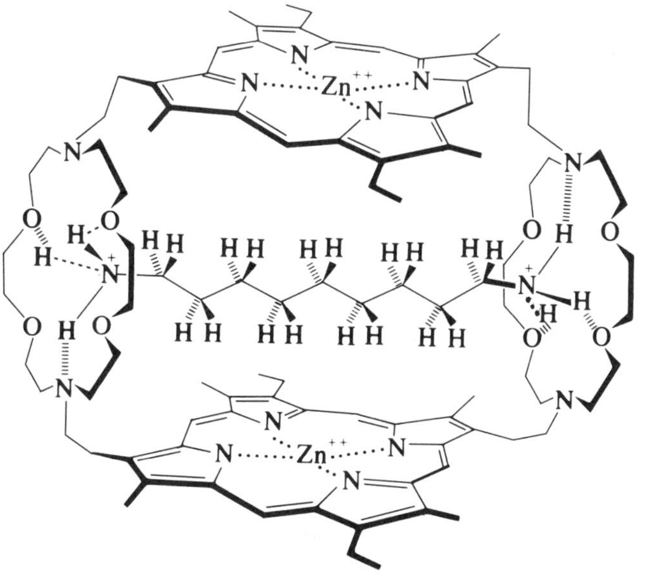

In the same series, a substance having porphyrin residues instead of biphenyl groups has been obtained (8,302). This heterotopic metallore-ceptor contains substrate-selective binding subunits for the complexation of both metal ions (Zn^{2+} in this case) and organic species (a C_9 diammonium) within the same superstructure. This mixed-substrate supermolecule can form a complex with several metal ions to yield a polynuclear cryptate. Many variations are conceivable involving the subunits as well as the overall macropolycyclic architecture. It could also mimic essential features of metalloenzymes. For instance, activation of the internally bound organic substrate by the metal-porphyrin sites may be envisioned. Also, by binding an effector species, the macrocyclic unit could act as regulation sites for external interaction with the metal-porphyrin centers, to the point of exerting allosteric control and cooperativity.

The combination of polar binding subunits with more or less rigid, apolar shaping components provides amphiphilic, macropolycyclic coreceptors of cryptand type, termed *speleands*, that form *speleates* by substrate inclusion (8). The following speleand combines two tartaric acid units with two diphenylmethane groups.

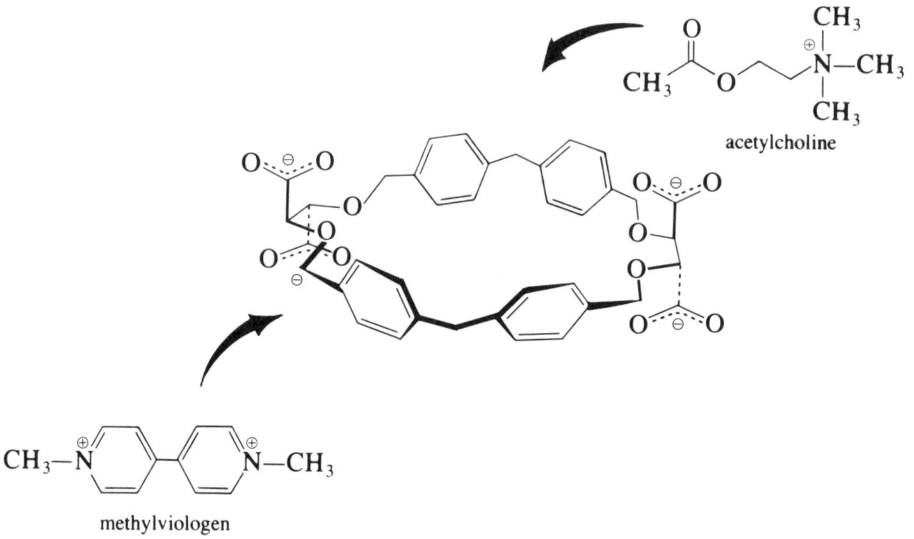

acetylcholine

methylviologen

This new type of receptor now has the ability to bind a range of molecular cations by electrostatic and hydrophobic effects. Not only are the complexes with primary ammonium ions more stable than with common polyether macrocycles, but the receptor also strongly binds tertiary and quaternary ammonium substrates. The complexation of acetylcholine is of special interest because it can provide an answer to the general question of how this neurotransmitter binds to its biological receptors (303).

Numerous other combinations of polar binding units and apolar architectural components may be imagined, making speleands attractive for the design of novel, efficient molecular receptors. For example, a macropolycyclic speleand combining an [18]-N_3O_3 macrocyclic binding unit with a cyclotriveratylene shaping component has been synthesized. Its tight intramolecular cavity allows inclusion of methylammonium ion (304).

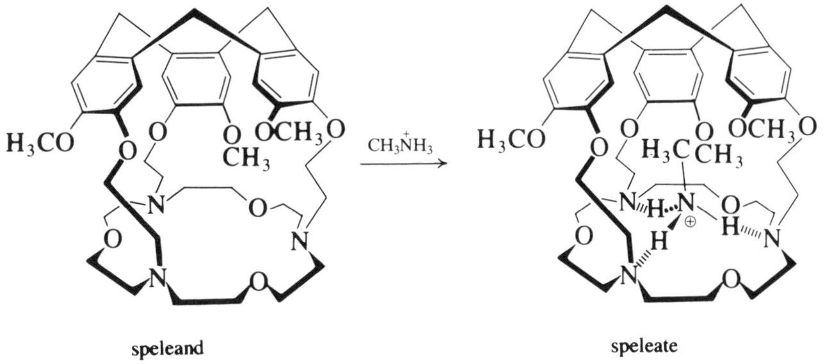

speleand

speleate

F.P. Schmidtchen (University of Munich) has developed an interesting series of new abiotic ditopic receptor molecules, based on aza-crown chemistry (305,306). One of them has a particularly good affinity for the amino carboxylate neurotransmitter GABA.

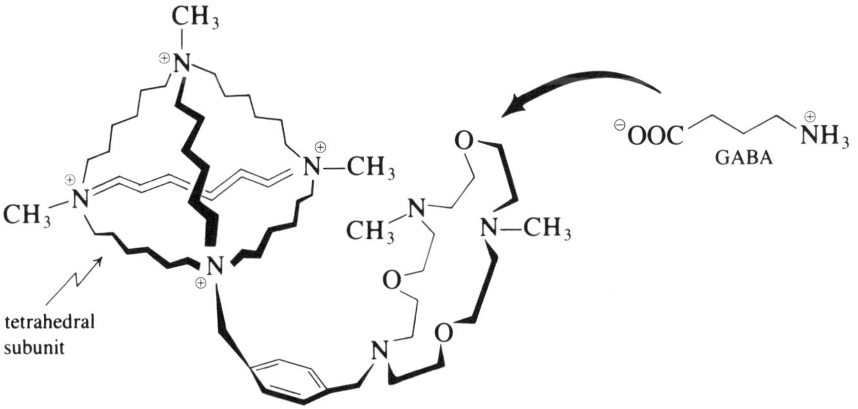

tetrahedral subunit

GABA

Charged polyamines open the way to the development of receptors for anions. In this effort, Lehn's team has used polyamine macrocycles to selectively bind anions, particularly ATP. This work led to the study of

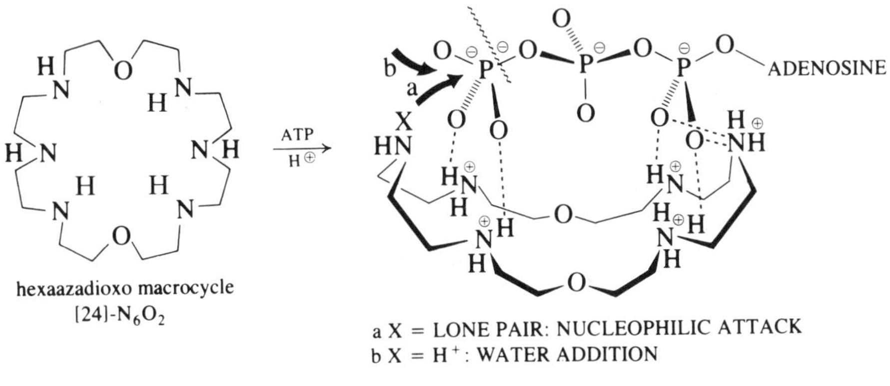

hexaazadioxo macrocycle
[24]-N_6O_2

a X = LONE PAIR: NUCLEOPHILIC ATTACK
b X = H^+: WATER ADDITION

ATP binding and hydrolysis by the polyamine [24]-N_6O_2. The binding study showed a marked kinetic of acceleration between pH 2.5 and 8.5 and the reaction resulted in the formation of ADP and eventually AMP. At pH greater than 6.5 a covalent intermediate seems to be formed between the macrocycle and the substrate (307). A ^{31}P-NMR study indicates the formation of a phosphoramidate intermediate (308) when

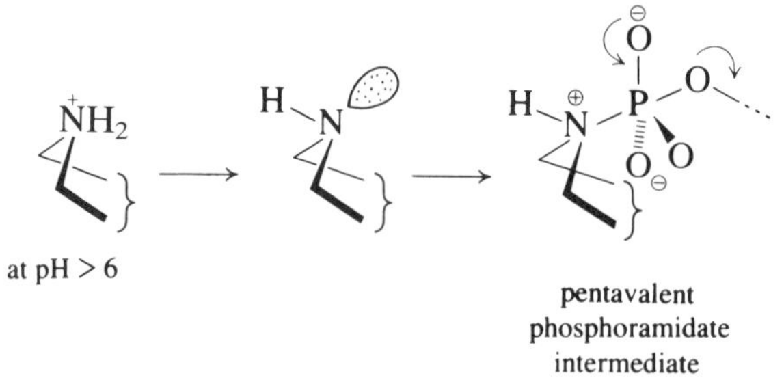

at pH > 6

pentavalent phosphoramidate intermediate

the reaction is carried out at high pH. It is not observed at low pH values. Interestingly, addition of Ca^{2+} causes a significant rate acceleration in ATP hydrolysis with the appearance of pyrophosphate, presumably via a phosphoryl transfer to a phosphate ion (309).

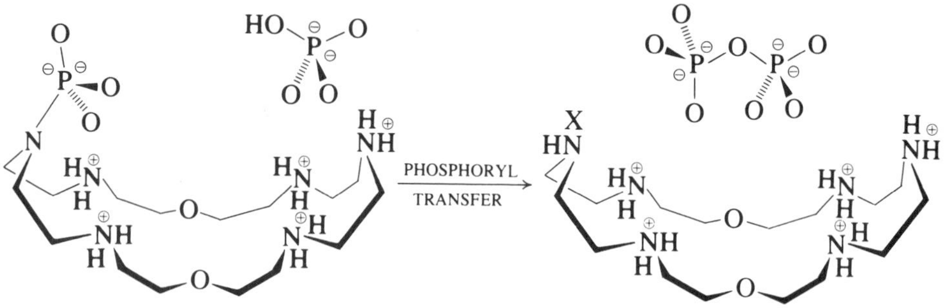

PHOSPHORYL TRANSFER

This same protonated macrocyclic polyamine [24]-N_6O_2 can catalyze the hydrolysis of acetyl phosphate to orthophosphate via the phosphoramidate covalent intermediate (310). This is schematically represented above and the aza-crown ether thus behaves as a ditopic receptor molecule and performs a process of *cocatalysis* in which both subunits cooperate for bringing together reagents and inducing bond formation. In this supramolecular reactivity and catalysis, both amine basicity (depending on pH) and favorable geometry effects operate.

This type of process lies in the development and design of other systems capable of inducing *bond formation*, thus effecting synthetic reactions as opposed to bond cleavage and degradation reactions. A step towards mimicking single synthetic reactors. The subject of polyammonium macrocycles as catalysts for phosphoryl transfer reactions has been reviewed (311).

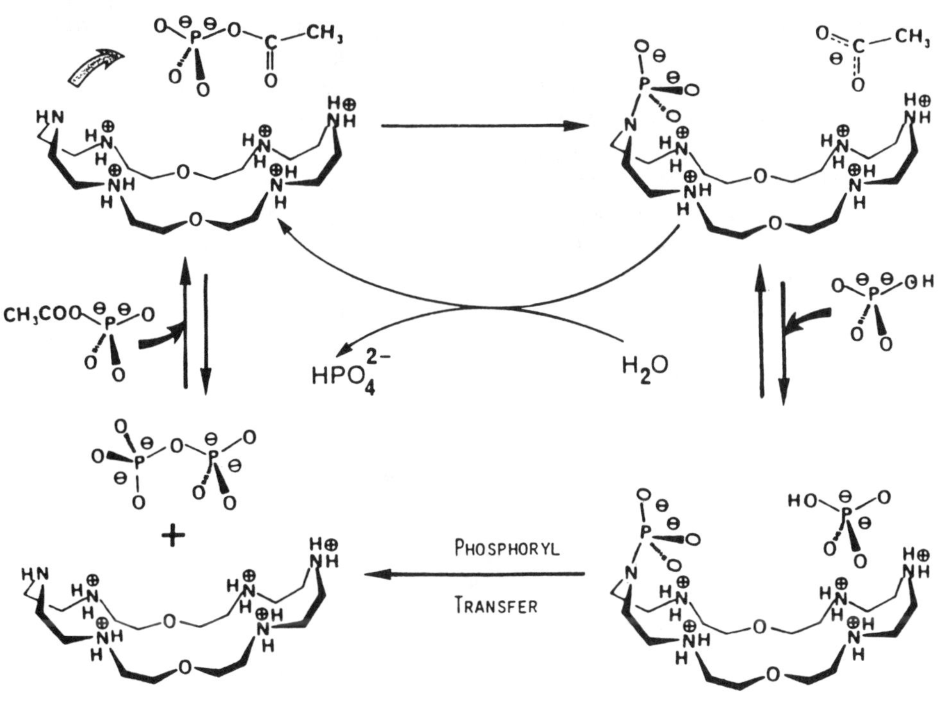

Further study with this supramolecular catalyst can provide entries to the direct cleavage of nucleic acid, related, for instance, to the self-splicing RNA process. It is also conceivable to design molecular catalysts capable of efficient phosphorylation of biosubstrates.

On a final note, aza-crown ethers and macrocyclic polyamines have both been used to anchor polypeptides. A cyclic tetrameric cluster of the chemotactic peptide *N*-formyl-Met-Leu-Phe has been prepared (312).

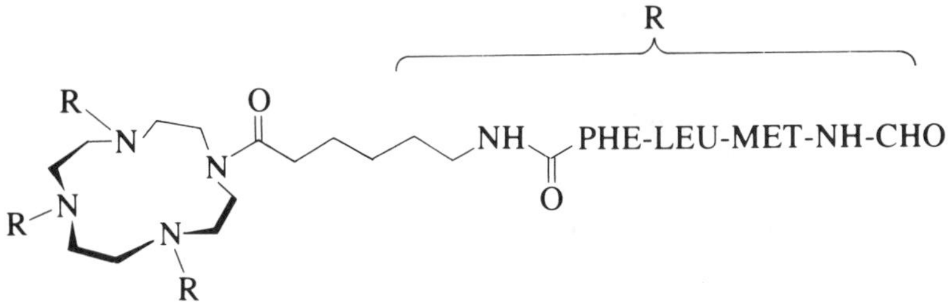

Its activity for the release of lysozyme from human neutrophils was compared to the monomeric analog. The cyclic clusters proved to be superactive by a factor of over 700 times and in a manner suggestive of a cooperative response of the membrane receptors. Therefore, the covalent binding of drugs (a peptide in this case) enhances its potency, suggesting a clustering effect of recognition sites on the membrane. This finding will promote a new approach to drug design with therapeutic applications in mind, such as the multivalent nature of antigens in immunochemistry or the topology of the neuropeptide receptors. Functionalized crown ethers are matrices of choice in this methodology.

Another promising domain is the use of aza-crown ethers with one or two peptide-based side chains for the binding of ions and the mimicking of ionophores like valinomycin (313). These pendants are called lariats and their utilities will be the subject of the next section. In this case, an X-ray study of the K^+ complex with N,N′-bis(methylglycidylglycidino)-4,13-diaza-18-crown-6 showed a twofold symmetry (314) with an *anti*-disposition of the two peptide chains.

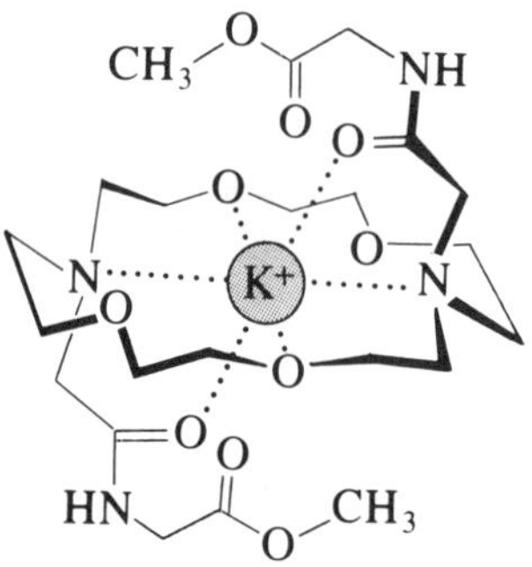

5.2.2 Lariat Crown Ethers

In 1980, G.W. Gokel (now at Washington University, School of Medicine) introduced a new class of crown ether, called *lariat crown ethers.** These compounds are characterized by a macrocyclic polyether ring and a flexible arm extending from the cycle to which are attached electron donor groups (315). This family of compounds has now grown to include single-armed species having side chains attached at a carbon atom of the macrocycle (i.e., *carbon-pivot lariat ethers*), side chains attached at a nitrogen atom of aza-crown ethers (i.e., *nitrogen-pivot lariat ethers*), and two-armed systems (*bibracchial lariat ethers*, BiBLEs) in which two arms are linked to the nitrogen atoms.

In principle, the presence of a ligating arm on a macrocyclic polyether should enhance cation-binding properties. This was not always the case because of a solvent dependence on the binding constants of these macrocycles bearing secondary donor groups on the flexible arms. In polar solvents, the binding constants are considerably diminished.

* The molecules were called lariats because of their resemblance on molecular models to a lasso and their ability to "rope" and "tie" cations.

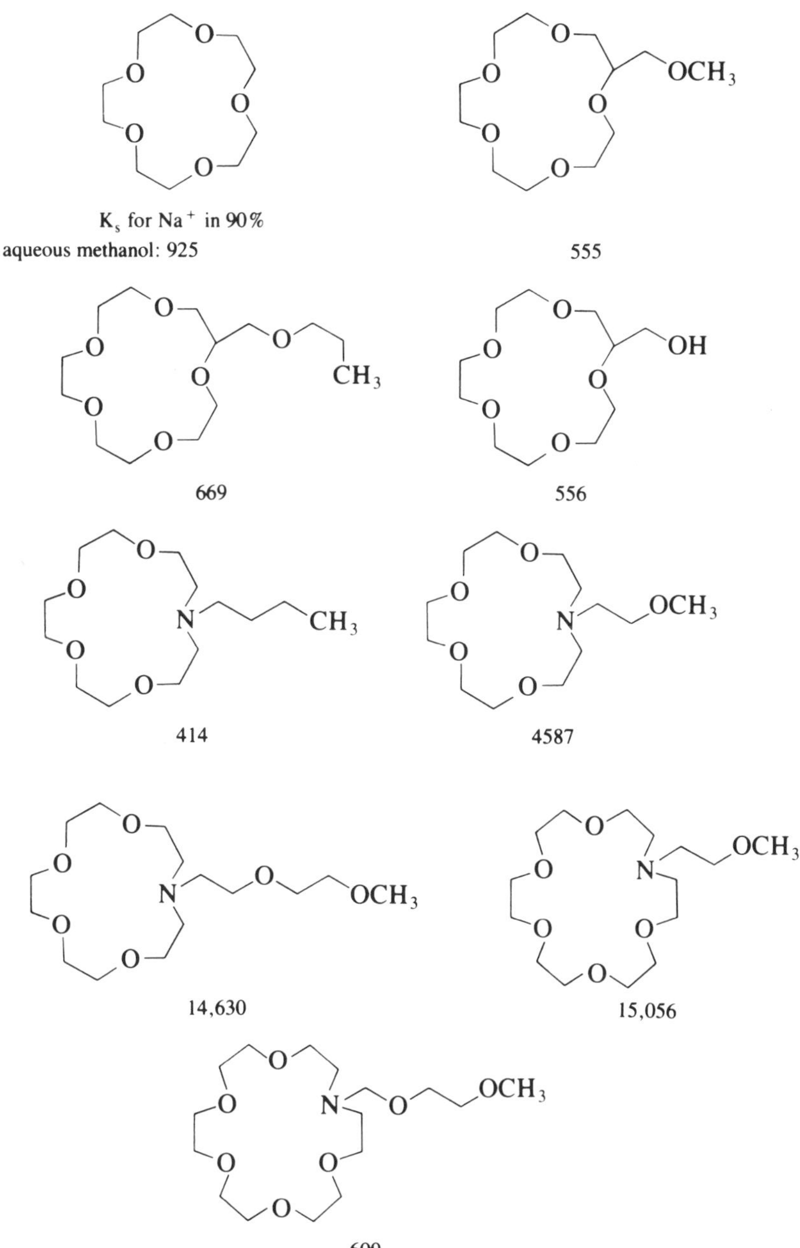

In fact, compounds having arms joined by a methyleneoxy group appear to be relatively inflexible and exhibit both relatively weak solvent-dependent cation binding. In contrast, when the secondary donor arm is attached to a nitrogen pivot rather than a carbon, binding constants are found to be substantially increased over simple monocyclic systems.

Gokel reasoned that the facile inversion about the nitrogen atom does reduce the potential problem of sidedness present in the former examples and may explain these differences. It is especially interesting to note that the high binding constant of the 15-crown-5 with two oxygens in the side arm is comparable to a 18-crown-6 system with only one donor oxygen in the side arm (316).

A series of *N*,*N*-disubstituted derivatives of 4,13-diaza-18-crown-6 (or BiBLEs) have also been prepared by Gokel. Some of these compounds exhibit Ca^{2+} over Na^+ or K^+ selectivity (317). These compounds have been prepared by a novel, one-step reaction of aliphatic primary amines with triethyleneglycol diiodide.

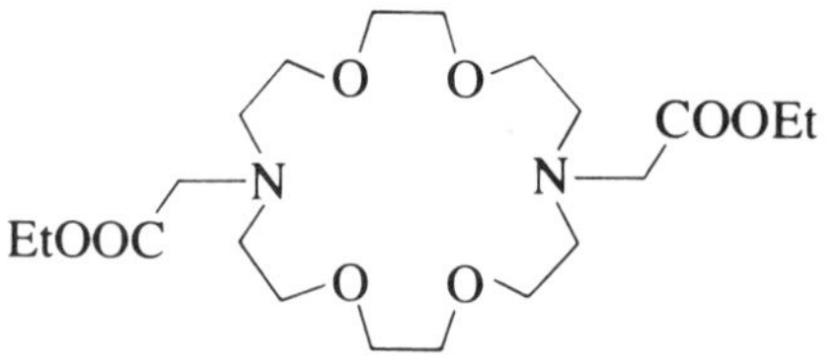

$\log K_s(Ca^{2+}) = 6.78$

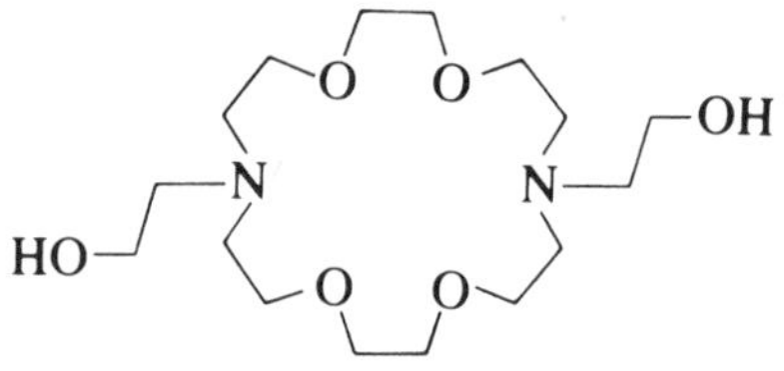

$\log K_s(Ca^{2+}) = 6.02$

The proximity in space of the two carboxylate groups accounts for the high selectivity for Ca^{2+} ions over monovalent cations. A similar situation has been encountered with a *syn*-diacid 18-crown-6 where $\log K_s$ for Ca^{2+} ions was 9.7 (318).

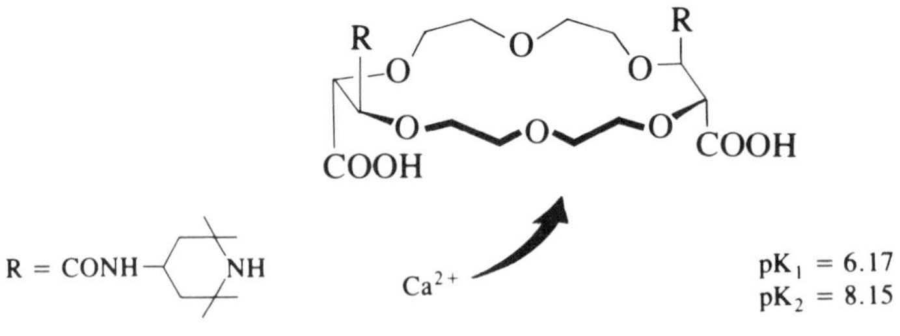

R = CONH–(piperidinyl)NH

Ca^{2+}

$pK_1 = 6.17$
$pK_2 = 8.15$

This *syn*-diacid crown ether is thus a particularly good Ca^{2+} ligand. The *syn* arrangement is unique in the sense that evaluation of the ionization of the carboxyl groups gave a value of about 2 for ΔpK, a phenomenon already observed with Rebek's molecular cleft (see Section 4.10).

The presence of carboxylate or anionic groups close to a crown ether cavity could have a profound effect on the stability of the corresponding ion complex, as well as on the ionophoric properties of such a macrocyclic compound. An interesting situation is the following phosphonate crown ether where the participation of the anionic side group in the coordination of Na^+ to the crown ring was demonstrated by NMR spectroscopy. The magnetic nonequivalence of the benzylic protons is a nice demonstration

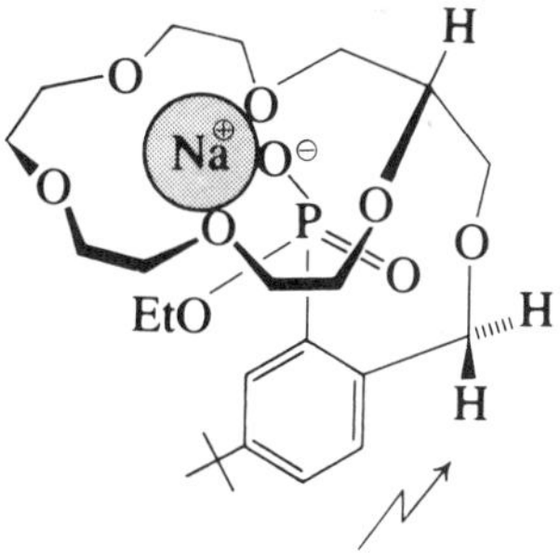

AB quartet of the benzylic protons

of anionic side group participation in complexation of a polyether-bound alkali metal cation in solution (319).

Finally, one of the most promising studies using aza-crown ether is the formation of a novel macrobicyclic cryptand formed by hydrogen bonding between two BiBLE precursors having adenine-terminated and thymine-terminated side arms (320). The presence of dodecane-1,12-diammonium bishydrochloride seems to enhance, in water, the formation of a self-assembled "molecular box." This should represent an interesting entry to the mimic of Watson–Crick base pairs common to DNA and RNA duplexes.

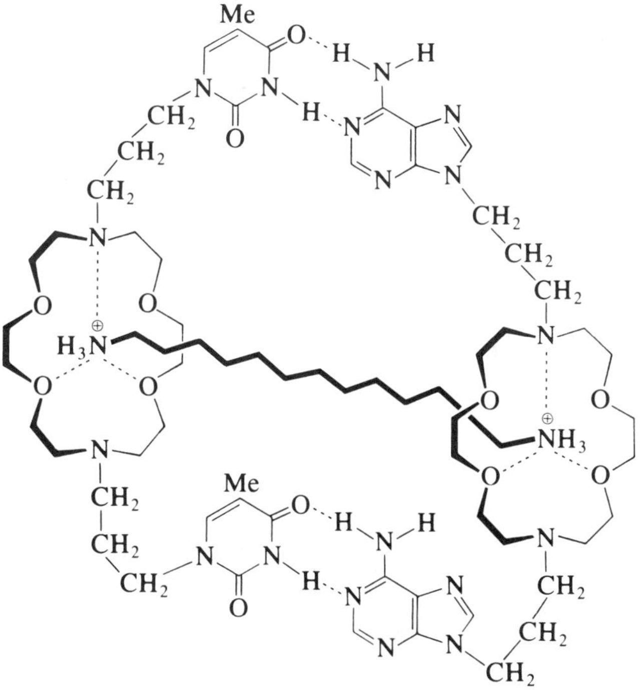

Proposed structure for the complex induced by addition of diammonium alkyl chain (320)

This work has now been extended to probe Hoogsteen vs. Watson–Crick hydrogen bonding and other base–base interactions in self-assembly processes (321). The authors report evidence for ternary induced-fit receptor complexes with such molecular boxes.

5.2.3 pH Regulation and Ion Selectivity

In Section 5.1.1 we showed examples where the selectivity of a crown ether for a given ion has often been intuitively attributed to a matching of the cavity size and the metal ion diameter. However, more recent investigation (322) has revealed that such a simple "cavity-size" selectivity concept is not always applicable, in particular when the macrocyclic ring is quite flexible. Consequently, the hole-size relationship probably plays its greatest role when the ligands are relatively inflexible, as in the case of monobenzo- and dibenzo-crown ethers. The special stability of the $K^+ \cdot$ 18-crown-6 complex may be due to a favorable combination of enthalpic and ligand symmetry factors (322).

A brief presentation of the effect of pH on ion selectivity and transport by macrocyclic carrier molecules is appropriate before going into the development of bis-crown ethers as ion carriers (next section).

R.A. Bartsch's group (Texas Tech University) has constructed a series of mono- and dibenzo-crown ether with an oxyacetic acid chain (323). In competitive extractions of alkali-metal cations, this crown ether carboxylic acid was found to be selective for Na^+ at high pH but selective for K^+ at $pH < 6$. The association constant is also greater for Na^+ than K^+ when the ligand is deprotonated.

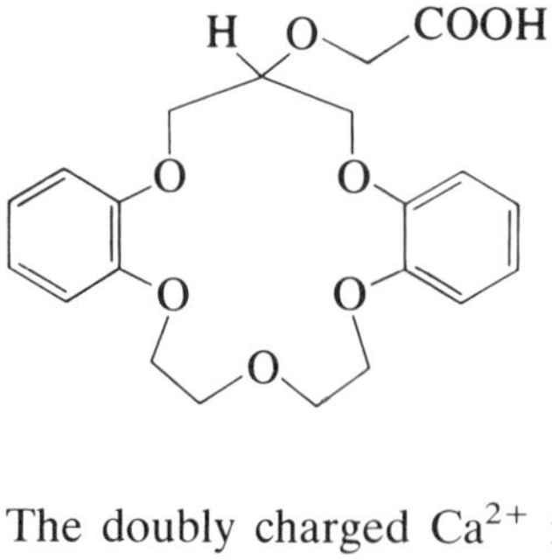

	basic pH	*acidic pH*
log K (Na^+)	4.02	2.75
log K (K^+)	3.71	2.78
log K (Ca^{2+})	4.10	2.34

The doubly charged Ca^{2+} ions bind quite well to the anionic ligand. The effect is even reversed as compared to low pH values. Thus, the ligand selectivity toward ion binding can be tuned by simply adjusting the pH of the medium. As pointed out by Bartsch, this property may be very important in biological ion transport through membranes with ionizable carboxylic ionophores such as monensin and nigericin.

High lithium ion selectivity in competitive alkali-metal solvent extractions was obtained with lipophilic crown carboxylic acids. The following compound is a good example (324).

selectivity order in competitive extraction:
$Li^+ \gg Na^+$ (no K^+, Rb^+, Cs^+)
Li^+/Na^+ ratio = 20

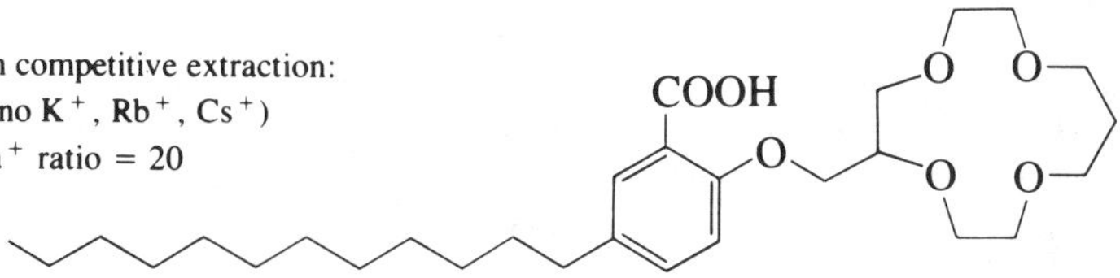

COOH

Incorporating an easily ionizable nitrophenol substituent on a lipophilic 14-crown-4 also produces an excellent ionophore for proton-driven cation transport (325). The electronically neutral carrier can transport the Li^+ ion across a microporous polypropylene film with extremely high Li^+ selectivity over Na^+. Because of the importance of lithium salts in the treatment of manic depression, this system or a variant of it should have wide applicability.

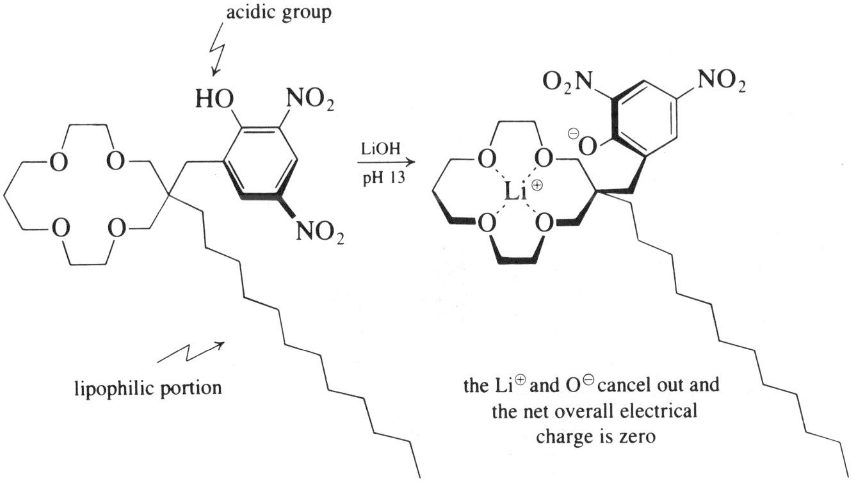

acidic group

HO NO$_2$ NO$_2$

LiOH pH 13

O_2N NO_2 $Li^{\oplus}$

lipophilic portion

the $Li^{\oplus}$ and $O^{\ominus}$ cancel out and the net overall electrical charge is zero

In this direction, chemists at Allied-Signal, New Jersey, have developed two sensitive and selective fluorescent cryptands for alkali metals. The [2,2,2]cryptand showed a selectivity for K^+ while the other showed great enhancement of fluorescence in Li^+ solutions. Therefore, these compounds may find uses in clinical analyzers with fiber optics to evaluate serum levels of lithium in patients suffering from depression and treated with lithium salts. On the other hand, the level of potassium could be determined the same way for patients treated for high blood pressure.

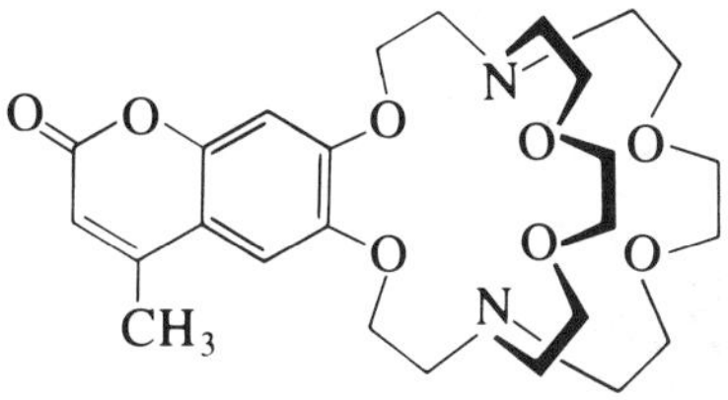

CH_3

4-methylcoumaro-[2,2,2]cryptand
$K^{\oplus}$ selective

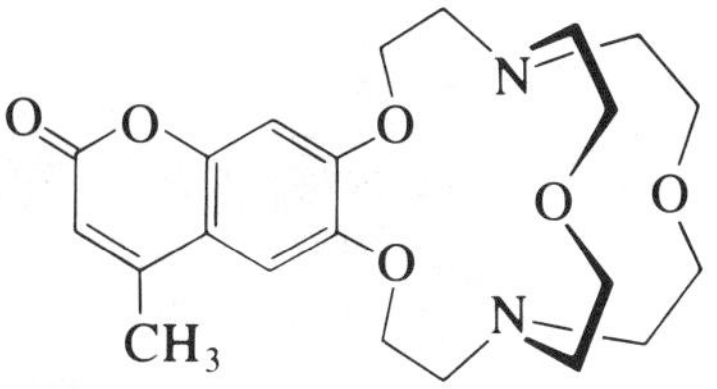

CH_3

4-methylcoumaro-[2,1,1]cryptand
$Li^{\oplus}$ selective

Divalent/monovalent (Ca^{2+}/K^+) cation transport selectivity has also been examined with a lipophilic dicarboxylic acid-dicarboxamide 18-crown-6 macrocycle (326). The carrier exhibits a pH regulation of Ca^{2+}/K^+ transport selectivity, coupled to proton *anti*-transport. At low pH, preferential K^+ transport occurs while preferential Ca^{2+} transport takes place at higher pH values. These results demonstrate how carrier design allows control of the rate and selectivity of divalent versus monovalent cation transport across an organic liquid phase.

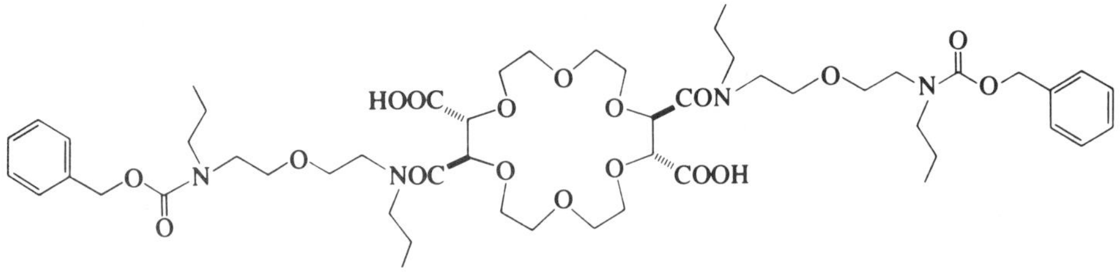

Ammonium tail lariat crown ethers have also been synthesized. Some of these new types of synthetic ionophores having both an 18-crown-6 ring and a primary amino group in the same molecule display excellent K^+/Na^+ selective and active transport across a liquid membrane in systems like the one in Fig. 5.4. The selectivity in ion transport is based on the reversible intramolecular complexation of the polyether ring and the ammonium ion of the side arm (327). Thus, the ionophore complexes with K^+ in the basic phase and transfers it with the counter ion to the acidic phase across the organic (CH_2Cl_2) membrane. In the acidic phase, the amino group becomes protonated and the primary ammonium ion formed in situ competes with K^+ for complexation like a "tail-biting" process. As a result, the system shows an excellent K^+/Na^+ selectivity over 48 hr; 60% of the K^+ is transferred compared to only 6% of Na^+. The transport cycle is completed by deprotonation of the ammonium ion in the basic phase.

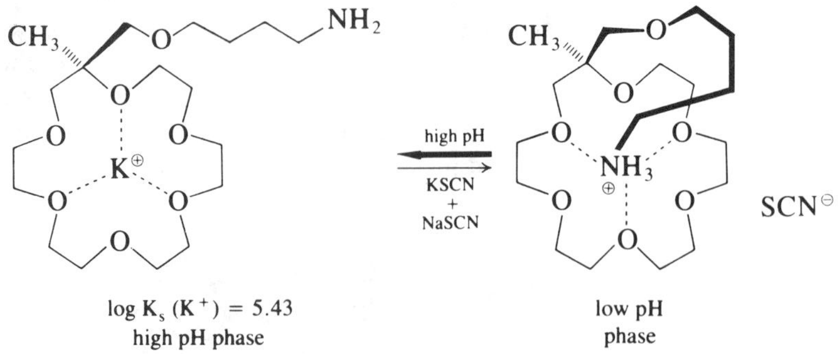

log K_s (K^+) = 5.43
high pH phase

low pH
phase

In a similar approach, Gokel's group made an anilinium cation lariat ether complex (328). The diazotization of the amine group gave the corresponding diazonium which underwent intramolecular crown complexation, as judged by IR spectroscopy, to form what they called, by analogy, an "ostrich molecule" complex.

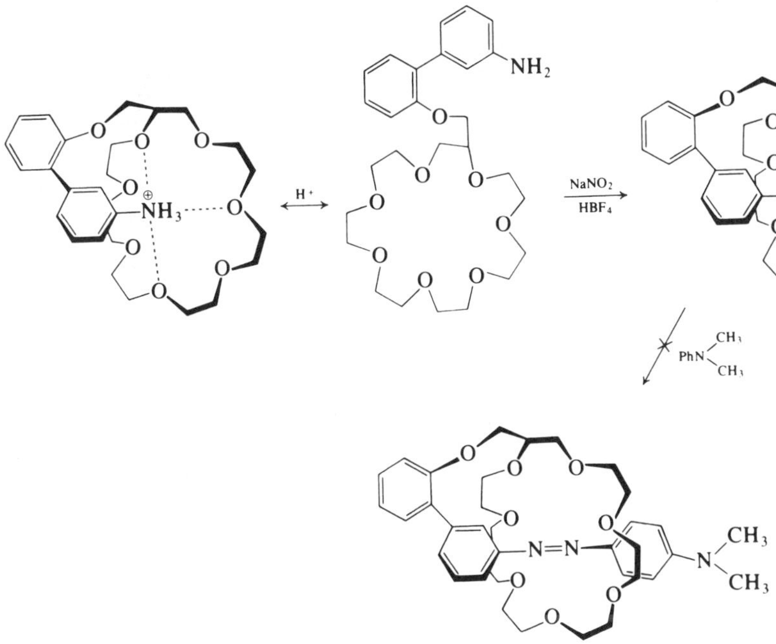

This morphological similarity is reminiscent of threading the "eye of the needle." Unfortunately, attempts to make a through-hole azo-coupling reaction with dimethylaniline were not successful, and only the normal nonthreaded azo structure was obtained.

5.2.4 Bis-Crown Ethers

A way to impose high selectivity in alkali metal cation complexation is to form an intramolecular complex in which the cation would be "sandwiched" between two adjacent crown ether rings. By using this concept, a series of malonate-type bis-crown ethers were prepared (329). The molecule with two 12-crown-4 rings exhibited a cation extraction selectivity order of Li^+/Na^+ selectivity of 37.

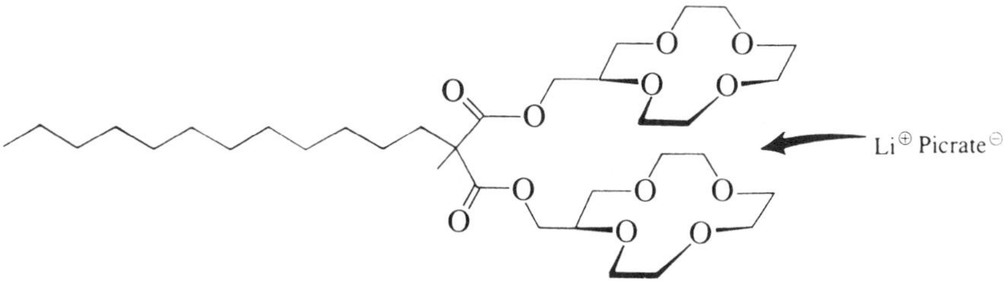

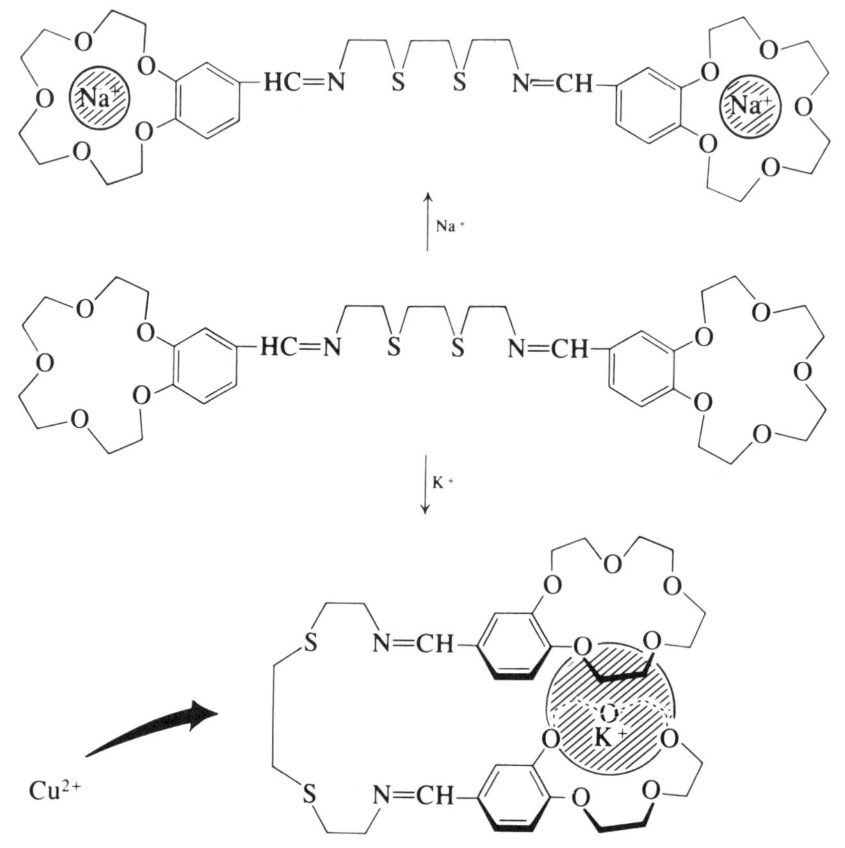

An interesting bis-crown ether was developed by P.D. Beer (University of Birmingham). This novel Schiff base bis-crown ether ligand contains recognition sites for alkali and transition metal guest cations (330). This multiple receptor may exhibit *allosteric* properties by binding sequentially two or more ions in close proximity to one another.

Its stoichiometry was found to be 1:2 for Na^+ and 1:1 for K^+ ions. Therefore, for K^+ ion, the two benzo-15-crown-5 units are lying cofacial to one another and are acting cooperatively to form an intramolecular sandwich complex.

Interestingly, in the presence of Ag^+ the complexation with K^+ is no longer of the 1:1 intramolecular sandwich complex type. It is believed that Ag^+ forms a tetrahedral coordination with the Schiff base and the sulfur sites.

To test this rationalization further, the preparation of a Cu^{2+} complex was undertaken. A 1:1 stoichiometry with K^+ was observed. This corresponds to the previous complex with a square planar coordination of the bound transition metal between the two sulfurs and the two nitrogen atoms.

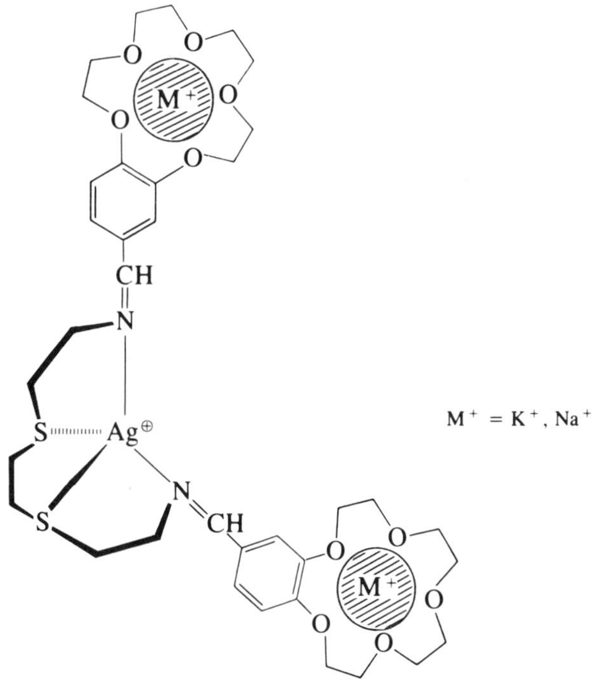

$M^+ = K^+, Na^+$

In a study aimed at taming a system with a quite different type of complexity, N. Nakashima's group (University of Kyushu) have been manipulating the ordering of molecules in a novel bilayer system that is responsive to chemical signals, in this case Na^+ ions. The bilayer was made of a chiral cationic lipid and a lipophilic bis-crown ether (331).

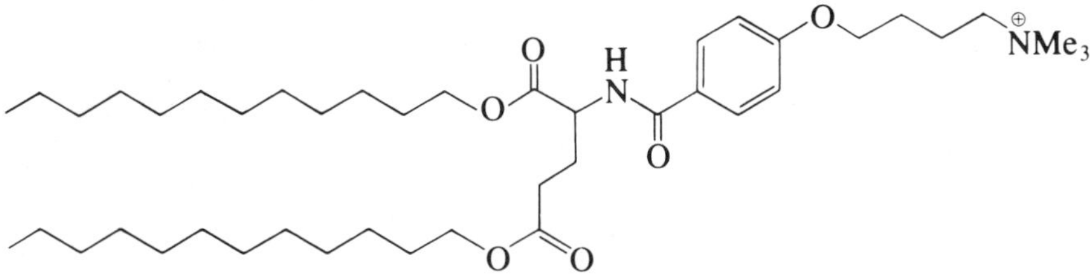

cationic lipid

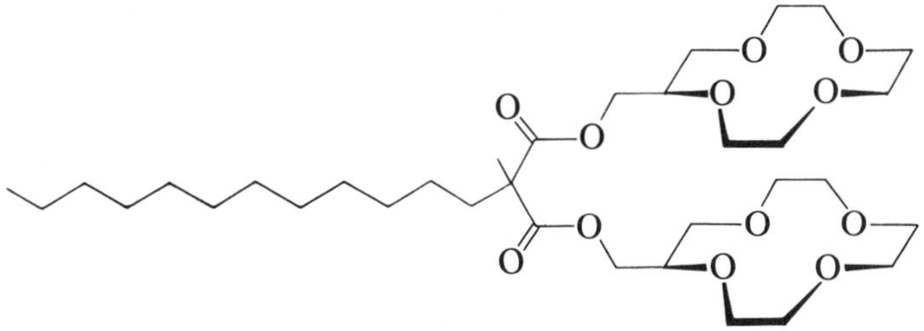

hydrophobic bis-crown ether

This mixed bilayer is thus composed of a chromophore-containing chiral amphiphile and crown ether chains where the crown ring and the quaternary ammonium head are at the surface of the bilayers. By using circular dichroism, they were able to show that chiroptical properties of the system are influenced more by Na^+ ions than by K^+ ions. Further work in this direction should be important in the development of artificial sensory systems to mimic biological membranes.

In the field of ion transport across membranes, the author became interested in developing a macrocyclic crown ether system that would be a simple model of the (Na^+,K^+)-ATPase pump for selective transport of ions (332). Recall that (Na^+,K^+)-ATPase is a transport enzyme that pumps sodium antiport to potassium coupled to ATP hydrolysis.

The approach was based on the design of the following bis-crown ether with two macrocyclic rings of different sizes: one with an affinity mainly for Na^+ ions and the other for K^+ ions but with the possibility of competing for the side chain ammonium group. This architecture has some similarity with the "tail-biting" or the "ostrich molecule" complexes presented in the preceding section.

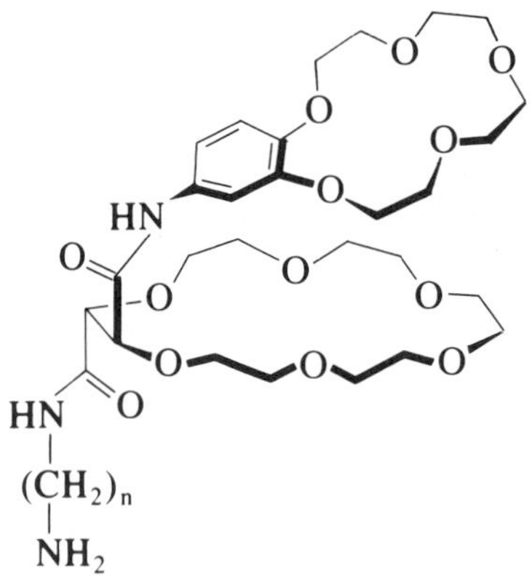

		Transport rates (μmol/h)	
		pH 6	*pH 10*
n = 2	Na^+	7.08	11.19
	K^+	5.74	7.78
n = 4	Na^+	8.67	13.61
	K^+	4.70	11.86

This bis-crown ether ionophore behaves as a pH-modulated ion transport. Indeed, it was observed that at low pH, where the ammonium tail is protonated, the carrier transports the Na^+ ion faster than the K^+. At higher pH values, both ions are transported by the bis-crown ether molecule. However, the differences in transport selectivity remain small. One of the reasons for this is that the amide bond of the ammonium tail is, as expected, mainly in the *trans* configuration and in such a geometry, very little of the charged ammonium tail could bind adequately to the crown ether cavity. In fact, the situation is already better for $n = 4$, compared to $n = 2$. To overcome this unfavorable energy barrier, the amide bond should eventually be substituted for by an ether function, therefore eliminating the undesired *cis* $\rightleftharpoons$ *trans* equilibrium of the former amide bond.

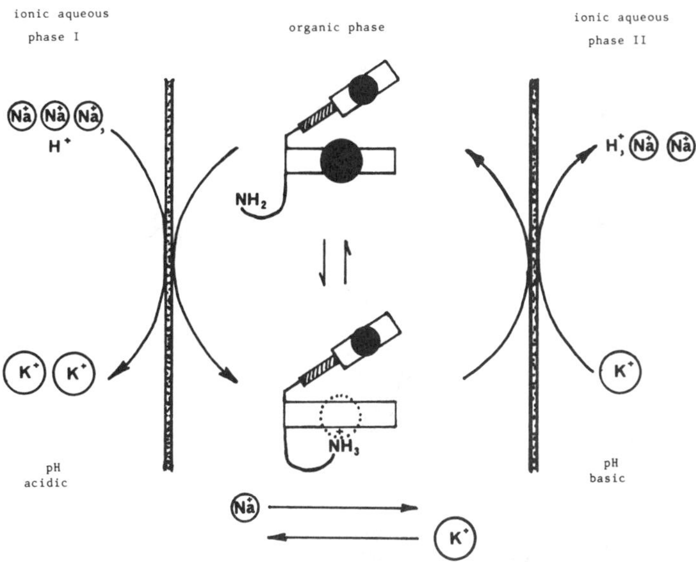

Fig. 5.7. Ion transporting machine.

These new compounds should then be tested for Na^+/K^+ transport selectivity with a system comparable to Fig. 5.7. The net result should be the flow of Na^+ from phase I to phase II with a reversed flow for K^+ ions. This simple mechanism has some similarity with natural, albeit more complex, systems of selective ion transport.

An important review article on bis-benzocrown ethers appeared in 1994 and should be consulted for more details on the synthetic implications (333).

5.2.5 Photoresponsive Crown Ethers

The group of S. Shinkai (University of Kyushu) made a remarkable contribution to the development of new mono- and bis-crown ethers. He designed crown ether molecules having a photosensitive group to promote photocontrolled extraction and transport of ions across artificial membranes (334). Aza-bis-benzocrown ethers are among the compounds synthesized. Irradiation by UV light of the stable *trans*-aza-isomer promotes a *trans* $\rightleftharpoons$ *cis* isomerization, which is thermally reversible. On molecular models this reversible interconversion resembles the motion of a butterfly.

The *cis*-isomer forms a stable 1:1 sandwich-type ion complex with large alkali metal cations. It can thus be used as a useful "phototweezer" in solvent extraction and ion transport systems (335). This is illustrated in Fig. 5.8. The isomerization occurs in the organic phase but at the

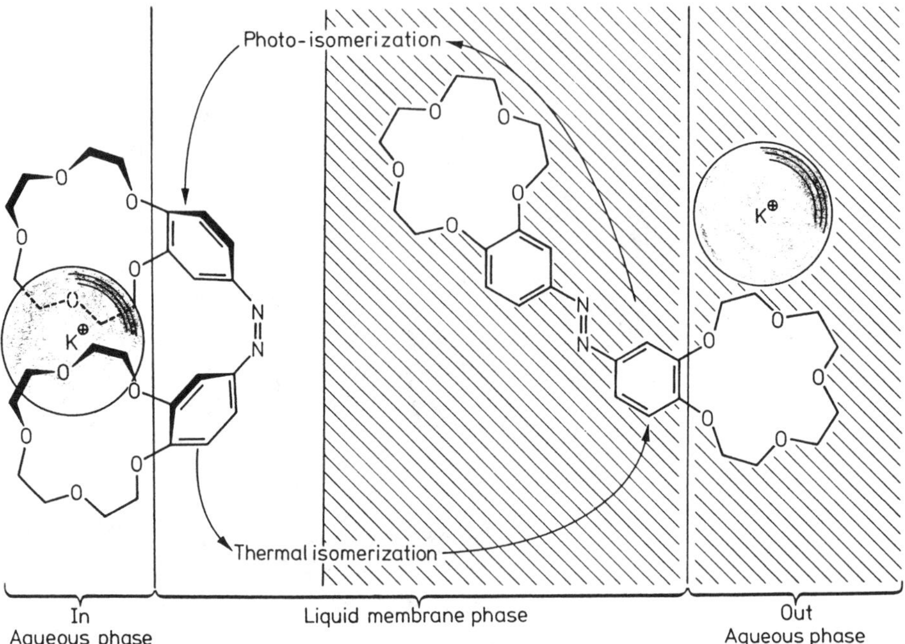

Fig. 5.8. Light-driven K^+ transport mediated by a photoresponsive crown ether (334). Reproduced with permission. Copyright © 1987 by Blackwell Scientific Publications Ltd.

trans-isomer

possibility of *cis* ⇌ *trans* isomerization by light

interface with the aqueous phases. It corresponds to a photoinduced rate enhancement in K^+ transport from the IN aqueous phase to the OUT aqueous phase across a liquid (*o*-dichlorobenzene) membrane in a U-tube. The counter ion is the hydrophobic picrate anion. The photostationary state of *cis*-isomer is markedly enhanced by addition of Rb^+ or Cs^+ ions.

This type of light-driven ion transport can also be facilitated by a crown ether with a photoresponsive anionic cap. In this particular case, markedly improved extractibities for Na^+ and Ca^{2+} were observed (336).

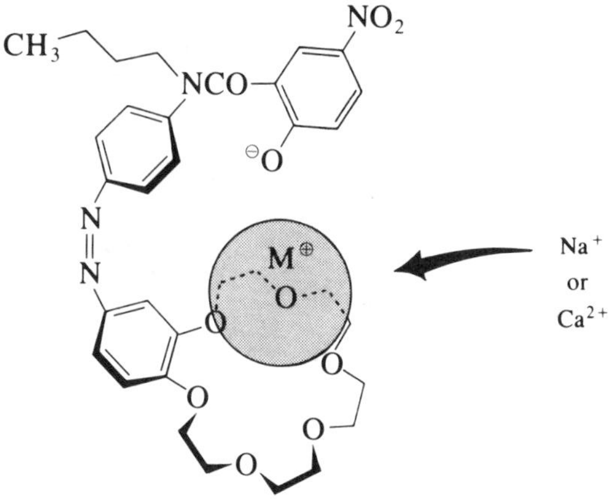

sandwich-type complex

This enhanced property is due to the participation of the phenolate ion, forming a sandwich-type complex with the crown ring. This was the first example of a light-driven Ca^{2+} transport across a liquid membrane.

In the development of photofunctional systems that change their chemical and physical properties in response to photoirradiation, a series of crown rings with an ammonio-alkyl lariat were prepared (337). The

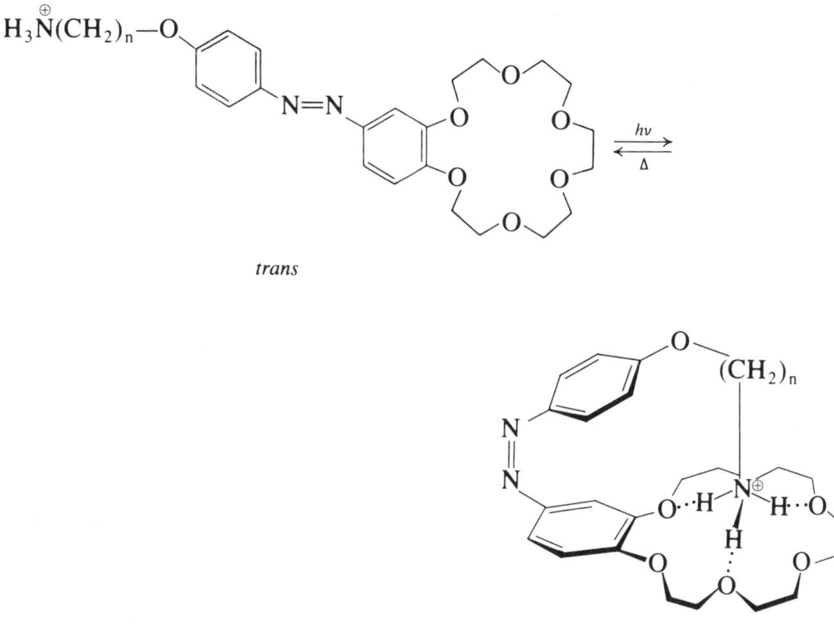

trans

cis

n = 4, 6 and 10

molecules have been designed so that intramolecular "tail-biting" of the ammonium group to the crown ring would occur upon photoisomerization, and would exist as discrete monomers. The *cis*-isomer with n = 4, however, cannot form the intramolecular "tail-biting" complex because of the short tetramethylene spacer, and exists as a pseudocyclic dimer.

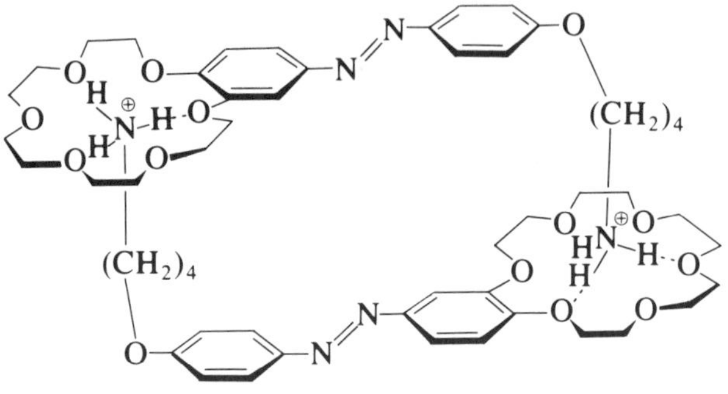

pseudo-cyclic dimer

This is well reflected in the electric conductance (for $n = 6$ and 10) which increases with *trans* to *cis* photoisomerization. Figure 5.9 shows how the photocontrol-reversible association–dissociation phenomenon of the ammonium tail-end with the crown ring mediates K^+ transport.

A different example of such association–dissociation process of the ammonium tail, modulated by pH, has already been presented in Section 5.2.4.

Regulation of membrane transport phenomena by an on-off type switch plays a fundamental role in biological processes. Such interactions

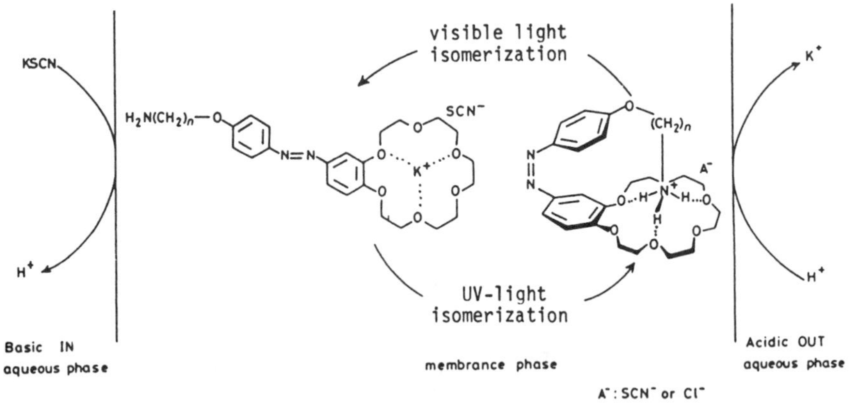

Fig. 5.9. Light-driven active transport of K^+ from the basic IN aqueous phase to the acidic OUT aqueous phase across an organic liquid membrane (334). Reproduced with permission. Copyright © 1987 by Blackwell Scientific Publications Ltd.

are regarded as allosteric in nature when binding at one site induces conformational changes that alter the receptivity at a remote site. In vision, the isomerization of retinal and the conformational changes on rhodopin coupled with Na^+ and Ca^{2+} transport is a good example.

The design of new switch-functionalized systems in biomimetic chemistry could lead to a better understanding of many biological stimuli modulated not only by pH and light but also by redox potentials and hormones.

In this respect, thiol groups are useful to give a redox-type switch-function to the crown ether family. The cryptand (oxidized) form binds K^+, Rb^+, and Cs^+ more efficiently than the reduced crown form. But they both have the same affinity for Na^+ ions (338). This difference was rationalized by assuming that K^+, Rb^+, and Cs^+ "perch on" the crown ring whereas Na^+ can "nest" in it. This illustrates the possibility of constructing redox-switched ionophores and eventually using the redox energy for their application in ion and drug transport across membranes.

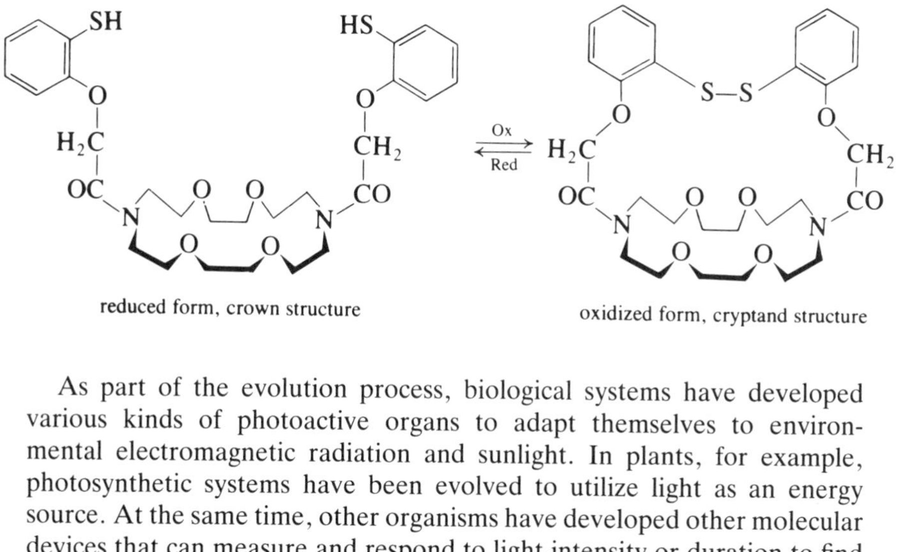

SH HS O O H_2C CH_2 OC CO N O O N O O Ox Red

reduced form, crown structure

S—S O O H_2C CH_2 OC CO N O O N O O

oxidized form, cryptand structure

As part of the evolution process, biological systems have developed various kinds of photoactive organs to adapt themselves to environmental electromagnetic radiation and sunlight. In plants, for example, photosynthetic systems have been evolved to utilize light as an energy source. At the same time, other organisms have developed other molecular devices that can measure and respond to light intensity or duration to find favorable conditions for survival (339). In some cases, light is used as information. In a similar manner to biological systems, light can be used in organic chemistry not only as an energy source for chemical synthesis but also as an information source or a trigger for subsequent events.

It is with this objective in mind that Irie and Kato (339) developed a new concept to control ion-binding ability: *photoresponsive molecular tweezers*. For this, the *trans-cis* photoisomerizable chromophore of thioindigo having polyether side chains was exploited for the photomodulation of metal ion binding by the polyethers.

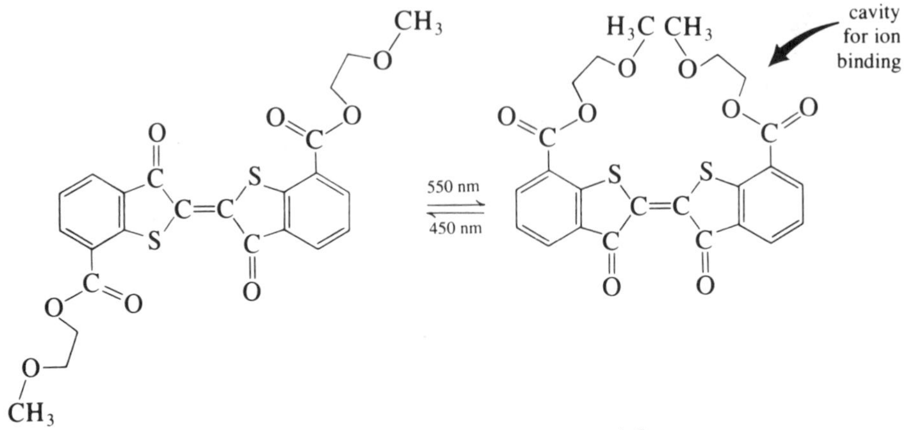

cavity for ion binding

550 nm

450 nm

trans form *cis* form

Thioindigo is known to change its absorption spectrum when the configuration changes from the *trans* to the *cis* form. The absorption maximum at 542 nm with the *trans* form is replaced by a maximum at 481 nm with the *cis* form upon irradiation at a wavelength of 550 nm. Solvent extraction of metal ions revealed that the *trans* form had no binding ability to any of the metal ions, whereas K^+, Rb^+, and Na^+ were selectively extracted ($Na^+ < Rb^+ < K^+$) by the photogenerated *cis* form. The *cis* form, in addition, had high binding ability to soft metal ions such as Ag^+, Hg^+, Hg^{2+}, and Cu^{2+} in comparison to alkali-metal ions. The association constant of Ag^+ to the *cis* form gave a value of over $10^6\ M^{-1}$.

Finally, ion transport experiments in a U-tube across a 1,2-dichloroethane liquid membrane showed that repeating cycles of alternate photoirradiation of light at 529 nm and 488 nm, which caused *trans* (488 nm) $\rightleftharpoons$ *cis* (529 nm) interconversion, resulted in the photoregulated Ag^+ transport from one aqueous phase ($AgNO_3$ + picric acid) to the other (only water).

This approach asserts an entry to the more and more exploited use of chromo- and fluoroionophoric colored crown ethers as dye reagents or probes for ion binding (340). As an example:

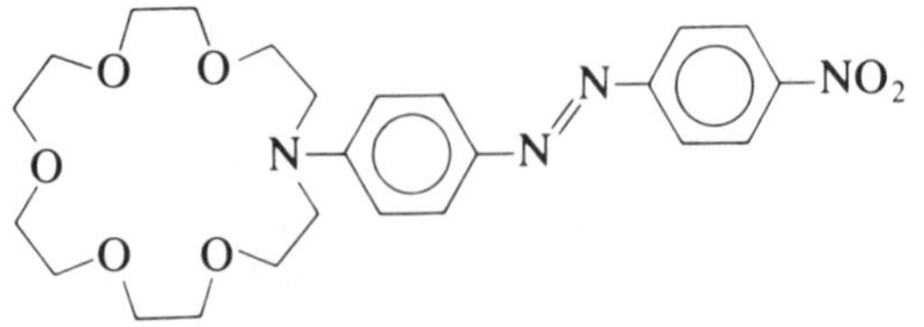

max = 477 nm, but moves to 450 nm in the presence of K^+ or Ca^{2+} and to 357 nm in the presence of Ba^{2+} (in acetonitrile)

5.2.6 Cavitands, Spherands and Calixarenes

In host–guest complexation chemistry, complexes are structured by contacts at multiple binding sites between host and guest. Contacts at several sites between hosts and guests depend on complementary placements of binding sites, and the binding energy at each single contact site is in the order of few kilocalories per mole. The crystal structures of crown ethers (*chorand*) and cryptands showed that they do not readily possess a cavity because one or two methylene groups are turned inward, toward the center of the molecule. The crystal structures of the corresponding complex, however, showed that the molecules now contain a cavity formed and filled by the ion.

D.J. Cram designed and synthesized new organic host molecules that contain enforced cavities large enough to complex and even surround simple inorganic or organic guest compounds. These new classes of *cavitands* were named *spherands*. What makes these ligands remarkable is that unlike the chorands and cryptands, whose cavities are developed by the complexing guest, the spherands already contain cavities in their uncomplexed state that become filled upon complexation (341). Therefore, these new systems of ligands are *organized prior to complexation* so

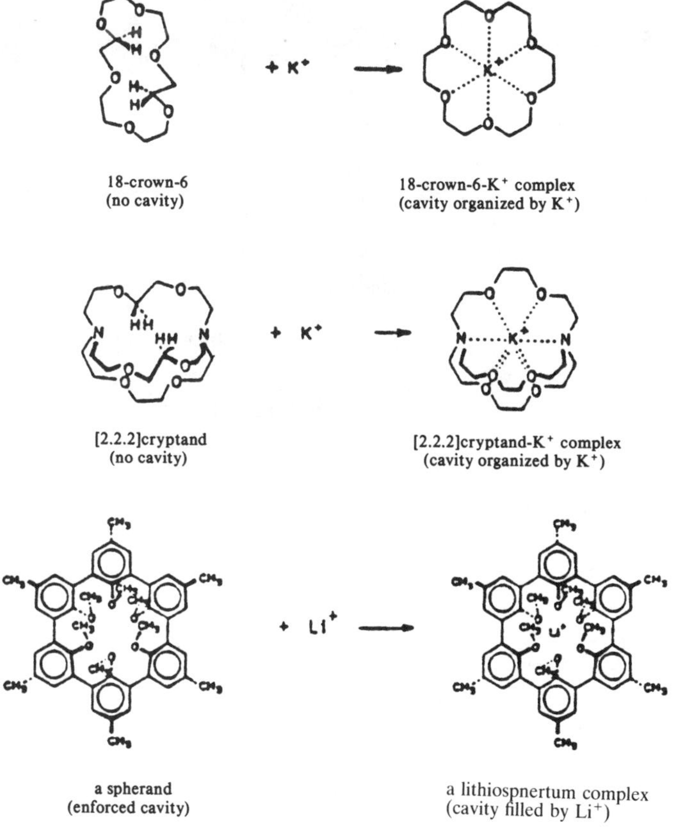

Fig. 5.10. Comparison between crown ether, cryptand, and spherand (341).

that the orbitals of unshared electron pairs of the binding sites shape a roughly spherical cavity reinforced by a support structure of covalent bonds (342).

The preorganized networks are thus determined during their synthesis rather than during their complexation. This results in a dramatic increase in the association constants for the binding of ions. For a lithium-spherand complex like the one depicted in Fig. 5.10, binding energy as high as $-23\,\mathrm{kcal/mol}$ has been measured which corresponds to an association constant of the order of 10^{10} to $10^{11}\,\mathrm{L \cdot mol^{-1}}$. However, the prototype just mentioned extracts Li^+ very slowly from water when in chloroform solution. The high proportion of carbon and hydrogen atoms in these molecules makes them quite insoluble in water. To improve this inherent insolubility problem and the efficiency of extraction, some anisole units have been replaced by cyclic urea units. The urea oxygen becomes less sterically hindered and, at the same time, is intrinsically a much better hydrogen-bonding site than an anisyl oxygen.

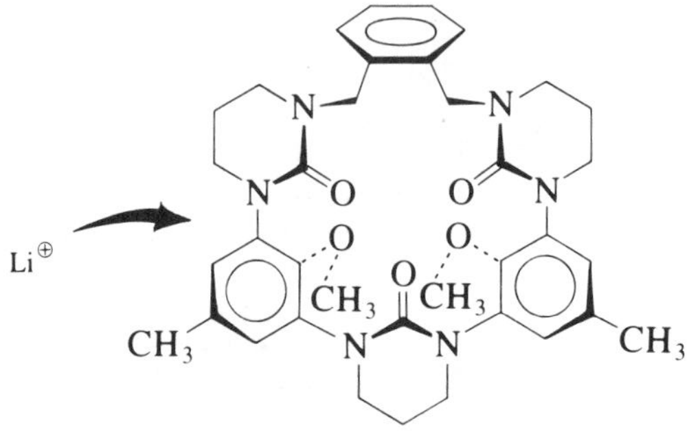

spherand with cyclic urea units

The synthesis of spherands and other organic bowl-shaped and calix-shaped (343) compounds with enforced cavities was in a way a tour de force in organic chemistry. Among other structures are *hemispherands*, hosts with a cavity half preorganized for binding, and *podands*, noncyclic hosts with an array of binding sites. These highly structured ligands of spherical form behave like "molecular prisons" by trapping ions permanently within them. One of these structures has even been named *carcerand* from the latin word for "prison;" in a way, its symmetry resembles industrial zolites.

These molecules are designed to bind not only ions but also small organic molecules such as CO_2, CH_4, and N_2. Projects at aiming to convert CO_2 to sugar or N_2 to NH_3 within these cavities to mimic photosynthesis or nitrogen fixation by plants are not far from being realized in the coming years. Another objective is to build cavitands with molecular pores for the passage of small molecules, a process well documented in polymeric films and membranes. In this context, Nature continually gives

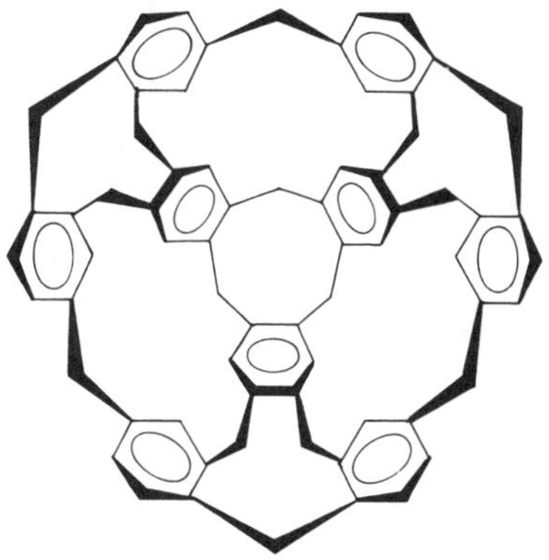

bowl-shaped calixarene

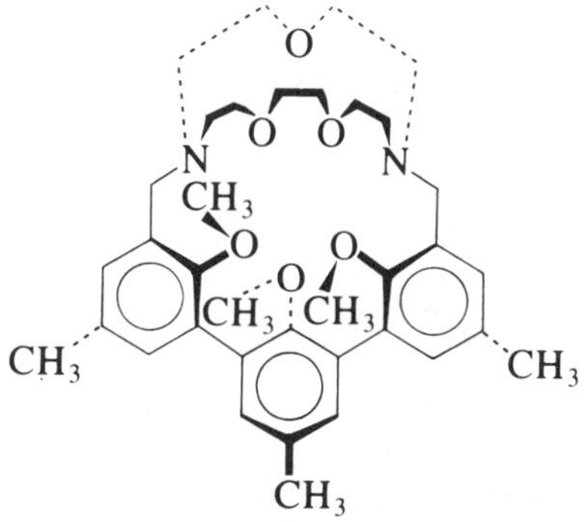

crypta-spherand

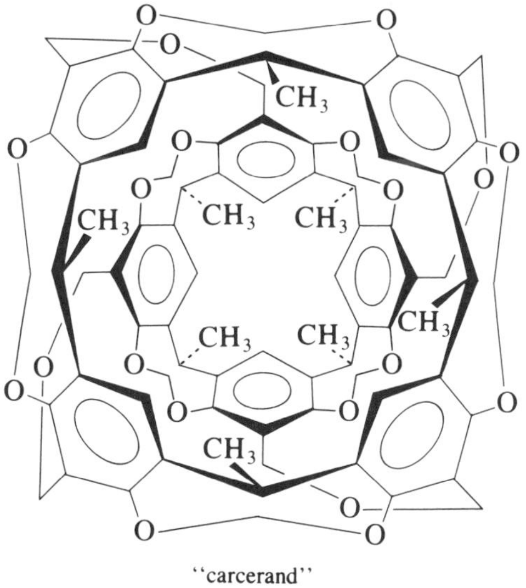

"carcerand"

us lessons. In its simplest form, the beautiful radiolarian marine organism *Trochodiscus longispinus* (Fig. 5.11) could represent an idealized but stimulating objective in molecular architecture that bioorganic chemists could humbly try to achieve.

The well-defined and organized structure of those synthetic molecular vessels were the beginning of a series of efforts by Cram's group to design and synthesize an enzyme-mimicking host compound. This challenge was oriented toward the mimic of serine proteases such as α-chymotrypsin (344). The target catalyst for the transacylase partial mimic possesses binding and catalytic groups preorganized to be complementary to salts of amino ester or amino amide substrates.

Fig. 5.11. Transmission electron microscopy of the marine organism *Trochodiscus longispinus*. Reproduced with the permission of J.I. Goldstein. Picture taken from Goldstein *et al.*, *Scanning Electron Microscopy and X-Ray Microanalysis*, Plenum Press, 1974

mimic Ser-195

nucleophile

mimic His-57

proton transfer catalyst

complexing site

mimic Asp-102 of α-chymotrypsin

target catalyst

The retrosynthesis analysis can be summarized as follows:

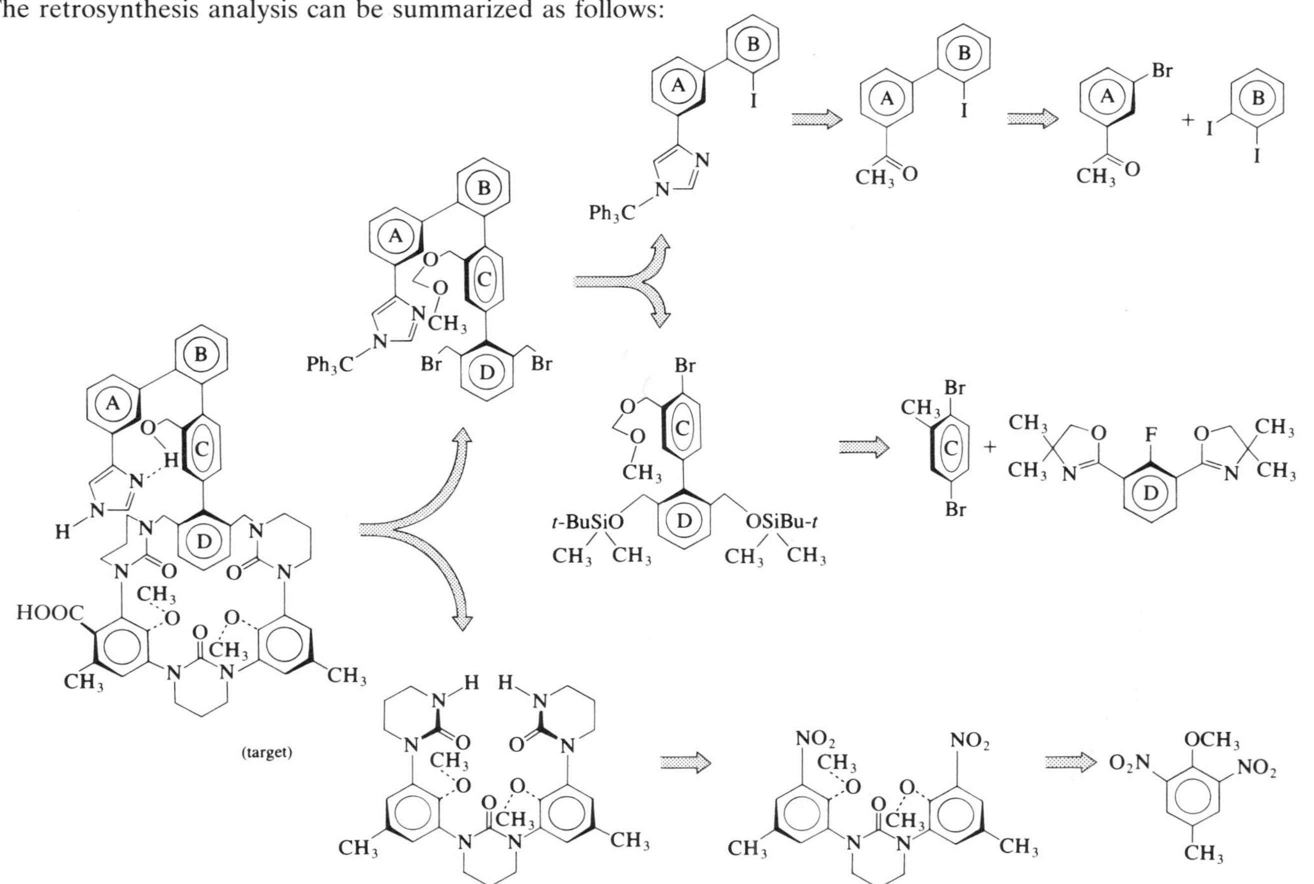

Because of the complexity of the problem, model compounds have been prepared in an incremental approach to serine transacylase mimics. One of these model catalysts has shown to direct acylation of an alanine ester by a factor of at least 100 billion times faster than when that guest is mixed with a comparable nucleophile lacking the ability to form a complex with that ester of alanine (345). The deacylation step, however, has not yet been mimicked. Nonetheless, such phenomenal rate enhancement in acylation surpasses by far the effect of ordinary organic catalysts and begins to rival the enzymatic efficiency whose properties Cram is seeking to duplicate.

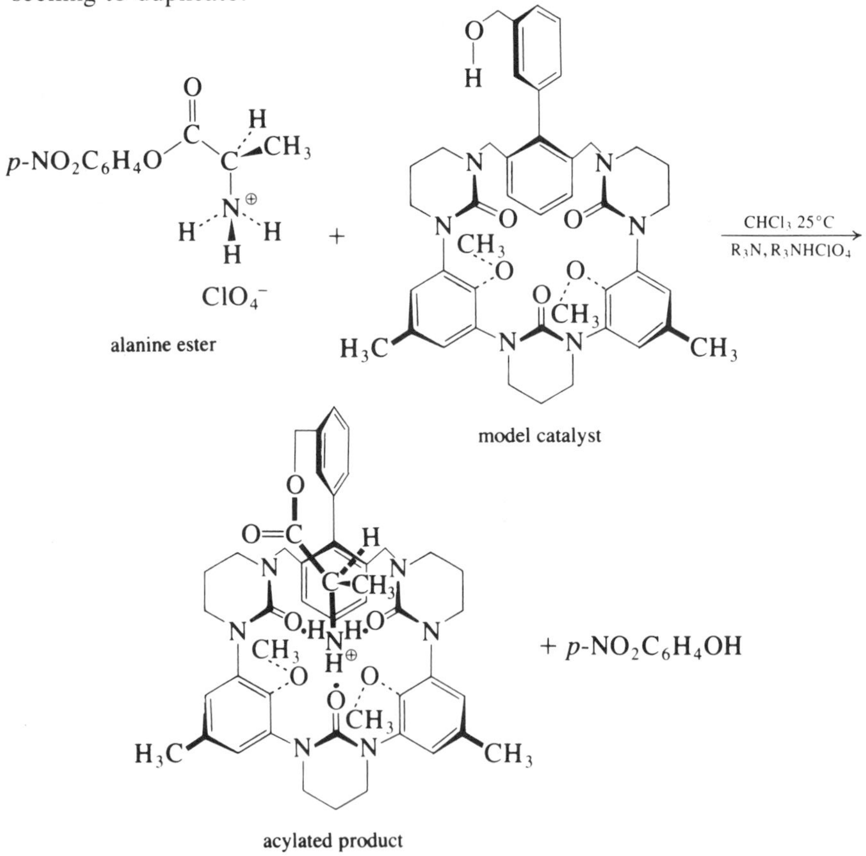

alanine ester + model catalyst $\xrightarrow[R_3N,\ R_3NHClO_4]{CHCl_3,\ 25°C}$ acylated product + p-$NO_2C_6H_4OH$

Other interesting molecular cavities worth mentioning are the *calixarenes*. A calixarene is a cavity-shaped, cyclic molecule made up of benzene units. These compounds were named first by C.O. Gutsche (Texas Christian University) because of their resemblance in shape to a calix-type Greek vase. Functionalized calixarenes can act as a new class of calalysts, ligands, and host molecules (346). One of the leaders in this field of research is S. Shinkai from Kyushu University. His group recently developed a series of so-called 1,3-alternate calix[4]arenes for the selective complexation of cations (347). Indeed, selective recognition of alkali metal cations are one of the central research interests in biological systems. During the last decade, it has been demonstrated that ionophoric cavities constructed from calix[4]arenes show Na^+ selectivity that is much higher than observed from crown ethers. The higher Na^+ selectivity is attributed to the size of the ionophoric cavity to fit Na^+ and to the rather rigid calix[4]arene skeleton serving as a platform. Furthermore, it became possible to freeze the rotation of phenyls units by introducing bulky substituents into OH groups or by cross-linking two phenyl units. Four possible calix[4]arene conformers that have been synthesized and isolated are represented below:

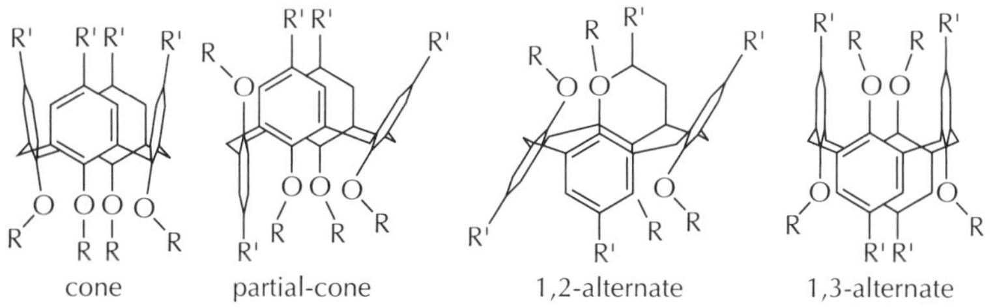

cone | partial-cone | 1,2-alternate | 1,3-alternate

Different conformations of calix[4]arenes
R = H,R′ = Pr^n

Conformationally rigidified calix[4]arenes are exceptionally useful in the design of ionophoric cavities with the desired size and preorganization of ligand groups (347).

The 1,3-alternate calix[4]arene conformer possesses several interesting structural characteristics that the other three conformers do not. For example, two independent ionophoric sites exist at both edges of the cavity; each ionophoric site is composed of two phenolic oxygens and two benzene rings; and the two ionophoric sites are connected by a "hole" surrounded by benzene rings. Careful examination of X-ray crystallographic pictures of a series of 1,3-alternate conformers reveals that the size of the ionophoric sites is comparable with the size of K^+ ion (2.6 to 2.7 Å). In fact, the K^+ complexes observed are stabilized not only by the interaction with the two phenolic oxygens, but also by the π-donor participation of two benzene rings. Shinkai's group found that the 1,3-alternate conformers show not only an unusually high affinity for K^+, but also for Ag^+, and support the π-donor participation in the metal binding. This is illustrated on page 310.

They also have evidence for metal tunneling through the calix[4]arene cavity. Indeed, dynamic H^1-NMR spectroscopy, especially at low temperatures (−50° to −85°) showed that the Ag^+ ion alternates intramolecularly between the two binding sites through a π-basic hole of the

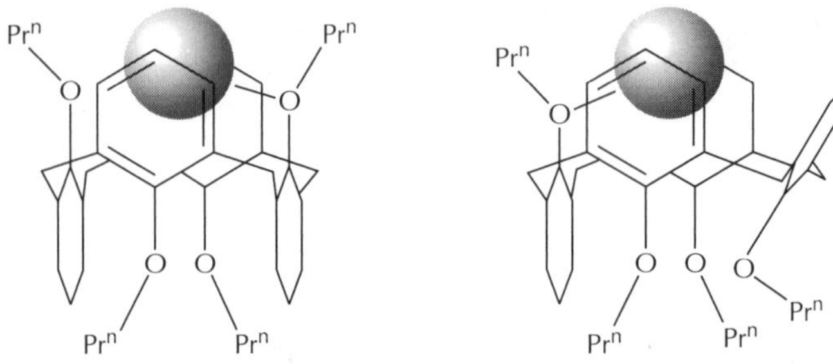

(a) (b)

Preferred binding mode of 1,3-alternate calix[4]arene conformer for K^+ in (A) and the partial-cone conformer for Ag^+ in (B) (347).

1,3-alternate calix[4]arene conformer. This represents the first example for Ag^+ tunneling across an aromatic cavity and has important implications with regard to the "metal cation-π interaction" expected for metal transport through ion channel, metal inclusion in fullerenes, intercalation of metal cavities into graphites, and so on.

5.2.7 Paracyclophanes, Polymer Crown Ethers, and Miscellaneous

Like the crown ethers, paracyclophanes (also a calixarene) are also macrocyclic molecules, but the backbone is entirely made of carbon and hydrogen atoms. The cavity is therefore much more hydrophobic in nature, and one has to rely on functional pendants in order to use those compounds as enzyme models.

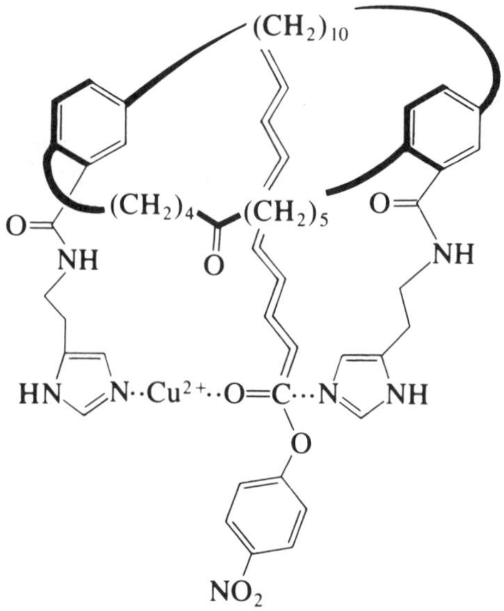

The previous structure is an instructive example because it combines hydrophobic effects to bind the substrate and imidazole chemistry coupled to a metal ion for performing a hydrolytic reaction on an aliphatic ester (348).

An acylated intermediate is formed but does not deacylate readily with water. The acceleration rate is primarily attributed to reduction of molecularity, going from a bimolecular to a pseudounimolecular process. However, the outcome is stoichiometric rather than catalytic.

Water-soluble cyclophane possessing a chiral hydrophobic cavity would have the advantage of stereoselective recognition. The first chiral cyclophane was made of two diphenylmethane units coupled to two chiral C_4-chains derived from L-tartaric acid (349). In water, this optically active paracyclophane forms diastereomeric complexes with small hosts like mandelic acid, lactic acid, and phenylglycine. The complexation can be followed by ^{1}H-NMR spectroscopy where the signals of (*R*)- and (*S*)-guests shifted to a different degree.

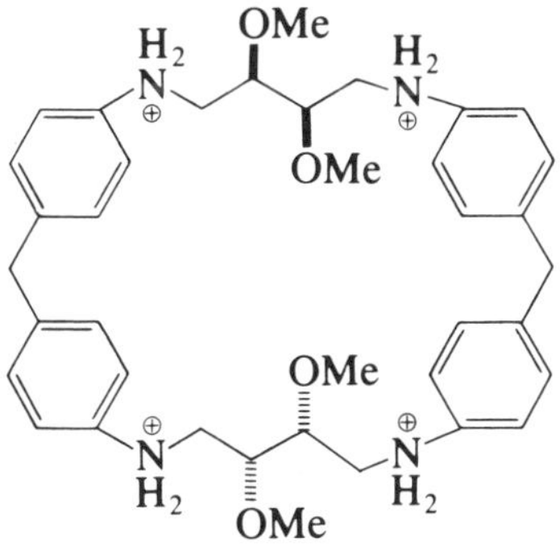

protonated form

Another interesting example is the synthesis of an optically active macrocyclic host incorporating an unnatural isoquinoline alkaloidlike element as the chiral building block (350). This novel chiral cyclophane

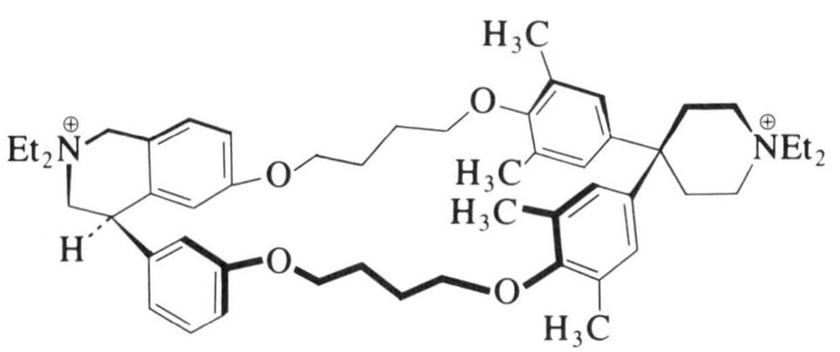

$2Cl^{\ominus}$

(+)-isomer

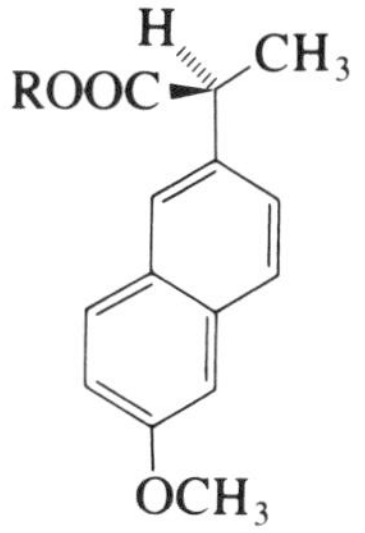

S-isomer of Naproxen

R = H
R = CH_3

forms diastereomeric complexes in D_2O/CD_3OD (60:40) with the anti-inflammatory drug Naproxen. Complexation between the (+)-host and the methyl ester ($R = CH_3$) of Naproxen is significantly stronger than the complexation with Naproxen itself (R = H). Furthermore, differences in proton chemical shifts of the aromatic signals between the (*R*)- and (*S*)-drugs are observed upon complexation of racemic Naproxen.

In the design of new molecular cavities, a receptor with six tentacular chains was synthesized from cyclotriveratrylene (351). By analogy with the marine animal octopus, it was labeled "hexapus." The molecule has a hydrophobic backbone with six tails ended by negative charges. It resembles an array of fatty acid chains but does exhibit much less surface tension activity than single-chained fatty acid salts. It is, however, a new complexing agent for a variety of organic molecules and can bind, for instance, to *p*-nitrophenyl butyrate and produce a base-catalyzed hydrolysis.

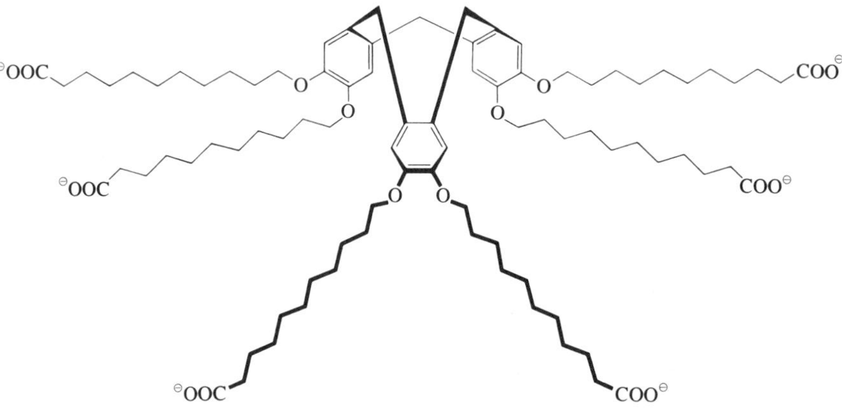

The "hexapus" with each six chains ending by a negative charge

Although this molecule is quite flexible, it allows inclusion of organic hosts in the cavity and eventual modulation of its tertiary structure. By convergence of the terminal carboxylates into efficient catalytically active groups, this model compound should make an attractive enzyme mimic.

Even more exciting is the recent development by Stoddart's group of "molecular belts" (Fig. 5.12), accomplished by employing repetitive Diels–Alder reactions on readily available bisdiene and bisdienophile (352). This molecule opens the way into a fascinating new range of molecular architectures. It can intercalate $CHCl_3$ molecules in a layerlike manner. Furthermore, the X-ray crystal structure reveals a certain structural elegance about this highly symmetrical novel molecule with all six oxygen atoms distributed around the outer surface of the stiff molecular collar.

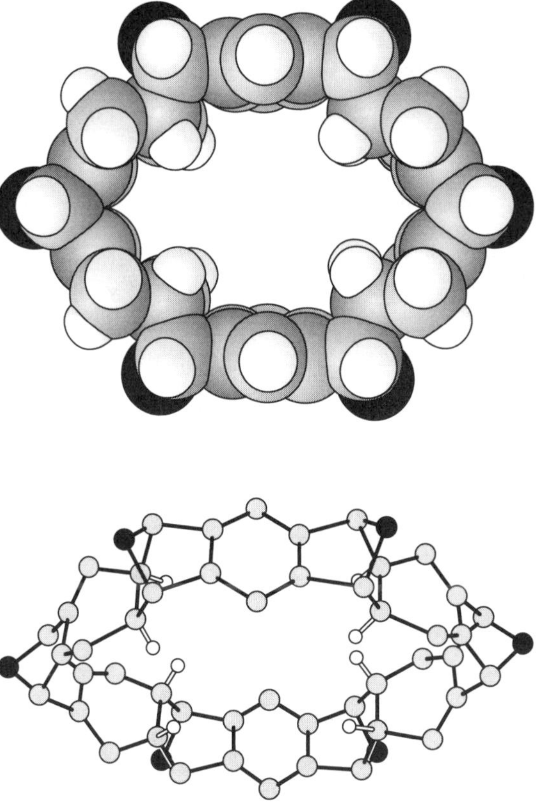

Fig. 5.12. X-ray structure of Stoddart's "molecular beit" compound, just described (352). The oxygen atoms are in black. Reproduced with permission. Copyright © 1987 by Verlag Chemie GmBH.

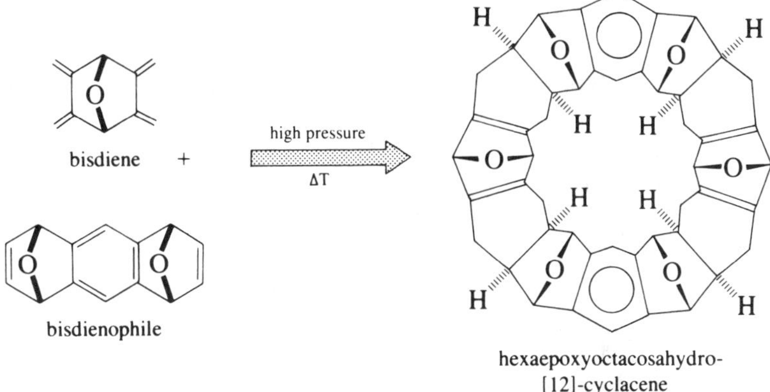

As pointed out by the conceptor, the path is now clear for the production of a vast array of other macropolycyclic derivatives with wedge-shaped clefts and grooves. A more complete account of this work is available (353).

Fullerene molecules have earned tremendous interest in recent years because of their growing application in industrial projects. The first monoaminated fullerene derivative, a C_{60}-azacrown ether, has now been reported (354). The computer-minimized structure is illustrated in Fig. 5.13. One can envisage the possibility that such a molecule could interact with sodium or potassium metal surfaces. As a result of ionization, the resulting cation would be trapped by the crown ether ring while the electron would be captured by the fullerene moiety.

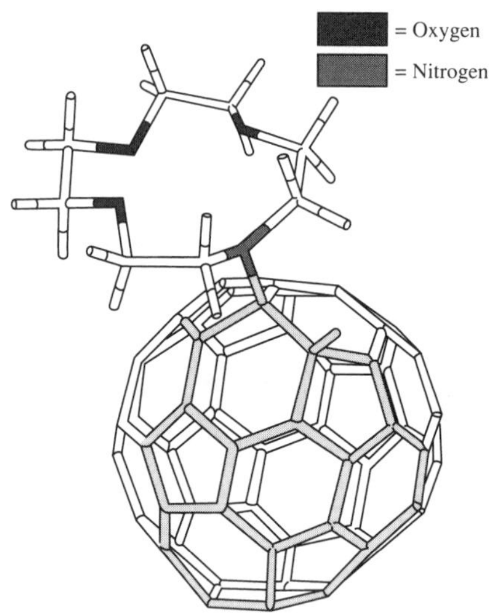

Fig. 5.13. Semiempirical minimized structure of the lowest energy isomer of C_{60}-amino-12-crown-4 (354). Reproduced with permission. Copyright © 1994 by the Royal Society of Chemistry.

This section ends with a few words about the importance of crown ether network polymers. Indeed, many active research programs are aiming at developing highly selective membrane carriers for alkali cations based on the polymerization of functionalized crown ethers (355). A few examples are illustrated below. The polymeric backbone could be derived from a polystyrene or from a polyamine or polypeptide. J. Smid (State University of New York) has contributed greatly to this venture (356). A large number of these crown ether polymers were used as efficient resins in ion selective chromatography.

A future development in this field concerns the construction of membrane with an artificial channel-type of ionophore and the prepara-

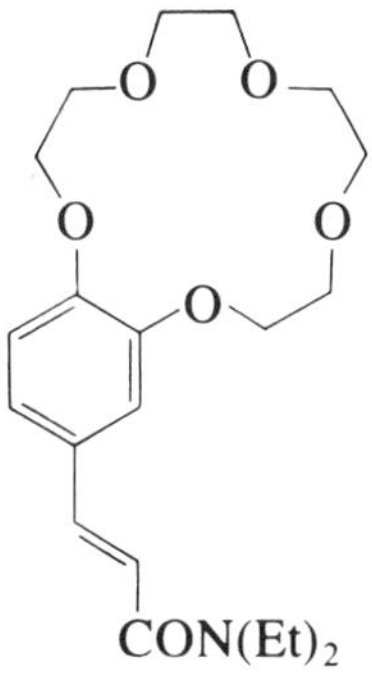

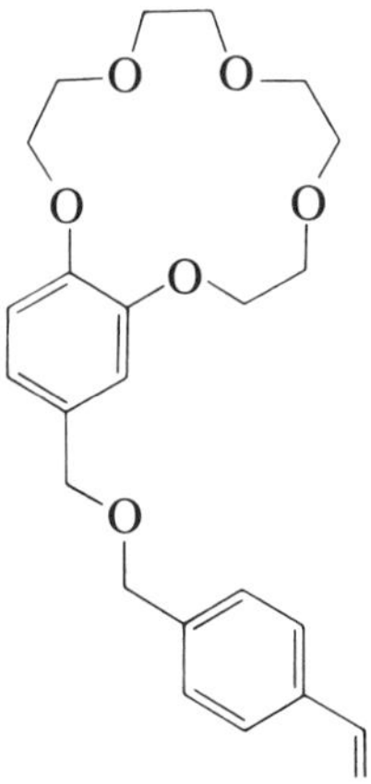

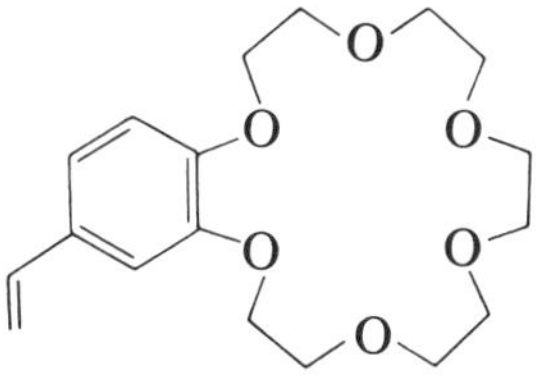

$CON(Et)_2$

tion of new polymeric materials. In this context, J. Simonet (University of Rennes) developed a bidimensional network of polymeric crown ethers by electrochemical polymerization of the aromatic system, tris(15-crown-5) triphenylene (357). The polymer matrix is composed of a large number of planar macrocycles piled up in a regular way. This molecular arrange-

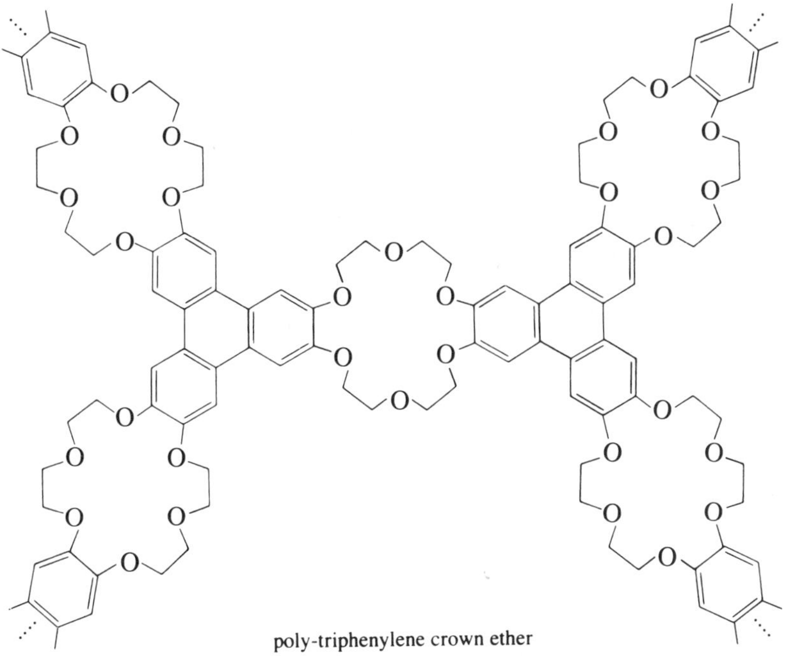

poly-triphenylene crown ether

ment of crown ethers, one above the other, facilitates the circulation of ions like K^+ via channels perpendicular to the planes of the polymer network. In a way this novel material mimics biological membranes and will permit the development of new ion selective and porous membrane-like organic ion conductors.

Accordingly, J.L. Dye (Michigan State University) has obtained a new kind of crystalline matter made of crown ether molecules organized in a sandwichlike manner (358). This new class of crystalline materials has been called *electrides*. They can trap electrons in their cavities leading to unusual optical, magnetic, and electronic properties.

W.C. Still's group has made constant efforts over the years to develop efficient and stereoselective receptors of amino acids and peptides (359). An illustration of this is the synthesis of this cuplike-shaped receptor for binding and selective recognition of (L)X-(L)Pro-(L)X tripeptides. With naphthlene rings, the cavity has a diameter of ~8 Å as compared to the original receptor (with benzene rings) with a cavity of ~6 Å. As a consequence, the new receptor (**N**) can now interact with the internal residue (R_2) of tripeptides, whereas the smaller original receptor (**O**) could interact primarily with small substituents of peptides. This is illustrated in the following diagrams.

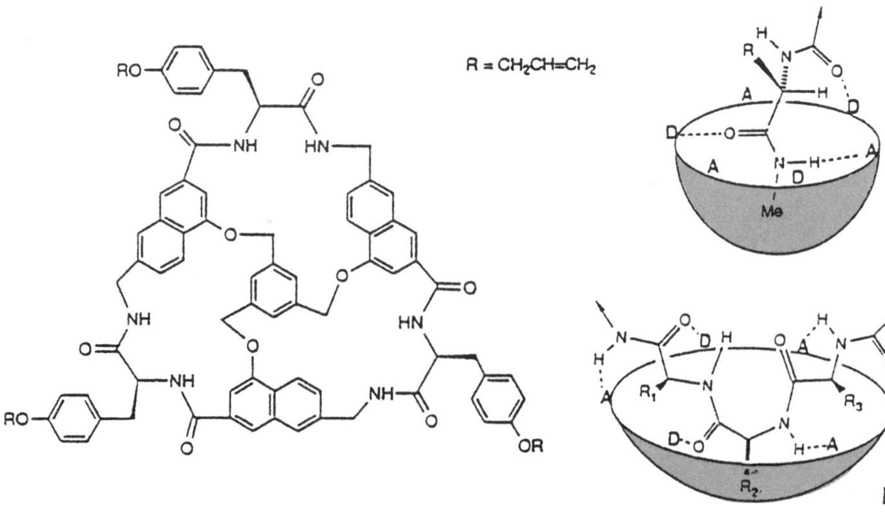

The structure of the peptide receptor and two cup-shaped diagrams illustrating the original small (**O**) cavity and the new wider (**N**) cavity. The new cavity has the naphthalene rings and can form as many as six hydrogen bonds (359).

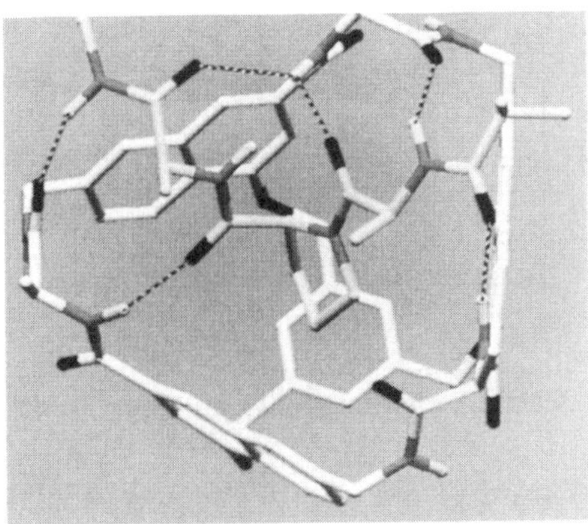

Fig. 5.14. A presentation of the global minimum of the complexation of iPrCO-(L)Ala-(L)Pro(L)Ala-NHMe (in blue-green) in the peptide receptor cavity (359). The dotted lines show the hydrogen bondings. Reproduced with permission. Copyright © 1994 by the American Chemical Society. See color print.

This receptor shows a sequence-selective tripeptide binding preference for the *L*-configuration and in particular for iPrCO-(L)Ala-(L)Pro-(L) Ala-NHMe. A computer simulation of the complex is illustrated in Fig. 5.14.

This result illustrates how known principles of receptor design and the increasing capacity of computer library screening can be combined to create a powerful approach to problems in molecular recognition. This model can be used to design receptors for other tripeptide sequences (359).

5.3 Membrane Chemistry and Micelles

What makes lipidic membranes unique as a molecular entity is the remarkable combination of their intrinsically high fluidity and, at the same time, the possibility of compartmentalization. This "molecular paradox" is related to the amphiphilic nature of the noncovalent lipid components plus the natural property to assemble and to organize themselves into an orderly tridimensional network. Depending on the nature and vocation of the molecular assembly, membranes develop specific functions as diverse as secretion, transport, regulation, and translocation, or other activities such as, for example, endocytosis of drugs and metabolites like cholesterol bound to lipoprotein.

Surfactants are amphiphilic molecules; that is, they have both pronounced hydrophobic and hydrophilic properties. A detergent is an excellent example. In solution, such low molecular weight electrolytes form ion pairs with the counterions. By increasing the monomeric concentration, clusters and then low molecular weight aggregates are formed. Finally, larger aggregates called *micelles* are produced (360–365). Therefore, micellization of monomeric surfactants is observed when the surfactant concentration exceeds the so-called *critical micelle concentration* (cmc). In general, the cmc varies from 10^{-2} to $10^{-4}\,M$ and the conductance of the solution changes sharply above this concentration.

Most often micelles are spherical. In a polar solvent such as water, the hydrophobic hydrocarbon chains of the surfactants are directed toward the interior while the polar or ionic head groups are distributed on the surface of the sphere facing the counterions in the aqueous solution. It is the coming together of hydrophobic groups in water that leads to a favorable entropy change because of the liberation of water molecules from the aqueous–apolar interfaces where the hydrophobic groups appear to have considerable freedom of movement in the micelle. It is this gain of entropy that leads to the favorable free-energy change on micellization.

Figure 5.15 gives an idealized spherical model of a micelle. Micellization of a surfactant such as dodecyltrimethyl ammonium bromide creates positively charged surfaces composed of cationic "heads." Coulombic attraction gathers the bromide ions into the vicinity of the quaternary nitrogens. This region forms the Stern layer and the most interesting micellar chemistry occurs in this region. Very little water is present in the interior of a micelle, which is hydrocarbonlike, and it is this difference in polarity between the interior and the surface that makes micelles resemble globular proteins. In effect, the polarities of micelle surfaces are generally similar to those of proteins and intermediate between that of water and

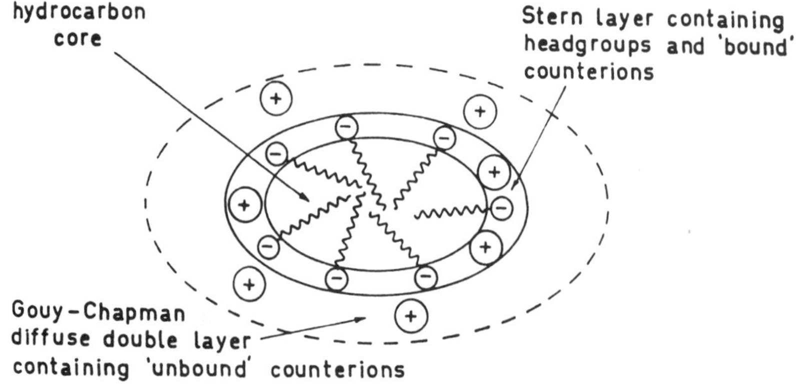

Fig. 5.15. Elliptical cross section of an idealized spherical model of an ionic micelle (363). Reprinted with permission. Copyright © 1977 by The Chemical Society.

ethanol. That the active site of an enzyme is apparently quite apolar, even though the enzyme as a whole is water soluble, makes micelle studies very pertinent (362,363).

It is well known that organic compounds, particularly nonpolar ones, can absorb onto or in micelles, increasing their solubility relative to that in pure water and often altering their chemical reactivity. At the same time, it is the micelles, rather than individual surfactant molecules, that are responsible for altering the rate of organic reactions in aqueous solution of surfactants. Therefore, a proper choice of surfactant can lead to rate increases of 5- to 1000-fold compared to the same reaction in the absence of surfactants. Depending on the type of micelle used, this results in a large concentration of H^+ or OH^- ions, which are gathered in the Stern layer and are responsible for the increase in the rate of the reaction. Other basic groups or nucleophiles in the micelle should also have an effect on catalysis. There is a much weaker interaction between the micelle and the counterions in the wider Gouy–Chapman layer, which extends for several hundred angstroms from the micelle surface and causes a gradual ion gradient.

Rate enhancements have been observed with cationic, anionic, and nonionic micelles.

For instance, bis-2,4-dinitrophenyl phosphate is rapidly hydrolyzed at pH 8 by cationic micelles. Anionic micelles will inhibit the reaction because of repulsions with the negatively charged product and competition with the negatively charged OH^- which causes the reaction. On the other hand, zwitterionic micelles, which are usually relatively ineffec-

$$\left(O_2N\text{–}C_6H_3(NO_2)\text{–}O\right)_2PO_2^{\ominus} \longrightarrow O_2N\text{–}C_6H_3(NO_2)\text{–}O^{\ominus} + O_2N\text{–}C_6H_3(NO_2)\text{–}O\text{–}P(=O)(O^{\ominus})\text{–}O^{\ominus}$$

tive in catalysis, are effective in the following decarboxylation reaction (361).

$$C_6H_5\text{–}CH(CN)(COO^{\ominus}) \longrightarrow C_6H_5\text{–}CH^{\ominus}(CN) + CO_2 \xrightarrow[\text{fast}]{H^{\oplus}} C_6H_5\text{–}CH_2\text{–}CN$$

The rationale is based on favorable coulombic interactions as follows:

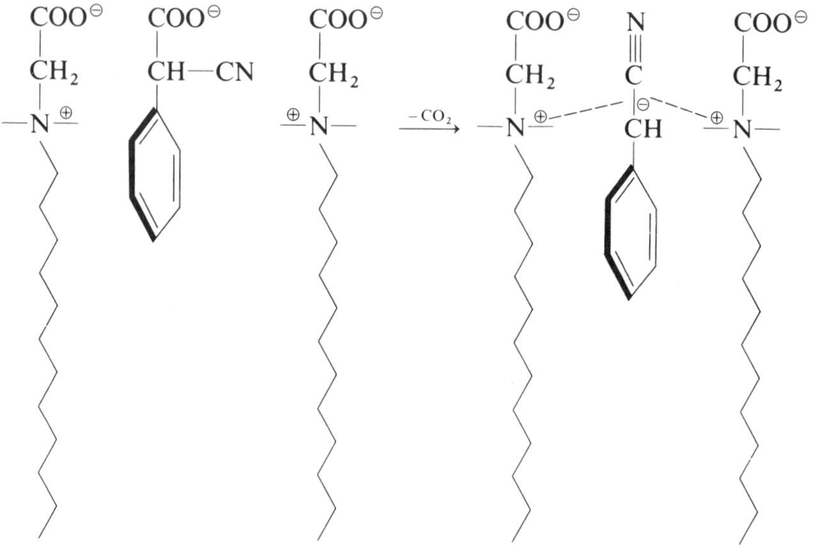

Consequently, many features of kinetics in micellar systems are related to reactions in monolayers and polyelectrolytes surfaces.

Sodium dodecyl sulfate (SDS) [$CH_3(CH_2)_{11}SO_3^-Na^+$], a well-known surfactant, forms spheres containing 50 to 100 molecules. The potential between bulk and micellar phases is about 50 to 100 mV, and electrostatic and hydrophobic interaction forces are important factors for maintaining the stability of the micelles. SDS is often used to denature proteins where similar forces are present in their tertiary structure.*

Examples of micellar systems are given below where the surfactant provides the medium for catalysis but does not directly participate in the reaction. For instance, *N*-acetyl histidine can bring about the hydrolysis of *p*-nitrophenyl esters via an acyl-imidazole intermediate.

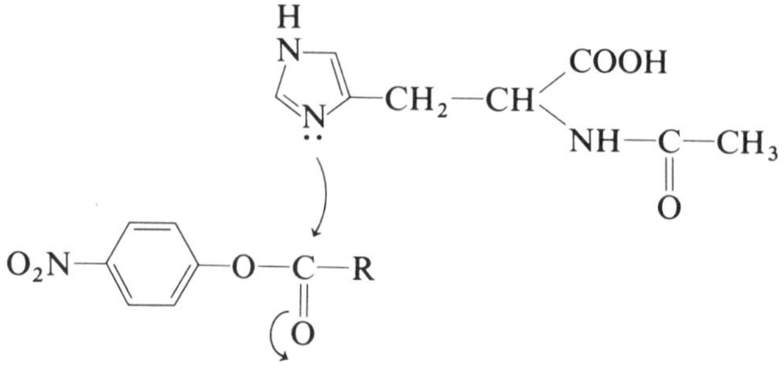

* That is, the final folded form of the polypeptide chain or active form of a protein.

The nucleophilicity of the imidazole ring toward the ester function is enhanced when the system is in a micellar form, that is, in the presence of SDS, as a result of a favorable high concentration of catalyst and substrate in the micelle.

Cyanide ions are known to react with *n*-alkyl pyridinium salts:

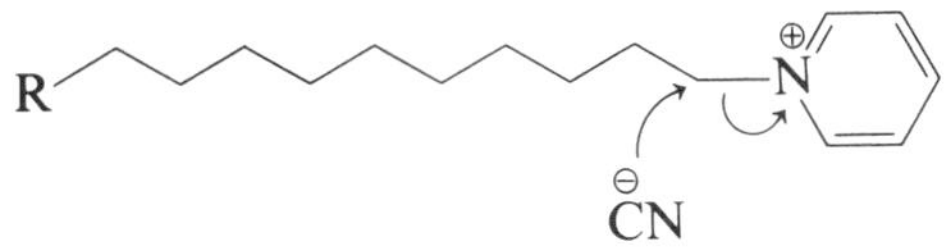

R N ⊕ $\overset{\ominus}{C}N$

However, the rate is markedly increased by cationic surfactants. The longer the alkyl chain, the faster the rate. This shows that hydrophobic binding is largely responsible for the increase in rate of substitution in this reaction.

Cyanide ions can also be added to 3-carbamoyl pyridinium bromide (NAD^+ analog):

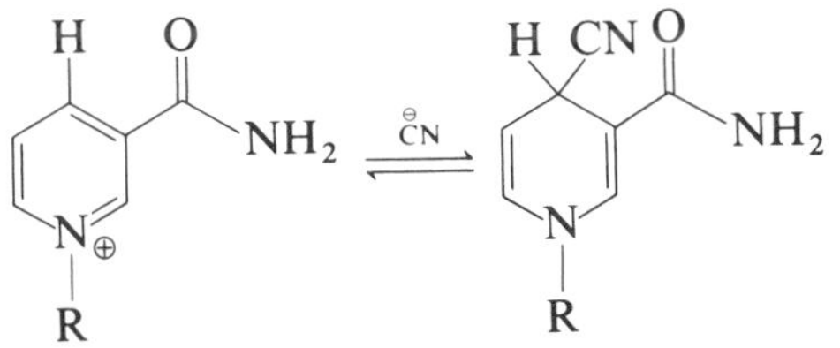

H O NH$_2$ N⊕ R ⇌ ($\overset{\ominus}{C}N$) H CN O NH$_2$ N R

$R = C_{16}H_{33}$

The presence of $0.02\,M$ cetyltrimethyl ammonium $[CH_3\text{-}(CH_2)_{15}-\overset{+}{N}(CH_3)_3]$ salts increases the rate constant by 950-fold for the addition of CN^- to the *N*-hexadecyl substrate ($R = C_{16}H_{33}$) and increases the corresponding association constant about 25,000-fold (366). Notice that in this transformation, the charged substrate becomes neutral after reaction, so the product of the reaction will be reoriented within the micelles and will be pulled inside. The hydrophobic interactions destabilize the reactant with respect to product, and it has been suggested that the presence of the surfactant is responsible for the large acceleration in the rate of attack by the CN^- ion. As a consequence, the proportion of [substrate] versus [product] of the reaction is displaced to the right in the micellar system.

Mention should be made of *reversed* or *inversed micelles* (367). Sulfosuccinate surfactants form reversed micelles where a remarkable amount of water (50 moles/mole of solute) can be incorporated inside the micelle in octane solutions.

The concept of "water pools" was introduced by F.M. Menger in 1973 to describe the nature of the cavity inside reversed micelles (367).

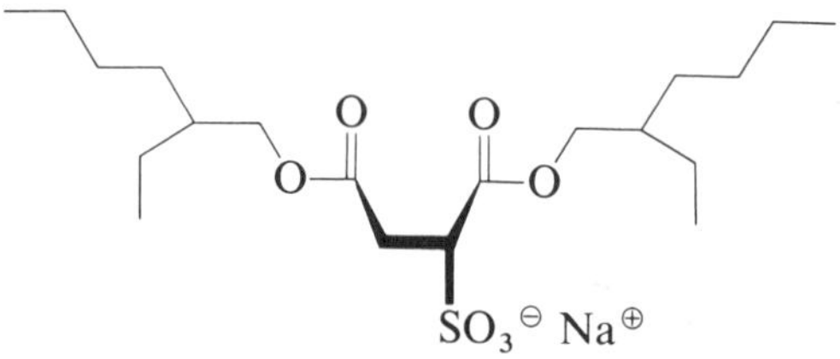

$SO_3^{\ominus}$ $Na^{\oplus}$

di-2-ethylhexyl sodium sulfosuccinate

Addition of *p*-nitrophenyl acetate in the presence of imidazole to this micellar system results in a 53-fold increase of hydrolysis of the acetate as compared to bulk water. Clearly, imidazole in the micelle is able to come very close to the substrate and to catalyze its hydrolysis. Therefore, remarkable rate enhancements in reversed micelles have been ascribed to favorable substrate orientation in the interior of the reversed micelles, where bond breaking may be assisted by proton transfer.

Dodecylammonium propionate (DAP) also forms reversed micelles at 0.10 *M* concentration in benzene and is able to entrap 0.55 *M* of water. This system was used by J.H. Fendler to investigate the protonation of pyrene-1-carboxylic acid using a nanosecond time-resolved fluorescence technique (368). The most striking feature of the data obtained is the extraordinarily large rate constant for the protonation of the carboxylate group in the surfactant-solubilized water pool; a value of the order of $10^{12}\,M^{-1}s^{-1}$ was observed! Figure 5.16 gives a model of this ultrafast proton transfer system in the reversed micelle which is feasible only if the donor and acceptor are in close proximity. A fraction of DAP is hydrolyzed to propionic acid which is pulled into the water pool region and proton transfer occurs within the hydration shell of the surfactant.

Charge and proton relay through hydrogen bonds have been proposed to contribute to the catalytic efficiency of enzymes, and in this sense reversed micelles provide an appropriate model to delineate the importance of such factors at the enzyme active site. Micellar surfaces also provide a convenient means for the reduction in dimensionality, an important factor in enhancing reaction rates. They also serve as good models to demonstrate the feasibility of ultrafast proton transfer when the reactants are localized in a suitable environment such as membrane surfaces and other complex biomacromolecules.

Micelles with attached catalytic groups are also known and can act as catalysts. For example, esters and carbonates can be hydrolyzed by long-chain *N*-acyl histidine surfactants (369).

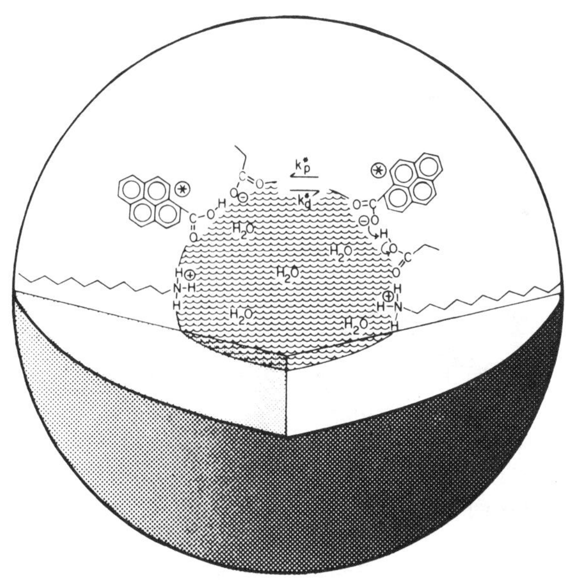

Fig. 5.16. A proposed model for the ultrafast proton transfer at the hydration shell of the surfactant head groups in reversed micellar DAP in benzene. Since the concentration of surfactant is in a very large excess over the probe, proton transfer must occur from dodecylammonium propionic acid to pyrene-1-carboxylate. For the sake of clarity, two pyrene moieties are drawn in the aggregate shown. In reality, there is much less than one probe per aggregate. The shaded area indicates the extent of water that hydrates the surfactant head groups (368). Reprinted with permission. Copyright © 1978 by the American Chemical Society.

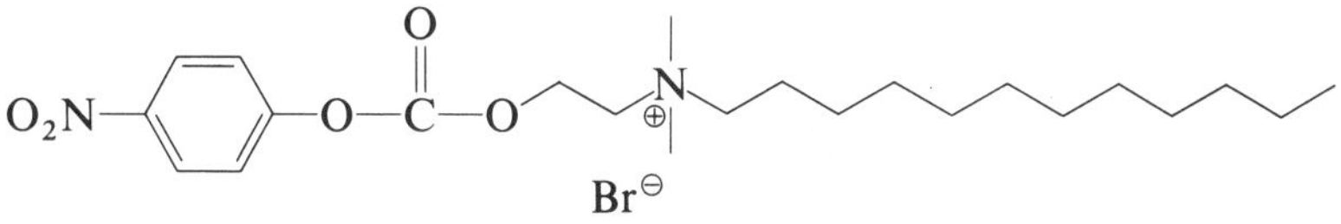

substrate (carbonate)

N-Dodecyl-*N'*,*N'*-dimethyl aminoethyl carbonate ion is hydrolyzed 2240 times faster with the following surfactant than with *N*-acetyl histidine.

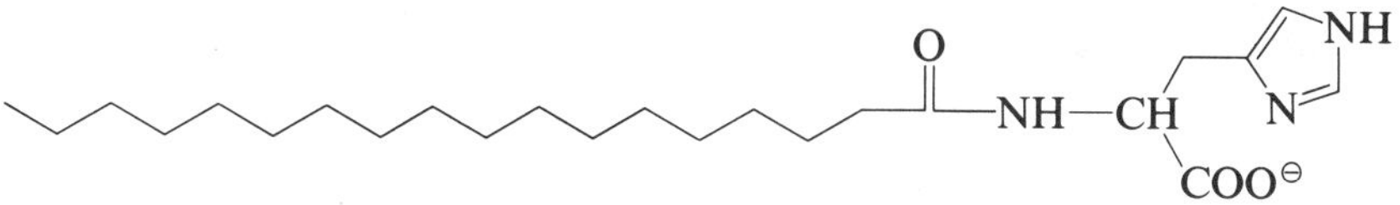

catalyst (*N*-acyl histidine)

In this context, many micellar enzyme analogs of serine proteases have been prepared (270). Cysteine proteases, however, such as papain and ficin, have been modeled only recently (370).

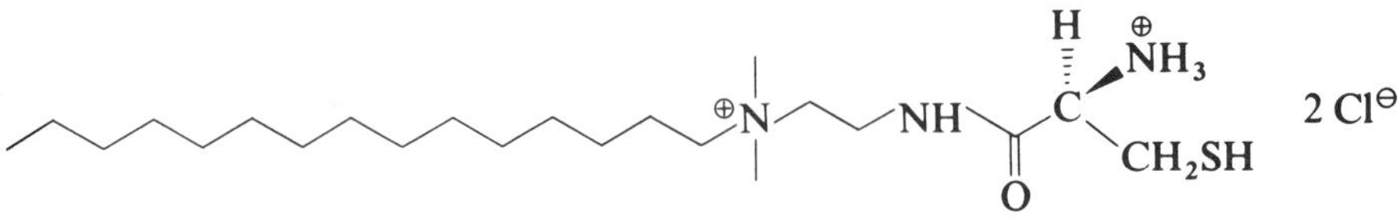

self-contained thiol-functionalized surfactant

The above cysteine-containing long hydrocarbon chain forms micelles between 0.003 and 0.05 *M* and cleaves *p*-nitrophenyl acetate with a pseudo-first-order rate constant. The surfactant is 180 times more reactive than cetyltrimethyl ammonium chloride, a micellar system without a functional group present.

An imaginative example of micellar catalysis (371) is in the acyloin (benzoin) condensation in the presence of *N*-lauryl thiazolium bromide (see Section 7.4).

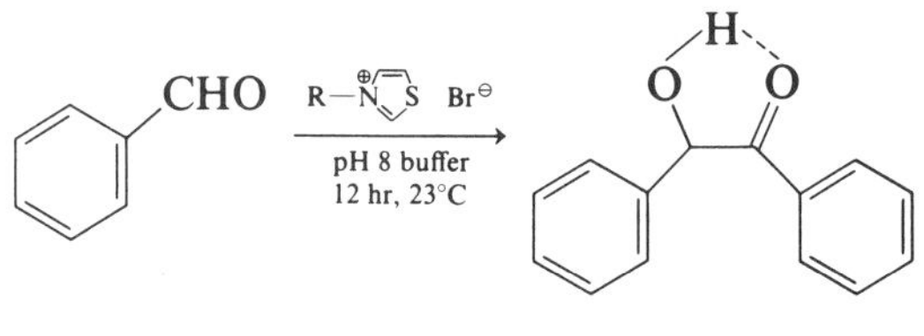

When R = butyl, the reaction does not work; if R = dodecyl, micelles are formed, benzaldehyde molecules intercalate, and the yield reaches up to 95% conversion.

In summary, impressive catalytic effects are obtained by incorporating reactants in micelles, thereby increasing their effective concentration and reducing the entropy loss in the transition state by providing an effective medium for the reaction. The other advantages of a micellar system are:

(a) favorable hydrophobic interactions;
(b) model to some degree the behavior of enzymes and membranes (phospholipid vesicules are catalysts);
(c) the forces that hold micelles are similar to those for tertiary structure of proteins;
(d) increases local concentration of ions responsible for the catalysis;
(e) rate enhancement up to 1000-fold have been observed.

However, it is difficult to have a greater increase in rate of reaction with micelles because of the inherent uncertainty as to the structure of the reaction site. Therefore, the limitations are:

(a) the structure of the micelle is not well defined;
(b) the structure depends on surfactant and substrate concentrations;
(c) rigid orientational effects are not expected unless immobilized surfactants on a polymer support can be achieved;
(d) nothing is known about the relative orientation of reactive groups.

Thus a micelle remains a very crude enzyme model.

5.3.1 Stereochemical Recognition

How about stereochemical recognition? So far, only few micelles of optically active surfactants have been used as catalysts in a number of reactions with chiral substrates, but in general the effects are small. Two examples will be given here where chiral micelles could stereoselectively catalyze the hydrolysis of chiral esters.

Cationic surfactants derived from D(−)-ephedrine analogs show different catalytic efficiencies in hydrolyses of *p*-nitrophenyl esters of D- and L-mandelic acid (372). Hydrolysis of the racemic mixture is slower than its enantiomers with D(−)-surfactant, suggesting that more than one substrate molecule is incorporated into each micelle. Therefore, an enantiomeric substrate molecule perturbs the micellar structure so that the resulting complex then exhibits markedly different activities toward the two enantiomers.

C_6H_5—CH(—OCH_3)—COO—C_6H_4—NO_2

D(−)-mandelic ester
L(+)-mandelic ester
substrates

C_6H_5—CH(OH)—CH(CH_3)—$N^{\oplus}(CH_3)_2$—R $Br^{\ominus}$

R = $C_{10}H_{21}$
= $C_{12}H_{25}$
D(−)-surfactant

The liberation of *p*-nitrophenol can be followed spectrometrically at 25°C, pH 9.0 in 0.01 *M* borate buffer plus 0.5% dioxane using a substrate concentration of 10^{-3} *M*. The results are presented in Fig. 5.17.

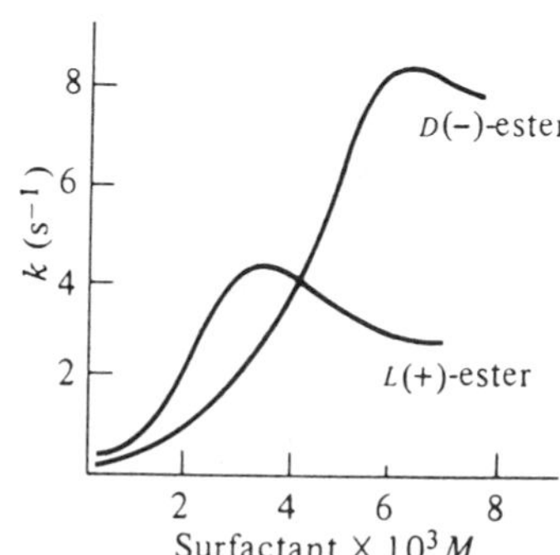

Fig. 5.17. Selective hydrolysis of mandelic ester by a chiral micelle (372).

A stereoselective hydrolysis of D(−)-mandelic ester over L(+)-enantiomer is taking place. At high surfactant concentration, some inhibition is observed, probably from the presence of surfactant counterions (salt effects).

A more dramatic example, again from C.A. Bunton's work (373), is the use of an optically active L-histidyl-cationic micelle. It shows a larger degree of stereoselective control (approximately threefold) for hydrolysis of amino acid derivatives.

The synthesis of the micelle is outlined below:

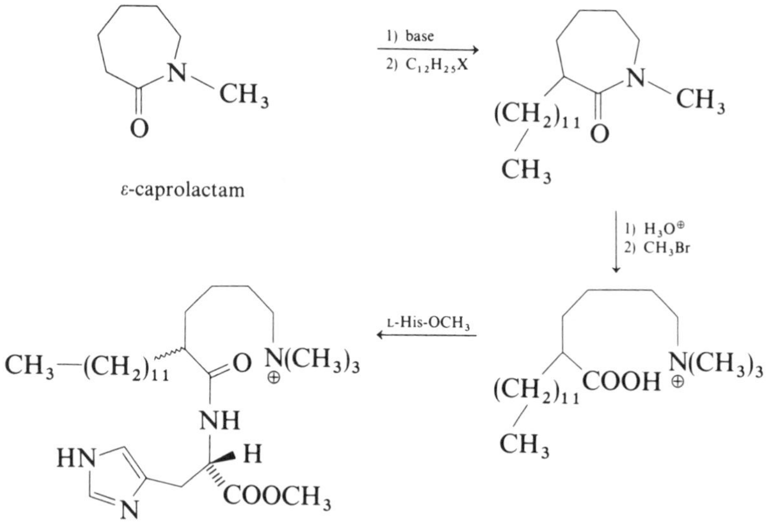

ε-caprolactam

1) base
2) $C_{12}H_{25}X$

1) $H_3O^{\oplus}$
2) CH_3Br

L-His-OCH$_3$

This surfactant is obtained in 33% excess of one pure enantiomer and acts as a powerful catalyst in 0.02 *M* phosphate buffer, pH 7.4, 25°C, for the deacylation of *p*-nitrophenyl-2-phenyl propionate.

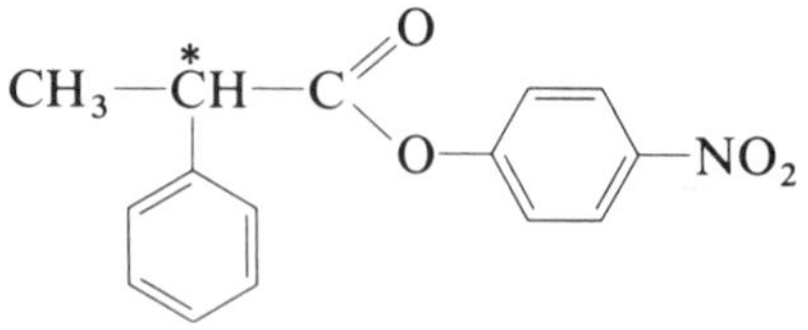

The rate of the reaction is pH dependent and shows that the group participating in the catalysis has a p*K* of 6.4 to 7.5. As compared to bulk buffer, rate enhancements of 260 and 283 for *R*- and *S*-isomers are obtained, respectively, but the binding constants are the same for both isomers.

However, a better stereoselectivity is obtained with *N*-acetyl-phenylalanine esters.

The *S*-isomer is deacylated faster with an enantiomeric reaction rate ratio of 3:1. The binding constants were again found to be the same for both enantiomers, so the difference in rates is probably due to ΔG differences in the transition state for the enantiomeric amino acid ester. The results are presented in Fig. 5.18.

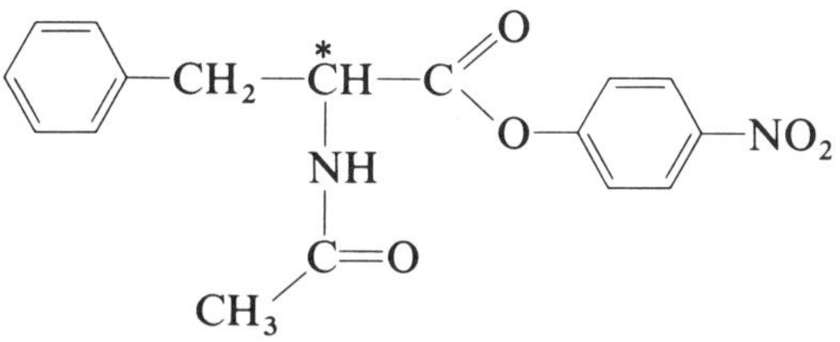

The initial rates are similar, confirming that the stereospecificity of the reaction depends on the transition state rather than initial state interactions. Addition of a competing surfactant such as cetyl ammonium salts still yields an enantiomeric rate ratio of 2:2.5 for the *S*-isomer depending on the concentration. Consequently, it is the presence of a functional group in the micelle rather than the individual surfactant molecules that is responsible for the selectivity observed, the imidazolyl residue functioning as the nucleophilic catalyst.

A hypothetical transition state intermediate between the chiral micelle and the *S*-isomer can then he presented:

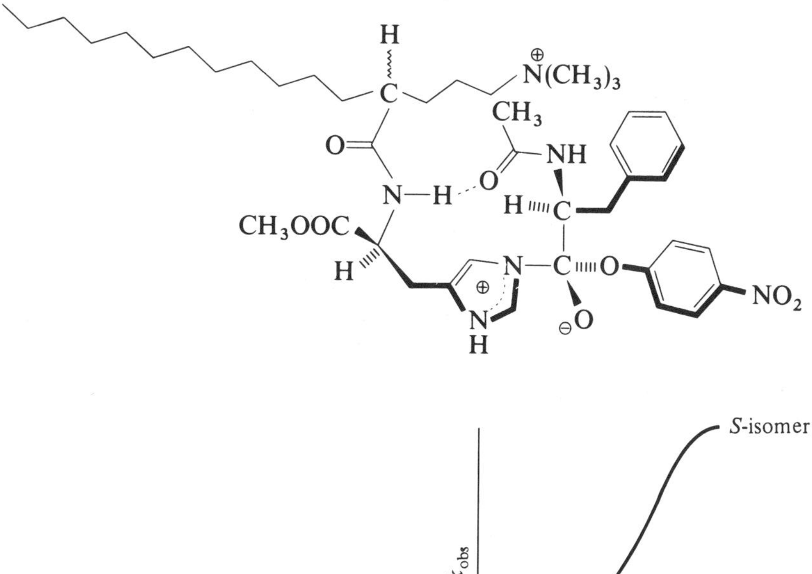

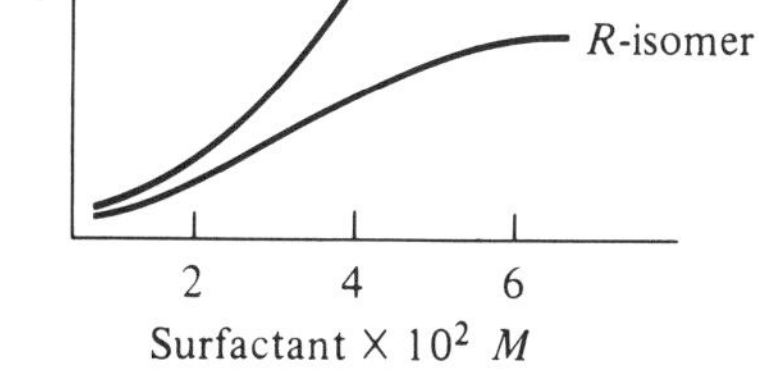

Fig. 5.18. Selective hydrolysis of *N*-acetyl-Phe-ester by the L-histidyl-cationic micelle (373).

The *R*-isomer, however, would suffer from a severe imidazole-phenyl ring repulsion.

This approach of micellar stereoselectivity by functional surfactants has been extended to cleavage of dipeptide diastereomeric substrates (374).

In conclusion, solvation changes and loss of some of the translational entropy in forming a transition state are two important factors responsible for catalysis and rate enhancements observed with micellar systems. In this respect they resemble enzymes. Another formal similarity between enzymatic and micellar catalysis is the strong hydrophobic binding with the substrate. However, the fact that micelles are of limited rigidity results in poor specificity in catalysis and only moderate rate enhancements are obtained.

Nevertheless, micellar systems have found applications in pharmacology and in industry, particularly in emulsion polymerization. Furthermore, synthetic organic chemists are frequently faced with the problem of reacting a water-insoluble organic compound with a water-soluble reagent (OH^-, MnO_4^-, IO_4^-, OCl^-, etc.). The use of surfactants can now alleviate this problem. Tow-phase reactions may be undertaken where the surfactants disperse organic liquid in water, generating higher yields and shorter reaction times (364).

5.3.2 New Developments in Membrane Mimic

Making artificial membrane systems, either via planar lipid bilayers or by lipid vesicules, to study the mechanisms of biological ion transport by ion carriers and ion channels has been the concern of many bioorganic chemists and biochemists for many years (375).

Many exciting new systems have been developed. One aspect concerns the development of vesicules made from polymerizable surfactants embedded with semiconducting particules (CdS) and the possible utilization in artificial photosynthesis (376) or in systems to convert water to hydrogen molecules (377). Active polymerized vesicules can be made, for instance, by sonicating a surfactant possessing a redox-active group

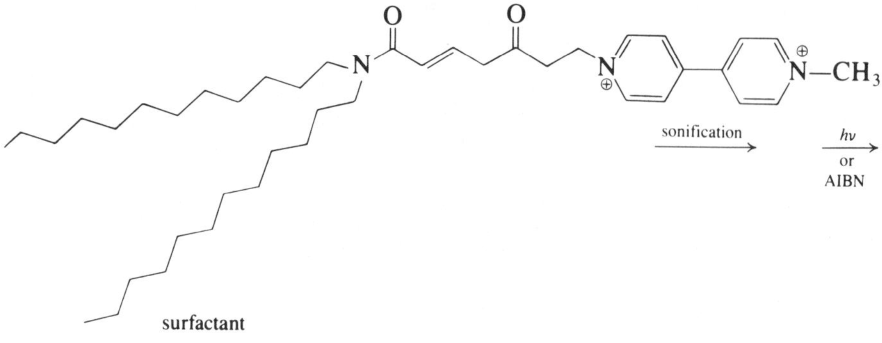

O O
N
N⊕
⊕N—CH$_3$
sonification
$h\nu$
or
AIBN
surfactant

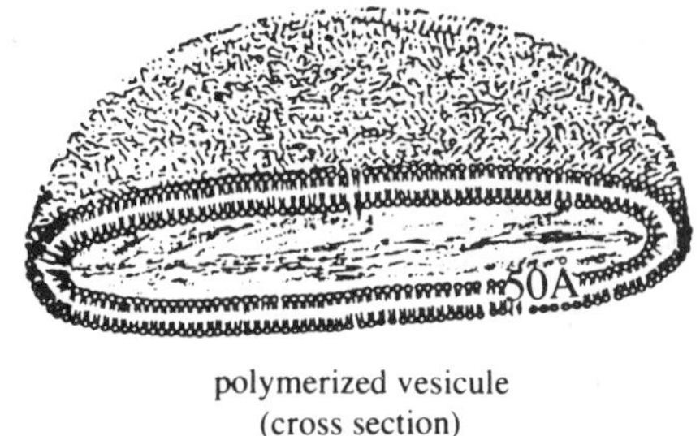

polymerized vesicule
(cross section)

like a viologen molecule and a residual double bond. The sonication forms bilayers that upon irradiation or in the presence of a promotor of polymerization (AIBN) are converted into vesicules polymerized across bilayers. Polymerized membranes made from polymerizable surfactants could ultimately be part of an artificial system that could catalytically split water molecules with hydrogen and oxygen formed and harvested in compartments separated by the bilayer. A model is presented in Fig. 5.19.

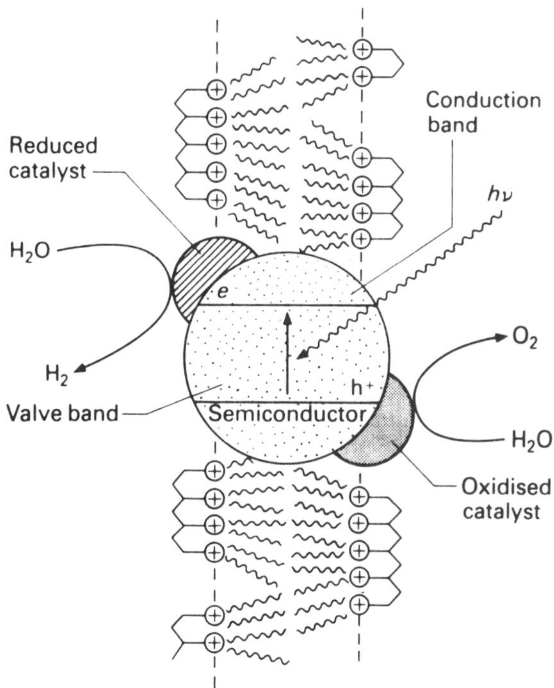

Fig. 5.19. A hypothetical model for a cyclic water-splitting system based on a semiconductor particle immobilized in polymerized membrane (377). Reproduced with permission. Copyright Royal Society of Chemistry.

Other types of polymerizable lipids have been made and some structures are represented below. They can be used as spectroscopic bilayer membrane probes (378).

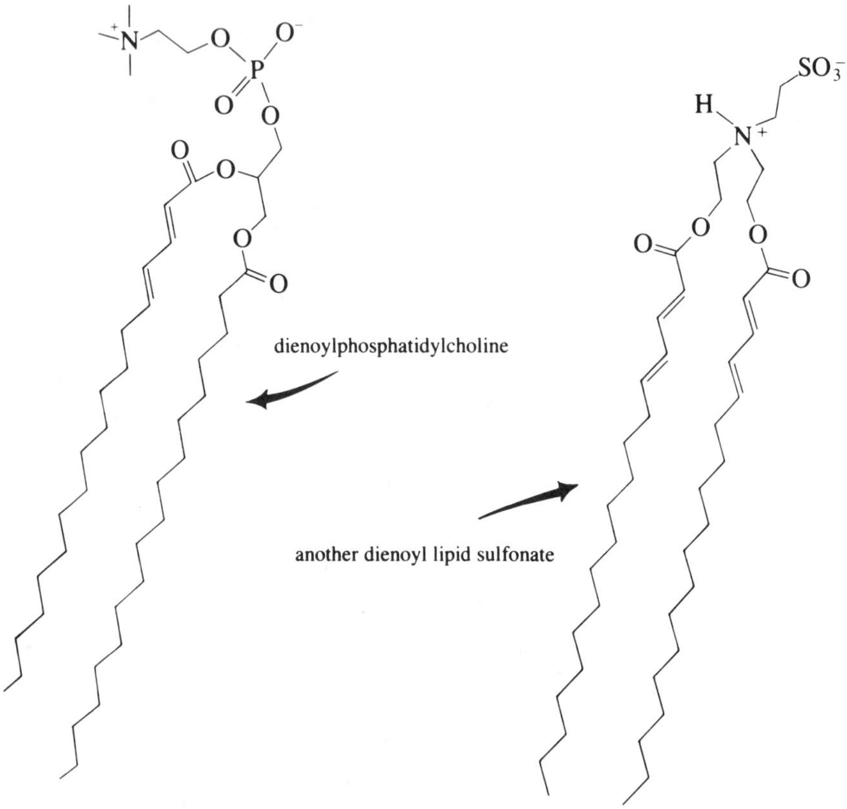

dienoylphosphatidylcholine

another dienoyl lipid sulfonate

Some novel globular aggregates, composed of synthetic peptide lipids, have also been made (379). The polar peptidic segments orient themselves like the ionic head group of phospholipids.

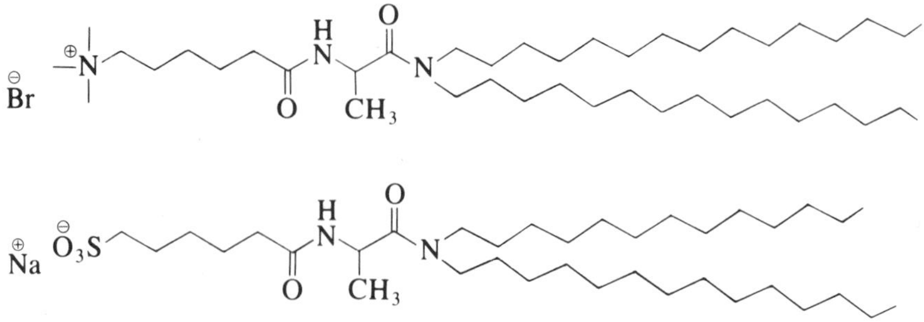

an equimolar mixture of these two lipids forms spherical aggregates when dispersed in water

Lipidic chains attached to aza-crown ethers have also been used by Shinkai to prepare lamella and rodlike aggregates (380).

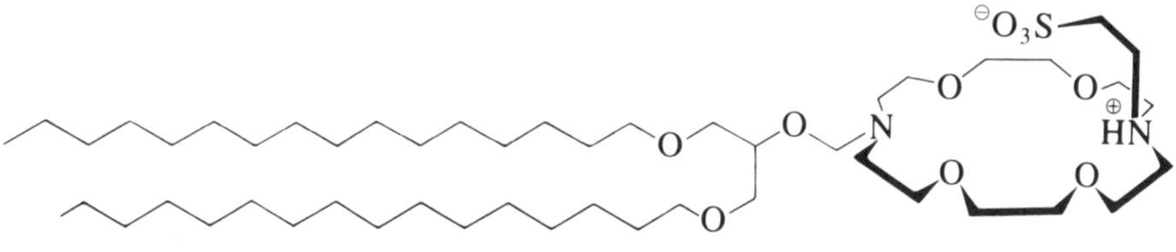

anionic-capped crown ether ring

Generally the critical micelle concentration of ionic surfactants is lowered by the addition of salts by a salting-out effect. In the present case the reverse effect (salting-ion effect) is observed by addition of K^+ ion. Consequently the crown ether rings of the membrane become more hydrophilic through complexation with metal cations.

Amphiphilic molecules like phospholipids but with head groups at both ends of a hydrophobic core have been named *bolaform amphiphiles* or *bolaamphiphiles* (381). The expression "bipolar lipids" has also been used by biochemists who found them in *archaebacteria*. A large variety of these molecules has been synthesized and used to form unsymmetrical vesicular membrane, a starting point for the preparation of charge-separating systems (382).

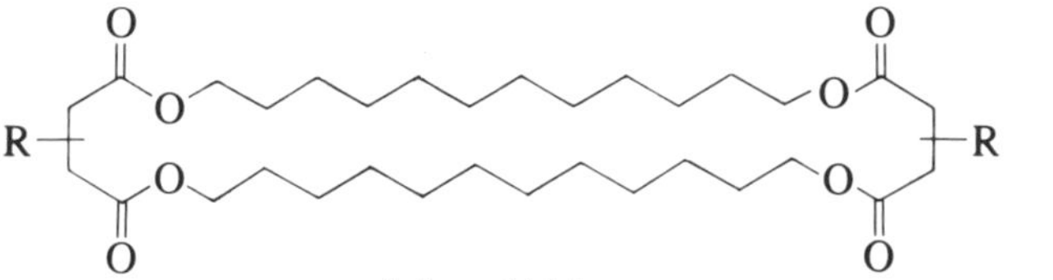

bolaamphiphiles

$R = SO_3^{\ominus}Na^{\oplus}$
$R = S{-}CH_2{-}COO^{\ominus}Na^{\oplus}$
$R = S{-}$ (sugar: OH, OH, OH, OH)

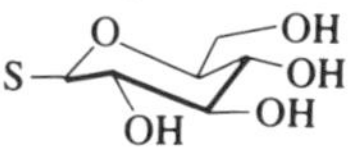

In particular, they produce in water extremely thin (~20 Å) monolayer lipid membrane vesicules as compared to the usual lipidic bilayers, which are in the range of 50 to 80 Å thick in the hydrophobic part. However, the bolaform monolayers are relatively unstable and collapse at relatively low pressures. On the other hand, they offer the possibility of spanning across the membrane.

A general strategy for the synthesis of membrane-spanning bipolar phospholipids equipped with a molecular probe in the hydrophobic membrane environment has recently been described (383). Photoaffinity label or spin label probes would be worthy candidates here to investigate lipid bilayer activities and drugs and ion transport.

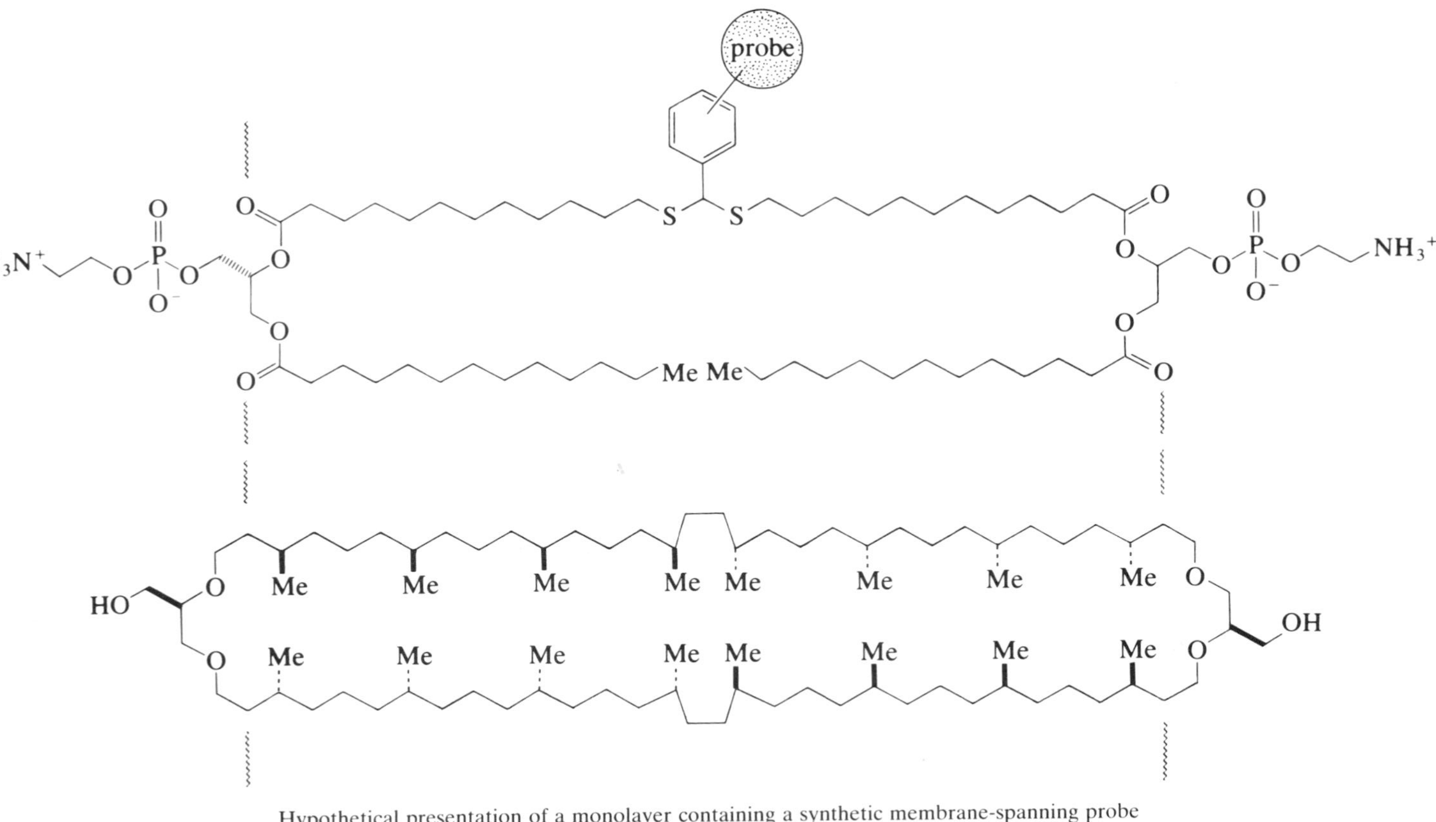

Hypothetical presentation of a monolayer containing a synthetic membrane-spanning probe in a naturally occurring transmembrane lipid found in *archaebacteria* (383).

In biomimetic model systems of ion-transport processes, two main molecular mechanisms can be discerned: transport mediated by *carriers* and transport mediated by *pores* or *channels*. So far, emphasis from the synthesis of such models has been on the former mechanism, despite the fact that the latter one appears to be more general among natural systems (384).

Using a crown ether-modified bolaamphiphile, T.M. Fyles's group (University of Victoria) developed a synthetic transporter that can mimic an ion channel in a bilayer membrane (385). A diagram of the transporter with the bilayer membrane is sketched in Fig. 5.20.

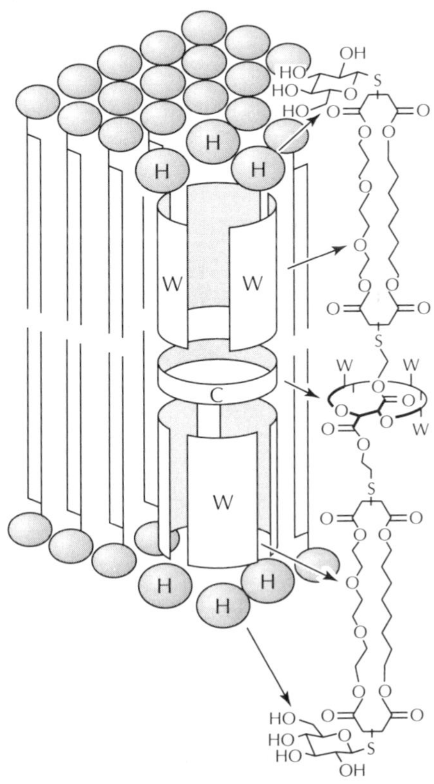

Fig. 5.20. Sketch of the ion-channel mimic in a bilayer membrane (385). A crown ether core (C, an 18-crown-6 hexaacid) acts as a framework to support the channel walls (W, formed from macrocyclic tetra esters) bearing polar head groups (H, derived from 1-mercapto-β-*D*-glucose in this example), which project toward the faces of the bilayer. The structure would be maintained across the bilayer by the cooperative interaction of the head groups in contact with water, hydrophobic contacts with the membrane lipids, the rigidity of the wall units, and the crown ether framework. Reproduced with permission. Copyright © 1990 by the National Research Council of Canada.

The influx of cations into these vesicles by the synthetic transporter was coupled to proton efflux with a selectivity for K^+ ion. Taken together, the available data were most consistent with the synthetic transporter acting via a channel mechanism, similar to gramicidin, than via a know carrier such as valinomycin.

In a continuing effort to design new structurally organized and functionally integrated chemical systems built into supramolecular architectures, Lehn and collaborators (386) described the first approach to the synthesis of what they called a *molecular wire*. The first molecules synthesized toward this goal are *caroviologens*, compounds that combine the features of both carotenoids and of the viologens. An example is the following bispyridine polyene:

Such structure can be incorporated into a membrane and behaves as a *membrane-spanning electron channel*. The length of the caroviologen is sufficient to span the bilayer membrane with the pyridinium sites being close to the negatively charged outer and inner surfaces of vesicules made of sodium dihexadecylphosphate. In principle, a redox active mobile carrier molecule could have terminal redox groups, such as quinones, that can be reduced by accepting an electron from one side of the membrane and transferring it to the other side via the polyene backbone of the molecular device.

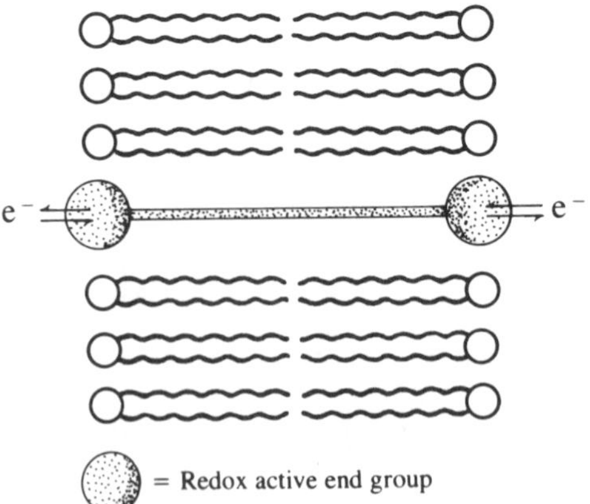

Schematic representation of a molecular wire as transmembrane electron channel (386). Each end of the molecular wire has a redox active group allowing electron flow to travel across the bilayer.

Further developments could involve the coupling with photoactive groups to yield *photoresponsive electron channels* and charge-separation devices by means of photoinduced electron transfer across membrane. Such structural elements would be of much interest for studies in artificial photosynthesis. Furthermore, the push-pull element in carotenoids could present interesting electronic and optical properties.

Thus, an intriguing new field of chemistry may develop, which Lehn calls *chemionics:* the design and operation of components, devices, circuitry, and systems for signal and information treatment at the molecular level (386).

Based on *molecular electronics*, molecular assembly of particular chemical elements (inorganic or organic) for the design of specific functions, characteristic of the chemical properties of the elements within, will be increasingly used in new microdevices. The molecular information and amplification signal to be obtained will be reduced to a single-molecule kind of function.

Because of their capability to host several types of molecules, reversed micelles have been used to solubilize enzymes and carry synthesis of peptides (387). The reversed micelles used were formed by the anionic surfactant bis(2-ethylhexyl) sodium sulfosuccinate (AOT, see p. 322). In a way, reversed micelles can be viewed as micro-reactors whose physical properties can be continuously modulated by the environment. Another important aspect of their structure is the difference between the polar core and the hydrophobic nature of the outside surface. It is this difference in microenvironment that was advantageously utilized by carrying the synthesis of a water insoluble tripeptide with α-chymotrypsin solubilized inside the micelles.

In this particular case the reaction chosen was:

$$\underset{\text{water insoluble dipeptide}}{\text{Cbz-Ala-Phe-OMe}} + \underset{\text{water soluble amino acid}}{\text{H-Leu-NH}_2} \overset{\alpha\text{-CT}}{\rightleftharpoons} \underset{\text{water insoluble tripeptide}}{\text{Cbz-Ala-Phe-Leu-NH}_2} + \text{MeOH}$$

The product of the reaction is expelled out of the reversed micelles. Typically, the reactions are carried out in 0.1 *M* borate buffer, pH 10, 20°C, and the yields vary between 40 to 60% in tripeptide formation.

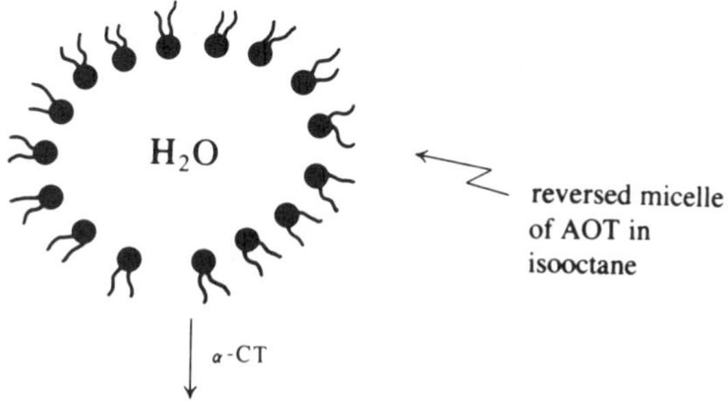

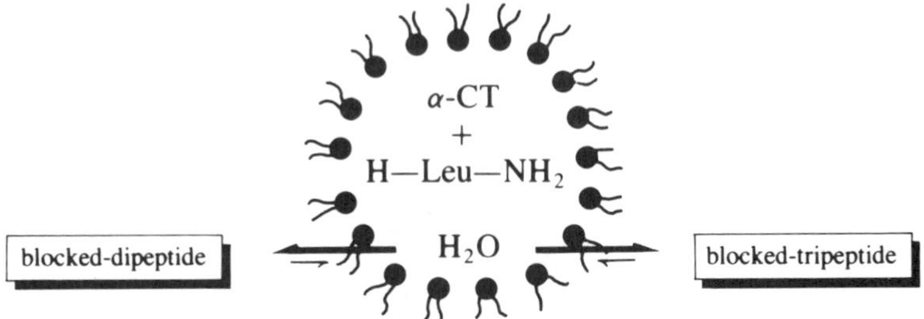

It was the first successful enzymatic peptide synthesis reported in a hydrocarbon micellar solution. A variant of this type of "organic" enzyme reactor could have potential industrial applications. In a different approach, Whitesides' group has developed a simple and practical technique for manipulating enzymes in organic synthesis by carrying out reactions with the enzyme enclosed in commercially available cellulose acetate dialysis membranes (388). This *membrane-enclosed enzymatic catalysis* (MEEC) method has been tested in a number of representative enzyme-catalyzed reactions. The chemistry of these transformations have already been presented in Chapter 4 (Section 4.8).

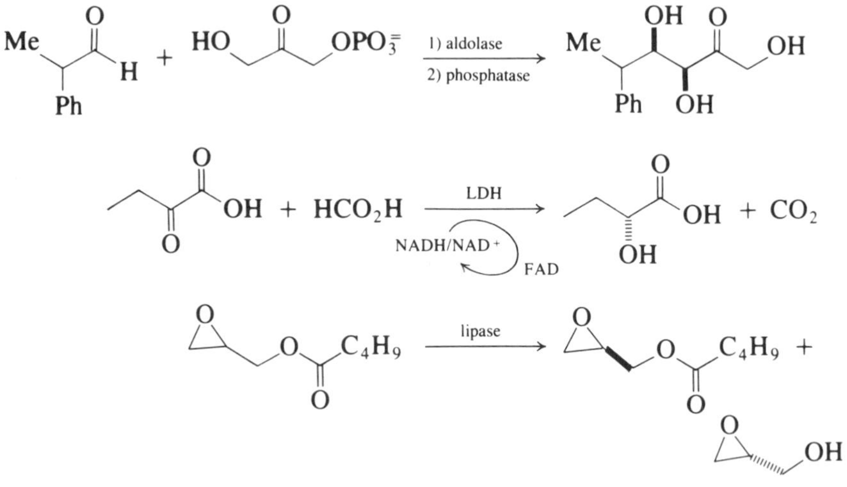

1) aldolase; 2) phosphatase

$+ \ HCO_2H \xrightarrow{LDH} \ + \ CO_2$ (NADH/NAD⁺, FAD)

lipase

In practice, the enzyme-containing bag is submerged in the aqueous phase of a two-phase water/hexane system. After the reaction has taken place in the aqueous phase inside the dialysis bag, the product diffuses out into the overlying organic phase. Because it is trapped in the membrane bag, the enzyme is protected from deactivation by contact with the organic solvent and can be recovered and reused.

The major advantage of MEEC technique relative to immobilized enzymes (Section 4.7) is that it is operationally more convenient and simpler with less loss of enzyme activity because of protease contaminants. Some reactions do proceed more slowly and yet globally, the advantages of MEEC technology surpass the disadvantages.

5.4 Polymers

Enzymes are copolymers omposed of various amino acid monomers. It is then easy to understand why the utilization of synthetic organic polymers to change the reactivities of low molecular weight substances has received more and more attention lately (389). These reactions can serve as models for more complex enzymatic processes. Although polymeric catalysts are considerably less efficient than enzymes, several analogies between natural and synthetic macromolecular systems have been revealed. In particular, a polymer with charged groups will tend to concentrate and/or repel low molecular weight ionic reactants and products in its vicinity and, consequently, will function as either an inhibitor or an accelerator of the reaction between two species. However, if catalytically active functions are added to a polymer that contains charged groups, the polymer itself, and not its counterions, will take part in the catalysis (390,391).

We will describe in this section examples of the charged group type of polymer, in particular imidazole-containing polymers, which have esterolytic properties and in many ways resemble serine proteases (390).

Early investigations with poly(methacrylic acid) showed that this polymer can catalyze the nucleophilic displacement of bromine ion from

$$\text{polymer}-COO^{\ominus} + Br-CH_2CONH_2 \longrightarrow \text{polymer}-COOCH_2(CONH_2) \xrightarrow{H_2O} \text{polymer}-COOH + HO-CH_2CONH_2$$

α-bromoacetamide. However, if the degree of ionization of this polyacid increases, the catalytic power of the polymer decreases markedly.

$$\text{methacrylic acid (COOH)} \xrightarrow{\text{polymerization}} \text{poly(methacrylic acid)}-(COOH)_n \underset{H^{\oplus}}{\overset{OH^{\ominus}}{\rightleftharpoons}} -(COO^{\ominus})_n$$

methacrylic acid poly(methacrylic acid)

Poly(vinyl-4-pyridine) has the catalytic capacity to accelerate the solvolysis of 2,4-dinitrophenyl acetate, but the rate constant of the reaction increases as the fraction of neutral pyridine residues increases. In fact, both free and charged species are probably needed for catalysis; the charged residues providing electrostatic binding.

From these two examples, it can be concluded that a balance of neutral and charged functions must be important for the polymer to act as an efficient catalyst. The above two polymers can be classified respectively as anionic and cationic polymers. Ion-exchange resins are among these categories. R.L. Letsinger was among the first to find applications of

vinyl-4-pyridine — polymerization → poly(vinyl-4-pyridine) $\underset{OH^{\ominus}}{\overset{H^{\oplus}}{\rightleftharpoons}}$

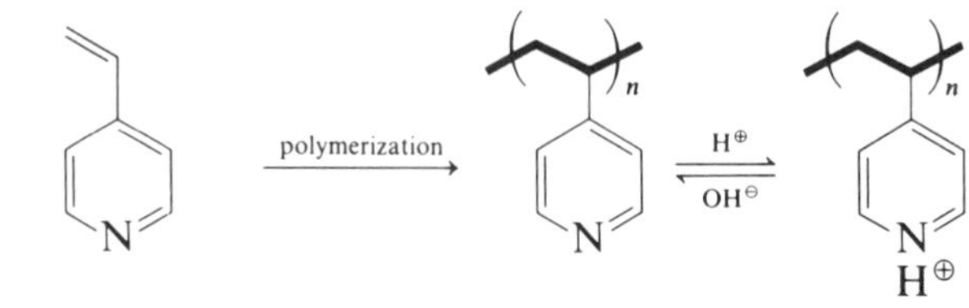

substrate binding to a polymeric catalyst using the concept of electrostatic interaction.

In 1965 C.G. Overberger and colleagues showed that vinyl polymers, containing imidazole and benzimidazole, have cooperative multifunctional interactions that also lead to enhanced catalytic action of the polymers in comparison to low molecular weight precursors (391).

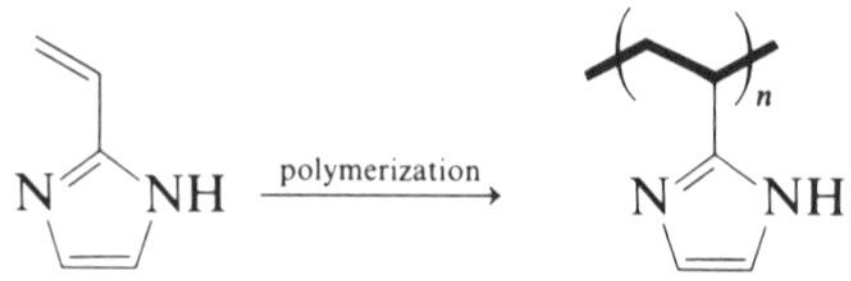

polymerization → poly(vinylimidazole)

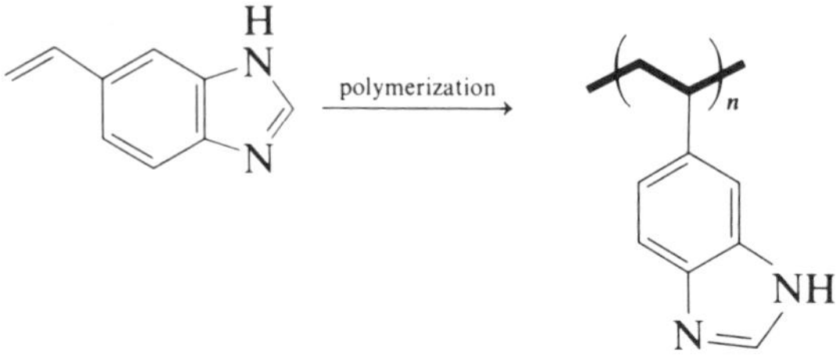

polymerization → poly(vinylbenzimidazole)

However, these polymers have a peculiar behavior. For example, the rate of solvolysis of *p*-nitrophenyl acetate (PNPA) in 28% EtOH—H_2O by poly(vinylimidazole) is presented in Fig. 5.21 and is compared to imidazole alone. The α value represents the fraction of the functional group in the nonionized (neutral) form. At low pH values, the imidazole ring is protonated and the α value is small.

It is clear that protonated imidazole ring (low α value) does not participate in catalysis. The upward curvature for the polymer shows that it is less efficient than imidazole itself at $\alpha < 0.8$, but more efficient at $\alpha > 0.8$. However, as the pK for the formation of anionic imidazole is ~14, it is impossible with this polymer to study the catalytic system as a function of (total) dissociation in a hydroxylic system. However, if one uses a poly(vinylbenzimidazole), which has a smaller pK value of ~12.2, better results are obtained. The rate of hydrolysis of the same substrate does

Fig. 5.21. Solvolysis of PNPA catalyzed by poly(vinylimidazole) (○) and imidazole (△) (28.5% ethanol-water, ionic strength 0.02, 26°C) (391).

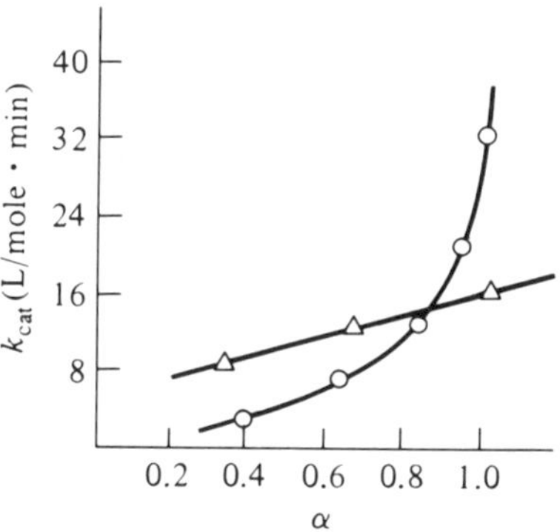

increase dramatically as a function of alkaline pH. Interestingly, a polymer of *N*-vinylimidazole, which cannot form the corresponding anion, is a much less efficient catalyst.

If anionic substrates such as 4-acetoxy-3-nitrobenzoic acid (NABA) and 4-acetoxy-3-nitrobenzenesulfonate (NABS) are used, a different rate profile is observed (Fig. 5.22).

Bell-shaped curves are obtained for both cases with maximal activity at 75% neutrality ($\alpha = 0.75$) with poly(vinylimidazole). These results are best explained if we assume that enough cationic sites (25%) have to be present on the polymer for substrate binding and also that a large portion of the polymer residues must be neutral (75%). It is these neutral imidazole rings that are probably responsible for the hydrolysis of the ester substrates.

If neutral–neutral imidazole interaction or bifunctional catalysis is involved at neutral or near neutral pH, then three mechanisms can be proposed to describe the interactions between the polymer catalyst and the substrate.

The first mechanism is a general-base nucleophilic type of catalysis but it cannot operate if the pH is too low because the first imidazole ring would be protonated and not act as a nucleophile.

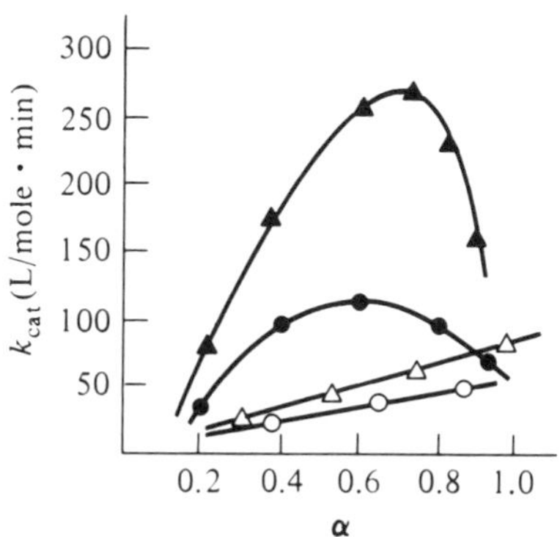

Fig. 5.22. Solvolyses of NABA and NABS catalyzed by poly(vinylimidazole) (●, ▲) and imidazole (○, △), respectively (28.5% ethanol-water, ionic strength 0.02, 26°C) (391).

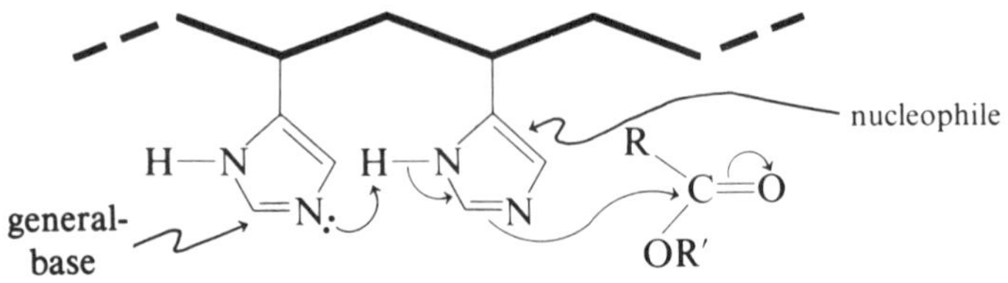

The second possibility is general-acid nucleophilic type catalysis, which seems unlikely since, in general, RO^- would not deprotonate an imidazole.

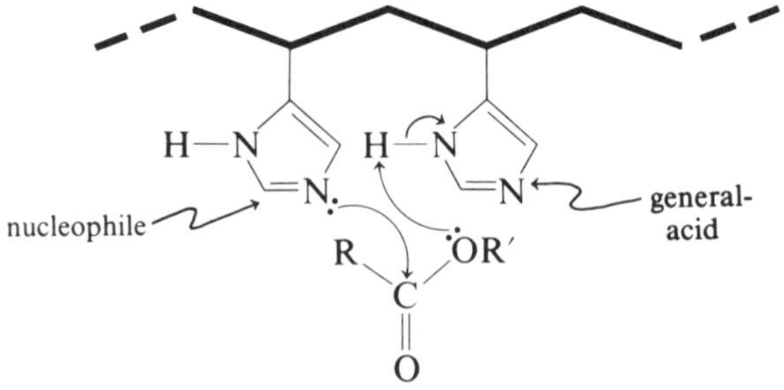

Finally, the third possible interaction is via stabilization of the tetrahedral intermediate:

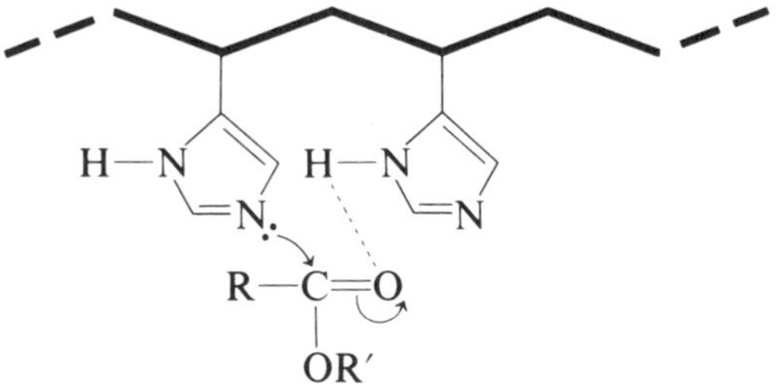

No evidence has been presented to rule out these proposed mechanisms. However, to explain the catalytic role of the polymer, two factors must be taken into consideration. First, increasing protonation of imidazole leads to a decrease in hydrophobic interactions with neutral and charged substrates but an increase in ionic interactions with oppositely charged substrates. Second, a polyion would be expected to be in a more extended conformation as the degree of protonation increases because of charge repulsions. This would render difficult interactions of two imidazole functions. Consequently, the efficiency of catalysis is believed to take place via the cooperativeness of two imidazole rings and not just one anionic ring. Furthermore, in the solvent of high polarity used, water molecules are important and may be involved in some way to help the imidazole rings to work in a cooperative fashion. In effects, the k_{cat} of the reaction is increased in water as opposed to a medium containing 30% alcohol.

Synthetic copolymers have also been shown to have catalytic power comparable to enzymes. The following poly(vinylimidazole)-*co*-poly(vinyl alcohol) has been prepared to verify if a cooperative interaction of imidazole and hydroxyl groups is possible. This situation is reminiscent of the enzyme α-chymotrypsin. However, the polymer is only slightly more active than poly(vinylimidazole) in esterolytic reactions.

poly(vinylimidazole)-*co*-poly(vinylalcohol)

Perhaps a more dramatic indication of a bifunctional participation is in the solvolysis of a positively charged substrate, 3-acetoxy-*N*-trimethylanilinium iodide (ANTI), by poly(vinylimidazole)-*co*-poly(acrylic acid). Figure 5.23 indicates that at high imidazole content in the copolymer, there are insufficient anionic sites to bind the positively charged substrate. On the other hand, at low imidazole content, the polymer begins to behave as a polyanion. As expected, the polymer was much less efficient with neutral substrates.

ANTI

poly(vinylimidazole)-*co*-poly(acrylic acid)

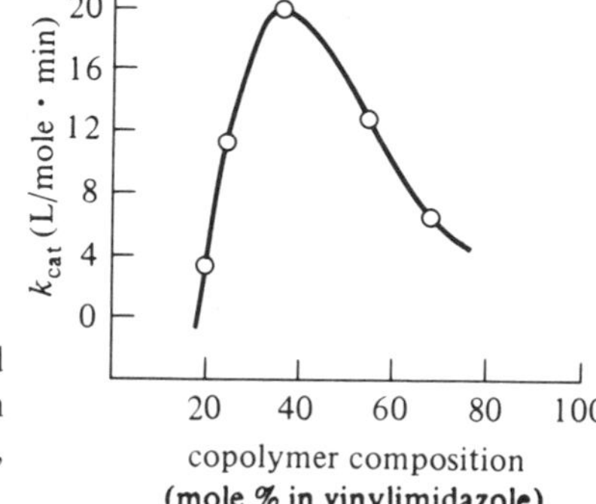

Fig. 5.23. Solvolysis of ANTI catalyzed by copolymers of vinylimidazole with acrylic acid (pH 9, 28% ethanol-water, ionic strenght 0.02, 26°C) (391).

The efficiency and selectivity of the copolymer for positively charged substrate is then rationalized by the electrostatic attraction of the substrate with the anionic carboxylate groups in the polymer which accumulates the substrate in a high local concentration of imidazole nucleophiles. This type of cooperative effect could serve as a model for the nervous system enzyme acetylcholinesterase. The enzyme catalyzes the hydrolysis of its positively charged substrate, acetylcholine.

Recently, high rate enhancements were observed with a synthetic polymer having long alkyl chains (10 residue mole %) attached to it. The group of I.M. Klotz (392), Northwestern University, developed the system and was able to attach dodecyl chains to a small cross-linked water-soluble poly (ethylenimine) matrix ($\overline{DP} \simeq 600$).*

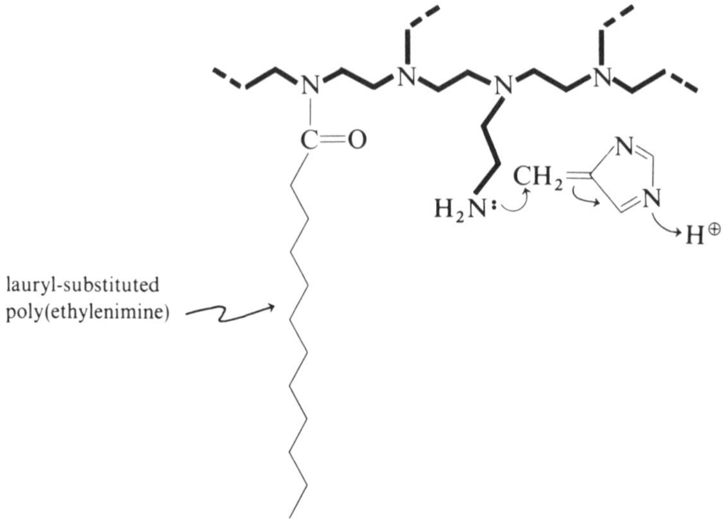

lauryl-substituted poly(ethylenimine)

Reaction of this polymer with methylene-imidazole (or chloromethyl-imidazole) leads to a 15% incorporation of imidazole ring on the polymer backbone. Hydrolysis of phenolic sulfate esters (catechol sulfate) was studied and accelerations of 10^{12}-fold, compared to unbound imidazole, were obtained! This remarkable macromolecular catalyst, possessing

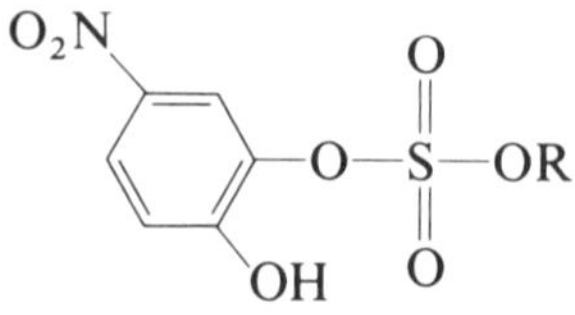

4-nitrocatechol sulfate

* $\overline{DP}$ means degree of polymerization.

a high local concentration of binding and catalytic groups, approaches catalytic constant values observed for the hydrolysis of nitrophenyl esters by α-chymotrypsin. Furthermore, the rates observed with sulfate esters are making this true polymer catalyst 10^2 times more effective than the type IIA aryl-sulfatase enzyme, although the substrates are not physiological.

This rigid macromolecular matrix possessing catalytic imidazole groups and micellar hydrophobic regions is the closest enzymelike synthetic polymer made to date. It has been called "synzyme" (synthetic enzyme) by Klotz since its reactivity is purported to be of the same order of magnitude as that of an enzyme (392).

Of course, not all polymeric catalysts have comparable reaction rates, but enzyme models have progressed significantly in the last decade. In the years to come, considerable progress will be expected in this field until we have enzyme models that will show both the speed and specificity of enzymes.

Another elegant example of the imitation of the properties of biopolymers by synthetic polymers comes from the school of E. Bayer of Tübingen (393). They have prepared chiral polysiloxane polymers for resolution of optical antipodes. The prochiral polymeric backbone was a copolymer of poly[(2-carboxypropyl)methylsiloxane], octamethylcyclotetrasiloxane, and hexamethyldisiloxane. Amino acids or small peptides were covalently linked to this polymer in order to introduce a chiral surface. For this, the free carboxyl function of the polymer was reacted with the L-amino acid in the presence of DCC. The individual chiral centers (amino acids) on the polymer surface were separated by siloxane chains of specified length in order to achieve optimum interaction with the substrate and polymer viscosity. An example of great value for optical resolution is the polymer designated "chirasil-Val," containing 0.86 mmole of *N-tert*-butyl-L-valinamide per gram of polymer (Fig. 5.24).

The polymer–substrate interaction has been studied by gas chromatography and has been used for determining the degree of racemization of amino acids and natural substances. In all cases, the D-enantiomer of a racemic mixture of amino acids is eluted from an L-amino acid phase (polymeric) before the L-form. Figure 5.24 shows the preferential interaction of one enantiomer. No such favorable stacking of the "receptor" polymer surface and the substrate is possible when the substrate has the D-configuration. The importance of the dimethylsiloxane units is also apparent since they keep the L-valinamide units at a distance and prevent formation of intramolecular hydrogen bonds which would give the polymer a quasi-crystalline structure.

Other applications include the separation of optical antipodes of drugs whose enantiomeric composition can be established conveniently and rapidly by gas chromatography. The effort of Bayer and his colleagues bring organic chemists one step closer to the synthesis of "tailor-made"

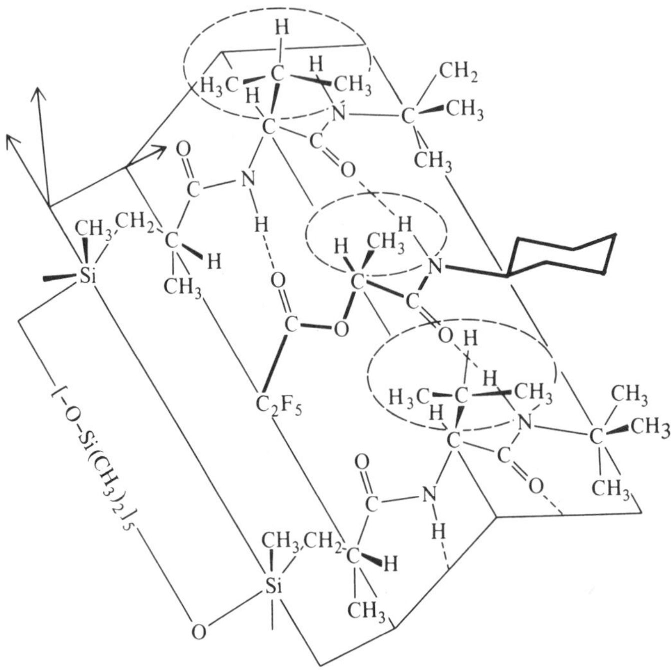

Fig. 5.24. Chirasil-Val, diasteriomeric association complex with *N*-cyclohexyl-*O*-pentafluoropropionyl-L-lactamide (393). Reprinted with permission. Copyright © 1978 by Verlag Chemie GmBH.

chiral matrices. Like proteins they can undergo selective interaction with enantiomers from a wide variety of substances.

In summary, the potential features of functionalized polymers as catalysts are:

(a) the possibility to achieve high effective concentration of catalytic groups on a polymer backbone;
(b) the possibility to generate a micellelike binding region by adding hydrophobic side chains to the polymer;
(c) the possibility to have electrostatic binding with high-charge density by attaching ionic side chains to the polymer.

On the other hand, the main disadvantages are their limited solubility in water and the random arrangement of the polymer chains. Furthermore, the kinetic profile is more complex because of the multisite nature of the polymer as compared to an enzyme which normally has only one active site region.

5.5 Cyclodextrins

α-Cyclodextrin is a naturally occurring host molecule composed of six D-glucose units linked head to tail in a 1α,4-relationship to form a ring called cyclohexaamylose (Fig. 5.25). It has a relatively inflexible doughnut-shaped structure where the top of the molecule has twelve hydroxyl groups from positions 2 and 3 of the glucose units and the bottom has the six primary hydroxyl groups from position 6. So the outside of the α-cylcodextrin molecules has hydrophilic hydroxyl groups, while the cavity features mostly C—H, C—C, and C—O bonds and is rather hydrophobic in nature. This situation is the reverse of that encountered with crown ethers which have rather hydrophilic cavities (394). α-Cyclodextrin can form insoluble, crystalline *inclusion complexes* with a variety of guest molecules. Usually a 1:1 ratio between host and guest molecules is observed, and the size of the guest is the determining factor for the formation of the complex. The inner diameter of the cavity of α-cyclodextrin (6-glucose units) is 0.45 nm. A benzene ring is small enough to penetrate 6-, 7-, and 8-glucose units in cycloamyloses. With anthracene,

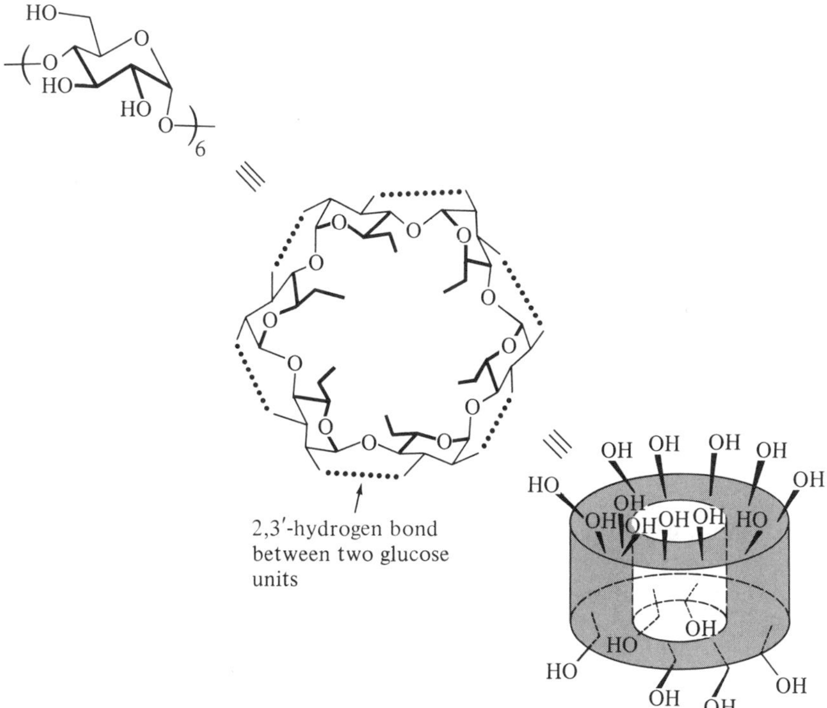

Fig. 5.25. Structural representations of α-cyclodextrin.

however, only 8-glucose units can accommodate this aromatic system. Hydrophobic interactions seem to be the most probable driving force for inclusion complex formation. In addition, hydrogen bonding, van der Waals, and London dispersion forces may also play a role.

The application of cyclodextrins to biomimetic chemistry was originated by F. Cramer in 1965, followed by R.L. Letsinger, H. Morawetz, and M.L. Bender, and further extended by R. Breslow and I. Tabushi. R. Breslow, leader of a group at Columbia University, was the first to show that selective aromatic substitution can take place with the α-cyclodextrin system (395). He found that treatment of anisole ($10^{-4}\,M$) in water at room temperature with HOCl ($10^{-2}\,M$) in the presence of an excess of α-cyclodextrin resulted in 96% chlorination at the *para* position of the anisole ring.

The results indicate not only that the cyclodextrin blocks all but one aromatic ring position to substitution, but also that it actively catalyzes substitution at the unblocked position. The schematic representation above shows an anisole molecule in the cavity of cyclohexaamylose. One or more hydroxyl groups can be converted into a hypochorite to explain the increased rate of chlorination in the complex. This illustrates a noncovalent catalysis (classical Michaelis–Menten binding) in which the host provides the cavity for the reaction without formation of a covalent intermediate.

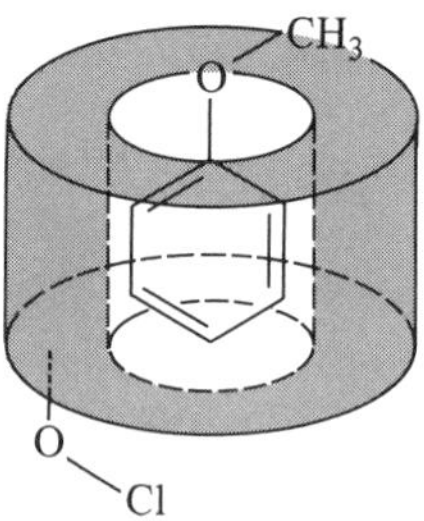

Similarly, aromatic esters are rapidly hydrolyzed by α-cyclodextrin. The secondary hydroxyl groups are believed to be involved in the catalysis but it is not known which one and how many. An intermediate is formed where the acyl group is transferred to the cyclodextrin host molecule. This situation is formally analogous to the mechanism for hydrolytic enzymes such as serine proteases and serves as an enzyme model because a complex with the substrate is formed prior to reaction. This transformation is classified in the category of covalent intermediate catalysis. It should be noted, however, that the second step in ester hydrolysis is very slow for cyclodextrins, so they are not true catalysts for these reactions.

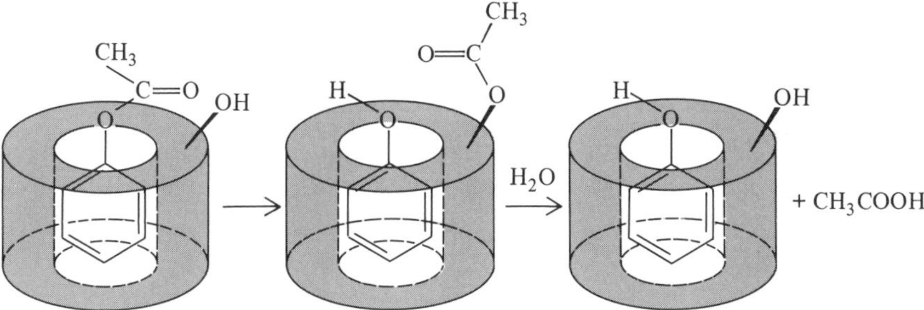

M.L. Bender's group at Northwestern University studied the hydrolysis of *m-tert*-butylphenyl acetate in the presence of 2-benzimidazoleacetic acid (396) with α-cyclodextrin in order to probe the charge–relay system of the enzymatic mechanism in serine proteases. The results showed that in the presence of catalyst a 12-fold acceleration of ester cleavage takes place after complex formation between the substrate and α-cyclodextrin. This observation is consistent with the formation of the substrate–enzyme complex in enzymatic reactions. However, the spacial disposition of the imidazole, carboxyl, and alkoxyl groups is apparently different from those shown for serine proteases. Serine proteases function by a nucleophilic attack of an alkoxide ion, whereas here the model shows nucleophilic participation by the imidazole group. Improvement in this system has been possible by selective modification of one of the secondary hydroxyl groups to a histamine residue. Unfortunately, one serious inconvenience in using cyclodextrin molecules is that they are not very active at neutral pH and prefer basic conditions, so that the kinetic data at pH 13 cannot be compared directly to those obtained, for instance, with α-chymotrypsin at pH 8. The host molecule is stable to alkaline solution but such a condition can be detrimental to the structure of the substrate. On the other hand, one should keep in mind the host's susceptibility to acidic media.

Note also that one of the chief problems is the binding step. The binding cavity is not apolar enough and is open at both ends, and the dissociation constants are larger than with enzymes. Immobilization of the substrate is not always adequate, high concentration of cyclodextrin is usually required, and the conformation of lowest energy of the substrate may not be the one for optimal catalysis. To overcome this difficulty, Breslow (397) attached a bulky group to the more reactive primary hydroxyl groups at one end of β-cyclodextrin (cycloheptaamylose with an inner diameter of 0.7 nm) (Fig. 5.26).

In this way the bottom of the host is "capped" and the cavity is more hydrophobic and shallow. It corresponds to the construction of acylclodextrin with a "floor" across one end of the doughnut's hole that forces shallower binding of the substrate. The following transformation was then examined:

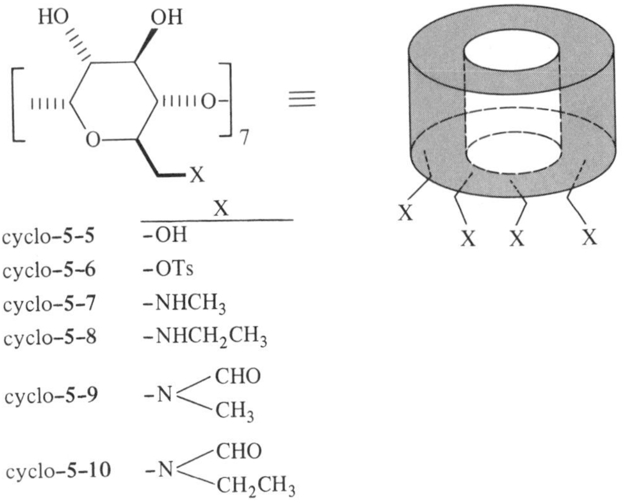

	X
cyclo-5-5	-OH
cyclo-5-6	-OTs
cyclo-5-7	$-NHCH_3$
cyclo-5-8	$-NHCH_2CH_3$
cyclo-5-9	$-N(CHO)(CH_3)$
cyclo-5-10	$-N(CHO)(CH_2CH_3)$

Fig. 5.26. Preparation of "capped" cyclodextrin molecules (397).

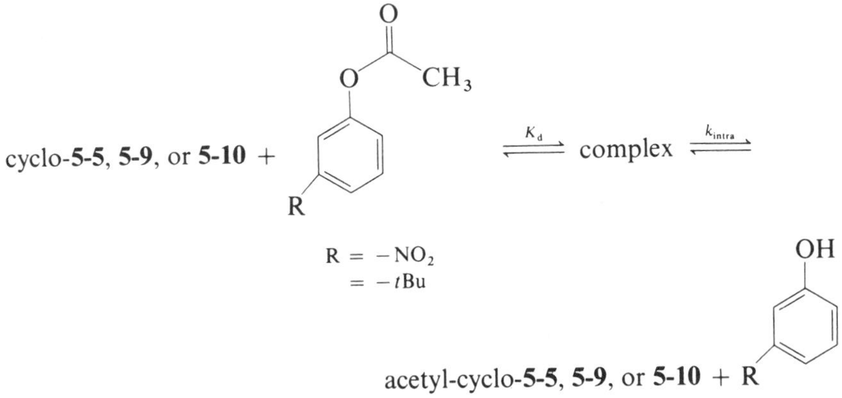

cyclo-**5-5**, **5-9**, or **5-10** + (m-R-phenyl acetate) $\underset{}{\overset{K_d}{\rightleftharpoons}}$ complex $\overset{k_{intra}}{\rightleftharpoons}$ acetyl-cyclo-**5-5**, **5-9**, or **5-10** + (m-R-phenol)

$R = -NO_2$
$= -tBu$

The results of this acyl-transfer process are presented in Table 5.1.

Table 5.1. Rate and Dissociation Constants for Reactions of Cycloheptaamylose Derivatives with *m*-Nitro and *m-tert*-Butylphenyl Acetate (397)

Substrate R	cyclo	$10^3 k_{intra}$ (sec^{-1})	k_{intra}/k_{OH^-}	$10^4 K_d$ (M)
$-NO_2$	**5-5**	11.9 ± 0.05	64	57 ± 7
$-NO_2$	**5-9**	123 ± 5	660	51 ± 7
$-NO_2$	**5-10**	210 ± 40	1140	260 ± 50
$-t$Bu	**5-5**	4.13 ± 0.25	365	1.9 ± 0.2
$-t$Bu	**5-10**	37 ± 5	3300	4.6 ± 0.9

The reactions show Michaelis–Menten kinetics. Cyclo-**5-9** (capped-Me) gives a 10-fold increase in rate and cyclo-**5-10** (capped-Et) an even larger increase. The rates are compared with the reaction (k_{OH^-}) in absence of host. Formation of a Michaelis–Menten complex is one of the reasons to justify the utilization of cyclodextrin-catalyzed reactions as models of hydrolytic enzyme reactions. It is interesting to observe that the presence of a *tert*-butyl group on the substrate dramatically reduces the dissociation constant, which means that it is better bound; the *tert*-butylphenyl group fills the cycloheptaamylose cavity. It is also worth noting that as the substrate is pushed higher in the cavity by the "floor," the dissociation constant generally goes up with the rate. The position for best binding in the absence of the "floor" is too low for good catalysis. Adamantane carboxylic acid, which properly sits on the surface of the cavity of β-cyclodextrin is an excellent competitive inhibitor of complex formation.

adamantane carboxylic acid

In the studies of the hydrolyses of substituted phenyl acetates by α- or β-cyclodextrins, it was observed that *meta*-substituted phenyl esters were hydrolyzed more rapidly than the corresponding *para*-isomers, a phenomenon termed "*meta*-selectivity." This observation indicates that the binding mode is probably asymmetric. This effect is apparently dependent on the depth of the cavity, and Fujita and co-workers (398) showed that appropriate simple modifications of β-cyclodextrin such as "capping" the host, for instance, can alter this selectivity and leads to conversion of the well-established *meta*-selectivity to *para*-selectivity.

In 1978, Komiyama and Bender gave further experimental evidence for the importance of hydrophobic bonding in complex formation of α- and β-cyclodextrins with 1-adamantanecarboxylate (399). Hydrophobic (nonpolar) bonding is characterized by a favorable entropy change that is attributed to a transfer of the guest molecule from aqueous medium to a more apolar medium such as the cavity of a cyclodextrin molecule. This transfer requires breakdown of structural water around the guest, resulting in a large favorable ΔS and a small unfavorable ΔH change. The importance of hydrophobic bonding in the complexation of cyclodextrins is also consistent with the finding of a stronger binding of "capped" cyclodextrin relative to native cyclodextrin with guests. In conclusion, Bender and co-workers showed that a favorable entropy change resulting from hydrophobic bonding is largely responsible for the stabilization of complexes of cyclodextrins with apolar guest compounds (399).

Further, complex formation of cyclodextrins with guest compounds such as drugs and insecticides introduced new physicochemical features to these compounds. This leads to interesting practical usages and reinforces the view that cyclodextrins are suitable models of enzymatic binding as well as enzymatic reactions.

In this regard, Breslow's group (400) synthesized a β-cyclodextrinyl-bisimidazole molecule to model ribonuclease A (RNase A) (see Chapter 3). The approach is based on the preparation of a "capped" disulfonate derivative made earlier by I. Tabushi and co-workers of Kyoto University (401,402).

The model hydrolyzes a cyclic phosphate substrate derived from 4-*tert*-butylcatechol in a selective manner with cooperative catalysis by a neutral imidazole and an imidazolium ion.

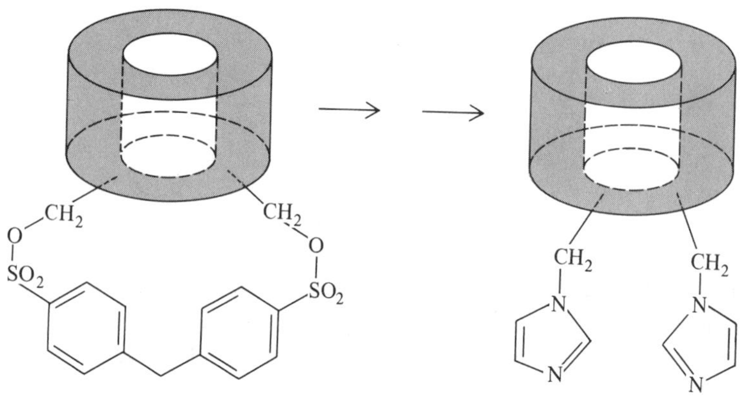

A normal chemical hydrolysis would produce a random mixture of two products, whereas hydrolysis by the "artificial enzyme" leads to the production of only the *m*-phosphate isomer.

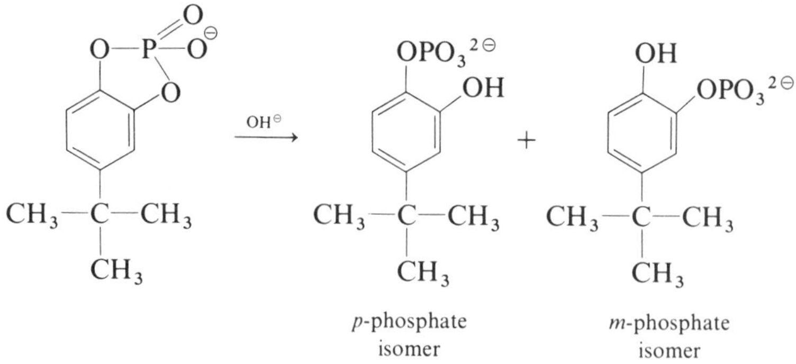

p-phosphate isomer

m-phosphate isomer

The reaction is much slower than with RNase (17-fold), but the selectivity is in accordance with an in-line mechanism without *pseudorotation* as is observed with the enzyme (refer to Section 3.3 for details). As in the case of *para*-chlorination of anisole (p. 346), this example of cyclodextrin reaction gives only one of two possible products.

On treatment with potassium iodide, the "capped" disulfonate β-cyclodextrin discussed above could easily be converted to the corresponding diiodide β-cyclodextrin. With appropriate nucleophiles (imidazole, histamine) a new route to bis(*N*-imidazolyl)-β-cyclodextrin and bis(*N*-histamino)-β-cyclodextrin was developed by Tabushi's team (403). In the presence of Zn(II) ion, both regiospecifically bifunctionalized cyclodextrins hydrate CO_2 and are the first successful carbonic anhydrase models. The Zn(II) ion binds to the imidazole rings located in the edge of the cyclodextrin pocket, and the presence of an additional basic group, as with bis(histamino)-cyclodextrin-Zn(II), enhances the activity. Therefore, the present models show that all three factors, Zn(II)-imidazole, hydrophobic environment, and a base seem to help to generate the carbonic anhydrase activity (403). The chemistry of this enzyme is further discussed in Section 6.2, page 390.

Also of interest is the model developed by Breslow and Overman (404) where a metal ion is introduced into a cyclodextrin–substrate complex. With *p*-nitrophenyl acetate as substrate, the presence of Ni(II) in the cyclodextrin–substrate complex results in a further increase in rate of hydrolysis by a 1000-fold.

This novel complex can be prepared in the following way:

α-cyclodextrin
2,5-pyridine dicarboxylic
ester formation on the cyclodextrin
Ni(II) +
complex

The cyclodextrin "cage" holds the ester while the metal ion positions other groups for attack. If cyclohexanol is added in the solution, it competes with the substrate and the efficiency of the system falls to 60%. Most of the catalytic power of the system is attributed to the binding of the substrate by the cyclodextrin moiety of the complex. In the absence of cyclodextrin, the rate enhancement is 350-fold. Thus, by combining the properties of cyclodextrin with those of a metal ion, a much more efficient catalytic system can be obtained that mimics enzyme features.

In a search for well-characterized polyfunctionalized α-cyclodextrins, Knowles's group (405,406) developed in 1979 the elegant approach shown below:

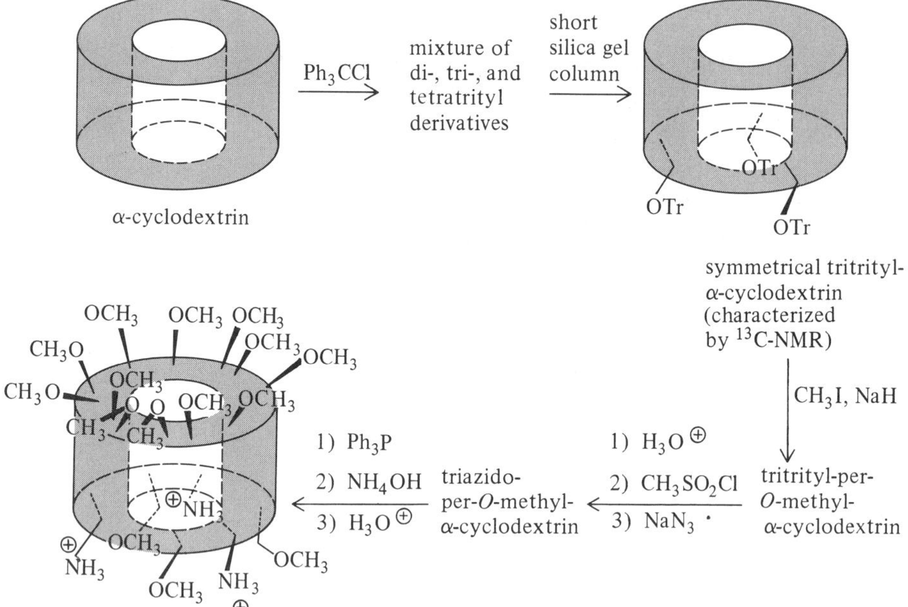

This procedure provides access to a wide variety of cyclodextrin derivatives and the rational synthesis of even more sophisticated model systems employing the regiospecifically disposed functionality on one side of the cyclodextrin cavity and possible additional functionalities on the other side.

This symmetrical multifunctional host has the ability to bind complementary guests. Figure 5.27 shows the binding of benzyl phosphate in the cavity.

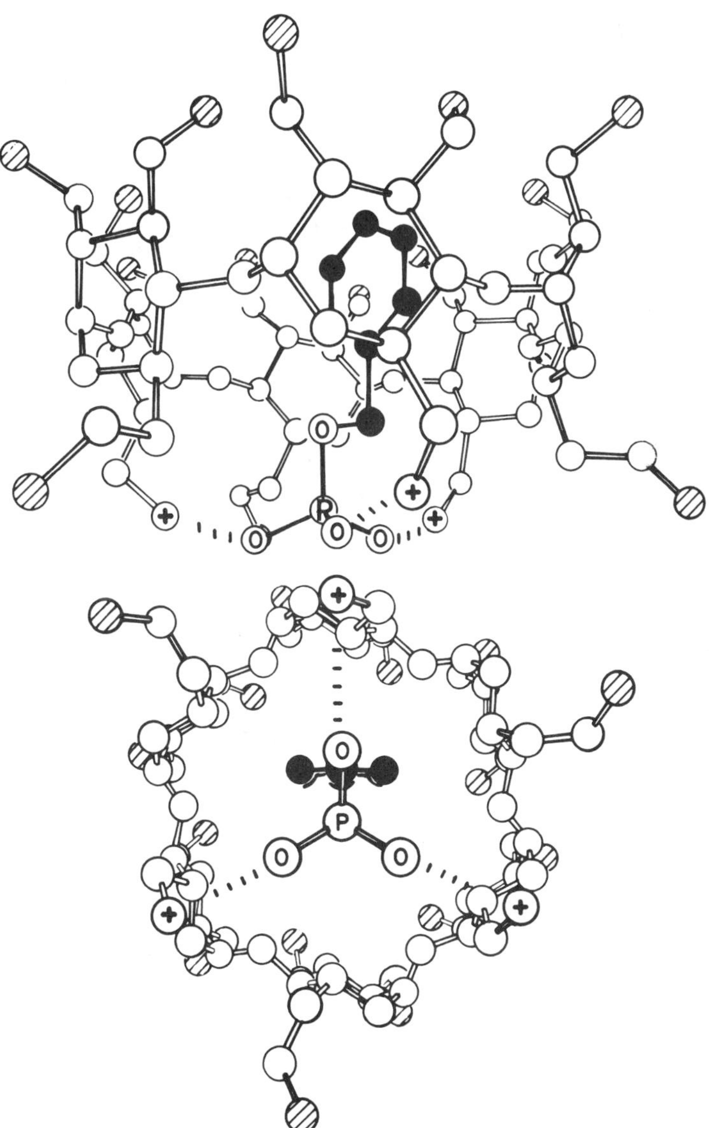

Fig. 5.27. Representation of the symmetrical triammonio-per-*O*-methyl-α-cyclodextrin complexed with benzyl phosphate. Each of the three ammonium ions is shown by the symbol ⊕. The 15 methyl groups are shaded. All hydrogen atoms are omitted and the size of the atoms is arbitrary (406). Reprinted with permission. Copyright © 1979 by the American Chemical Society.

Potentiometric titration of the ammonium groups of the host gave apparent pK_a values of 7.42, 8.02, and 8.79. These pK_a values are rather low for primary ammonium groups but could reflect their nonpolar environment. At pH 7, the host is primarily in the fully protonated species and shows a dissociation constant for benzyl phosphate (0.031 mM) which is almost three orders of magnitude smaller than for benzyl alcohol (24.3 mM). It is thus a properly designed host for benzyl phosphate. Furthermore, the dissociation constant for the complexation of benzyl phosphate is pH dependent, showing that the guest binds in the cyclodextrin cavity with the phosphate oxygen interacting (at pH 7) strongly with the three symmetrically disposed ammonium ions. Interestingly, inorganic phosphate itself does not bind to this cyclodextrin derivative and does not expel benzyl phosphate from the cavity, showing again the importance of hydrophobic binding between host and guest.

Therefore, both hydrophobic and electrostatic interactions in aqueous solution are exploited in this specific host–guest complexation, and this provides a model for multiple recognition sites common in biological systems.

Being formed of D-glucose units, cyclodextrins are chiral and chiral induction on substrates have been observed with cyclodextrin reactions (394). A recent application of this is the preparation by Breslow's team (407) of a covalently linked coenzyme-cyclodextrin "artificial enzyme." It consists of a β-cyclodextrin-pyridoxamine (refer to Section 7.2 for the chemistry of this coenzyme) that can selectively transaminate phenylpyruvic acid to phenylalanine with a 52% excess of the natural L-enantiomer.

W. Saenger's group (408) from the Max Planck Institute has data to show that cyclodextrins can be even better models for enzymes than was hitherto assumed. They have demonstrated that a conformational change takes place in the host when substrates are included in it. This situation is analogous to enzyme which exhibits "induced fit" in their interaction with substrates (see Section 6.2 for an example).

Other interesting systems involving cyclodextrin chemistry have been applied as enzyme models by I. Tabushi. They are specific allylation-oxidation of hydroquinone (409), cyclodextrin having an amino group attached to retinal as a rhodopsin model (410), and specific inclusion catalysis by β-cyclodextrin in the one-step preparation of vitamin K_1 and K_2 analogs (411).

For all the enzyme models that we have examined so far, the catalytic efficiency appears to be understandable in terms of three main effects:

(a) hydrophobic interactions to provide an efficient binding;
(b) a medium (or cavity) effect where proper polarity provides a large driving force for the reaction;
(c) an orientation effect that leads to restriction of conformation and provides a large rate acceleration.

Two recent review articles appeared on cyclodextrins as building blocks for supramolecular structures and functional units (412,413). They should be consulted by those interested in other elaborated examples of molecular architectures based on supramolecular positioning.

5.5.1 New Biomodels

The hydrophobic nature of the cyclodextrin cavity and the success of making molecular inclusion complexes are well illustrated by the possibility of performing intramolecular Diels–Alder reactions in water. This observation was first noted in 1980 by Breslow. For instance, in the following transformation:

OH S S O ΔT → HO O S S + O HO S S

isomer-A isomer-B

When the furan molecule is heated in water at 89°C for 6 hr, a 20% yield of an epimeric mixture A/B (1:2) of Diels–Alder adducts is formed. The same reaction in the presence of an equivalent of β-cyclodextrin enhances the yield to 91% with an epimeric ratio of 1:1.5 (414). The authors suggested that for the intramolecular reaction to occur, both diene and dienophile are simultaneously complexed within the β-cyclodextrin cavity. They do not exclude, however, the possibility that complexing the dithiane portion of the molecule would also result in juxtapositioning the two reactive ends of the molecule. In the presence of α-cyclodextrin, no improvement in reaction yield was observed.

The catalytic role played by β-cyclodextrin in intramolecular cyclization reactions of organic molecules in water media reinforces the importance of hydrophobic interactions in the cavity. The addition of a less polar co-solvent could also be used to influence the stereoselectivity in Diels–Alder reactions (415).

In the previous section, various examples were presented to show how cyclodextrin has been used by many investigators to mimic enzyme activity, particularly serine proteases. In most cases, *p*-nitrophenyl esters of substrates are used, yet serine proteases naturally hydrolyze amides and not esters. Because of the popularity of *p*-nitrophenyl esters, labeled "*p*-nitrophenyl ester syndrome," Menger tried to understand the origin of rate accelerations found in previous enzyme models and to throw light on the principle governing mimic behavior (416).

The argument is based on a study of the hydrolysis of a series of ferrocenylacrylate esters complexed with β-cyclodextrin. Breslow had

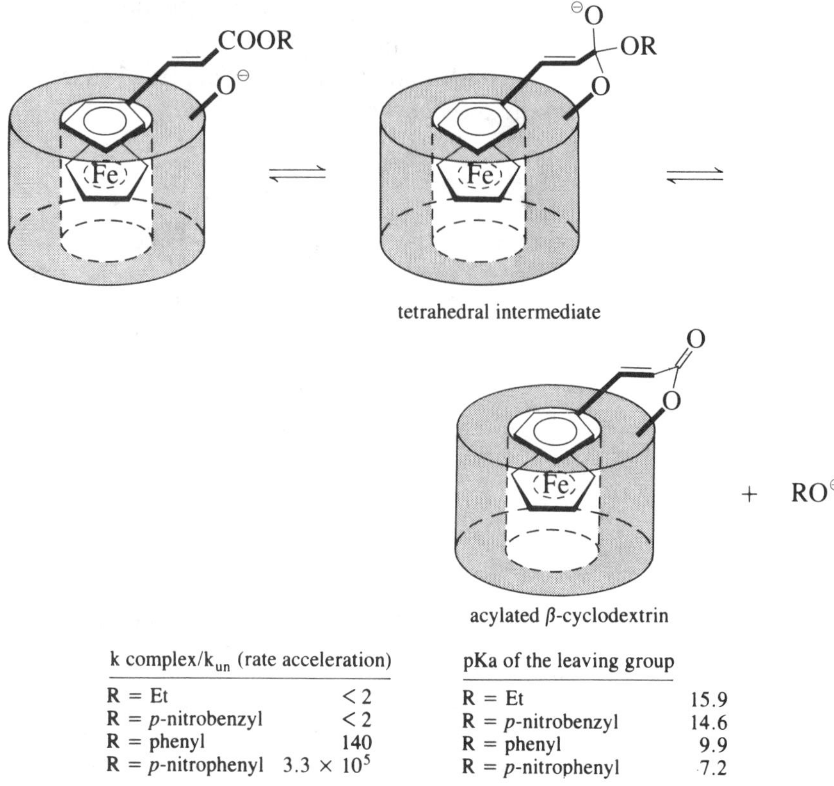

k complex/k_{un} (rate acceleration)		pKa of the leaving group	
R = Et	< 2	R = Et	15.9
R = *p*-nitrobenzyl	< 2	R = *p*-nitrobenzyl	14.6
R = phenyl	140	R = phenyl	9.9
R = *p*-nitrophenyl	3.3×10^5	R = *p*-nitrophenyl	7.2

pKa of cyclodextrin OH = 12

already observed that the ferrocenylacrylate structural element fits nicely into the cavity and the corresponding *p*-nitrophenyl ester gave a remarkable 3.3×10^5 acceleration in hydrolysis rate. The reaction with the ethyl ester is extremely slow considering that it is only 56 times less reactive than the *p*-nitrophenyl ester toward basic hydrolysis. The absence of a significant rate acceleration with the ethyl ester is not caused by insufficient binding since $K_{ass} = 24\,M^{-1}$ as compared to $133\,M^{-1}$ reported for the *p*-nitrophenyl ester. The geometry of the complex could be distorted by the ferrocene unit rather than by its ester appendage. However, the *p*-nitrobenzyl ester, which is only 31 times less reactive toward OH^- than the *p*-nitrophenyl ester, is as unreactive as the ethyl ester under the same complexation conditions.

The situation is already better with a series of phenyl esters and consequently the rate of acceleration seems to manifest itself when the pK_a of the leaving group drops to that of phenol. It is conceivable that the preferential movement of the tetrahedral intermediate toward product requires leaving groups with $pK_a < 9$. This value is 3 units below the pK_a of the reactive β-cyclodextrin hydroxyl group. Surprisingly, the β-cyclodextrin oxyanion behaves as a much better leaving group that one might have expected from its pK_a value. This behavior is the pivotal point for the understanding of the "*p*-nitrophenyl ester syndrome." Considering that the acylated β-cyclodextrin possesses a high-energy s-*cis* configuration, the tetrahedral intermediate would tend to eject the β-cyclodextrin oxyanion and revert to unreacted complex with simple esters. Furthermore, additional contribution to rate acceleration could be the critical distance between the OH nucleophile and the carbonyl ester junction within the complex. A productive alignment could operate for the *p*-nitrophenyl ester but not for the ethyl ester.

In the preceding section (p. 350) a model of ribonuclease A was presented. A β-cyclodextrinyl-6,6′-bisimidazole could cleave with a modest rate a cyclic phosphate of 4-*tert*-butylcatechol with excessive formation of the *meta*-phosphate isomer. The kinetics of the reaction indicates cooperative catalysis by a basic imidazole group and an acidic imidazolium group, a situation analogous to the nuclease enzyme itself.

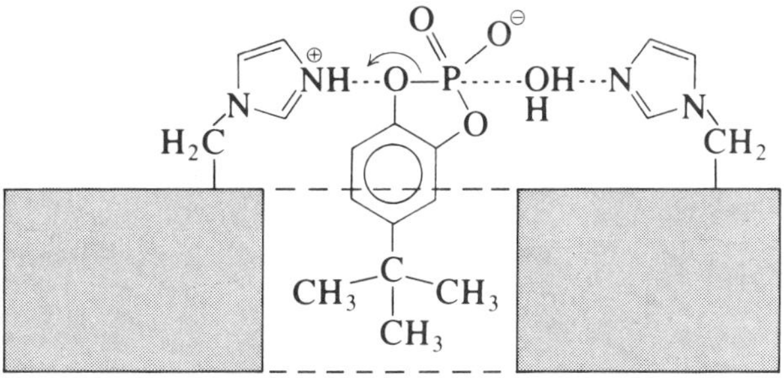

schematic representation of the cyclic phosphate cleavage with formation of the *meta*-phosphate isomer, consistent with an "in-line" displacement at phosphorus (388)

In a continuous effort to improve this supermolecular assembly, Breslow designed a bis-imidazole-cyclodextrin similar to the previous one but with an extra -CH_2-S- linkage (417). Interestingly, in this geometrically different arrangement, the new catalyst still cleaves the cyclic phosphate with an "in-line" orientation, but the outcome of the reaction give predominantly the *para*-phosphate isomer over the other.

The new catalyst, having different geometric requirements, completely reverses the direction of catalyzed cleavage in the substrate. The regioselective cleavage seems to reflect the direction of approach of the water molecule in the process. An H_2O/D_2O isotopic effect of 2.5 is consistent with the general-base delivery of water. This more flexible catalyst shows less rate acceleration (8-fold) than the previous model (17-fold, p. 350). The bell-shaped pH-rate profile indicates, however, a cooperative catalysis by two imidazole groups.

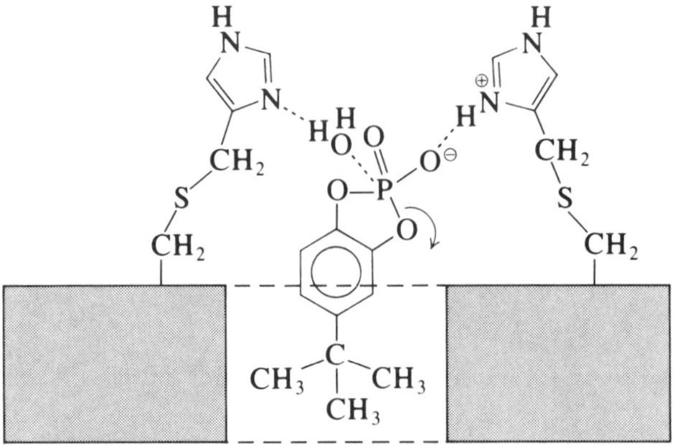

schematic representation of the specific cleavage of the cyclic phosphate with a different bis-imidazole β-cyclodextrin catalyst (417)

If a substrate ester molecule is included too deeply in the cyclodextrin cavity, favorable positioning between the carbonyl group of the ester and the hydroxyl groups at the surface of the cyclodextrin is not met and the efficiency of catalysis is dramatically reduced. To prevent this, capped cyclodextrins have been prepared (see p. 348). They offer the advantage of blocking the access of one face of the cyclodextrin ring and at the same time preventing the substrate, which enters into the cavity from the other open end, from going too deeply inside the hydrophobic core of the cyclodextrin structure.

An interesting example that illustrates this concept is the preparation of a cyclodextrin capped with a photosensitive function. For this, the following azobenzene-capped β-cyclodextrin was synthesized and showed that the rate of hydrolysis of *p*-nitrophenyl acetate is accelerated by photoirradiation (418). Indeed, in the photoinduced *cis*-isomer, the binding ability of the substrate is increased whereas the cavity of the *trans*-isomer is too shallow to allow the formation of a stable complex. Although the k_{cat} of the *trans*-isomer is larger ($1.6\,s^{-1}$) as compared to the *cis*-isomer ($0.7\,s^{-1}$), the Michaelis constant, K_m, is more favorable for the *cis*-isomer ($0.2 \times 10^{-1}\,M$) compared to the *trans*-isomer ($2.3 \times 10^{-2}M$). As a result, the k_c/K_m ratio for the *cis*-isomer is 5 times larger, favoring this form for more efficient catalysis.

An inherent problem facing the construction of capped cyclodextrins is the constant formation of two or more transannular isomers. The possibility, however, of making two cooperating functional groups in appropriate spacial arrangements on a cyclodextrin is crucial for the preparation of more refined and sophisticated enzyme models. It is to this regiospecific difficulty that the group of I. Tabushi has consecrated many rewarding efforts.

trans-isomer
"shallower cavity"

hν / Δ

cis-isomer
"deeper cavity"

β-Cyclodextrin, being nonsymmetrical, poses a real challenge. An exhaustive study showed that the nature of the capping disulfonate compounds used can in some cases govern the outcome of the distribution of positional isomers (419). This is summarized in the following presentation, viewed from the top of the cavity.

β-CD + Cl–▬–Cl ⟶ AD isomer + AC isomer

capping disulfonate	**AD** isomer	**AC** isomer
ClO_2S–C₆H₄–CH=CH–C₆H₄–SO_2Cl	nearly exclusive	very small portion
ClO_2S–C₆H₄–C(=O)–C₆H₄–SO_2Cl	small portion	nearly exclusive
ClO_2S–C₆H₄–X–C₆H₄–SO_2Cl, X = O, CH_2	non selective	non selective

A planar *transoid* geometry of the reagent seems to favor the **AD**-isomer while a planar *cisoid* arrangement with a small distance between the two ends of the molecule favors the other isomer.

Furthermore, it seems that the major determining factor of the regio-specific **AC** and **AD** capping is the formation of the first functionalization site which determines the site of the second functionalization at the "best fit" position. The distance between the two reacting groups, the strain in the transition state, and the direction of approach of an entering group must all be considered. This mechanism of multifunctionalization is probably applicable to other models of enzymes.

In this same context, Tabushi has prepared a bifunctionalized β-cyclodextrin as an artificial receptor for amino acids (420). An **AB**-capped *meta*-benzene disulfonate was first made and eventually converted to a 6**A**-amino-6**B**-carboxy-β-cyclodextrin.

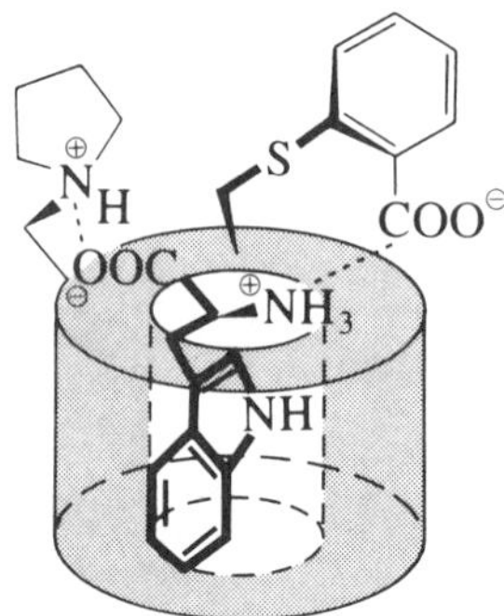

D-Trp complex

The example above illustrates the favorable polar interaction with the amino acid at the surface of the cyclodextrin and the hydrophobic contact between the rest of the amino acid in the cyclodextrin cavity. For tryptophane, the D-isomer has an association constant of $54\,M^{-1}$ and the L-isomer of $42\,M^{-1}$ for the receptor.

Values for ΔH^0 and ΔS^0 for this complexation suggest a significant contribution of hydrophobic environment for enhancement of polar interaction.

Aza-crown ether has also been attached to β-cyclodextrin (421). The crown-ether-capped cyclodextrin provides a molecular assembly with two recognition receptor sites, which cooperate in the association of alkali-metal *p*-nitrophenolates as substrates. In the case of the Na^+ salt, the association constant is increased 70-fold relative to the nonfunctionalized cyclodextrin. For the Li^+ or K^+ salt, the association constants are enhanced by factors of only 6 to 10. This observation is rather surprising since the binding properties of alkali ions to aza-crown ether is normally 10-fold less efficient for Na^+ than K^+ ions. Therefore, other factors such as steric constraints must be involved with the association of the cation in the proximity of the phenolate anion. The best balance between association of the cation to the crown ether ring and effective electrostatic interactions seems to be obtained with Na^+ ions. This result of cooperative

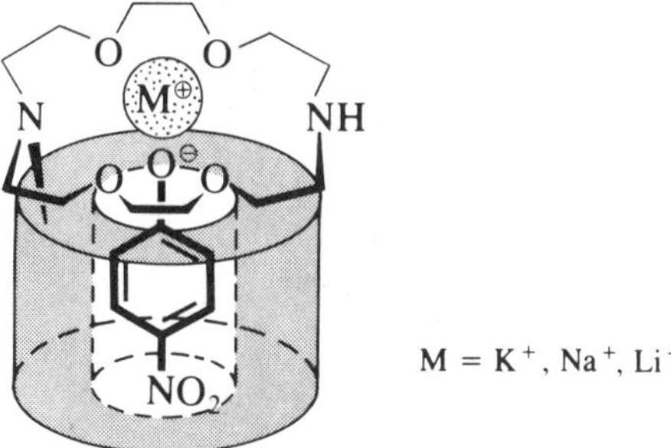

function represents a mimic for the participation of remote functional groups of the peptidic backbone of natural enzymes to induce strong association of substrates.

Lastly, mention should be made of the efforts of H. Ogino (Tohoku University) toward the synthesis of a compound consisting of a cyclodextrin molecule threaded by an α,ω-diaminoalkane coordinated to cobalt(III) complexes (422). Such a molecular arrangement of a ring threaded by a chain having large end groups that cannot be extruded from the ring is called *rotaxane*. This represents the first example of a rotaxane containing a chiral ring between two metal-containing complexes.

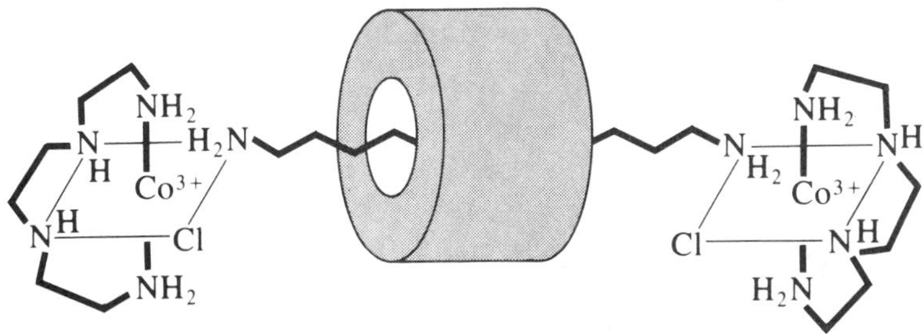

rotaxane-based inclusion complex

More elaborated molecular constructions using cyclodextrins, rotaxanes, or catenanes have been developed (423).

5.6 Enzyme Design Using Steroid Template

Long alkyl chains with a catalytic group at one end can provide enough hydrophobic binding sites with fatty ester substrates to produce a rate enhancement. In the following case, a 10-fold increase in rate is observed at pH 8 in Tris buffer, 25°C. Micelles are formed and there is no case yet of a large rate enhancement for a 1:1 association of alkyl chains with monomeric binding site. Therefore, a serious problem is the fact that these molecules are rather insoluble in water and thus have a tendency to aggregate or form micelles. Work has to be done at concentration well below the cmc level which is about $10^{-4}\,M$ for this system. Furthermore, the conformation of the chain varies even when extended. Since the alkyl chains are very flexible and a strong hydrophobic area is never produced, a search for a more oriented network was needed.

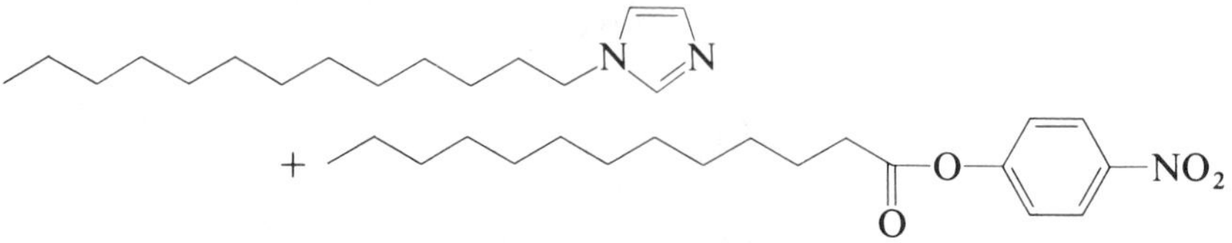

To overcome this difficulty, J.P. Guthrie, from the University of Western Ontario, suggested the use of a planar, large, and rigid hydrophobic backbone, a steroid skeleton (424).

For this, he synthesized the following steroid molecule, androstane 3β, 11β-diamino-17β(4)-imidazole.

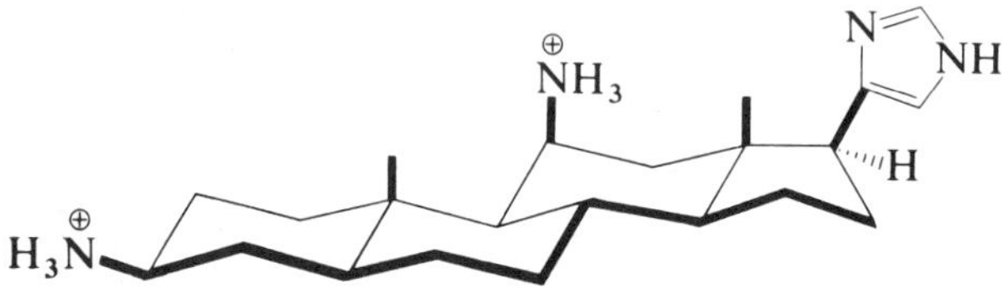

The presence of two ammonium ions makes the molecule water soluble enough to inhibit micelle formation under the experimental conditions. It should be recalled that bile acids are water soluble but form micelles.

As compared to cyclodextrin which is a three-dimensional system, this one is a two-dimensional hydrophobic surface. In this regard it is a useful model to evaluate hydrophobic and electrostatic interactions of well-defined geometries. The presence of an imidazole group at position 17β is responsible for the catalytic effect observed during the hydrolysis of esters.

To illustrate this, a series of aromatic esters, with increasing degree of hydrophobicity have been tested. The relative rate is doubled for $n = 2$ or 3 as compared to $n = 1$:

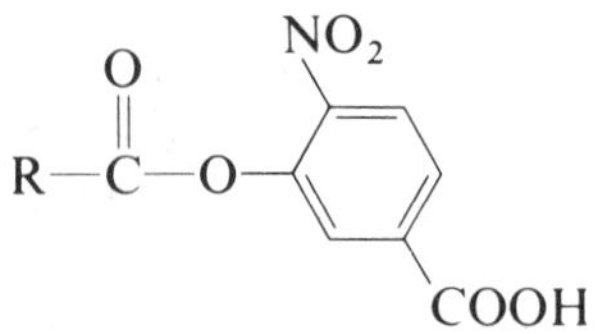

R
CH_3

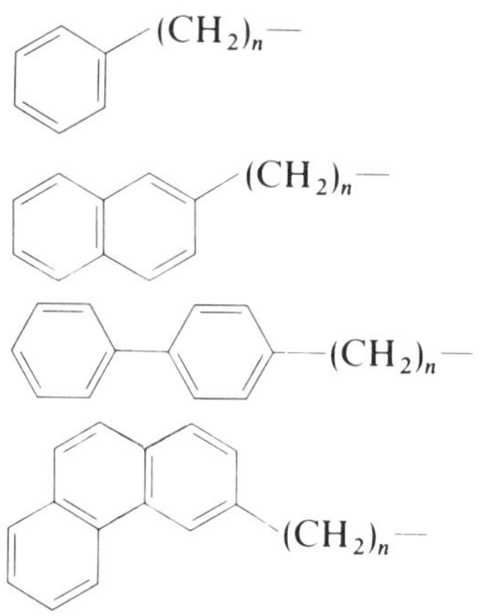

It was found that as the size of the hydrophobic group in the ester increased, the rate of catalyzed hydrolysis also increased. However, these esters showed essentially identical reactivity toward imidazole alone as nucleophile. Space-filling models then suggested that two CH_2 groups are required between the aromatic ring system and the ester group in order to permit enough hydrophobic contact with the α-surface of the steroid molecule. It is concluded that imidazole-catalyzed hydrolysis of the ester is facilitated by hydrophobic interactions between substrate and catalyst.

A true second-order rate is observed, indicating a 1:1 stoichiometry. Because of solubility problems, no saturation kinetics are measured, only the second-order rate constants are evaluated but not k_{cat} nor the K_m binding constants. A productive binding representation of a tetrahedral intermediate formed indicates that efficient hydrophobic binding is possible between the substrate and the steroid rings. The advantage of this system is that very little flexibility in either substrate or catalyst is allowed.

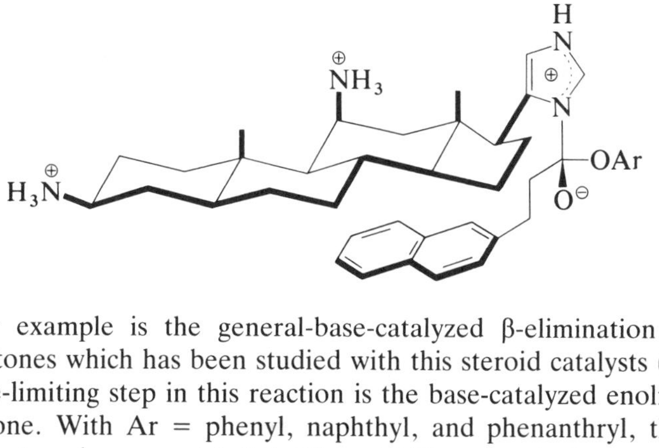

Another example is the general-base-catalyzed β-elimination of β-acetoxy-ketones which has been studied with this steroid catalysts (425).

The rate-limiting step in this reaction is the base-catalyzed enolization of the ketone. With Ar = phenyl, naphthyl, and phenanthryl, the hydrophobic rate enhancement is 9.0, 27, and 110, respectively.

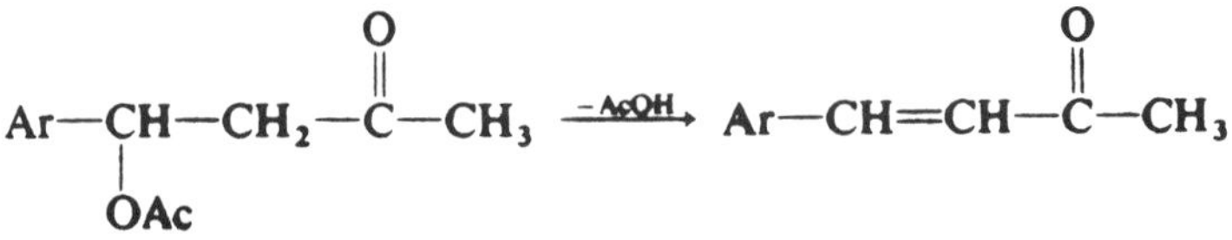

$$\mathrm{Ar{-}CH(OAc){-}CH_2{-}C(=O){-}CH_3 \xrightarrow{-AcOH} Ar{-}CH{=}CH{-}C(=O){-}CH_3}$$

Since the pK of imidazole is 7.1 and the pK of the corresponding steroid–imidazole is 7.2, the catalyst basicity is not likely to be the factor responsible for the rate enhancement. The results show that the three substrates have comparable kinetic behaviors. In fact, the steroid–imidazole complex should, if anything, retard the reaction. Therefore, the size of the aryl group must make the difference for proper fit. In other words, some favorable noncovalent interactions with the rest of the steroid molecule and the substrate must be involved. In effect, the conformation of the transition state must be such that the rings of the steroid and the substrate interact in such a way that the bottom face of the steroid skeleton is shielded from water. Consequently, strong and favorable hydrophobic bindings are the main reason for the rate enhancement observed with the phenanthryl derivative. This is illustrated below:

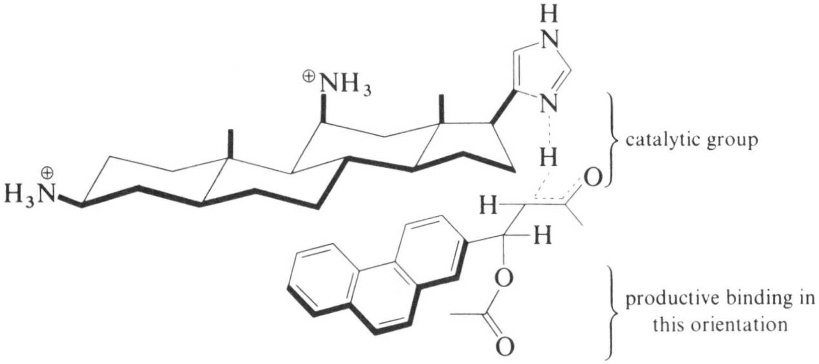

H
N
⊕NH3
N
catalytic group
H
⊕
H3N
O
H
H
O
productive binding in this orientation
O

Some studies have also been carried out with anionic and cationic acetyl substrates (426). The general picture that emerges is that rate retardation is observed with cationic species while rate enhancement is obtained with anionic substrates. However, a *meta* anionic group shows a larger rate enhancement than a *para* anionic orientation.

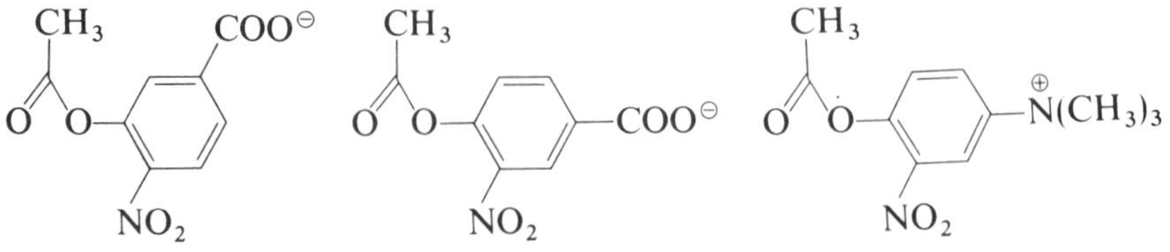

CH3 COO⊖ CH3 CH3
O O O O —COO⊖ O O ⊕N(CH3)3
NO2 NO2 NO2

This is the first clear demonstration of catalysis resulting from electrostatic interaction of charged groups remote from the reaction center with reasonably well-specified geometry.

The following structures show that the *m*-COO$^-$ and $-\overset{+}{N}H_3$ groups in the tetrahedral intermediate can come into close contact without strain. The contact between the 11-ammonium ion and *p*-carboxylate of the substrate is possible but only at the expense of desolvating the ionic groups without formation of a compensating hydrogen bond, whereas the 11-ammonium and *m*-carboxylate can come into contact, without intervening water, and form a good hydrogen bond. In other words, close contact is necessary for a large electrostatic affect but is possible only if loss of solvating water is compensated for by formation of a new hydrogen bond. The presence of the C-18 methyl group is not a steric barrier.

In conclusion, the steroid system is a useful enzyme model because there is binding of catalyst and substrate and some selectivity for some substrates over others. Stronger binding affinities within the steroid–imidazole system is needed and should be developed to constrain the substrate to bind in only one mode.

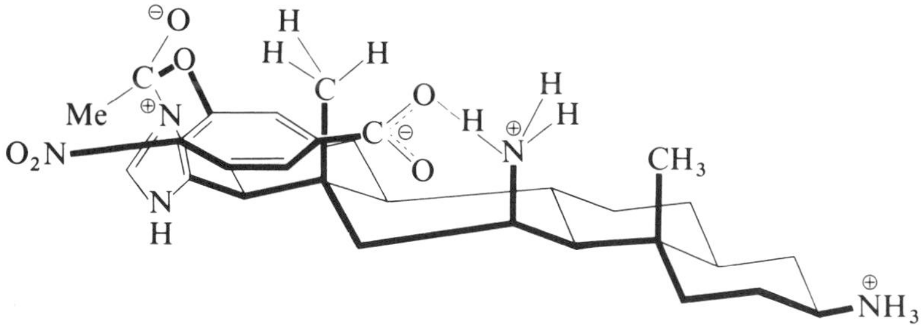

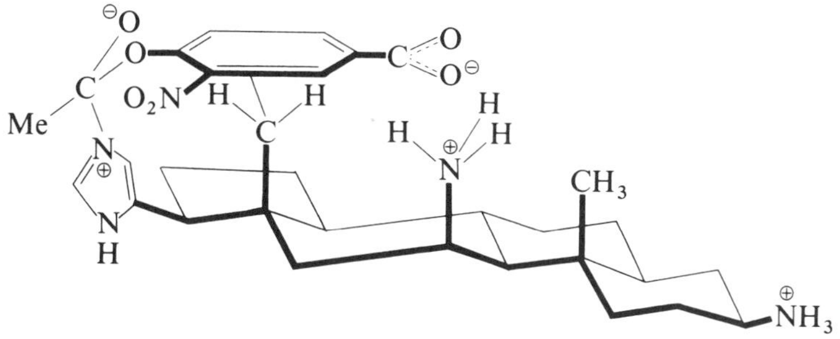

For this, Guthrie's group has planned to synthesize a steroid–imidazole dimer by combining two ketone analogs with an aromatic diamine.

This way, a three-dimensional frame is built and the substrate is expected to be trapped in the "jaw" of the dimeric catalyst where hydrophobic binding now comes from both faces of the substrate reaction center. So far, the bis(11)-keto derivative has been prepared and increased catalysis by 3000-fold over that of simple imidazole has been observed (427).

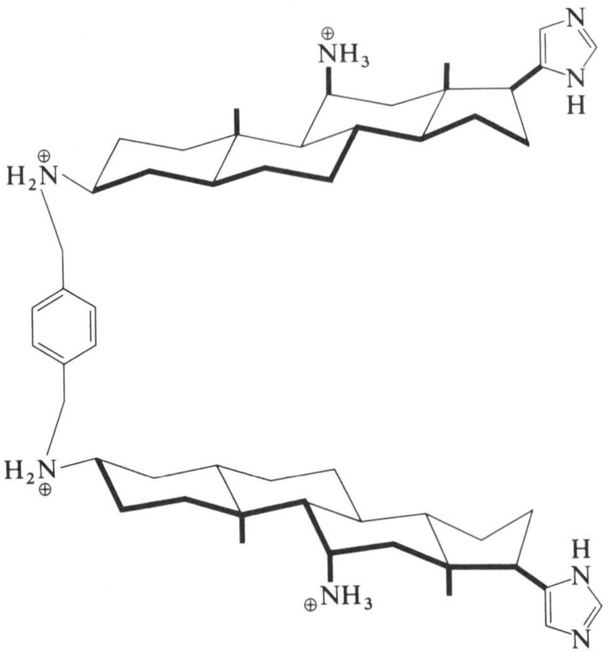

With the water-soluble bis(11)-ammonium derivative, hydrolysis of arylpropionate esters with good leaving groups showed that rate enhancements, relative to imidazole, of up to 5500 were obtained, and analysis of the data indicated that the potential rate enhancement is over 10^5 (428).

5.7 Remote Functionalization Reactions

Nature's ability to carry out selective functionalization of simple substrates utilizes a principle of great power that has not been applied by chemists until recently (429). For example, enzymatic systems such as desaturases can oxidize a single unactivated carbon–hydrogen bond at a specific region on the alkyl chain of stearic acid and convert it to oleic acid, possessing only a *cis* geometry.

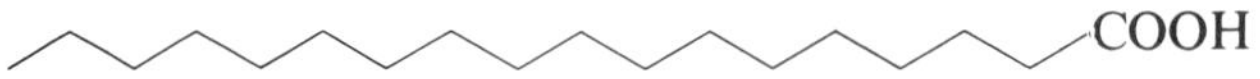

stearic acid

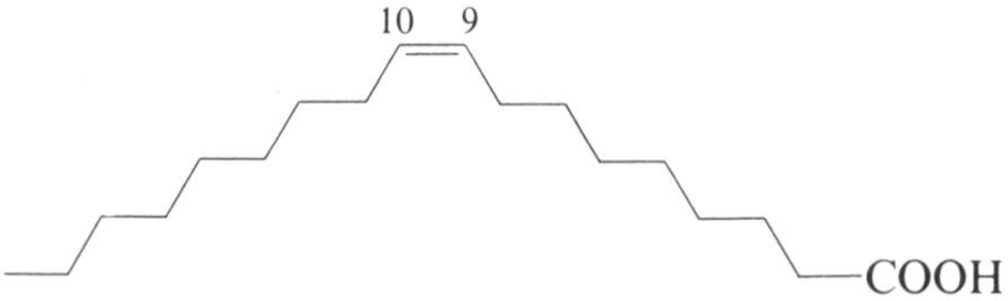

oleic acid

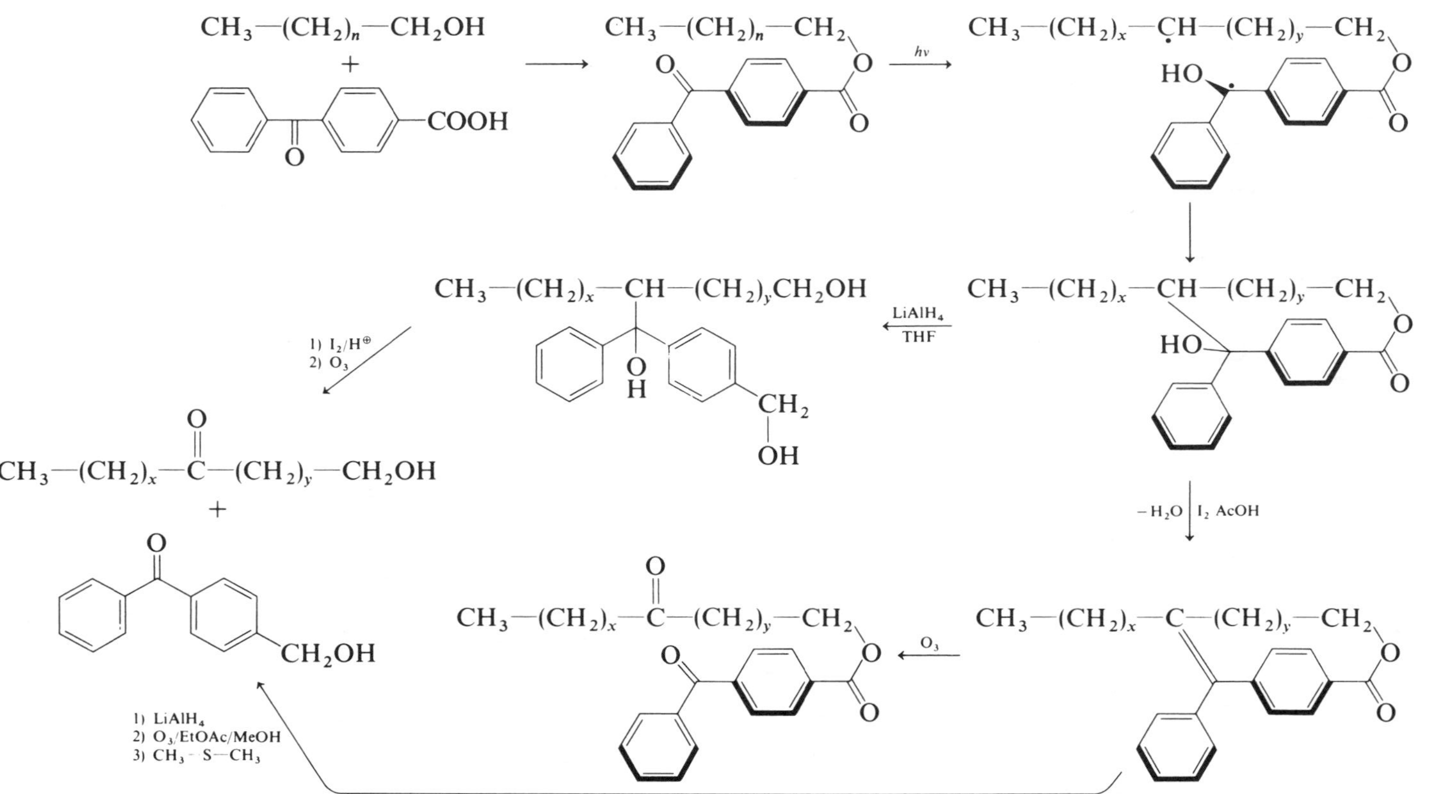

Fig. 5.28. Breslow's strategies to remote functionalization reactions.

Therefore, high rate enhancement and the production of only one product from a reaction where many are possible are important characteristics of enzymatic transformations. In this context, the team of R. Breslow has produced the first chemical example of the application of this principle of *remote functionalization* by undertaking selective oxidation of alkyl chains (429,430). As the reagent to attack an unactivated CH_2 group, they selected the photochemical activation of benzophenone triplet (Yang triplet reaction) and studied the intramolecular insertion of benzophenone carboxylic ester of alkanols (Fig. 5.28).

The radical reaction proceeds with the formation of a covalent bond intermediate. Benzophenone is regenerated after ozonolysis and the covalently attached alkyl chain (substrate) is selectively oxidized with yields up to 66% (Table 5.2).

Table 5.2. Observed Degree of Oxidation of the Alkyl Chain Given in Percent of Oxidation (429)

Oxidation site	Number of carbons in ester alkyl chain[a]			
	C-14	C-16	C-18	C-20
C-8	—	—	—	—
C-9	—	—	—	—
C-10	—	8	8	5
C-11	11	10 to 20	17	15
C-12	49	8 to 13	21	20
C-13	22	3 to 10	18	19
C-14	—	56 to 66	12	19
C-15	—	7 to 10	5	13
C-16	—	—	13	8
C-17	—	—	6	—

[a] A dash means that no product was detected.

Fatty acid esters of benzophenone-4-carboxylate can be incorporated into a micelle such as that from sodium dodecyl sulfate (SDS) or lecithin and can be used as photochemical probes for model membrane structures (431).

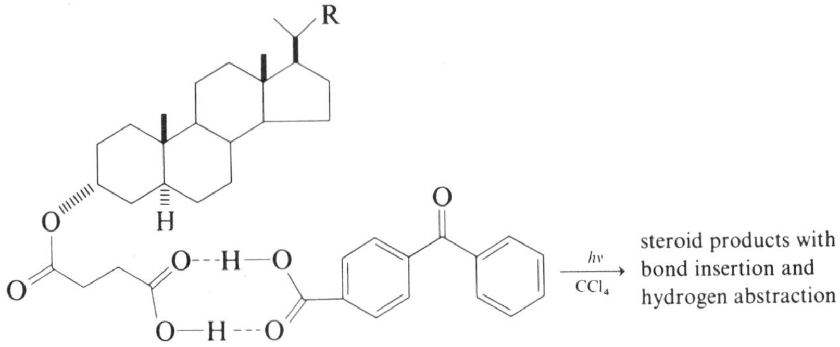

So the possibility of developing a variety of specific functionalization reactions utilizing this concept is at hand. In particular, properly designed rigid models such as steroid templates were used to develop new approaches toward simpler syntheses of hormones where greater selectivity could be achieved.

For example, by attaching the benzophenone chromophore to position 3α of a steroid molecule (267), a series of selectively functionalized derivatives have been obtained under various experimental conditions (Fig. 5.29).

It was also demonstrated that a covalent bond between substrate and "catalyst" is not always necessary. In the example below, hydrogen bondings between the two substances are sufficient to bring the two molecules close enough together for bond insertion and hydrogen abstraction.

A variant of this approach of remote functionalization is the use of a *p*-dichloroiodobenzoate ester to provoke intramolecular halogenation and elimination reactions (432):

R

O

H

O R′

R′ = CH_2—⟨phenyl⟩—ICl_2 (para)

$\xrightarrow[PhCl]{h\nu}$

R

O

R′

H

53%

R′ = CH_2—⟨phenyl⟩—ICl_2 (meta)

$\xrightarrow[C_6H_6]{h\nu}$

R

O

R′

H

43%

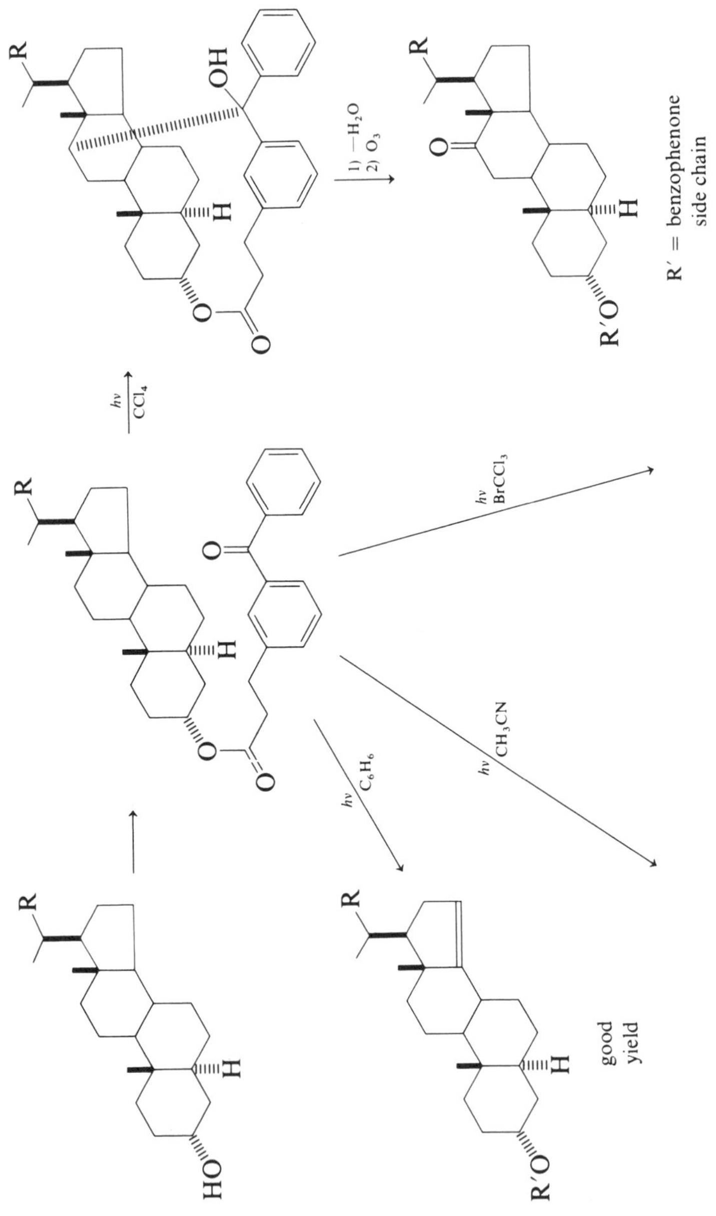

$h\nu$ / CCl_4

1) $-H_2O$
2) O_3

$R' =$ benzophenone side chain

$h\nu$ / $BrCCl_3$

$h\nu$ / C_6H_6

$h\nu$ / CH_3CN

good yield

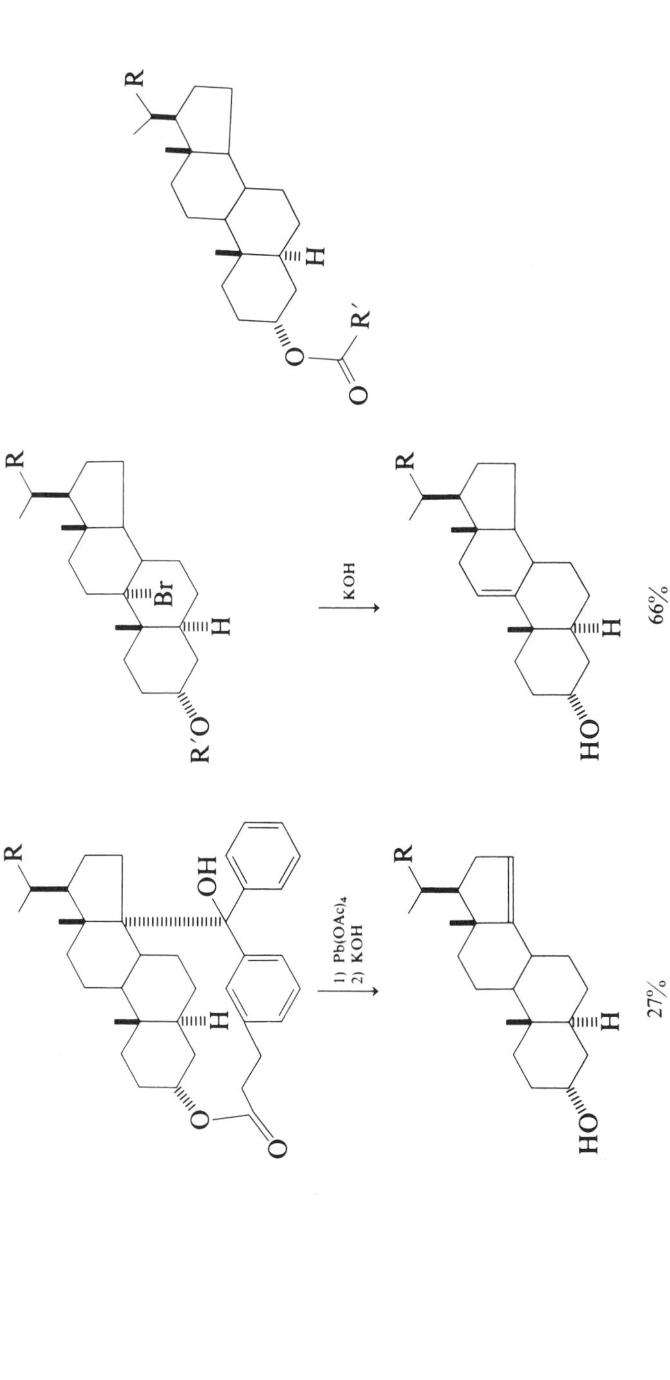

Fig. 5.29. Remote functionalization of steroids.

This method has been applied to a new synthesis of the adrenal hormone cortisone (433):

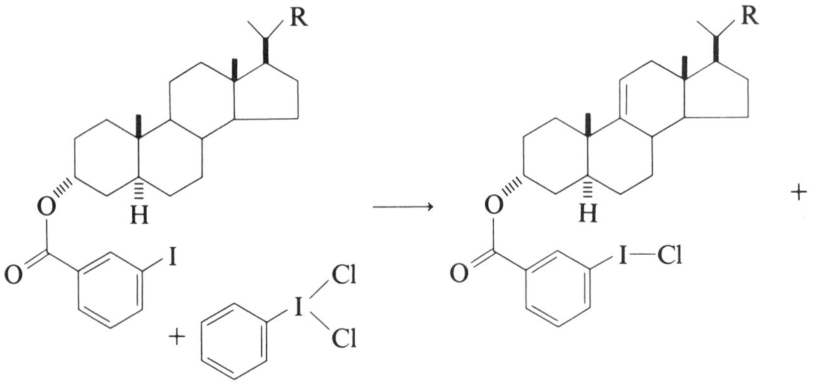

By using different templates, different positions on the steroid are attacked. This is a case of enzymelike geometric control of the site of functionalization rather than the usual control by reactivity seen in chemical reactions. These template-directed halogenation reactions show both catalysis and specificity and are indeed biomimetic (434).

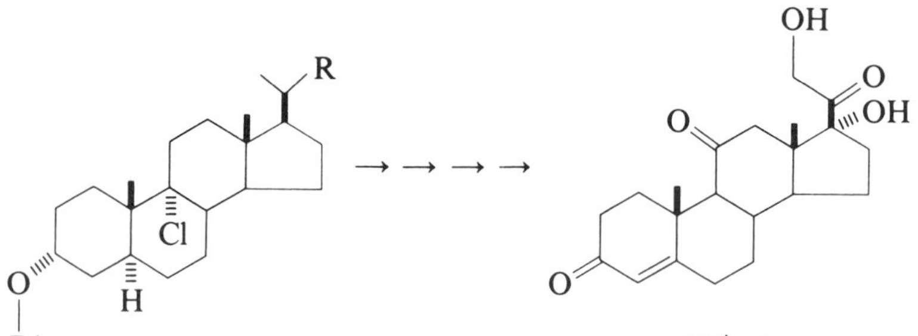

cortisone

More recent findings include the remote functionalization of the steroid β-face with 6-β-benzophenone-alkanoate esters. Photolysis in CCl_4 gave in 20% yield products with carbonyl insertion at C-15, C-16, and on the angular C-18 methyl group (435). Depending on the length of the attacking chain, the steroid side chain could also be functionalized.

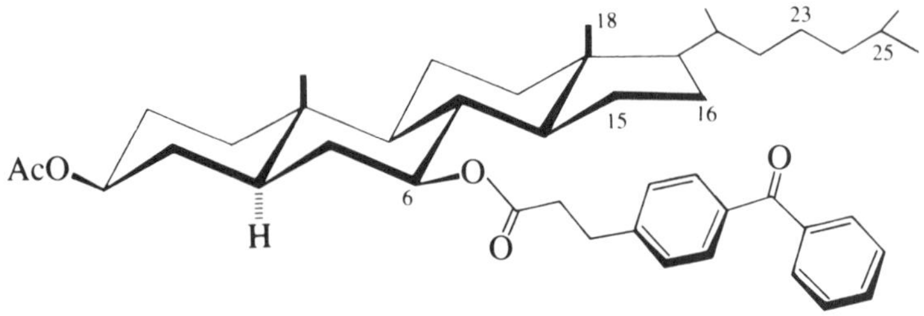

Steroid sulfates could also be catalytically chlorinated in the presence of *m*-chloroiodophenyltrimethylammonium cation. In the following example, positions C_9 and C_{14} are chlorinated to the extent of 57% and 10% respectively. This direct steroid halogenation involves an ion-paired template intermediate on the α-face of the steroid nucleus (436).

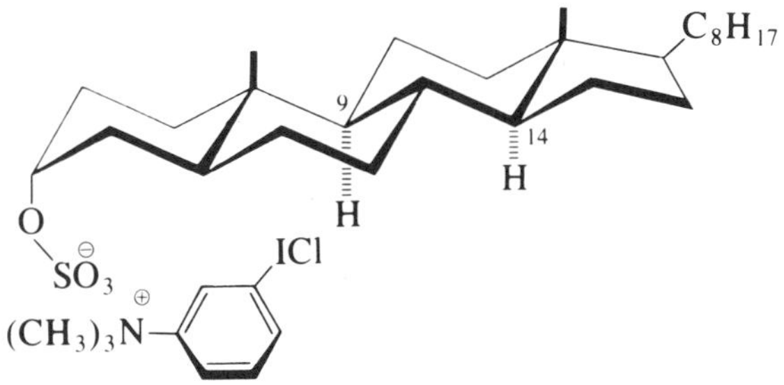

C_8H_{17} 9 14 H H O $SO_3^{\ominus}$ ICl $(CH_3)_3N^{\oplus}$

5.8 Biomimetic Polyene Cyclizations

Biomimetic-type synthesis may be defined as the design and execution of laboratory reactions based upon established or presumed biochemical transformations. This implies the development of chemical transformations new to the nonbiological area and the elaboration of elegant total synthesis of various natural product precursors. Efforts in this direction have been conducted by two schools: one by E.E. van Tamelen (437) and the other by W.S. Johnson (438,439), both at Stanford University.

5.8.1 From Squalene to Lanosterol

Squalene is the precursor of sterols and polycyclic triterpenes. In the 1950s, G. Stork (Columbia University) and A. Eschenmoser (E.T.H., Zurich) proposed that the biogenetic conversion of squalene to lanosterol involved a synchronous oxidative cyclization pathway. The transformation is acid catalyzed and proceeds through a series of carbonium ions to allow the closure of all four rings. There is now ample evidence that the first step is the selective epoxidation of the $\Delta^{2,3}$-double bond to form 2,3-oxidosqualene (Fig. 5.30).

The earlier work of van Tamelen showed that squalene can be converted chemically to the 2,3-bromohydrin with hypobromous acid in aqueous glyme (440). In addition, *N*-bromosuccinimide (NBS) treatment of squalene in water selectively gives the desired terminal bromohydrin. Treatment with ethanolic base results in the racemic 2,3-oxidosqualene. Using rat liver homogenate under standardized aerobic conditions, racemic 2,3-oxidosqualene finally gives rise to sterol fractions that can be purified by chromatography.

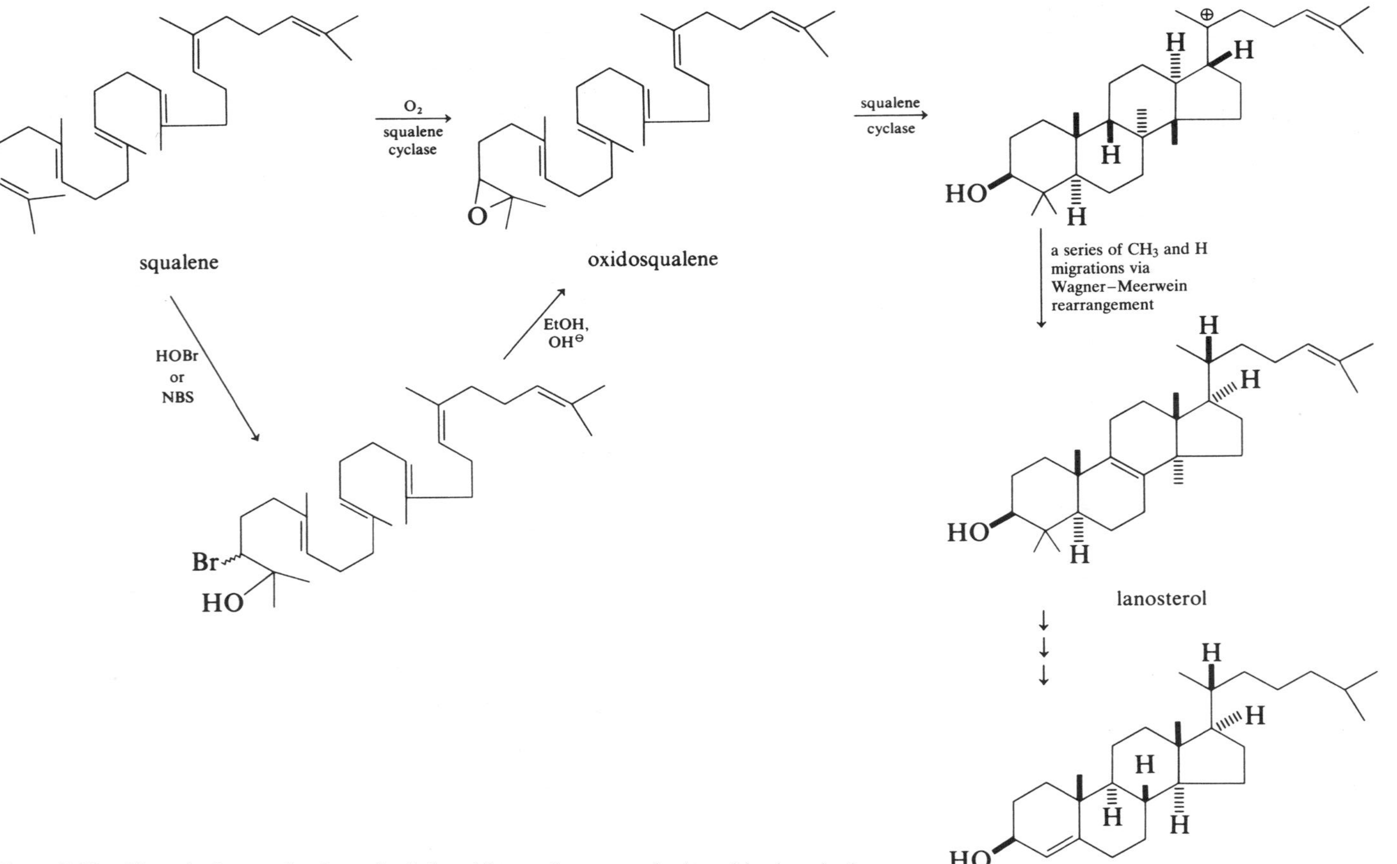

Fig. 5.30. Chemical synthesis of 2,3-oxidosqualene and its biochemical transformation to cholesterol.

Furthermore, it was demonstrated that 2,3-oxidosqualene is synthesized directly from squalene in the sterol-forming rat liver system. The former is a precursor of sterols and is far more efficiently incorporated than squalene under anaerobic conditions. Therefore, it seems very likely that the intermediate 2,3-oxidosqualene is cyclized by an enzymatic mechanism that leads to lanosterol, the precursor of cholesterol and other steroid hormones. Of course, in the enzymatic process only one isomer, the 3-*S*-isomer, of the epoxide is formed.

Closer examination of the 2,3-oxidosqualene skeleton reveals the presence of three distinct π-electron systems designated α, β, and γ-regions (440,441).

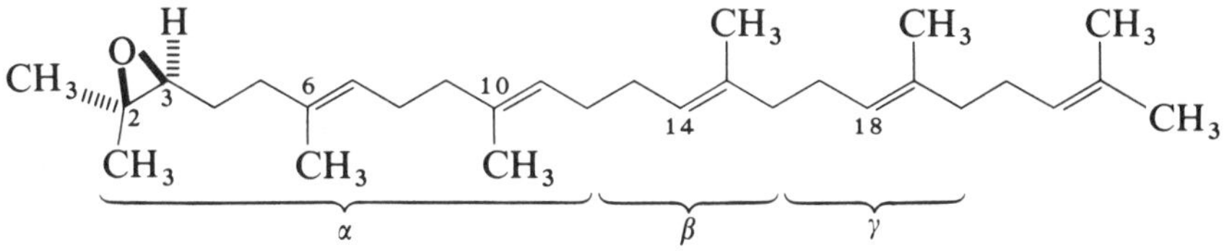

For the cyclization to occur, the α-region is essential for the enzyme cyclase. A second enzymatic control is involved in the β-region for the Δ^{14}-double bond to orient properly on the incoming carbonium ion. The reaction is therefore believed to proceed nonstop until a tetracyclic system is obtained. The role of the squalene oxide cyclase is to maintain the carbon skeleton in place to maximize orbital overlap which allows generation of the σ-bonds in the sterol molecule. This so-called Stork–Eschenmoser hypothesis is represented below:

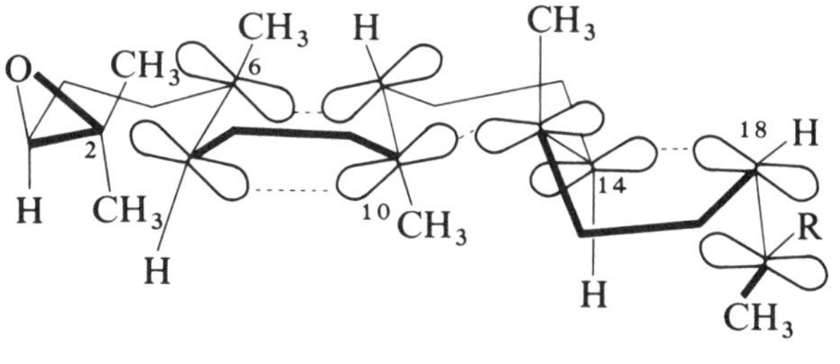

This three-dimensional representation of the skeleton shows the three important orbital overlaps. First the epoxy-Δ^6 bond system where a S_N2 reaction occurs at C_2, then the Δ^6–Δ^{10} overlap which is maximized in a boat form, and finally the Δ^{10}–Δ^{14} and Δ^{14}–Δ^{18} systems where the orientation of the π-planes are perpendicular. Note the boat conformation of the B-ring. Obviously, the biological cyclization of squalene can be rationalized on stereoelectronic grounds. Such a "helixlike" conformation in the transition state very likely favors the concerted cyclization process. This process of cyclization is quite complex and the group of E.J. Corey was among the first to be concerned with the question of how the cyclization of squalene is initiated (442). In this process only one of six possible double bonds is selectively activated.

In animals, squalene adopts the chair–boat–chair folding shown above to generate a chair–boat–chair tricyclic structure, which then undergoes a series of methyl-hydrogen migrations and finally a proton loss. In plant sterols, however, an all-chair folding mechanism takes place:

Particularly interesting is the mechanism by which hydrogen and methyl shifts occur in the Wagner–Meerwein rearrangement to lanosterol. A priori, two types of methyl migration are possible: a single 1,3-shift or two 1,2-shift of methyls.

Bloch suggested in 1958 a very imaginative experiment to solve this ambiguity (443). Starting from two selectively enriched ^{13}C-labeled (•) ψ-ionones, they obtained four differently labeled squalene molecules by condensation with a double Wittig reagent.

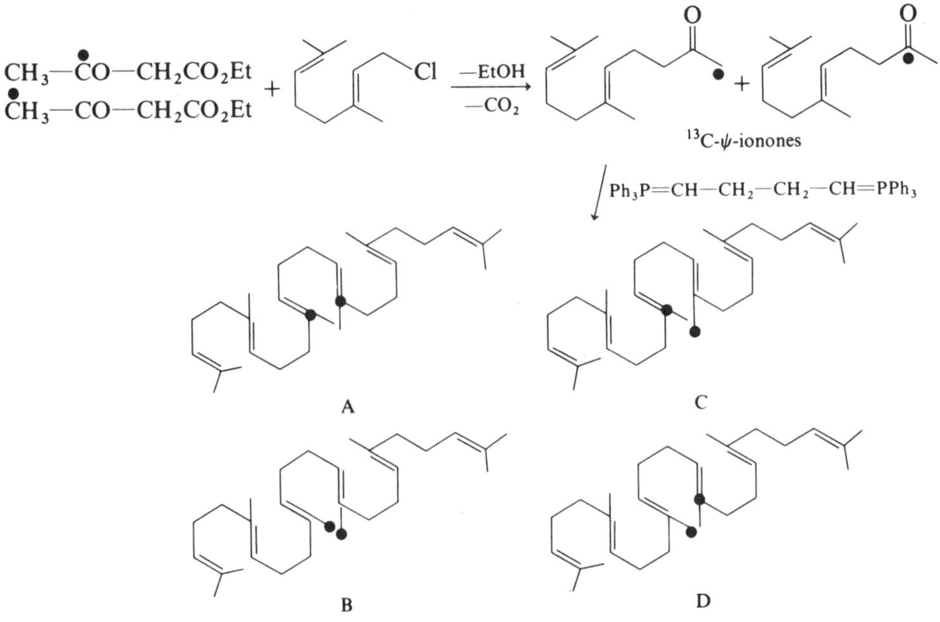

CH_3—C̊O—CH_2CO_2Et
C̊H_3—CO—CH_2CO_2Et + [Cl] $\xrightarrow[-CO_2]{-EtOH}$

^{13}C-ψ-ionones

$Ph_3P{=}CH{-}CH_2{-}CH_2{-}CH{=}PPh_3$

A C

B D

Enzymatic cyclization of these labeled squalenes A, B, C, and D led to A′, B′, C′, and D′ lanosterol partial structures shown in Fig. 5.31.

Lanosterol prepared from the labeled squalene mixture was degraded to acetate by Kuhn–Roth oxidation (sulfochromic treatment) and, in turn, the acetate was degraded to ethylene for mass spectrometric analysis. Consideration of the labeling patterns reveals that only structure D can give rise to ethylene containing ^{13}C at *both* carbon atoms. Since such doubly labeled ethylene was actually detected, it was concluded that the process of two 1,2-methyl migrations actually occurs.

With this brief overview of the enzymatic conversion of squalene to lanosterol, we can now envisage strategies for the total synthesis of polycyclic natural products in the laboratory. This field of bioorganic chemistry has been called *biomimetic polyene cyclizations* by Johnson.

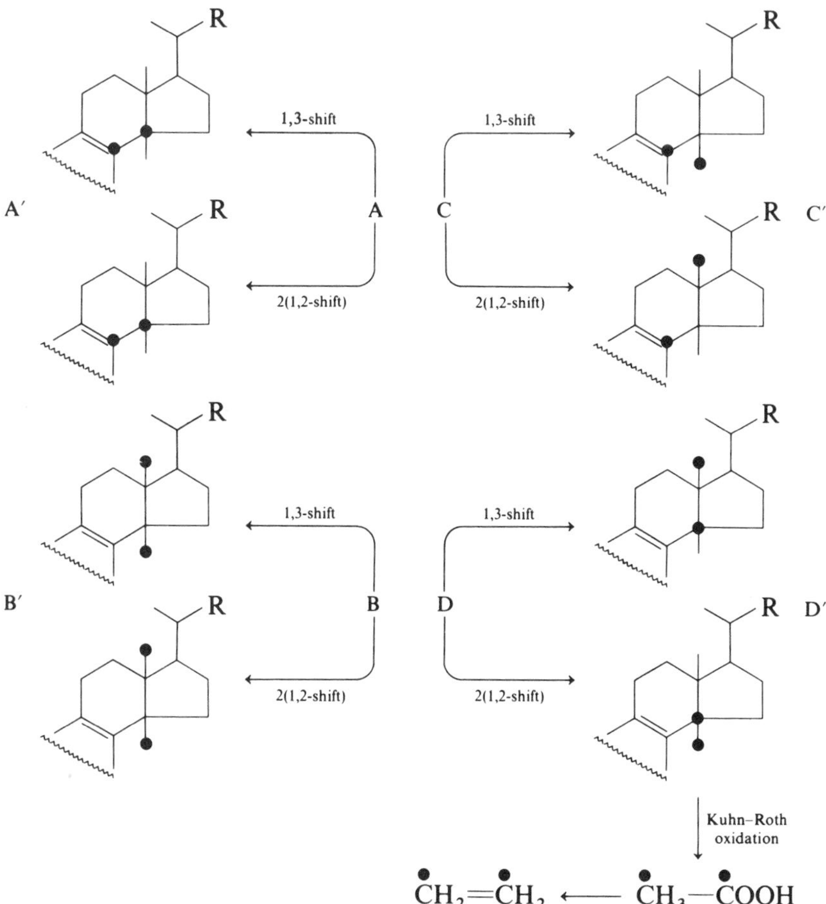

Fig. 5.31. Bloch's proof of two 1,2-methyl migrations in the biosynthesis of lanosterol.

5.8.2 The Biomimetic Approach

How much of the biosynthetic operation can one expect to simulate in the organic laboratory? The first studies directed toward nonenzymatic biogenetic-like polyene cyclization were not promising. However, van Tamelen found that if the D-ring is preformed, the isoeuphenol system can be obtained in good yield by treatment of the epoxide precursor with a Lewis acid (437).

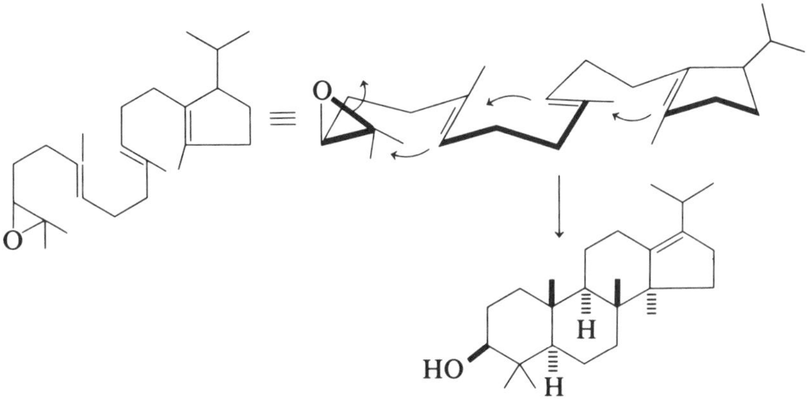

isoeuphenol skeleton
35%

In this cyclization, five asymmetric carbons are formed at one time, based on an all-chain cyclization mechanisms. Of course, the precursor had to be prepared and its synthesis is outlined in Fig. 5.32.

Van Tamelen argues that by selecting such a key intermediate, the preformed D-ring acts as an "insulator" to deter involvement of the side chain π-bond with any carbonium ion developing during cyclization (437). It also prevents the C-ring from becoming five-membered, which was one of the previous difficulties in such cyclization.

The biomimetic approach of Johnson, like van Tamelen, also envisages the production of a number of rings stereospecifically in a single step by the ring closure of an acyclic chain having oppositely placed *trans* olefinic bonds (438). The approach was from the beginning very systematic; *trans* olefinic bond precursors of growing complexity were used to study the annealation of one, two, and three ring products with the natural all-*trans* configuration.

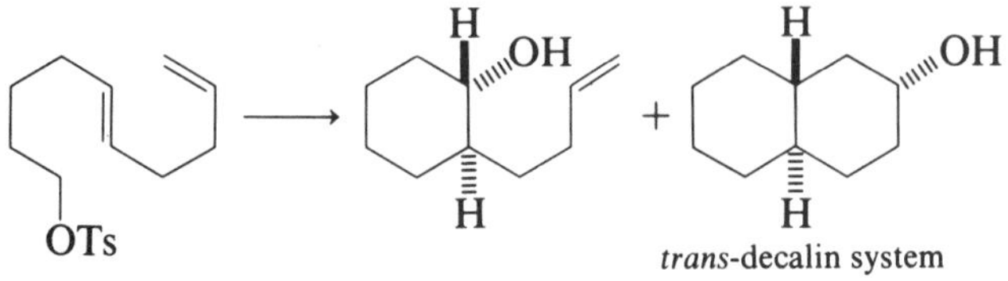

trans-decalin system

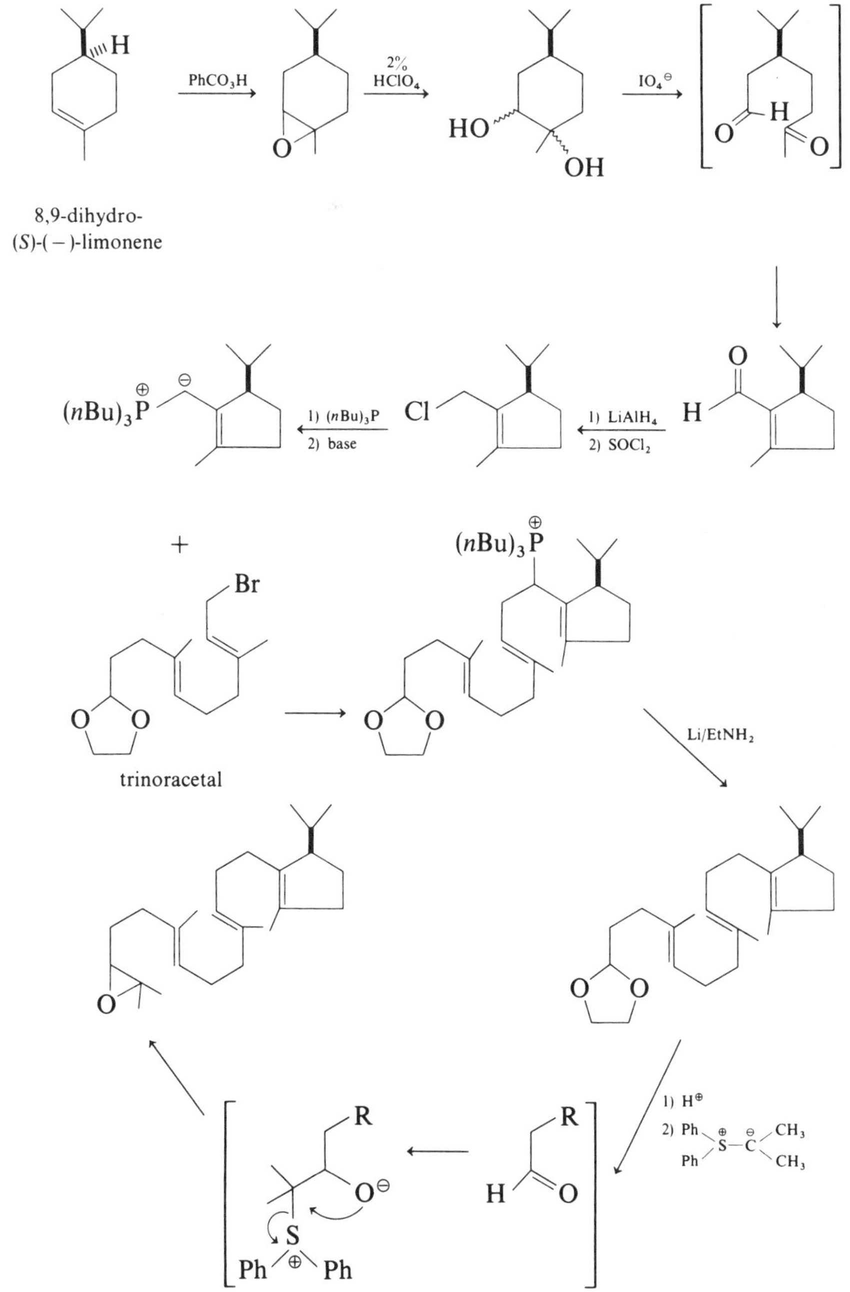

PhCO$_3$H

2% HClO$_4$

HO OH

IO$_4^{\ominus}$

O H O

8,9-dihydro-(S)-(−)-limonene

$(nBu)_3\overset{\oplus}{P}$ $\ominus$

1) $(nBu)_3P$ 2) base

Cl

1) $LiAlH_4$ 2) $SOCl_2$

O H

+ Br

O O

trinoracetal

$(nBu)_3\overset{\oplus}{P}$

O O

Li/EtNH$_2$

O

O O

1) $H^{\oplus}$ 2) $Ph_2\overset{\oplus}{S}-\overset{\ominus}{C}(CH_3)_2$

R O$^{\ominus}$ S Ph $\oplus$ Ph

R H O

Fig. 5.32. Van Tamelen's synthesis of the epoxide precursor of the isoeuphenol system (437).

For example, sulfonated diene esters in 1,5-relationship undergo cyclization–solvolysis to yield a *trans*-decalin system. The major bicyclic component is *trans-syn*-2-decalol.

Although the yield is low, the stereospecificity of the reaction represented the first simple example of a system that followed the theoretical predictions of a stereoelectronically controlled synchronous cyclization.

Much better yields are obtained when acetal as well as allylic alcohol functions are used as initiators for these cyclizations. For instance, the following *trans*-dienic acetal gives quantitative conversion to a *trans*-bicyclic material:

CH_3NO_2, $SnCl_4$, 0 C, 80 sec

The isomeric *cis*-internal olefinic precursor yielded only a *cis*-decalin derivative, in accordance with the predictions of orbital orientation and overlap. It is important to note that the products of all these biomimetic cyclizations are racemic. However, it is interesting to observe that when an optically active dienic acetal is used, a very high degree of asymmetric induction is realized:

(*R*, *R*) C_6H_6, $SnCl_4$

(+)-hydrindanone 92% 8%

For this case, an enantiomeric ratio of 92:8 is observed and the major product was converted to (+)-hydrindanone and its optical activity correlated by optical rotatory dispersion (ORD). The reason for the high selectivity is not clear.

The next objective was to examine the possibility of forming three rings from trienic acetal precursors. Some successful examples are given below:

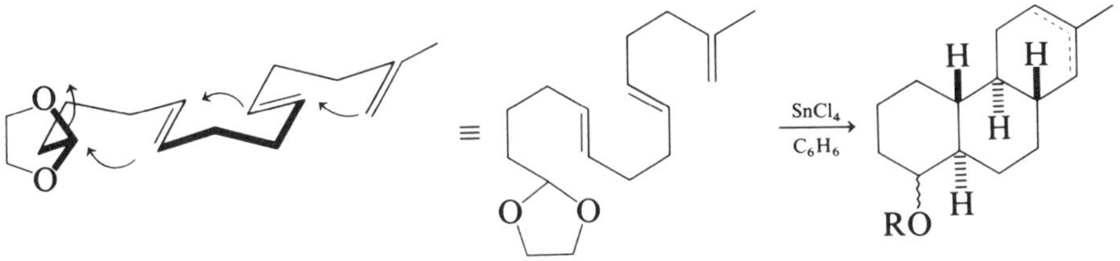
SnCl4
C6H6
H
H
H
H
RO

50%

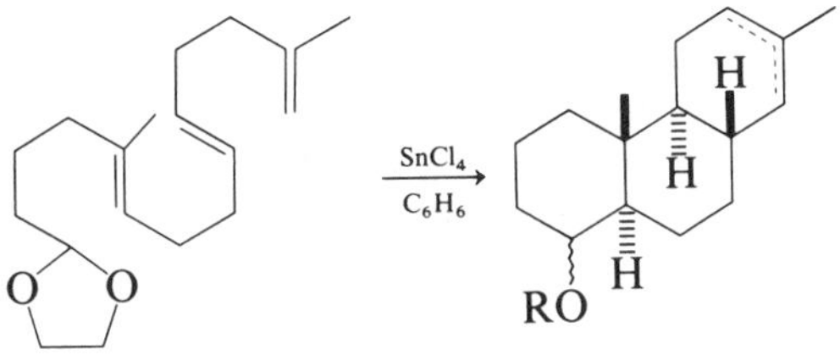
SnCl4
C6H6
H
H
H
RO

63%
(faster cyclization)

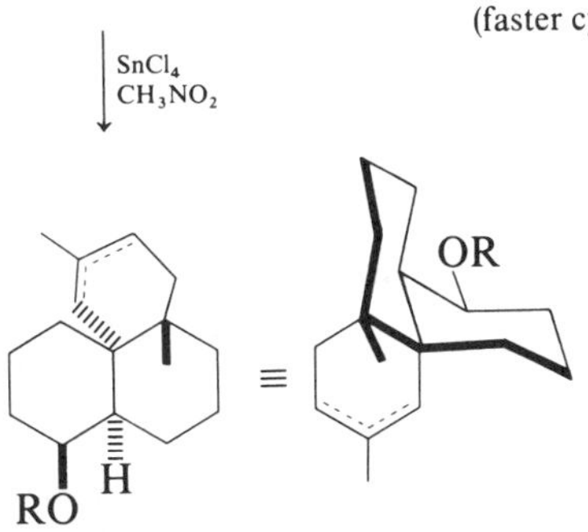
SnCl4
CH3NO2
OR
H
RO

44%
abnormal course

Surprisingly, when the acetal is cyclized in the presence of stannic chloride in nitromethane (conditions that had been so successful with the dienic acetal) the reaction takes a completely abnormal course involving rearrangement. The main product is a tricyclic substance. This product could have arrived through a consecutive 1,2-hydride and methyl shifts of a bicyclic cationic intermediate:

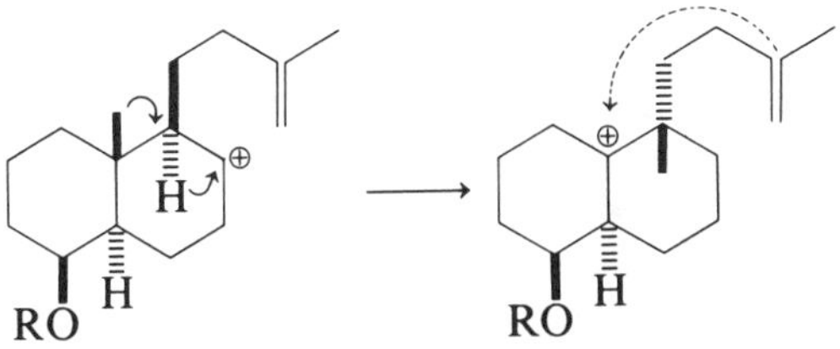
H
H
RO
H
RO

It seems that for this particular case the cyclization is not concerted and a stepwise process is involved leading to mixtures.

Finally, extension to a four-ring system gives D-homosteroidal epimers in the all-*trans* stereochemical series in 30% yield.

$SnCl_4$, C_5H_{12}

Note that seven asymmetric centers are produced in this cyclization with a remarkable stereoselectivity, yielding only 2 out of 64 possible racemates. This concerted conversion is the closest nonenzymatic approach realized thus far with all-*trans* configuration. Undoubtedly, the conversion of an open-chain tetraolefinic acetal having no chiral centers into a tetracyclic compound having 7 such centers and producing only 2 out of a possible 64 racemates is a striking tribute to Johnson's method of biomimetic cyclization.

Because the five-membered ring widely occurs in natural products, particularly in the D-ring of steroids, it was of special interest to search for systems that would give five-membered ring closure in biomimetic processes. After many trials, Johnson and co-workers found that the introduction of a methylacetylenic end group is a useful method to induce five-membered ring closure:

CF_3COOH, CF_2HCH_3, ethylene carbonate, $-30°C$

H_2O

1) t-$BuOCrO_3H$
2) DDQ($-H_2$)
3) H_2,$Rh(PPh_3)_3$

progesterone

DDQ = 2,3-dichloro-5,6-dicyano-1,4-benzoquinone, an oxidizing agent

In this example, the A ring is preformed and the cyclization gives an A/B *cis* junction. The product was then readily transformed into the naturally occurring hormone, progesterone. Thus, acetylenic groups are particularly useful terminators of polycyclization.

In a different approach, a styryl terminating group was also found to favor five-membered ring formation by virtue of the relative stability of the allylic carbonium ion formed.

Ph ⊕ Ph H *trans* 6/5 junction Ph ⊕ H *trans* 6/6 junction

This reaction proceeds in 70% yield in CF_3COOH/CH_2Cl_2 at −50°C with exclusive formation of a *trans* 6/5 ring junction.

In summary, certain polyenic substances having *trans* olefinic bonds in a 1,5-relationship and, possessing effective initiator and terminator functions, can be induced to undergo stereospecific, nonenzymatic, cationic cyclizations. These give polycyclic products of natural interest with an all-*trans* configuration (438,439).

The mechanism of these biomimetic cyclizations and that of their enzymatic counterparts is still unknown, but the majority of the evidence favors a synchronous process over a stepwise one. The study of these nonenzymatic simulations of sterol biosynthesis is of fundamental importance because it could clarify and explain processes that could have been operative since the beginning of life forms on Earth.

5.8.3 Recent Findings

The Stork–Eschenmoser proposal for the stereospecific and synchronous cyclization of epoxy-squalene (see p. 375) has been reexamined by Nishizawa's group with model compounds. The objective was to find evidence for deciding whether the mechanism of biomimetic olefin cyclization operates in a "stepwise" or a "synchronous" fashion (444). If a stepwise mechanism takes place, a series of conformationally rigid cationic intermediates should be involved.

Using his very efficient olefin cyclization agent, mercury(II) triflate/*N*,*N*-dimethylaniline complex, Nishizawa found that in nitromethane/water, (*E*,*E*,*E*)-geranylgeranyl acetate is converted into a mixture of mono-,

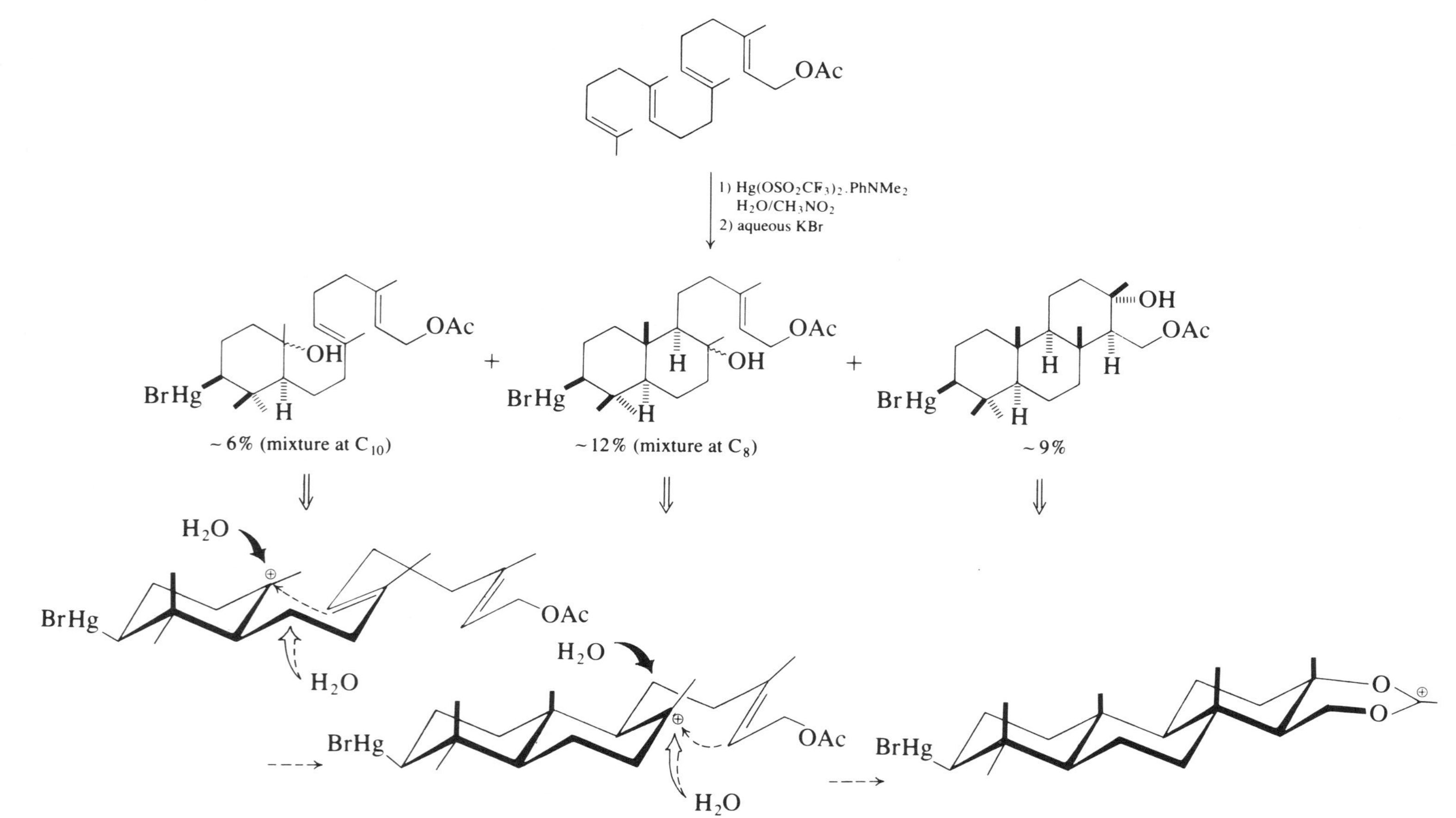
OAc
1) $Hg(OSO_2CF_3)_2 . PhNMe_2$
H_2O/CH_3NO_2
2) aqueous KBr
BrHg
OH
H
OAc
+
BrHg
H
H
OH
OAc
+
BrHg
H
H
H
OH
OAc
~6% (mixture at C_{10})
~12% (mixture at C_8)
~9%
H_2O
BrHg
OAc
H_2O
H_2O
BrHg
OAc
H_2O
BrHg
O
O

bis-, and tricyclic tertiary alcohols via trapping of the cationic intermediate of each step. See page 384.

The results obtained by this cyclization clearly suggest the existence of the three cationic intermediates just mentioned. In the mono- and bicyclic compounds, a predominant formation of α-equatorial alcohols is observed. This indicates that the α-side of the intermediates is solvated, a situation that would not predominate if the conformation of the incipient cationic species were already fixed by π-complexation with the double bond close by to form the next ring.

In the biosynthesis of lanosterol from squalene (Fig. 5.30), the third ring (C-ring) is a six-membered ring. This is rather surprising considering that if during the stereocontrolled π-bond participation the more stable trisubstituted carbonium ion were to be formed, the resulting C-ring would be a five-membered ring. This is what is expected from classical organic chemical concepts. Puzzled by this problem, van Tamelen wondered what nature is doing to alter the normal organic pathway and produce exclusively the six-carbon ring in lanosterol.

To answer this problem, the Stanford chemists tried the reaction on a different substrate, 15′-nor-18, 19-dihydro-2,3-oxidosqualene. Contrary to lanosterol, with this precursor the D-ring cannot form and the intermediate carbonium ion at C_{10} could equally well react with C_{14} or C_{15} double bond to give a five- or a six-membered C-ring. The tricyclic compound obtained from this enzymatic transformation contained only a six-membered C-ring, and a $C_{17}-C_{18}$ double bond as a result of a hydrogen transfer from C_{17} or C_{18} to the carbonium ion on C_{14} (445).

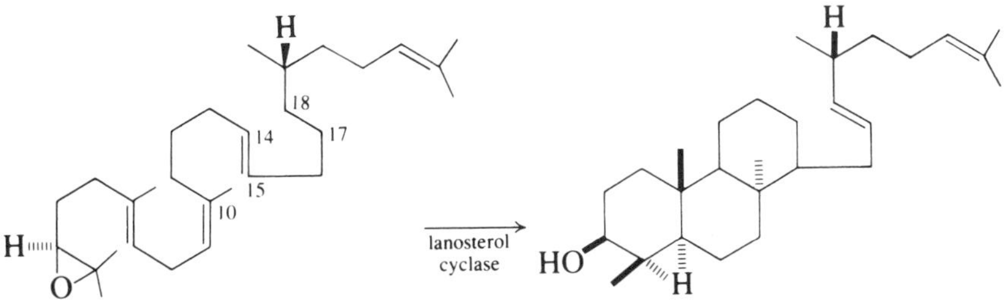

lanosterol cyclase

Van Tamelen reasoned that unless the enzyme holds these carbons close together, such a transfer would be unlikely. For the same reason, it is the conformation imposed by lanosterol cyclase that forces the formation of exclusively a six-membered C-ring.

It is also conceivable that the enzyme may stabilize each cationic center through complementary negative charges during the conversion of epoxy-squalene to lanosterol. If this occurs enzymatically, Johnson reasoned that by preparing polyenes with cation-stabilizing groups, one should improve biomimetic cyclizations. This turned out to be the case. An isobutenyl group at position C_8 was chosen as the cationic-stabilizing auxiliary (446).

$SnCl_4$

$\overset{\ominus}{O}SnCl_4$

polyene with cation-stabilizing isobutenyl group

stabilized intermediate

77% combined yield of tetracyclic products

tricyclization

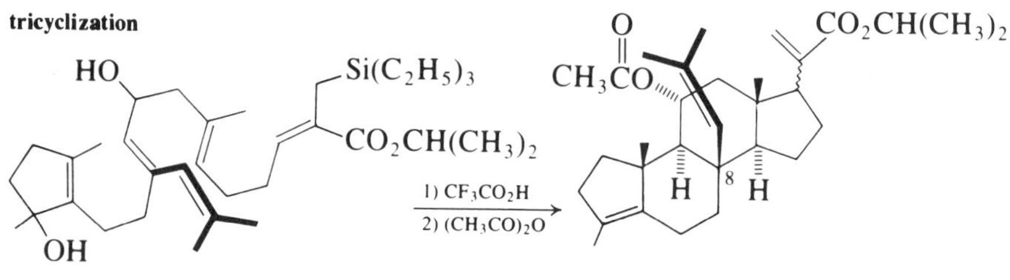

polyene with cation-stabilizing isobutenyl group

83% yield of tricycle

These tetracyclic compounds were obtained in a single step in an improved yield of 77%, compared to 30% for the same polyene without the cation-stabilizing isobutenyl group. Tricyclization was also obtained in high yield. By a process of "fine tuning" the interior of linear polyenes, Johnson's group is currently looking at other stabilizing groups, including heteroatoms.

Nature may use the same sort of molecular device in the enzymatic cyclization of squalene by using specific amino acids of its tertiary structure to stabilize the developing cationic intermediates in the transition state of the cyclizing substrate. Constructing molecules that are transition state analogs (see Section 2.7) of the intermediates formed during the cyclization and seeing if they inhibit the enzyme are among the dreams that Johnson is contemplating.

The first example of nonenzymatic, biomimetic polyene pentacyclization has been reported by Fish and Johnson (447). For this, a series of closely related (*E*,*E*,*Z*,*Z*)-polyenes, with the same tetraene carbon backbone but with differing initiator groups, have been synthesized and cyclized. The work was then focused toward the synthesis of the natural pentacyclic triterpenoid sophoradiol.

In the development of a new strategy for the construction of steroid skeleton, Deslongchamps has recently demonstrated the application of transannular Diels–Alder reactions on macrocycles. For instance, heating at 300°C a well-designed macrocyclic *trans-cis-cis*-trienone gave a tricyclic ketone (448).

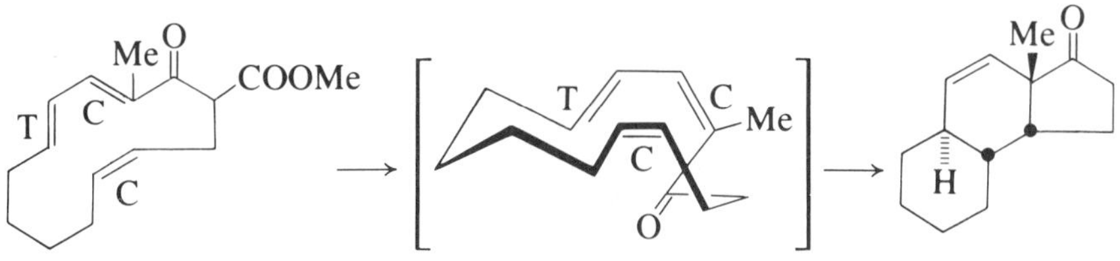

Me O COOMe T C C → T C Me C O → Me O H

hypothetical transition-state

The potential of this single but powerful transformation has been extended to open-chain starting materials for the construction of tricycles bearing alcohol functions. By this approach, well-constructed intermediates in polyene cyclization could eventually be obtained.

Chapter 6
Metal Ions

"Les belles Actions cachées sont les plus estimables."

Pascal

This chapter will provide some basic concepts concerning the reactivity of biological systems utilizing metal ions. Although N, S, O, P, C, and H are the basic elements used to construct the building blocks of biological compounds, certain metal ions are essential to the organisms. It will be seen that the interactions of metal ions with biological molecules are generally of a coordinate nature and are used primarily for maintaining charge neutrality. Also, they are often involved in catalytic processes. Thus, the subjects developed in this chapter interphase organic and inorganic chemical principles.

6.1 Metal Ions in Proteins and Biological Molecules

Many metal ions are essential to living cells. They are Na, K, Mg, Ca, Mn, Fe, Co, Cu, Mo, and Zn and constitute about 3% of human body weight. Na(I), K(I), and Ca(II) are particularly important in the so-called ion pump mechanism where active transport of metabolites and energetic processes are taking place. Transitions metals such as Zn(II) and Co(II) are found in various metalloenzymes where they coordinate with amino acids and enhance catalysis at the active site (449). They act as *super acid* catalysts having a directional or template effect. On the other hand,

Fe(II) and Cu(II) ions prefer to bind to porphyrin-type prosthetic groups and are involved in many electron transport systems.

In contrast, other metal ions such as Hg, Cd, and the metalloid As are strong chelating agents for thiol groups. Consequently, they are efficient enzyme inhibitors. Toxic metals differ from organic poisons in that they cannot be transformed to harmless metabolites. For example, a very potent poisonous gas is the organo-arsenic compound named *Lewisite*. It is capable of acting in the lungs, the skin, or any other of the inner or outer surfaces of the body (449). Within the last thirty years, the development of specific competitive chelating agents has been responsible for significant advances in the treatment of most types of heavy metal poisoning. One of these antidotes for arsenic poisoning is 2,3-dimercapto-1-propanol or BAL (British antiLewisite), developed by Sir Rudolph Peters during World War II. EDTA and some of its analogs are also important and are used to remove, for instance, plutonium in victims of nuclear accidents.

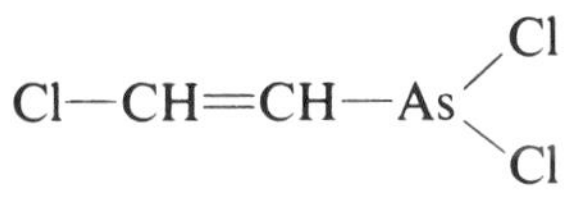

$$Cl-CH=CH-As\begin{smallmatrix} \diagup Cl \\ \diagdown Cl \end{smallmatrix}$$

Lewisite

$$\begin{array}{ccccc} CH_2 & - & CH & - & CH_2 \\ | & & | & & | \\ SH & & SH & & OH \end{array}$$

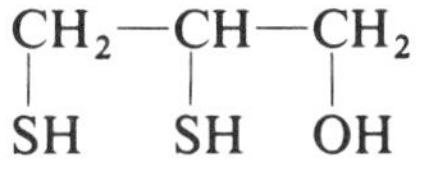

BAL

The development of new chelating agents, through a better understanding of the role and mode of action of metal ions in biosystems, will lead in the near future, it is hoped, to far more selective and effective agents for the therapeutic control of both toxic and essential metal ions (450,451).

In recent years there has been a rapid development in the understanding of how metalloenzymes function. Specifically, the position and the sequence of reactions at the reactive site, and some clues as to the mechanism, have been forthcoming. Important in these processes is the hydrolysis (or hydration) of carbonyl- and phosphoryl-type substrate such as CO_2, carbon esters, phosphate esters and anhydrides, and peptides. It is perhaps not so surprising that a significant number of these systems require a divalent metal ion to function. More surprisingly, however, is the finding that this usually is Zn(II) or Mg(II) (e.g., in enzymes that act on DNA, RNA, cAMP, or cGMP). Thus, the high concentration of zinc in mammals (2.4 g per 70 kg in humans) is exceeded only by that of iron (5.4 g per 70 kg), and most of it is necessary for enzyme function (450).

Zinc ions are important for a number of diverse biological functions. For example, zinc ions serve as a stabilizer for insulin, the hormone that facilitates the passage of glucose into the cell; they are vital components of spermatozoa; they are important in speeding up the healing of wounds; they are involved in the breakdown of protein food in the intestinal tract. Zinc ions are also involved in one of the key reactions in vision.

6.2 Carboxypeptidase A and the Role of Zinc

Enzymes that hydrolyze amide and ester bonds may be divided into three classes: (1) those requiring a thiol group for activity, such as papain, ficin, and other plant enzymes; (2) those inhibited by diisopropylphosphorofluoridate (DFP), such as α-chymotrypsin, trypsin, subtilisin, cholinesterase, and thrombin; (3) those that require a metal ion for activity. This last class includes dipeptidases, and exopeptidases such as carboxypeptidase and leucine aminopeptidase. The metal ion is involved in the stabilization of the tetrahedral intermediate (refer to Section 4.4.1).

Another important metalloenzyme is carbonic anhydrase which catalyzes the conversion of carbon dioxide into bicarbonate ion. It is a zinc-dependent enzyme that is necessary for the absorption of CO_2 into

$$CO_2 + H_2O \rightleftharpoons HCO_3^{\ominus} + H^{\oplus}$$

the blood and subsequent discharge at the lungs. It is believed that a coordinated hydroxide ion facilitates the hydration of CO_2, and it was suggested that an ordered icelike water structure at the active site induces ionization (452).

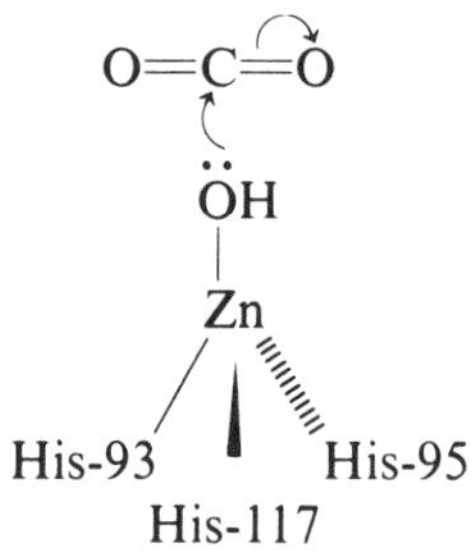

O=C=O
ÖH
Zn
His-93 His-95
His-117

bound to three histidine residues

X-Ray studies of carbonic anhydrase show that the active center is composed of three imidazole ligands that have distorted tetrahedral coordination to the Zn(II) ion. Molecular models suggested that a similar geometry could be attained with a tris(imidazolyl) methane derivation. For this reason, R. Breslow's team (453) in 1978 synthesized

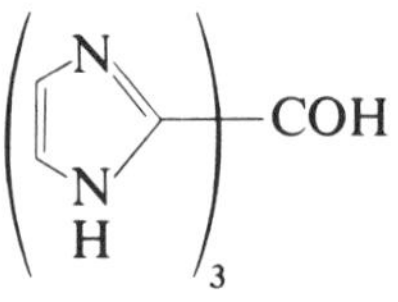

2-TIC 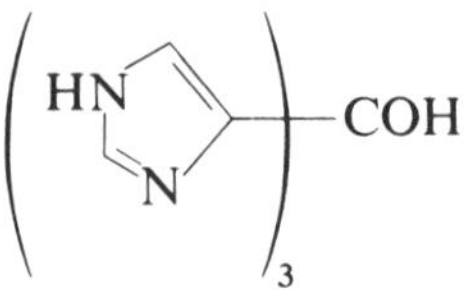 4-TIC 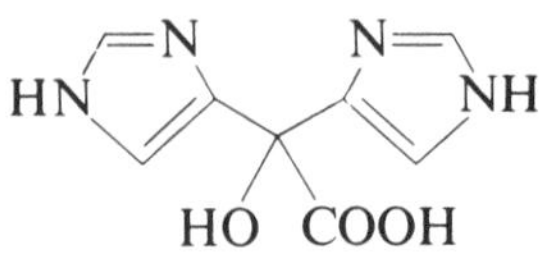 4-BIG

tris[4(5)imidazolyl] carbinol (4-TIC) and tris(2-imidazolyl) carbinol (2-TIC) as models for the zinc binding site of carbonic anhydrase and alkaline phosphatase. Similarly, bis[4(5)-imidazolyl] glycolic acid (4-BIG) has been synthesized to mimic the zinc binding site of carboxypeptidases and thermolysin.

Comparison between the different models clearly showed that 4-TIC is a tridentate ligand capable of using all three imidazole rings for coordination to Zn(II), Co(II), or Ni(II).

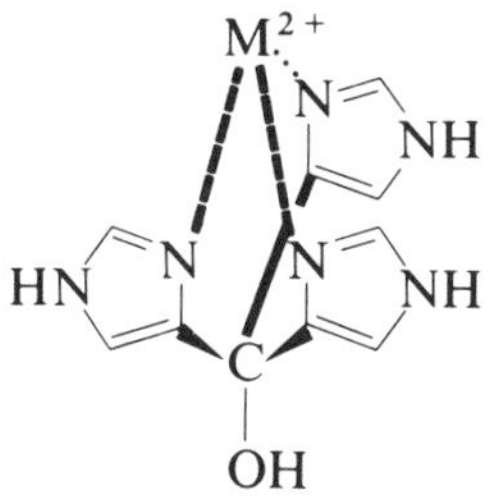

M^{2+} NH HN N N NH C OH

4-TIC metal complex (M = Zn, Co, or Ni ions)

However, the Co(II) complex with 4-TIC or with 4-BIG in nonbasic aqueous solution does not have the blue color characteristic of tetracoordinate Co(II), as in carbonic anhydrase or carboxypeptidase complexes of Co(II). The models are probably too small, so that octahedral complexing is favored. Consequently, spectral and binding studies suggest that the geometry of 4-TIC is not quite right for a good mimic of carbonic anhydrase. It is hoped that a somewhat larger ligand related to 4-TIC might mimic better the spectroscopic behavior of carbonic anhydrase and its extraordinary Zn(II) affinity.

In connection with the chemical approach using model compounds, P. Woolley (454) tried to calculate the activity expected of a zinc complex of suitable geometry, taking explicit account of (1) acceleration by immobilization of substrate, (2) acceleration by activation of substrate, and (3) acceleration by activation of co-factors and prosthetic groups (here, the zinc ion). The result of the calculation came very close to the activity of native carbonic anhydrase. The smallness and symmetry of the CO_2 substrate indicates that CO_2 would turn up at the active site about as frequently as water molecules.

In this section we will orient the rest of the discussion on carboxypeptidase A. This is a zinc-dependent enzyme that catalyzes the hydrolysis of C-terminal amino acid residues of peptides and proteins and the hydrolysis of the corresponding esters. This exopeptidase has a molecular weight of 34,600 and a specificity for aromatic amino acid in the L-configuration.

Carboxypeptidase A is one of those enzymes for which most of the detailed structural information has been obtained by X-ray crystallographic studies, through the efforts of W.N. Lipscomb's group at Harvard University (455,456). The only zinc atom is situated near the center of the molecule and is coordinated by His-69, Glu-72, His-196, and a water molecule. X-Ray crystallographic studies have also been undertaken in the presence of the substrate Gly-Tyr, bound to the active site. The main features of the binding mode show that the C-terminal carboxylate group of the substrate interacts electrostatically with the guanidium group of Arg-145. The carbonyl oxygen of the substrate amide bond displaces the water molecule coordinated to the Zn(II) ion. There is also an electrostatic interaction between the NH_3^+-terminal group of the substrate and Glu-245. Most important is the movement of Tyr-248 which is displaced about 1.2 nm from its original position upon addition of the substrate, thus placing the phenolic OH group about 0.3 nm from the NH of the peptide bond to be cleaved. This is an indication of the *induced-fit* effect, proposed by Koshland, upon enzyme–substrate complex formation.

However, the Tyr-248 hydroxyl is not implicated in the catalysis since site-directed mutagenesis was recently used to replace this group by a phenylalanine group and the enzyme retained its catalytic ability to hydrolyze peptides or related esters (457). Finally, the carboxylate group

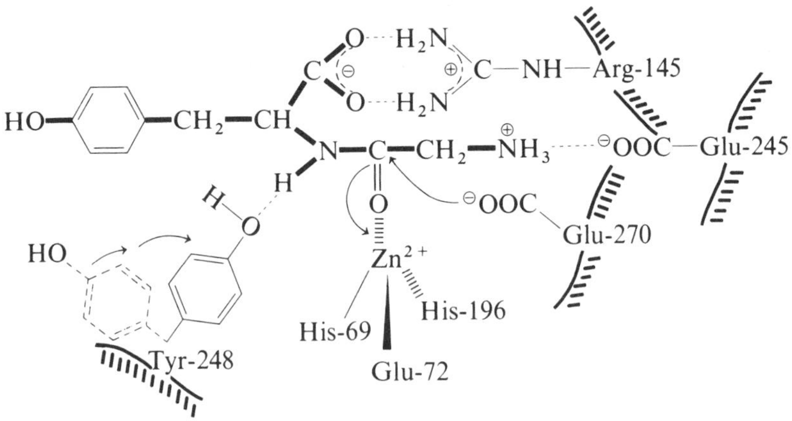

of Glu-270 is believed to be implicated in the catalysis by acting as a nucleophile, forming an anhydride intermediate:

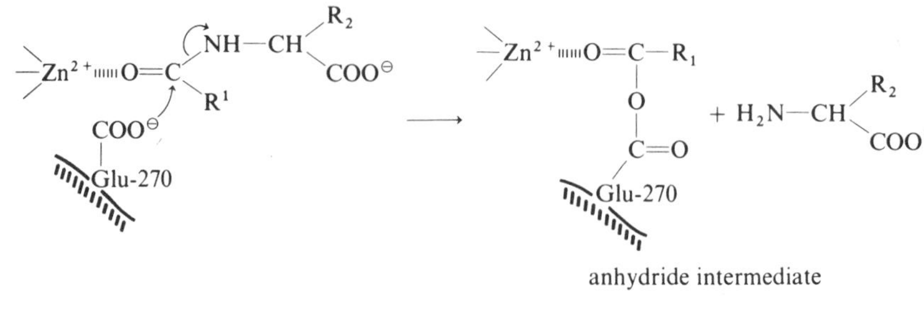

$$\downarrow 2H_2O \text{ or a peptide (resynthesis)}$$

$$\rangle Zn^{2+} \cdots OH_2 + R_1{-}COO^{\ominus}$$

$$+ \quad COO^{\ominus}{-}\text{Glu-270}$$

The mechanism shown above involves two steps and an anhydride (acyl-enzyme) intermediate. In the first step, Zn(II) of the enzyme electrophilically activates the substrate carbonyl toward nucleophilic attack by a glutamate residue. Departure of an alkoxyl group (with ester substrates) or an amino group (with peptide substrates) results in the production of an anhydride between the enzyme glutamate residue and the scissile carboxyl group. In the second step, the hydrolysis of this anhydride can be catalyzed by the Zn(II) ions, the only remaining catalytic group. Although the reaction in $H_2{}^{18}O$ results in the incorporation of ^{18}O in the carboxylate group of the glutamate residue, this suggested mechanism is ambiguous because the anhydride intermediate has never been trapped by a nucleophile such as hydroxylamine.

Therefore, one could wonder if this mechanism is correct. To approach such a problem, three general questions concerning the mechanism of the action of carboxypeptidase and metalloenzymes in the broad sense must be examined.

(a) How do metal ions affect ester and amide hydrolysis?
(b) Under what conditions will a neighboring carboxylate group participate in ester and amide hydrolysis and what is the mechanism of such participation?
(c) How will a metal ion affect intramolecular carboxylate group participation in ester and amide hydrolysis?

These questions should be answerable from model studies. Considerable information concerning questions (a) and (b) are available, but there are experimental difficulties in attempts to answer (c) because of the difficulty of finding a system where carboxylate group participation is possible but where the metal ion does not bind to it.

To evaluate these points, Breslow and his group developed in 1975 and 1976 interesting and more precise models for carboxypeptidase (458,459). They examined the hydrolysis of the following anhydride models with and without Zn(II) ions. The reaction is pH dependent only in the presence of Zn(II) ions and pseudo-first-order hydrolysis rate constants at pH 7.5 have been measured (Table 6.1). Therefore, in the presence of Zn(II), the rate falls in the range of carboxypeptidase A and cleavage of anhydride must be a reasonable step. As with carboxypeptidase A, the Zn(II) ion coordinates to two nitrogen atoms and one carboxylate group. Studies of

Table 6.1. Pseudo-First-Order Rate Constants at pH 7.5 for the Hydrolysis of the Anhydride Model Compounds (458)

Model compound		k_{obsd} (sec^{-1})	k_{rel}
with X = COOH	−Zn(II)	2.7×10^{-3}	1.0
	+Zn(II)	3.0	10^3
with X = H	−Zn(II)	5.5×10^{-3}	2.0
	+Zn(II)	1.5	5×10^2

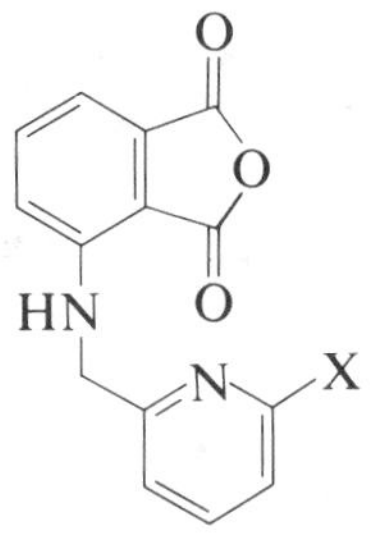

X = COOH, H

the opening of the anhydride (with X = COOH) by hydroxylamine showed that in the absence of Zn(II), hydroxylamine is an effective nucleophile for attack on the anhydride function, but the attack is not catalyzed by Zn(II). Rather, the pH-dependence indicates that Zn(II) catalyzes the attack of an OH^- ion on the anhydride instead of a water molecule and this process is much faster than the uncatalyzed attack by hydroxylamine. Consequently, binding of Zn(II) and OH^- (from H_2O) allows OH^- ion to be an extremely efficient nucleophile, so that a good nucleophile such as hydroxylamine cannot compete. Remember that in the first step of carboxypeptidase action, the Zn(II) ion catalyzes the anhydride formation between Glu-270 and the substrate by acting as a

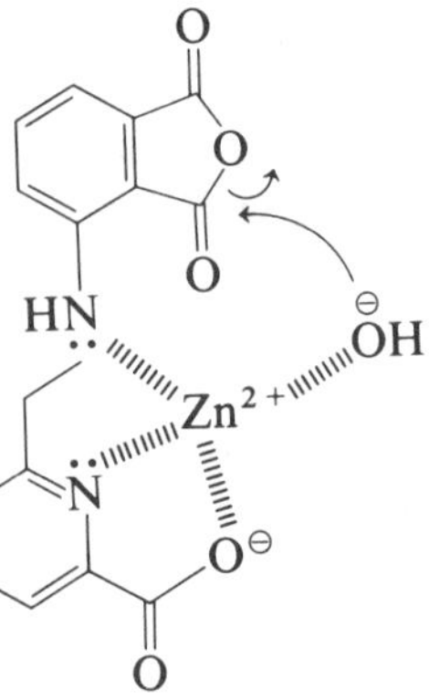

Lewis acid. The present model shows that it can also catalyze the second step by delivering a specific nucleophile to the anhydride intermediate. Nonetheless, the two-step mechanism proposed for carboxypeptidase A remains one of the most attractive explanations of all the data.

Zinc ions and other ions in matalloproteins could be considered as having a *dual role*. The first is an orientation or template effect while the second is a concentration (of the nucleophile) effect at the site of reaction.

Based on his study on model compounds, Breslow suggested a second mechanism for peptide hydrolysis by carboxypeptidase A which does not involve the formation of an acyl-enzyme intermediate (459,460). Essentially, in the hydrolysis of a peptide bond there is participation of a zinc ion, a carboxylate ion, and a tyrosine hydroxyl group. The Zn(II) ion still plays the role of a Lewis acid to coordinate the carbonyl oxygen but the carboxylate group acts rather as a general base. The argument is based on the fact that in the presence of CH_3OH (to substitute for water), methanolysis of a peptide substrate cannot be directly observed because the equilibrium constant is unfavorable. Thus, the enzyme cannot incorporate methanol in the transition state of the reaction (catalyzed in either direction) for either ester or peptide substrates. This suggests that removal of *both* protons of water is required in the transition state for hydrolysis.

In brief, with this mechanism the carboxylate group of Glu-270 acts as a general-base to deliver nucleophilic water to the carbonyl. In the presence of methanol the first step would simply reverse. Consequently, only a second deprotonation could drive the reaction in the forward direction, and this proton transfer might well involve the hydroxyl group of Tyr-248 as a bridge between the OH and the N of the amide bond to be cleaved. This proposal has the advantage of giving a justification for the "induced-fit" effect mentioned earlier in this section (460,461).

Finally, much controversy has surrounded the question of whether Arg-145 or Zn(II) is the binding site for the C-terminal carboxyl group of the substrate. Breslow's mechanism suggests that both are true (462). Indeed, the Zn(II) ion binds the carboxylate group of the hydrolyzed product which becomes the substrate for the reverse or next reaction. One would thus expect that exopeptidases should have two alternate binding sites, separated by a distance corresponding to one amino acid residue in the substrate, as in this proposed mechanism.

From these studies of the mode of action of carboxypeptidase A, the following two points can be extracted: (a) the Zn(II) ion presumably complexes the carbonyl of ester and amide substrates, and (b) Glu-270 is also implicated as a participant, and both general-base and nucleophilic mechanisms have been proposed. There is also strong evidence that the mechanisms are different with esters and amides (461). However, another mechanistic possibility to be envisaged for carboxypeptidase A is to consider a nucleophilic attack of the substrate ester or amide bond by Zn(II) coordinated hydroxide ion. Such a possibility was scrutinized by

Zn^{2+} O R_2—C—NH—CH R_1 $COO^{\ominus}$····$\overset{\oplus}{Arg}$-145 OH H $COO^{\ominus}$ Glu-270

⇌

Zn^{2+} O R_2—C—NH—CH R_1 $COO^{\ominus}$····$\overset{\oplus}{Arg}$-145 O H H COOH O Glu-270 Tyr-248

⇌

Zn^{2+} O R_2—C—$\overset{\oplus}{N}H_2$—CH R_1 $COO^{\ominus}$····$\overset{\oplus}{Arg}$-145 $O^{\ominus}$ H O COOH Glu-270 Tyr-248

⇌

Zn^{2+} O R_2—C=O + H_2N—CH R_1 $CO\overset{\ominus}{O}$····$\overset{\oplus}{Arg}$-145 H O O=C Glu-270

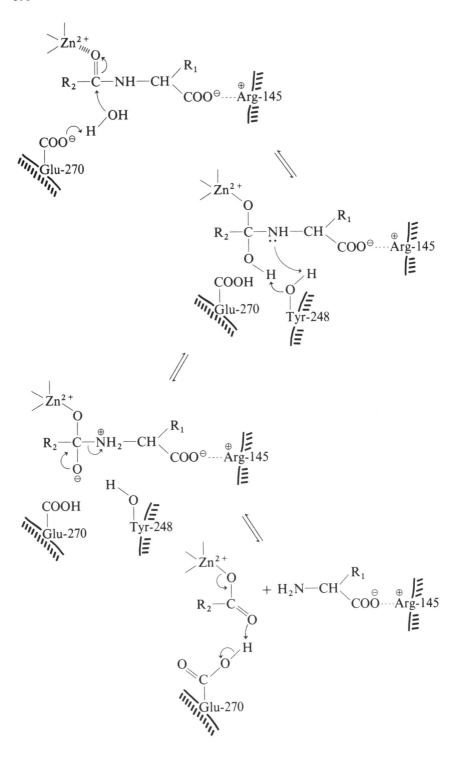

T.H. Fife and V.L. Squillacote (463). In particular, they examined the hydrolysis of the carboxyl-substituted ester 8-quinolyl hydrogen glutarate in the presence of Zn(II) ion.

Extensive kinetic studies showed that at pH values above 6, Zn(II) ion catalyzes the hydrolysis of the ester function through an intramolecular attack of the metal-bound hydroxide ion or by a metal-ion-promoted attack of an external OH^- ion. However, at pH below 6, the metal-ion-facilitated OH^- catalysis, although capable of large rate enhancements, cannot compete with the intramolecular attack of the carboxylate ion to eject the 8-hydroxyquinoline bound to the zinc ion (463).

With regard to the mechanism of carboxypeptidase A, the question is whether a neighboring carboxyl group (Glu-270) functioning as a nucleophile can compete with metal-ion-promoted OH^- catalysis. The present model seems to indicate that with ester substrates, both mechanisms could be operative at appropriate pH values.

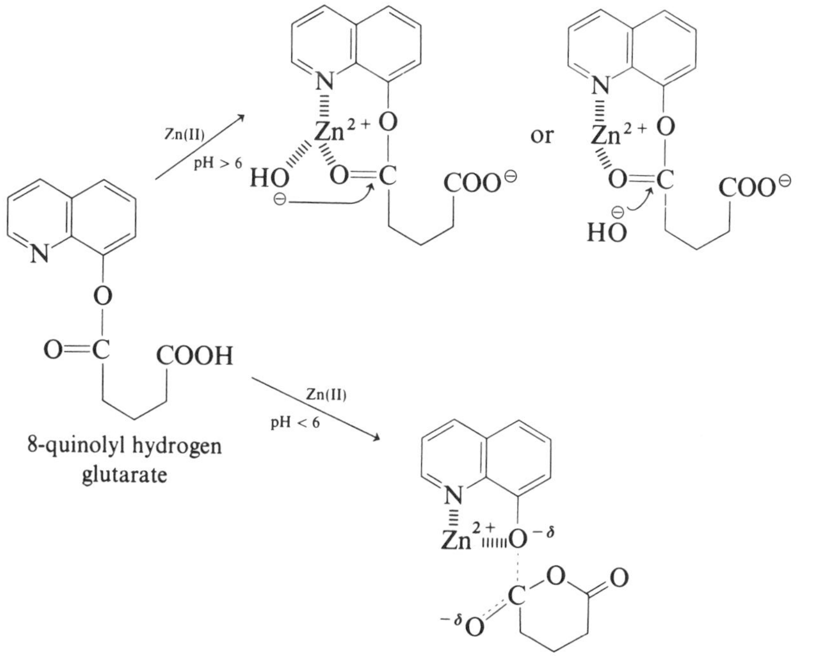

Zn(II), pH > 6; or; Zn(II), pH < 6

8-quinolyl hydrogen glutarate

Therefore, a key argument in regard to the proposed enzymatic mechanism of carboxypeptidase A is whether the carboxylate group of Glu-270 is sterically capable of participating efficiently in a nucleophilic reaction. In fact, such evidence has now been obtained by spectral characterization in the subzero temperature range (−60°C) study of a covalent acyl-enzyme intermediate obtained in the hydrolysis of the specific substrate *O*-(*trans-p*-chlorocinnamoyl)-L-β-phenyllacetate by carboxypeptidase A (464). Furthermore, the results indicate that deacylation of the mixed anhydride intermediate is catalyzed by a Zn-bound hydroxide group.

As part of a continuing X-ray crystallographic study of carboxypeptidase A and its interaction with inhibitors and substrates, Lipscomb's group prepared a ketonic substrate analog, 5-benzamido-2-benzyl-4-oxopentanoic acid (BOP) and examined the slow hydrolysis of the substrate *N*-benzoyl-L-phenylalanine (Bz-Phe) (465,466). Interestingly, BOP binds to the active site as a covalent hydrated product. It was

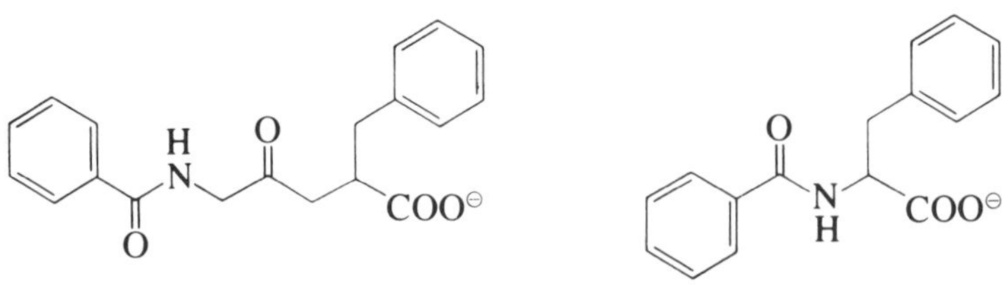

BOP

Bz-Phe

suggested that the zinc-bound water of the native enzyme adds to the ketone carbonyl function. This intermediate would be stabilized by hydrogen bonding with Glu-270 and Arg-127. With the substrate Bz-Phe, the presence of an ionic interaction between the carboxylate group and Arg-127 was also rewarding, and it is believed that both zinc ion and Arg-127 may serve to polarize the scissile carbonyl group of substrate molecules.

6.3 Hydrolysis of Amino Acid Esters and Amides and Peptides

Commonly, transition metals have d orbitals that are only partially filled with electrons. In solution, these positively charged metal ions can readily combine with negative ions or other small electron-donating chemical functions called *ligands* to form complex ions. The geometry of the ligand–metal complex depends on the nature of the metal ion and can be as varied as tetrahedral, square planar, trigonal bipyramidal, or octahedral. Two aspects must be considered in the evaluation of transition metal ion complexes with ligands: first, the nature of the ligand–metal bond involved, and second, the geometry of the complex formed. These factors will contribute to the stability of the ionic complexes.

In many metal ion catalytic reactions, the role of the metal ion is similar to that of a proton, but in a more efficient way because more than a single positive charge can be involved and the ion concentration may be high in aqueous solution. Experimentally, concentrations up to 0.1 M are possible whereas with H^+ only $10^{-7}\,M$ is obtained in a neutral medium. Furthermore, the progress is energy favored because the ion forms a complex with a substrate that is more stable than the free form since the complex has a lower ΔG, and a lower $\Delta G^{\ddagger}$ as well in the transition state.

For example, if a molecule such as oxaloacetic acid binds to Cu(II) ion, decarboxylation readily occurs:

H_2O Cu^{2+} OH_2 ⊖O O O O ⊖O ⟶ H_2O Cu^{2+} OH_2 ⊖O O⊖ O CH_2 + CO_2

The metal ion can form many bonds to the substrate and draw electron density from it. It thus behaves as a *super acid*. Furthermore, ligands such as AcO^- or OH^- relative to H_2O should enhance the positive character of the metal and increase its effectiveness. Also, Fe(III) is a better catalyst than Fe(II) because of the additional positive charge.

An enzyme, because of its well-defined three-dimensional structure, will form a catalytic site where catalysis is involved. A small peptide, however, has very little rigidity and will not possess a catalytic capacity. But it is interesting to realize that if a metal ion is bound to the peptide, amide bond hydrolysis can take place, analogous to that observed with hydrolytic enzymes. Thus, hydrolysis of amides (and esters) is susceptible to the catalytic action of a variety of metal ions because the α-amino and the carbonyl oxygen groups are two good potential ligands for complex ion formation. In other words, coordinated ligands (peptide) acquire remarkable reactivity as a result of the electron-withdrawing effect of positively charged metal ions.

Co(II) and Cu(II) ions can promote the hydrolysis of glycine ethyl esters at pH 7 to 8, at 25°C, conditions under which they are otherwise stable. Complexation takes place between the metal ion (M^{2+}) and the amino acid ester to form a five-membered metal chelate. Subsequently, catalysis occurs as a result of the coordination of the metal ion with the amino and ester functions of the amino acid. In either case, the metal ion can polarize the carbonyl group, thereby promoting attack of OH^-. The rate of hydrolysis increases with pH showing that OH^- ion participates in the mechanism. Thermodynamically, the hydrolysis occurs presumably because the carboxylate anion formed coordinates more strongly to the metal cation than the starting ester.

Experimentally, it is found that metal ions do not hydrolyze simple esters unless there is a second coordination site in the molecule in addition

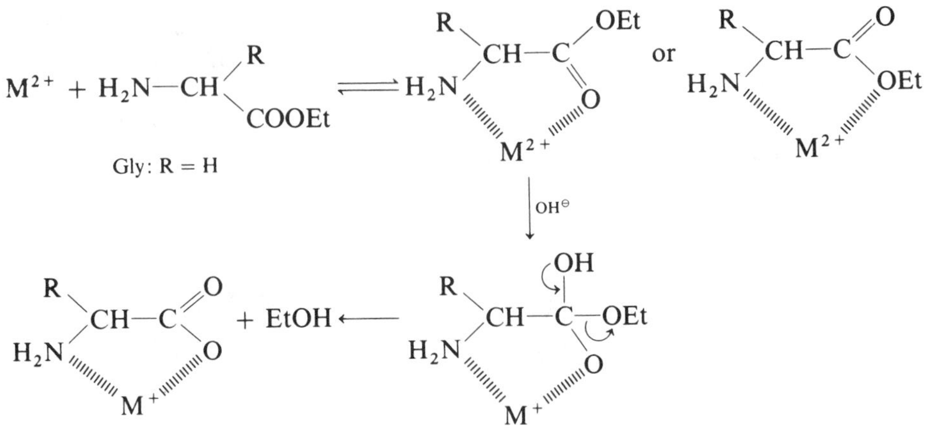

$$M^{2+} + H_2N{-}CH(R)COOEt \rightleftharpoons \text{[chelate: } H_2N{\cdots}M^{2+}{\cdots}O{=}C(OEt){-}CH(R)\text{] or [chelate: } H_2N{\cdots}M^{2+}{\cdots}OEt{-}C(=O){-}CH(R)\text{]}$$

Gly: R = H

$$\xrightarrow{OH^{\ominus}} \text{tetrahedral intermediate } H_2N{\cdots}M^{+}{\cdots}O{-}C(OH)(OEt){-}CH(R) \longrightarrow \text{[chelate: } H_2N{\cdots}M^{+}{\cdots}O{-}C(=O){-}CH(R)\text{]} + EtOH$$

to the carbonyl group. Hydrolysis of the usual types of ester is not catalyzed by metal ions, but hydrolysis of amino acid esters is subject to catalysis.

In a thermodynamic sense, the transition state energy of a metal–amino acid complex is lowered relative to the free transition state for amino acid hydrolysis principally because of charge stabilization. There is likely a lesser solvent reordering term in $\Delta S^{\ddagger}$ in the metal-catalyzed step as well. Consequently, a good template effect of the metal ion in binding the substrate is important. Metal ions will also catalyze the hydrolysis of a number of amides but the effect is not as large as with properly bound esters. The reason lies in the difference in the nature of the leaving group. The poorer amide leaving group makes breakdown of the tetrahedral intermediate rate-determining.

Let us examine a few models. The rate of hydrolysis of the ester function of the phenyl and salicyl esters of pyridine-2,6-dicarboxylic acid is significantly enhanced upon complexation with Ni(II) or Zn(II) ions (467).

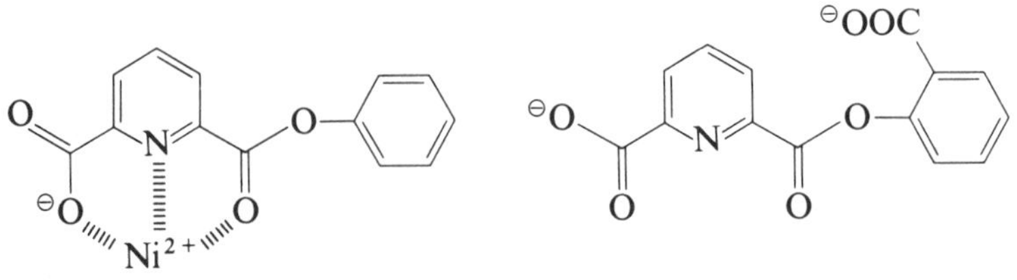

[Structures: Ni²⁺ complex of the pyridine-2,6-dicarboxylic acid monophenyl ester (⊖O, N, O coordinated to Ni^{2+}); salicyl ester with $^{\ominus}OOC$ on the phenyl ring and $^{\ominus}O$ carboxylate]

The relative rate with Ni(II) is 9300-fold with the phenyl ester and 3100-fold with the salicyl ester. In the latter case the rate is slower presumably because two anionic sites are in competition for the metal cation. But the reaction rate is almost twice as fast if the second carboxyl group is ionized, suggesting that carboxylate exerts a weak catalytic effect.

Pyridine carboxaldoxime-Zn(II) complex is an excellent catalyst for the deacylation of 8-acetoxyquinoline-5-sulfonate. Again the dual role of Zn(II) ion comes into play (468,469).

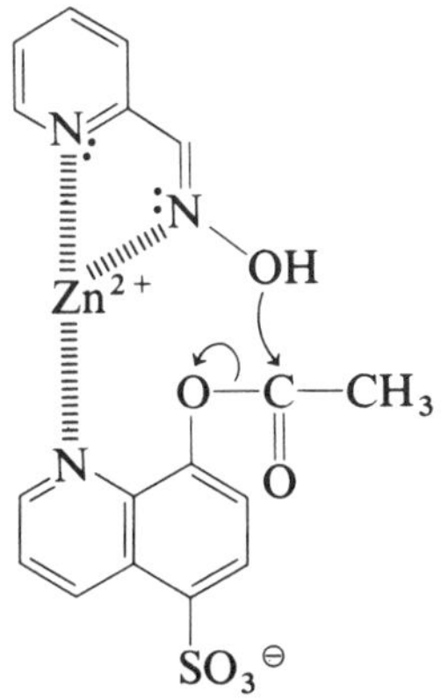

Similarly, phosphorylation of 2-hydroxymethyl phenanthroline by ATP is facilitated by Zn(II) complexation.

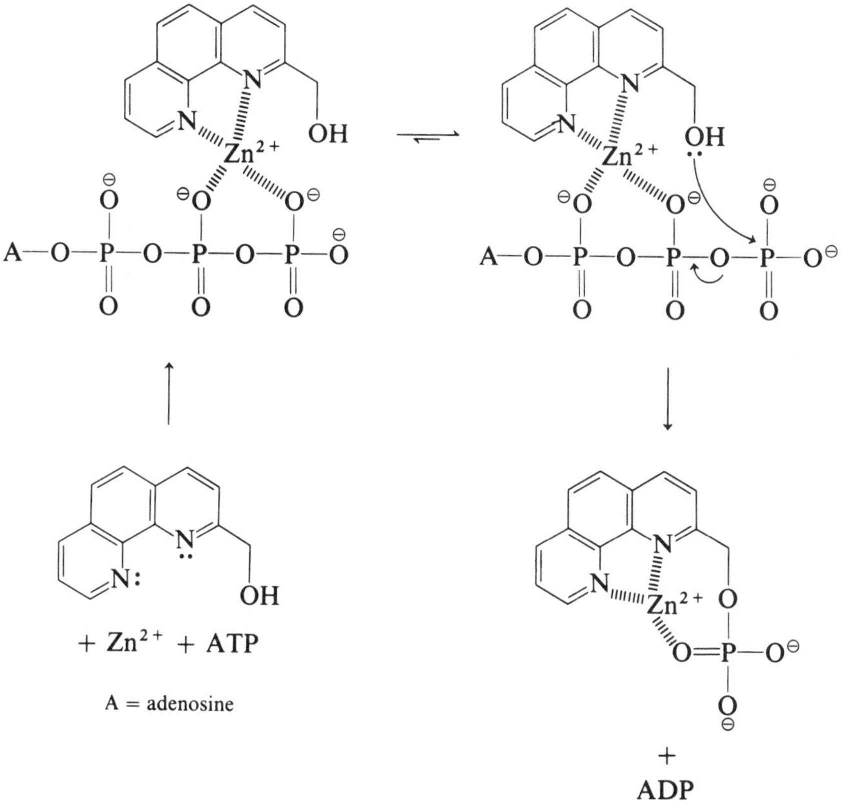

$+ Zn^{2+} + ATP$

A = adenosine

+ ADP

This example illustrates the catalytic influence of a chelated metal ion. The metal has a *directional effect* or *template effect* as is seen by the binding together of ATP and phenanthroline. In addition, it serves a *neutralization* function, thus reducing the electron repulsion between the incoming hydroxymethyl group and the phosphate of the ATP. Of course, the nature of the metal also is important to mediate the position of bond cleavage. The complex also lowers the pK_a of the alcohol function of the phenanthroline moiety to 7.5 so that an appreciable fraction is ionized at the experimental pH value. Further electron delocalization in the aromatic ring system upon complexation with the metal ion increases considerably the acidity of the alcohol group.

In summary, a metal ion facilitates hydrolysis or nucleophilic displacement by some other species in essentially two ways: direct polarization of the substrate and external attack, or by the generation (by ionization) of a particularly reactive (basic) reagent. One of the prime functions of a metal ion in any biological system is thus to provide a useful concentration of a potent nucleophile at a biologically acceptable pH.

Cobalt ions can also exert interesting catalytic properties on ester and amide hydrolysis. For instance, Co(III) is much more effective than Zn(II) at polarizing the carbonyl group of a peptide bond. However, it has been estimated that carboxypeptidase A catalyzes the hydrolysis of benzoylglycyl-L-phenylalanine about 10^4 faster than a Co(III) activation could afford. Therefore, the enzyme must also be exerting an additional effect.

The group of D.A. Buckingham (470,471), then at the Australian National University, has prepared a number of stable Co(III) ion complexes with amino acids and peptides. Ethylene diamine and triethylene tetramine are usually used as ligands. A first example is given here where Hg(II) ion drives the reaction.

With the complex ion catalyst *cis*-β-hydroxoaquatriethylenetetramine cobalt(III), abbreviated $[Co(trienH_2O)OH]^{2+}$, peptide bond hydrolysis of the dipeptide L-aspartylglycine takes place, but only in the productive binding mode.

$[Co(trien)(H_2O)OH]^{2+}$ + L-Asp-Gly

productive complex

nonproductive binding

+ Gly

hydrolyzed dipeptide

The rate-determining initial step involves the replacement of a coordinated water molecule by the terminal NH_2-group of the peptide. The remaining OH-ligand then acts as a nucleophile to promote the hydrolysis. Such intramolecular attack of a coordinated water or hydroxo group is very common in coordination chemistry.

The beauty of the Co(III)-trien systems is that water exchange and substitution in the coordination sphere of the metal ion is, in general, a very slow process ($t_{1/2} \simeq$ minutes to hours), which means that kinetic parameters can be easily evaluated. The slow exchange of ligands in aqueous solution has thus the advantage of using ^{18}O tracers (indicated as •) to follow the path of the coordinated aqua or hydroxo group and thereby allowing the possibility of distinguishing between the direct nucleophilic and general-base paths for hydrolysis. These advantages bring with them obvious problems when one considers their parallel (or lack of it) in enzymatic processes. For instance, the Co(III)-trien complexes–promoted reactions are stoichiometric rather than catalytic, with the product of hydrolysis or hydration remaining firmly bound to the metal center. For this reason, Co(III) complexes are not as useful as they could be for enzyme mimicking. However, because of a favorable decrease in $\Delta S^\ddagger$ ($\Delta H^\ddagger$ remains essentially unchanged) upon complexation with appropriate ligands, rates of $\sim 10^4$ have nevertheless been observed. In spite of that, the system still imitates reasonably well the metals in their ability to polarize adjacent substrate molecules and to activate coordinated nucleophilic groups.

Let us look at another example: the hydrolysis of glycyl-L-aspartic acid by $[Co(trienH_2O)OH]^{2+}$. The situation is slightly more complicated than the preceding one because three structures can be postulated:

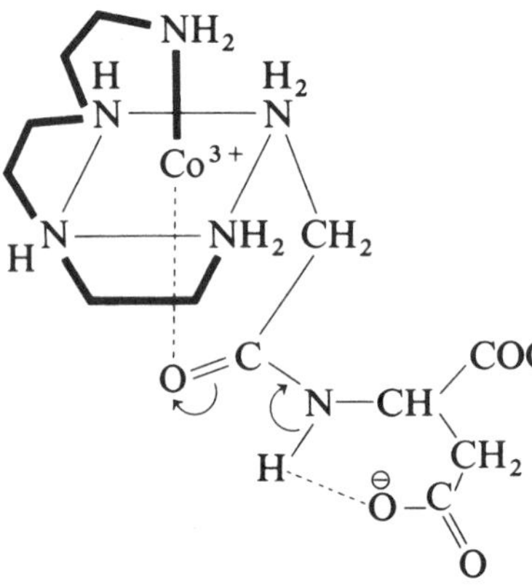

6-1 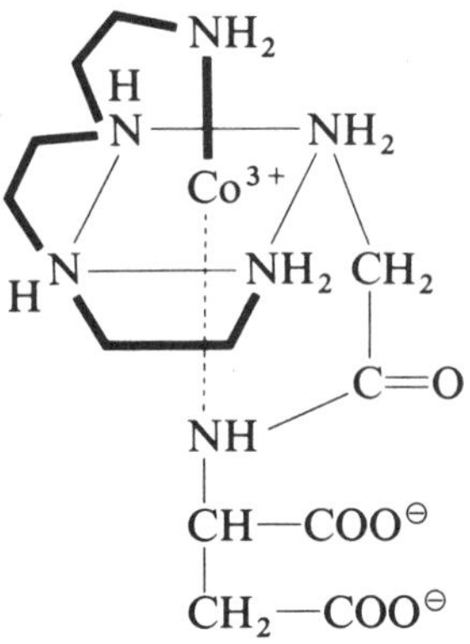 **6-2**

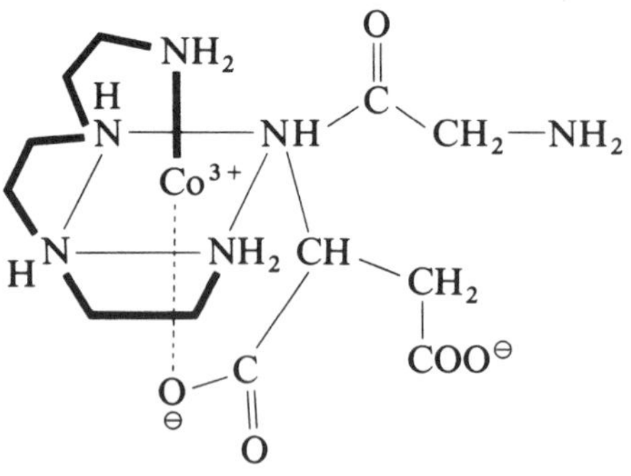

6-3

Structure **6-1** has been favored on the basis of NMR and proton–deuteron exchange studies and hydrolysis occurs in about 20% yield. Again, the N-terminal first binds to the cobalt ion followed by the carbonyl of the amide bond. The γ-carboxylate group of aspartic acid might be implicated in the hydrolysis by contributing to the stability of the complex.

In general, with a peptide not having polar side chains, two pathways are possible for the hydrolysis of the metal–substrate complex. An external water molecule activates the amide carbonyl or the adjacent coordinated OH group acts as a nucleophile on the amide function. This process corresponds to an intramolecular hydrolysis by a metal–ligand. In other words, process A involves a carbonyl coordination intermediate whereas in process B a carbonyl attack by the coordinated hydroxide takes place. These pathways can be followed with ^{18}O-enriched water. Therefore, water and OH are the two ligands replaced by the amino acid terminal

$$[Co(III)(trien)(H_2O)OH]^{2+} + H_2N-\underset{R}{CH}-\overset{O}{\overset{\|}{C}}-NH-\underset{R'}{CH}-\overset{O}{\overset{\|}{C}}-peptide$$

pink color

A ... B

pH > 7

$$H_2N-\underset{R}{CH}-COOH \xleftarrow[\text{(reduction for regeneration of } Co^{2+} \text{ complex)}]{H_2S \text{ or } HCl/Na_2S}$$

$$+ H_2N-\underset{R'}{CH}-CO-peptide$$

group in process A. In process B, the coordinated OH or H_2O is a potent nucleophile and has the proper geometry to attack the amide bond via a tetrahedral intermediate followed by the liberation of the peptide minus the last N-terminal amino acid residue. It is a particularly interesting case because of its selectivity for the hydrolysis of only the N-terminal amino acid of the peptide chain. The resulting Co(III)–amino acid complex has to be decomposed by a reducing agent. It should be realized that the metal ion is not a true catalyst in this hydrolysis since it must be regenerated. Its role is more of a *promoter* of the hydrolytic reaction.

With such Co(III) complexes, rate enhancements of 10^8 have been observed relative to alkaline hydrolysis of the amide function. Such acceleration of hydrolysis reactions is comparable to rates obtained with carboxypeptidase A for its substrates. Notice that in process A, which is believed to be favored over process B with Co(III) ion, the carbonyl group becomes polarized and susceptible to external water attack much more so than in the noncomplex form. Thus the metal ion plays again the role of a *super acid*. In other words, the direct polarization of a carbonyl function by a metal ion generates a more electrophilic center at the carbon atom. Of course, different metal ions have different abilities in this respect, depending largely on their overall charge, size, coordination number, and ease of displacement of (usually) a coordinated water molecule.

These findings have been applied to a stepwise sequential hydrolysis and analysis of peptides from the N-terminal end. This is comparable to the action of an aminopeptidase. With most amino acids the yield of hydrolysis varies between 30 to 50%. To improve the yield, the approach has been altered to include a solid phase support. In alkaline buffer at 60°C only the N-terminal residue remains bound to the resin support and unreactive materials can be washed and recycled (472).

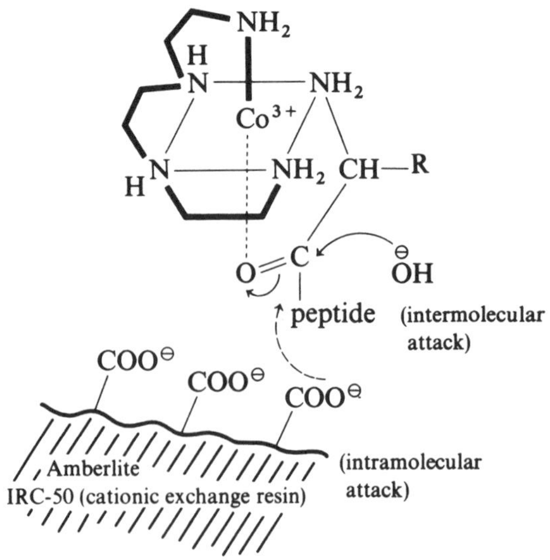

A brief review article appeared in 1994, listing a number of binuclear metal complexes that can act as efficient intermediates in biochemically relevant hydrolysis reactions (473).

6.4 Iron and Oxygen Transport

Iron functions as the principal electron carrier in biological oxidation–reduction reactions (474). Both Fe(II) and Fe(III) ions are present in human systems and when acting as an electron carrier, they cycle between the two oxidation states. This is illustrated by the cytochromes.* Iron ions also serve to transport and store molecular oxygen, a function that is essential to the life of all vertebrates. In this system only the Fe(II) form exists; Fe(III)-hemoglobin does not carry oxygen. In order to satisfy the metabolic requirement for oxygen, most animals have developed a circulating body fluid that transports oxygen from an external source to the mitochondria of tissues. Here it is needed in the respiratory chain to permit oxidative phosphorylation and production of ATP. However, the solubility of oxygen in water is too low to support respiration in active animals. To overcome this problem, nearly all bloods contain supplies of oxygen-carrying protein molecules. The muscles of many animals also contain proteins that reversibly bind oxygen. These molecules facilitate the uptake of oxygen through the muscle and may also provide a reservoir for the storage of oxygen.

The molecular architecture of oxygen-carrying proteins is fascinating, and in the course of biological evolution, nature has developed several alternative molecular devices to serve as oxygen carriers. Each being strikingly colored, these proteins fall into three major families: *hemoglobin*, the familiar red substance in the blood of humans and many other animals; *hemocyanin*, the blue pigment in the blood of many mollusks and arthropods; and *hemerythrin*,† the burgundy-colored protein in the body fluids of a few minor invertebrates. All are metalloproteins. Hemoglobins contain iron by virtue of the heme group; hemocyanins possess dimeric copper clusters (see Section 6.5), and hemerythrins have dimeric iron centers. Hemoglobin is the red protein of red blood cells which carries oxygen from the lungs to the tissues and corresponds to about three-fourths of all the iron in the human body (475).

The hemoglobin molecule is a tetramer composed of two similar globins (polypeptide chain) of unequal lengths. In the center of the protein lies a prosthetic group made of a porphyrin nucleus. This acts as a tetradentate ligand for the iron ion. The porphyrin–iron complex is called a *heme* group. The association of a protein molecule with a heme is referred to as a *hemoprotein*.

Hemoproteins, ubiquitous among plants and animals, perform at least four important functions related to oxygen and the production of energy:

---

* A cytochrome is an electron-transporting protein that contains a heme prosthetic group covalently attached to the peptide chain.

† The common prefix *hem-* in these names does not derive from the presence of heme (iron protoporphyrin IX), which occurs only in hemoglobin, but from the Greek word for blood.

(1) oxygen transport to tissue, (2) catalytic oxidation of organic compounds, (3) decomposition of hydrogen peroxide, and (4) electron transfer.

Hemoglobin reversibly binds oxygen, so that under conditions of high oxygen pressure, such as prevails in the lungs, oxygen will associate preferentially with the protein. Conversely, in the tissues where oxygen is required, the oxygen–hemoglobin complex will dissociate. The oxygen is then transferred to another oxygen-binding hemoprotein, myoglobin. This is a monomeric polypeptide chain. Hence, myoglobin facilitates transfer of oxygen from blood to muscle cells, which then store the oxygen as an energy source (476).

It should be realized that the ability to bind oxygen reversibly is a unique property found in nature only in iron–porphyrin proteins, iron proteins, and copper proteins. However, other small molecules such as CO, CO_2, or CN^- can also interact with these metalloproteins. In fact, CO binds to hemoglobin even more avidly than oxygen, producing a cellular oxygen deficiency, which is sometimes identified as "carbon monoxide poisoning."

In both proteins the heme is tightly bound to the protein part (globin) through about eighty hydrophobic interactions and a single coordinate bond between an imidazole ring of the so-called proximal histidine and the iron atom. In spite of numerous differences in their amino acid sequences, all myoglobin- and hemoglobin-heme units have very similar tertiary structure consisting of eight helical regions. The heme is wedged in a crevice between two helical regions; oxygen binds on one side of the porphyrin while the histidine residue is coordinating the other side. It is believed that the unique oxygen-binding properties of hemoglobin depends on structural features of the entire molecule of hemoglobin and myoglobin.

In deoxyhemoglobin the iron atom is in a high-spin Fe(II) state and lies slightly out of the porphyrin plane. On binding O_2, however, the iron becomes low-spin Fe(II) and moves into the plane. This apparently results in a motion of the proximal imidazole ring by 0.06 nm which causes conformational changes in the protein backbone, producing a higher O_2 affinity quaternary form of the protein molecule. This structural change is the basis of M.F. Perutz's molecular model to explain *cooperativity* in hemoglobin.

Indeed, the most significant property of hemoglobin is cooperative oxygen binding. That is, the oxygen affinity of the tetramer rises with increasing saturation. The concept of *allostery*, developed by J. Monod and co-workers at the Pasteur Institute, took origin with this protein. Cooperativity is required for the transfer of oxygen from the carrier hemoglobin to the receptor myoglobin as well as for responses to other physiological requirements.

Finally, it should be recalled that the iron bound to porphyrin (the heme) is in the ferrous state. Upon oxygen binding, hemoglobin exhibits a reversible behavior with 1:1 stoichiometry *per* iron atom without oxidation of Fe(II) to Fe(III). Scientists have been searching for ways to study this reversible binding of molecular oxygen by Fe(II) in heme. In fact, heme becomes capable of reversibly binding oxygen when it is incorporated into a large protein structure. However, when the heme is removed from any of the proteins and placed in solution at room temperature, molecular oxygen irreversibly oxidizes the iron to the ferric state.

In fact, at neutral pH, ferric ion is readily hydrolyzed to form insoluble hydrated oxides commonly known as rust. Of course, Nature had to avoid the formation of *rust* or insoluble iron oxide in porphyrin. For this, macrocyclic porphyrin ligand had to be encapsulated in a hydrophobic pocket formed by the polypeptide chains of hemoglobin and myoglobin.

Model studies have been undertaken in an attempt to understand the manner by which hemoglobin and myoglobin regulate oxygen affinity (477). The goal in making heme-model compounds is to answer the following questions: (1) How is oxidation to Fe(III) retarded during physical measurements of hemoglobin such as X-ray crystallography? (2) What is the detailed molecular geometry of the heme, heme–CO, and heme–O_2 complexes? (3) How are oxygen affinity and oxidation rate controlled in the hemoprotein?

In 1970 it was first shown that an organic iron complex absorbs oxygen from the air, taking a single oxygen atom into its molecular structure (478). The oxygen pickup occurs in a reaction of triphenylphosphine with $(Bu_4N)_2$ $\{Fe_2[S_2C_2(CF_3)_2]_4\}$. X-Ray works showed that molecular oxygen slips in between iron and phosphorus to give an oxide adduct, (Bu_4N) $\{(C_6H_5)_3P\ OFe[S_2C_2(CF_3)_2]_2\}$. In preparing this complex, E.F. Epstein and I. Bernal at Brookhaven National Laboratory and A.L. Balch at the University of California at Los Angeles found the compound much more stable than similar cobalt complexes previously made.

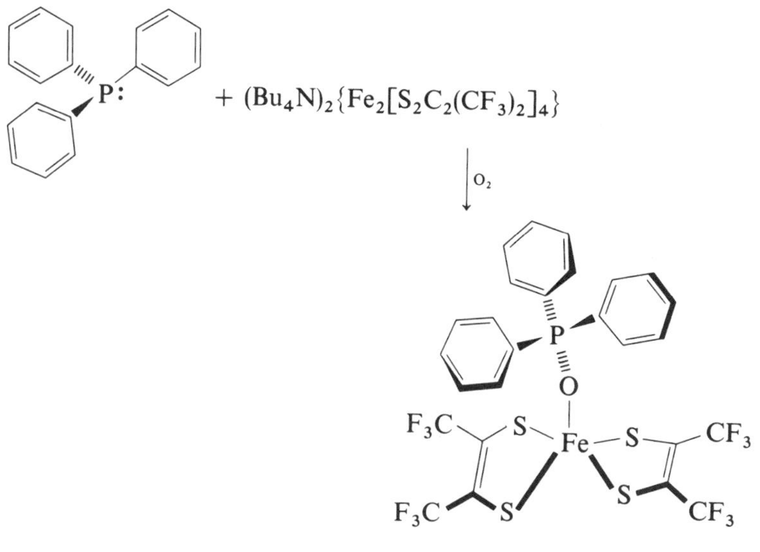

Of course the current direction is to synthesize complexes that will more closely mimic hemoglobin and myoglobin and to develop oxygen-activating species in general. In 1973 and 1974, several model systems were devised. We will thus be concerned here with the interaction of molecular oxygen (dioxygen) and carbon monoxide with matalloporphyrin complexes as models of biologically important hemoproteins.

J.E. Baldwin and co-workers (479) at the University of Oxford first synthesized a porphyrinlike structure surrounding an Fe(II) atom that reversibly bound oxygen in solution. However, this occurred only at −85°C. Later he developed a simple method to prepare a cagelike or a "capped" porphyrin molecule. This was done in the hope that upon O_2 binding, the irreversible Fe(II) → Fe(III) process will not occur and O_2 will bind reversibly to the model molecule at room temperature (480).

For this, they have examined the direct condensation of suitable tetraaldehydes with pyrrole, as a route to capped porphyrins. This is schematically represented as follows:

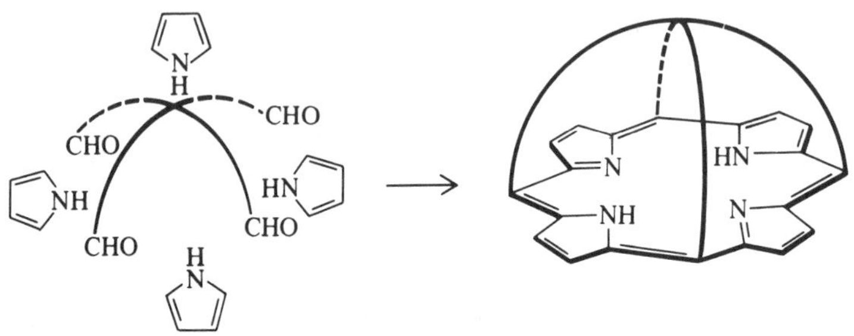

The suitable tetraaldehyde was prepared from salicylaldehyde and condensed with pyrrole (Fig. 6.1).

The corresponding "capped" porphyrin–Fe(III) complex was reduced with chromous bis(acetylacetonate) in benzene to a crystalline ferrous porphyrin that reversibly oxygenates in solution at room temperature. The kinetic stability of solutions containing this oxygen complex depends on many factors such as the temperature, the nature and coordination of the porphyrin and of the axial base (1-methylimidazole is much superior to pyridine), the partial pressure of oxygen, and the solvent polarity.

In a different approach, T.G. Traylor and his group (481, 482) at the University of California at San Diego introduced an Fe(II) atom within a modified porphyrin ring having the desired imidazole side chain properly designed to mimic the neighboring histidine groups in hemoproteins. This man-made active site of myoglobin now has the proximal histidine residue covalently attached to the porphyrin nucleus. It was obtained by the following sequence of chemical transformations, from natural chlorophyll *a* (Fig. 6.2). The strategically placed imidazole group creates both physical and chemical conditions that induce the synthetic molecule (protoheme) to bind oxygen reversibly (but only at −45°C) in approximately the same

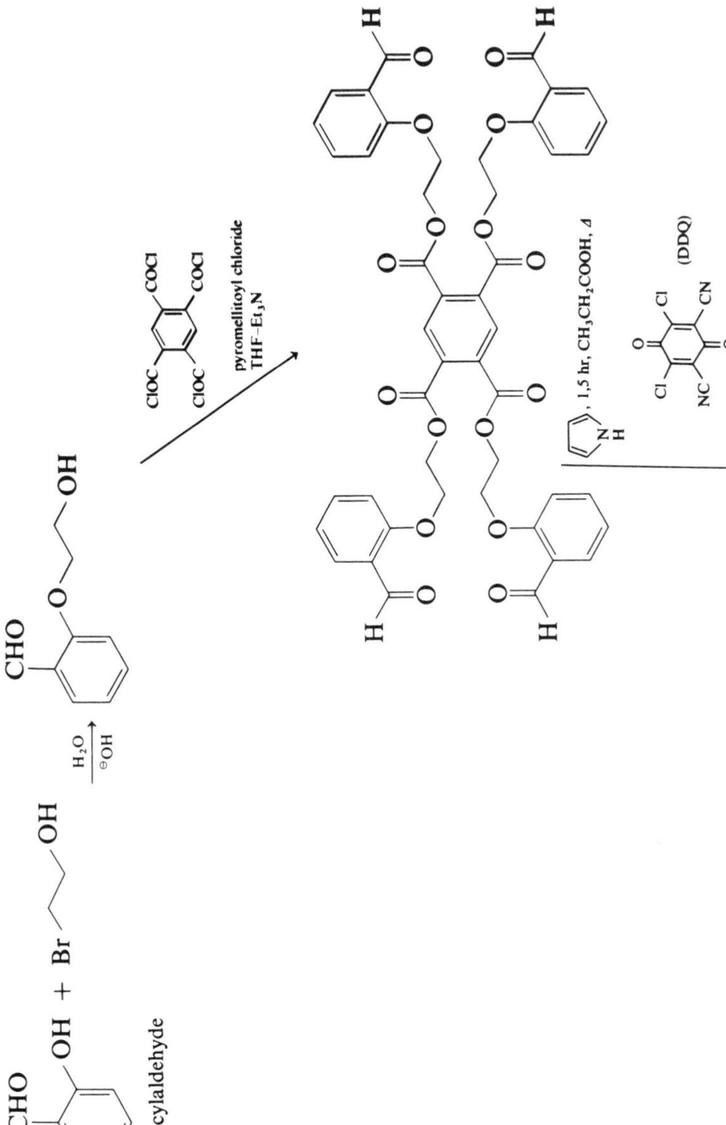

Fig. 6.1. Baldwin's synthesis of "capped" porphyrin (480). Reprinted with permission. Copyright © 1975 by the American Chemical Society.

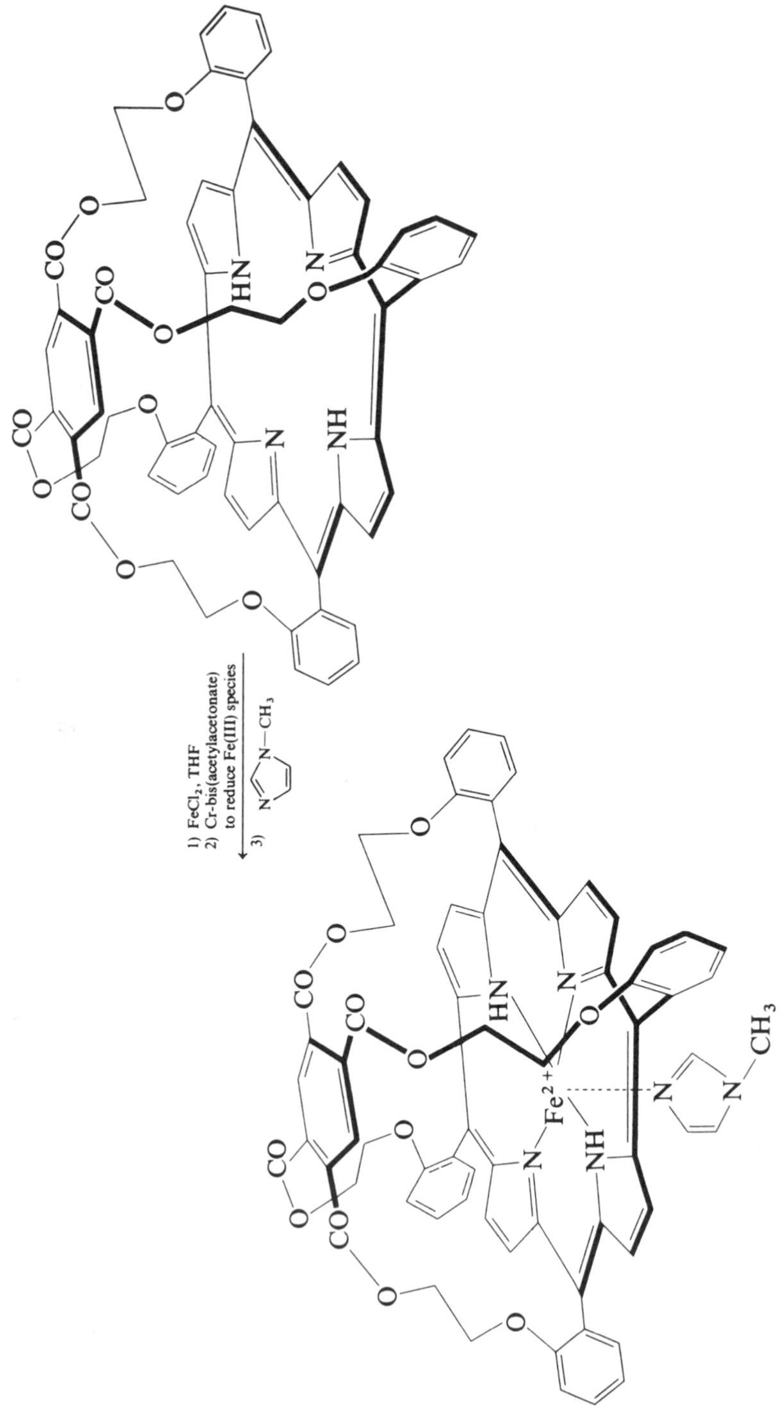

1) $FeCl_2$, THF
2) Cr-bis(acetylacetonate) to reduce Fe(III) species
3) N-methylimidazole (N–CH_3)

2% yield from starting aldehyde
(luster violet crystals)

Fig. 6.1. (continued)

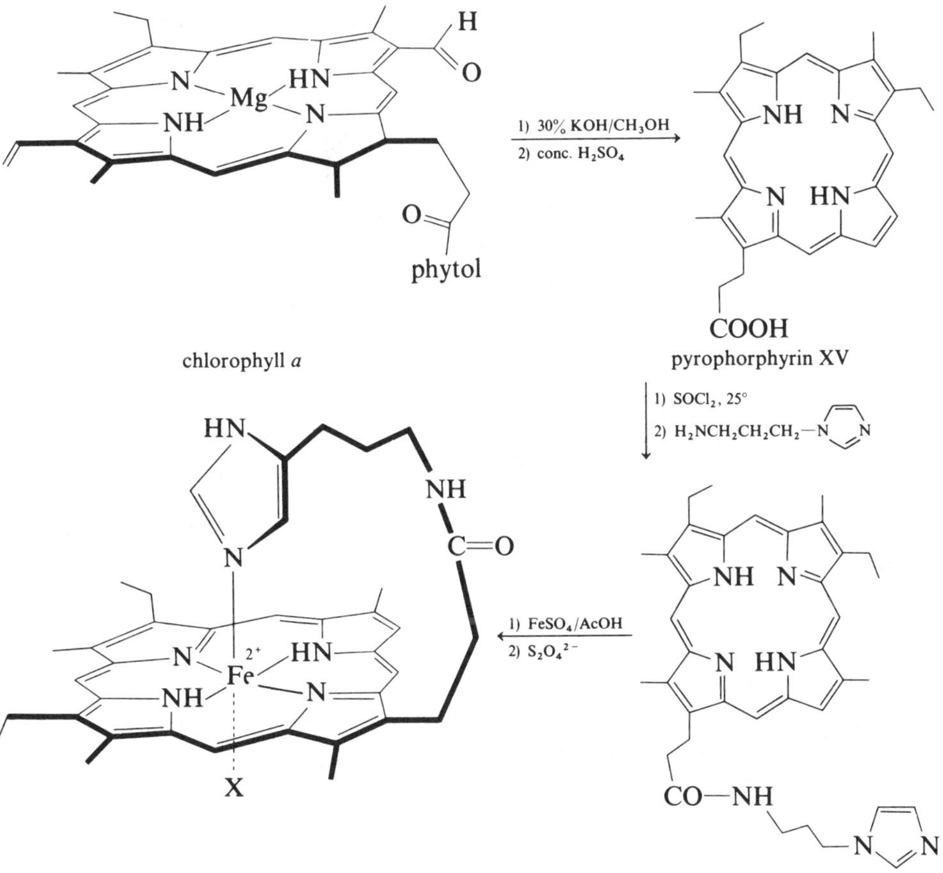

Fig. 6.2. Traylor's approach to man-made active site of myoglobin (477).

way as the natural molecule does. In CH_2Cl_2 at $-45°C$, the model shows almost the same binding constant with oxygen as does sperm whale myoglobin at room temperature. Interestingly, this result shows that the protein might not be as important to the active site as was formerly believed, although current theories explain hemoglobin's efficient binding of oxygen through some type of beneficial interaction among the four subunits. Notice that this ion selectivity by the porphyrin ring is analogous with host–guest designed chemistry, discussed in Chapter 5.

If the imidazole group is replaced by a pyridine molecule, the compound does not bind oxygen but is oxidized to the ferric state. So the basic nature of the imidazole ring and its geometry relationship to the Fe(II) ion in the heme results in a unique binding power for oxygen. The model binds CO much more strongly than it binds O_2.

The binding efficiency for oxygen is also present in the frozen solid state and when the model compound is dissolved in a film of polystyrene. Practically, it is now possible to make materials similar to a synthetic active site that can be built into plastic film to facilitate oxygen transfer in artificial lung machines. Of course, such technology can become really practical only with the development of new models that can bind oxygen more efficiently at room temperature without being oxidized. Before 1975 reversible oxygenation of model complexes containing heme was observed only at low temperatures or in nonphysiological media.

The close proximity of the porphyrin system to certain residues of the peptide chain in hemoglobin can create steric hindrance toward CO or O_2 ligation to the Fe(II) ion. In 1979 Traylor's group developed two variants of the previous model to evaluate this steric effect (483). Figure 6.3 outlines the synthesis of these molecules that are referred to as cyclophane porphyrins.

With these two new and more sophisticated heme models, a large distal side steric effect is introduced because the distance between the anthracene ring and the porphyrin ring is smaller than 0.5 nm. By converting the anthracene system to the "pagoda" system via a Diels–Alder reaction, further restriction and a tighter pocket are achieved. Addition of 1-methylimidazole in methylene chloride to the models give molecules showing a five-coordinated visible spectra as previously observed with "capped" heme. The corresponding CO complex has also been prepared in dry benzene and corresponds to pure monocarbonyl heme. Therefore, the anthracene ring above the porphyrin ring has greatly reduced the affinity of the second CO molecule. The binding is reduced by 10^3-fold as compared with chelated hemes without the steric effect. The presence of an additional steric factor in the "pagoda" molecule results in a further decrease by 10-fold the rate of CO association. These dynamic models thus illustrate that the dramatic reduction in CO association rates can be attributed to the steric hindrance in the synthetic heme pocket (483).

In 1977, E. Bayer's group in Tübingen synthesized for the first time a heme group embedded in a polymer network (484). The structure is shown in Fig. 6.4 and was obtained from poly[ethylene glycol bis(glycine ester)] and polyurethanes from polyethylene glycols and diisocyanates as the basic polymers, using the procedure of *liquid-phase peptide synthesis*.* At the end of the synthesis the iron is present in the Fe(III) state and must be reduced with sodium dithionite prior to oxygenation.

The polymer shares the following properties with hemoglobin and myoglobin: (a) good solubility in water in order to achieve high concen-

* Liquid-phase peptide synthesis resembles the solid phase method in that the synthesis takes place on a polymer backbone (i.e., polyethylene glycol). However, the polymer is soluble in the solvent system used, so that a homogeneous reaction results. This methodology has found far less use than the analogous solid phase synthesis.

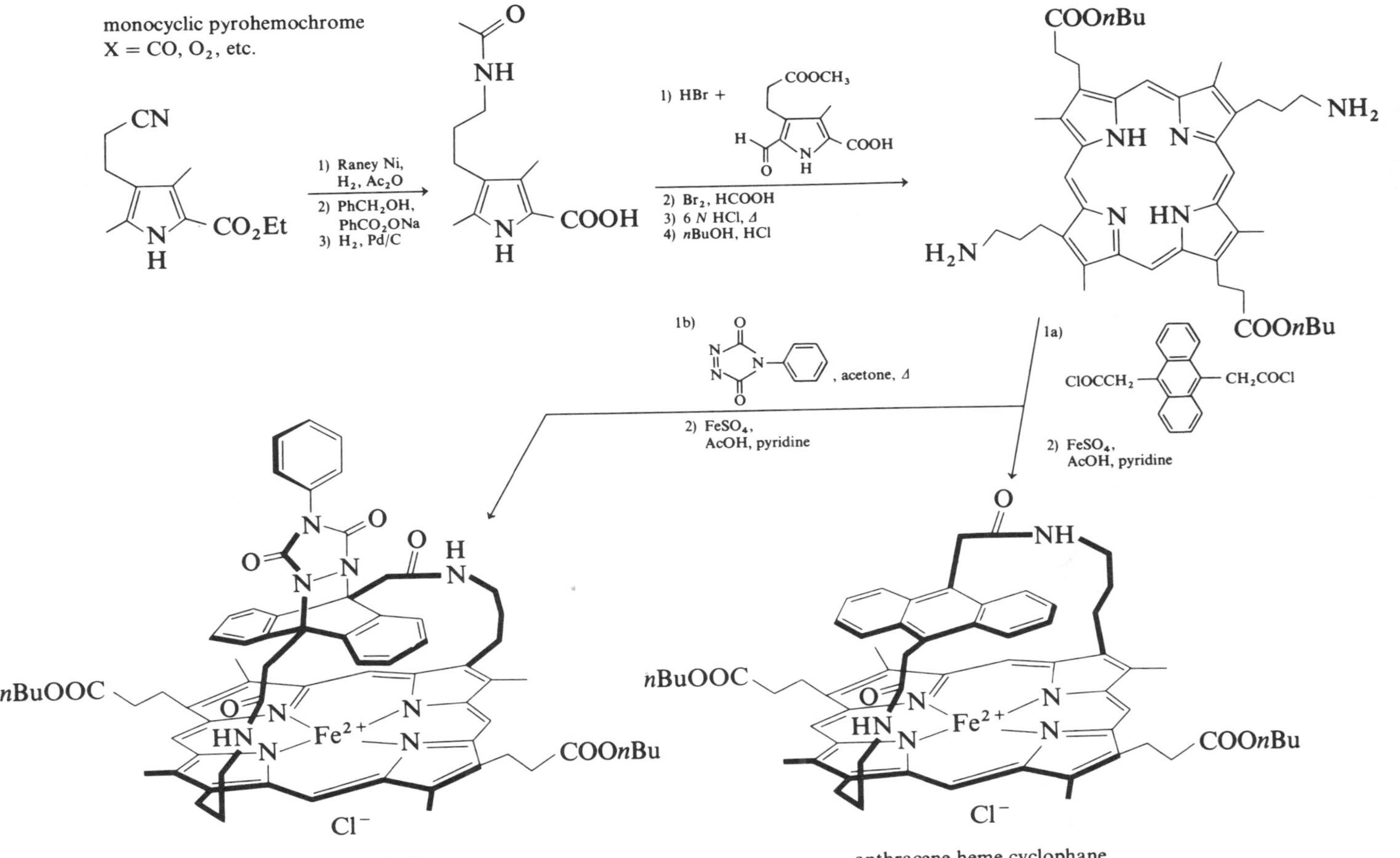

Fig. 6.3. Synthesis of anthracene-heme [6,6]cyclophane and of "pagoda" porphyrin (483).

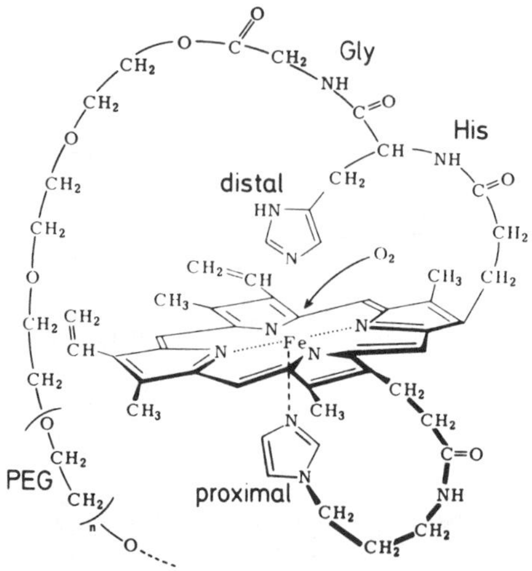

Fig. 6.4. Bayer's hemopolymer (484). Reproduced with permission. Copyright © 1977 by Verlag Chemie GmBH.

tration of O_2; (b) hindrance of irreversible oxidation of the oxygen complex by the functionalized polymer; and (c) imitation of a distal imidazole.

Furthermore, a second imidazole ligand corresponding to the proximal histidine of natural oxygen carriers was introduced by modification of the propionic acid side chain of the porphyrin ring. Interestingly, the structural units of the active site of myoglobin or hemoglobin which are essential for binding oxygen are all present, albeit in a different arrangement (484). In particular, the oxygen coordination site is shielded both by histidine and by the synthetic polymer. The oxygen absorption curve for the complex at 25°C is in good agreement with the values for myoglobin (0.2 to 1.2 torr). Furthermore, the sigmoid shape of the curve is indicative of a cooperative effect in oxygen binding, as in the case of hemoglobin.

Consequently, the synthetic hemopolymer has properties closely resembling those of natural oxygen carriers, although the basic polymer is completely different. This demonstrates that specificity and reactivity are by no means privileges of naturally occurring polymers. Such findings could have been obtained only from model studies.

Another leader in this field of bioorganic models of oxygen-binding site in hemoproteins is J.P. Collman (485,486) of Stanford University. He used an approach that modified the molecular geometry of the porphyrin such that the oxidation process of iron is sterically hindered and diminished. The effort consisted in constructing a porphyrin ring with four bulky *o*-pivalamidophenyl groups projecting from one side of the ring. The other side is left unencumbered with the hope that it could be protected by a bulky axial imidazole group (Fig. 6.5). This new com-

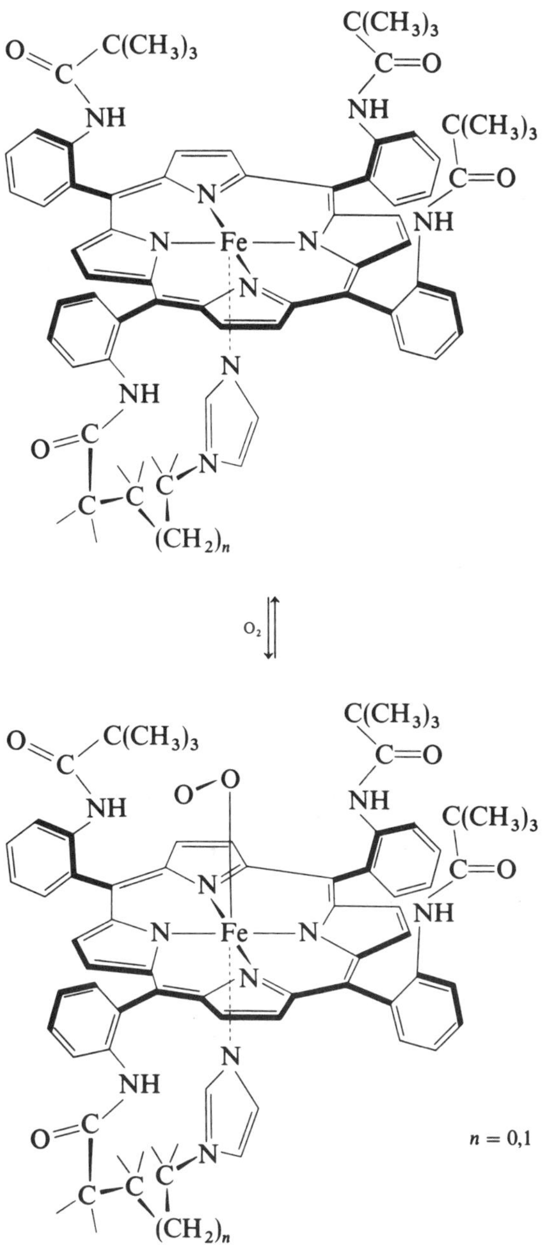

$n = 0,1$

Fig. 6.5. Collman's "picket fence" ferrous porphyrin model (486). Reprinted with permission. Copyright © 1977 by the American Chemical Society.

pound, although less of a model for myoglobin than protoheme itself, prevents heme–heme approach and Fe—OO—Fe formation (482). Collman called his model "picket-fence" porphyrin. It does bind molecular oxygen reversibly at room temperature and the dioxygen complex is crystalline. This model system shows significant cooperativity in its binding of O_2 but only in the solid state so far. At low O_2 pressures a low O_2 affinity form exists, and at high O_2 pressures a higher O_2 affinity form develops. This ability to show high affinity at high partial pressures of O_2 (as in the lungs) and low affinity at low O_2 pressures (as in the muscles and tissues) is critical to efficient O_2 transport. In this regard, the model mimics hemoglobin quantitatively (486).

By making precise measurements of the physical properties of the oxygen–iron bond in their model, Collman's group found that the oxygen molecule binds in an end-on fashion (Pauling's model) to the iron of the protoheme, rather than in the sideways manner proposed by some theories (Griffith's model). In fact, the way in which molecular oxygen binds to certain iron-containing biological molecules has been the subject of great theoretical debate for the past thirty years. Finally, X-ray crystallographic studies of Collman's model prove that oxygen binds in the end-on fashion, a suggestion first proposed by L. Pauling in 1948!

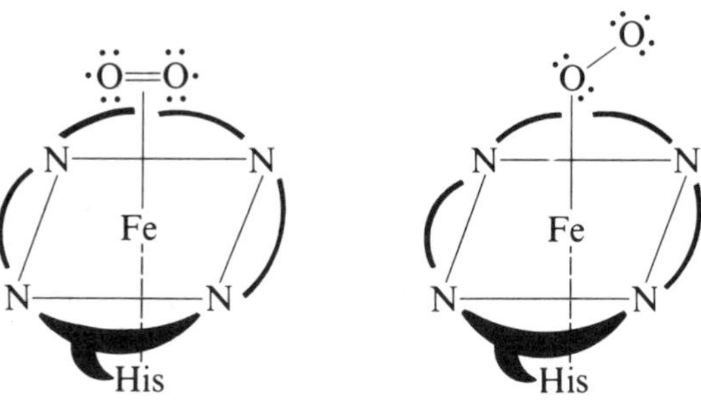

Griffith's model Pauling's model

In 1978, a new approach by the team of D.H. Dolphin (487) at the University of British Columbia was taken in devising molecular models that mimic the reversible binding of molecular oxygen by heme. They successfully incorporated ruthenium(II) into *meso*-tetraphenylporphyrin and octaethylporphyrin systems. They found that these porphyrin systems complexed with two molecules of the solvent but only *one* molecule of oxygen per ruthenium atom is absorbed reversibly at 25°C and at 1 atm in DMF. They also made monocarbonyl and dinitrogen (one N_2 molecule) species reversibly in polar aprotic solvents. Therefore, these new systems parallel that of corresponding Fe(II) systems but at more convenient temperatures.

Artificial hemoglobins containing zinc, manganese, copper, and nickel have also been reconstituted, and their properties compared with those of the native iron proteins. These artificial systems, however, are incapable of reversible oxygenation. In contrast, cobalt-substituted hemoglobin and myoglobin are functional, although their oxygen affinities are 10 to 100 times less than those of the native oxygen carriers. In fact, it has been known since the 1930s that several synthesized Co(II) complexes bind molecular oxygen reversibly. Cobalt is adjacent to iron in the periodic table, and it was hoped that the elucidation of the behavior of cobalt complexes might shed some light on that of hemoglobin and the other naturally occurring oxygen carriers. In other words, the Co(II) complexes were recognized as possible models for hemoglobin and other natural O_2 carriers. A second reason for selecting cobalt is that more extensive thermodynamic data are available for related cobalt systems.

A recent synthetic model of cobalt-substituted hemoglobin is shown in Scheme 6.1 (488). The long side chain facilitates the coordination of the pyridine ring to the central cobalt atom. The Co(II) complex of this so-called looping-over porphyrin reacts reversibly with molecular oxygen at low temperatures (−30°C to −60°C) but the side chain has a negligible enhancement effect on the oxygen affinity for such models, as compared to iron–porphyrin systems.

Collman's group also examined oxygen binding to "picket fence" porphyrin–cobalt complexes of *meso*-tetra-(α,α,α,α-*o*-pivalamidophenyl) porphyrinato-cobalt(II)-1-methylimidazole and 1,2-dimethylimidazole (489).

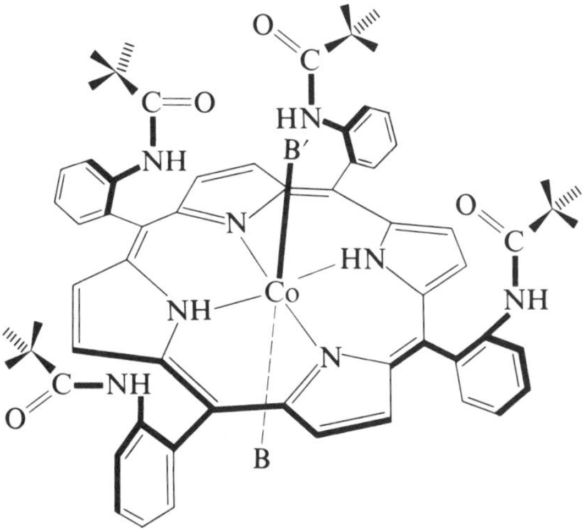

B, a base such as *N*-methylimidazole or 1,2-dimethylimidazole

B′, the position where O_2 can bind

Interestingly, these complexes bind oxygen with the same affinity as cobalt-substituted myoglobin and hemoglobin in the solid state and in toluene. More elaborated models await the synthesis of "picket fence" porphyrins with different "pickets."

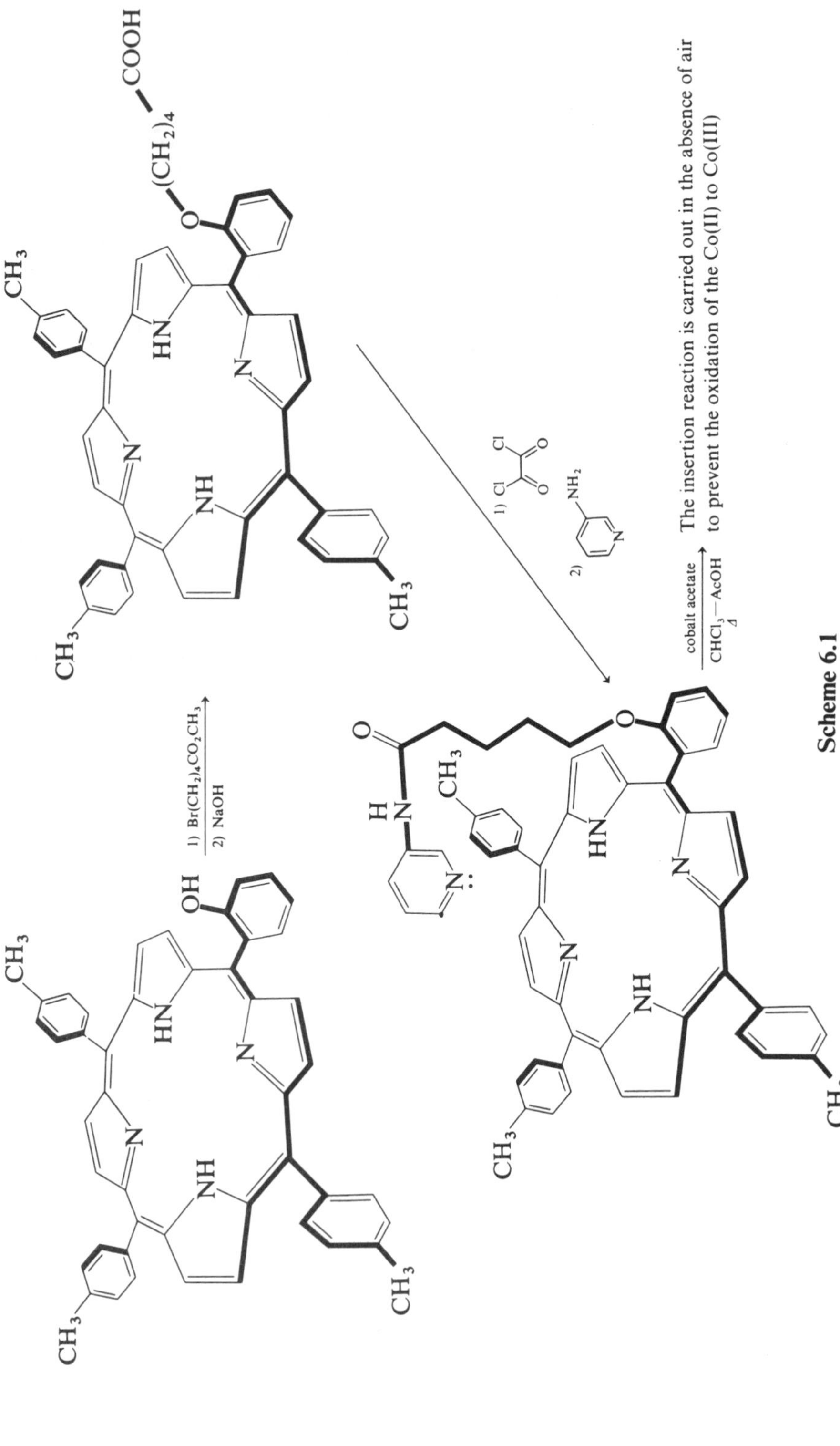

1) $Br(CH_2)_4CO_2CH_3$
2) NaOH

$(CH_2)_4$—COOH

1) Cl–CO–CO–Cl
2) 3-aminopyridine (NH_2)

cobalt acetate, $CHCl_3$—AcOH, Δ

The insertion reaction is carried out in the absence of air to prevent the oxidation of the Co(II) to Co(III)

Scheme 6.1

6.4.1 Other Models

For oxygen-"carrying" molecules (myoglobin, hemoglobin) and oxygen-"consuming" molecules (cytochrome, catalase, etc.) many questions still remain unanswered regarding the detailed fine tuning process of O_2 binding to heme groups. For example:

1. How does hemoglobin recognize O_2 and at the same time try to discriminate the toxic CO molecule?
2. Why myoglobin developed the capacity to transfer O_2 molecules while cytochrome b_5 transfers electrons?
3. Do we completely understand the role of the porphyrin nucleus?
4. Why imidazole is considered one of nature's best bases?

More elaborated biomodels ought to find some explanation to these questions. As far as the orientation of the O_2 molecule relative to the porphyrin plane is concerned, many model studies have shown that it is angularly oriented (see p. 418). With the CO molecule (with two electrons less) however, the fixation is believed to be parallel and does not seem to depend on the presence of a basic imidazole.

In this regard, Traylor's group has developed two new "strapped" porphyrin complexes, one using a 1,3-adamantane unit and the other made with a 3,5-pyridine arm (490). In the solid state, the adamantane bridge has an offset orientation relative to the plane of the porphyrin nucleus, and short intramolecular contacts leave no free cavity between the molecular strap and the porphyrin. In solution, however, the NMR results suggest a rapid flip of the adamantane ring between two asymmetric orientations. The affinity for both CO and O_2 is greatly reduced in the (dicyclohexylimidazole) Fe(II) complex as compared to open heme compounds. The decrease in affinity was somewhat more

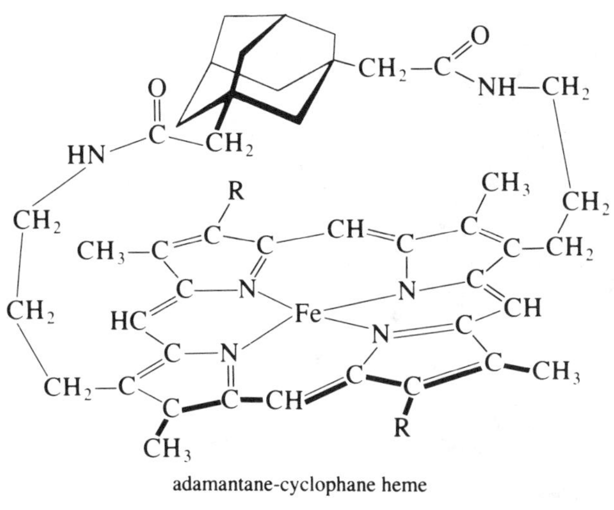

adamantane-cyclophane heme

pyridine-cyclophane heme
$R = CH_2CH_2CO_2CH_2Ph$

pronounced for CO than for O_2 molecules. Besides steric interference, a "local polar effect" was evoked to account for this difference. This is based on the observations that electron donation to the heme, increased basicity in the base B, or increased polarity of the solvent increases O_2 affinities but has little effect on CO affinities. One of the remarkable

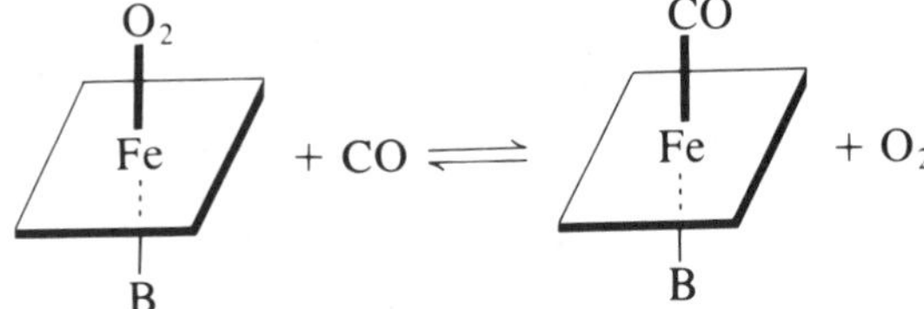

properties of the shorter strapped pyridine-cyclophane heme in the ferrous form is the impossibility of giving either an internal or external Fe–pyridine complex.

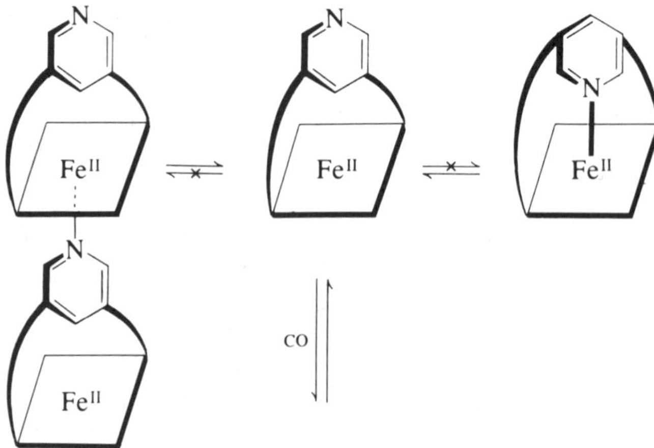

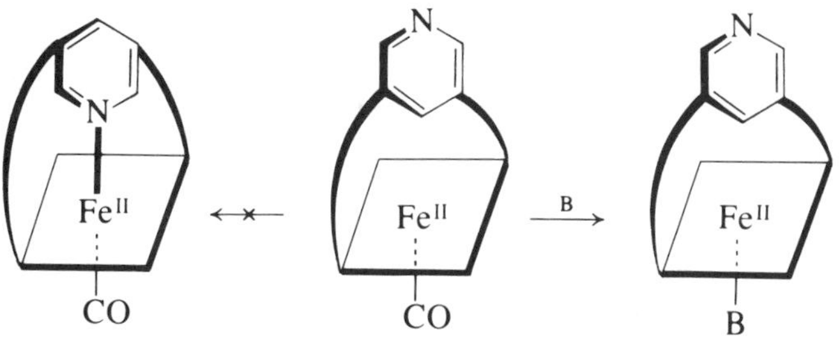

B = 1,5-dicyclohexylimidazole

Furthermore, an important discovery was the observation of a rather spectacular decrease in the ratio of CO/O_2 affinity. From about 20,000 for single-type hemes, the value drops here to 14 in toluene and 5 in *o*-dichlorobenzene. These findings appear to confirm earlier speculations that solvent or local polar effects are the principal factors governing the differentiation of O_2 and CO binding. Such results are particularly interesting since the heme pocket in both hemoglobin and myoglobin has been referred to as being hydrophobic and sterically congested. However, one should imagine, rather, that the presence of an amino acid side chain imidazole group, which is very polar in an environment of rather low dielectric constant, presents a very "polar pocket effect" for O_2 binding. This observation with this model suggests ways of designing new artificial oxygen-binding materials that will preferentially bind dioxygen molecules.

In both hemoglobin and myoglobin an important function of the heme pocket is to prevent dimeric interactions between hemes. In accordance with this requirement, various new sterically protected or "picket fence" porphyrins have been synthesized in recent years (452–453). Two examples are presented below.

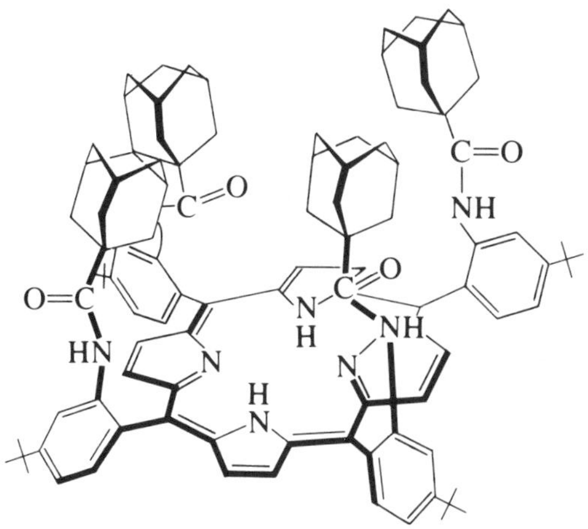

"tulip garden" porphyrin (ref. 467)

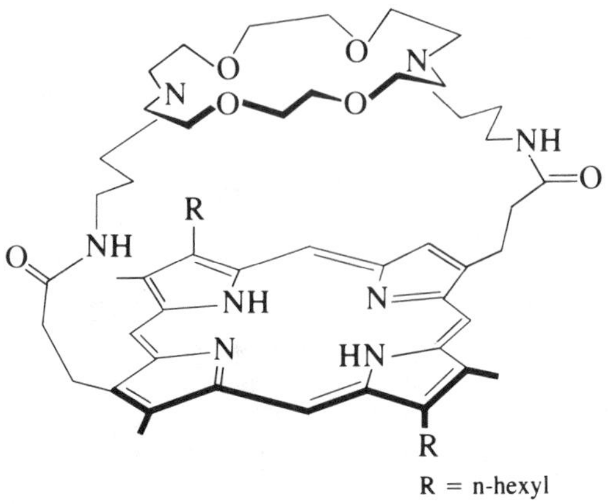

R = n-hexyl

"crowned" porphyrin (ref. 491)

These superstructures have a hydrophobic pocket on one face of the porphyrin nucleus and help to inhibit irreversible autooxidation of the Fe(II) ion. At room temperature they bind O_2 reversibly and mimic myoglobin. The bulkiness of the adamantyl pendants in the "tulip garden" porphyrin allow the Co(II) complex to show almost the same O_2 affinity as coboglobins.

A.R. Battersby has synthesized a doubly bridged Fe(II)-porphyrin having a cavity formed on one face of the porphyrin by an anthracene-containing bridge and a looser imidazole ligand strapped across the other face. The compound binds O_2 reversibly at ambient temperature in DMSO and is stable for almost two days. Only 20% of the molecule undergoes irreversible oxidation (494).

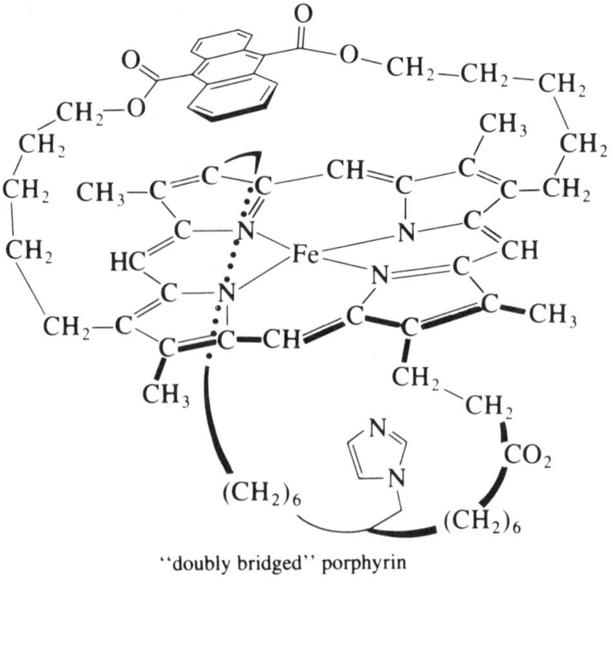

"doubly bridged" porphyrin

Another concern in hemoglobin chemistry is a better understanding of the cooperative O_2 binding. For this, Tabushi has developed a new *artificial allosteric system* made of *m*-bis (*meso*-triphenyl porphyrinyl)-benzene. This so-called *gable porphyrin*, with a fixed hinge, can form *bis*-metal (Zn, Co, or Fe) complexes (495). Dimeric (bridging) ligands

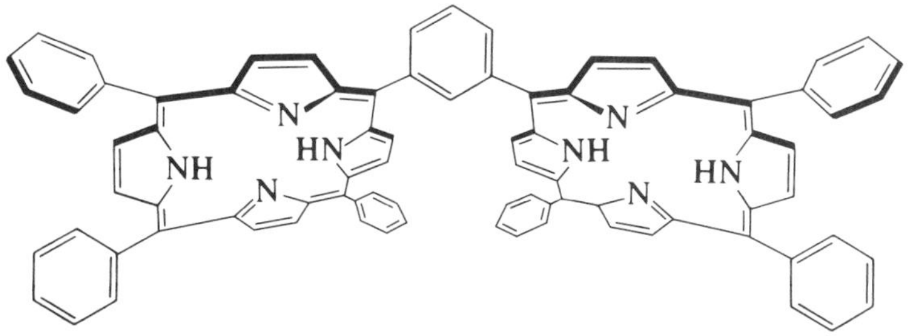
N
NH
HN
N
N
NH
HN
N

"gable" porphyrin

such as γ,γ′-dipyridylmethane (DPM) bind to monometal-gable complexes in a noncooperative fashion and mainly from the *exo* side. For bismetal-gable complexes, the same dimeric ligand binds preferentially to the *endo* side and much more strongly than monomeric ligands like 1-methylimidazole. A remarkable allosteric behavior is observed when

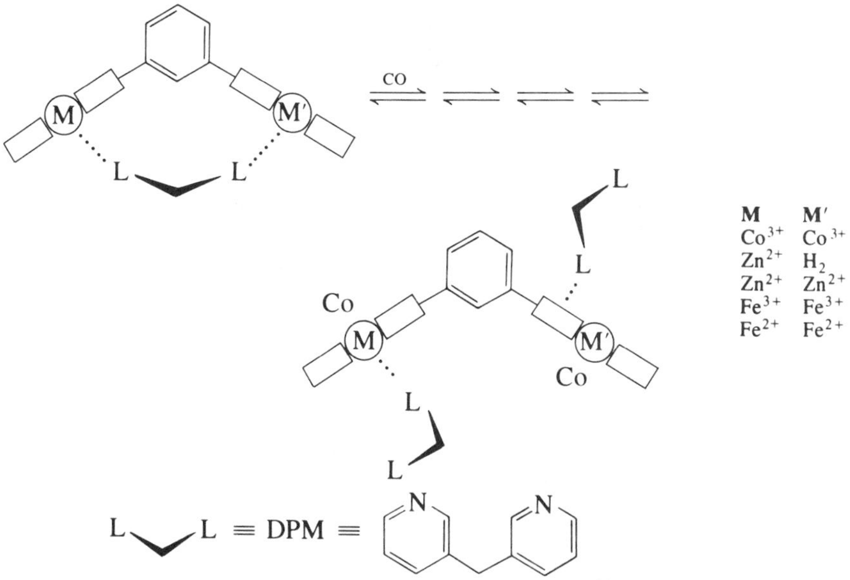
M
M′
L
L
CO
L
L
Co
M
M′
Co
L
L
M M′
Co3+ Co3+
Zn2+ H2
Zn2+ Zn2+
Fe3+ Fe3+
Fe2+ Fe2+
L L ≡ DPM ≡
N
N

1-methylimidazole is added to allosteric system gable porphyrin-Zn_2-DPM. The enhanced binding of this base is interpreted as a favorable local structural change induced by the presence of the first dimeric base: a situation modeling perturbation between subunits in the oxygenation of deoxyhemoglobin.

Dimeric "face-to-face" porphyrin compounds have been synthesized by Collman's team (496). They are made of two porphyrin rings linked by two amide bridges located on opposite β-pyrrolic carbons, with the amide carbonyl groups attached to the porphyrin nucleus. In the case where the connecting bridges contain four atoms, the interporphyrin distance is sufficient to allow an O_2 molecule to fit in the cavity of the dimer. When adsorbed on edge-plane pyrolytic graphite electrodes, the dicobalt derivative can catalyze the four-electron reduction of O_2 to water molecules with a turnover rate of about 400 dioxygen molecules per second per catalyst site.

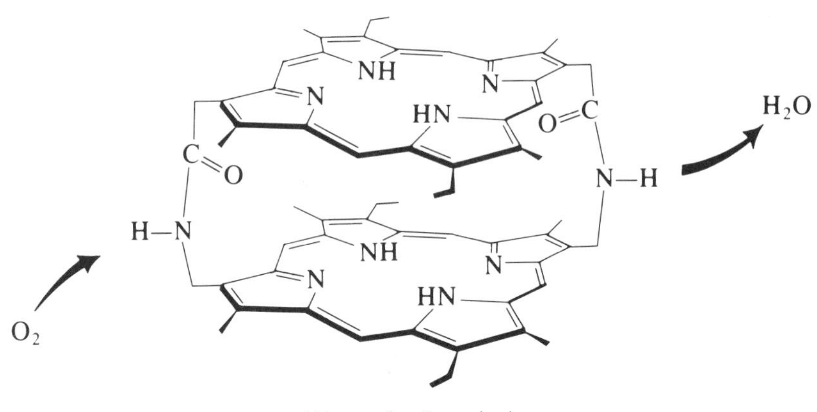

"face-to-face" porphyrin

This electrochemical reduction of dioxygen with model systems could have in the near future important industrial applications.

Depending on the type of molecules present at the center of the porphyrin rings in these covalently linked dimeric porphyrins, the parallel disposition of the porphyrin rings can be affected and such conformational changes have been examined (497). Knowledge of this relative orientation will be crucial in the design of photosynthetic reaction centers (see Section 6.6).

Finally, mention should be made of the interesting contribution of E. Tsuchida (Waseda University, Tokyo), who constructed a "picket fence" amphiphilic iron–porphyrin complex that reversibly binds O_2 when embedded in phospholipidic bilayer liposomes (498). The amphiphilic nature of the pendants on the porphyrin ring permits its incorporation into aqueous lipidic solution. Polymerized lipid liposome, poly[1-{9′-(*p*-vinylbenzoyl)nonanoly)}-2-*O*-octadecyol-*rac*-glycerol-3-phosphocholine], was used and the porphyrin-Fe(II) was complexed to 1-laurylimidazole within the liposome structure. The oxygen uptake and evolution in this system could be induced by changes in pH. See structures on pages 427–428.

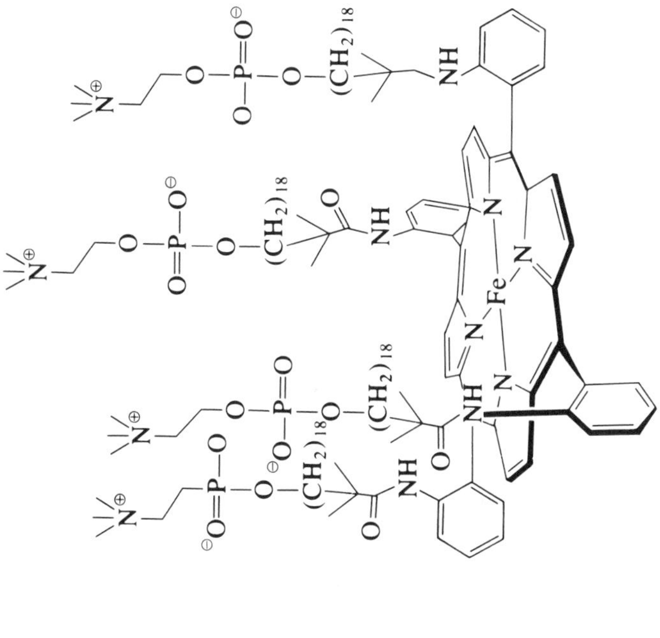

amphiphilic iron-porphyrin complex

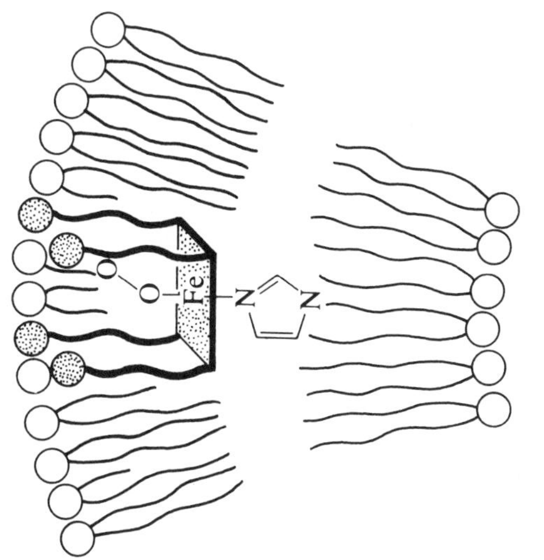

schematic representation when incorporated into a bilayer membrane

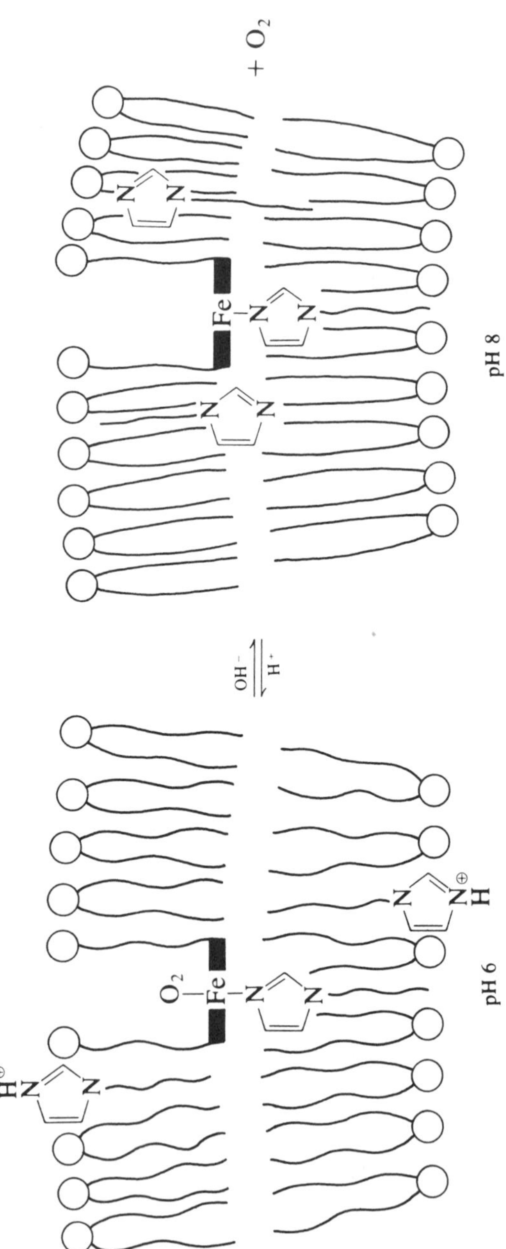
+ O2
Fe
pH 8
OH−
H+
O2
Fe
pH 6

This system tries to mimic the well-known *Bohr effect* in red blood cells where the presence of CO_2 decreases the pH of the medium and increases the release of oxygen, favoring the deoxyhemoglobin form.

However, in the present model system, an increase in pH promotes the evolution of oxygen at a rate of 3.6 mL per 100 mL of solution and the solution turns from dark red to brownish red. This behavior is in contrast to the anticipated Bohr effect. To explain this, one could invoke that the role played by the protonated imidazole is to increase the oxygen-binding affinity to the porphyrin nucleus and at the same time to destroy some of the packing structure of the bilayer by increasing the mobility of the phospholipids. This change would provide an environment where the heme complex would adopt a more relaxed structure showing higher oxygen-binding affinity.

The directionally controlled anchoring of myoglobin molecules on dipalmitoylphosphatidylcholine bilayers has been successfully conducted (499). Such self-organized, lipid-protein layers are of unquestionable importance in protein-based technology, particularly for signal peptides in membrane transport of protein and ion signal transduction.

A reconstituted myoblobin, with a ruthenium tris(2,2′-bipyridine) pendant, has also been developed recently (500). The following protoheme derivative bound covalently to $[Ru(bpy)_3]^{2+}$ was prepared and incorporated to apo-myoglobin.

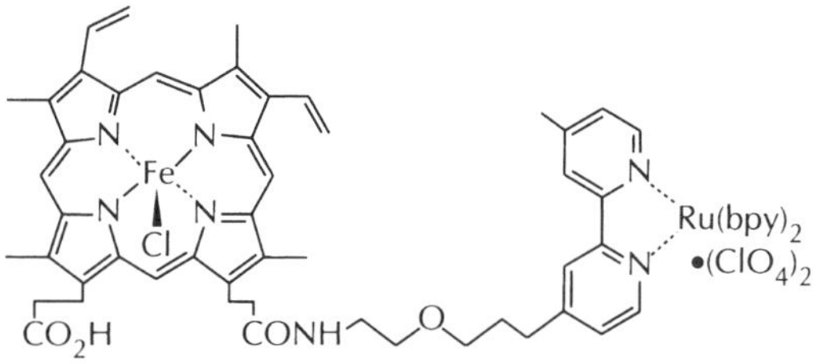

Ru(II) protoheme

This reconstituted myoglobin with a heme bearing a photosensitive Ru(II) pendant was effectively activated by visible light to function as a dioxygen storage protein. Such molecular environment of a protein with a covalent bond between Ru(II) in its active site may facilitate electron-transfer communication, as does an electron wire via electron tunneling pathways.

The chemical modification of proteins with nonnatural functional groups, such as photochromic or redox-active molecules, has gradually been proven one of the most promising approaches to the development of a general methodology for the regulation of proteins or enzymes that are activated by external stimuli. As pointed out by I. Hamachi (Kyushu University), this is essential for the design of sophisticated biocatalysts that have potential applications to a wide area of biotechnology and medicine (500).

6.4.2 Other Systems

In addition to the iron–porphyrin compounds, there are a large number of nonheme proteins implicated in many oxidation–reduction processes. They contain relatively large amounts of sulfur and iron and have thus been designated *iron–sulfur proteins* (501). They are widely distributed in nature, occurring in all living organisms, and their physiological function is that of electron transfer rather than catalysis of chemical transformations. Among these proteins are the *rubredoxins, adrenodoxins*, and *ferredoxins*. They are the most ubiquitous electron carriers in biology and have been directly associated with, or implicated in, a larger area of metabolic reactions, especially in photosynthetic processes and CO_2 and N_2 fixation in bacteria and plants. The first ferredoxin protein was isolated and purified in 1962 from *Clostridium pasteurianum* and spinach. This nonphotosynthetic anaerobic bacterium is one of the oldest bacteria present on Earth—more than 3.1 billion years ago.

The presence of a Fe–S spectroscopic chromophore and their relatively low molecular weight (about 6,000 to 12,000) make these proteins fascinating molecules for investigation. Little can be said at this time about their mechanism of action but they are known to function as strong reducing agents. They are also characterized by evolution of H_2S in acidic solution (acid-labile sulfur). It is believed that part of the primitive and reductive atmosphere prevailing at the origin of organic molecules on Earth could have come from such transformations. Hence ferredoxins might become an essential link in the comprehension of the origin of life.

These iron–sulfur redox proteins contain a cagelike cluster of iron and sulfur linked to cysteine residues of the peptide chain.

It is these metal–sulfur bridges that are responsible for the biological properties of these sites. The sulfur ligands make the Fe(II) state very unstable and account for their powerful reducing properties in various biological processes. Such cagelike clusters behave as one-electron redox catalysts where the Fe(II) $\rightleftharpoons$ Fe(III) oxidoreduction reactions occur within the cluster.

Only few synthetic analogs of the Fe—S clusters are known today and this particular chemistry is still in the development stage. In this line of thought, mention must be made of R.H. Holm (502) at Harvard University who made a series of synthetic analogs in an approach toward elucidation of the active sites of certain iron–sulfur proteins and enzymes (503).

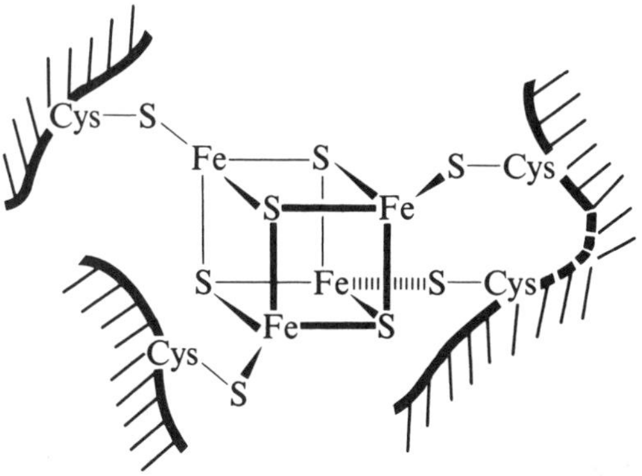

cagelike cluster proposed as active site of ferredoxin

As mentioned earlier, hemerythrin is a protein present in marine invertebrates. It contains an oxo-bridged dinuclear iron center bound to two carboxylate groups of amino acid side chains. Model components of

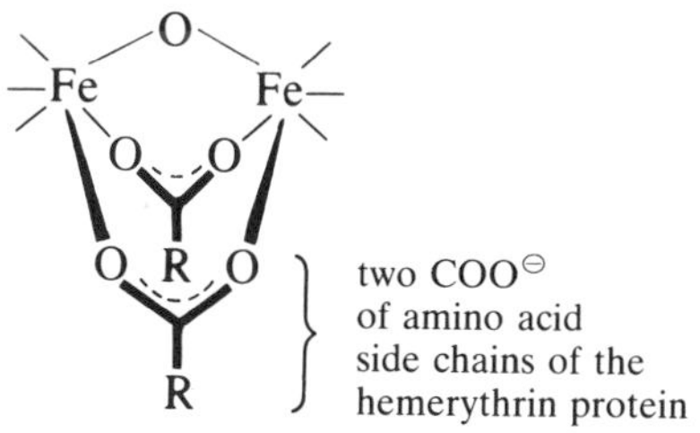

two $COO^{\ominus}$ of amino acid side chains of the hemerythrin protein

this which recognize iron-oxo bridge in biology and other non-heme iron(II) centers with dioxygen have been proposed by S.J. Lippard's group (504,505). The following structure is an example. The Fe—O bond length is 1.78 Å but increases to 1.96 Å upon oxygen protonation.

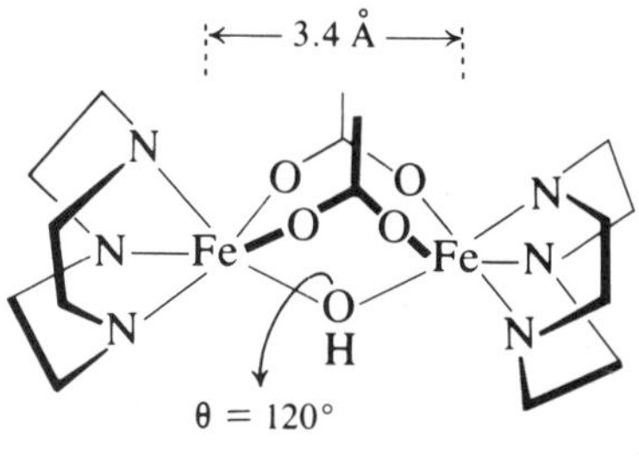

3.4 Å

θ = 120°

This corresponds to an appreciable energy change that might favorably influence a two-electron oxidative addition of molecular oxygen in the deoxy- to oxyhemerythrin equilibrium. This rich redox chemistry may have a real function in biology.

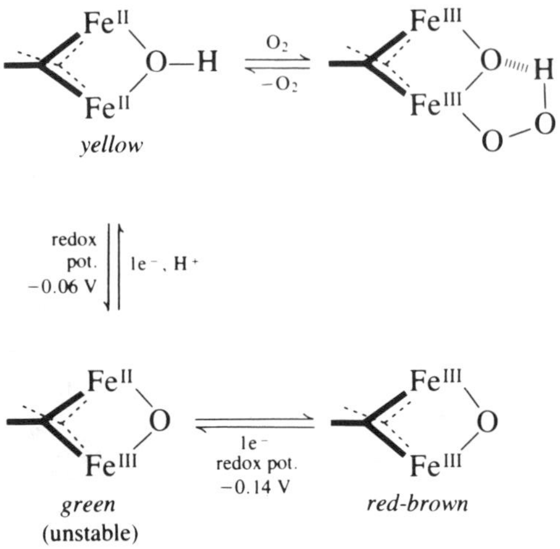

Fe^II
O—H
Fe^II
yellow
O₂
−O₂
Fe^III
O
H
Fe^III
O
O
redox
pot.
−0.06 V
1e⁻, H⁺
Fe^II
O
Fe^III
green
(unstable)
1e⁻
redox pot.
−0.14 V
Fe^III
O
Fe^III
red-brown

Among the strongest iron-binding agent in nature is enterobactin, found in enteric bacteria. It contains three symmetrically disposed catechol units on a macrocyclic ring. These microbes have evolved specialized iron ligands or *siderophores* to acquire and transport environmental iron into the cell. The siderophores bind Fe(III) by wrapping six oxygen-ligating atoms around the metal center in either a right-handed or left-handed coordination propeller shape (506).

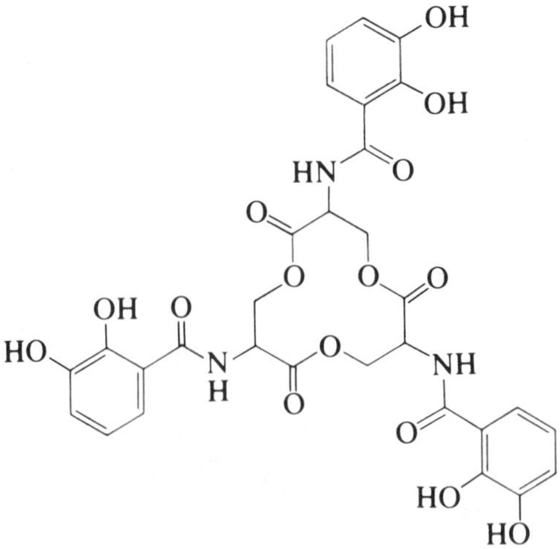

OH
OH
HN
O
NH
HO
HO
N
H

enterobactin

Few chiral analogs of these iron chelators have been prepared (506, 507). Large molecular cavities bearing siderophore-type functions have

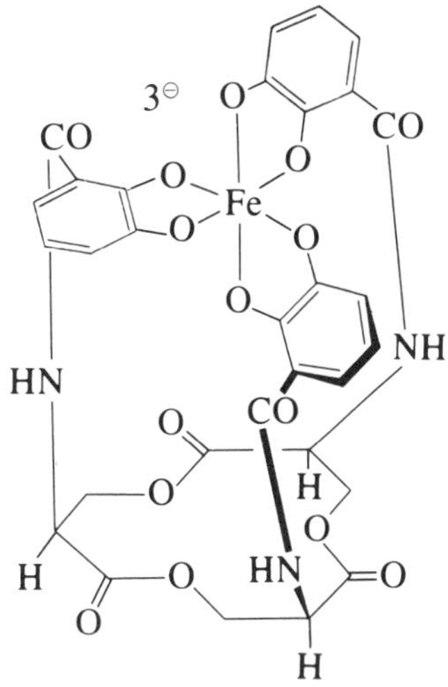

Δ-*cis* form (right-handed)

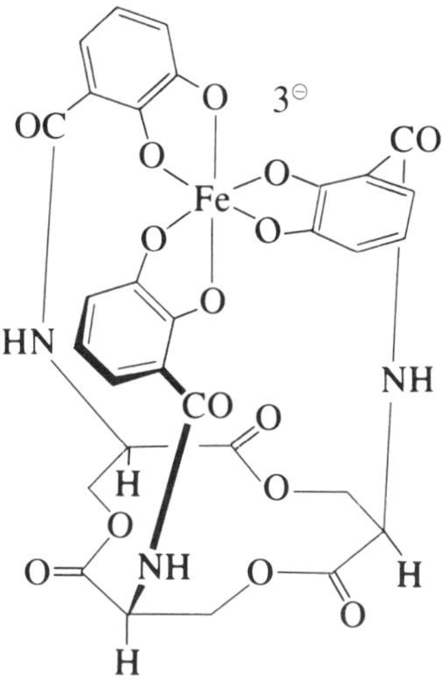

Δ-*cis* form (left-handed)

also been constructed by Vögtle's group (508). The following two molecules are an illustration:

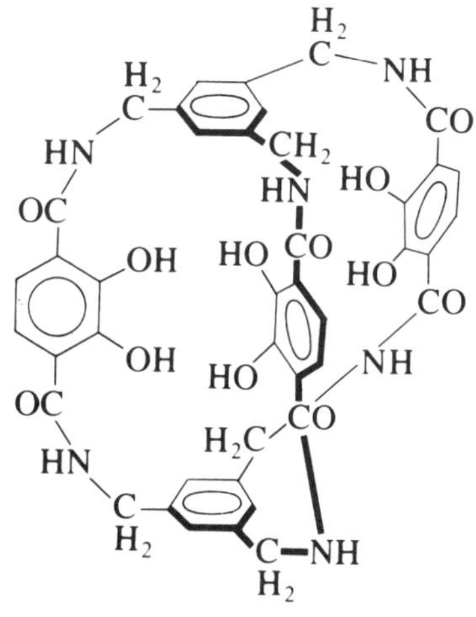

26-membered ring

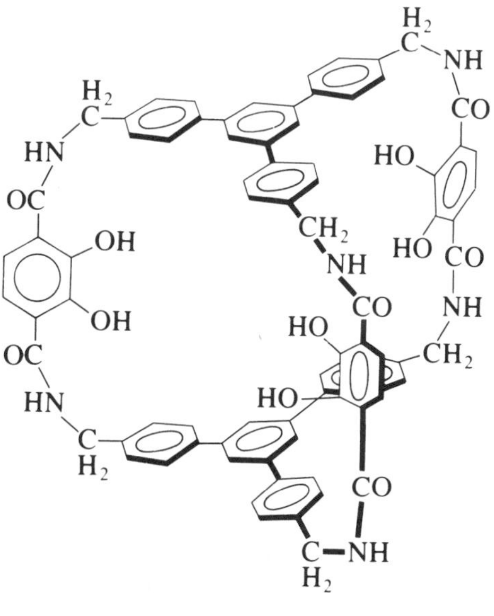

42-membered ring

The smaller receptor binds Fe(III) very strongly but the larger one does not. However, the 42-membered ring constitutes one of the largest molecular host cavities known so far that can potentially enclose in its cavity organic guest molecules. In addition, the molecule could allow further functionalization utilizing the OH groups.

6.4.3 Model Studies of Cytochrome P-450

The way that we human beings operate is basically by using enzymes to break down food; the resulting electrons pass through mitochondria and eventually react with molecular oxygen to form water (see Section 2.1). We have also seen that those exothermic reactions are used to produce high-energy phosphates (see Section 3.2). All these processes involve oxidation reactions and transfer of electrons by electron transfer proteins, the cytochromes. Nature operates this way by using, in a fascinating constellation, different proteins that are compartmentalized at different distances along this metabolic pathway.

One class of hemoproteins that is capable of metabolizing many compounds are the enzymes known collectively as *cytochrome P-450* (509). Cytochromes P-450 are ubiquitous in nature, being found in plants, animals, and bacteria. They participate in numerous metabolic pathways. These heme proteins are monooxygenases, catalyzing the hydroxylation of substrates at the expenes of molecular oxygen. The reductive cleavage of O_2 results in the incorporation of one oxygen into the substrate, the other being converted to water.

$$\underset{\text{substrate}}{\mathbf{R}\text{-H}} + O_2 \xrightarrow{2e^-\ 2H^+} \mathbf{R}\text{-OH} + H_2O$$

In mammalian systems, the roles played by cytochrome P-450 are as diverse as steroid biosynthesis and hydroxylation of lipidic substances, drugs, and hydrocarbons (510). A well-known example is the in vivo oxidation of polyaromatic compounds present in tobacco smoking and the resulting carcinogenicity.

The most important property attributed to cytochrome P-450 is its ability to form an extremely potent oxidizing agent from molecular oxygen. This contrasts sharply from the more familiar oxygen- or electron-transport functions of many other heme proteins. The active site of cytochrome P-450 also contains an iron bound to a prosthetic porphyrin nucleus, but in contrast to O_2-transporting proteins that have a basic imidazole group, it possesses a thiolate group from a cysteine residue. Furthermore, the iron in cytochrome P-450 is in the (V) oxidation state and forms an iron-oxo complex. Both Collman's and Battersby's groups have tried to model this oxidizing protein by synthesizing "mercaptantail" porphyrin compounds (510,511). Cytochrome P-450 exhibits an unusual "hyper"-Soret absorption spectrum ($\lambda_{max} \sim$ 380,450 nm) in the car-

Fe^{II} —B $\xleftarrow{O_2}$ H_2O Fe^{II} $\xrightarrow{O_2}$ $\xrightarrow{2e^-, 2H^+}$:O: Fe^{V} —S $+ H_2O$

O_2 carrier molecule (B = base)

"active" oxygen in P-450 (iron-oxo complex)

bonmonoxide adduct; giving a doublet associated to the S—Fe—CO bonding. The model compound reproduces quite well the same characteristic spectroscopic properties and suggests that it is indeed a viable model for further studies. It also shows that the iron remained in the ferrous state.

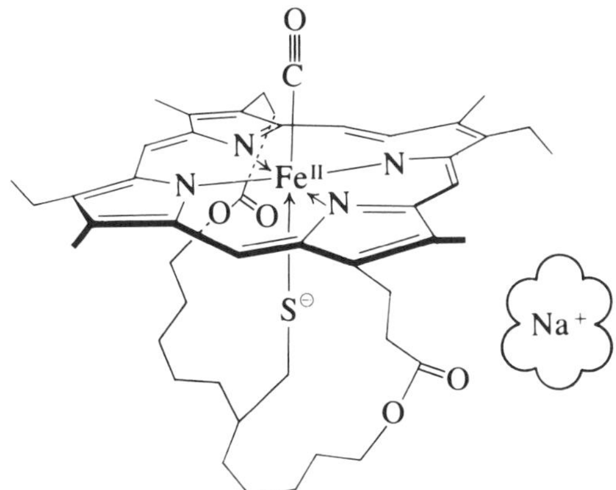

bound CO molecule to strapped thiolate porphyrin (511)

The importance of organometallic chemistry of iron porphyrin molecules is being more and more recognized, and the use of models may serve as an excellent basis for elucidating the mechanism of the catalytic reactions in which cytochromes P-450 play important roles (512).

In this direction, Collman and Tabushi have reported highly efficient synthetic model systems that use Mn(III)-porphyrins and a substituted imidazole ligand (B) to catalyze the transfer of an oxygen from O_2 or ClO^- to olefins (513,514). In some instances, regioselective monoepoxidation of polyolefins was observed using a colloidal platinum support, and recycling the catalyst was also possible (see page 436).

More elaborate electron-transfer assemblies, which are composed of a membrane-spanning *bis*-heme unit made of Zn(II)-porphyrin and Mn(III)-porphorin with four steroid side chains, have also been constructed by J.T. Groves and coworkers (Princeton University) (515).

O
+
Mn^{II}
$^{+}]Cl^{-}$
B
$ClO^{\ominus}$
Cl
O
Mn^{V}
$^{+}]Cl^{-}$
B
Mn^{III}
B

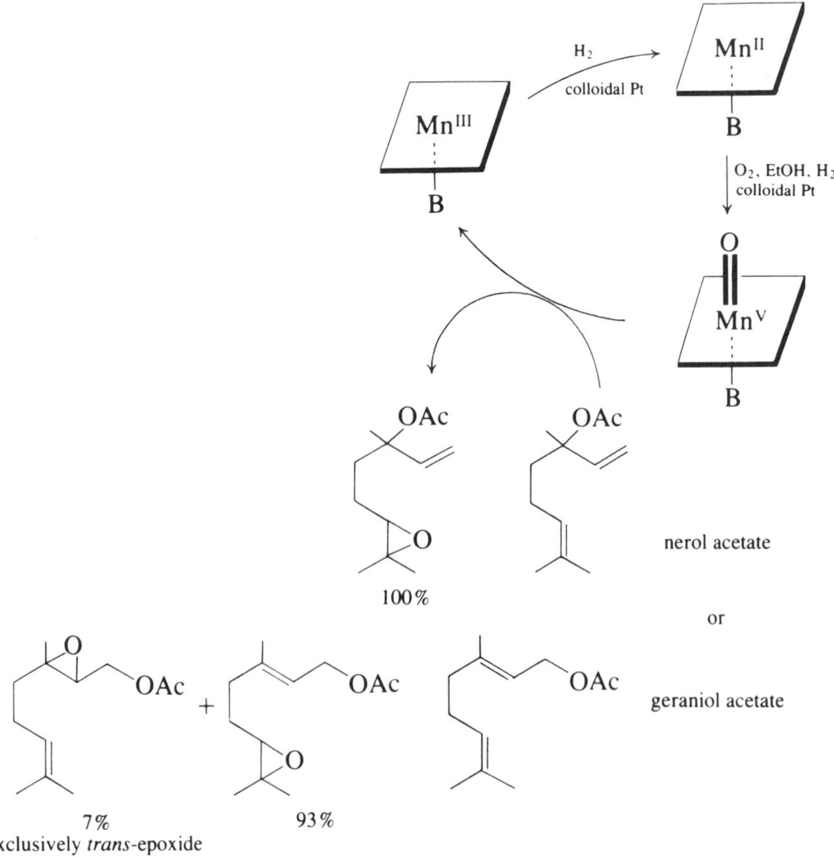

6.5 Copper Ion

Copper ions in the (I) and (II) oxidation state are biologically important. These ions appear to stabilize walls of certain blood vessels including the aorta and the sheath around the spinal cord. They are involved in the body's production of the color pigments of the skin, hair, and eyes and in the in vivo synthesis of hemoglobin (475,516). Binding and activation of molecular oxygen by copper complexes has been reviewed by K.D. Karlin (517).

In certain copper-containing proteins the copper appears to serve principally in electron transport with no evidence of $Cu-O_2$ interaction, such as in cytochrome oxidase. Of importance, however, is that many copper proteins and enzymes participate in reactions in which the oxygen molecule is directly or indirectly involved. An example is hemocyanin, the oxygen carrier in the blood of certain sea animals such as snails, octopuses, and crustacea. Oxygenated hemocyanin is blue and the cephalopods (crabs and lobsters) are literally the blue bloods of the animal kingdom. Hemocyanins are giant molecules of $MW > 10^6$ that occur free in solution.

Both hemocyanin and tyrosinase, the enzyme that activates molecular oxygen for the oxidation of tyrosine, rely on direct covalent interaction between Cu(I) and O_2, forming an observable dioxygen adduct. The mushroom *Gyroporus cyanescens* (Bluing Boletus), which turns blue instantly when bruised, also contains a copper protein or a "blue protein," as it is called.

Basically, three different types of copper centers are known (518). "Blue" or type I copper occurs in the blue electron-carrying proteins such as stellacyanin, plastocyanin, and azurin. The spectroscopic display of blue copper is very useful for characterizing the different ligand environments and coordination numbers around the copper atom. There is also a "nonblue" or type II copper and an electron paramagnetic resonance (EPR) "nondetectable" or type III copper center, apparently containing a pair of contiguous Cu atoms. Type II is usually found in combination with types I and III but it occurs alone in galactose oxidase. Type I is also found accompanied with type II and III, as in the blue oxidases, laccase, ceruloplasmin, and ascorbic oxidase. For instance, laccase is known to be a four-copper-atom protein that binds covalently with molecular oxygen.

Amino acid analyses of a variety of hemocyanins indicate that a large amount of histidine and methionine per copper pair is present as well as cysteine, although the number involved in disulfide bridges has not been determined. Intuitively, three types of donor atoms are likely to be involved in these protein complexes, namely, oxygen (carboxylate, phenolate, and water), nitrogen (amine, amide anion, and imidazole), and sulfur (thioether and thiolate). Furthermore, copper (II) can adopt square-planar, square-pyramidal, trigonal-bipyramidal, octahedral, and tetrahedral geometries.

So far, only a few biomodels of copper proteins have been made where the structure and ligand environments of the copper sites are based on the analysis of the electronic spectra.

B. Bosnich's group, now at the University of Chicago, designed a variety of ligands based on the above considerations (518). Some are presented in Fig. 6.6.

This extended series of ligands and their Cu(II) complexes have been prepared as spectroscopic models (observable in visible and near-IR regions) for determining the geometries and ligand coordination of copper proteins. From their results it was possible to propose the following structures for the copper coordination and geometry in (blue) type I copper proteins and for the copper site in oxyhemocyanin (518).

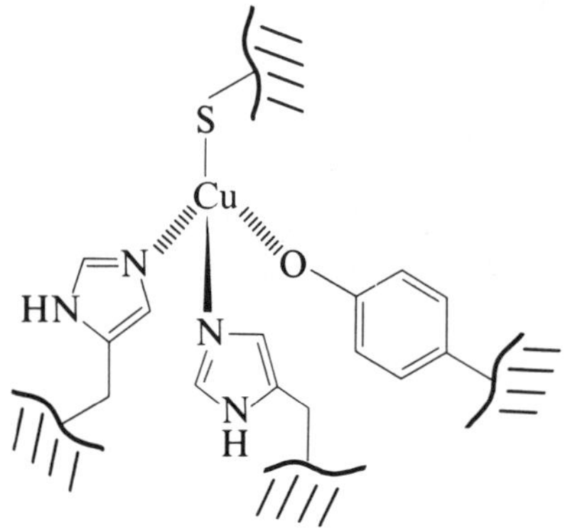

proposed structure and ligand coordination in the blue type I Cu proteins; the copper is bound at the active site via imidazole, phenolate and thioether ligands

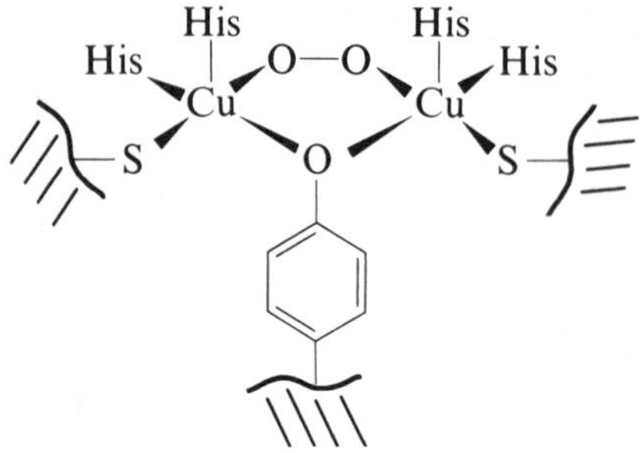

proposed environment of the Cu site in oxyhemocyanin; type III copper may also have a similar molecular architecture

With the same objective of elucidating the active site geometry and mechanism of copper proteins, the following model has also been made by other workers (519).

To inhibit binding of a second copper atom and to prevent possible hydrogen atom transfer reactions, the bridging oxime hydrogen was replaced with BF_2 via treatment with BF_3–etherate in dioxane. The Cu(I) complex reacts with monodentrate (i.e., CO, 1-methylimidazole,

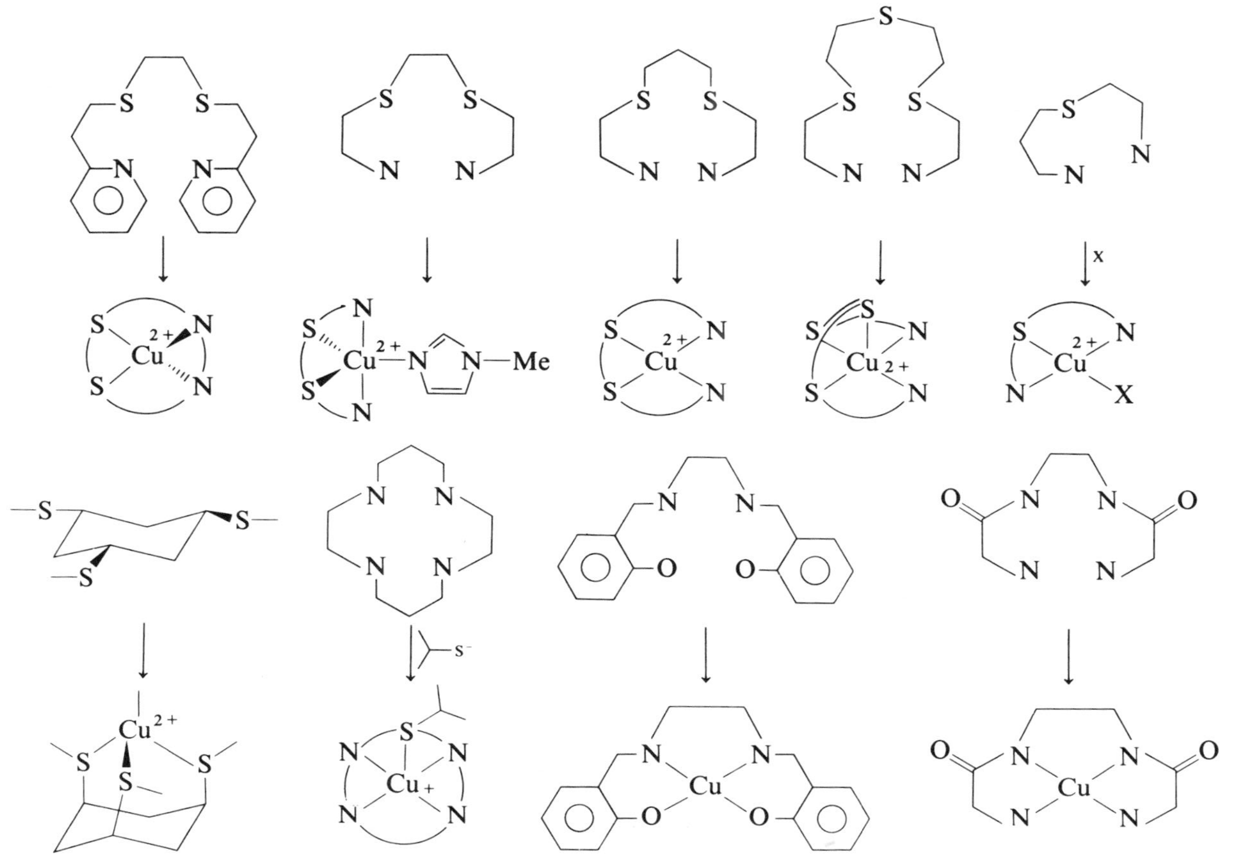

Fig. 6.6. Some copper-containing biomodels (518). Reprinted with permission. Copyright © 1977 by the American Chemical Society.

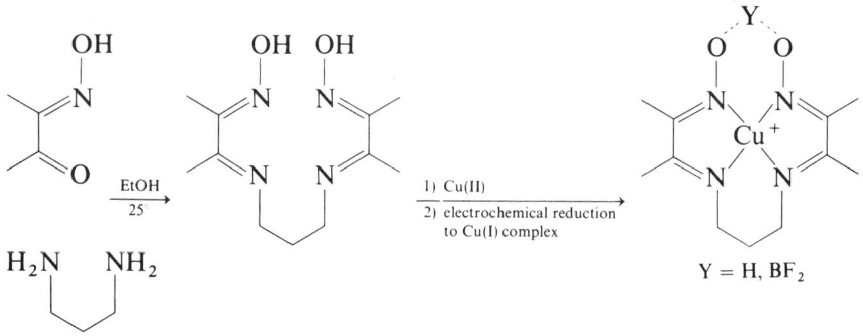

acetonitrile) yielding five-coordinate adducts. Of couse, such a model not an accurate mimic of the active sites of any copper protein but suggests that consideration of Cu(I) active site structures must include the possibility of five-coordination.

Since many enzymes have two metal ions in their active site, it would be interesting to mention the synthesis, in 1978, by Buckingham of a model (Fig. 6.7) for bismetallo active sites based on the "capped" or "picket fence" porphyrin concept (520). The structure of the model, shown below, consists of a tetrapyridine coordination arrangement joined via nonflexible links to a tetraphenylporphyrin. This arrangement, which is capable of considerable modification, allows the insertion of the same or different metal ions into the two different coordination sites.

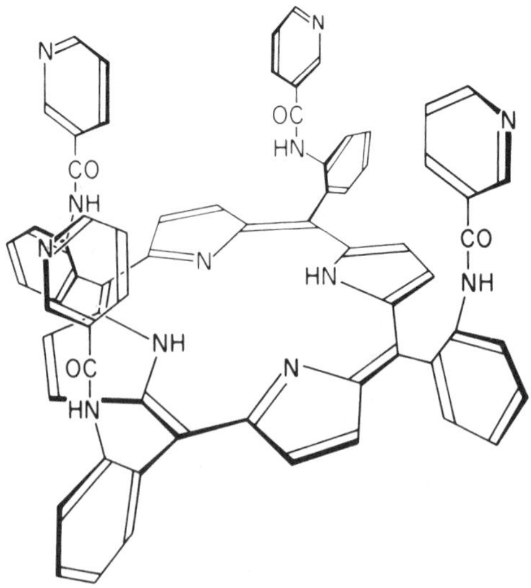

Fig. 6.7. Buckingham's "picket-fence" porphyrin (520). Reprinted with permission. Copyright © 1978 by the American Chemical Society.

Figure 6.8 illustrates this arrangement where M and M′ can be iron, copper, or nickel ions. Interconversions have been repeated several times without displacement of the metal ion from the porphyrin nucleus.

Electron paramagnetic resonance (EPR) spectra show strong interactions between the two metal centers and for the bis-Cu system, the Cu · · · Cu separation was estimated to be 0.59 nm. This very imaginative synthetic model appears capable of considerable extension.

In 1979, the groups of R. Weiss (521) and J.-M. Lehn (522) from Strasbourg in collaboration with S.J. Lippard from M.I.T. developed new bimetallic copper models based on macrocyclic complexes. The first group synthesized the following elongated macrocyclic ligand that can bind two Cu(I) or Cu(II) ions.

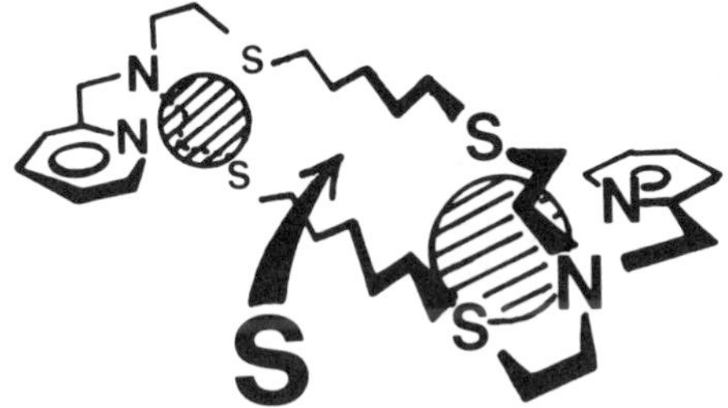

The spheres represent the space occupied by the copper ions and S the region where substrates can bind (521). Reprinted with permission. Copyright © 1979 by the American Chemical Society.

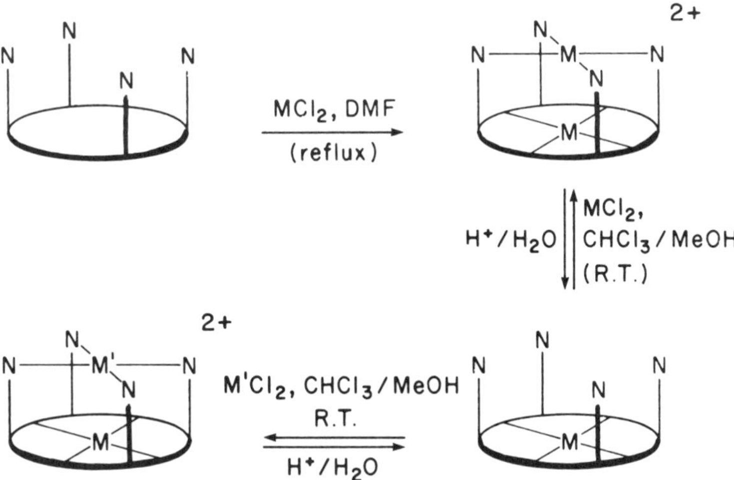

Fig. 6.8. The bismetallo active site concept of Buckingham (520). Reprinted with permission. Copyright © 1978 by the American Chemical Society.

Insertion of linear diatomic and triatomic substrate molecules such as CO, NO, O_2, or N_3^- are possible. The metal ions are located inside the macrocyclic ligand, each linked to a NS_2 ligand donor set as well as to the substrate (four molecules for the case of N_3^-) trapped in the middle. Since some metalloproteins use binuclear metal centers to perform catalytic function, such a model represents a mimic for type III copper pairs in copper enzymes. In the binuclear Cu(II) complex, the Cu · · · Cu distance was estimated to be 0.52 nm. Interestingly, this complex exhibits antiferromagnetism and is diamagnetic at room temperature. The groups of Lehn and Lippard (522) tackled the problem in a different way although conceptually the approach is quite similar. They synthesized an imidazole bridged dicopper (II) ion incorporated into a circular cryptate macrocycle (see Section 5.1.1).

The present approach is to prepare an imidazolate-bridged bimetallic center that would prevent the two metal ions from migrating away from each other when the bridge is broken. The preparation involves the synthesis of a dinuclear cryptate macrocycle containing two Cu(II) ions and where a substrate such as sodium imidazolate is able to bind.

macrocycle A

1) $Cu(NO_3)_2 \cdot 3H_2O$ in CH_3OH

2) sodium imidazolate ($N{:}^- Na^+$), 90% CH_3OH

3) $NaClO_4/CH_3OH$

(diamond-shaped blue platelets)

= Cu(II) ion

$[Cu_2(imH)_2(im) \subset A]^{3\oplus}$

symbol used for complexation with macrocycle A

X-Ray work showed that the $Cu_2(im)^{3+}$ ion is incorporated into the circular cryptate cavity by binding to the two diethylenetriamine units at each end of the macrocycle. Each Cu(II) ion is further coordinated to a neutral imidazole (imH) ligand, achieving overall pentacoordination. The synthesis of such a bimetallic complex offers the possibility of mimicking the 4-Cu(II) form of bovine erythrocyte superoxide dismutase and opens the way for the preparation of selective coreceptors as well as the development of bioinorganic models of metalloproteins in general.

6.5.1 Dinuclear Receptors

Since the earlier work of Lehn, many more dinuclear receptors, particularly with two copper atoms, have been made (523). Among the most representative dimetallic macropolycyclic inclusion complexes are the following *lateral* macrobicyclic dinuclear Cu(II) cryptates. The structure of the chelating unit should allow the possible stabilization of different oxidation states of the two copper ions. In fact, the reduction potentials

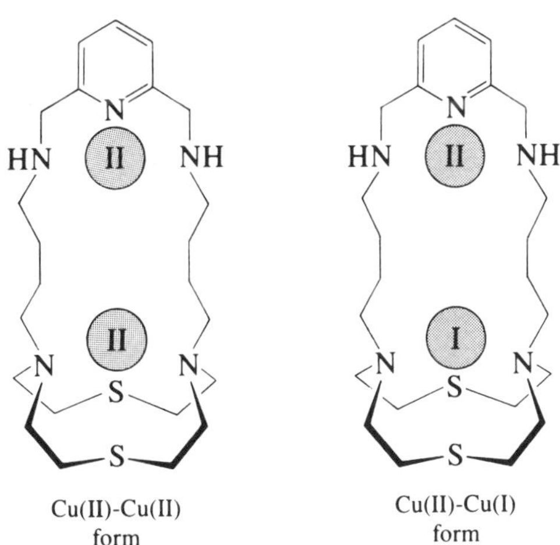

of each ion are markedly different, and the Cu(II)-Cu(II) form undergoes a facile one-electron reduction at +500 mV to give the Cu(II)-Cu(I) form. The second Cu(II) ion is reduced at +70 mV. The formation of a Cu(II)-Cu(I) mixed valence dinuclear cryptate suggests the possibility of the formation of heterometallic dinuclear systems. Not only does this provide a novel approach to studying *cation–cation interactions* at short distances, but it may also permit the *inclusion of substrates* imbricated between the cations in a supramolecular structure. Furthermore, such *co-systems* may display successive binding of different substrates yielding "*cascade*"-type complexes.

To model cytochrome c oxidase, M.J. Gunter and co-workers have synthesized a lateral macrobicyclic receptor where a porphyrin nucleus is capped by a pyridino bridging unit (524). A heterodinuclear complex was

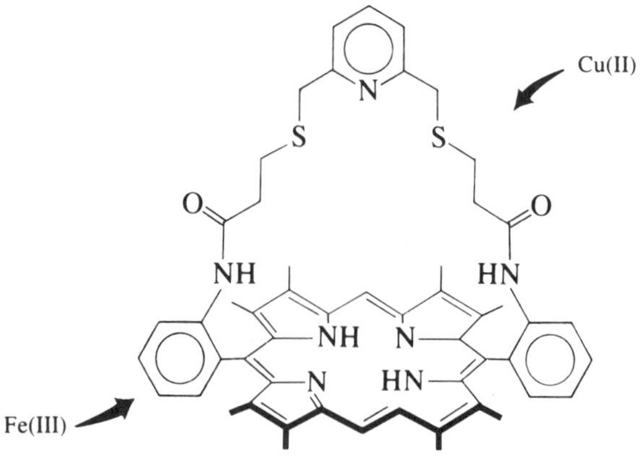

Cu(II)

Fe(III)

cytochrome c oxidase model

formed with an iron(III) ion in the porphyrin unit and a copper(II) ion bound to the lateral pyridino bridge.

Other heterometallic dinuclear complexes belonging to this general lateral macrobicyclic class of ligand might be envisaged as bioinorganic models of metalloproteins like superoxide dismutase.

"Face-to-face" linkage of two macrocycles by two bridges yields macrotricyclic molecules of the *cylindrical* type. Their architecture defines *three cavities*: two lateral circular cavities inside the macrocycles and a central cavity. Connecting two porphyrin nuclei gives "face-to-face" porphyrins (see Section 6.4.1) which are members of this class and form *homotopic* systems.

An impressive effort has been deployed by Collman's group to synthesize "face-to-face" porphyrin dimers and to exploit them as potential binuclear multielectron redox catalysts (525). Many synthetic strategies were elaborated before to arrive at an efficient and flexible synthetic scheme to generate isomerically pure porphyrin dimers. "Face-to-face" porphyrins with interporphyrin amide, amine, and hydrocarbon bridges of varying length have been prepared. The crystal structure of the dinuclear Cu(II) complex of one of the amide-linked dimers reveals a shear-like displacement of one porphyrin unit with respect to the other by an average distance of 4.95 Å. The Cu · · · Cu separation is 6.33 Å and the interplanar separation of the porphyrin rings is 3.87 Å.

These dinuclear copper and cobalt complexes have been used in the development of new catalysts for the direct, four-electron electrochemical reduction of O_2 or N_2 molecules to water or ammonia, a mimic of the nitrogenase in soil bacteria. Nitrogenase model compounds containing iron-sulfur centers have also been prepared (526).

The search for aesthetically new and attractive molecular architectures has been a constant concern among bioorganic chemists. One facet of beauty in molecular edifices rests in *molecular topology* and in the

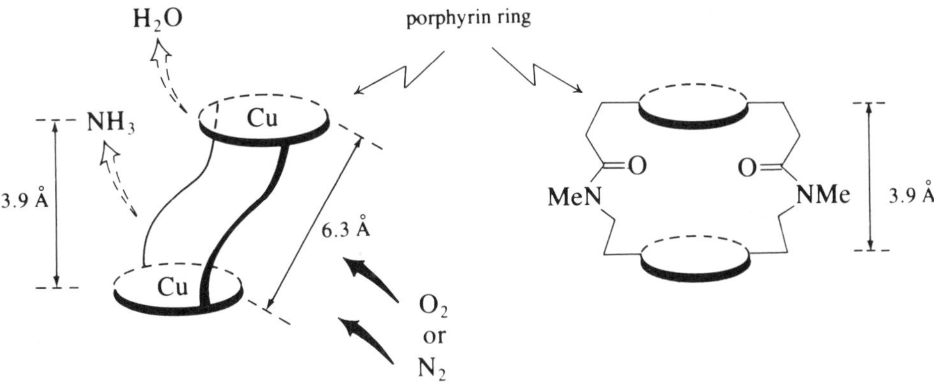

topological properties of the objects. Double-stranded circular DNA is a good example where supercoiled and interlocked (catenated) structures are known to yield "superstructures" or topological DNA conformations. A whole class of enzymes can even manipulate these super-DNA structures and effect topological transformations; they are called *topoisomerases*.

An important class of topological isomers are the *catenanes*. A catenane is composed of interlocked rings made of molecular threads. Constructing such a molecular edifice represents an achievement of considerable efforts where the imagination of the chemist is directly dependent on the power of new synthetic tools. J.P. Sauvage and his team (Institut de Chimie, Strasbourg) deserve all the credit for this impressive contribution (527).

Sauvage developed a novel family of molecules: the *metallo-catenanes* (528). A recent strategy consisted in a template-directed synthesis of three interlocked rings using two Cu(I) centers in a cyclodimerization process (529). The approach is schematized below.

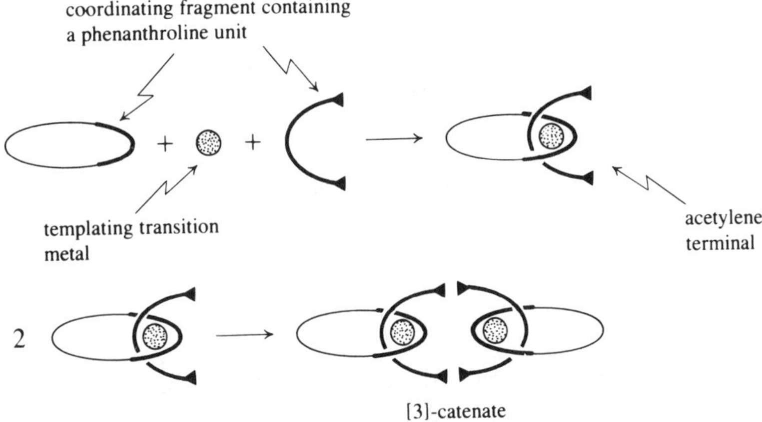

The cyclodimerization step is based on acetylenic oxidative coupling (Glaser reaction) where four reacting centers are implicated. Strikingly high yields of up to 60% allow gram scale preparation of those compounds. The following example illustrates this.

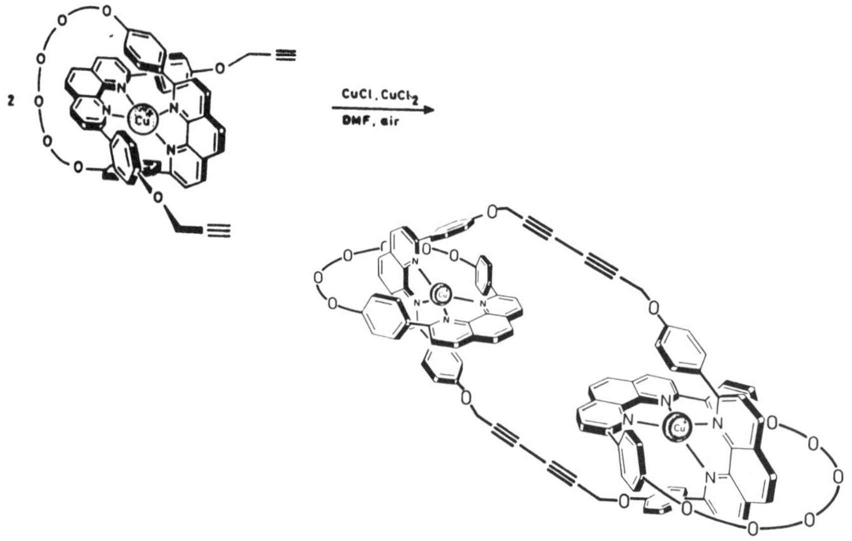

2 CuCl, $CuCl_2$ / DMF, air →

Demetallation of the bis-Cu(I) catenate is performed with KCN, affording the free ligand. The resulting new coordinating system is relatively flexible and adaptable owing to its catenane nature, but some rigidity in the central 44-membered ring is ensured by the presence of the two diyne spacers.

Both crystallographic and NMR solution studies demonstrated that the geometry of this catenate is dramatically different from that implied by the structure shown above (530). In fact, the molecular complex exists in a curled-up conformation of the central ring, bringing its two phenanthroline groups and the other two 30-membered rings in a stacking disposition. This is illustrated in Fig. 6.9. The copper atoms in this molecule are only 8 Å apart.

The structure of this "molecular bracelet" is stabilized by π—π interactions between the aromatic rings belonging to the two copper complex subunits. This "tertiary structure" is reminiscent of other reported systems consisting of a cationic noble-metal complex linked to a large macrocyclic polyether which has to fold up in order to adjust its geometry to the shape of the bound transition-metal complex (531). The disposition of the aromatic nucleus resembles also the stacking disposition of nucleic acid bases in DNA duplex. More complex molecular trefoil knots have also been synthesized (532). With the same Cu(I)-based template strategy, Sauvage and his colleagues have also made a rotaxane-like molecule consisting of two rigidly held porphyrin rings as molecular stoppers (533).

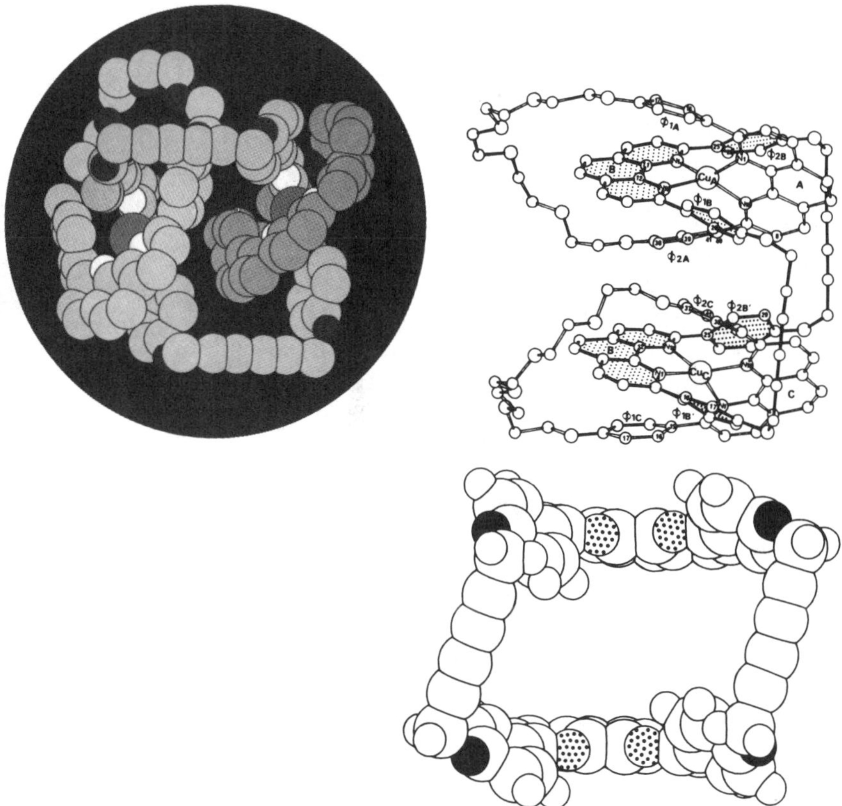

Fig. 6.9. Computer drawing and crystal structure of di-Cu(I) [3]-catenate (530). Reproduced with permission. Copyright © 1987 by Verlag Chemie GmBH. See color print of figure in upper left.

This new field of chemistry at the edge between bioorganic and bioinorganic chemistry will offer in the future many new challenges of a conceptual and practical nature with applications of technological importance.

6.6 Biomodels of Photosynthesis and Energy Transfer

One of the most complex and fascinating transformations in biology is the photoinduced "hydration" of CO_2 and C–C bond formation to chiral sugar molecules. The complex event of transformation is summarized in Fig. 6.10. This accepted zigzag scheme represents a two quantum per electron photoreduction system in chloroplasts. The highly mobile electron carriers are composed of plastoquinone, plastocyanin, ferrodoxin, and

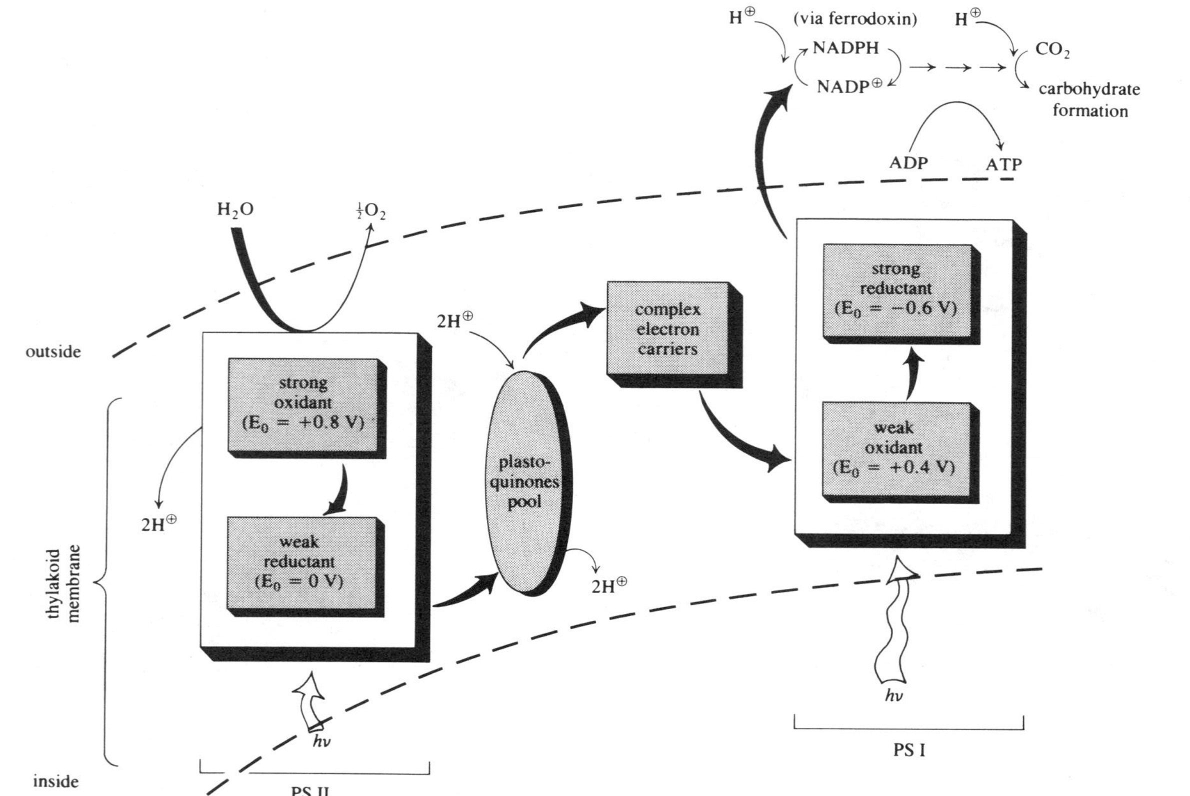

Fig. 6.10. Oversimplified zigzag scheme for natural photosynthesis. The energy of the process is harvested by two distinct photosystems (PSI, PSII) that operate in series. Two photons are absorbed for every electron liberated from water. The electron flow is eventually coupled to $NADP^+ \rightleftharpoons NADPH$ cycle with formation of ATP and reduction of CO_2 to carbohydrates. Adapted from (535). Reproduced with permission. Copyright © 1987 by the American Chemical Society.

cytochrome b_6-f complex. The essential features of the photosynthetic events are as follows (534):

1. Light trapped by chlorophyll molecules in an extensive antenna system is channelled, via singlet excitation transfer, to a special chlorophyll center.
2. At this stage, part of the photon energy has been converted into the energy of the excited singlet state of the special pigment center.
3. An electron then leaves the chlorophyll center and, residing momentarily on a number of intermediate electron acceptors, in a time period of a few hundred picoseconds, reaches an electron acceptor. In photosynthetic bacteria the acceptor molecule is a ubiquinone. In the PSI of plants it is a membrane-bound ferrodoxin center while in PSII a plastoquinone is implicated.

To sum up, photosynthesis could be regarded as a process that converts light into chemical potential energy in the form of long-lived charge intermediates within and across a bilayer thylakoid membrane. Recombination of any charge-separated species is prevented by the large distance separating the final electron donor and acceptor site. In fact, the reaction centers employ a series of electron-transfer steps, each of which occurs over a short distance with high quantum yield because the yield of a single long-range electron transfer across the membrane bilayer would be vanishingly small.

Because of the importance of quinone chemistry in the process, many model studies will incorporate this functional group in their structures.

Efficiently mimicking such a process will be a monumental achievement in bioorganic chemistry. Many attempts have been directed toward

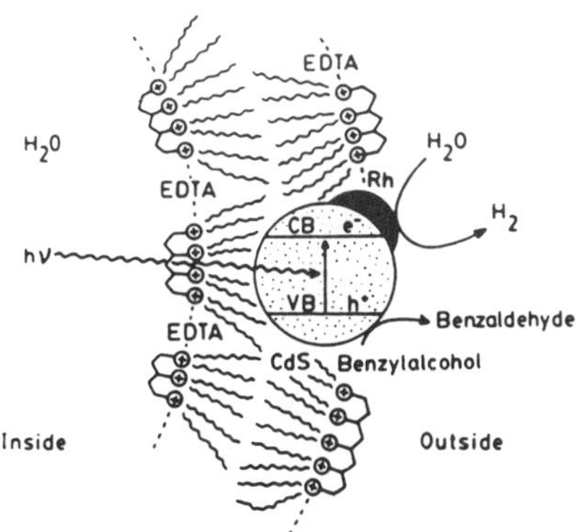

Fig. 6.11. Reduction of water with hydrogen formation by irradiation of semiconducting particules of CdS/Rh embedded in an artificial polymerized micelle (535). Reproduced with permission. Copyright © 1987 by the American Chemical Society.

such a goal. In one mimetic system, water rather than CO_2 has been successfully reduced in the reduction half-cycle to hydrogen at the expense of an organic substrate, benzyl alcohol. This is depicted in Fig. 6.11, page 449.

The ultimate goal is to tap light energy to generate hydrogen fuel from water in a redox reaction. Formation of oxygen, however, proved more difficult to realize. J.-M. Lehn tackled this problem using a mixed system made of three components. The first one is composed of a photosensitive ruthenium complex such as Ru(2,2′-bipyridine)$_2$Cl$_2$. The second one includes a cobalt(III) complex, typically Co(NH$_3$)$_5$Cl·Cl$_2$. The last module is ruthenium oxide. In this assemblage, water oxidation is photochemically driven.

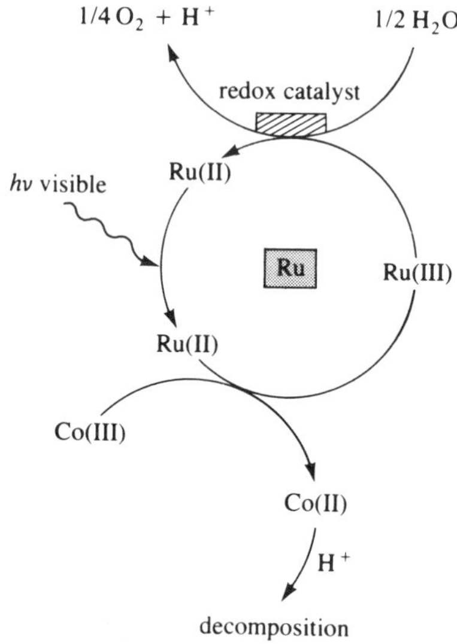

The next step in complexity will be to seek a means of combining the two half-systems (oxidation and reduction) to generate both hydrogen and oxygen simultaneously.

In the light of recent findings, synthetic biomimetic models of special pair bacteriochlorophyll *a* have also been prepared (536,537). The reason is that, in the molecular organization of chlorophyll in the photoreaction centers of both green plants and photosynthetic bacteria, it is believed that special pairs of chlorophyll molecules are oxidized in the primary light conversion event in photosynthesis. Dimeric chlorophyll derivatives such as the one in Fig. 6.12, in which the porphyrin macrocycles are bound by a simple covalent link, exhibit several photochemical properties that mimic in vivo special pair chlorophyll.

Like the cofacial chlorophyll-based dimer just mentioned, hinged and singly linked chlorophyll dimers have also been synthesized and the spectroscopic properties examined (538). Being more flexible, they may

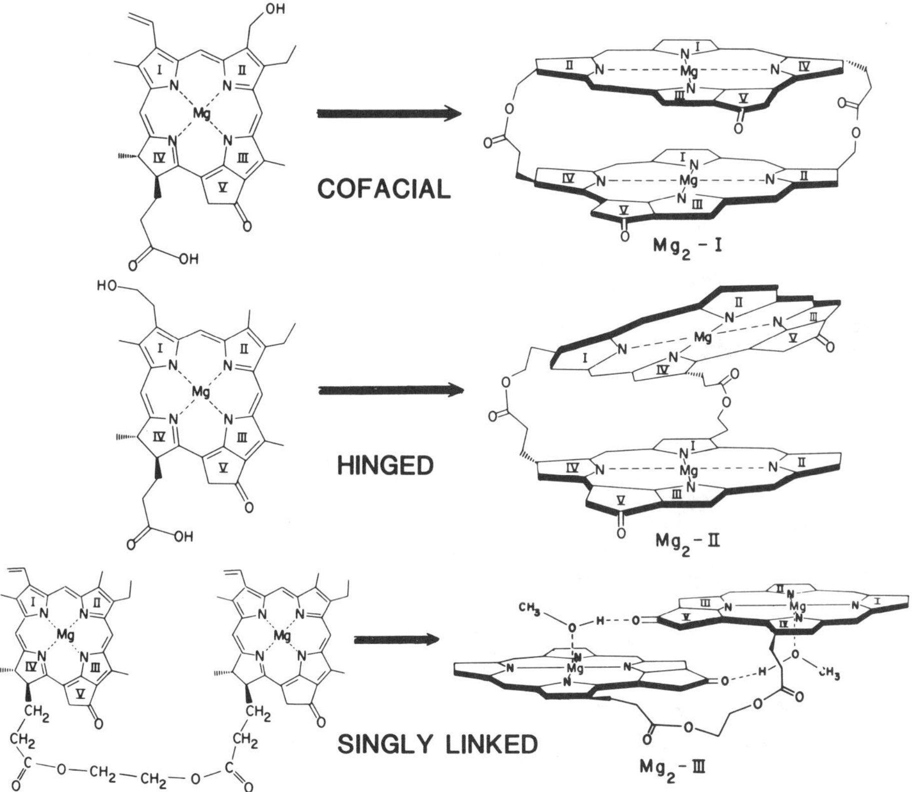

Fig. 6.12. Biomimetic models of paired chlorophyll (538). Reproduced with permission. Copyright © 1982 by the American Chemical Society.

be better representatives of the primary electron donor in green plant (PSI) and bacterial photosynthetic systems. These models exhibit a large red shift to about 700 nm and provide further evidence for interchromophore resonance (exciton) interactions. They also support electron spin resonance (ESR) data and the presence of a cationic radical character of the electron donor in bacterial systems.

Other imaginative chlorophyll dimers have been prepared, and one of them using a bis-pyridylethylene spacer has been developed at Gas Research Institute by R.V. Serauskas and K. Krist. The strategy consisted in linking a chlorophyll nucleus in an electron-acceptor form (vinyl group present) to another one in an electron-donor form (vinyl group and carbonyl absent). The resulting di-Zn(II) complex showed enhanced photochemical properties (539). The geometry of the molecular assembly can be varied by changing substituents and/or ligands. For efficient photoconversion however, a 14 Å separation between the chlorophyll rings has been found as crucial. The objectives behind this venture are

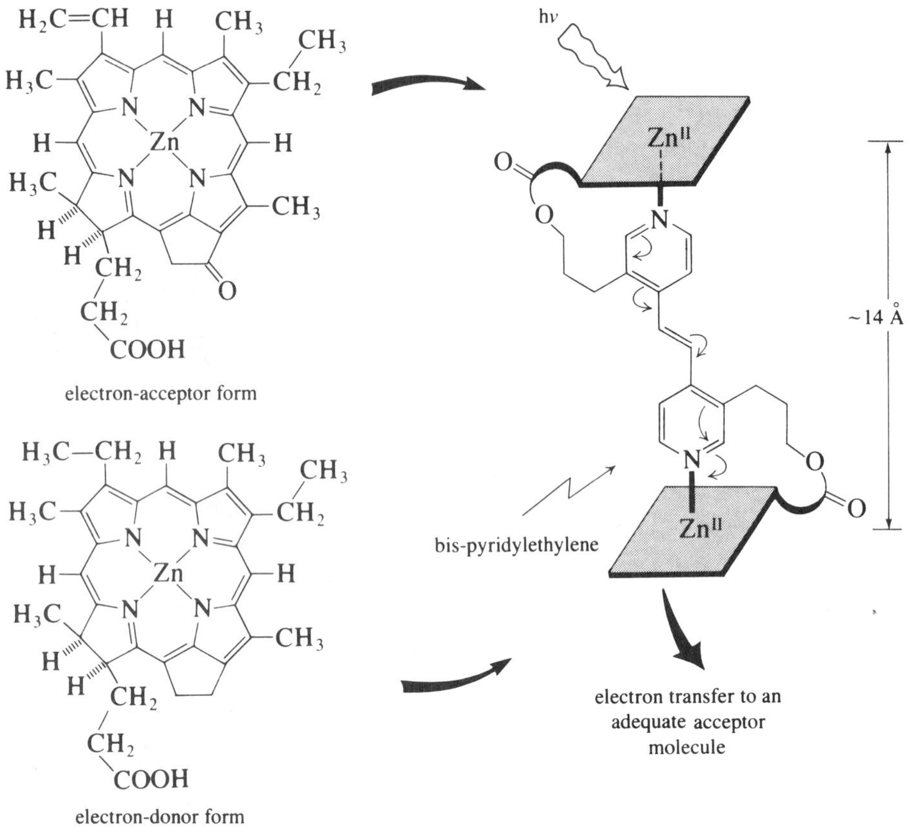

electron-acceptor form

electron-donor form

bis-pyridylethylene

electron transfer to an adequate acceptor molecule

multiple. For instance, this new "molecular toy" or a variant of it could be used as a catalyst for water splitting ($H_2O \rightarrow H_2 + O_2$) or for the aqueous reduction of carbon dioxide ($CO_2 + H^+ \rightarrow CH_4$), either photochemically or electrochemically. This way, new sources of gaseous fuels (H_2 and CH_4) might be obtained at costs and scales more accessible than conventional photovoltaic devices.

Many recent biomimetic models have been constructed aiming for artificial photosynthesis. A "photo-driven charge separation" is the key step in the photosynthesis process where solar energy is converted to chemical potential energy. An efficient model should have such a molecular device built in. The light-excited chlorophyll can transfer electrons to an intermediate acceptor to generate high-energy molecules that later assist in the conversion of CO_2 into sugar molecules. In the meantime, the positive charge left on the chlorophyll is replenished with electrons coming from the oxidation of nearby water molecules to produce oxygen. However, nature ensures that at all times the light-induced positive and negative centers stay separated from one another. This reduces loss of energy through charge recombination. Any biomimetic models of photoelectron acceptor site

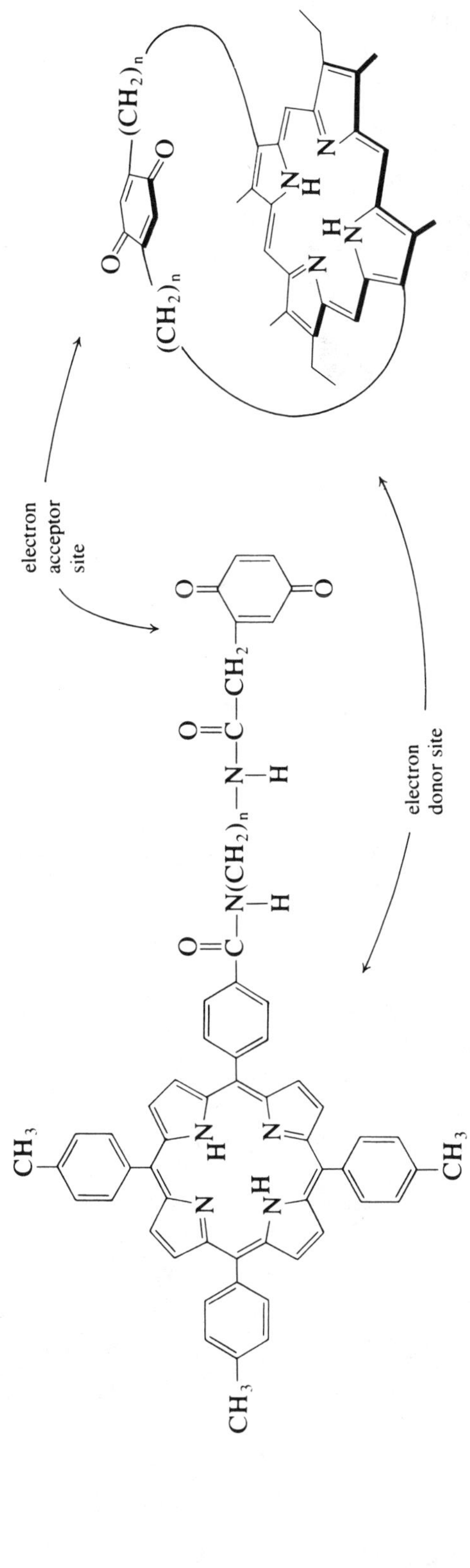

electron donor site

Bolton's model (534)

Dolphin's model (540)

in analogous molecules with a single flexible chain, intramolecular charge transfer reactions are obscure by competing intermolecular reactions

Lindsey's model (541)

an acceptor to snatch the electron coming from the phorphyrin

trapping light center

the β-carotene unit serves to "neutralize" the positive "hole" left on the porphyrin

Moore's model (542)

synthesis should have some sort of a similar molecular device built in to be efficient. The most representative models are shown on page 453. After the photoexcitation and before the electron has a chance to jump back on the porphyrin, the β-carotene shares an electron to the porphyrin ring. The net result is a charge-separated molecule that has a relatively long life of 3 μs, compared to 3 ps for simple porphyrin–quinone models. The polyolefinic carotene chain has a second function in protecting the supermolecule from photodamage by singlet oxygen and also absorbs shorter wavelengths of light that the porphyrin cannot. This group thus serves the same function as in the natural system, but with only 10% efficiency.

The recent X-ray analysis of a bacterial photosynthetic reaction center from *Rhodopseudomonas viridis* dramatically revealed the presence of the intriguing arrangement of six interacting porphyrin nuclei at the active site (543). This imbricated hexameric unit has approximately 2-fold rotational symmetry, and at its center is a "special pair" of overlapping bacteriochlorophyll molecules that seem to function as the primary electron donor (544). Such molecular disposition of porphyrin rings suggests the possibility of two electron-transfer pathways, one of them to a nearby quinone species.

Furthermore, these types of molecules may provide a mechanism for designing well-characterized molecular antennae arrays that could transfer light energy to the central core porphyrin nucleus. These transfers may operate by electron tunnelling or charge delocalization, but this remains to be determined.

In an effort toward constructing a synthetic model of this photosynthetic reaction center, A.D. Hamilton (University of Pittsburgh) with the help of G.M. Dubowchik devised a way to build a structure containing a

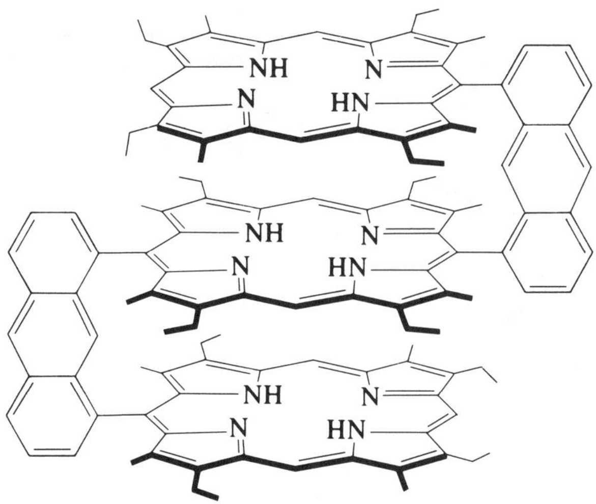

triple-deckered triporphyrin

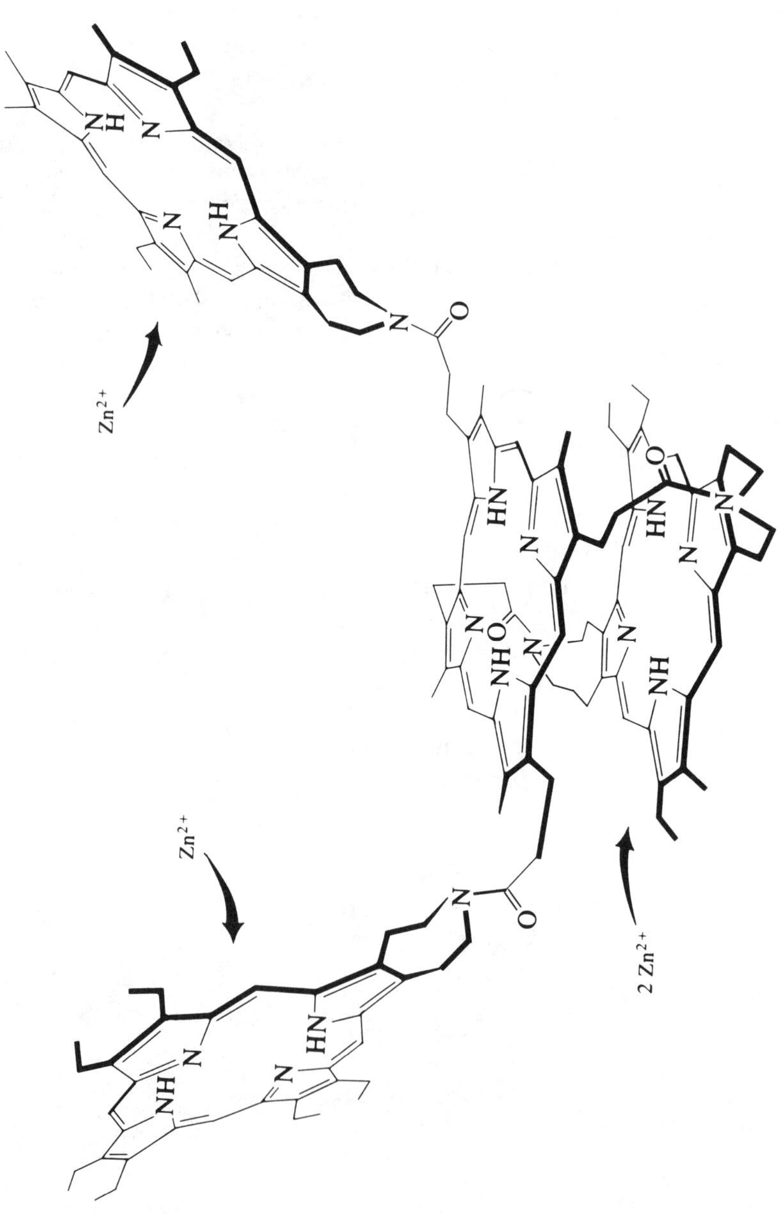
Zn2+
Zn2+
2 Zn2+

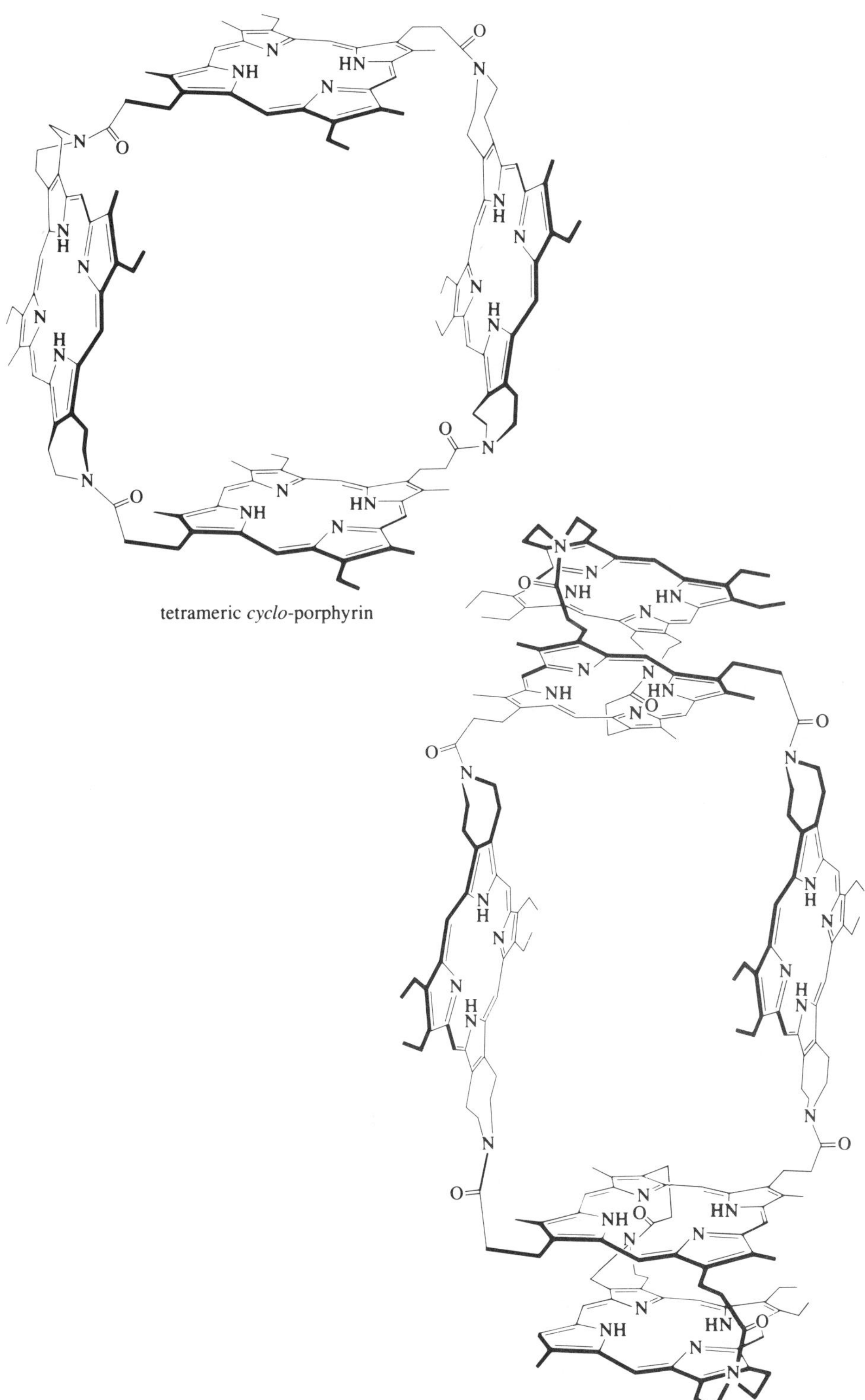

tetrameric *cyclo*-porphyrin

hexameric *cyclo*-porphyrin

dimeric tetrapyrrole covalently linked to two other adjacent tetrapyrroles (544). Reaction of this tetramer with zinc acetate in dichloromethane-methanolic solution afforded the tetra-Zn(II) complex. These structures show two Soret bands corresponding to the two porphyrin environments. Preliminary fluorescence emission spectroscopy and lifetime measurements provide evidence for electron- or energy-transfer interactions between dimeric and monomeric chromophores. See page 456.

In a continuing effort to develop enforced and controlled aggregates of multiple tetrapyrroles, the same team recently disclosed the synthesis of tetrameric and hexameric *cyclo*-porphyrins (545). (See page 457.) Again, they serve as models or the biochemical aggregates present in bacterial photosynthetic active centers. They represent interesting photoactive ligands capable of binding four- and six-metal atoms, respectively, within a single molecular unit. With a different synthetic strategy, but with the same objective in mind, a triple-deckered triporphyrin was assembled (546). This highly structured empilement of porphyrin nucleus is designed to mimic the primary electron-donor-acceptor complex in photosynthetic reaction centers as well as the molecular events involved in the electron-transport processes.

In this three-story molecular edifice each floor is made of well-ordered and stacked porphyrin planes (see p. 455). The ensemble represents another attempt to model the light-driven charge separation step in photosynthesis using prophyrin and chlorophyll analog molecules. All these constrained molecules represent real synthetic challenges.

The group of Staab (Max-Planck-Institut, Heidelberg) has synthesized and studied the properties of vertically stacked porphyrin(1)-porphyrin(2)-quinone cyclophane (547,548). An example is illustrated below:

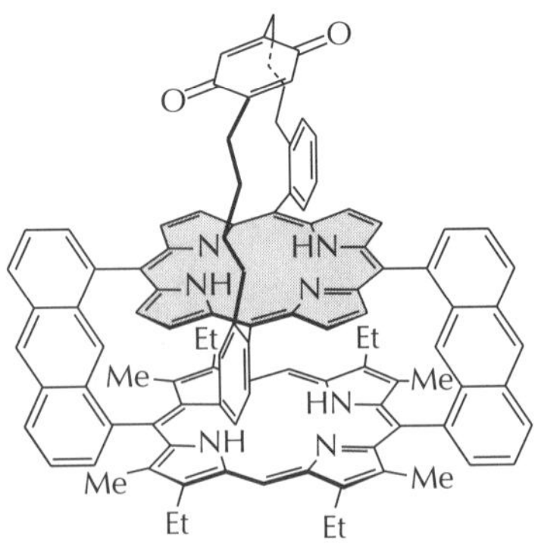

A vertically stacked porphyrin(1)-porphyrin(2)-quinone cyclophane (548).

The two porphyrin units are aligned parallel to each other and are separated by approximately 5 Å. The porphyrin that is the weaker donor (the one in gray) is flanked by the quinone, which functions as an electron acceptor. Such a molecule offers the possibility of photoinduced electron transfer across several successive porphyrin and quinone units and has been increasingly studied in the context of the primary process in biological photosynthesis.

In 1984 Lehn developed *photoactive cryptands* containing bipyridine and phenanthroline ligands (549). Their metal-ion complexes may have attractive features by combining the cation inclusion nature of the cryptand with the photoactivity of bipyridine and phenanthroline groups. The corresponding cryptates could be the start of the development of new systems capable of performing photoinduced processes related to solar energy conversion and artificial photosynthesis.

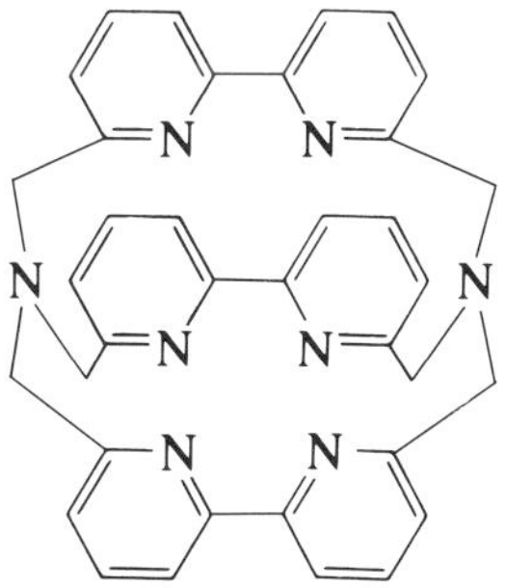

[bpy]$_3$-cryptand

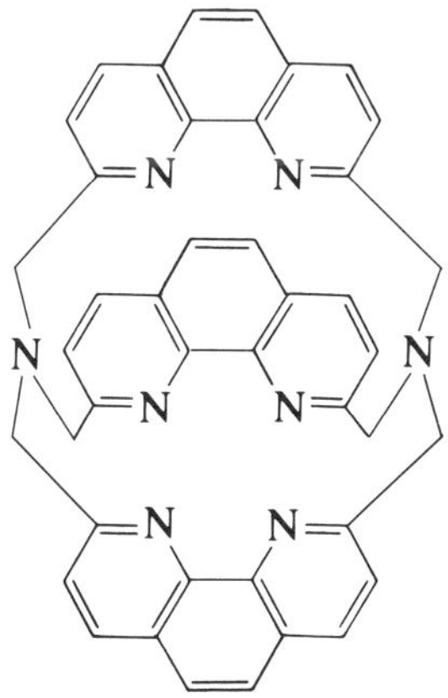

[phen]$_3$-cryptand

In fact, europium cryptates have been found to present a notable emission in aqueous solution at room temperature. In the present case, strong red and strong green emissions, respectively, are readily obtained for the Eu(III) and Tb(III) complexed to [bpy]$_3$-cryptand. The light absorbed by the bipyridine groups is reemitted as visible light by the complexed lanthanoid ions (see Fig. 6.13).

An intramolecular energy transfer from the π,π^* excited states of the ligand groups to the excited levels of the Eu(III) ion may be involved. The efficiency of conversion is close to unity. This photoactive Eu(III) cryptate can be considered as an efficient luminophore that operates as a **A-ET-E** light conversion molecular device, transforming UV light absorbed by the aromatic rings into visible lanthanoid emission via intramolecular energy transfer (550).

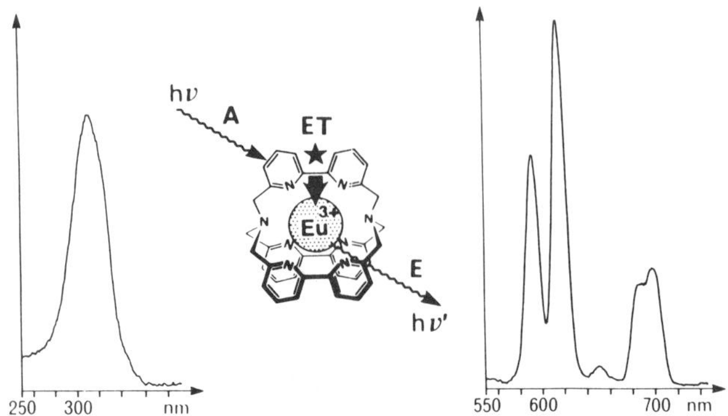

Fig. 6.13. Illustration of the absorption-energy transfer-emission **A-ET-E** light conversion process by an Eu(III) cryptate. On the left, the excitation spectrum (emission at 700 nm); on the right, the emission spectrum (excitation at 320 nm, 10^{-6} *M* in water at 20°C (550). Reproduced with permission. Copyright © 1987 by Verlag Chemie GmBH.

Systems performing photoinduced electron transfer processes represent components for *light-to-electron conversion devices* and, in a broader perspective, such investigations may lead to the developments of *photo-active molecular devices*. A number of potential uses may come to mind like the development of luminescent materials and photolabels in immunoassays.

6.7 Cobalt and Vitamin B_{12} Action

Although cobalt ions are found in both the (II) and (III) oxidation states, the most important biological compound of cobalt is vitamin B_{12} or cobalamin where the Co(III) form is present (551) (Fig. 6.14). Cobalamin or related substances are important biological compounds that are involved in a great variety of activities, particularly in bacteria. Vitamin B_{12} is also necessary in the nutrition of humans and probably of most animal and plant species. It is of critical importance in the reactions by which residues from carbohydrates, fats, and proteins are used to produce energy in living cells. Pernicious anemia is a severe disease in elderly people. This disease is usually accompanied in mammals by the increased excretion of methylmalonic acid in the urine. Today it is effectively controlled by a 100 μg injection of vitamin B_{12}.

Vitamin B_{12} (cyanocobalamin, R = CN) is one of the most complicated naturally occurring coordination compounds and certainly the most complex nonpolymeric compound found in nature. In the 1950s, H.A. Barker

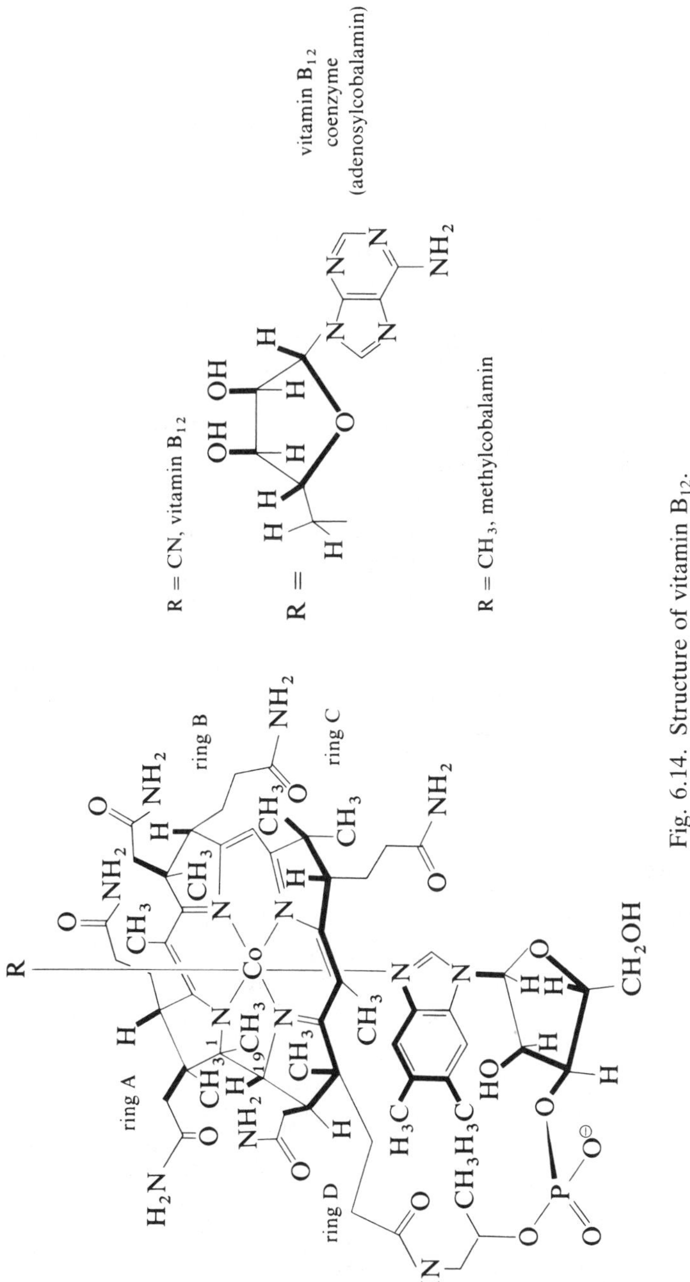

Fig. 6.14. Structure of vitamin B$_{12}$.

demonstrated the involvement of a derivative of vitamin B_{12} in the following reaction:

HOOC–CH(NH$_2$)–CH$_2$–CH$_2$–COOH $\underset{\text{mutase}}{\overset{\text{glutamate}}{\rightleftharpoons}}$ HOOC–CH(NH$_2$)–CH(CH$_3$)–COOH

glutamic acid β-methylaspartic acid

It was then realized that the biochemically active form of the cobalamin is the coenzyme B_{12}, containing an adenosyl moiety bound through its 5′-carbon to form a covalent bond to the central cobalt atom of vitamin B_{12}. Although the conversion of vitamin B_{12} (cyanocobalamin) to adenosyl-B_{12} (the coenzyme) appears to involve the replacement of a CN^- group by an adenosyl group from ATP, the enzymatic conversion has been found to be fairly complex.

One of the remarkable properties of B_{12} is its ability to form alkyl derivatives (551). Before the discovery of vitamin B_{12} by Barker, it was thought that a $Co-C_\sigma$ bond would be unstable, if capable of existence at all. This is the first and only example of a water-stable organometallic compound occurring in nature. Its complete structure was revealed in 1956 by the crystallographic work of D.C. Hodgkin, aided by the earlier chemical studies of Lord A.R. Todd and A.W. Johnson. Its full synthesis was completed in the early 1970s by a team effort of the late professor R.B. Woodward (Harvard) and A. Eschenmoser (E.T.H., Zurich).

The core of the molecule resembles an iron–porphyrin system, but here the ring is called a *corrin* and has two of the four modified pyrrole rings directly bonded (C_1-C_{19} link). All the side chains are made of acetamide and/or propionamide groups. One of them is linked to an isopropanol phosphate residue attached to a ribose and finally to a dimethylbenzimidazole ring to the Co(III) ion.*

Surprisingly, the cobalt ion can have the $+1(B_{12s})$, $+2(B_{12r})$, or $+3(B_{12a})$ oxidation state. Indeed, one of the remarkable properties of alkyl cobalamins is that three routes for cleavage of the Co–C bond are possible:

$$R-Co(III) \rightarrow R^{\cdot} + \dot{C}o(II) \text{ (called } B_{12r} \text{ form)}$$
$$R-Co(III) \rightarrow R^{\oplus} + \ddot{C}o(I) \text{ (called } B_{12s} \text{ form)}$$
$$R-Co(III) \rightarrow R^{\ominus} + Co(III) \text{ (called } B_{12a} \text{ form)}$$

* Naturally occurring vitamin B_{12} analogs have been found where the benzimidazole ring is replaced with purines (adenine, guanine, etc.).

The B_{12s} form is a potent nucleophile (also a reducing agent) and coenzyme B_{12} is believed to be formed via a nucleophilic attack on ATP. In the presence of diazomethane, the B_{12s} form is converted to methylcobalamin ($R = CH_3$ in the general structure). Therefore, an alkyl radical, a carbocation, and a carbanion can all be produced in B_{12} chemistry. In both bacteria and in liver, the 5′-deoxyadenosyl coenzyme is the most abundant form of vitamin B_{12}, but lesser amounts of methylcobalamin also exist.

The rearrangement described above by Barker using the coenzyme B_{12}-dependent enzyme glutamate mutase is a most remarkable reaction. Until very recently, no analogous chemical reaction was known. In fact, elucidation of the structure of the coenzyme form of vitamin B_{12} did not clarify its mechanism. Besides this transformation, nine distinct enzymatic reactions requiring coenzyme B_{12} as cofactor are known. Most are without precedent in terms of organic reactions. They are listed in Fig. 6.15. In choosing vitamin B_{12} derivatives as coenzymes, enzymes appear to have reached a peak of chemical sophistication that would be difficult to mimic by the chemist.

Interestingly, all the reactions mentioned above can be generalized as the migration of a hydrogen atom from one carbon to an adjacent with concomitant migration of a group X from the adjacent carbon to the one where the hydrogen was originally bound.

$$-\overset{|}{\underset{\underset{X}{|}}{C_1}}-\overset{\overset{H^*}{|}}{\underset{\underset{H}{|}}{C_2}}-Y \rightleftharpoons -\overset{|}{\underset{\underset{H^*}{|}}{C_1}}-\overset{\overset{H}{|}}{\underset{\underset{X}{|}}{C_2}}-Y$$

This unique carbon skeleton rearrangement has been studied in some detail with most of the B_{12}-dependent enzymes. Of course, the crucial question concerning the mechanism is: *How does the 1,2-shift occur?* Before examining the details of the mechanism, let us examine some useful experimental observations.

It was observed that for methylmalonyl-CoA (reaction 6-2), the X group (—COSCoA) is transferred intramolecularly and the hydrogen that is transferred does not exchange with water during the process. The same observation applies for the dehydration of propanediol (reaction 6-4) and the conversion of glutamic acid to methylaspartic acid (reaction 6-1). However, for these last two cases, inversion of configuration takes place at the carbon atom to which the hydrogen migrates. But for the methylmalonyl-CoA, the hydrogen that migrates from the methyl group to the C_3 position of succinyl-CoA occupies the same steric position as previously occupied by the —COSCoA group. The migration occurs with retention of configuration.

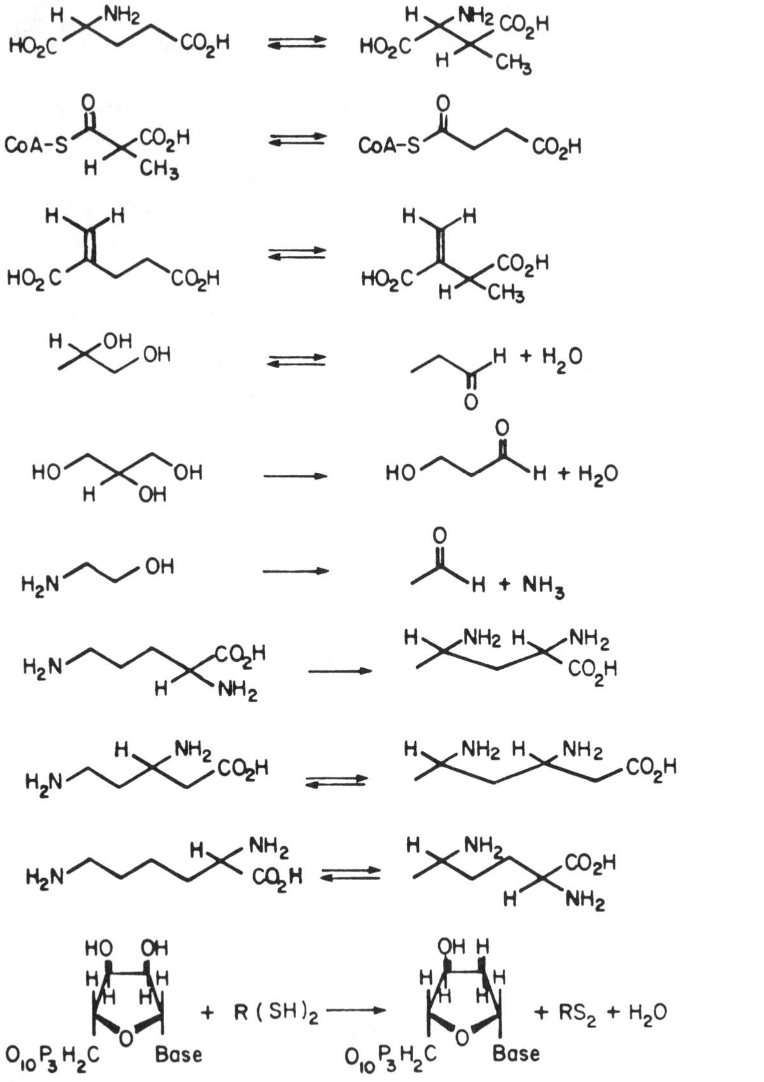

(6-1)

(6-2)

(6-3)

(6-4)

(6-5)

(6-6)

(6-7)

(6-8)

(6-9)

(6-10)

Fig. 6.15. The ten different reactions that require coenzyme B_{12}.

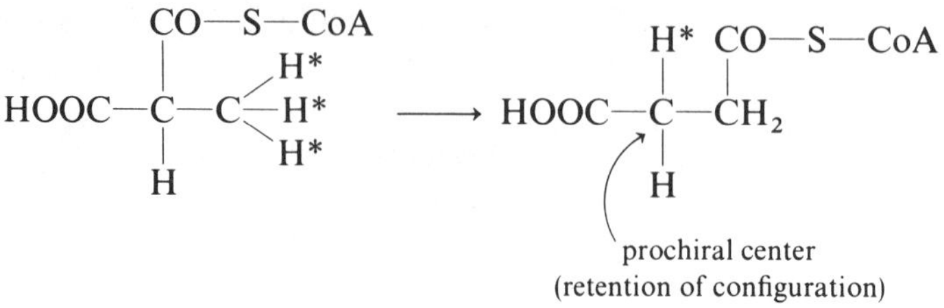

$$\mathrm{HOOC{-}C(H)(CO{-}S{-}CoA){-}CH^*_3} \longrightarrow \mathrm{HOOC{-}C(H)(H^*){-}CH_2{-}CO{-}S{-}CoA} \tag{6-11}$$

prochiral center (retention of configuration)

$$HOOC-CH_2-CH^*H-CH(COO^{\ominus})(NH_3^{\oplus}) \longrightarrow CH_2H^*-CH(COOH)-CH(COO^{\ominus})(NH_3^{\oplus}) \quad (6\text{-}12)$$

(inversion of configuration at this center)

$$CH_3-CH(OH)-CH(OH)H^* \longrightarrow \left[CH_3-CHH^*-CH(OH)OH\right] \longrightarrow CH_3-CHH^*-CHO$$

prochiral center
(inversion of configuration)

(6-13)

In this last example, it has been shown that the migrating hydrogen atom does not exchange with the solvent but exchanges with similar hydrogen atoms of other substrate molecules. The transformation corresponds to an internal redox reaction but is not exclusively intramolecular. Hence:

$$CH_3-CH(HO)-CD_2(OH) \longrightarrow$$

$$CH_3-CH_2-C(=O)D,\ CH_3-CH_2-C(=O)H,\ CH_3-CHD-C(=O)D \quad (6\text{-}14)$$

$$CH_3-CT(HO)-CT_2-OH + CH_2(HO)-CH_2(OH) \longrightarrow CH_3-C(=O)T + CH_3-CH_2-C(=O)T \quad (6\text{-}15)$$

Thus, the stereochemical course of vitamin B_{12}-dependent enzymes is a fascinating subject for the chemist to study (551). For instance, the enzyme propanediol dehydrase catalyzes also the dehydration of deuterated analogs of 1,2-propanediol (reactions 6-16 and 6-17). Migration of the deuterium atom is observed only with one isomer. These initial experiments showed that the dehydration proceeds by way of a 1,2-shift.

(6-16)

(1R,2S)-[1-²H]propanediol

H migrates in this isomer to produce [1-^{2}H]propionaldehyde

(6-17)

(1R,2R)-isomer

^{2}H migrates in this isomer to produce [2-^{2}H]propionaldehyde of 2S-configuration

Arigoni and colleagues (552) used 2*S*- and 2*R*-1-[^{18}O]-1,2-propanediol to bring an elegant and additional proof for the concomitant 1,2-shift of an OH from C-2 to C-1 as the hydrogen migrates from C-1 to C-2 in this process.

More interesting is the finding by R.H. Abeles of Brandeis University that the C-5′ hydrogen of the coenzyme is also replaced by tritium when 1,2-[1-^{3}H] propanediol is used as substrate and the tritium could be transferred back to the product. In his view, the three nonequivalent hydrogen atoms of the 5′-deoxyadenosine moiety have equally a one-third probability of transfer back to the substrate. Consequently, the enzyme does not distinguish between the two prochiral hydrogens at C-5′ and shows that the role of the coenzyme B_{12} is basically to act as carrier. This apparent lack of stereoselectivity involved a homolytic cleavage of a cobalt–carbon bond (see below).

The experimental evidence suggests that a coenzyme-bound intermediate must exist in the initial event of some B_{12}-catalyzed rearrangements. It can be pictured in the following manner, using ethylene glycol as substrate:

CH_2—CH_2 + [CH_2R–Co] ⇌ [HOCH—CH_2OH–Co] + CH_3—R
HO OH

[CH_2R–Co] + CH_3CHO

remains tighly bound to the apoenzyme and does not exchange with solvent

abbreviates corrin system or a biomodel analogue (R = adenosyl)

The detailed mechanism of this fascinating biochemical transformation was elucidated mainly through the efforts of the groups of R.H. Abeles (551), D. Arigoni (552), D. Dolphin (559–561), G.N. Schrauzer (553–555) and P. Dowd (556). For simplification, various biomodels of the corrin system were used. Some are presented in Fig. 6.16. The most commonly used is the bis-(dimethylglyoximato)–Co complex (called cobaloxime) which shows many properties of the cobalt atom in the corrin ring.

If a cobaloxime is coordinated with a suitable basic group (B) in one of the axial positions, it behaves in many ways like vitamin B_{12}. When

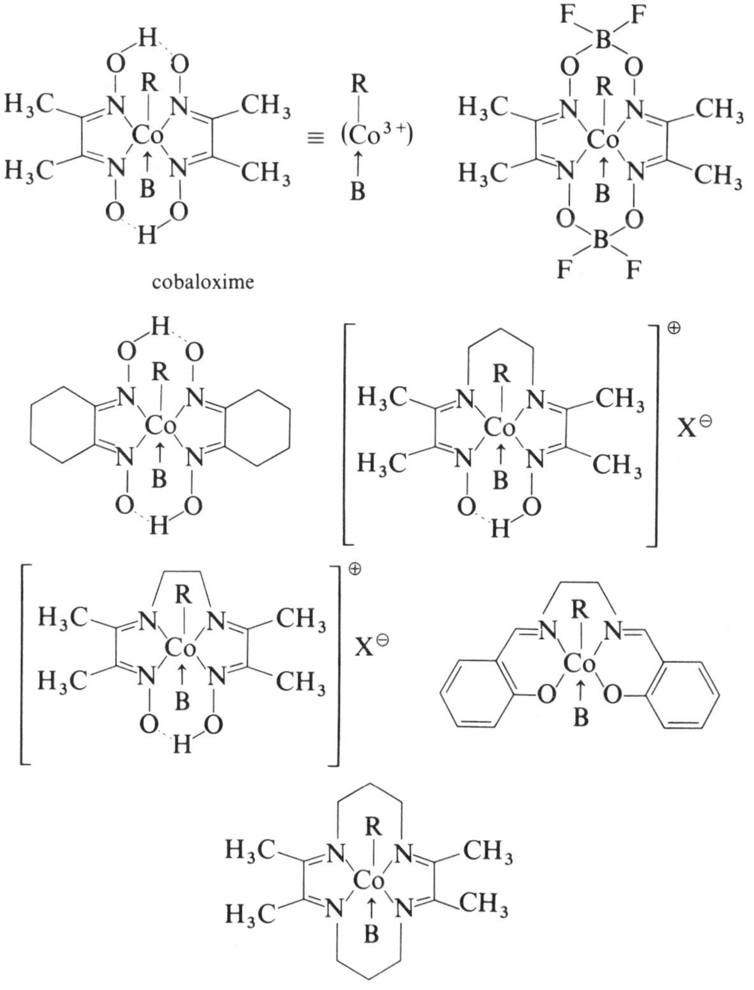

cobaloxime

$\equiv$ R–(Co^{3+})←B

Fig. 6.16. Some biomodels of the corrin system (554). Reproduced with permission. Copyright © 1976 by Verlag Chemie GmBH.

reduced to the Co(II) state, it reacts with alkyl halides to form alkyl cobaloxime, analogous in many properties and chemical reactivity to vitamin B_{12} coenzyme.

As seen previously, breaking the Co—C bond of the coenzyme B_{12} is a necessary step in the catalytic cycle. The evidence also suggests that the cleavage is *homolytic*, producing B_{12r} radical [Co(II) species and a C-5′-methylene radical of deoxyadenosine]. This cleavage can also be induced nonenzymatically both thermally and photochemically with chemically related cobaloximes.

$$\underset{\text{HO}\quad\;\text{OH}}{CH_2{-}CH_2} + \overset{CH_2R}{|}\,Co^{3+} \rightleftharpoons \underset{\text{HO}\quad\;\text{OH}}{CH_2{-}CH_2} + \overset{\cdot CH_2R}{}\;Co^{2+}$$

This homolytic cleavage results in the formation of a reactive intermediate that can abstract a hydrogen atom from the substrate to give the CH_3—C-5′-adenosyl intermediate (CH_3—R) and a substrate radical:

$$\overset{\cdot CH_2R}{\dot{C}o^{2+}} \rightleftharpoons \overset{HOCH_2{-}\dot{C}HOH \quad CH_3R}{\underset{B_{12r}}{\dot{C}o^{2+}}} \rightleftharpoons \overset{HOCH_2{-}CH(OH)\quad CH_3R}{\overset{|}{Co^{3+}}}$$

The combination of this radical with the B_{12r} species generates a new alkyl cobalamin with the substrate as ligand. We have thus accomplished a transalkylation of the cobalt atom. How does the Co—C bond become activated toward homolytic cleavage? It is believed that the presence and proper orientation of the propionamide side chains on the corrin ring are responsible for the ease of the enzymatic system, possibly by some distortion of the corrin (557). Support for this hypothesis comes from the fact that hydrolysis of a side chain to the corresponding acid results in an inactive coenzyme B_{12} molecule.* It is not known why nature proceeds by homolytic fission, a unique situation in coenzyme chemistry! Abeles was the first one to recognize the free radical nature of vitamin B_{12} chemistry.

However, J. Halpern (University of Chicago) argues that the strength of the cobalt–carbon bond and the forces acting to weaken it are of prime

---

* In a personal communication, N.T. Anh of Orsay, France, indicated that X-ray work on coenzyme B_{12} showed that the angle of the Co—CH_2—CH bond of the methylene 5′-adenosyl moiety is unusually wide. The peculiar hybridization of this carbon could account for the ease of breaking of the Co—C bond.

importance for such understanding. He obtains a value of 26 kcal/mol for this organometallic bond and believes that part of the bond strength depends on the basicity of ligand underneath the corrin structure (558). The weaker the basicity of a ligand in that position, the weaker the bond will be. As such, coenzyme B_{12} may act as a reversible carrier of free radicals, in the same sense that other organometallics like myoglobin is a reversible carrier of dioxygen.

Now the crucial point. How does the following rearrangement occur?

$$HO-\overset{\beta}{C}H_2-\overset{\alpha}{C}H(OH)-Co^{3+} \quad CH_3R \rightleftharpoons \overset{\beta}{C}H_2(-Co^{3+})-\overset{\alpha}{C}H{=}O \quad CH_3R + H_2O$$

Since OH^- has to be a leaving group, three extreme electronic forms of the resulting intermediate can be envisioned.

$$\overset{\beta\oplus}{C}H_2-\overset{\alpha}{C}H(OH)-Co^{3+} \qquad \overset{\beta}{C}H_2-\overset{\alpha}{C}H(OH)-Co^{3+}\ (\oplus) \qquad \overset{\beta}{C}H_2-\overset{\alpha}{C}H(OH)\ (\oplus\ Co^{3+})$$

primary carbonium ion — delocalized carbonium ion — π-complex

Model studies indicated that solvolysis of ^{13}C-labeled 2-acetoxy-ethyl(pyridine) cobaloxime in methanol gives equal amounts of the two ^{13}C-labeled products.

$$\text{Pyr}-Co^{3+}-CH_2-{}^{13}CH_2-O-\overset{O}{\overset{\|}{C}}-CH_3 \xrightarrow{CH_3OH} \text{Pyr}-Co^{3+}-CH_2-{}^{13}CH_2-OCH_3 + \text{Pyr}-Co^{3+}-{}^{13}CH_2-CH_2-OCH_3$$

This suggests that a π-complex intermediate is the simplest pathway for the rearrangement. Consequently:

enol intermediate ⇌ diol intermediate $\xrightleftharpoons{-H_2O}$ (aldehyde σ-complex)

Since the hydroxyl group on the α-carbon stabilizes the developing positive charge, a nucleophile (H_2O, or OH^-) will more likely attack this position to generate a new σ-complex on the cobalt ion with the β-carbon of the substrate.

In summary, the sequence of events taking place are, first, loss of the β-substituent from the cobalt σ-complex, followed by formation of a π-complex, and finally readdition of the leaving group on the π-complex to generate a new cobalt σ-complex with the substrate.

The transformation is completed by a second transalkylation that will regenerate the original coenzyme B_{12} and the dehydrated product.

$\dot{C}H_2$—CHO / $\dot{C}o^{2+}$ / CH_3R ⇌ $\dot{C}H_2R$ / $\dot{C}o^{2+}$ + CH_3—CHO ⇌ CH_2R—Co^{3+} (B_{12} coenzyme)

Therefore, the critical intermediate in B_{12} chemistry is a Co(III)-olefin π-complex. The formation of this intermediate is chemically reasonable because an enol is an electron-rich species whereas a metal (trivalent Co) is electron deficient.

An analogy to such a process is found in the heterogeneous polymerization of olefins (Section 4.1) where the existence of a σ–π complex between a metal ($TiCl_3$) and a double bond is well documented.

Further evidence for the σ–π rearrangement with the model system cobaloxime comes from Dolphin's work with electron-rich olefins, such as vinyl ethers (559,560). As expected, the quenching of the π-complex by a nucleophile occurs on the most positive carbon (Fig. 6.17).

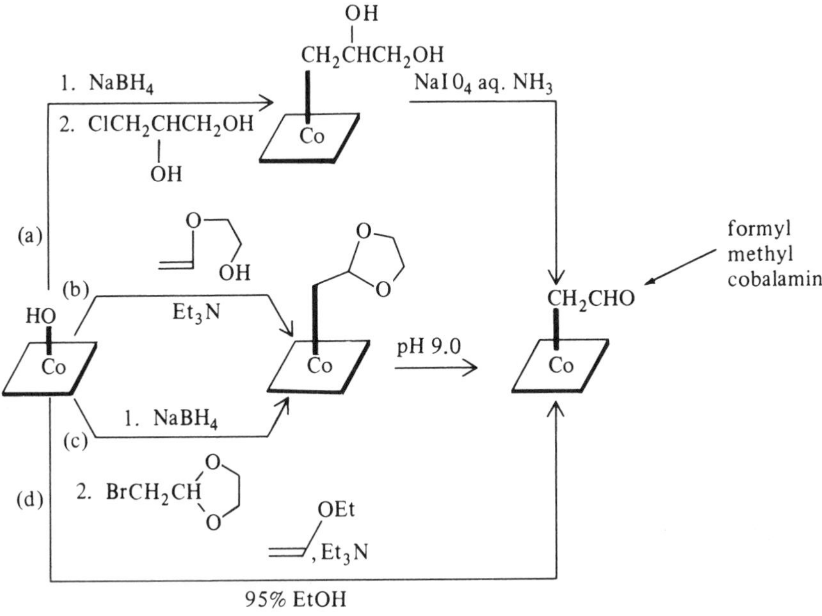

Fig. 6.17. Four different routes for the synthesis of formyl methyl cobalamin (561). Reprinted with permission. Copyright © 1974 by the American Chemical Society.

In these sequences, formyl methyl cobalamin, the intermediate in the enzymatic conversion of ethylene glycol to acetaldehyde, is synthesized in four different ways.

With the more elaborate systems, β-methylaspartate to glutamate (reaction 6-18) and methylmalonyl-CoA to succinyl-CoA (reaction 6-19), it has been proved that the largest group is the one to migrate, as illustrated below:

$$HOOC{-}CH(NH_2){-}CH(CH_3){-}COOH \rightleftharpoons HOOC{-}CH(NH_2){-}CH_2{-}CH_2{-}COOH \tag{6-18}$$

$$CoA\text{-}S{-}C(=O){-}CH(CH_3){-}COOH \rightleftharpoons CoA\text{-}S{-}C(=O){-}CH_2{-}CH_2{-}COOH \tag{6-19}$$

$$\text{HOOC}-C(=CH_2)-CH(CH_3)-\text{COOH} \rightleftharpoons \text{HOOC}-C(=CH_2)-CH_2-CH_2-\text{COOH} \quad (6\text{-}20)$$

In the first case the migration occurs with inversion whereas in the second case it occurs with retention of configuration.

For the system methylitaconate to α-methyleneglutarate (reaction 6-20), P. Dowd's group at the University of Pittsburgh made an interesting model to demonstrate the migration of the acrylate fragment in the carbon-skeleton rearrangement leading to α-methyleneglutaric acid (556).

The coenzyme model intermediate was synthesized by the reaction of vitamin B_{12s} [Co(I)] with bis(tetrahydropyranyl) bromomethylitaconate. On standing in aqueous solution the model yields rearranged α-methyleneglutaric acid together with unrearranged methylitaconic acid and butadiene-2,3-dicarboxylic acid. Since the model reaction provides no role for deoxyadenosine, its 5′-methylene being the instrument of

COOTHP
COOTHP
CH_2
Br
THP = tetrahydropyranyl
B_{12s}
COOTHP
COOTHP
CH_2
Co^{3+}
Zn/AcOD
^2H_2O
25°, dark
N_2 atom, 300 hr, pH 5–9

COOH
COOH
CH_2
butadiene-2,3-dicarboxylic acid (50%)

\+

COOH
COOH
CH_2
D
(nonrearranged product) (25%)

\+

COOH
CH_2
COOH
D
α-methyleneglutaric acid (rearranged product) (25%)

Zn/AcOD

COOH
COOH
Br
(used to prove the structure of rearranged product)

hydrogen transfer in all coenzyme B_{12}-dependent carbon-skeleton rearrangement reactions, it was important to learn the source of the hydrogen introduced into the products. This was accomplished by performing the reaction in ^2H_2O and analyzing the product by NMR and mass spectrometry. Both rearranged and nonrearranged products contained deuterium but not the butadiene-2,3-dicarboxylic acid. The position of labeling was further proved by Zn/AcOD reduction of the starting material.

It thus establishes that the C—Co bond in the model series is hydrolyzed by proton transfer from the solvent water. It should be recalled that in the corresponding enzymatic process, exchange with solvent water does not occur in the course of the rearrangement because of the presence of the 5′-deoxyadenosine moiety. However, if the assumption is made that the position of the deuterium is indicative of the position of the cobalt, prior to hydrolysis, then the presence of deuterium exclusively at the γ-carbon in this model requires that the acrylic acid moiety be the migrating group in the rearrangement reaction (556). Thus, this model system leads one to expect that in the enzyme-catalyzed rearrangement, it is the more complex group that migrates.

Whether the real coenzyme actually takes opportunity of this mechanism remains to be proven. The group of L. Salem (562) of the University of Paris-Sud proposed, on theoretical grounds, that in coenzyme B_{12} rearrangements the more electronegative group of the substrate should migrate preferentially. This is in accordance with Dowd's finding (556).

A further proof supporting Dowd's proposal came by studying the same system with ^{13}C-enriched butadiene-2,3-dicarboxylic acid (563). This investigation became necessary because if the carboxy group were the migrating group and if carboxy migration were followed by an interchange of hydrogen with cobalt, then the deuterium-labeling result would have been the same.

The ^{13}C-model compound was eventually decomposed, in aqueous solution, in the dark, at room temperature and at pH 8.3, and the rearrangement product α-methyleneglutaric acid was isolated. The proton-decoupled ^{13}C-NMR spectrum of the desired product showed a vinyl methylene carbon singlet at 126 ppm and a β-methylene carbon singlet at 28 ppm with no resonance at 33 ppm for the γ-position.

^{13}C * COOH COOH $\xrightarrow[AcOH]{HBr}$ * COOH COOH Br + COOH * COOH Br

↓

↓

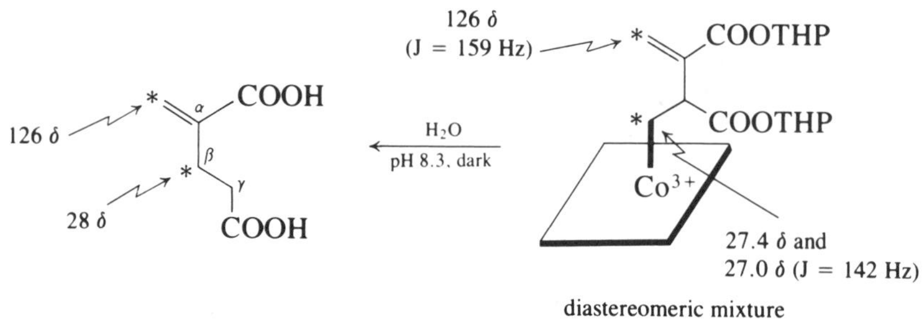

This result demonstrates beyond question that the acrylate group is indeed the migrating group in the model, as it is in the enzyme-catalyzed rearrangement.

It is also known that a malonic ester substituted in the β-position with cobalt can rearrange into the corresponding succinic ester. To test the possible role of the metal and to better simulate the situation of the enzyme, the group of J. Rétey (564) synthesized the following cobalt complex in which the "substrate" is anchored covalently by two methylene bridges to planar cobaloxime.

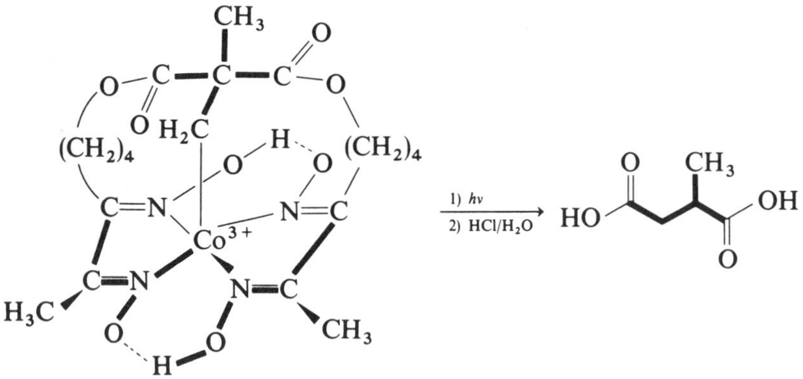

Irradiation in methanolic solution, followed by acid workup, gave the rearranged product in good yield. This observation strongly suggests that the homolytic rearrangement of the bound substrate to the product is promoted by the cobalt atom.

These extensive efforts on the biochemical reaction catalyzed by coenzyme B_{12} led to the following unified picture. First, coenzyme B_{12} inserts into an unactivated C—H bond of the substrate. The result is that the adenyl group of the coenzyme now carries a hydrogen atom of the substrate, while the substrate methylene group (in the case of methylmalonyl-CoA to succinyl-CoA) becomes bonded to the cobalt of B_{12}, replacing the original adenyl-B_{12} C—Co bond. Second, a rearrangement occurs within the B_{12}-substrate molecule to produce a B_{12}-product molecule. Finally, this undergoes the equivalent of a reverse of the first reaction in which the adenyl-cobalt bond of coenzyme B_{12} is again formed, and the reaction product is released carrying one of the hydrogen atoms of the 5′-deoxyadenosine methyl group.

In 1976, Breslow (565) examined various potential model systems of B_{12}-homolytic cleavage and found simple chemical transformations that furnish good precedent for the step involving insertion into an unactivated carbon. He made a biphenyl-cobaloxime compound that by irradiation generates the expected benzylic radical.

CH_3 CD_2 CH_3 Co^{3+} $\xrightarrow{h\nu}$ CH_3 $CD_2^{\cdot}$ CH_3 $\not\rightarrow$ CH_3 CD_2H $CH_2^{\cdot}$

However, the resulting radical dimerized without hydrogen transfer from the nearby methyl group. Breslow thus turned to a cyclodocecyl radical, a system in which intramolecular hydrogen transfer to a carbon-free radical had already been demonstrated. Again deuterium was used as a marker.

Since B_{12s} [Co(I)] is a powerful nucleophile, it reacts rapidly with the alkyl iodide to give a cyclododecyl-cobalamin. Upon bromination this affords cyclododecyl bromide with the deuterium marker distributed over several ring carbons.

D X

X = OTs, I $\xrightarrow{B_{12s}}$ D Co^{3+} $\longrightarrow$ D $\xrightarrow{Br_2}$ D Br

Transannular hydrogen transfer thus occurs in the overall sequence. Since the reaction of cyclododecyl iodide with B_{12s} is certainly not a simple displacement at carbon but also involves transannular hydrogen transfer via homolytic fragmentation, the insertion reactions common to all coenzyme B_{12} catalyses now seem to have a good chemical precedent (565).

However, convincing direct evidence for the intermediacy of carbon radicals in natural coenzyme B_{12} rearrangements is still lacking, and for this reason E.J. Corey proposed in 1977 an interesting and different mechanism, consistent with current knowledge of organometallic reactions (566). A key feature of the proposal is an electrocyclic opening of the corrin ring of the coenzyme, cleaving the unique direct chemical bond that joins rings A and D (see Fig. 6.14) and thus allows a role for the side chain of the corrin system. Corey argues that nature's construction of such direct linkage in a corrin system is both nonaccidental and essential. A rational stepwise mechanistic interpretation of the rearrangement taking place involves a cobalt–carbene complex with the substrate.

This interesting proposal is beyond the scope of the present discussion but puts into reevaluation the existence of a Co π-complex in coenzyme B_{12} rearrangements.

Because of the complexity of the enzymatic systems involved in coenzyme B_{12} chemistry, there are several reports on the purification of B_{12}-dependent enzymes or B_{12}-binding proteins by vitamin B_{12} affinity adsorbents. In fact, for purification of enzymes or proteins, *affinity chromatography* has been widely used as one of the most attractive methods (567). For that purpose, the synthesis of a cobalamin–Sepharose insoluble support has been prepared and applied to the purification of N^5-methyltetrahydrofolate-homocysteine cobalamin methyltransferase from *E. coli.* The scheme for the synthesis of the solid support is summarized in Fig. 6.18.

Here the apoenzyme* of methionine synthetase binds to the adsorbent and can be obtained as the holoenzyme by elution after the cleavage of the carbon–cobalt bond by light irradiation:

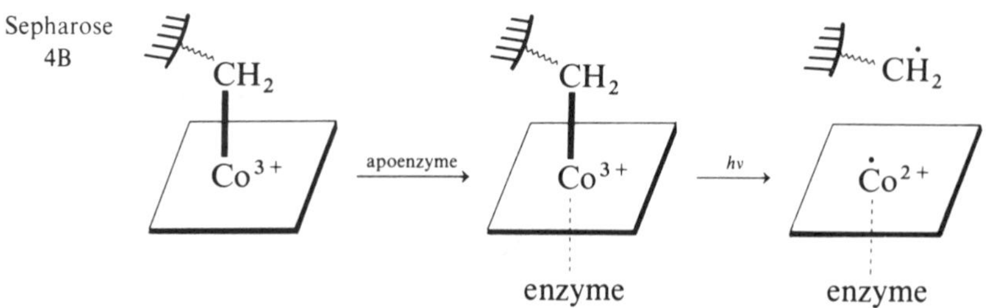

* In the classical terminology of enzymology, the complete enzyme–cofactor complex is termed *holoenzyme*, and the protein component minus its cofactor is termed an *apoenzyme*. The apoenzyme is generally inactive as catalyst (see Chapter 7).

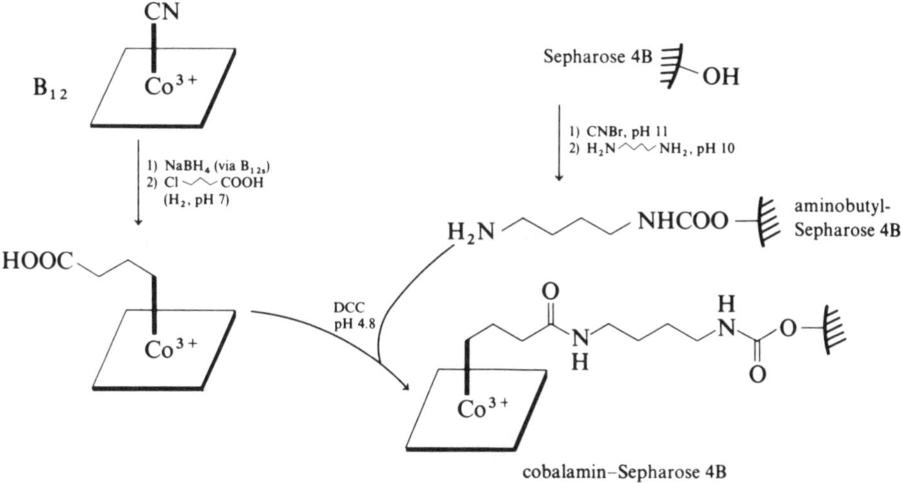

Fig. 6.18. Preparation of a cobalamin—Sepharose insoluble support for purification of B_{12}-dependent enzymes by affinity chromatography (531).

So far the affinity adsorbent is not reusable, but if a method is developed of eluting the protein without photolysis so that the adsorbent would be reusable, it might open the way for the wide application of this type of adsorption in the purification of B_{12}-dependent enzymes.

Hydrophobic vitamin B_{12} derivatives, which have long chain ester groups in place of the peripheral amide moieties of the naturally occuring vitamin B_{12}, have been prepared by Y. Murakami and coworkers (Kyushu University). The derivatives have been incorporated into vesicles in aqueous media to study the 1,2-carbon-skeleton rearrangement of vitamin B_{12} with these new model complexes (568,569). The migratory aptitude of the electron-withdrawing groups was found to follow the sequence: CN $\sim CO_2C_2H_5 < COCH_3$ (570).

Finally, because of the growing importance of mercury in the environment, a few words should be said about its methylation and its relationship with vitamin B_2. J.M. Wood, a biochemist then at the University of Illinois, was one of the first to study in detail the mechanism for methylation of mercury (571). The research of Wood proved that the biosynthesis of methylmercury could occur under anaerobic conditions. Historically, it was the discovery that inorganic mercury could be alkylated to methylmercuric ion in natural systems that pointed out the present dimension of the mercury pollution problem. Mercuric ion is an extremely strong methyl acceptor and the monomethyl derivative is stable in aqueous solution. Methylmercuric ion in aqueous solution can then be absorbed by unicellular organisms, thereby entering an aquatic food chain. The concentration builds up along the chain, reaching highly toxic levels in fish, and most strongly affecting those species that eat fish. Wood reasoned that B_{12}-containing enzyme systems should be capable of methylmercury synthesis in biological systems via *transmethylation*. This refers to the transfer of a methyl group from one substrate to another, usually as part of biochemical reaction. This process is also referred to as *biological methylation*. Three major coenzymes are available in nature to carry such methyl transfer reactions in biological systems: *S*-adenosylmethionine (SAM), N^5-methyltetrahydrofolate derivatives (N^5-CH_3-THF), and methylcorrinoid derivatives (methylcobalamin). The first two involve methyl group transfer as a carbonium ion. For the third one, Wood argued that methylcorrinoids are particularly efficient methyl transfer agents since they are capable of transferring methyl group as carbanions, carbonium ions, or radicals (see page 462):

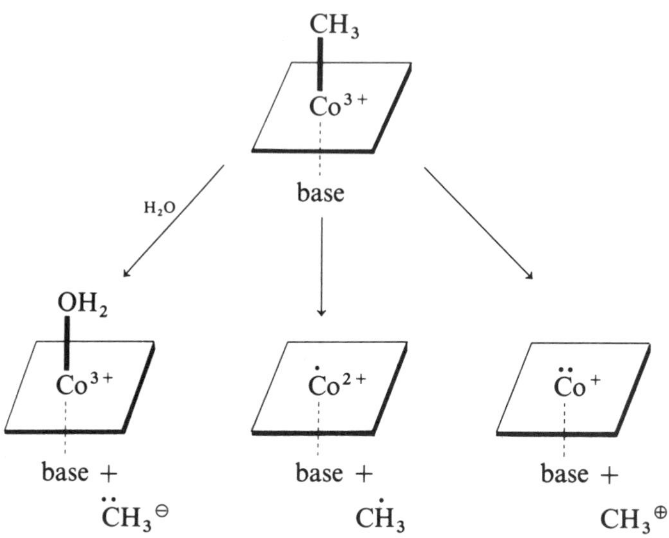

Among the enzymes implicated in mercury methylation are methionine synthetase, acetate synthetase, and methane synthetase. The last one is a very common enzyme in anaerobic ecosystems, such as lake and river sediments. These enzymatic processes can generate CH_3^+ and CH_3^- species where vitamin B_{12} derivatives act as mediator to methylate Hg(II) ions.

Furthermore, it has also been shown on models that when methylcorrinoids are photolyzed under anaerobic conditions, homolytic cleavage of the Co—C bond can also occur to give Co(II) and methyl radicals:

$$\underset{\text{(planar)}}{\overset{CH_3}{|}\;Co^{3+}} \xrightarrow[\text{anaerobic condition}]{h\nu} \dot{C}H_3 + \dot{C}o^{2+} \xrightarrow[H_2O]{O_2} \overset{OH_2}{|}\;Co^{3+}$$

$$\downarrow Hg^0$$

$$\dot{C}o^{2+} + CH_3{-}\dot{H}g \xrightarrow{CH_3} CH_3{-}Hg{-}CH_3$$

Dimethylmercury is then believed to be synthesized also by methyl radical addition to metallic mercury. Indeed, it seems feasible in certain organisms that Hg(II) is transported across cell membranes, reduced to metallic mercury, and then methylated. Dimethylmercury, being volatile, would readily diffuse out of the microbial cells and be released to the water. If the pH is acidic, it would be converted to monomethylmercury and methane:

$$\underset{\text{volatile}}{CH_3{-}Hg{-}CH_3} + H^{\oplus} \longrightarrow \underset{\text{soluble}}{CH_3\overset{\oplus}{Hg}} + CH_4$$

$$\downarrow h\nu$$

$$C_2H_6 + CH_4 + Hg(0)$$

Apparently, volatile methylmercury in the atmosphere is degraded to ethane, methane, and Hg(0) by irradiation. The metallic mercury then goes back to the soil and lakes and the "environmental mercury cycle" just referred to is repeated again (572,573).

Vitamin B_{12} and its derivatives are thus remarkable catalysts of versatile functions and are implicated in numerous unusual biochemical transformations in various organisms. The transformations of mercury in the environment is only one of the many facets of vitamin B_{12} action.

In concluding this chapter, it should be noted that we have presented various models of enzyme mechanisms in which metal ions participate. We have seen that reactions catalyzed by metalloenzymes or enzymes activated by metal ions span a remarkably broad spectrum of reaction types. Of course, many facets such as the remarkable speed and specificity of enzyme catalysis have not received complete explanation on the basis of model system chemistry. However, the missing link may be found at the point where biological molecules deviate from the model systems. Here, perhaps the most interesting chemical features will be found (553,554).

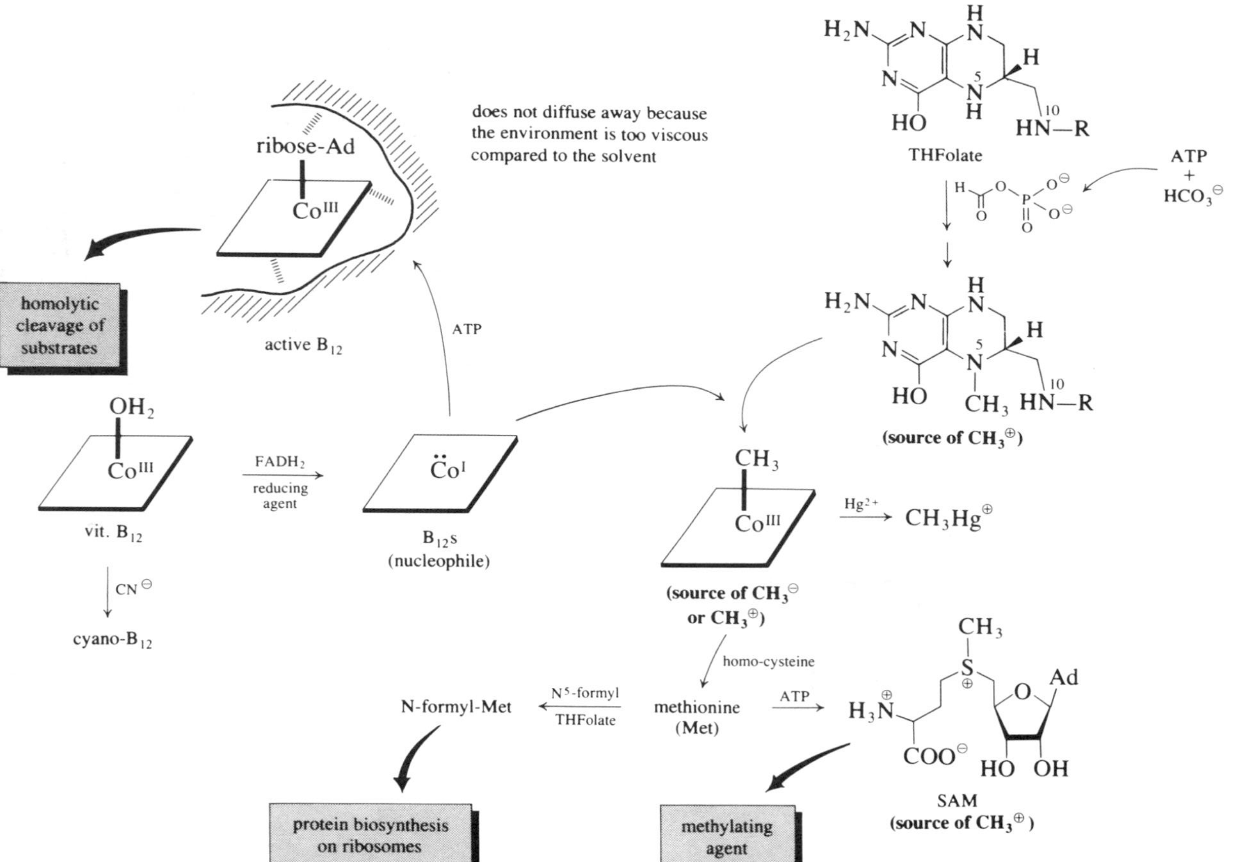

Fig. 6.19. Interrelation between vitamin B_{12} and other methylating agents in biology.

Finally, Fig. 6.19 on page 480 gives a global picture of the interconnection between other biologically active methyl donors and acceptors and vitamin B_{12}.

The tetrahydrofolate cofactor (THFolate) is considered as a "carrier for one-carbon units" and is composed of a 2-amino-4-hydroxy-pteridine nucleus, often called *pterin*.

Chapter 7
Coenzyme Chemistry

". . . it is by logic that we prove, but by intuition that we discover."

J.H. Poincaré

Cellular metabolism is under enzymatic control and often the enzymes involved need a substance or cofactor in order to express their catalytic activity. In these systems the protein portion of the enzyme is designated the *apoenzyme* and is usually catalytically inactive. The cofactor is a metal ion or a nonprotein organic substance. Many enzymes even require both cofactors. A firmly bound cofactor is called a *prosthetic group*. If, however, the organic cofactor is brought into play during the catalytic mechanism, it is referred to as a *coenzyme*. The complex formed by the addition of the coenzyme to the apoenzyme is referred to as a *holoenzyme* (or enzyme, for short).

Some coenzymes serve as carriers of chemical groups, hydrogen atoms, or electrons. Others such as ATP function in energy coupling reactions within the cell and are often regarded as a substrate rather than a coenzyme. Other coenzymes have more complex structures and are derivatives of *vitamins*. They act at the active site of the enzyme by combining with the substrate in a way that permits the reaction to proceed more readily. By definition, vitamins cannot be synthesized by the host and must therefore be provided by the diet. Their presence is thus required for normal growth and health and their absence causes specific or "vitamin deficiency" diseases.

As is often the case, coenzyme biochemistry leads to unconventional organic chemistry. In this regard, coenzymes are nature's special reagents and their well-defined structures make them ideal molecules to use for developing the concept of structure–function relationships by bioorganic approaches (574). This chapter will thus be concerned with this aspect and special attention will be devoted to model design of coenzyme action.

7.1 Oxidoreduction

Enzymatic oxidoreduction sequences lie at the heart of cellular energy metabolism. The energy released in oxidation of reduced organic or inorganic compounds is captured with varying efficiencies in useful forms such as ATP, membrane potentials, or reduced coenzymes. Because of their physiological role, the mechanism of action of enzymes catalyzing electron transfer processes has been actively studied.

It should be recalled that redox enzymes or *oxidoreductases* are divided into four main classes.

(a) *Dehydrogenases* are enzymes that catalyze the transfer of a hydrogen from one substrate to another. This reversible system normally needs a nicotinamide coenzyme such as NAD^+. Chiral alcohols are made from the corresponding ketones or aldehydes this way. Some dehydrogenases contain an essential atom of Zn(II) (575).

(b) *Oxidases* are enzymes that catalyze the transfer of a hydrogen from a substrate to molecular oxygen. A flavin coenzyme, FAD, is usually required. Water or hydrogen peroxide may be the end product, depending on the enzyme specificity.

(c) *Oxygenases* catalyze the incorporation of oxygen, from molecular oxygen, into a substrate. *Monooxygenases* incorporate one atom of molecular oxygen into the substrate; the other appears as H_2O_2 or H_2O. *Dioxygenases* insert both oxygen atoms into the substrate.

(d) *Peroxidases and catalases* both catalyze reactions involving hydrogen peroxide as oxidant. Water, the reduced form, is produced. These enzymes require ferric and/or cupric ions for catalysis.

7.1.1 NAD^+, $NADP^+$

Nicotinamide adenine dinucleotide (NAD^+) and nicotinamide adenine dinucleotide phosphate ($NADP^+$) are pyridine nucleotides, first identified by O. Warburg in 1935. NAD^+ is the coenzyme involved in dehydrogenase reactions and is reduced to NADH during the process.

The reaction is stereospecific and only one isomer (the α- or *R*-isomer in the above example) of NADH is produced. Other dehydrogenases with a different specificity are known to yield the β-isomer stereospecifically rather than the α-isomer.

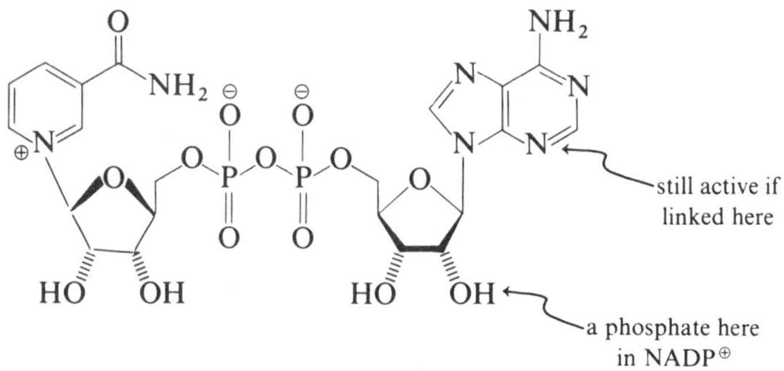

still active if linked here

a phosphate here in $NADP^{\oplus}$

$NAD^{\oplus}$

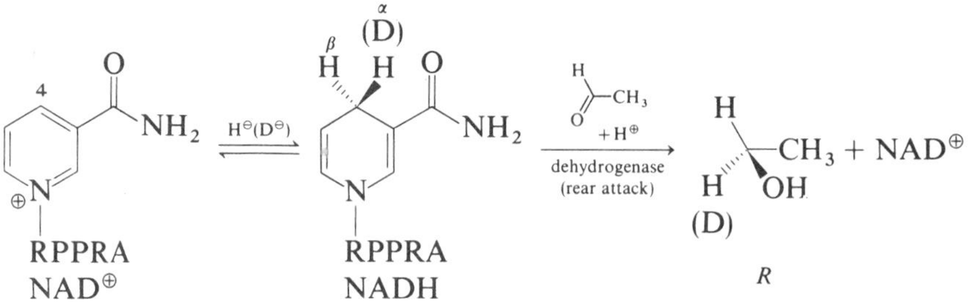

$H^{\ominus}(D^{\ominus})$

dehydrogenase (rear attack)

RPPRA NAD$^{\oplus}$ RPPRA NADH

RPPRA = ribose-phosphate-phosphate-ribose-adenine

Theoretical calculations of charge distribution on the nicotinamide ring by B. Pullman in Paris agree with a hydride attack at position C-4. Experimental evidence comes from reaction with a deuterated alcohol

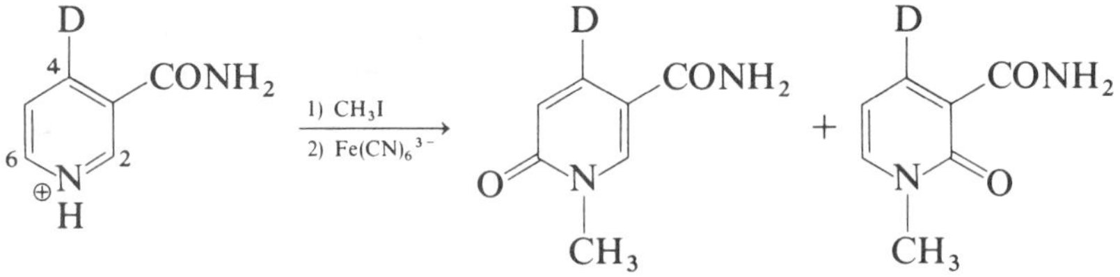

1) CH_3I
2) $Fe(CN)_6^{3-}$

and isolation of the pyridinium salt. Methylation and oxidation take place without loss of deuterium, proving that the labeling occurs only at C-4.

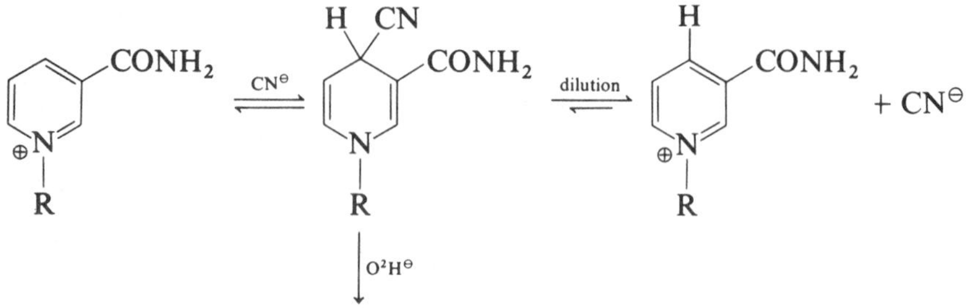

$CN^{\ominus}$ dilution $+ CN^{\ominus}$

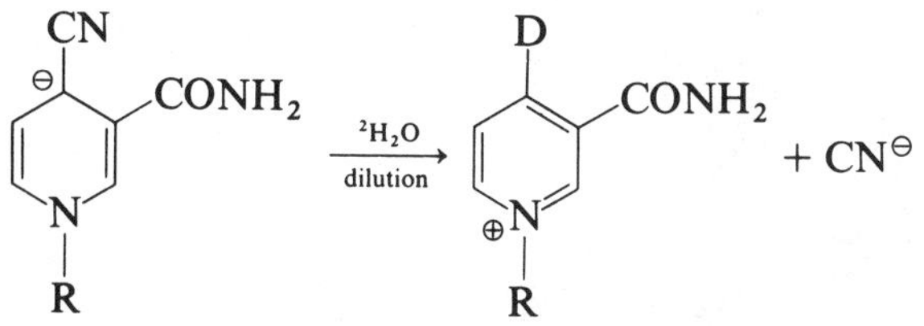

Additional reactions show that a good nucleophile such as cyanide ion can add to NAD^+. It can then be lost by dilution to give the rearomatized product. Basic treatment in ^2H_2O affords the 4-*d* analog of NAD^+.

The specific preparation of the deuterated analog demonstrates the electrophilic character of position C-4. In this regard, bisulfite addition at pH 7 is easily obtained and a dithionite treatment of NAD^+ leads to a reduction reaction at C-4.

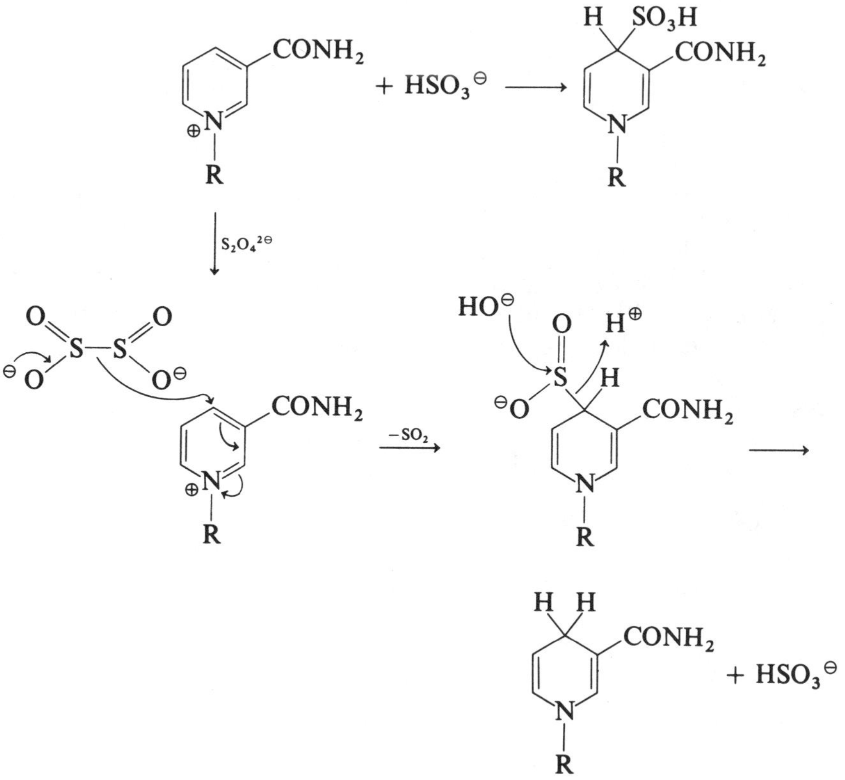

Nucleophilic addition to NAD^+ was also demonstrated in the case of pyruvate at alkaline pH. Addition presumably occurs via the enol (or the corresponding enolate) to give the reduced and oxidized adducts (576).

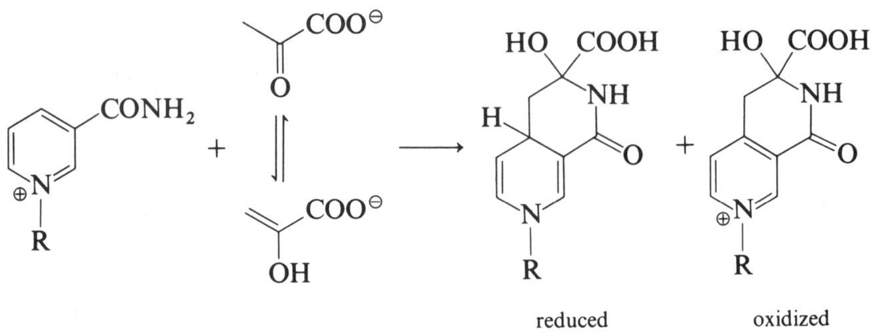

reduced oxidized

These data suggest the following equilibrium:

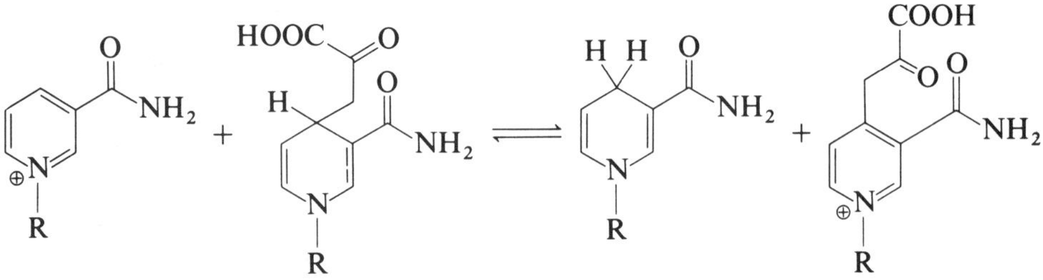

In the NADH $\rightleftharpoons$ NAD$^+$ + H$^-$ reaction, what is the evidence for the direct transfer of a hydride ion?

D.J. Creighton and D.S. Sigman (577) made, in 1971, the first model for an alcohol dehydrogenase. They used a zinc ion to catalyze the reduction of 1,10-phenanthroline-2-carboxaldehyde by *N*-propyl-1,4-dihydronicotinamide, a simple model compound of NADH.

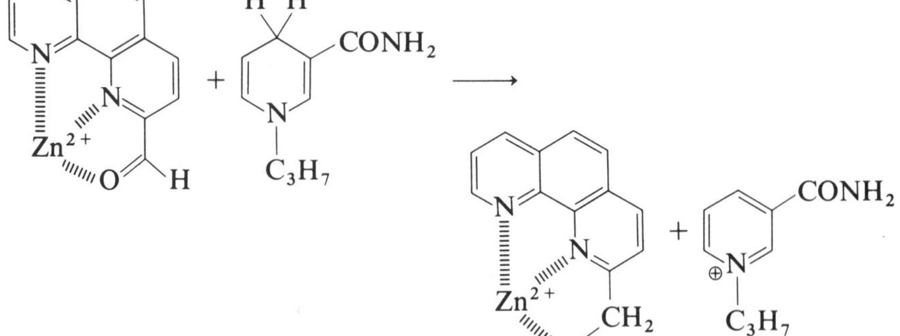

Interestingly, in acetonitrile at 25°C the presence of Zn(II) was found essential for an efficient catalysis of this reaction. Coordination of the metal ion to the substrate is thus a driving force for the reduction to take place. This was the first example of a reduction of an aldehyde by an

NADH analog in a nonenzymatic system. If 1-propyl-4,4-dideuterionicotinamide is used, monodeuterated 1,10-phenanthroline-2-carbinol is produced. This result demonstrates that the product is formed by direct hydrogen transfer from the reduced coenzyme analog. It strengthens the view that coordination or proximity of the carbonyl to the zinc ion is probably important in the enzymatic catalysis. Zn(II) ion is known to be essential for the catalytic activity of horse liver alcohol dehydrogenase (575).

Surprisingly, if the proximity model below is used to mimic NAD^+ reduction, no transfer occurs.

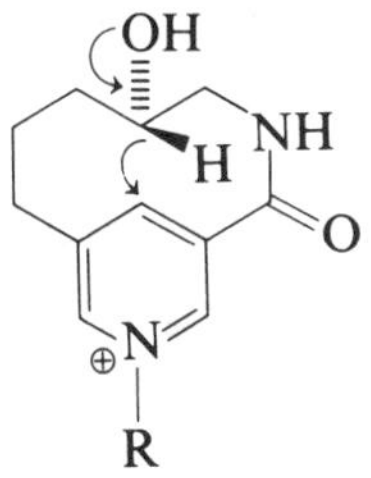

Although favorable factors are present, the system prefers to remain aromatic. Hence, the formation of NADH in the enzymatic system could be driven by conformational changes that shift the equilibrium toward the nonaromatic species. However, in 1978, a German group (578) observed an *intra*molecular hydride transfer in the presence of pig heart lactate dehydrogenase using a coenzyme–substrate covalent analog composed of lactate and NAD^+.

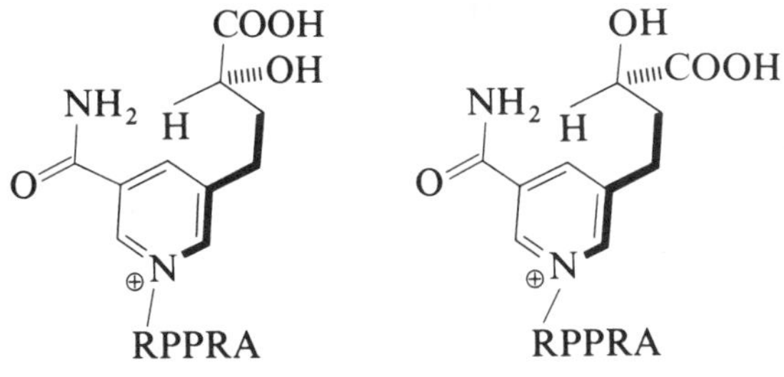

S-isomer

R-isomer

The two synthetic diastereomeric nicotinamide adenine dinucleotide derivatives are attached via a methylene spacer at position 5 of the nicotinamide ring. Only the *S*-isomer undergoes the intramolecular hydride transfer, forming the corresponding pyruvate-nicotinamide analog and NADH. Two (*R*)-lactate specific dehydrogenases, however, do not catalyze a similar reaction with either one of the two diastereoisomers. Consequently a possible arrangement of the substrates (lactate and pyruvate) at the active centers of these enzymes can be proposed:

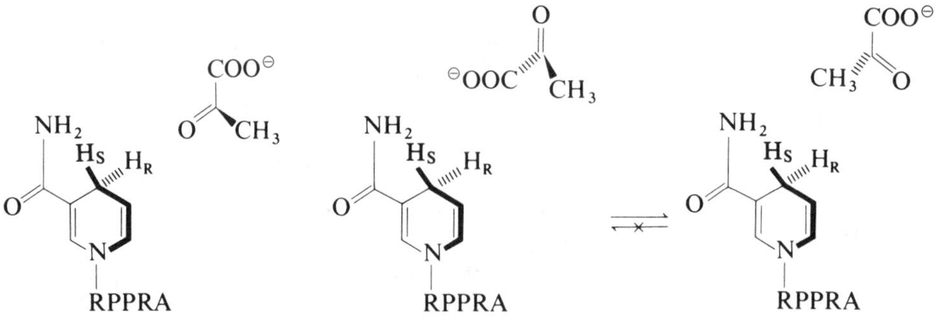

possible arrangement of pyruvate in the active center of (*S*)-lactate-specific dehydrogenase allowing for the transfer of the *pro-R* hydrogen from the dihydropyridine ring; this orientation agrees with X-ray data for ternary inhibitor complexes

this arrangement is not favored since no hydride transfer occurred with (*R*)-lactate-specific dehydrogenase

C_2–C_3 rotation gives this new orientation of pyruvate; this arrangement is postulated for (*R*)-lactate-specific dehydrogenase

Models for the reverse reaction (NADH → NAD^+) have also been prepared. An interesting example is the following crown ether NAD(P)H mimic (579).

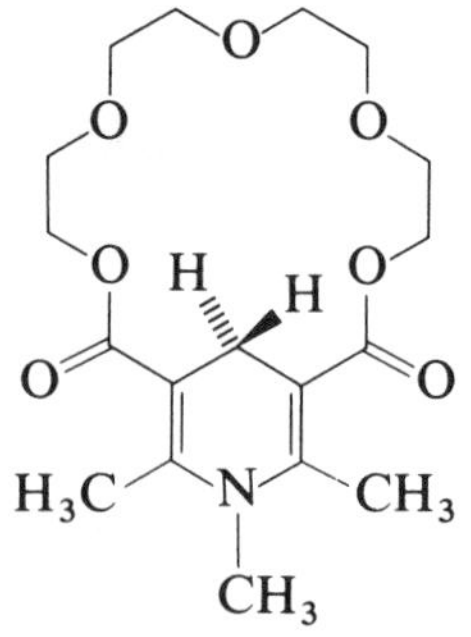

In the presence of sulfonium salts, transfer of hydride occurs 2700 times faster than with the corresponding analog without the crown ether ring.

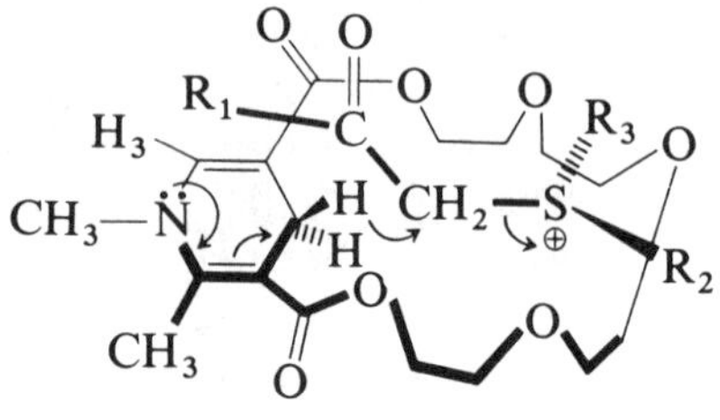

Obviously, the presence of the crown ether allows proper complexation of the cation.

This work has been extended to chiral crown ether analogs (580). The presence of the two isopropyl groups forces the phenyl group of ketoester substrates to be oriented over the dihydropyridine moiety of the receptor. In this orientation, the *re*-face of the carbonyl is the one properly exposed for acceptance of a hydride transfer, resulting in an *S*-alcohol in 86% ee. The presence of a Mg^{2+} ion helps to organize the structure of the macrocycle even if it complexes only relatively weakly.

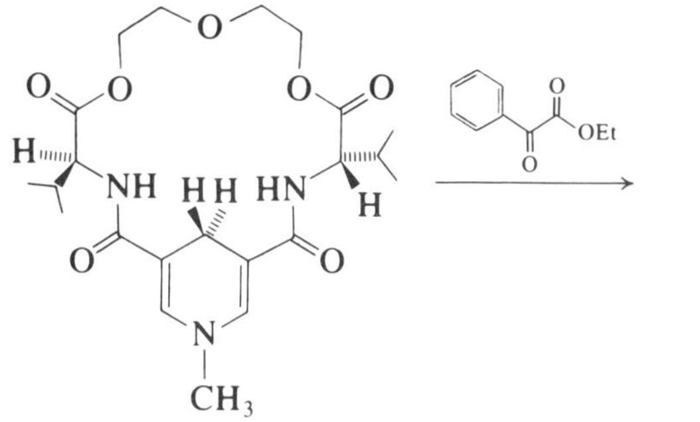

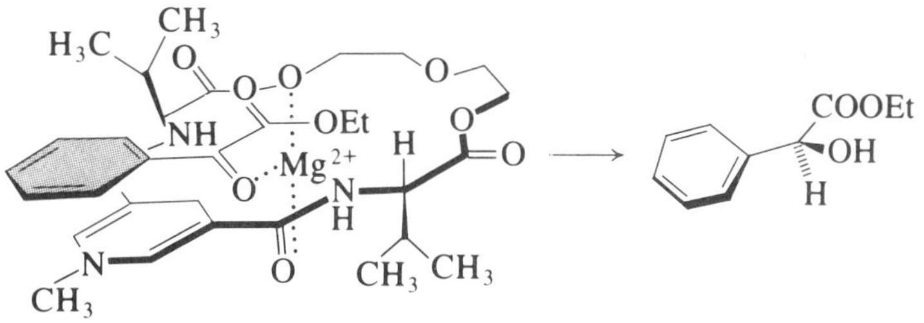

A nonenzymatic stereospecific intramolecular reduction by an NADH mimic containing a covalently bound substrate has been carried out by Meyers's group (Colorado State University). It also offers an *intramolecular* version of the Mg^{2+}-ion-mediated reduction. In the presence of $Mg(ClO_4)_2$ in THF, the benzoylformate ester of the dihydropyridine carbinol model compound gave after four days the corresponding *S*-alcohol with a stereocenter transfer greater than 99% (581).

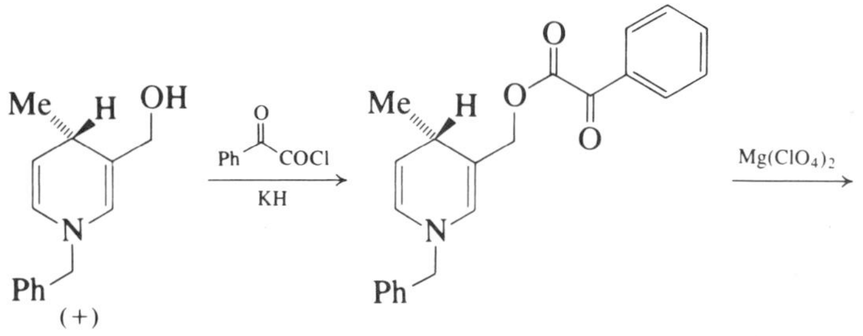

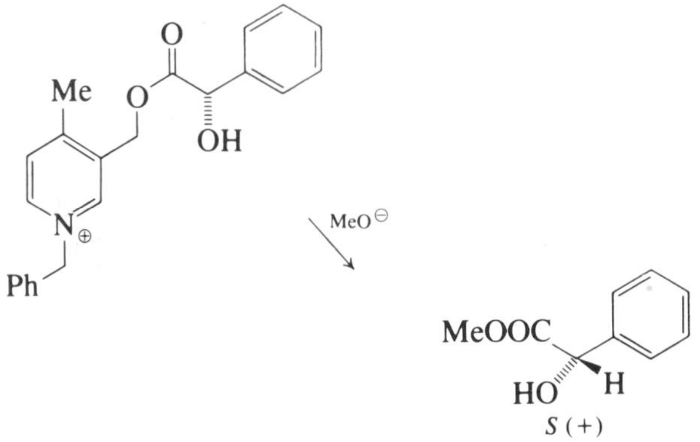

MeO⊖

MeOOC HO H S(+)

This stereochemical outcome can be rationalized by assuming the presence of the following pretransition-state intermediate. Its structure is reasonably strain free, incorporates the metal ion comfortably, and allows the pyridine ring to pucker as it is known to exist. Furthermore, in this form electron donation to the hexa-coordinated magnesium comes from both electron-rich carbonyl groups and the lone pair of electrons of the nitrogen atom. The other ligands (ClO_4^- and solvent molecules) are not shown on the structure.

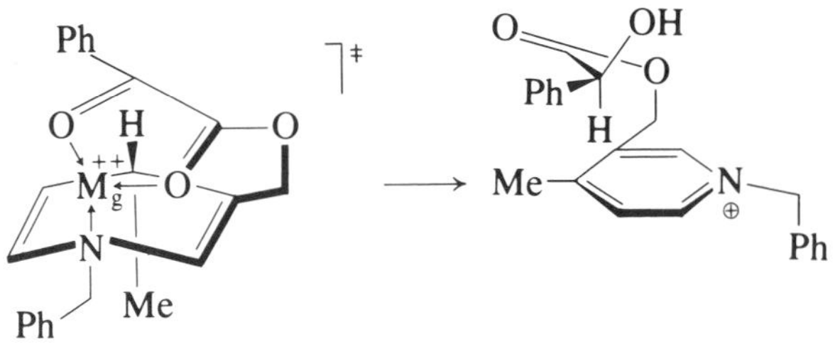

Another model was developed, by the group of J.-M. Lehn (582) of Strasbourg. They prepared a complex between a pyridinium substrate and a chiral crown ether receptor molecule having four dihydronicotinamide derivatives as the side chain (Fig. 7.1).

In this complex, enhanced rate of H-transfer to the bound pyridinium salt substrate is observed. This represents the first example of accelerated 1,4-dihydropyridine to pyridinium H-transfer (transreduction) in a synthetic molecular macrocyclic receptor–substrate complex. Therefore, such a synthetic catalyst displays some of the characteristic features that molecular catalysts should possess. It provides both a receptor site for substrate binding and a reactive site for transformation of the bound substrate. Consequently it is of interest as both an enzyme model and as a new type of efficient and selective chemical reagent (582).

Models incorporating a cyclodextrin unit have also been exploited. In one instance Kojima prepared two β-cyclodextrin-nicotinamide derivatives, an axial and an equatorial isomer on secondary hydroxyl sites

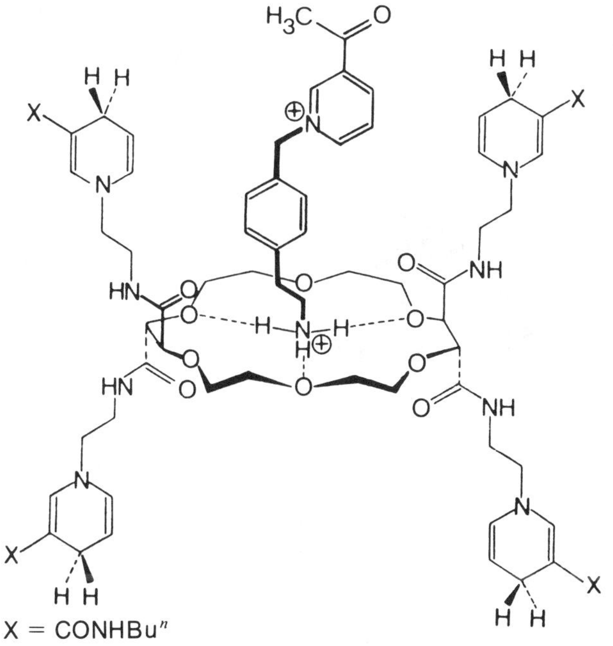

X = CONHBu^n

Fig. 7.1. Lehn's crown ether model of NADH (582). Reprinted with permission. Copyright © 1978 by the Chemical Society.

(583). The corresponding reduced forms can reduce the aromatic ketone ninhydrin with rate enhancements up to 60-fold relative to NADH. Enzymelike saturation kinetics indicate that the reaction involves the formation of a host–guest complexation.

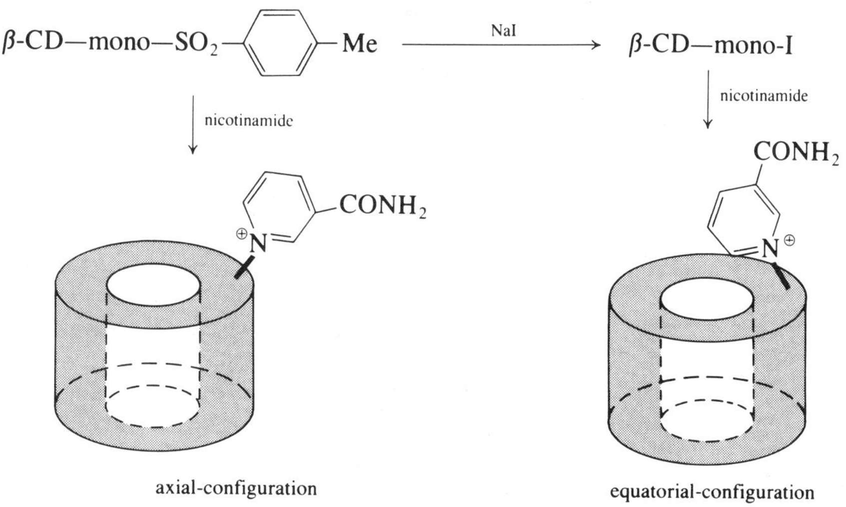

β-CD—mono—SO_2—C_6H_4—Me —NaI→ β-CD—mono-I

nicotinamide

$CONH_2$

axial-configuration

equatorial-configuration

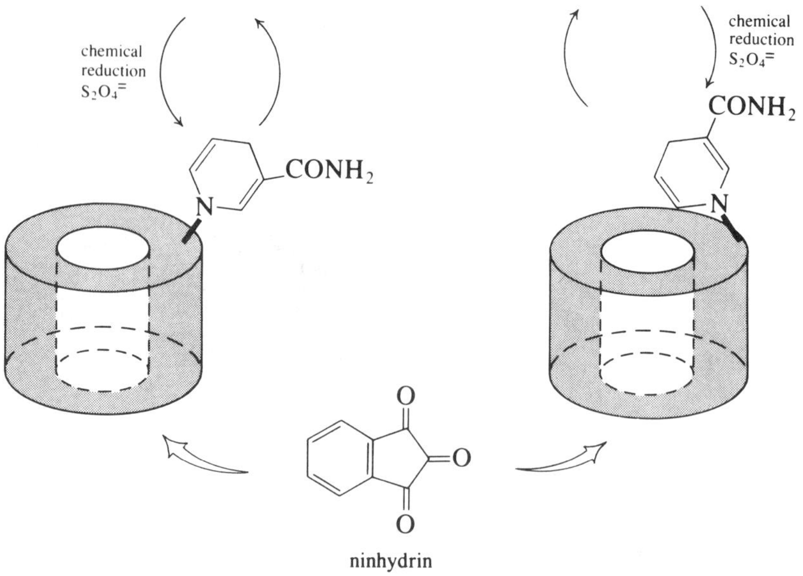

chemical reduction $S_2O_4^{=}$

$CONH_2$

N

chemical reduction $S_2O_4^{=}$

$CONH_2$

N

O

O

O

ninhydrin

When designing models of coenzymes it is important to examine the structure of the corresponding enzymes. The three-dimensional structure of horse liver alcohol dehydrogenase was resolved at 0.24 nm resolution by the group of C.I. Brändén from Uppsala (575) and was correlated to a number of physical and chemical studies in solution. The active enzyme has a molecular weight of 80,000 and is a dimer of two identical subunits. Each subunit is composed of a single polypeptide chain of 374 amino acids. Furthermore, each subunit is divided into two separate domains, each associated with a particular function. Three is a coenzyme (NAD^+)–binding domain where the adenine part of NAD^+ is oriented in a hydrophobic pocket and there is a catalytic domain that binds different substrate molecules. Three residues, Cys-46, His-67, and Cys-174, within this domain provide ligands to the catalytic zinc atom. Interestingly, a second zinc atom is present in this region and is liganded to four other cysteine residues. The function of this extra zinc atom remains unknown, but it has been suggested that it might be essential for the structural stability of the enzyme. The similarities with ferredoxin (Section 6.4.2) also suggest that this region might be a catalytic center for a redox process possibly with one or two of the cysteine ligands as catalytic group (575). Finally, the two domains of the subunit are separated by a crevice that contains a wide and deep hydrophobic pocket with the catalytic zinc atom at the bottom of this pocket.

Obviously, the system is more complex than previously thought since the coenzyme and the substrate are bound in different domains on the enzyme. Future models of NAD^+-dependent enzymes will thus have to take these observations into consideration.

Last, let us consider the possibility of a mechanism other than a hydride transfer in NAD^+ chemistry. Indeed, G.A. Hamilton argued that if a direct hydride transfer process occurs in dehydrogenase reactions, it is unique in biology since proton transfer would be more favorable (584). However, it is not a simple task to distinguish between these two possibilities. Generally, it is simpler to say that the reduction reaction is analogous to a transfer of two electrons rather than postulating a hydride ion. More will be said on this subject in Section 7.1.3 on flavin coenzyme.

7.1.2 Nonenzymatic Recycling of Coenzymes and Some Applications in Organic Synthesis

In recent years, the requirements of synthetic chemists for reagents capable of effecting selective or asymmetric transformations have increased dramatically. Enzymes present unique opportunities in this regard, and the exploration of their properties as chiral catalysts is now receiving considerable attention. Of course, one of the great synthetic attractions of an enzyme is that the various facets of its specificity can endow it with the potential for effecting highly controlled and selective transformations in a single step (585).

The oxidoreductase of most current utility is horse liver alcohol dehydrogenase (HLADH), and its application in the stereoselective oxidoreduction of ketones and alcohols has been explored. It requires an NAD^+ coenzyme which becomes expensive if reactions have to be performed on a preparative (up to 5 g) scale. In fact, this cofactor sells for as much as $250,000 per mole!

To overcome this problem, methods of recycling this coenzyme were developed. One of these is a nonenzymatic method, studied by J.B. Jones (586–588) of the University of Toronto, for regenerating catalytic amounts of the nicotinamide coenzyme continuously. (It should be recalled that Section 4.7, devoted to enzyme technology and its application to chemistry and medicine, has been presented.)

The method of Jones uses sodium dithionite to regenerate NADH from catalytic NAD^+. It has been shown to be preparatively viable for HLADH-catalyzed reductions of a broad range of aldehydes and ketones in high yields.

ketone $R_2C{=}O$ → alcohol $R_2C(OH)H$ — enzymatic (HLADH): NADH → $NAD^{\oplus}$

$NAD^{\oplus}$ → NADH — chemical ($S_2O_4^{2\ominus}$): $S_2O_4^{2\ominus} + OH_2$ → $HSO_3^{\ominus} + SO_2$

O_2 — $FMNH_2$ ← → $NAD^{\oplus}$ — R,R C(OH)(H) alcohol

chemical (flavin) — enzymatic (HLADH)

H_2O_2 ← → FMN — NADH ← → $R_2C{=}O$ ketone

For the opposite mode (oxidation of alcohol to ketone) the system needs to be coupled to a flavin cofactor ($FMNH_2 \rightleftharpoons FMN$) where oxygen is reduced to H_2O_2. The practicability of applying HLADH as a chiral oxidoreduction catalyst was demonstrated in many instances. Two examples are as follows:

(±)-2-nonbornanone —HLADH, 20°C, pH 7; NADH recycling; 46% reduction, 5 hr→ (+)-(1*S*,2*R*,4*R*)-endo exclusively, 64% opt. pure, 39% yield + (−)-(1*R*,4*S*) 47% opt. pure, 31% yield

(±)-bicyclo[3.2.1]-2-octanone —HLADH, 20°C, pH 7; 25% reduction, 11 hr→ (−)-(1*S*,2*S*,5*S*)-exo (difficult to get otherwise) 83% opt. pure, 36% yield + (−)-(1*R*,5*R*) 17% opt. pure, 58% yield

Both reactions are performed with a high degree of stereospecificity. These stereochemical results are interpretable in terms of the diamond lattice section approach of V. Prelog (589), using the composite model shown in Fig. 7.2.

By fitting a cyclohexanone in the diamond lattice, Prelog developed a step-by-step analysis of the HLADH-catalyzed reduction. The positions marked ● (A to D) are "forbidden;" oxidoreduction will not take place if binding of a potential substrate places a group in one of these locations. Positions marked ○ (E to G) are "undesirable," although their occupation by part of a substrate does not necessarily preclude the oxidoreduction. The rate of reaction will be very slow. The positions under the lattice (U) are also in this category. The location ○ (I) is a newly identified "unsatis-

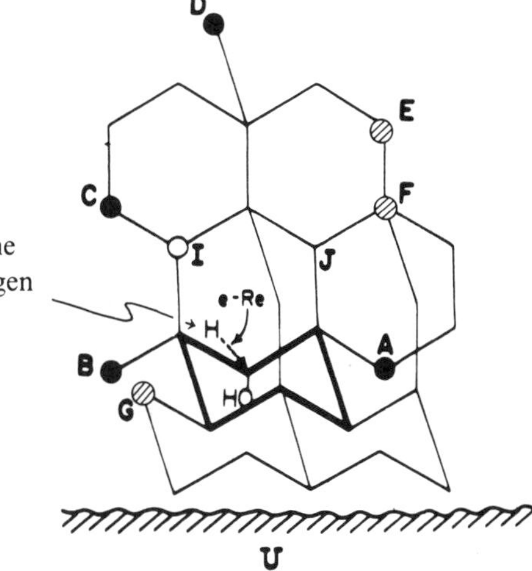

Fig. 7.2. Diamond lattice section for HLADH, as developed originally by Prelog (586). Reproduced with permission. Copyright © 1978 by the American Chemical Society.

factory" position. Placement of a group here is to be avoided if possible, but slow reaction will still take place if it is occupied.

Returning to the previous two examples, orientations of the enantiomers in their preferred "flat" positions within the diamond lattice section of HLADH allows an easy interpretation of the observed stereospecificity. This is illustrated in Figs. 7.3 and 7.4.

Jones also examined the oxidation of a dihydroxycyclopentene and found that HLADH has the ability to retain its enantioselectivity while effecting regiospecific oxidation of only one of two unhindered hydroxyl groups within the same molecule.

OH, OH (±) —HLADH, 20°C, pH 9; $NAD^{\oplus}$ recycling; 40% oxidized, 4 hr→ O, =O + OH, OH

(+)-(1*R*,2*S*) 49% opt. pure 48% yield

(−)-(1*S*,2*R*) 23% opt. pure 35% yield

This provides a synthetically useful combination of properties that cannot be duplicated in a single step by traditional chemical oxidation methods. Figure 7.5 (page 498) shows that the regiospecificity observed is as predicted by the diamond lattice section of the active site.

The use of rigid molecular models is thus considered essential when analyzing or interpreting substrate behavior in diamond lattice terms. It

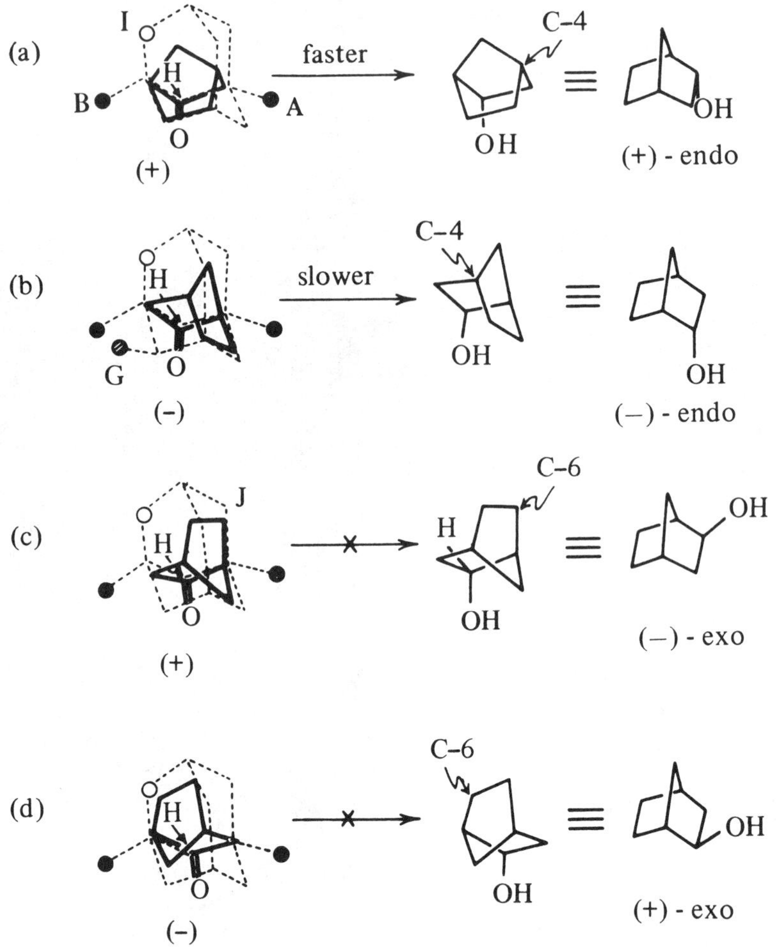

(a) I B A H O (+) faster C-4 OH OH (+) - endo

(b) H G O (-) slower C-4 OH OH (−) - endo

(c) J H O (+) C-6 H OH OH (−) - exo

(d) H O (-) C-6 OH OH (+) - exo

Fig. 7.3. Diamond lattice section analysis of the stereochemical course of HLADH-catalyzed reduction of the 2-norbornanone enantiomers. The orientations shown are considered to be those of the "alcohollike" transition states that would be involved, with H being delivered from the e-*re* direction of Fig. 7.2. There are no unfavorable lattice interactions when (+)- or (−)-isomers are positioned as shown in (a) and (b), respectively. Reductions to the endo-alcohols (+)- and (−)- are thus permitted processes. The predominant formation of the (+)-enantiomer is thought to be due to the preference of C-4 for its unhindered location in (a) over its position in (b) in which it approaches the forbidden *A,I,C* region of the lattice. The discrimination of HLADH against exo-alcohol formation is accounted for by the unfavorable interactions of C-6 with lattice positions J and I, as represented in (c) and (d), respectively (586). Reproduced with permission. Copyright © 1978 by the American Chemical Society.

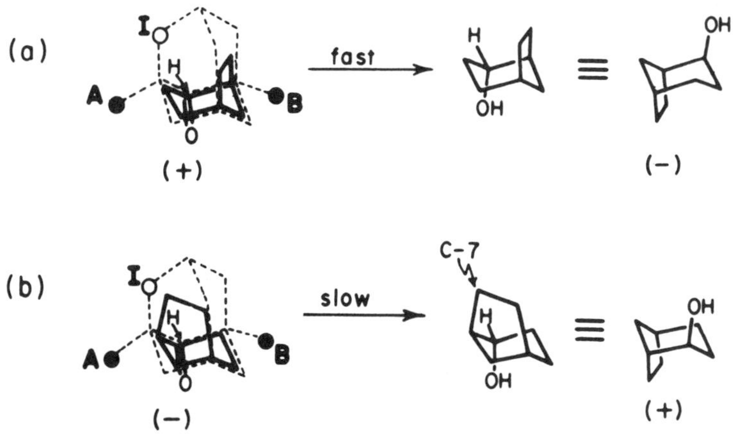

Fig. 7.4. Schematic representation of the orientations of the enantiomers of bicyclo[3.2.1]-2-octanone in their preferred "flat" positions within the diamond lattice section of HLADH. Delivery of H to the carbonyl group from the e-*re* direction ensures the formation of an exo-alcohol. In (a) orientation of (+)-isomer as shown does not place any substituents at undesirable positions. Reduction to the observed (−)-product is thus facile. The corresponding orientation of the enantiomeric (−)-ketone is shown in (b). Here, C-7 is required to locate close to the unsatisfactory position I and reduction in this mode is not favored. Formation of (+)-product is thus a relatively slow process (586). Reproduced with permission. Copyright © 1978 by the American Chemical Society.

is hoped that this approach and others with different enzymes and coenzymes will attract more organic chemists to use enzymes in difficult steps of a complicated organic synthesis.

In 1982, Jones introduced a new and more reliable cubic-space section model for predicting the specificity of HLADH-catalyzed oxidoreductions of organic ketones, based on new available kinetic and X-ray data of the enzyme (590). The reader is also invited to consult Section 4.8 for other examples of the utilization of enzymes in synthetic organic chemistry.

The field of synthetically useful enzymes was reviewed in 1993 (591).

7.1.3 Flavin Chemistry

Many hydrogenation–dehydrogenation processes are mediated by a flavin adeninedinucleotide (FAD) coenzyme where two electrons (equivalent to a hydride ion) form NADH are carried over to the respiratory chain (592). Flavins are the means of coupling one and two electron redox systems to cytochromes in the respiratory chain where oxygen is ultimately reduced to water. Like NAD^+, FAD is a coenzyme that transfers electrons from a substrate, but in a much more versatile way. The structure of

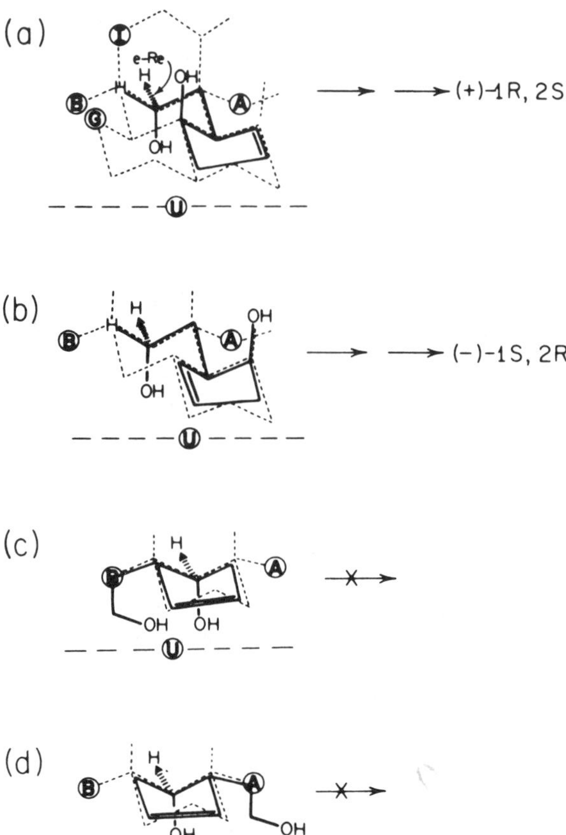

Fig. 7.5. Diamond lattice analysis of the regiospecificity of the cyclopentene diol. The relevant portions of the lattice, including the forbidden or undesirable positions A, B, G, I, and U are indicated by the dashed lines. When the hydroxyethyl groups of (+)- or (−)-compounds are oriented such that removal of the *pro-R* hydrogen from the e-*re* direction can occur as required by the model, unfavorable interactions with the lattice are completely avoidable. Substrate placements meeting these requirements and leading to the observed products are depicted in (a) for (+)-isomer and (b) for (−)-isomer. In contrast, positioning of the secondary alcohol function of (+)- or (−)-starting materials in a manner permitting hydride removal from the e-*re* direction would compel a methylene group of the hydroxyethyl function to violate position B or A, respectively, as shown in (c) for (+)-isomer and (d) for (−)-isomer. Oxidation of the C-1 hydroxyl groups is thus precluded for both enantiomers (587). Reproduced with permission. Copyright © 1977 by the American Chemical Society.

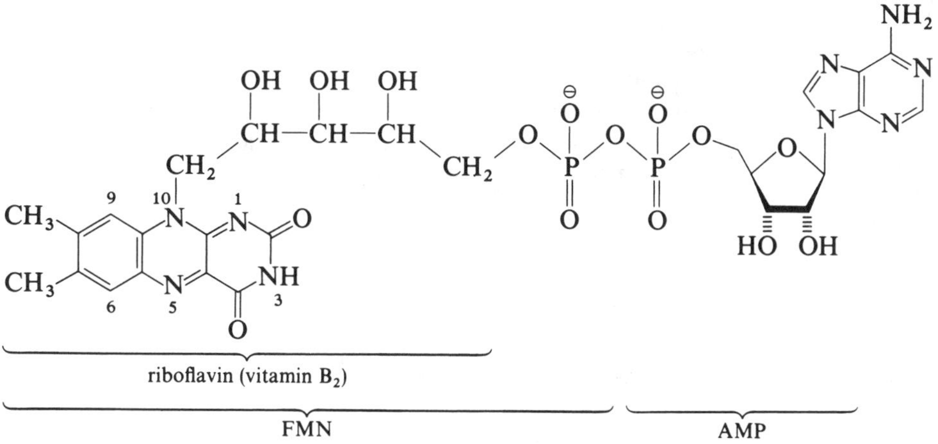

riboflavin (vitamin B_2)

FMN AMP

FAD

FAD was determined in 1935 by R. Kuhn. The first mechanism of action was proposed in 1938 by O. Warburg and its total synthesis was accomplished in 1954 by Lord Todd's group. The hydrophilic nature of the substituent at N-10 is responsible for its solubility in an aqueous protein solution.

The site of fixation of a hydride ion from NADH (or two electrons) was suggested to be at the electrophilic nitrogen N-5 of the flavin nucleus.

$$\text{NADH} + \text{FAD} \xrightleftharpoons{H^{\oplus}} \text{NAD}^{\oplus} + \text{FADH}_2$$

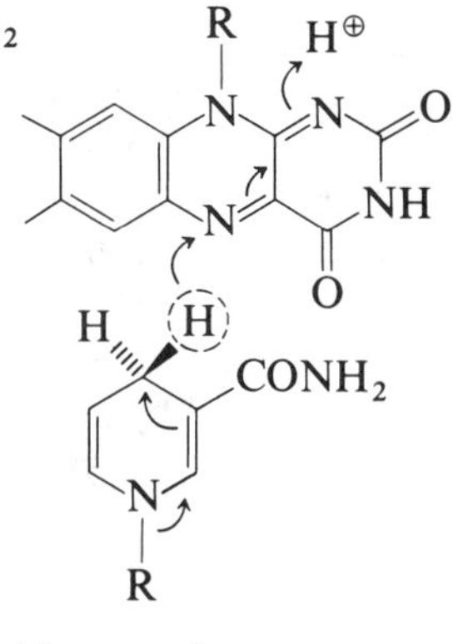

In 1973 Sigman showed that direct hydride transfer can occur from NADH (or *d*-4-NADH) by using an aromatic analog; *N*-methyl acridinium salt (593).

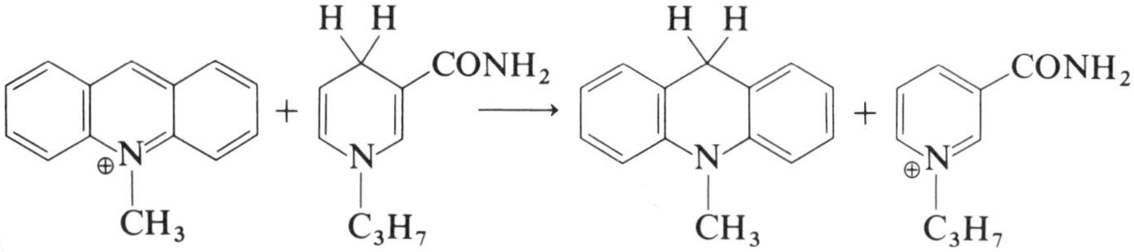

The reaction was also performed with a reduced nicotinamide molecule having a propyl side chain. The reverse reaction, however, does not take place. Free radical quenching agents had no effect on the rate of the reaction, suggesting an ionic rather than a radical mechanism. However, isotope effects with deuterated NADH show that under nonenzymatic conditions the reaction with FAD is not a simple bimolecular process.

	H H, $CONH_2$, N–C_3H_7	H D, $CONH_2$, N–C_3H_7	D D, $CONH_2$, N–C_3H_7
$k(M^{-1}s^{-1})$ for loss of hydride at C-4:	2040	1620	1398

From the above results a secondary isotope effect of 0.74 can be calculated. Since it is well known that conversion of an sp^3-carbon to sp^2 hybridization should give a secondary deuterium effect of greater than 1, a simple bimolecular process is impossible. Consequently, a kinetically important intermediate must be postulated to account for the experimental facts.

Still, no one general mechanism for flavin reactions exist. A charged-transfer complex cannot be excluded, but a free radical intermediate is also possible. It should be recalled that usually metal ions are required in flavin enzymes, and they may have an important role in the mechanism. In fact, the central position which flavoenzymes occupy in many biological pathways, after nicotinamide coenzymes (two electrons process) and before cytochromes (one electron process) in the respiratory chain,* may be caused by the versatility of the flavin structure, allowing both ionic and free radical mechanisms.

Free radical reactions are possible because the system can easily be oxidized or reduced by a one electron step via a stable intermediate called *semiquinone* (stabilized by ten resonance forms). In these planar structures, one ring always remains fully aromatic. On the other hand, if radicals are not intermediates, reactions must involve a hydride transfer or proceed by PPC or PPM mechanisms. These two mechanistic classes were suggested by G.A. Hamilton, University of Pittsburgh (584): *PPC* (proton, proton, covalent bond compound) for systems that do not require a metal ion and *PPM* (proton, proton, metal) where a metal ion is involved. The flavin nucleus is thus structurally and electronically

---

* Cytochromes are electron carriers coupled to oxidative phosphorylation, a process in which ATP is formed as electrons are transferred from NADH or $FADH_2$ to molecular oxygen. The overall process consists of the oxidation of a substrate (glucose, for instance) where the flow of electrons cascade through a respiratory assembly (the cytochromes) to reach O_2 which is finally reduced to water. Ubiquinones are used to bridge flavin two-electron donor chemistry to cytochromes' one-electron carriers.

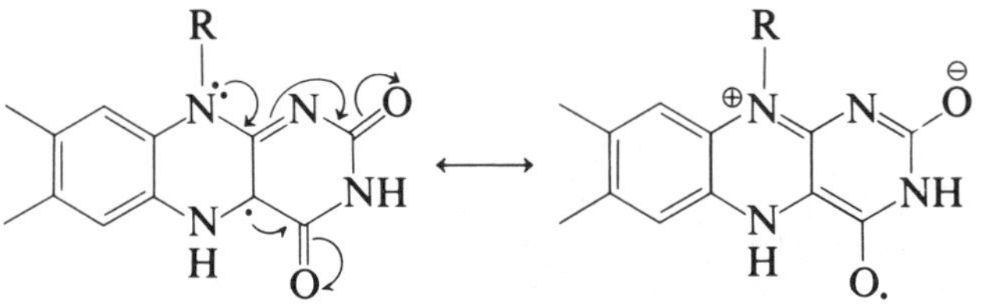

flavin radical semiquinone radical

designed to stabilize both one- and two-electron transfers in redox reactions.

For the PPC mechanism (Hamilton mechanism) to be operative, a covalent compound must be formed between the flavin coenzyme and the substrate, so that some mechanism is available for transmitting electrons from one molecule to the other. Since this reaction is ionic, there must be an electrophilic site on the flavin that can be attacked readily by the nucleophilic substrate. Examination of the flavin nucleus reveals four potential sites of attack, among which position 4a is believed to be the most reactive electrophilic site. The reactivity is in fact increased somewhat by the inductive effect of the adjacent amide and amidine functions.

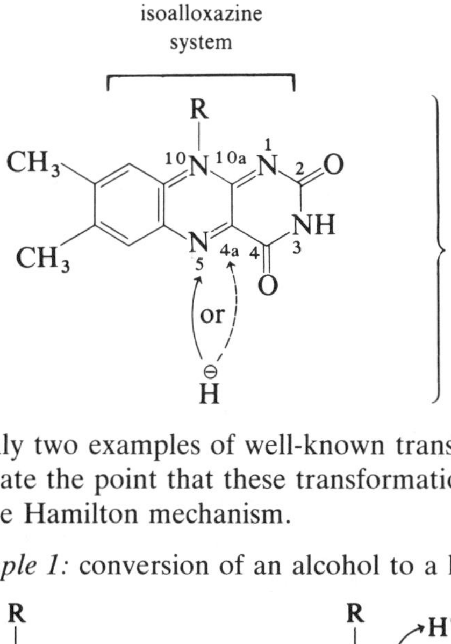

isoalloxazine system

possible electrophilic sites

position 2 (urea type)
position 4 (amide type)
position 10a (amidine type)
position 4a (Schiff base carbon)

Only two examples of well-known transformations are given to simply illustrate the point that these transformations are mechanistically feasible via the Hamilton mechanism.

Example 1: conversion of an alcohol to a ketone

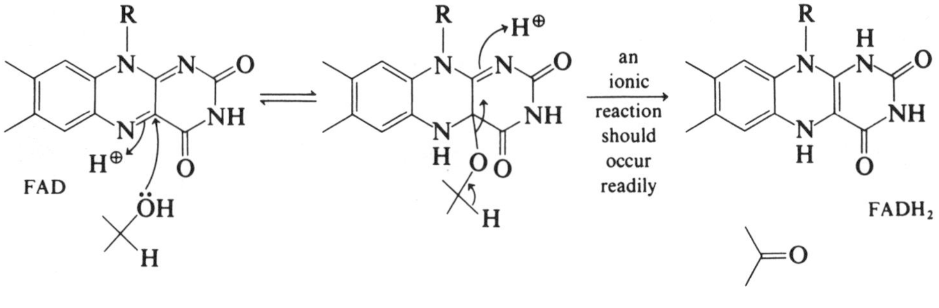

FAD — an ionic reaction should occur readily → $FADH_2$

Example 2: conversion of NADH to NAD^+

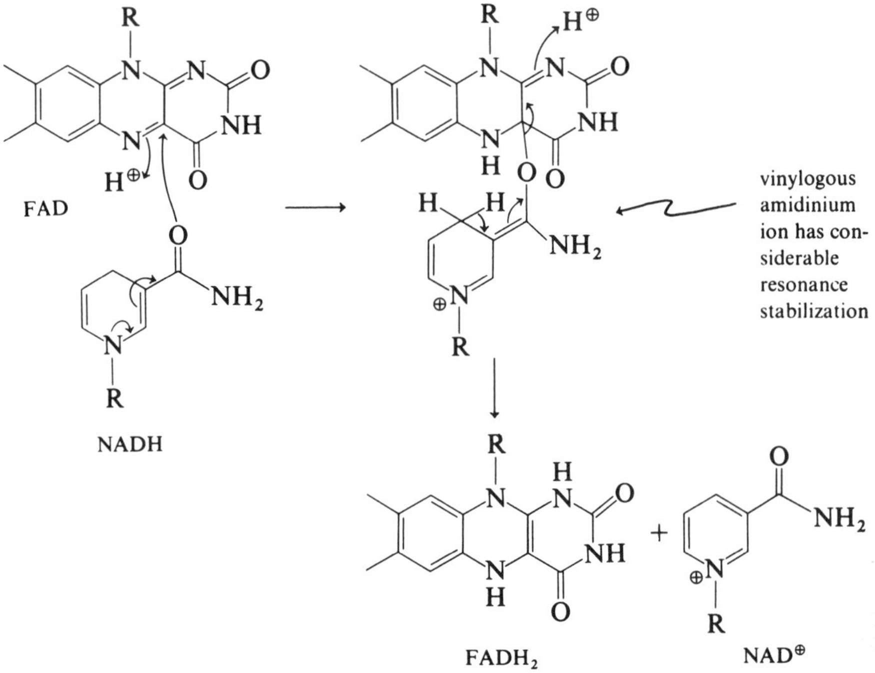

FAD

NADH

vinylogous amidinium ion has considerable resonance stabilization

$FADH_2$ + $NAD^{\oplus}$

Clearly, no hydride transfer is involved, only proton shifts via the formation of a covalent intermediate at position 4a. Hamilton argues that biological oxidoreduction (dehydration) reactions rarely, if at all, involve hydride ions because protons are not shielded by electrons and thus travel much faster and more efficiently in biological media (544).

Here again, model studies are necessary to clarify the situation.

Model 1

The first evidence of a 4a-adduct on a model compound appeared in 1967 (594).

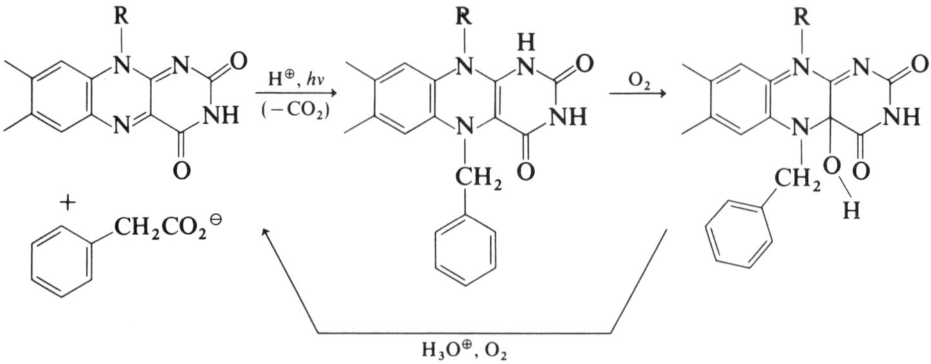

$H^{\oplus}$, $h\nu$ $(-CO_2)$

O_2

$H_3O^{\oplus}$, O_2

The compound was obtained by a photo-induced benzylation of flavin by phenylacetate. The resulting product reacts slowly with air and suggests that the reactivity of N-5 may be of importance in flavoprotein catalysis. It also indicates that an oxygen can add readily to position 4a if position N-5 is substituted, or protonated as it could be on the enzyme.

Although many models show reactivities at positions 4a and N-5, the question as to what is the exact electrophilic site in enzymatic flavin chemistry has not been satisfactorily answered.

Model 2

The formation of a covalent bond between NADH and FAD has been suggested, and previous studies indicated that a preequilibrium complex is probably formed with flavins. Second, theoretical calculations have shown that position N-5 should be a good electrophilic center. Therefore, Bruice (595) argued that changing that position for a carbon atom should also yield a strong electrophilic center with the advantage that the new C—H bond formed will be stable and not exchange with the solvent.

He synthesized the following flavin model compound and submitted it to the action of NADH in ^2H_2O for three days in the dark, under an argon atmosphere.

CH3 9 N N O N—CH3 6 H O

NADH ^2H_2O 3 days

CH3 D N N O N—CH3 H H O

no deuterium incorporated

He obtained a reduced product with no deuterium incorporated at position 5. The results thus clearly show that there is a direct hydrogen (or two-electron) transfer to position 5, and that in the preequilibrium complex, the NADH probably does not occupy the area adjacent to positions 1, 9, and 10. This result is a strong argument against the Hamilton proposal where this FAD analog acts as a hydride acceptor.

Model 3

Similarly, a 5-deazariboflavin was synthesized and reacted in the presence of a NADH:FAD oxidoreductase (596). C. Walsh, at Harvard Medical School, showed that this 5-deaza-analog functions coenzymatically, undergoing reduction by direct hydrogen transfer from NADH.

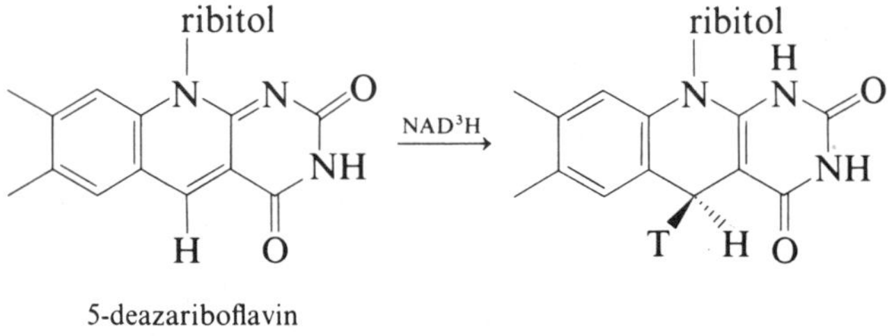

$\xrightarrow{NAD^3H}$

5-deazariboflavin

The analog was reduced at 0.3% the rate of riboflavin. However, the reaction was not absolutely stereospecific but the oxidoreductase showed a preference for the *R*-isomer of [5-^{3}H]NADH with attack on the *re*-face of 5-deazariboflavin. It turns out that a 5-deaza-FAM derivative (called factor F_{420}) has now been found as a natural cofactor in the anaerobic bacteria that produce CH_4 from CO_2 (597) and makes the study of analogous model systems even more significant.

Model 4

In 1979, Sayer and co-workers (598) used 1,3-dimethyl-5(*p*-nitrophenyl-imino) barbituric acid as a flavin model and were able to isolate a covalent intermediate in the reduction of this highly activated imine substrate by a thiol (methylthioglycolate or mercaptoethanol).

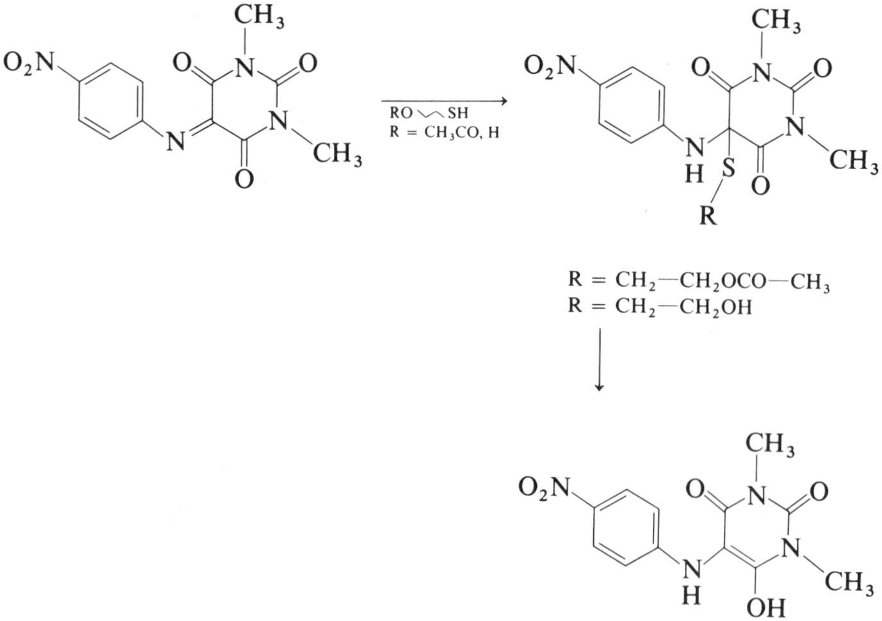

RO⌒SH
$R = CH_3CO, H$

$R = CH_2{-}CH_2OCO{-}CH_3$
$R = CH_2{-}CH_2OH$

The thiol addition product was detectable spectrophotometrically and the intermediate with R = CH_2CH_2OH was isolated and its structure proved by ^{1}H- and ^{13}C-NMR. The intermediate could then be converted to the dihydro compound showing that the addition must have occurred at position C-4a and not N-5. This work thus provides evidence for the existence of a thiol addition intermediate involving the C(4a)—N(5) bond in nonenzymatic reductions of flavins and analogs by thiols.

Among the more recent coenzyme models of flavin, two deserve special attention. The first one involves the "remote control" of flavin reactivities by an intramolecular crown ether ring serving as a metal-binding site (599,600):

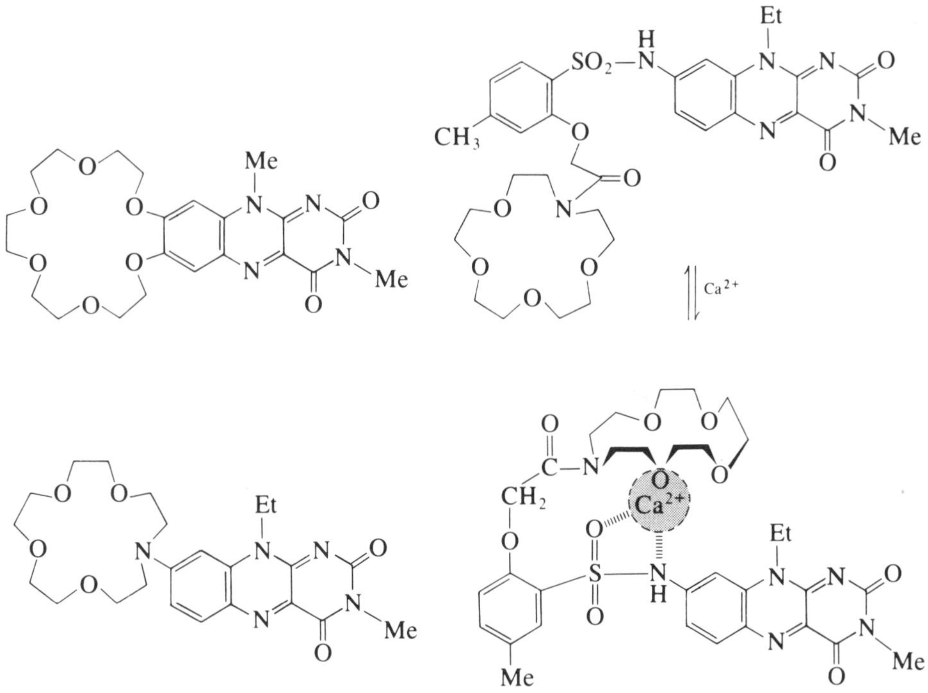

The presence of a crown ether could be used to mimic allosteric functions by binding an ion or an effector at a site remote from the flavin chemistry. In fact, allosteric effects by which some catalytic activities of coenzyme-mediated enzymes may be regulated are quite interesting from the bioorganic standpoint. The oxidizing ability of the flavin nucleus could be influenced by resonance contributions coming from substituents at C_7 and C_8 positions. Spectral changes can monitor these variations. An intriguing situation is with the 8-sulfonamide derivative. The compound gives a new absorption band at 452 nm upon addition of $Ca(ClO_4)_2$ and the band increases with increasing Ca^{2+} concentration. The large association constant ($4.3 \times 10^4\ M^{-1}$) was attributed to a synergistic effect of the crown ring and the sulfonamide group. It is equivalent to a "lariat effect" by the sulfonamide chain capping the Ca^{2+} ion bound to the crown ether cavity.

More important is the result of a photooxidation study of benzyl alcohol by this coenzyme model. The quantum yield in aqueous solution was markedly greater (up to 640-fold) than the corresponding $PhSO_2NH$-flavin species and increases with increasing Ca^{2+} concentration. The presence of a crown ether moiety thus serves as an allosteric effector site to induce a change in the catalytic activity of the flavin moiety.

In a different approach, M. Hirobe's group synthesized and studied the properties of a novel flavin-linked porphyrin molecule (601,602). It was observed that in the manganese Mn(III)-complex form, the flavoporphyrin is efficiently reduced by *N*-benzyl-1,4-dihydronicotinamide with a rate enhancement of up to 8 as compared to similar intermolecular electron transfer. A smaller k_{intra}/k_{inter} value of 2 was obtained with the molecule linked to position N_{10}. This variation may be due to a difference in the conformation of the transient ternary complex.

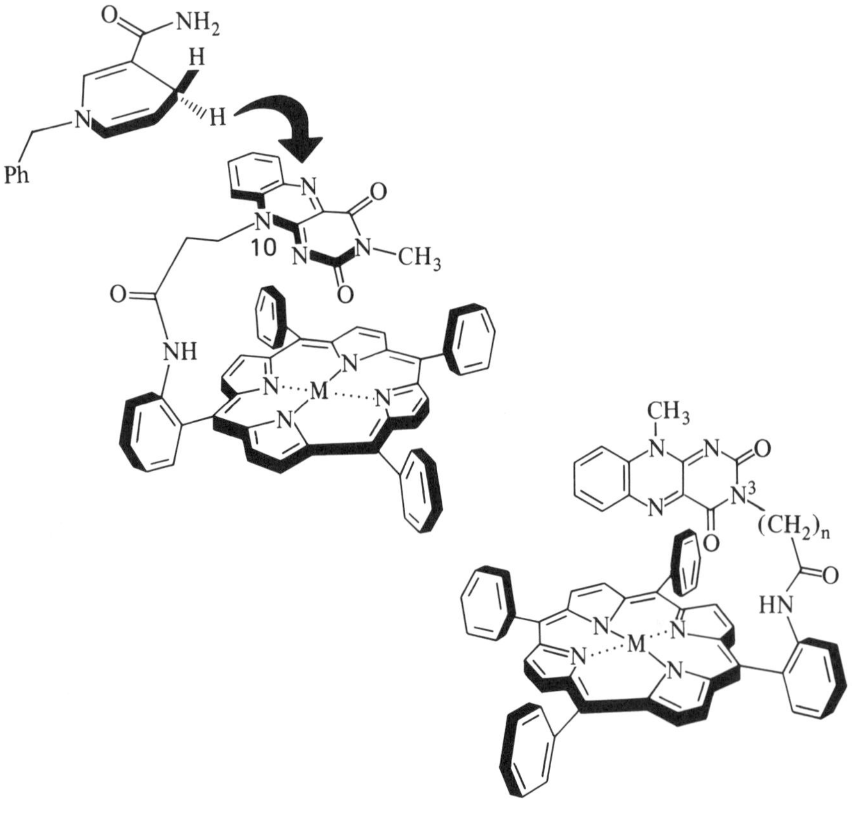

M = Mn(III) C*l*

Such model compounds serve as two-electron/one-electron transfer carriers.

7.1.4 Oxene Reactions

Many oxidases (monooxygenases) are FAD dependent and catalyze the oxidation of various substrates using molecular oxygen. FAD coenzymes are thus able to transfer oxygen to an organic substrate, another versatility in this cofactor not present in NAD^+.

The flavin-dependent monooxygenases bind and activate molecular oxygen, ultimately transferring one oxygen atom to substrate and releasing the second as water. Among the reactions catalyzed by the flavin monooxygenases are the hydroxylation of *p*-hydroxybenzoate, the oxidative decarboxylation of salicylate, and a variety of amine oxidations mediated by the mixed-function aminooxidase system (603). Apparently these reactions involve an "oxene" mechanism. This term is used because it is analogous to carbene and nitrene reactions, two other electrophilic species.

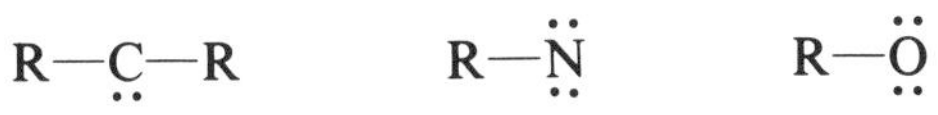

R—C̈—R (carbene) R—N̈: (nitrene) R—Ö: (oxene)

Some enzymes in these reactions require a transition metal ion while others do not. For simplicity, we will focus only on mechanisms not involving a metal ion.

One group of monooxygenases for which it is quite clear no metal ion is necessary requires a flavin as cofactor. Not requiring a metal ion also means that the reaction with oxygen could have a free radical character at some stage of the mechanism. Since substrate radicals are very unstable, it seems more likely that oxygen reacts with the reduced form of flavin to give an intermediate which then reacts by an ionic mechanism with the substrate. This way spin conservation is preserved. The oxygenation cycle for flavin monooxygenases is presented in Fig. 7.6.

In these aerobic flavin-linked monooxygenases the reduced flavin $FADH_2$ first reacts directly with oxygen to give a $FADH_2O_2$ adduct. This adduct is believed to rearrange to a flavin 4a-hydroperoxide, a good oxidizing agent, and a stronger oxidant than H_2O_2.

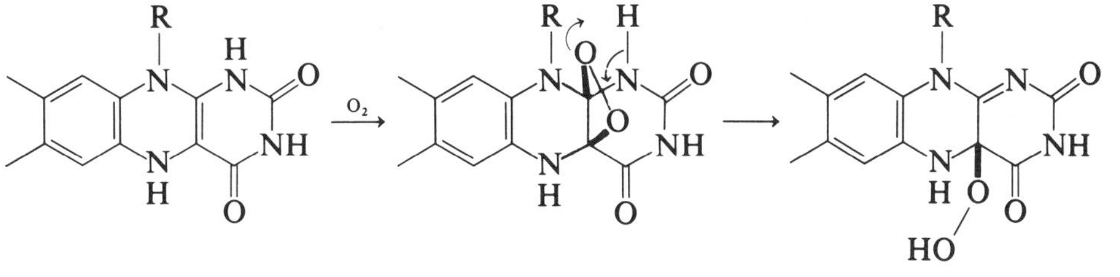

$FADH_2$ $\xrightarrow{O_2}$ $FADH_2O_2$ $\longrightarrow$ flavin 4a-hydroperoxide

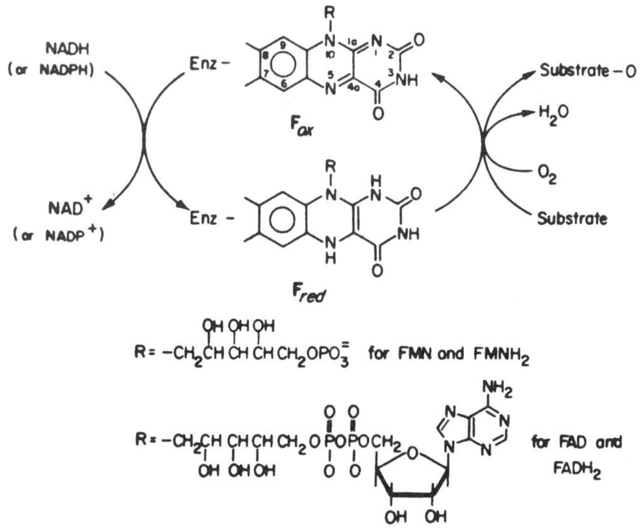

Fig. 7.6. Oxygenation cycle for flavin monooxygenases (604). Reproduced with permission of D. Dolphin.

Since flavomono- and flavodioxygenase reactions are the only non-metal-ion-requiring oxygen activation reactions in biochemistry, considerable attention has been paid to understanding the mechanism by which molecular oxygen reacts with 1,5-dihydroisoalloxazine nucleus (605).

How does a flavin molecule activate molecular oxygen? It should be realized that in the transformation above, oxygen reacts as a ground state triplet molecule whereas organic molecules (flavin) are usually in the singlet state. However, the reaction of a singlet with a triplet to give a singlet product is a spin-forbidden process! Nevertheless, it is possible to obtain an ionic reaction of oxygen without requiring the formation of singlet oxygen if it is complexed to a transition metal ion that itself has unpaired electrons. Since many oxidases do not require a metal ion to function, the situation is still not clear, to say the least, but a radical process with flavin is a possibility. In fact, the addition of oxygen to reduced flavin is similar to the reaction of oxygen with a substituted tetraaminoethylene, a strong electron-donating double bond.

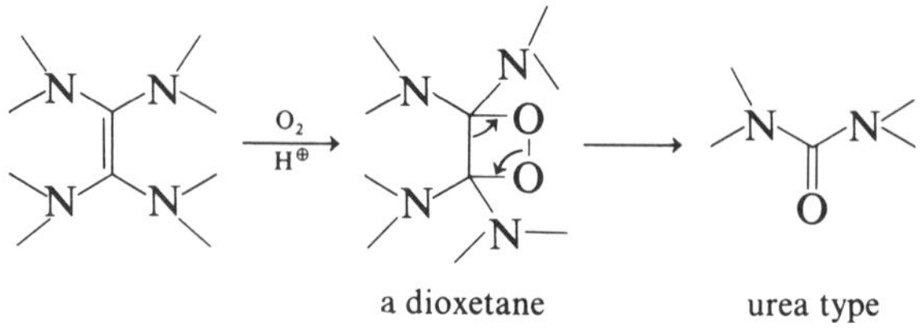

a dioxetane

urea type

Furthermore, aromatic compounds (phenol) can be hydroxylated by oxygen in a nonezymatic system when various reduced flavins are present. It is suggested that the hydroxylating agent is the flavin hydroperoxide which rearranges by a proton transfer to give the actual hydroxylating agent. It is a carbonyl oxide. Similar intermediates with the reaction of ozone on double bonds are known.

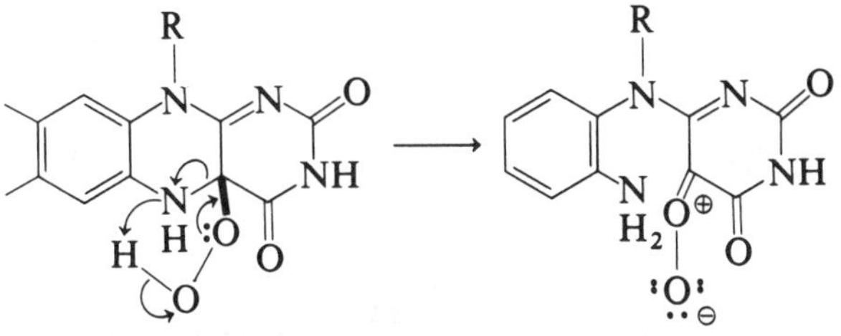

flavin hydroperoxide

carbonyl oxide

Since carbene and nitrene are electrophilic agents, the terminal oxygen will have to become more electropositive by stabilizing a negative charge on the rest of the molecule.

This mechanism was first proposed by Hamilton (584). It has, however, the disadvantage of involving an open form of the flavin ring.

Other ionic mechanisms can also be operative. The transfer of oxygen can occur via a radical type transfer rather than ionic mechanism (see page 514). At this point we could ask why nature selected flavin and not nicotinamide coenzyme to transfer oxygen to a substrate? The answer

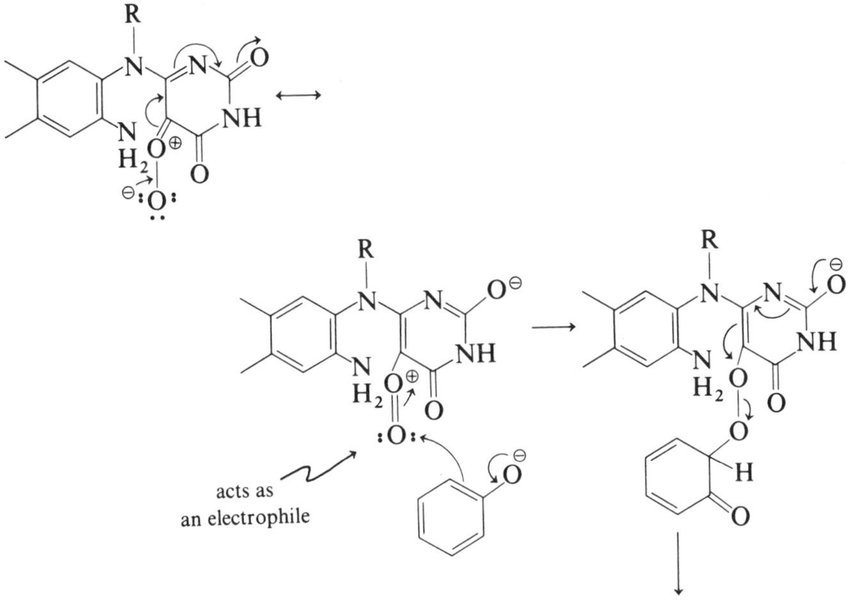

acts as an electrophile

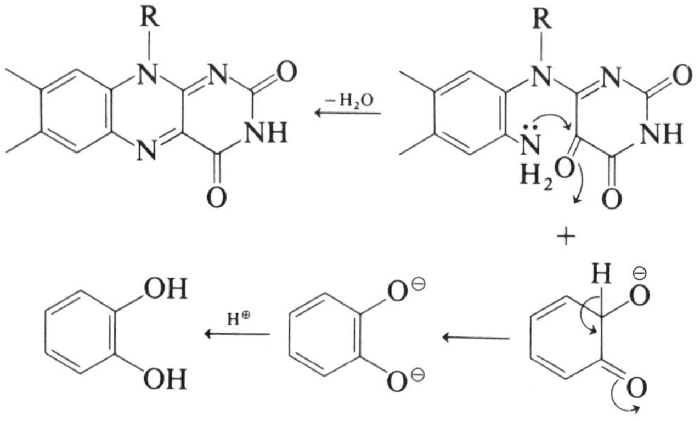

R R
N N O N N O
$-H_2O$
NH NH
N N
H_2O
O O
+
OH $O^{\ominus}$ H $\ominus$ O
$H^{\oplus}$
OH $O^{\ominus}$ O

probably lies in the fact that NAD^+ can transfer electrons only via a hydride ion whereas FAD has the chemical versatility to be converted to an oxygen-adduct intermediate that can then react with the substrate so that a direct transfer of oxygen to the substrate does not take place.

To overcome the difficulty encountered in Hamilton's mechanism, Dolphin in 1974 proposed an interesting alternative (604) where no open-ring product intermediate is formed. His mechanism involves a reactive oxaziridine-4a,5-flavin, obtainable from flavin hydroperoxide.*

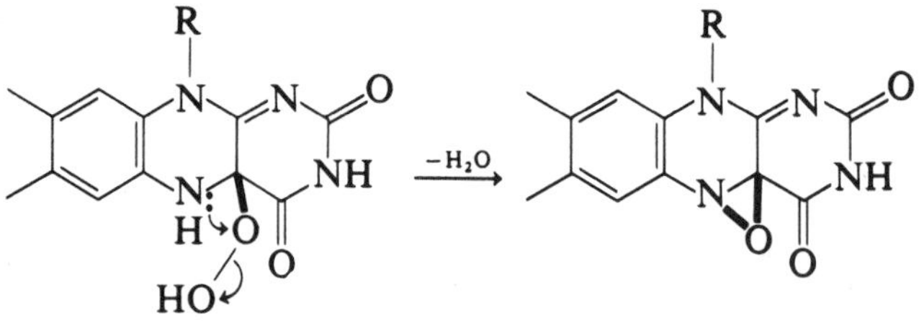

R R
N N O N N O
$-H_2O$
NH NH
N N
H O O O
HO O

oxaziridine

This postulate is based on numerous chemical precedents, among which is the following that shows the electrophilic character of the oxaziridine system:

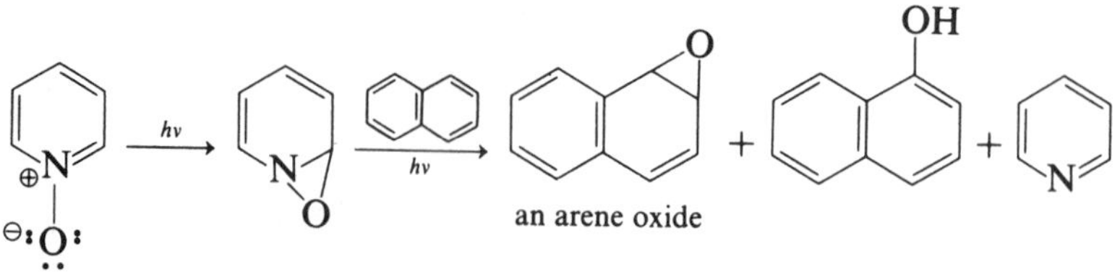

OH
O
$h\nu$ $h\nu$ + +
$\oplus$N N N
$\ominus$:O: O
an arene oxide

---

* As mentioned earlier (page 503), 5-deazaflavin analogs are more susceptible to nucleophilic attack which typically occurs at position 5 to yield 1,5-dihydro adducts. In this respect, the formation of a 4a,5-epoxy-5-deazaflavin derivative by the reaction of 5-deaza-isoalloxazine with H_2O_2, *tert*-butyl hydroperoxide, or *m*-chloroperbenzoic acid has been reported recently (606).

So the following rational mechanism can be written for the hydroxylation of phenol:

stabilized anion

stabilized cation

FAD

+

$H^{\oplus}$

$-H^{\oplus}$

Interestingly, transfer of oxygen from the oxaziridine reagent leads to a benzene epoxide intermediate.

B. Witkop and his group (607) from N.I.H. already suggested in 1969 that epoxides are intermediate in a number of monooxygenase-catalyzed oxidations of aromatic compounds. For example:

salicylate hydroxylase, $FADH_2O_2$; $+ CO_2$

Once again, a flavin cofactor is the agent of epoxidation.

Flavin hydroperoxide is also implicated in oxidative decarboxylation:

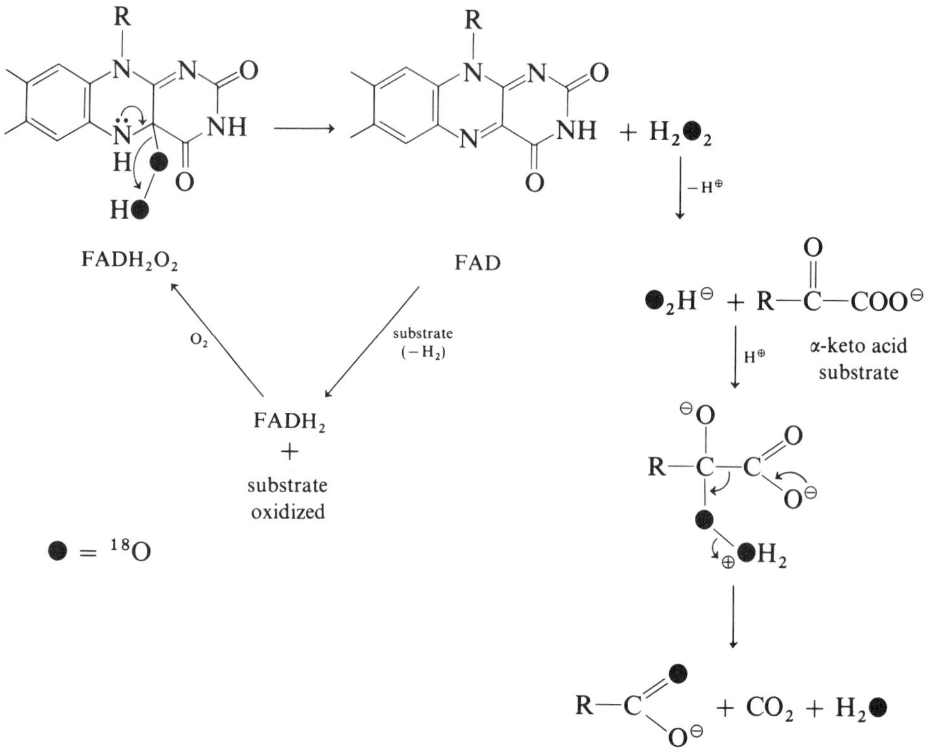

FADH$_2$O$_2$ FAD + H$_2$●$_2$

O$_2$ substrate ($-$H$_2$) $-$H$^{\oplus}$

FADH$_2$ + substrate oxidized

●$_2$H$^{\ominus}$ + R—C(=O)—COO$^{\ominus}$ α-keto acid substrate

H$^{\oplus}$

● = ^{18}O

R—C(=●)—O$^{\ominus}$ + CO$_2$ + H$_2$●

It was shown by using $^{18}O_2$ that only one of the two oxygen atoms ends up in the carbonyl group. An interesting example of this operation is the conversion of *p*-hydroxyphenylpyruvic acid to homogentisic acid by a dioxygenase:

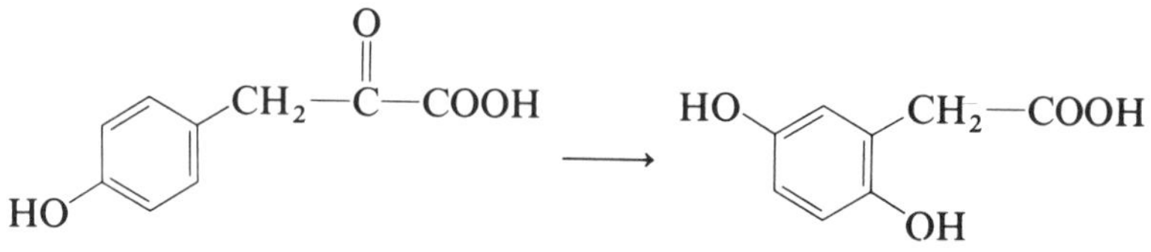

p-hydroxyphenylpyruvic acid → homogentisic acid

Witkop first proposed the following mechanism:

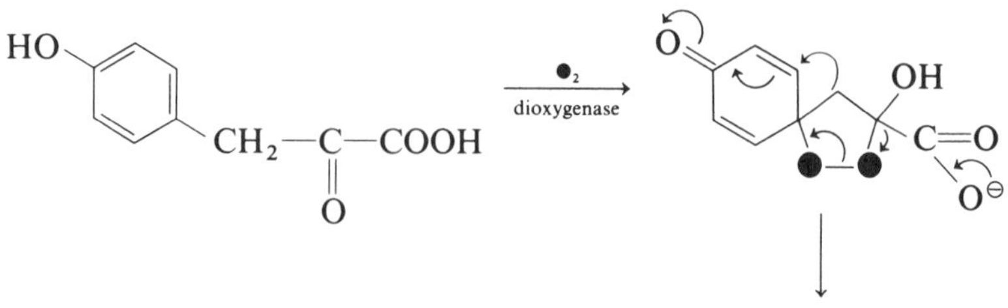

●$_2$ dioxygenase

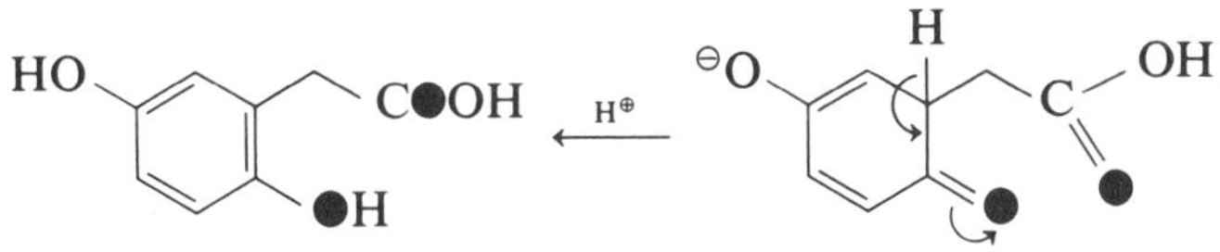

A side chain migration to the *ortho* position takes place in this rearrangement and this has been referred to as the *NIH shift* (608). However, this mechanism was questioned because the reaction still worked with phenylpyruvic acid (no *p*-OH). A more likely mechanism would be the involvement of a transition metal ion to form first a peracid that would then undergo an "oxene" mechanism via an epoxide intermediate:

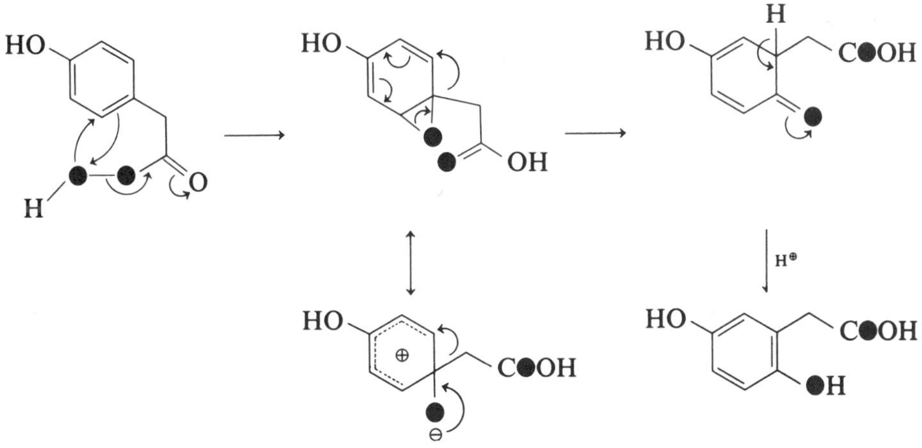

The proposal by Dolphin of an oxaziridine mechanism for flavin-mediated oxidations has been criticized by Rastetter's group (603) since oxygen insertion in enzymatic oxidations is not induced photochemically. What they have shown, however, is that nonenzymatic oxidations can be effected by a flavin N(5)-oxide (nitrone). Upon irradiation in the presence of a phenol, a flavin nitroxyl radical and a phenoxy radical are generated, as detected by EPR spectroscopy. The coupling of these two species leads to flavin and hydroquinone.

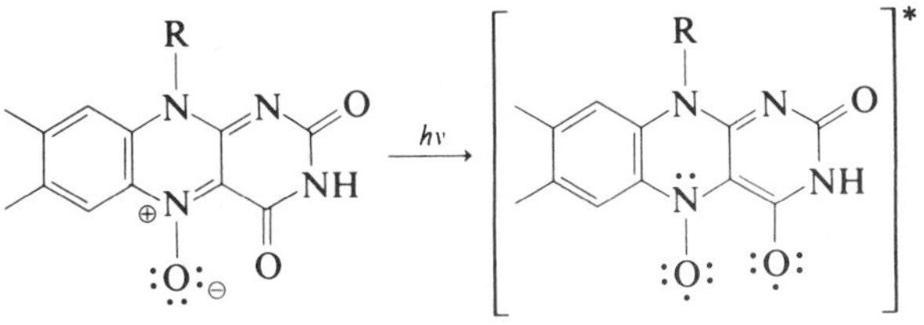

flavin N(5)-oxide
(nitrone)

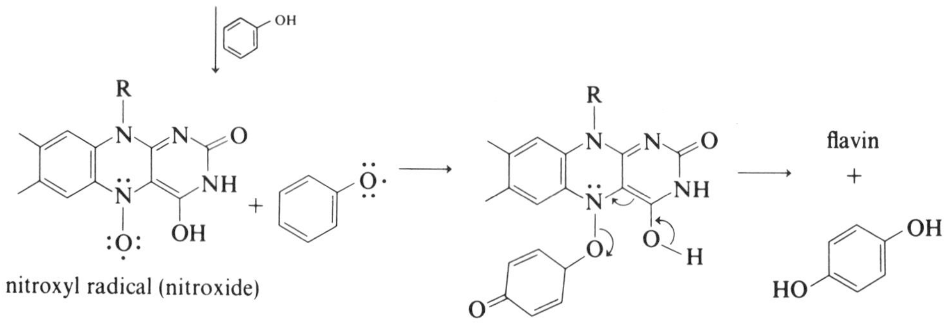

nitroxyl radical (nitroxide)

flavin +

The authors argue that flavin nitroxyl radical may play a role in flavo-enzyme chemistry and suggest that the nitroxyl radical intermediate may be derived from enzyme-bound flavin 4a-hydroperoxide.

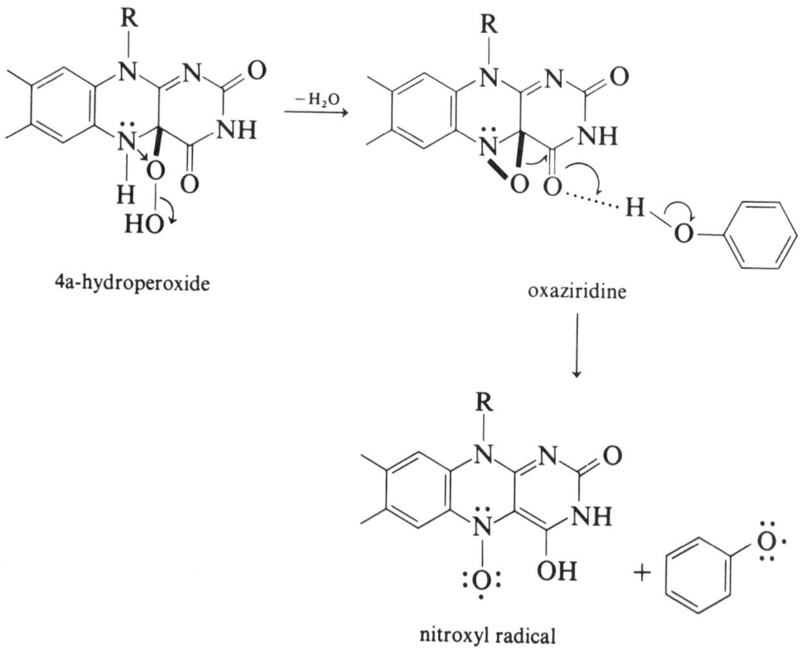

$-H_2O$

4a-hydroperoxide

oxaziridine

nitroxyl radical

Such a radical-pair mechanism avoids the need to invoke an "oxene" intermediate during the oxygen transfer by the flavoenzyme, and at the same time releases the ring strain of the oxaziridine species.

In 1978, Caltech's W.A. Goddard proposed a different mechanism for the hydroxylation of phenolic compounds and attempted to show how flavin coenzymes carry out such oxidations. It is a theoretical proposal based on wave functions and quantum mechanics using generalized valence bond theory, applied to biological problems (609). An example is shown in Fig. 7.7 for the oxidation of phenol to catechol.

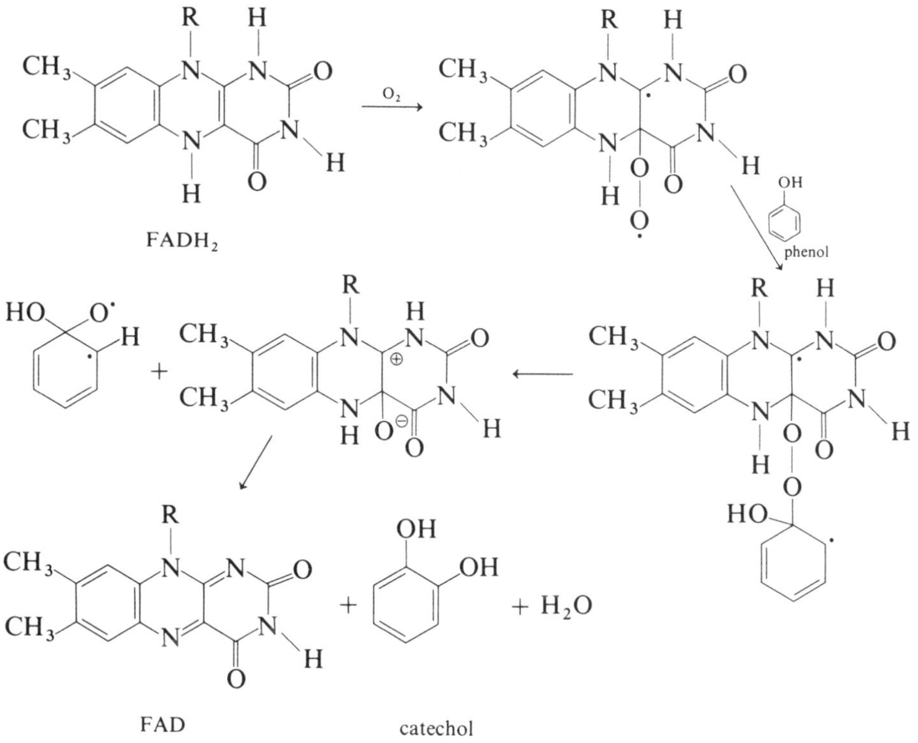

FADH$_2$

FAD

catechol

Fig. 7.7. Goddard's proposal of how flavin coenzyme carries out the oxidation of phenol (609).

In the first step of this oxidation process, triplet oxygen attacks flavin, forming a biradical. In a way flavin helps to activate oxygen and for this, one of the unpaired electrons of the biradical is stabilized by the flavin ring, and the other remains associated with the oxygen atom that reacts with the phenol in the second step. This leads to an intermediate in which the flavin and phenol are joined through the two oxygen atoms of molecular oxygen. Then, the flavin molecule helps to drive the reaction to the next step where the lone-pair electrons on the nitrogen atoms at positions 5 and 1 stabilize the positive charge of the intermediate. According to Goddard, the nitrogen atoms in flavin assist in breaking the oxygen–oxygen bonds, an uphill (energetically) step in the reaction.

Interestingly, the mechanism postulated involves an attack on phenol by an oxygen atom at the carbon bearing the hydroxyl group. That radical intermediate is believed to be lower in energy than the one with oxygen on the adjacent ring carbon, according to bond energy calculations. The oxygen later migrates in a rearrangement reaction. At this point, the catechol product has been made, but the flavin still contains an extra oxygen. It is believed to be removed from the flavin by picking up protons and forming water. The process would certainly be enzyme mediated. The proof of this proposal awaits model studies.

Among the miscellaneous redox reactions in nature, the emission of visible light by living beings is one of the most fascinating of natural phenomena. Luminescent bacteria, worms, and the amazing firefly have all been the objects of the biochemists' curiosity. A natural question is: What kind of chemical reaction can lead to emission of energy in the visible region of the spectrum? Although an AMP-intermediate is found, the process requires far too much energy to be provided by the hydrolysis of ATP itself. A clue comes from the fact that chemiluminescence is very common when O_2 is used as an oxidant, and extraction of luminous materials from organisms showed that one of the active principles in the firefly system is luciferin. This compound is decarboxylated by the enzyme luciferase (610).

firefly luciferin

ATP, Mg^{2+}, luciferase → + PP_i

O_2 + basic residue on luciferase

α-peroxylactone

$-CO_2$

→ light

electronically excited oxyluciferin

After activation by ATP, luciferin is believed to be converted to a α-peroxylactone by dehydrative cyclization. This intermediate serves as a chemienergizer for the formation of an electronically excited oxyluciferin by decarboxylation.

This enzymatic oxygenation by carbanion attack on O_2 followed by decarboxylation occurs without conjugated-cofactor assistance and is thus different from the mode of action of bacterial luciferase, a flavin-dependent monooxygenase (611). In the bacterial system, a 4a-flavin hydroperoxide undergoes a chemiluminescent reaction in the presence of an aldehyde.

A simple model compound for the oxidation–decarboxylation sequences is now known (612). Dehydration of the following very reactive α-peroxyacids by DCC yields the corresponding α-peroxylactone (dioxetanone).

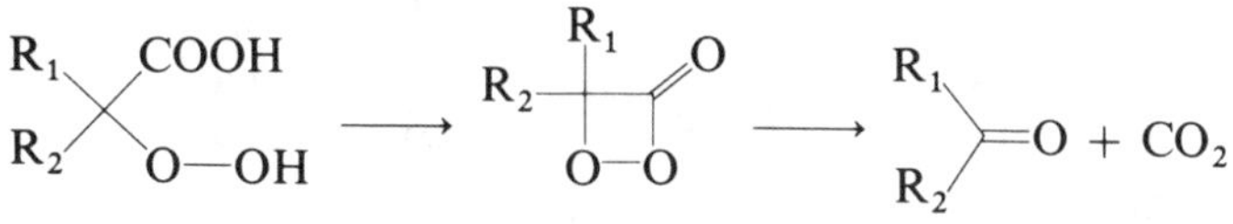

$R_1 = R_2 = CH_3$
$R_1 = R_2 = t\text{Bu}$
$R_1 = t\text{Bu}, R_2 = CH_3$

$\rightarrow$ $R_1R_2C{=}O + CO_2$

Thermolysis of the α-peroxylactone in CH_2Cl_2 at 25°C leads to the quantitative generation of the ketone and to light emission. The decarboxylation is believed to take place in a first step by a one-electron transfer from a catalytic amount of an aromatic activator present in the solution (612).

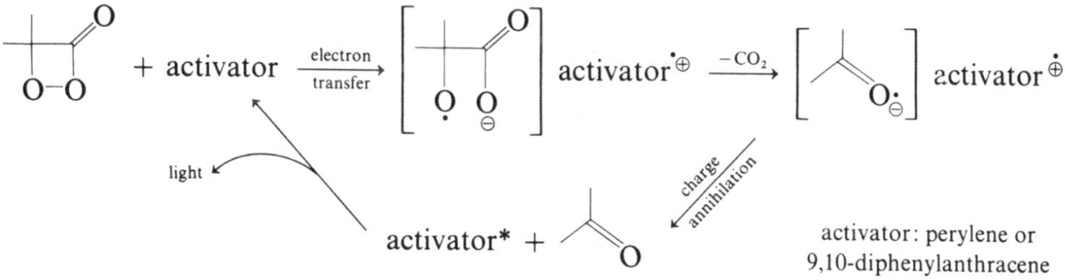

+ activator $\xrightarrow[\text{transfer}]{\text{electron}}$ [ ] activator$^{\cdot\oplus}$ $\xrightarrow{-CO_2}$ [ ] activator$^{\cdot\oplus}$

charge annihilation

activator* +

light

activator: perylene or 9,10-diphenylanthracene

The next step along the chemiluminescence path is the rapid loss of CO_2 from the reduced dioxetanone. The final step in the sequence is light emission from the excited activator, which is detected as chemiluminescence. It seems that the central feature of chemiluminescent reactions is radical anion formation.

Finally, mention should be made of a bacterial flavin-containing cyclohexanone oxygenase that can oxygenate cyclohexanone by catalyzing a four-electron reduction of dioxygen with two electrons derived from reduced nicotinamide cofactor and two electrons derived from substrate (613). This is a good illustration of the well-known Baeyer–Villiger reaction in organic chemistry.

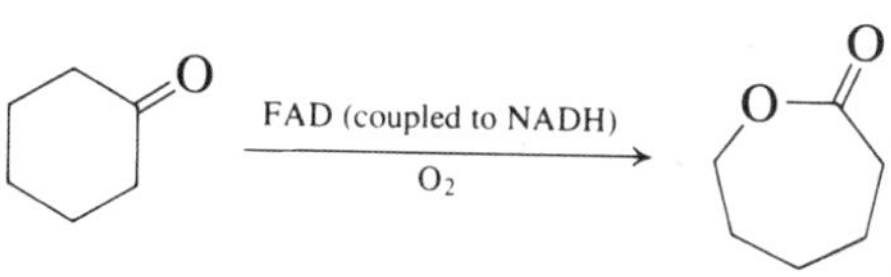

$\xrightarrow[O_2]{\text{FAD (coupled to NADH)}}$

7.1.5 Lipoic Acid

Lipoic acid (1,2-dithiolane-3-valeric acid) is widely distributed among microorganisms, plants, and animals. It belongs to the group of cofactors containing sulfur and in nature it is coupled to thiamine pyrophosphate (see Section 7.4). However, lipoic acid basically belongs to another class of electron transfer cofactors where the net oxidoreduction function is to produce ATP. The cofactor is needed in fatty acid synthesis and in the metabolism of carbohydrates.

Indeed, it has been recognized for a long time that lipoic acid is an interesting growth factor found in a number of microorganisms. Chemically, it can be reduced and the reduced from can be readily reoxidized to lipoic acid. Dihydrolipoic acid is an efficient reducing agent

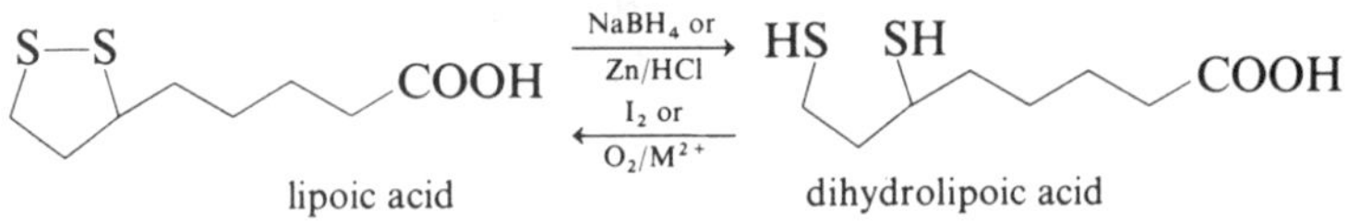

S—S COOH $\underset{I_2 \text{ or } O_2/M^{2+}}{\overset{NaBH_4 \text{ or } Zn/HCl}{\rightleftharpoons}}$ HS SH COOH

lipoic acid dihydrolipoic acid

of sulfate ion to sulfite. The sulfate is first activated as an adenyl sulfate, also called adenosine 5′-phosphosulfate (APS). The formation of lipoic acid is an entropy-favored process because both sulfur atoms are on the same molecule.

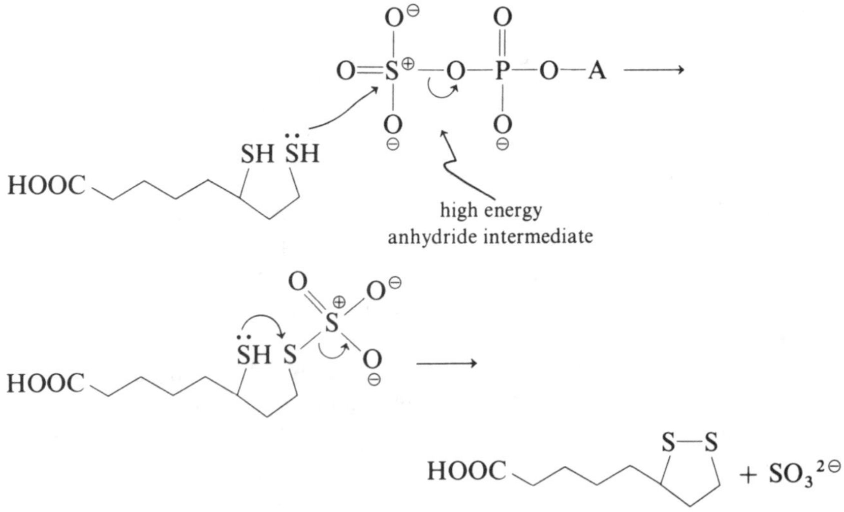

HOOC SH SH O=S⊕(O⊖)(O⊖)—O—P(=O)(O⊖)—O—A ⟶

high energy anhydride intermediate

HOOC SH S—S⊕(=O)(O⊖)(O⊖) ⟶

HOOC S—S $+ SO_3^{2\ominus}$

One should realize that the five-membered ring of lipoic acid is not planar but has a C—S—S—C dihedral angle of 26°. Normally, disulfides are colorless but lipoic acid is yellow. This property is attributed to the ring strain. This ring strain is caused in part by the repulsion of electron pairs on adjacent sulfurs and makes lipoic acid a better oxidizing agent than a less strained six-membered ring analog. There is thus a structure–function relationship to its biological role (614).

Finally, on the multienzyme complex that performs the oxidative decarboxylation of an α-keto acid, lipoic acid is not found free but is covalently linked to a lysine residue, through an amide linkage (Fig. 7.8).

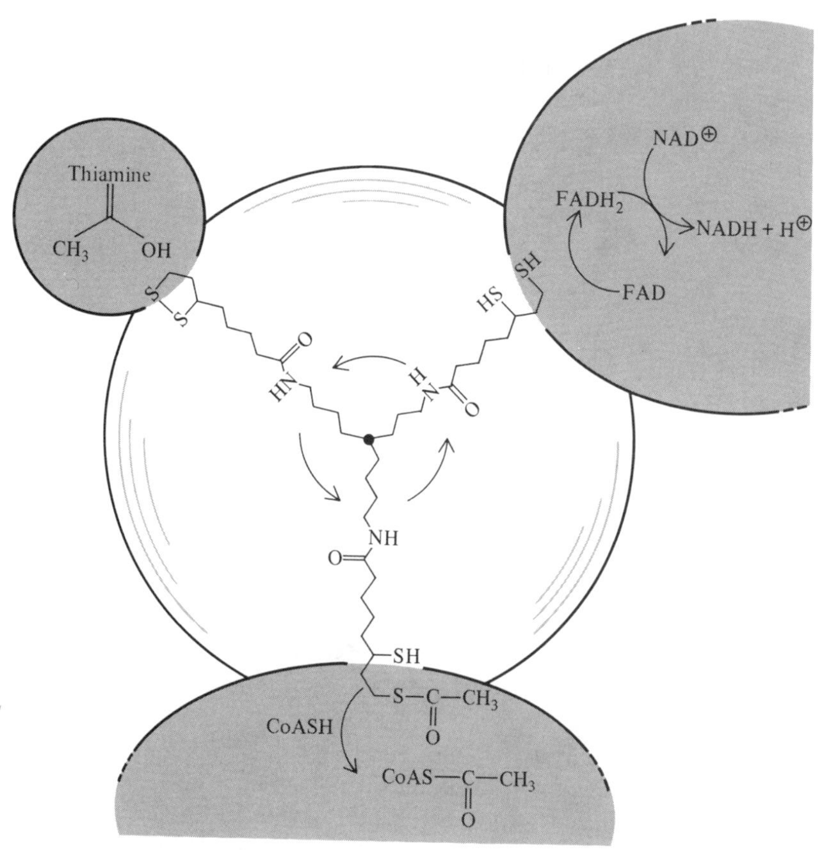

Fig. 7.8. Diagram of the enzyme complex in the oxidative decarboxylation of pyruvic acid.

The swinging arm of this prosthetic group coordinates other subsites of the complex so that the efficiency of the process is related to this unique chemical organization at the subcellular level.

In the cell the thioester of dihydrolipoic acid is not stored but is coupled to another sulfur-containing coenzyme: *coenzyme A*, CoA–SH, the universal acyl-transfer agent.

$$HS{-}CH_2CH_2NHC(=O)CH_2CH_2NHCOCH(OH){-}C(CH_3)_2{-}O{-}P(=O)(O^-){-}O{-}P(=O)(O^-){-}O{-}\text{(3'-phosphoadenosine)}$$

pantothenic acid

CoA—SH

Acetyl-CoA is the real form of "activated acetate" where 36.9 kJ/mol (8.8 kcal/mol) is liberated upon hydrolysis (see Chapter 2). Reduced lipoic acid is reoxidized with an FAD coenzyme present on the complex.

$$CH_3{-}C(=O){-}S\ \ SH\text{(dihydrolipoyl-R)} + CoA{-}SH \rightleftharpoons CH_3{-}C(=O){-}S{-}CoA + HS\ \ SH\text{(dihydrolipoyl-R)}$$

$$\xrightarrow{FAD} \text{S—S (lipoyl-R)} + FADH_2$$

7.2 Pyridoxal Phosphate

Pyridoxine or vitamin B_6 is an important dietary requirement. The aldehyde form is called pyridoxal and its phosphate ester is implicated in many enzyme-catalyzed reactions of amino acids and amines. The reactions are numerous and pyridoxal phosphate (pyridoxal-P) is surely one of nature's most versatile catalysts or coenzymes. The chemistry that will be emphasized here is one of proton transfer. In transamination (equation

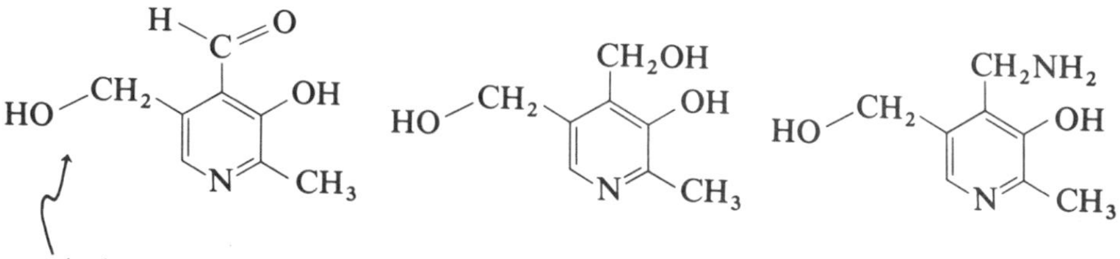

$$\text{pyridoxal (4-CHO, 3-OH, 2-CH}_3\text{, 5-CH}_2\text{OH pyridine)} \qquad \text{pyridoxine (pyridoxol) (4-CH}_2\text{OH)} \qquad \text{pyridoxamine (4-CH}_2\text{NH}_2\text{)}$$

in the active cofactor this hydroxyl is phosphorylated

pyridoxal

pyridoxine (pyridoxol)

pyridoxamine

Structure of vitamin B_6 compounds

7-1), a process of central importance in nitrogen metabolism, it is converted to pyridoxamine.

In fact, pyridoxal-P coenzyme catalyzes at least seven very different reactions where acid–base chemistry and tautomerism is fully exploited.

$$R-\underset{\substack{\|\\O}}{C}-CO_2^{\ominus} + R'-\underset{\substack{|\\ {}^{\oplus}NH_3}}{CH}-CO_2^{\ominus} \rightleftharpoons R-\underset{\substack{|\\ {}^{\oplus}NH_3}}{CH}-CO_2^{\ominus} + R'-\underset{\substack{\|\\O}}{C}-CO_2^{\ominus} \tag{7-1}$$

transamination

$$HOCH_2-\underset{\substack{|\\ {}^{\oplus}NH_3}}{CH}-CO_2^{\ominus} \longrightarrow CH_3-\underset{\substack{\|\\O}}{C}-CO_2^{\ominus} \tag{7-2}$$

serine pyruvate

elimination–hydration

$$\overset{2\ominus}{HO_3PO}-CH_2CH_2-\underset{\substack{|\\ {}^{\oplus}NH_3}}{CH}-CO_2^{\ominus} \longrightarrow CH_3-\overset{\substack{OH\\|}}{CH}-\underset{\substack{|\\ {}^{\oplus}NH_3}}{CH}-CO_2^{\ominus} \tag{7-3}$$

homoserine phosphate threonine

elimination–hydration

$$HO_2C-CH_2CH_2-\underset{\substack{|\\ {}^{\oplus}NH_3}}{CH}-CO_2^{\ominus} \longrightarrow HO_2C-CH_2CH_2-CH_2-NH_2 + CO_2 \tag{7-4}$$

glutamic acid γ-aminobutyric acid (GABA)

decarboxylation

$$HOCH_2—CH(NH_3^{\oplus})—CO_2^{\ominus} \longrightarrow H_2C{=}O + CH_2(NH_3^{\oplus})—CO_2^{\ominus} \quad (7\text{-}5)$$

serine glycine

reverse condensation

$$R—C(H)(NH_3^{\oplus})—CO_2^{\ominus} \rightleftharpoons R—C(H)(NH_3^{\oplus})—CO_2^{\ominus} + R—C(CO_2^{\ominus})(NH_3^{\oplus})—H \quad (7\text{-}6)$$

racemization

$$\text{indole} + CH_2(OH)—CH(NH_3^{\oplus})—CO_2^{\ominus} \longrightarrow \text{tryptophan} \quad (7\text{-}7)$$

indole serine tryptophan

tryptophan synthesis

While it may be surprising that the above diverse reactions require the same cofactor, this will be readily understood when it is realized that these reactions have certain common features. All require imine (Schiff base) formation between the aldehyde carbonyl of the cofactor and the amino group of the substrate. The pyridoxal phosphate becomes an electrophilic catalyst or "electron sink," since electrons may be delocalized from the amino acid into the ring structure. It is the direction of this delocalization that dictates that reaction type and in model systems more than one reaction pathway is often observed. Thus the enzyme both enhances the rate of reaction and gives direction to that reaction (see page 529).

process a: leads to decarboxylation
process b: leads to transamination
process c: leads to aldol-type condensation and retrocondensation

In the transamination process, the pyridoxal coenzyme is transferred from the enzyme–imine intermediate to the substrate–imine. The evidence of an imine function comes from reduction by borohydride which does not yield pyridoxine but shows that a covalent bond is formed with a lysine residue of the enzyme. The protonation of the pyridine ring is also essential for catalysis.

What advantage, if any, is served by formation of a Schiff base between the enzyme and coenzyme? Enzyme-bound imine must provide a more rapid pathway than the substrate-bound imine (614). There is thus a structural reason linked with the greater reactivity of imines as compared to the corresponding aldehydes. A more basic nitrogen forms a stronger hydrogen bond (i.e., with an appropriate hydrogen bond donor on the enzyme surface) and is protonated to a much greater extent than an

$^{-2}O_3PO$ H⊕ N OH CHO

NH_2 Enz

$^{-2}O_3PO$ ⊕H N OH N Enz

1) $BH_4^{\ominus}$ 2) $H^{\oplus}$, Δ

$^{-2}O_3PO$ ⊕H N OH NH $(CH_2)_4$—CH $CO_2^{\ominus}$ $\overset{\oplus}{N}H_3$

Lys derivative

R – CH $CO_2^{\ominus}$ $\overset{\oplus}{N}H_3$

$^{-2}O_3PO$ ⊕H N OH N CH $^{\ominus}O_2C$ R

NH_2 Enz

oxygen. Furthermore, an imine carbon is more electrophilic than a carbonyl carbon; hence it is attacked more readily by nucleophiles.

Thus, an enzyme–imine intermediate facilitates a more rapid formation of a covalent intermediate between substrate and coenzyme.

The role of the phosphate is to bind the coenzyme to the corresponding apoenzyme. Lowe and Ingraham (614) and D. Arigoni of E.T.H., Zurich, proposed an attractive hypothesis (Fig. 7.9) where the phosphate and the methyl groups form an axis about which pyridoxal could pivot between enzyme–imine and substrate–imine covalent structures.

As pointed out by the authors, if everything in nature exists for a reason, this may indeed be the case (614).

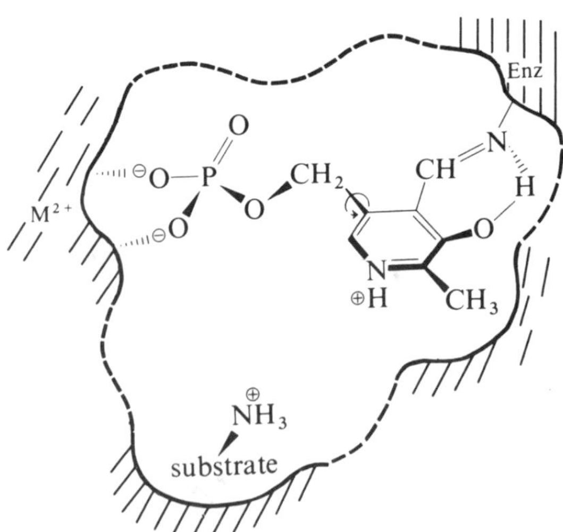

Fig. 7.9. Hypothetical mode of action of pyridoxal phosphate. Adapted from Lowe and Ingraham (614) with permission.

7.2.1 Biological Role

Pyridoxal phosphate has several features that make it an excellent catalyst for transamination reactions. In particular, the hydroxyl group is ideally located to provide general-acid and -base catalysis. Being intramolecular it is particularly effective. Second, the positively charged nitrogen of the pyridine ring acts as an electron sink to lower the free energy of C—H bond tautomerization. Finally, the groups are positioned in just the right geometry on the enzyme surface.

Addition of a metal ion such as Al(III) to the nonenzymatic system considerably enhances the catalytic effect (615). It complexes with the imine and acts as a general-acid catalyst.

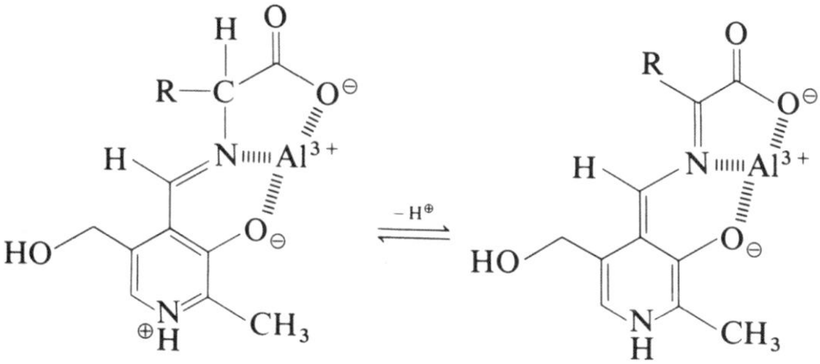

H O R—C O⊖ H N Al^{3+} O⊖ HO N ⊕H CH$_3$ $\underset{}{\overset{-H^{\oplus}}{\rightleftharpoons}}$ O R O⊖ H N Al^{3+} O⊖ HO N H CH$_3$

It has been demonstrated that Cu(II) and Fe(III) can also greatly accelerate the rate of reaction and actual chelate intermediates have been isolated. In model systems, all reactions catalyzed by enzymes except decarboxylation have been reproduced: transamination, oxidative deamination, elimination of β- and γ-substituents, and the like. Many substitutes for pyridoxal have been synthesized and examined in the biological system. The following compounds were observed to be ineffective:

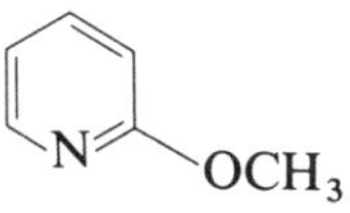

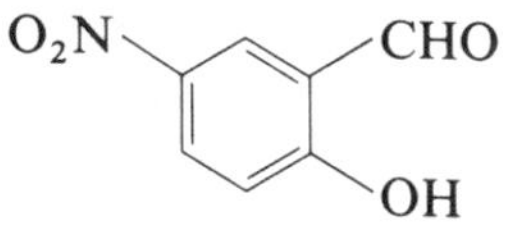

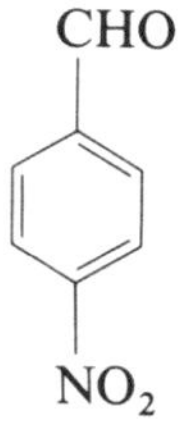

N OCH$_3$ O$_2$N CHO OH CHO NO$_2$

whereas the following were effective catalysts:

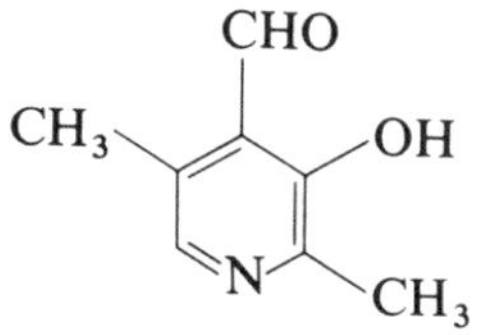

CHO CH$_3$ OH N CH$_3$

5′-deoxypyridoxal analog

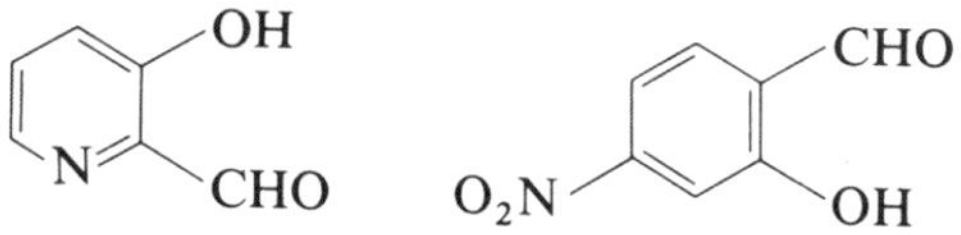

OH N CHO

CHO O$_2$N OH

Proper electron delocalization is thus important. Furthermore, the presence of a hydroxyl group for chelation with metal ions appears essential. Surprisingly, with α-phenyl-α-aminomalonic acid in the presence of Cu(II) and appropriate model compounds (see below), an oxidation reaction takes place that has no biological precedent (616).

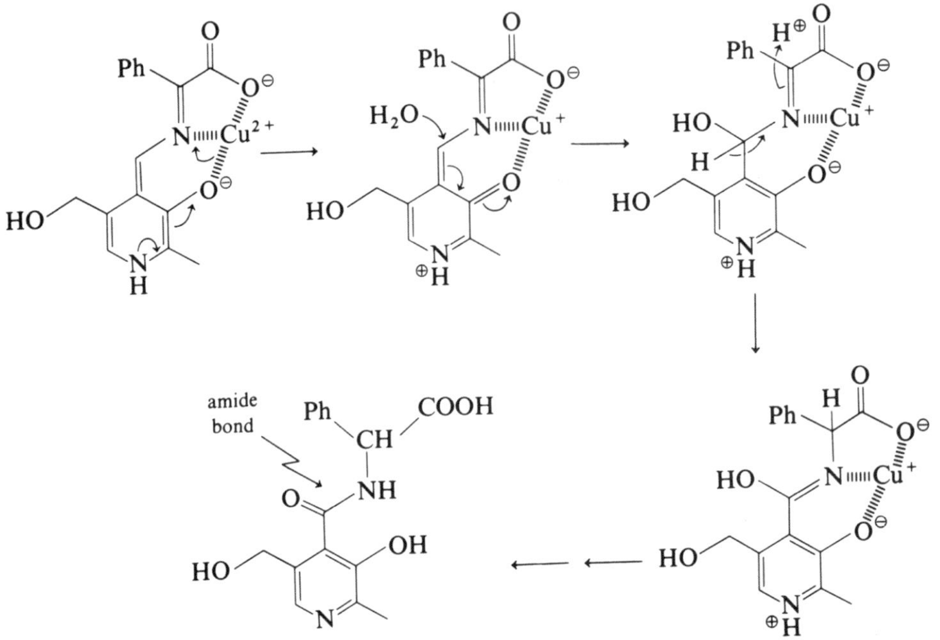

It should be noted that the role of the metal ion has often been overestimated in pyridoxal-P-dependent enzymes because most of these enzymes are not metalloenzymes. Rather, it is the enzyme itself that plays a template role, as the metal ion.

To better understand the mode of action of pyridoxal-P, let us examine in detail equation 7-3; the conversion of homosereine phosphate to threonine. It is an elimination–hydration transformation. The first process (Fig. 7.10) involves an aldimine ⇌ ketimine tautomerization which is subjected to general-acid catalysis intramolecularly by the hydroxyl group, followed by the slow breaking of a C—H bond. The latter is the rate-determining step.

The presence of a negative charge on the β-carbon can be shown by trapping the anion with *N*-methylmaleimide (NMM) via a Michael addition. Interestingly, pyridoxal phosphate helps stabilize a negative charge (anionic form) on α-, β-, and γ-carbon positions. Conjugation with the ring nitrogen stabilizes the negative charge on α and γ positions while the imine nitrogen stabilizes the negative charge on the β position. In the presence of ^2H_2O, positions α and γ can be deuterated. This particularly clear cut sequence of protonations shows that the role of pyridoxal phos-

(NMM)

aldimine

ketimine

enamine

$-H^{\oplus}$

$-H^{\oplus}$

$-PO_4^{3\ominus}$

One fascinating aspect of this Schiff-base chemistry is that, depending on the need of the coenzyme, all three positions (α, β, and γ) of the bound substrate can have either carbanionic or carbocationic character

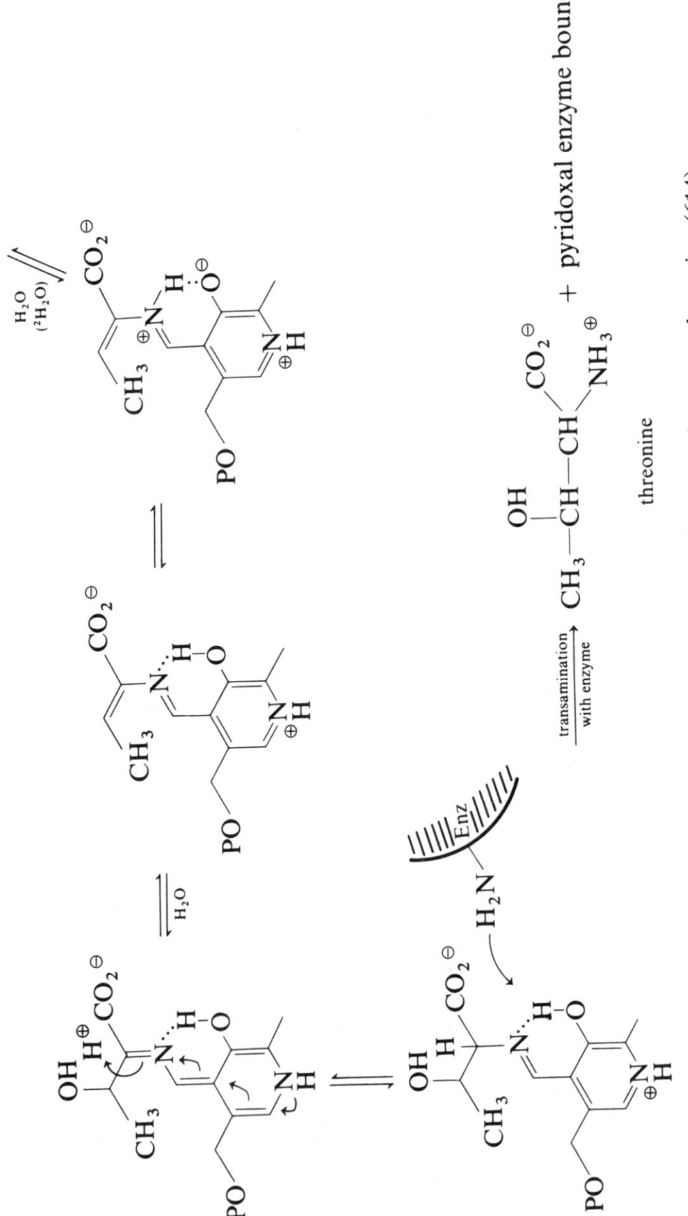

Fig. 7.10. The conversion of homoserine phosphate to L-threonine (614).

phate is to stabilize carbanionic intermediates by acting as a delocalizing electron sink for the extra electron density (617).

It should be realized that since pyridoxal is converted to pyridoxamine or the corresponding enzyme–imine intermediate, it is not really a true catalyst! Rather, it is a reactant.

One fascinating aspect of this Schiff-base chemistry is that, depending on the need of the coenzyme, all three positions (α, β, and γ) of the bound substrate can have either carbanionic or carbocationic character.

The most important central problem in transamination processes is a consideration of stereochemistry. The enzyme–coenzyme complex, depending on the reaction and enzyme type, can remove from the substrate amino acid the R group, the carboxylate group, or the hydrogen of the α-carbon. What structural features determine which bonds are to be broken? This depends on the enzyme, as does the reaction rate. The critical factor is the lowest energy pathway of the transition state of the covalent intermediate. In other words, the correct conformation of the coenzyme-bound substrate on the enzyme must have a dominant influence (614).

transaminase aldolase decarboxylase

Orientational dependency of pyridoxal-phosphate-containing enzyme-catalyzed reactions

The labile bond is always the one perpendicular to the pyridine ring and combined ionic, polar, and hydrophobic interactions on the enzyme determine which conformer predominates. This is easily seen, for example, in the Newman projection of enzymatic decarboxylation. The con-

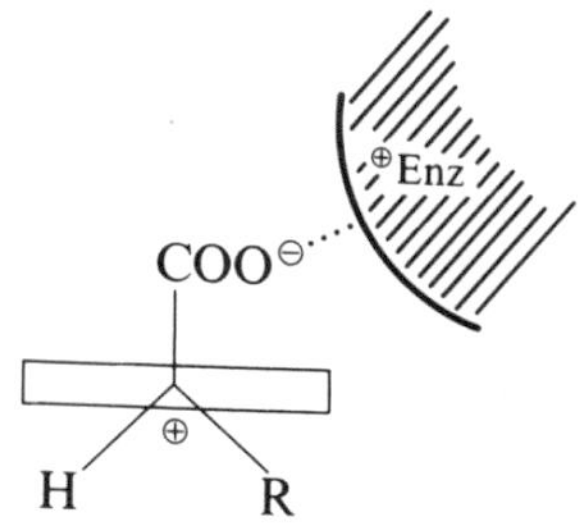

Newman projection for the transition state of a pyridoxal phosphate enzyme catalyzed decarboxylation

formation required for decarboxylation places the carboxyl group substantially out of the plane of the conjugated system. Consequently, the specificity of the reaction is manifested principally at this stage. For instance, enzymatic decarboxylation of amino acids occurs with retention of configuration and thus allows the preparation of optically pure α-deuterated amines if the reaction is carried out in heavy water (618).

Metal binding, if it occurs, can give different pathways. An interesting example of this effect is observed with L-serine hydroxymethyltransferase which can also catalyze the transamination of D-serine. The same binding site is implicated but different products are obtained:

M^{2+} ··· $^{\ominus}O_2C$–C(H)(CH_2OH)–N= ; $\xrightarrow{-H^{\oplus}}$ M^{2+} ··· $^{\ominus}O_2C$–C(CH_2OH)=N– ; $\xrightarrow{H_2O}$ $^{\ominus}O_2C$–C(=O)–CH_2OH

if L-Ser is bound to the coenzyme

M^{2+} ··· $^{\ominus}O_2C$–C(CH_2OH)(H)–N= ; $\xrightarrow{-\overset{\oplus}{C}H_2OH}$ M^{2+} ··· $^{\ominus}O_2C$–C(H)=N– ; $\xrightarrow{H_2O}$ $^{\ominus}O_2C$–CH(H)–NH_2

Gly

if D-Ser is bound to the coenzyme

The orientation of the carboxyl group is thus determined here by the structure of the binding site (614). It shows also the important role of vitamins because excess glycine and serine in the system can be toxic but not if enough vitamin B_6 is administered.

Clearly, besides proton transfer, pyridoxal phosphate is also implicated in mechanisms where carbanion chemistry is involved. By leaving a negative charge on the α-carbon of an amino acid substrate, a new problem is introduced; one of stereochemistry. On the enzyme, does this carbanion (negative charge) finally get protonated by proton exchange with the medium or by a tautomeric form of the coenzyme? What kind of models can one select to best imitate such processes?

7.2.2 Model Systems

The following results on models are evidence of the existence of carbanion intermediates in pyridoxal enzymes. For instance, NMR studies by Gansow and Holm (615) in 1969 on pyridoxamine in the presence of Zn(II) or Al(III) ions in ^2H_2O, pH 1 to 13 range, showed the condensation of substrate (pyruvate) with a B_6 vitamin (pyridoxamine). The rate of exchange of hydrogen atoms increases with pH.

$CH_3{-}C(=O){-}CO_2^{\ominus}$ + pyridoxamine $\underset{}{\overset{M^{+m}}{\rightleftharpoons}}$ [metal complex] $\overset{^2H_2O}{\rightleftharpoons}$ [deuterated complex]

1) further ^2H_2O exchange
2) H_2O hydrolysis

$CD_3{-}C(=O){-}CO_2^{\ominus}$
+
pyridoxamine

Earlier it was observed that the complex of pyridoxamine, ethyl pyruvate, and $Al(NO_3)_3$ in methanol gives a species absorbing at 488 nm: This visible band corresponds to the presence of a carbanion intermediate that disappears after a few hours to give transaminated products.

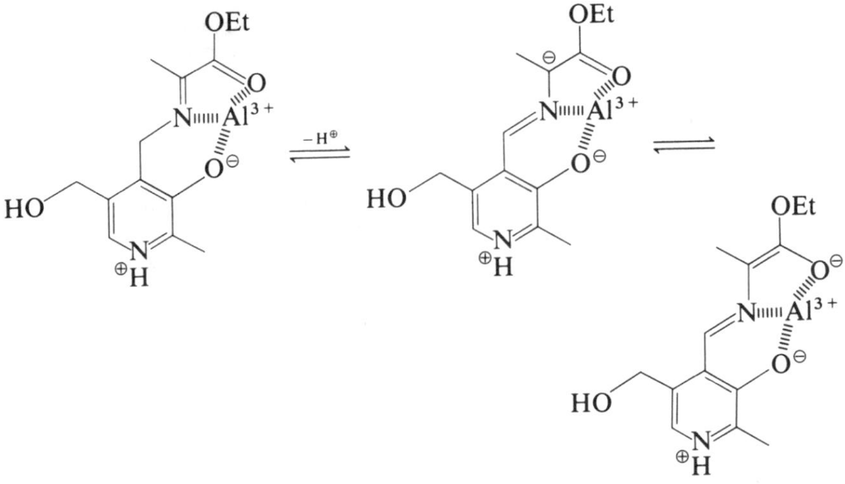

Other data should also be examined. For example, pyridoxamine in the presence of an α-keto acid and appropriate enzyme gives pyridoxal plus the corresponding L-amino acid. If the reaction is undertaken in ^2H_2O, half of the methylene hydrogens of pyridoxamine are exchanged and only one monodeuterated pyridoxamine isomer is formed.

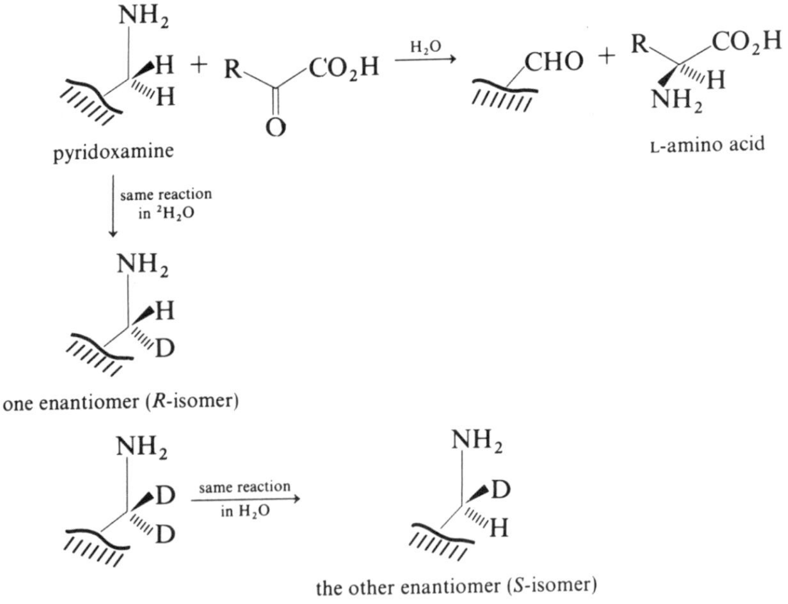

pyridoxamine

L-amino acid

same reaction in ^2H_2O

one enantiomer (*R*-isomer)

same reaction in H_2O

the other enantiomer (*S*-isomer)

If double-deuterated pyridoxamine is used for the same reaction in water, half of the deuterium is lost and the other monodeuterated pyridoxamine enantiomer is obtained. Furthermore, if an α-deuterated

L-amino acid is used, the deuterium is transferred stereospecifically to give only one of the two possible stereoisomers of monodeuterated pyridoxamine.

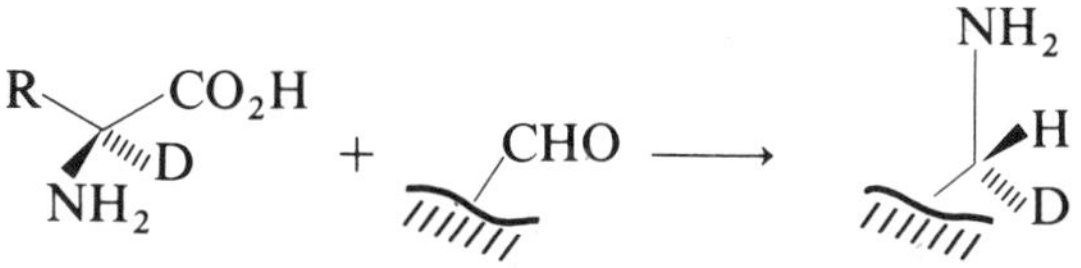
R
CO_2H
D
NH_2
CHO
NH_2
H
D

only the *R*-isomer

H.C. Dunathan and his coworkers (619,620) at Haverford College in Pennsylvania contributed information about these processes by studying the mechanism and stereochemistry of transamination reactions. They defined the problem by three stereochemical variables:

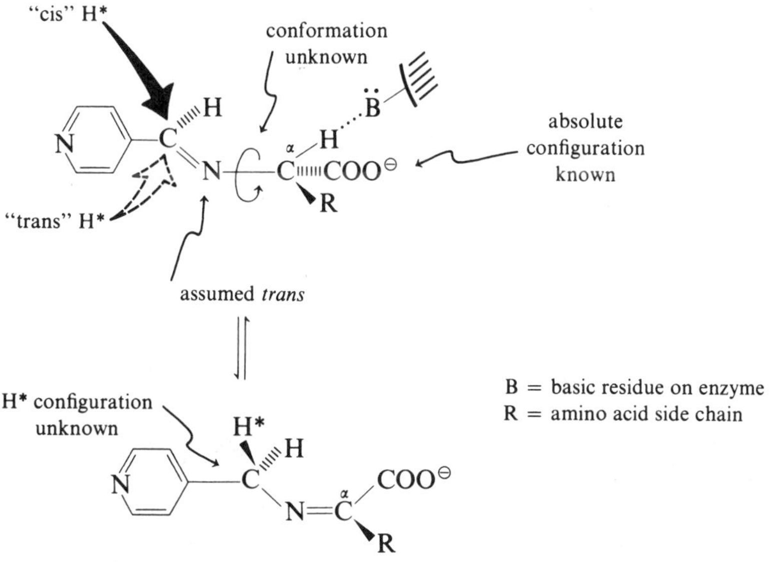
"cis" H*
conformation unknown
absolute configuration known
"trans" H*
assumed *trans*
H* configuration unknown
B = basic residue on enzyme
R = amino acid side chain

Therefore, the tautomerization process can be pictured as either an intramolecular (enzyme-assisted) *cis* transfer or a *trans* transfer of hydrogen. This rearrangement is referred to as a tautomeric 1,3-prototropic shift and model studies have tried to solve this stereochemical ambiguity:

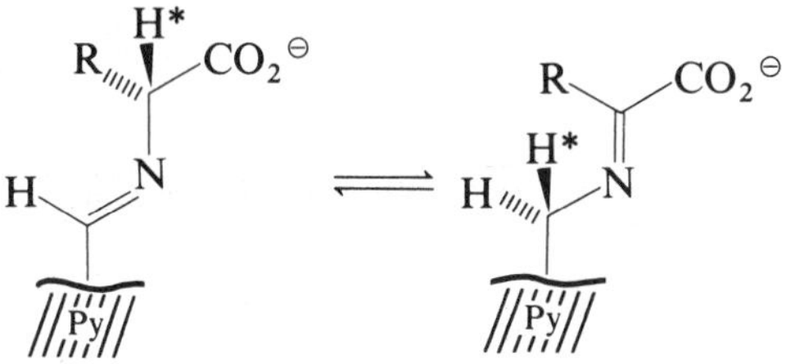
H*
R
$CO_2^{\ominus}$
N
H
Py
R
$CO_2^{\ominus}$
H*
N
H
Py

cis transfer
(suprafacial)

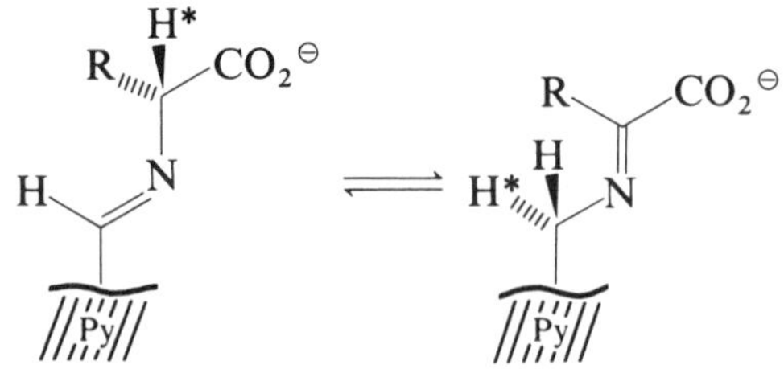

trans transfer (antarafacial)

First, the biochemical observations of Dunathan (620) and others led to the hypothesis that the prototropic shift is stereospecific with a *cis* internal transfer. If so, a basic residue on the apoenzyme must be implicated in the process where an azaallylic anion intermediate is formed (Fig. 7.11).

The group of D.J. Cram (621–624) studied in detail this base-catalyzed methylene–azamethine rearrangement.

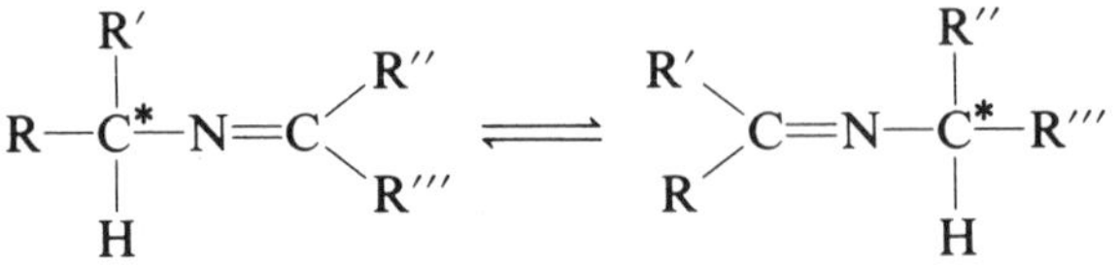

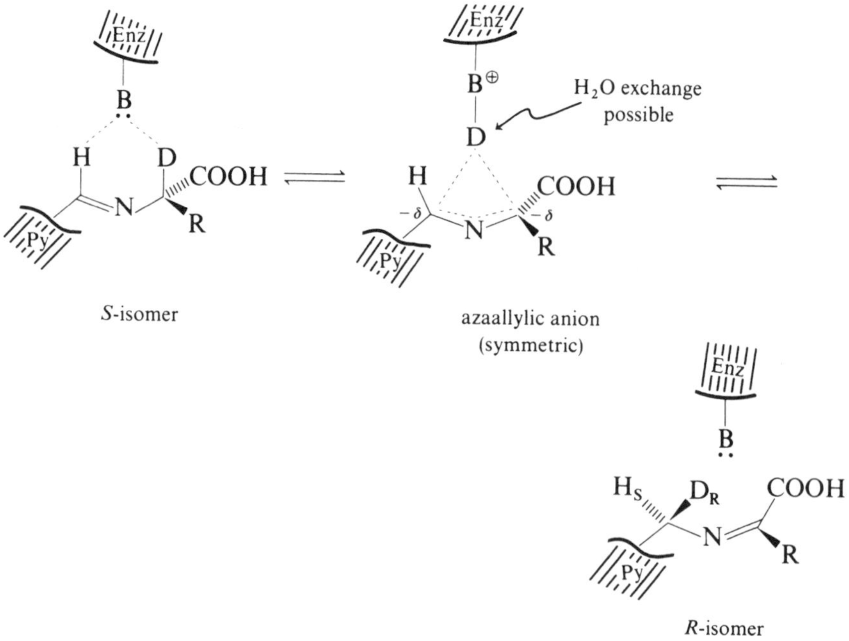

S-isomer

azaallylic anion (symmetric)

H_2O exchange possible

R-isomer

Fig. 7.11. The *cis* internal 1,3-prototropic shift involved in enzyme-mediated pyridoxal-phosphate-catalyzed reaction. Py, the rest of the coenzyme molecule.

For such carbanionic rearrangement, two mechanisms in EtO^{-}–EtOH can be proposed. One stage concerted:

$$\begin{array}{c} -\overset{|}{\underset{|}{C}}-N=\overset{|}{C}- \;\rightleftharpoons\; -\overset{|}{C}\cdots N-\overset{|}{C}- \;\rightleftharpoons\; -\overset{|}{C}=N-\overset{|}{\underset{|}{C}}- \\ \quad H \qquad\qquad RO-H \quad HO-R \qquad\qquad H \\ RO^{\ominus} \quad HO-R \qquad\qquad\qquad\qquad R-OH \quad {}^{\ominus}OR \end{array}$$

transition state

or two stages of anionic mechanism:

$$\begin{array}{c} -\overset{|}{\underset{|}{C}}-N=\overset{|}{C}- \;\rightleftharpoons\; -\overset{|}{C}\overset{\cdots}{=}\underset{\ominus}{N}\overset{\cdots}{=}\overset{|}{C}- \;\rightleftharpoons\; -\overset{|}{C}=N-\overset{|}{\underset{|}{C}}- \\ H \qquad\qquad\qquad\qquad\qquad\qquad H \\ RO^{\ominus} \qquad\qquad ROH \qquad\qquad {}^{\ominus}OR \end{array}$$

azaallylic anion

Monitoring the isomerization process in deuterated solvent allows the determination of the mechanism operative in the model system. For instance, in the concerted mechanism, that proceeds through a symmetric transition state, for each deuteron incorporated from the solvent, an isomerization must occur. On the other hand, via the azaallylic intermediate, it is possible for a deuteron to be incorporated without a subsequent isomerization. It is this latter process that has been observed, thus strongly supporting the existence of the azaallylic intermediate (621,622).

Since the kinetics of this reaction were studied in $^{2}H_2O$ and in H_2O, a few comments should be made concerning H–D exchange. Starting with a compound having an acidic hydrogen at a chiral center, three different products are possible via a carbanion intermediate (621):

(a)(b)(c)C*—H	$\xrightarrow{ROD,\ RO^{\ominus}}$	(a)(b)(c)C—H	retention, no exchange
″	⟶	(a)(b)(c)C—D	retention, with exchange
″	⟶	D—C(a)(b)(c)	inversion, with exchange
″	⟶	H—C(a)(b)(c)	inversion, no exchange (*isoinversion*)

The process of inversion (racemization) without exchange is called *isoinversion*. This involves an ion-pair mechanism. Indeed, in the presence of a crown ether, to separate the ions, the percentage of racemate increases (621). In fact, the stereochemical course of many metal-alkoxide-catalyzed reactions in nonpolar solvents can be drastically modified by addition of catalytic amounts of crown ethers to the medium. For this reason, ion pairs, in low dielectric media, play a remarkable role as intermediates in reactions in which the negative ion is a carbanion. For example, Cram studied the rate of exchange (k_e) and racemization (k_α) as a function of the substituents present with deuterium-labeled 9-methylfluorene (623). The most interesting case was $k_e/k_\alpha < 0.5$, racemization without exchange. This happens for X = dimethylamide in the presence of tri-*n*-propylamine base and in *tert*-butanol-THF solvent-replacing methanol.

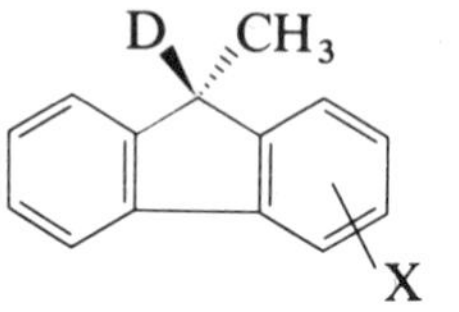

k_e/k_α	Reaction solvent-base
0.6 (exchange with inversion)	MeOH-N(Pr)$_3$
1 (exchange with racemization)	*t*BuOH-*t*BuO$^-$K$^+$
> 1 (exchange with retention)	C_6H_6-C_6H_5OH-$C_6H_5O^-K^+$
< 0.5 (isoinversion)	*t*BuOH-THF-N(Pr)$_3$

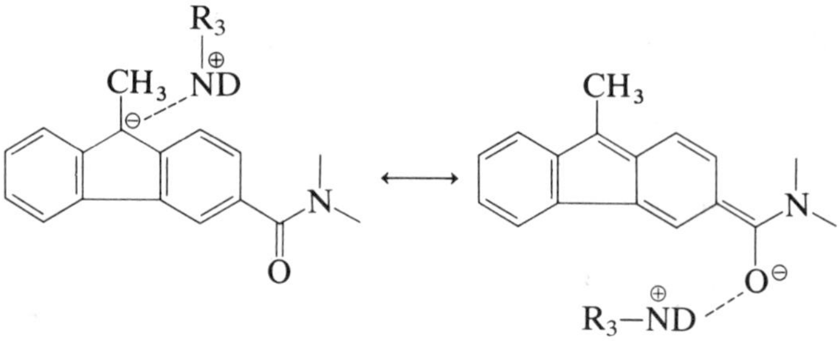

Cram reasoned that the deuterium is removed by the amine, and the ammonium ion formed remains paired with the carbanion via an ionic bond. The cation need not remain at the original site because the resonance of the ring system allows the negative charge to be delocalized all the way to the oxygen of the substituent. Then the ion pair which now lies in the plane of the ring may slide to the other side of the planar structure or return to the original position without exchanging its deuterium with the solvent protons. For this particular process, Cram coined the name *conducted-tour mechanism* (the base has taken a conducted tour) to explain the phenomenon of isoinversion. Notice that in methanol (which is a stronger acid than *tert*-butanol) the carbanion is more readily protonated and so does not have a long enough half-life to allow a conducted-tour process.

It should be noted that isoinversion can also occur without a conducted-tour mechanism. This has been observed again with 9-methylfluorene and *tert*-butanol solvent, but with tri-*n*-propylamine replaced by the base pentamethylguanidine. That an alternative mechanism is operative is indicated by the fact that a substituent such as dimethylamide no longer needs to be present. The base can provide a charge delocalization pathway for travel of the deuteron from one side of the planar ring system to the other without solvent exchange. The deuteron can thus add to either side of the plane, as is illustrated below:

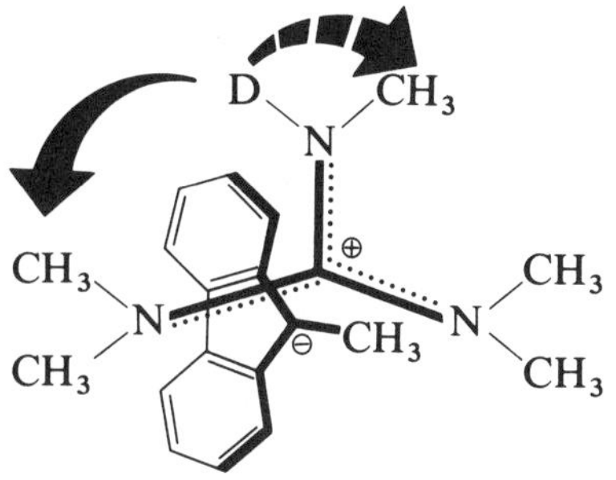

With this brief excursion into carbanion chemistry we can now examine the models that Cram and co-workers used to study the stereochemical outcome of the proton transfer stages of pyridoxal-mediated biological transaminations (624).

The following model systems were examined (625):

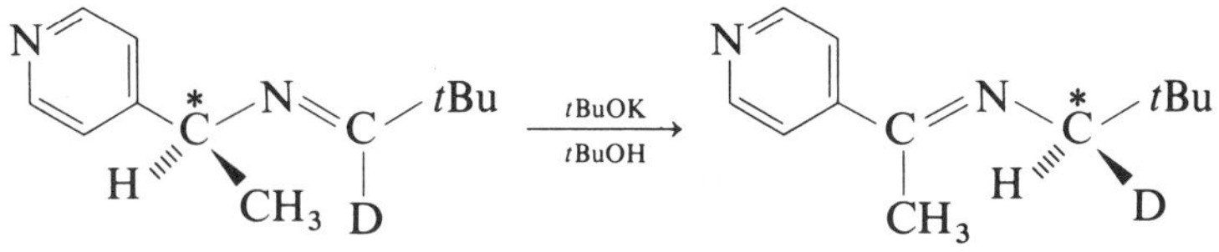

(*S*)-(−)-**7-1** $\xrightarrow[tBuOH]{tBuOK}$ (*R*)-(+)-**7-2**

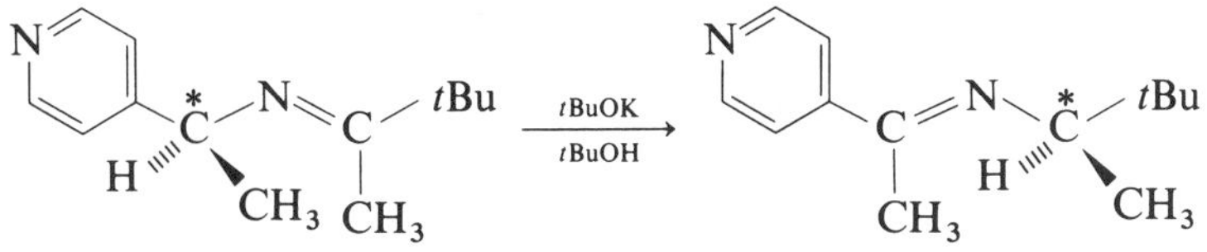

(*S*)-(−)-**7-3** $\xrightarrow[tBuOH]{tBuOK}$ (*R*)-(−)-**7-4**

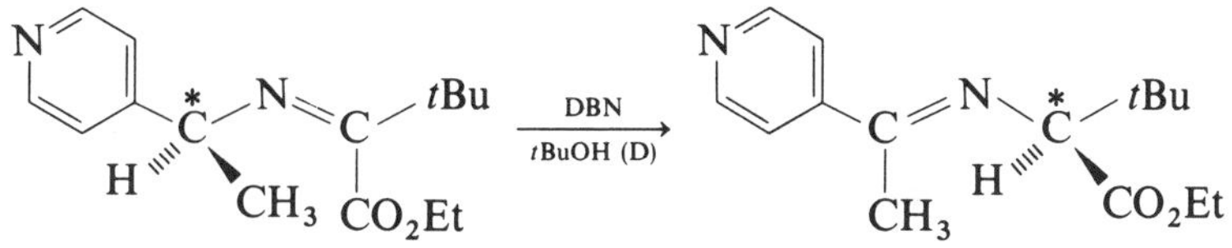

(*S*)-(−)-**7-5** $\xrightarrow[tBuOH\ (D)]{DBN}$ (*S*)-(−)-**7-6**

With all the processes, isomerizations were stereospecific as indicated. Reaction in *tert*-butyl alcohol-*O-d* allows isotopic exchange and helps to monitor the course of the reactions.

For instance, starting with pure (−)-(*S*)-**7-5**, in the presence of the base diazobicyclo [4.3.0] nonene (DBN), after 811 hr of equilibration, less than 0.5% of optically pure **7-5** remains and a 30 to 60% conversion to (−)-(*S*)-**7-6** is observed. This result indicates that the interconversion is stereospecific and that *suprafacial* (same side) dominates *antarafacial* (opposite side) isomerization of **7-5** → **7-6**. The large pyridine ring and the *tert*-butyl group enforce conformational homogeneity of the azaallylic system. The chiral center of **7-5** racemizes without exchange with the solvent. This corresponds to the phenomenon of isoinversion via a conducted-tour mechanism. The inversion is believed to occur via a bridge between DBN and the pyridine nucleus:

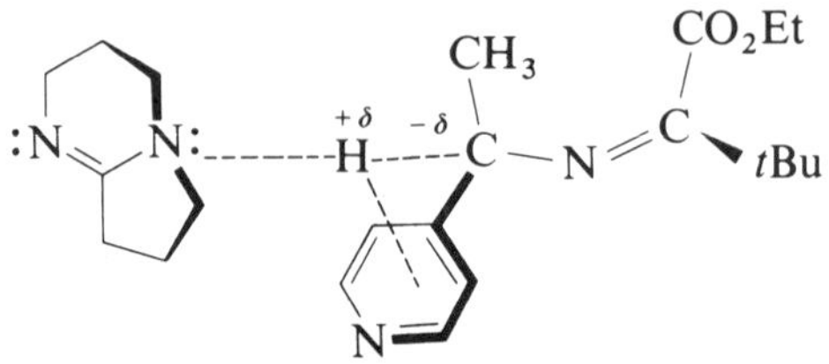

The chiral center of **7-6**, however, exchanges with retention of configuration; it cannot invert because it is too bulky.

In conclusion, the tautomerization is intramolecular and the suprafacial 1,3 proton shift occurs across an azaallylic anion. The model differs slightly, however, from the biological system by providing competing stereochemical and isotope-labeling reactions pathways. Therefore, coenzymes carry out stereospecific reactions because of their apoenzymes, while nonenzymatic model reactions are not as stereospecific (624).

This stereospecific transformation can be used advantageously in the case of a bulky amino acid such as L-threonine for the preferential asymmetric synthesis of a monodeuterated pyridoxamine:

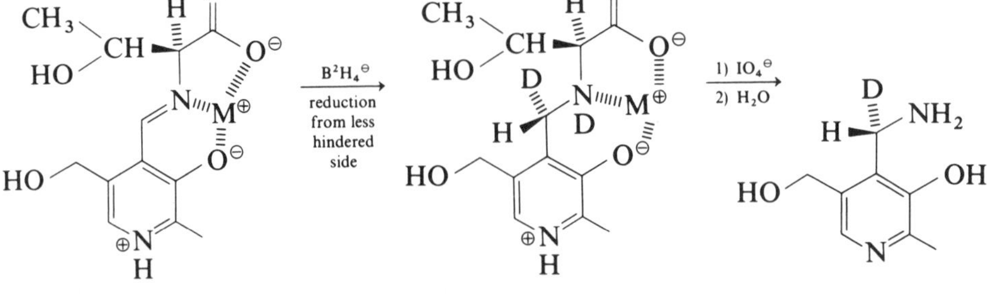

$B^2H_4^{\ominus}$ reduction from less hindered side; 1) $IO_4^{\ominus}$ 2) H_2O

L-amino acid

S-isomer

From D- and L-threonine, the two pyridoxamine products give different isotopic primary effects with glutamic-oxaloacetic transaminase.

Again, cyclodextrin and cyclophane cavities have been used as hosts to mimic some of the properties of pyridoxal phosphate and pyridoxamine phosphate. R. Breslow is considered one of the principal artisans in the development of artificial transaminase enzymes (626).

A β-cyclodextrin-pyridoxamine was used to mimic tryptophan synthase as well as the conversion of indolepyruvic acid to tryptophan. The chirality of β-cyclodextrin induces a selective transamination with 12% ee for the L-isomer. The product selectivity is lost with time because of rapid equilibration of keto acid with amino acid. In a similar approach to the synthesis of phenylalanine, chiral induction gave a 5:1 preference of L-Phe over the D-enantiomer. The rate of conversion was accelerated 18-fold as compared to the conversion of pyruvic acid to alanine. Surprisingly, with pyridoxamine bound to secondary hydroxyl groups of β-CD (at C_3), on the more open side of the cyclodextrin structure, very little enantiomeric preference was detected.

from β-CD-6-tosylate

β-CD-pyridoxamine

bound intermediate

With the following water-soluble macrocyclic cyclophane derivative, conversion of phenylpyruvic acid to phenylalanine or α-ketovaleric acid to norvaline was slower and the reaction showed no significant selectivity compared to β-CD-induced chirality.

cyclophane-pyridoxamine

In a joint effort, Breslow and Tabushi converted an **A**,**B**-capped cyclodextrin (see Section 5.5.1 for details) into a new derivatized pyridoxamine-aminoethylamine-β-CD.

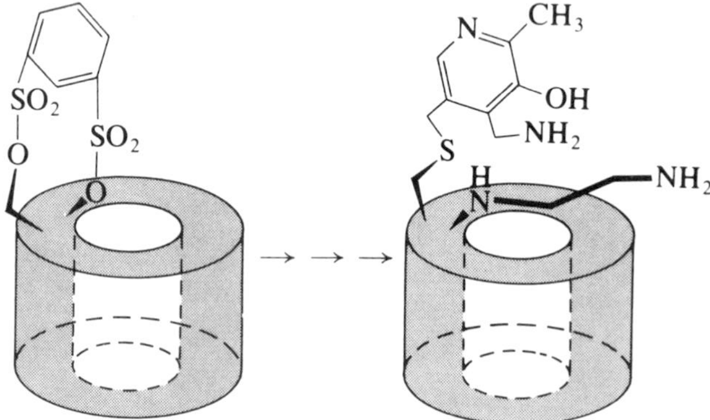

Conversion of indolepyruvic acid to tryptophan resulted in this case to a 95:5 formation of the L-isomer over the D-isomer (627). The observed high chiral induction may be understood on the basis of a stereospecific assistance of the terminal amine group of the neighboring chain toward the prochiral planar azaallylic intermediate during the protonation step.

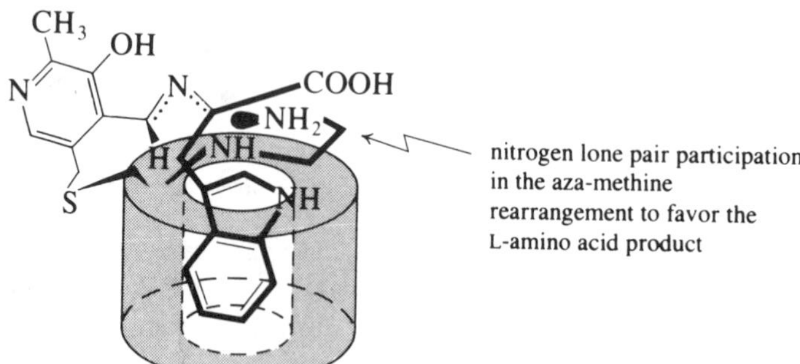

With similar objectives, a mono-(3-aminopropylamino)-β-cyclodextrin-Ni(II) complex was prepared. It bound very favorably with a tryptophan anion and exhibited a 10-fold enantioselectivity in favor of the L-Trp isomer.

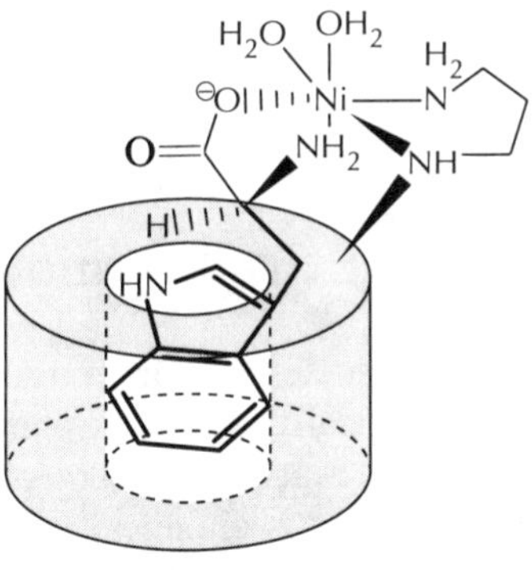

In this case, both ion binding and hydrophobic binding are used in the complexation and molecular recognition (628).

In nature, the enzyme uses a catalytic amino group from a lysine residue to remove the *pro-S* hydrogen from the 4′-methylene of the ketimine coenzyme–substrate intermediate (see p. 534). The resulting ammonium cation reprotonates the intermediate on the *si*-face to eventually generate the chiral α-carbon of the amino acid. In order to mimic the aldimine–ketimine proton transfer, Zimmerman and Breslow developed an appropriate system based on a tetrahydroquinoline backbone (629).

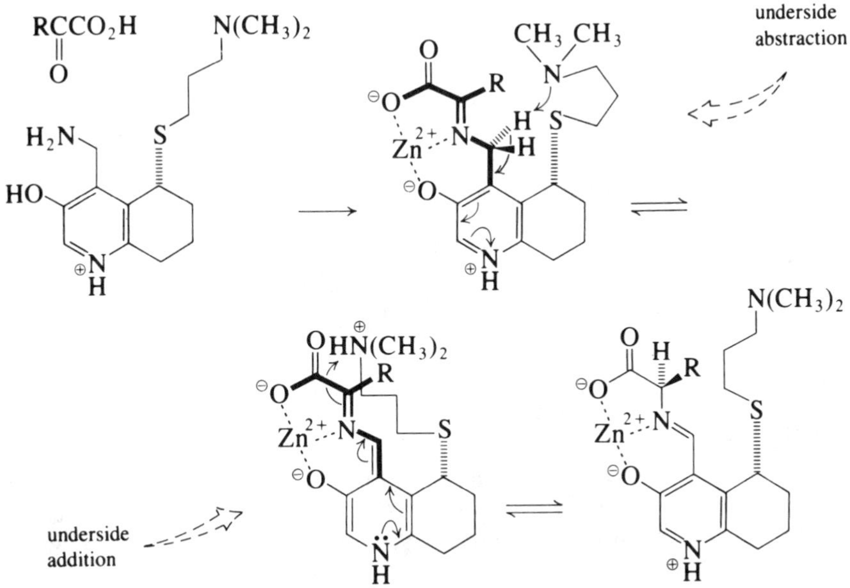

underside abstraction

underside addition

In this molecule, the rigidly mounted side arm is constrained to perform proton transfers on one face only of the transamination intermediate, as in the enzyme. Indeed, this constraint resulted in an extraordinary stereoselectivity of 93:7 in D- to L-isomer ratio with an 83% conversion of pyruvic acid to alanine. The strong D preference indicates that this general base–acid catalysis must promote proton transfer along the *si*-face of the ketimine C=N bond.

7.3 Suicide Enzyme Inactivators and Affinity Labels

In a broad sense, an enzyme is specifically inhibited when its active site is blocked physically and/or chemically without significant alteration of the rest of the molecule. For this, many types of covalent inhibitors have been developed. The desired goal is chemical modification of an active site amino acid residue of the enzyme and subsequent loss of catalytic activity. The most common approach has been the synthesis of structurally and chemically reactive analogs of a substrate of the target enzyme. Such inhibitors have been referred to as *active-site-directed irreversible inhibitors* or *affinity labels* (630,631). Generally, the affinity label has a reactive electrophilic substituent that can generate a stable covalent bond with an active site nucleophilic group. Such experiments may serve to identify a nucleophile important in catalysis.

The principles of the method can be illustrated by one of the first affinity-labeling experiments developed by E. Shaw, the reaction of *N*-tosyl-L-phenylalanine chloromethylketone (TPCK) with α-chymotrypsin (632).

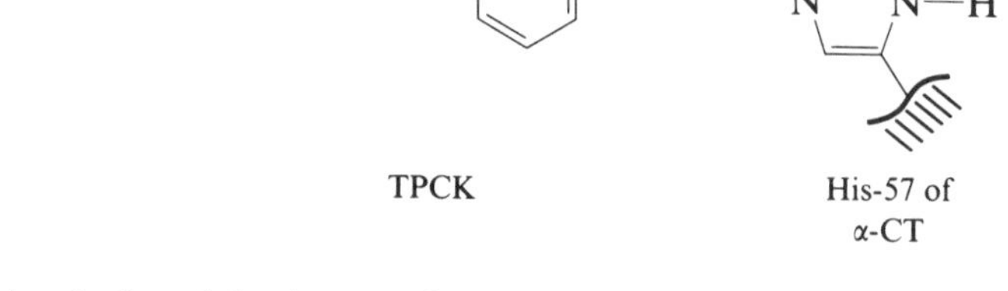

CH_3–C_6H_4–SO_2–NH–CH($CH_2C_6H_5$)–C(=O)–CH_2–Cl

TPCK

His-57 of α-CT

The tosyl-phenylalanine moiety mimics a substrate like *N*-tosyl-phenylalanine methyl ester, and the chloroketone function acts as an electrophile where the chloride ion is displaced by His-57 at the active site of the enzyme. Similarly, a chloromethylketone analog of lysine inhibits trypsin.

Diisopropylphosphorofluoridate (DFP, Section 4.4) is also an active site irreversible inhibitor that blocks the active serine residue of serine proteases. It is easy to show that the inhibition is irreversible since exhaustive dialysis still produces an inactive enzyme.

Ser-195 of α-CT

Alternatively, diazomethylketone substrate derivatives can be efficiently used as active site–directed inhibitors of thiol proteases. For instance, the carbobenzoxyphenylalanine analog reacts stoichiometrically at the active center cysteine residue of papain.

R = Cbz-Phe

active site residue of papain

However, active site–directed inhibitors have two disadvantages. First, they are intrinsically reactive molecules and a large portion is simply hydrolyzed by the aqueous medium. Second, they might react nonspecifically with other reactive moieties on the protein surface.

Many more examples of active site–directed inhibitors can be found in recent books and review articles (617,630,631). Besides chemical modification of enzyme and affinity labeling, a few new techniques have been developed in the past ten years. Although not directly relevant to bioorganic models of enzymes, these techniques should be mentioned since, in general, they have given useful information when applied to biological problems. Among those are *photoaffinity labeling* (633,634) and *fluorescence energy transfer* as a spectroscopic ruler (635). These recent methods involve mainly biophysical aid for a better understanding of biological processes. The information obtained can be a valuable guide for the planning and design of new bioorganic models of macromolecules of biological interest.

It became apparent in the 1970s that new kinds of inhibitors with increased selectivity were needed. For instance, it has long been recognized that carbonyl reagents such as hydroxylamine and hydrazine are interesting inhibitors (irreversible Schiff-base formation) of pyridoxal phosphate-dependent enzymes. A well-known example is isonicotinyl hydrazide (isoniazid), one of the most effective drugs against tuberculosis. Apparently, this drug competes with pyridoxal to form a hydrazone that blocks the corresponding kinase enzyme. The latter catalyzes the biosynthesis of pyridoxal phosphate from pyridoxal and ATP.

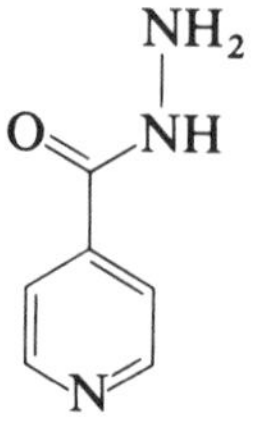

isoniazid

Many naturally occurring irreversible enzyme inhibitors also exist. Some are referred to as *toxins*. A well-known and intriguing toxin is the β,γ-unsaturated amino acid rhizobitoxine, produced by *Rhizobium japonicum*. This natural metabolite is a highly specific irreversible inhibitor of pyridoxal-linked β-cystathionase from bacteria and plants (636).

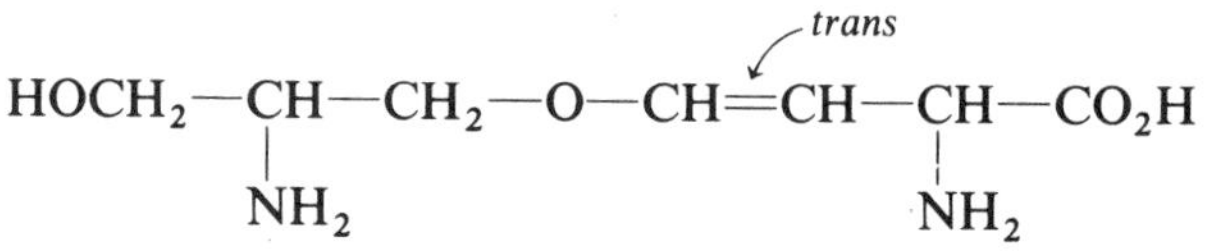

$$HOCH_2{-}CH(NH_2){-}CH_2{-}O{-}CH{=}CH{-}CH(NH_2){-}CO_2H$$

trans

rhizobitoxine

$$HO_2C{-}CH(NH_2){-}CH_2{-}S{-}CH_2{-}CH_2{-}CH(NH_2){-}CO_2H$$

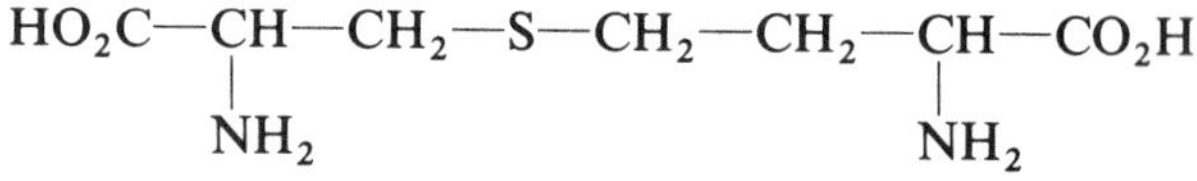

cystathionine

The role of β-cystathionase is to degrade cystathionine as follows:

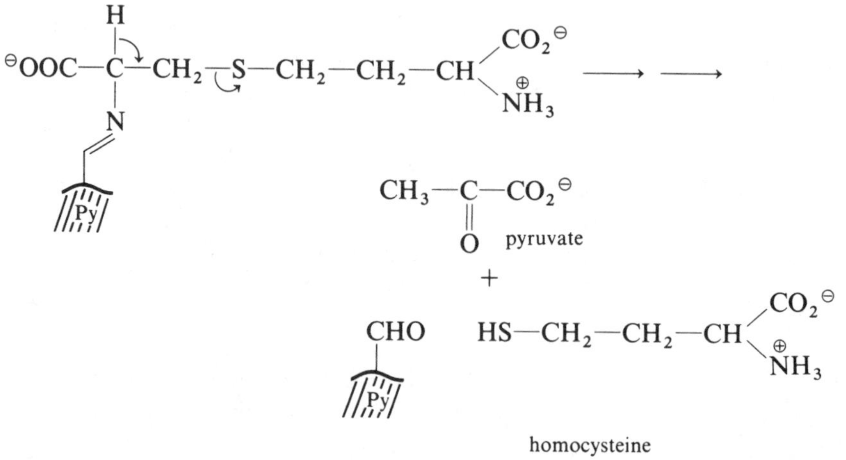

$$^{\ominus}OOC{-}C(H)(N{=}CH{-}Py){-}CH_2{-}S{-}CH_2{-}CH_2{-}CH(CO_2^{\ominus})(\overset{\oplus}{N}H_3) \longrightarrow \longrightarrow$$

$$CH_3{-}\underset{\underset{O}{\|}}{C}{-}CO_2^{\ominus}$$ pyruvate

+

$$Py{-}CHO \qquad HS{-}CH_2{-}CH_2{-}CH(CO_2^{\ominus})(\overset{\oplus}{N}H_3)$$

homocysteine

The inhibitor binds similarly but follows a different chemical pathway. The introduction of a conjugated system greatly facilitates attack by a neighboring nucleophile on the apoenzyme. This forms a new stable covalent bond resulting in an irreversible inhibition of the enzyme, preventing the system from normal metabolic functioning.

$$^{\ominus}OOC{-}C(H)(N{=}\text{Py}){-}CH{=}CH{-}O{-}CH_2{-}CH(CH_2OH)(\overset{\oplus}{N}H_3) \longrightarrow \quad H^{\oplus}$$

$$^{\ominus}OOC{-}C({=}N{-}\text{Py}){-}CH{=}CH{-}O{-}CH_2{-}CH(CH_2OH)(\overset{\oplus}{N}H_3) \quad H^{\oplus} \quad \ddot{N}u$$

It is important to realize that the natural toxin is so constructed that it requires chemical activation by the target enzyme. Upon activation, a chemical reaction ensues between the inhibitor and enzyme resulting in the irreversible inhibition of the latter. Thus, the enzyme by its specific mode of action catalyzes its own inactivation or "suicide."

In fact, many well-known drugs have inhibitory properties on one or another enzyme, but only a few were conceived with that goal in mind. The knowledge of the structure and mechanism of action of enzymes has progressed tremendously over the past four decades so that the search for specific enzyme inhibitors as potential new drugs has become more attractive. In order to do so, it is necessary to know as much as possible about the specificity of the enzyme and about the secondary binding sites, if any, near its active site. This has led in recent years to the development of a fascinating new group of irreversible enzyme inhibitors: the *enzyme-activated irreversible inhibitors*, also referred to as *suicide enzyme inactivators* (631,636,637). In a sense, the expression "suicide inactivator" is rather ambiguous but nonetheless useful.

To put it another way, the system behaves as a "molecular kamikaze." As these enzyme inhibitors owed their activity to the catalytic properties of the enzyme, Rando named them "k_{cat} inhibitors." Later, the expression "mechanism-based inhibitors" was adopted. Originally, it was the work of K. Bloch and coworkers with the enzyme β-hydroxydecanoyldehydrase that gave impetus to the recent interest in suicide inactivators.

There is considerable promise in this type of inhibitor because the potential reactive group may be rather innocuous in the presence of a number of enzymes or in vivo until it reaches and is activated by the target enzyme. It is really a question of exploiting the organic and physical chemistry of enzymes (617).

This concept of *suicide substrates* or *enzyme inactivators* has many potential implications whereby organic chemists can synthesize well-designed substrate analogs of specific enzymes. Of course, the design of new suicide enzyme inactivators involves higher and higher levels of chemical sophistication. Most of the work done with suicide inhibitors has been on nonproteolytic enzymes, especially pyridoxal- and flavin-dependent enzymes.

Nowadays, computer-assisted molecular modeling is a technique routinely used in the design and execution of these projects.

Pargyline is a potent irreversible inhibitor of a flavin-linked monoamine oxidase (MAO) and has found clinical application. The latter catalyzes the inactivation of biologically important catecholamines. It forms a covalent bond with the enzyme via the flavin cofactor and the mode of action is believed to be as shown in Scheme 7.1.

Alternatively, β,γ-unsaturated amino acids, γ-acetylenic amino acids, or amino acids having a leaving group at the β-position are potentially good inhibitors of pyridoxal-dependent enzymes involved in amino acid metabolism. The first two become activated either by carbanion formation

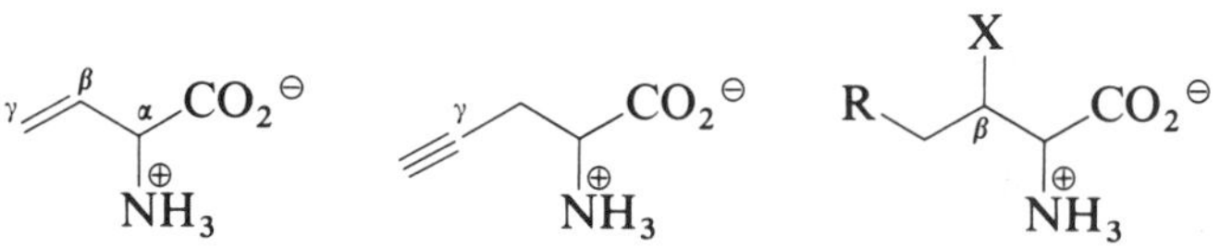

γ β α $CO_2^{\ominus}$ $\overset{\oplus}{N}H_3$ γ $CO_2^{\ominus}$ $\overset{\oplus}{N}H_3$ X R β $CO_2^{\ominus}$ $\overset{\oplus}{N}H_3$

adjacent to the unsaturated function or by two-electron oxidation to the bound conjugated ketimine. For the halogenated compounds, an enzyme-mediated loss of HX produces an aminoacrylate Schiff base. In all cases, the activated electrophiles are attacked by enzyme nucleophiles at or near the active site.

Needless to say, the design of such inhibitors is of therapeutic importance, and three factors are necessary for a successful design:

(a) The enzyme or the enzyme–coenzyme complex must convert a chemically unreactive molecule to a reactive one.
(b) The reactive molecule must be generated within bonding distance of a crucial active site residue.
(c) The activated species created must be designed in such a way that reaction with an active site residue can occur rather than with an external nucleophile.

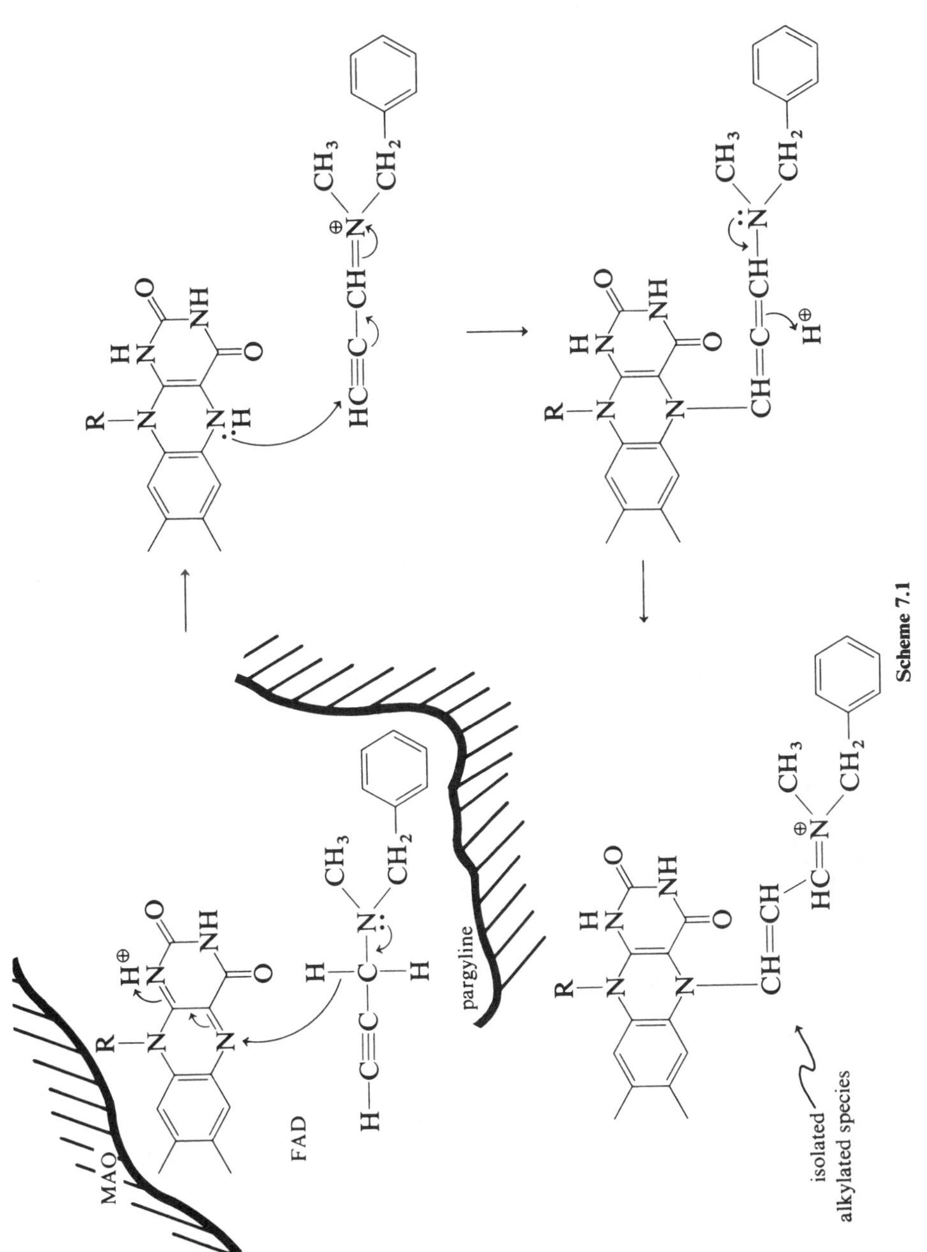

Scheme 7.1

In other words, a harmless compound must be converted into a potent inhibitor, the enzyme serving as the agent of its own destruction. A few examples will serve to illustrate this point.

Example 1:

Serine *O*-sulfate and β-chloroalanine are both inhibitors of aspartate aminotransferase (638) and aspartate-β-decarboxylase (639).

Normal process:

Schiff base of aspartate after 1,3-prototropic shift

In the presence of the inhibitors:

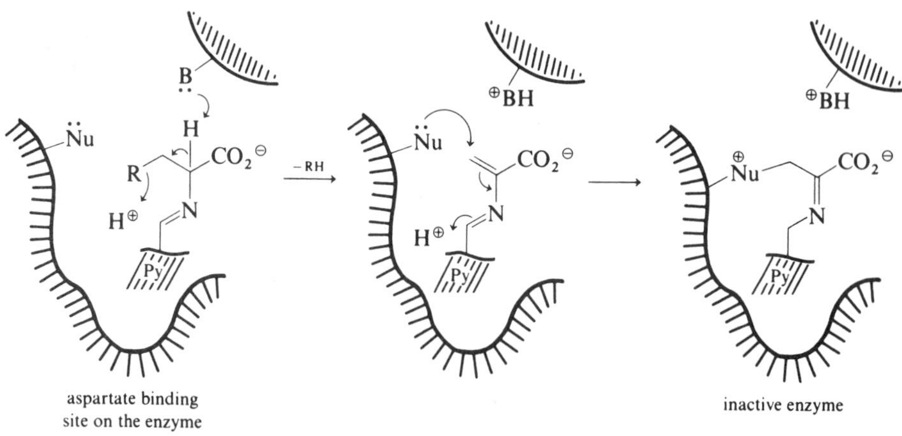

B = basic group on the enzyme
$\ddot{N}u$ = nucleophile on the enzyme
R = $-Cl$, $-SO_4^{2\ominus}$
Py = pyridoxal nucleus

Example 2:

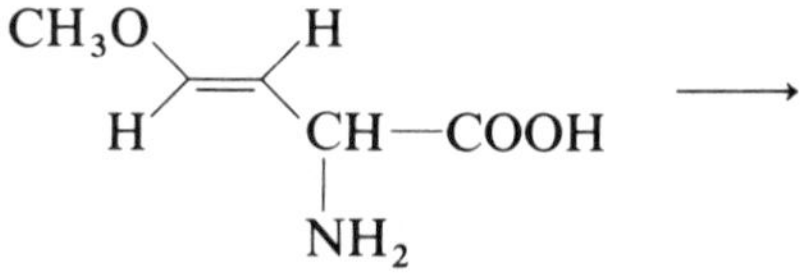

2-amino-3-methoxy-*trans*-3-butanoate (a substituted vinyl glycine), an antibiotic that inhibits aspartate transaminase (617)

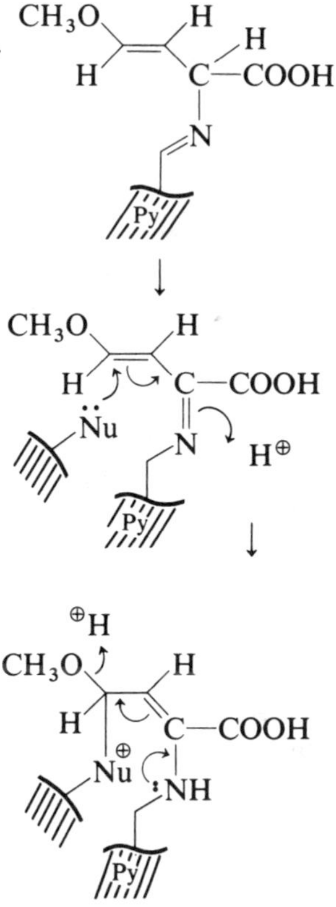

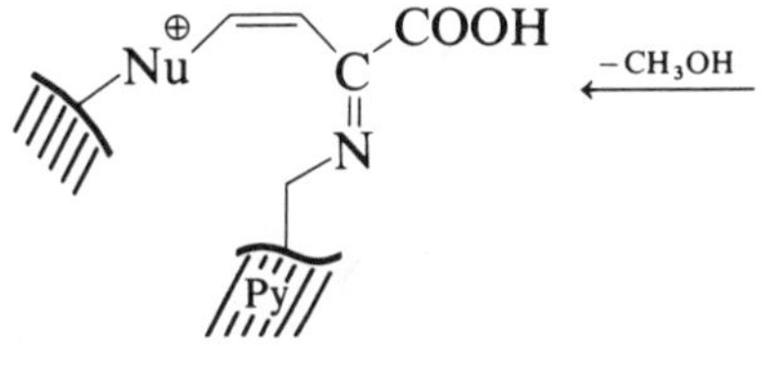

CH_3O H H COOH N Py CH_3O H C COOH Nu N $H^{\oplus}$ Py $^{\oplus}H$ CH_3O H C COOH Nu NH Py $-CH_3OH$ $\overset{\oplus}{Nu}$ COOH C N Py

Example 3:

H C C H—C $CO_2^{\ominus}$ $\overset{\oplus}{N}H_3$

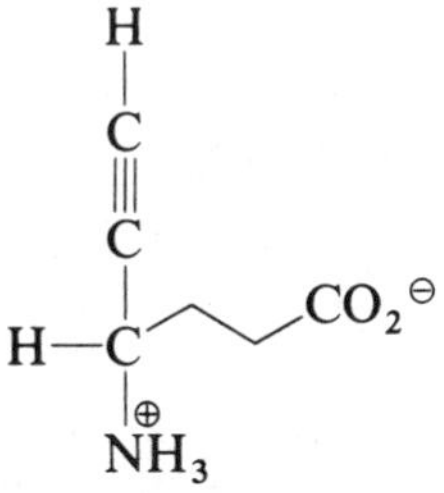

an γ-acetylenic analog of GABA that blocks GABA transaminase in brain (640)

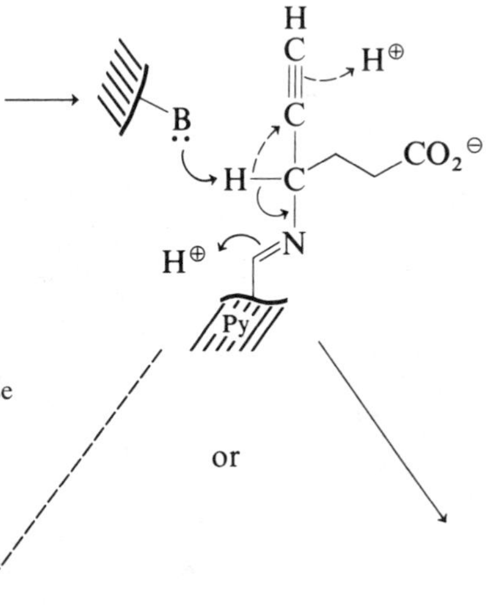

B H C C $H^{\oplus}$ H—C $CO_2^{\ominus}$ N $H^{\oplus}$ Py

or

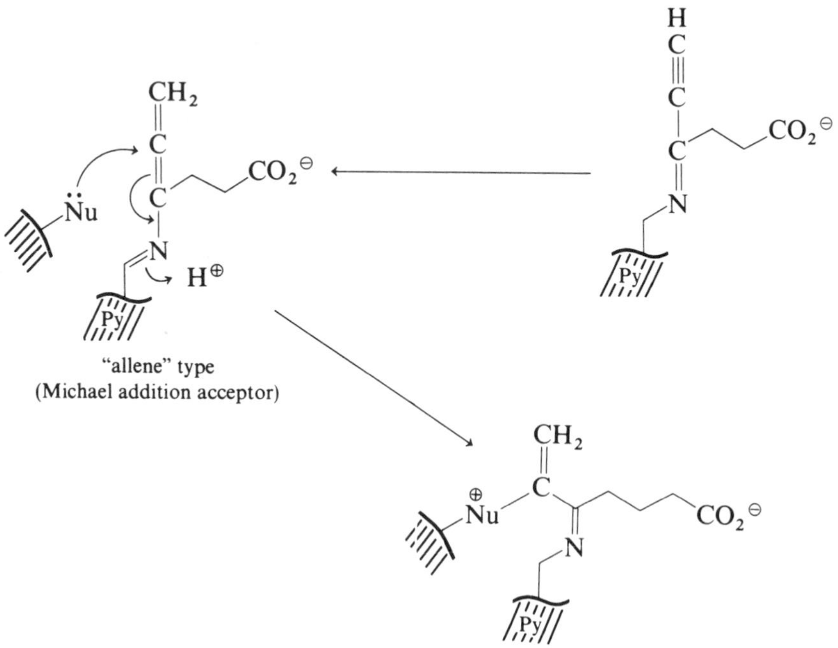

"allene" type
(Michael addition acceptor)

This analog of γ-aminobutryic acid (GABA) (640) appears to be a promising clinical candidate as an antiepileptic drug.

Example 4:

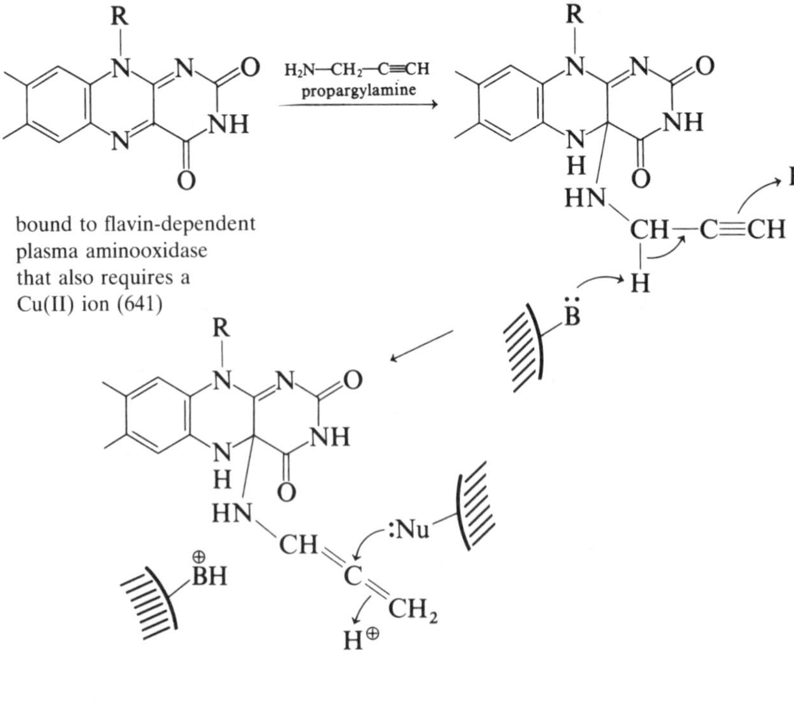

$H_2N-CH_2-C\equiv CH$
propargylamine

bound to flavin-dependent plasma aminooxidase that also requires a Cu(II) ion (641)

Example 5 (636):

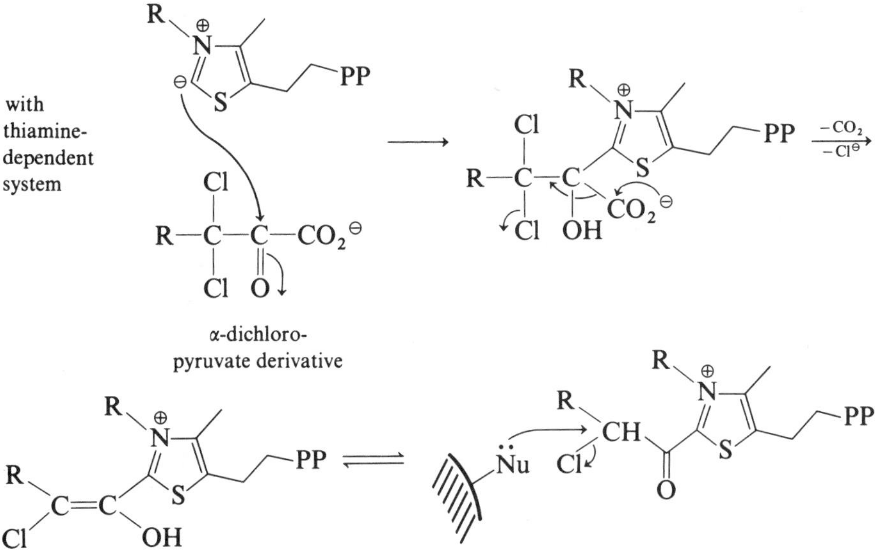

with thiamine-dependent system

α-dichloro-pyruvate derivative

Example 6:

Among stable halogenated amino acids, α-fluoromethyl-glutamic acid is a good example of a mechanism-based inactivation of pyridoxal-dependent enzymes (642).

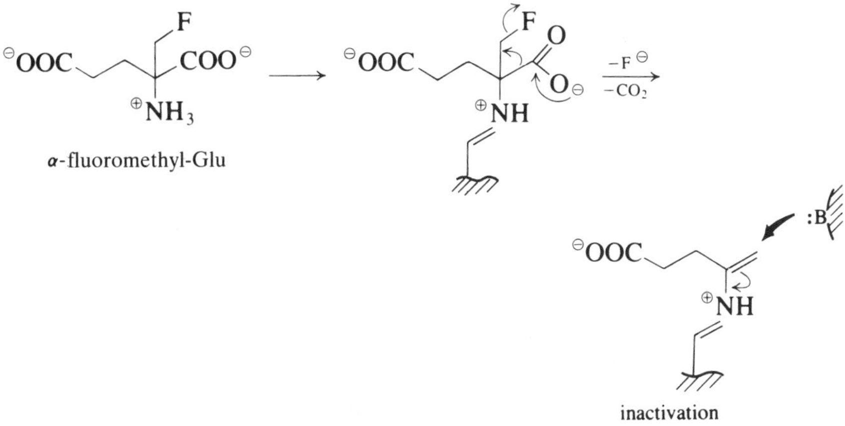

α-fluoromethyl-Glu

inactivation

In this respect, fluoromethyl-3,4-dihydroxyphenylalanine is used as a suicide inhibitor of L-DOPA decarboxylation enzyme. A product of therapeutic potential.

Example 7:

The fungal metabolite gabaculine is a very efficient killer of GABA transaminase (643).

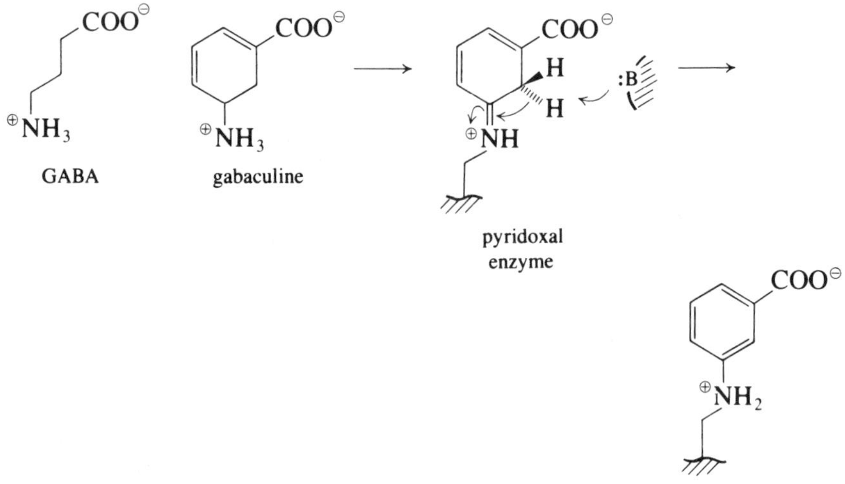

GABA gabaculine pyridoxal enzyme

No product molecules get away and gabaculine thus kills every enzyme molecule that takes part in the enzymatic transformation.

Example 8:

Proteases and phosphatases should be also amenable to targeting if a normal intermediate can be rerouted to mechanism-based auto-destruction. Katzenellenbogen came up with a few good illustrations of this possibility (644). Haloenol lactones inactivate α-chymotrypsin by an enzyme-mediated process.

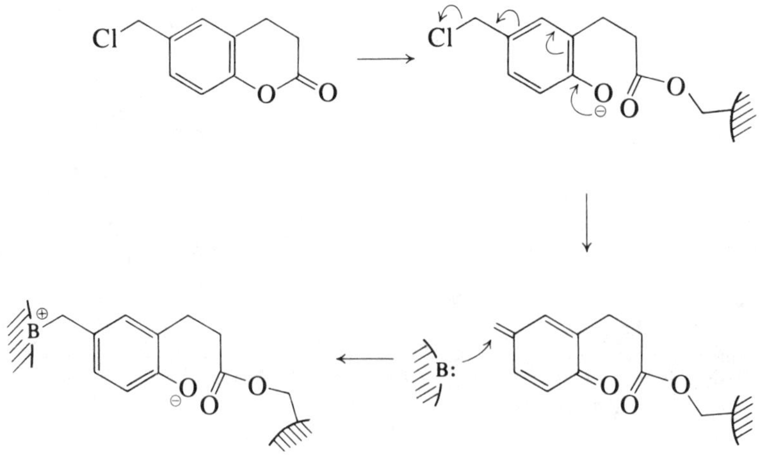

In a similar way:

bromoenol lactone

further hydrolysis (turnover) but remains inactive by covalent fixation

Example 9:

Abeles showed that benzylchloropyrones both inactivate α-chymotrypsin and are hydrolyzed by it. However, the benzyl group controls the orientation of the inactivator in the enzyme active site (645).

3-benzyl-6 chloropyrone

5-benzyl-6 chloropyrone

inactivated enzyme

inactivated enzyme by a site-directed inactivation rather than a suicide process

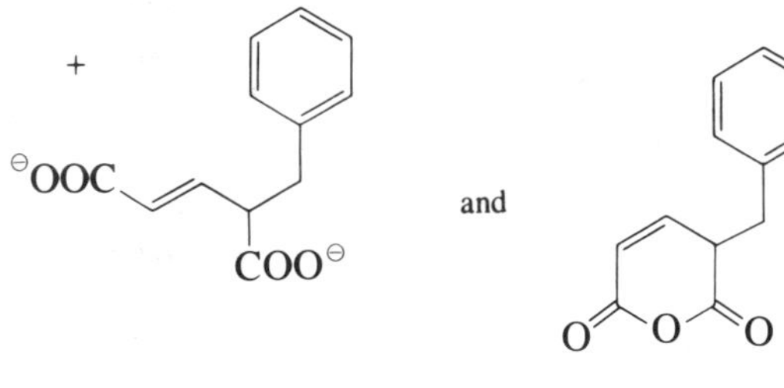

+

$^{\ominus}OOC$ ... $COO^{\ominus}$ and

hydrolysis products

Example 10:

Thromboxane A_2 (TXA_2) is a physiologically important metabolite of arachidonic acid and may play a significant role in the development of several diseases. Consequently, compounds that can inhibit its synthesis may become important therapeutic agents. Merrell's scientists have synthesized a diazoketone that is designed to inactivate the enzyme that converts the endoperoxide intermediate to TXA_2 (646).

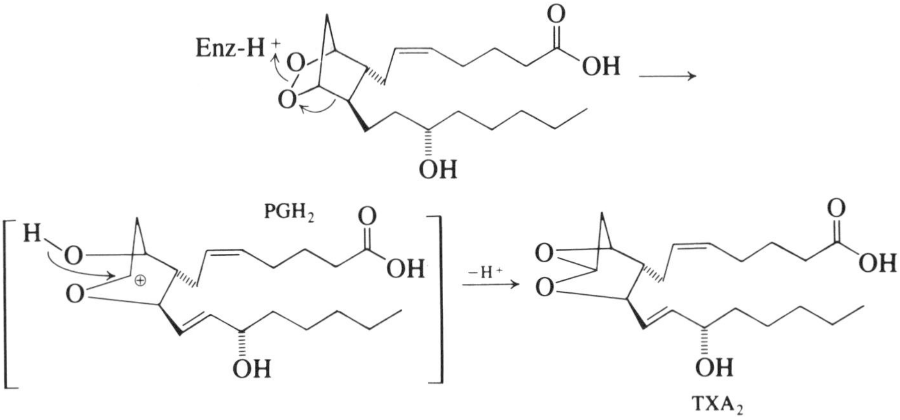

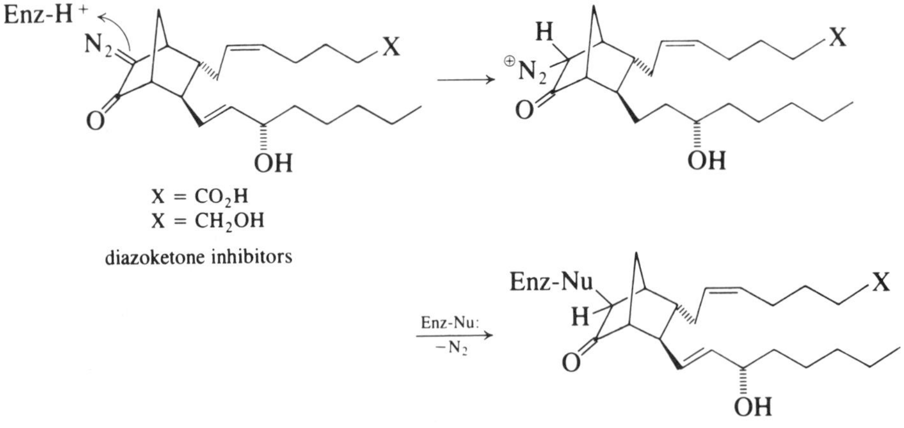

Enz-H$^+$ O O OH OH

PGH$_2$

H O O OH OH $-H^+$ O O OH OH

TXA_2

Enz-H$^+$ N_2 O X OH

$X = CO_2H$
$X = CH_2OH$

diazoketone inhibitors

H $^{\oplus}N_2$ O X OH

Enz-Nu: $-N_2$

Enz-Nu H O X OH

Example 11:

Silicon chemistry was at the heart of an ingenious way to inactivate the cytochrome P-450-mediated oxidation of cholesterol side chain.

cholesterol

hydroxylated intermediates

silicon-mediated suicide inhibitor

pregnenolone

formation of a covalent intermediate

This approach could also be applied to the development of a new class of monooxygenase mechanism-based inhibitors (647).

Example 12:

The last series of examples will focus on penicillin chemistry and the search for efficient β-lactamase inactivators. The workers at Beechman Pharmaceuticals put many fruitful efforts into this endeavor since the discovery of clavulanic acid. They marketed a drug called *Augmentin*. It is a mixture of clavulanic acid and amoxicillin. The role of clavulanic acid

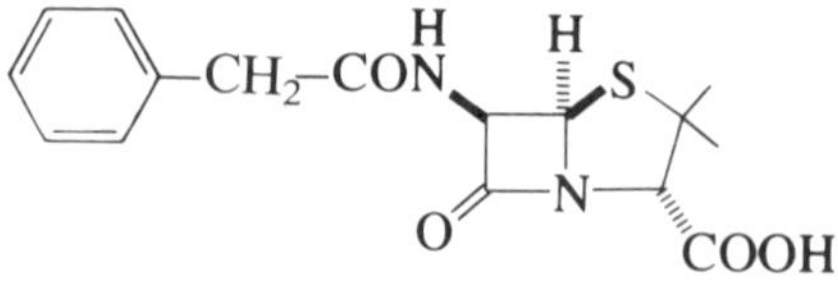

penicillin G

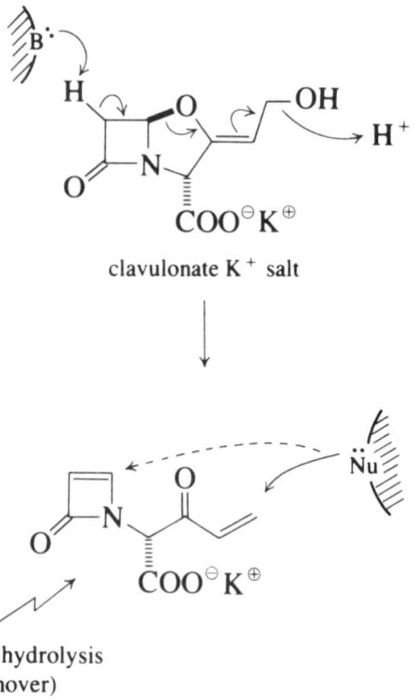

clavulonate K^+ salt

amoxicillin

possible hydrolysis (turnover)

covalent capture and inhibition

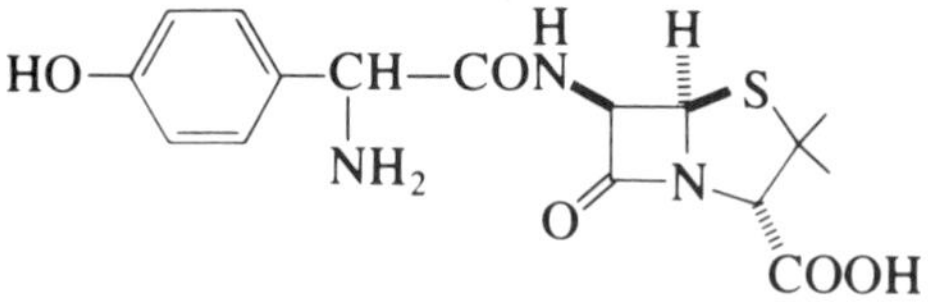

is to first trap the lactamase that the bacteria naturally manufacture to destroy the antibiotic. By doing so, the other component, the antibiotic amoxicillin, will act and bacteria known to be resistant to penicillin will in this way become sensitive and be killed. Similarly, the semisynthetic penicillanic sulfone is first acylated by the penicillinase and the acyl-enzyme intermediate is converted to a mechanism-based potent inhibitor that undergoes a Michael addition.

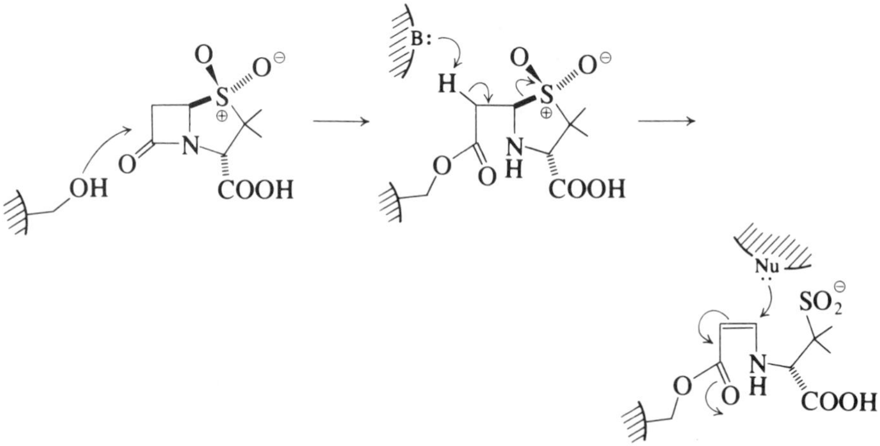

Most of these examples illustrate that proton abstraction is an important concept in suicide enzyme inactivation. For this, pyridoxal-dependent enzymes are obvious potential victims. In the future, one can expect more developments on mechanism-based inactivators of pyridoxal-dependent enzymes based on functional group activation achieved by way of carbanion chemistry of intermediates (617). It is very likely it was the design of more selective active site inhibitors that brought about the development of suicide enzyme inhibitors or inactivators. Compared to the active site–directed irreversible inhibitors evocated earlier, they have the advantage of being relatively unreactive but becoming so only after interaction with the enzyme active site residues. The reactive form depends on the specific catalytic capacities of the active site. It is thus an enzyme-catalyzed conversion. However, both types of inhibition allow labeling and identification of active site residues and functional groups on enzymes.

7.3.1 Structure-Based Design of an Inhibitor

Traditional approaches to the rational design of enzyme inhibitors have usually been founded on an understanding of the mechanism of the catalyzed reaction, as exemplified by the transition-state analogs (Section 2.7) and suicide inhibitor strategies (Section 7.3). Recently, advances in structural biology, computer graphics, and theory have stimulated other approaches to the design of biologically active compounds. These approaches are based on structural rather than mechanistic considerations. Two recent examples of structure-based and mechanism-based design inhibitors will be mentioned briefly.

The first one is the "classical" glutamate containing thymidylate-synthetase inhibitor presented below:

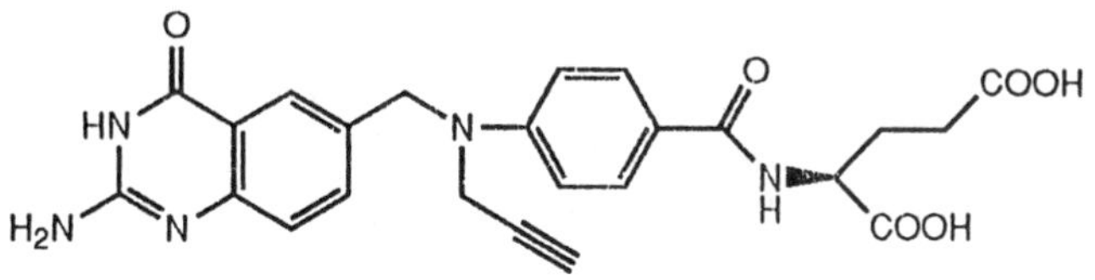

"Suicide" or mechanism-based inhibitor of thymidylate synthetase

Since then, the X-ray crystal-structure-based design, synthesis, and biological activity of a novel family of ben[*cd*]indole-containing lipophilic inhibitors for thymidylate synthetase have been described (648). Following this, another class of antifolate inhibitors of thymidylate synthetase, using the 6,7-imidazotetrahydroquiniline skeleton, has appeared (649).

The second example is in the development of drugs to fight AIDS. Based on transition-state mimic, a series of antiviral HIV-1 protease peptidomimetic inhibitors have been prepared (650).

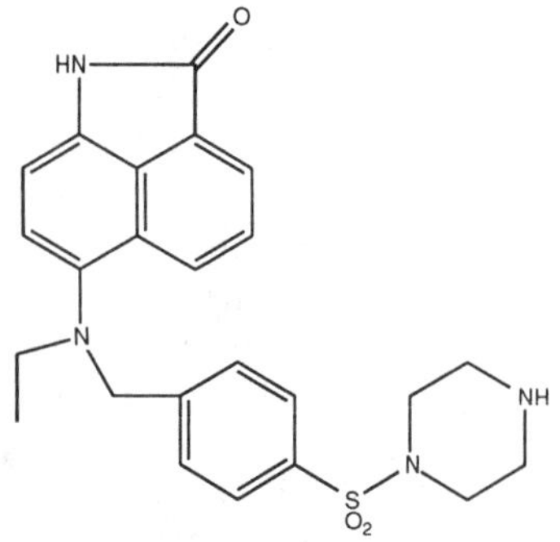

benz[*cd*]indole-containing inhibitor of thymidylate synthetase

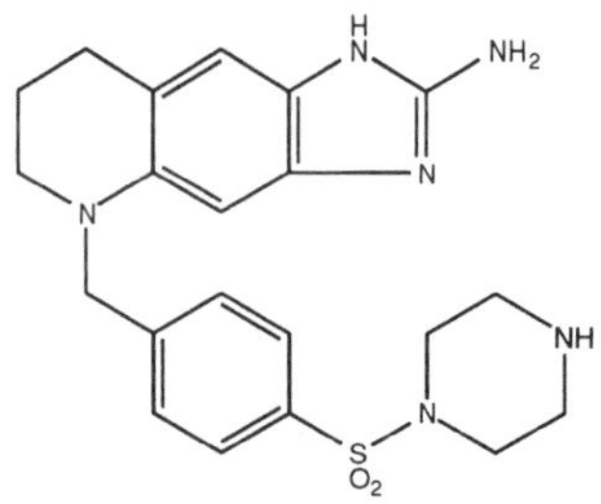

6,7-imidazotetrahydroquiniline-based inhibitor of thymidylate synthetase

The strategy of the structure-based design of inhibitors relies on the increasing availability of protein crystal structures and of NMR structures of bound and free ligands. It also builds on the growing body of information concerning the changes in binding energy that result from specific alterations in the structure or conformational properties of small molecule inhibitors or protein-binding sites (651). A central tenet in these strategies is that the binding affinity of a ligand will be improved if its conformational motion can be restricted to that of the bound state.

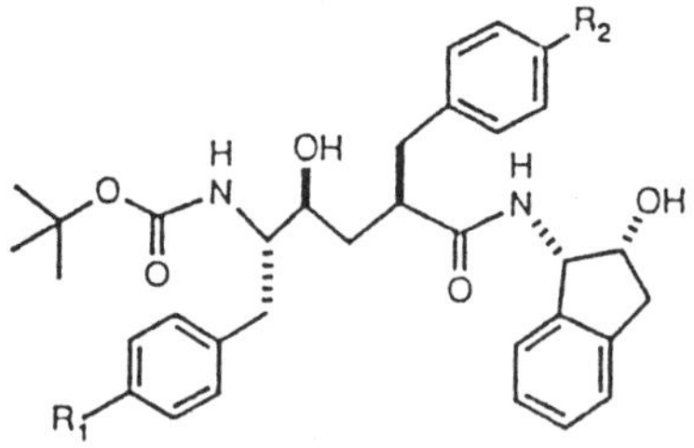

Transition-state peptide mimic for HIV protease inhibition

Aside from the inherent attraction of this novel approach, structure-based design offers potential solutions to some of the challenges encountered in optimizing the biological activity of peptides and peptide analogs (651). However, the use of peptides as lead compounds cannot progress without the development of methods for maximizing their binding affinity, altering their receptor selectivity, and reducing their susceptibility to metabolic degradation. The goals are frequently sought and sometimes achieved through the design of cyclic peptide analogs. This is the approach that P.A. Bartlett (University of California, Berkeley) has used to design an inhibitor of the zinc-dependent enzyme thermolysin (651).

Using the crystallographic structure of the complex of thermolysin with the known inhibitor Cbz-Gly^P-Leu-Leu (K_i = 9 nM; "Gly^P" = $NHCH_2PO_2^-$) as the starting point, his research group has designed an optically active macrocyclic (*S,S*)-phosphonamidate analog as a conformationally constrained derivative.

Cbz-Gly^P-Leu-Leu (*S,S*)-phosphonamidate cyclic analog

Of course, several control compounds were designed and synthesized to assess the success of the design and to gauge the binding enhancement it produced. The chroman linking unit was utilized to rigidify the structure

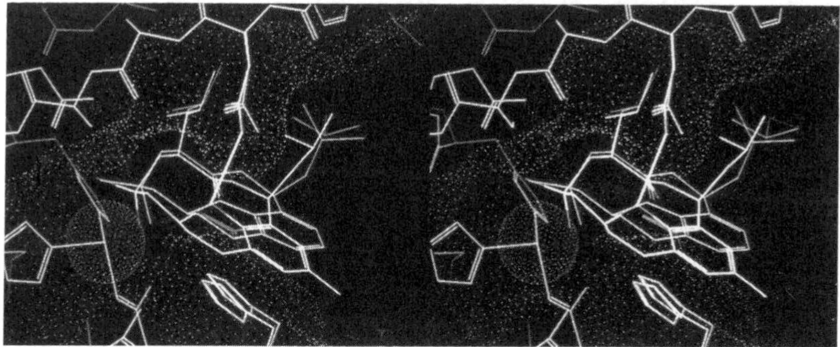

Fig. 7.12. A stereoview of a computer-assisted molecular modeling of a structure-based design of a semirigid peptidomimetic inhibitor of the zinc-dependent enzyme thermolysin. This figure compares modeled (orange) and observed (color-coded) conformations of the *S,S*-inhibitor in the active site of thermolysin (blue). The protein structure is taken from the complex with Cbz-Glu^P-Leu-Leu; color-coded representations of the side-chains of His-231 and Asn-112 (below and above the inhibitor, respectively) indicate the different orientations they adopt in the inhibitor complex (651). Reproduced with permission. Copyright © 1994 by the American Chemical Society. See color print.

while avoiding unfavorable steric interactions with the enzyme. The (*S,S*)-isomer had a K_i of 4 nM, was found more potent than other diastereoisomers, and showed a 50-fold increase in affinity as compared to some similar noncyclic analogs. This increase comes from the inherent affinity of the chroman unit itself and from its constraint in the peptide chain. An illustration of the binding of the inhibitor to the enzyme is shown in Fig. 7.12 and on the cover of this book.

This work illustrates clearly the importance to binding affinity of *dynamical complementarity* between ligand and receptor or enzyme active site; it may be an important consideration in structure-based design. However, the interpretation of the binding effects remains speculative, without structural characterization of the protein-ligand complexes themselves. Furthermore, structural information by molecular modeling provides a necessary foundation for further design enhancements. The successful design of this macrocyclic thermolysin inhibitor demonstrates the power of such a combined approach (651).

7.4 Thiamine Pyrophosphate

As seen in the first section of this chapter, a large percentage of the chemical reactions occurring in biological systems involve oxidation or reduction of one or more reactants. However, a particularly important type of reaction that apparently occurs in the majority of nonredox enzymatic reactions involves proton transfer that is aided by general-acid or -base catalysis. Of course, many of these enzymatic transformations are accomplished with the aid of a nonproteinic cofactor or coenzyme. In this category are some sulfur-containing coenzymes among which thiamine pyrophosphate (often referred to as vitamin B_1) is the most important. It will soon become apparent that thiamine pyrophosphate involves carbanion chemistry as an intermediate but with a degree of sophistication not yet encountered.

The coenzyme participates in reactions involving formation and breaking of carbon–carbon bonds immediately adjacent to a carbonyl group. Examples include nonoxidative and oxidative decarboxylations and aldol condensations. For instance, it is involved in the nonoxidative decarboxylation of pyruvic acid to acetaldehyde:

$$\underset{\text{pyruvic acid}}{CH_3-\underset{\underset{O}{\|}}{C}-COOH} \xrightarrow[\text{decarboxylase}]{\text{pyruvate}} \underset{\text{acetaldehyde}}{CH_3-C\!\begin{smallmatrix}\nearrow\!\!\!\!/ O\\ \searrow H\end{smallmatrix}} + CO_2$$

It also participates in the formation of acetyl phosphate:

xylulose 5-phosphate $+ P_i \rightleftharpoons$ glyceraldehyde 3-phosphate + acetyl phosphate

or in the following condensation:

xylulose 5-phosphate $\rightleftharpoons$ hydroxyacetaldehyde + glyceraldehyde 3-phosphate

In attempting to understand the mechanism of these condensations, the simplest analogy to be made by the chemist is that thiamine behaves like a cyanide ion in the catalysis of a benzoin-type condensation (614):

$R = C_6H_5$, benzoin

To explain this behavior, we have to examine its structure more carefully. The coenzyme thiamine pyrophosphate (thiamine-PP) contains a thiazolium ring system:

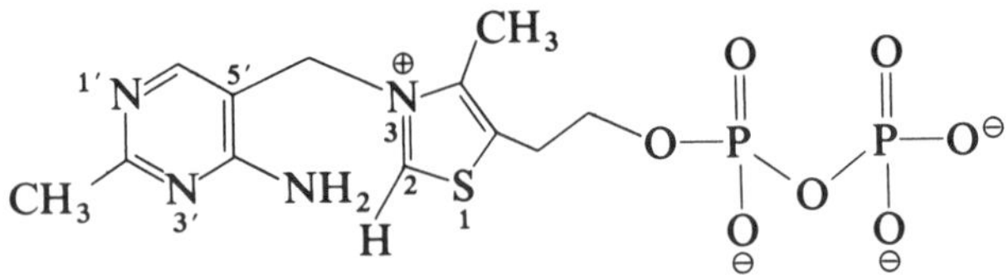

thiamine-PP (TPP)

R. Breslow (652) was the first to recognize, from NMR studies, that the thiazolium ion is acidic and the hydrogen at the C-2 position can be exchanged by deuterium in basic heavy water.

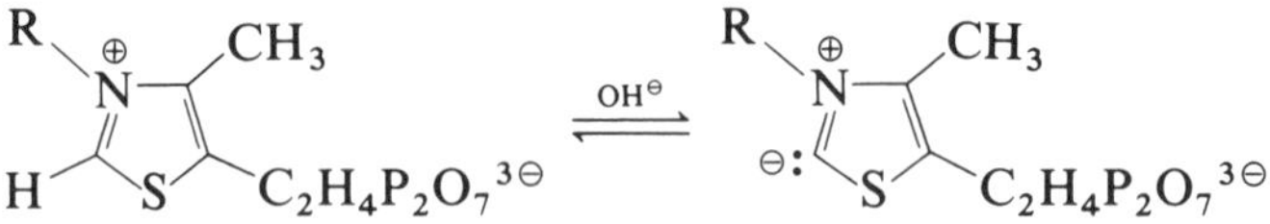

strong acid $\underset{}{\overset{OH^{\ominus}}{\rightleftharpoons}}$ ylid (dipolar ion)

What is the role of the pyrimidine portion of the coenzyme? One can only speculate that the C-4′ amino group can get close enough to the C-2 hydrogen and act as a weak base to facilitate the generation of the thiazolium dipolar structure. Recent ^{13}C-NMR work on thiamine salts has indicated that this is the case and X-ray diffraction studies show that, in the crystalline form, the two rings of thiamine are oriented in a manner to favor this function (653). This process could be enzyme mediated and protonation at N-1′ would facilitate the process. In fact, introduction of a methyl group onto the N-1′ pyrimidine position, to mimic the presence of a single charge on this ring, confers to thiamine catalytic properties superior to thiamine itself in enzymatic reactions requiring the coenzyme (654). This positive charge on the pyrimidine very likely accelerates ylid formation, decarboxylation, and acetoin formation relative to the natural vitamin. As pointed out by Metzler (574) the possibility that the enzyme

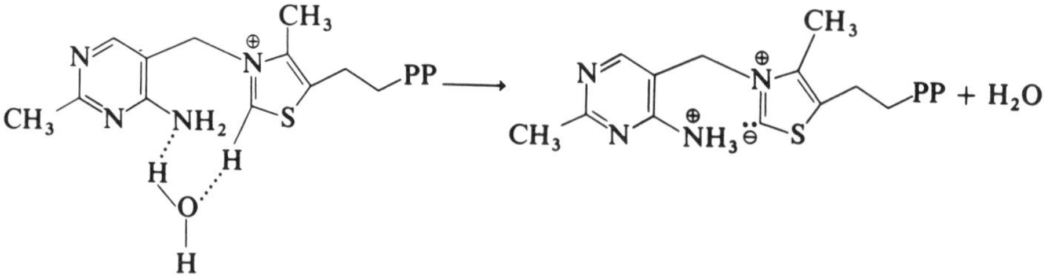

might stabilize a minor tautomeric form of the aminopyrimidine ring is attractive. It would constitute a charge–relay system in which the removal of the C-2 hydrogen of the thiazolium ring would be assisted by the amino group (in the imino form) of the pyrimidine ring.

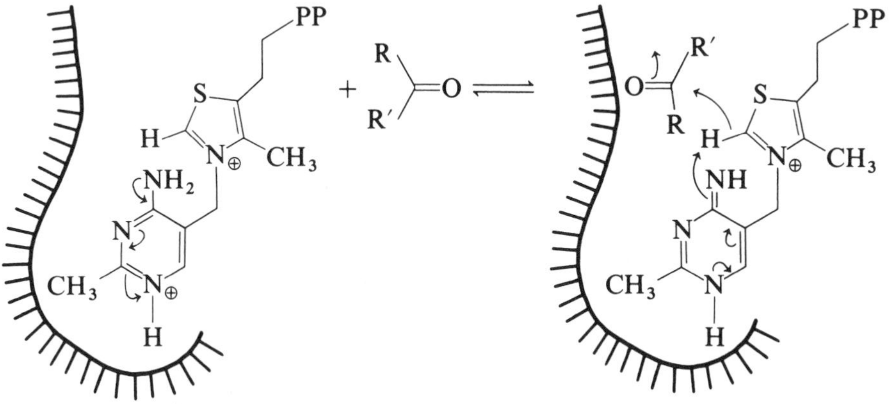

One of the remarkable facets of thiamine chemistry is that the thiazolium ring forms a covalent intermediate with the substrate and can act either as an electrophile or a nucleophile. It is thus a catalyst with dual functionality. In other words, the electron flow can occur from the bond to be broken into the structure of the coenzyme (toward the $=\overset{|}{N}^{+}-$group) and then back toward the new bond to be formed. For example, the ylid has a nucleophilic carbanion that can condense with pyruvic acid and other α-keto acids:

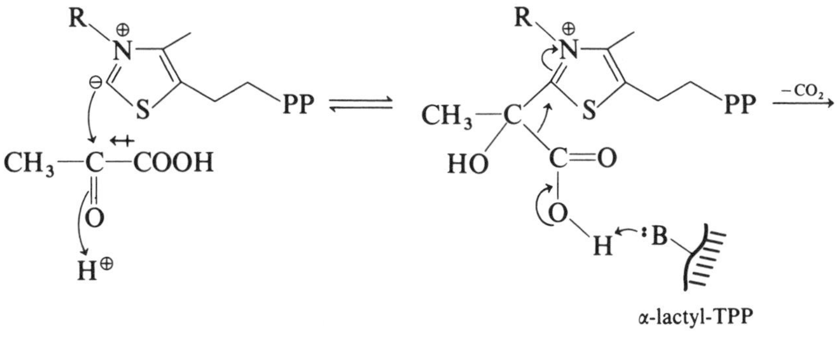

α-lactyl-TPP

The thiazolium ion then behaves as an "electron sink" or nucleophile and decarboxylation follows. The enolic intermediate, on the other hand, acts as a nucleophile that can be protonated. This intermediate has been isolated. Finally, acetaldelyde is formed and the coenzyme (ylid form) is regenerated at the same time. The liberation of acetaldehyde is the rate-limiting step in the pyruvate decarboxylase mechanism.

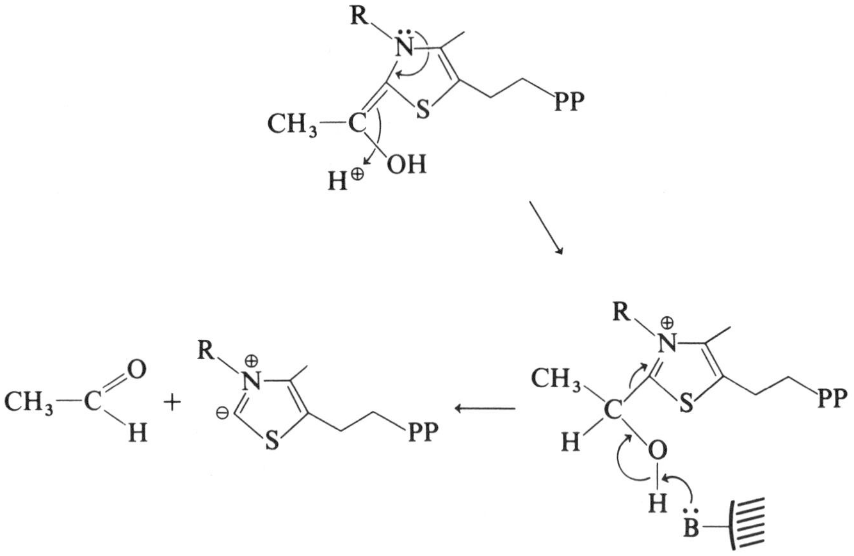

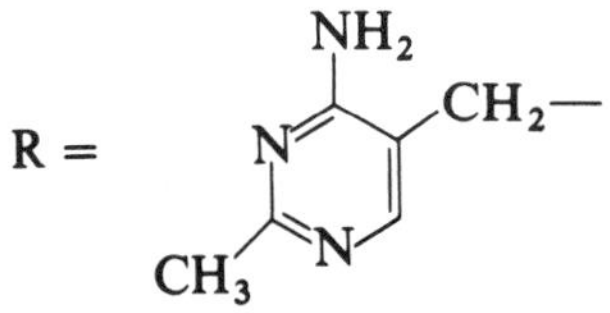

R =

It should be realized that the nucleophilic enolic intermediate (hydroxyethyl-thiamine pyrophosphate) is a form in which much of the coenzyme is found in vivo and can react with other electrophiles such as a molecule of acetaldehyde (aldol condensation):

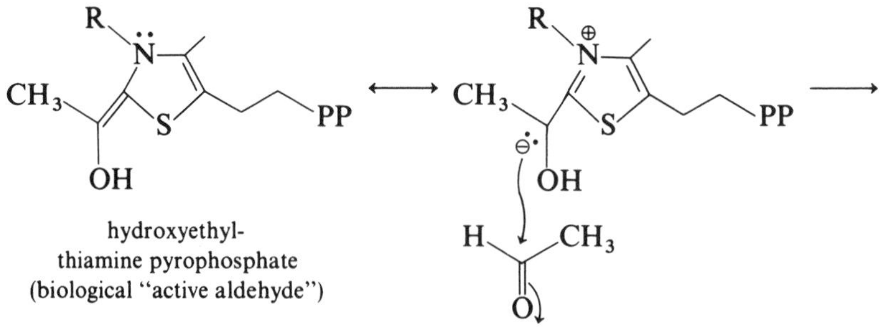

hydroxyethyl-
thiamine pyrophosphate
(biological "active aldehyde")

The intermediate is thus a potent nucleophile and adds to carbonyl compounds to form carbon–carbon bonds. The medium has to be basic

$$\longrightarrow CH_3-\underset{\underset{O}{\|}}{C}-CH(OH)CH_3 + TPP$$

acetoin

enough to ensure significant amount of the ylid, but not too basic or significant thiazolium ring opening will occur.

A second molecule of pyruvic acid can also add to the same intermediate to generate a molecule of acetoin after decarboxylation. With pyruvate decarboxylase only trace amounts of acetoin are reported in the enzymatic reaction.

Hydroxyethyl-thiamine pyrophosphate is also nucleophilic toward lipoic acid. In nature, lipoic acid operates in tandem with thiamine pyrophosphate.

$$CH_3-\underset{\underset{O}{\|}}{C}-CH(OH)CH_3 \xleftarrow[-CO_2]{\Delta} CH_3-\underset{\underset{O}{\|}}{C}-\underset{OH}{\overset{CH_3}{C}}-COO^{\ominus} + \text{TPP ylid}$$

acetoin

This transformation corresponds to a reductive acylation of lipoic acid and was formulated first by L.L. Ingraham (655).

An interesting aspect of the organic chemistry of these reactions is the transformation of pyruvic acid with an $R-\overset{\ominus}{C}=O$ character to an acetyl thioester of lipoic acid with an $R-\overset{\oplus}{C}=O$ character. This behavior is not very common in organic chemistry. The explanation comes from the fact that the sulfur atom of the thioester can expand its valence shell by using d orbitals.*

In connection with lipoic acid chemistry, mention should be made of arsenic compounds, some of the oldest and best-known poisons used throughout history. More recently, organic derivatives have been used as fungicides and insecticides. The more important arsenic compounds from a toxic standpoint are trivalent compounds. Arsenite ($O=As-O^{\ominus}$), for instance, is noted for its tendency to react rapidly with thiol groups, especially dithiols such as reduced lipoic acid. The result is that by blocking oxidative enzymes that require lipoic acid, arsenite causes the accumulation of pyruvate and other α-keto acids.

The thiamine transformations described above show clearly the analogy of thiamine with a cyanide ion where the resulting carbanion is relatively stable because the negative charge is partly shared by the nitrogen atom (Fig. 7.13).

* The synthon dithiane of Corey and Seebach is one of the closest organic analogs where a $>\overset{\oplus}{C}=O$ character is transformed to a $>\overset{\ominus}{C}=O$ character.

R(H)C=O ↔ R(H)C⊕–O⊖ + HS–(CH₂)₃–SH ⟶ R(H)C(S–(CH₂)₃–S) $\xrightarrow{R^{\ominus}Li^{\oplus}}$ R–C⊖(S–(CH₂)₃–S)

Other organic examples in the same trend are:

$\overset{\ominus}{RC}O$ character		$\overset{\oplus}{RC}O$ character
$R-CH(CN)-O\text{-}Si(CH_3)_3$	$\xrightarrow{F^{\ominus}}$	$R-CHO$
$H-CCl_3$	$\xrightarrow[H_2O]{Ag^{\oplus}}$	$H-COOH$
$CH_3-C(=O)-COOH$	$\xrightarrow{NaIO_4}$	CH_3-COOH

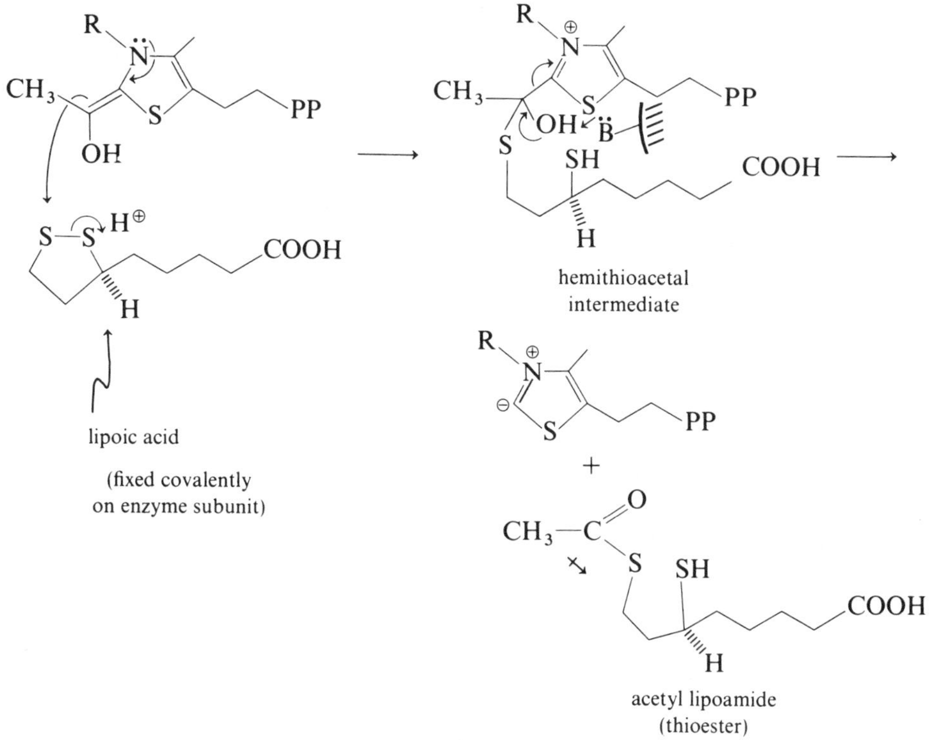

lipoic acid

(fixed covalently on enzyme subunit)

hemithioacetal intermediate

acetyl lipoamide (thioester)

Since the nitrogen atom in the thiazolium ring is already positively charged, it probably takes on an even greater share of the negative charge than in the cyanide ion. Because of this it can be called "biological cyanide" (614).

Furthermore, the adducts formed between thiamine-PP and substrates could be compared with a β-keto acid and a β-keto alcohol that readily undergoes β-cleavage:

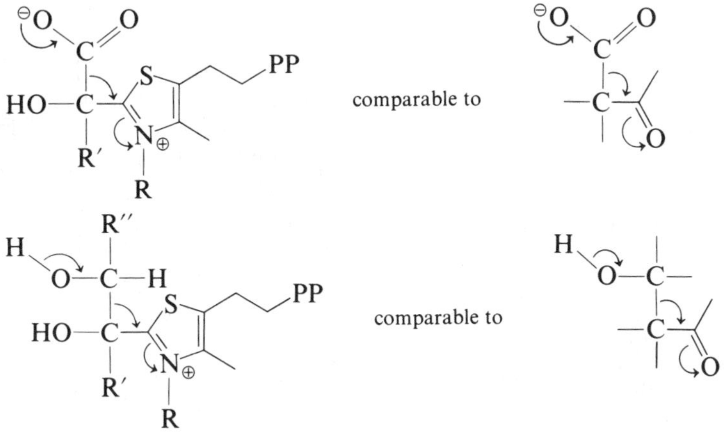

comparable to

comparable to

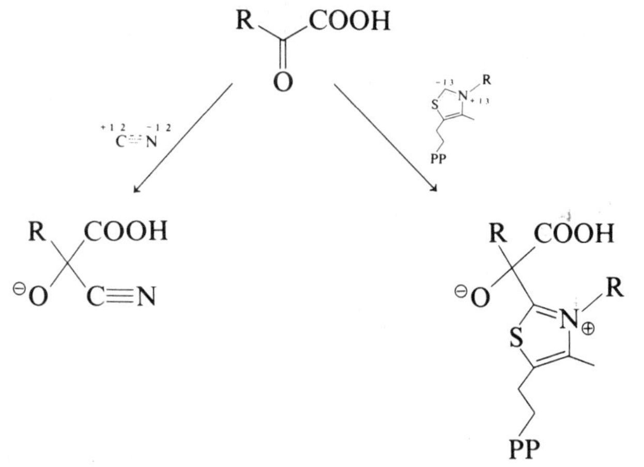

Fig. 7.13. Analogy between cyanide ion and thiazolium ion of thiamine-PP.

7.4.1 Model Design

Only a few biomodels of thiamine have been reported among which is *N*-benzyl thiazolium ion. At pH 8 and 25°C, it can readily decarboxylate pyruvic acid and modify other substrates:

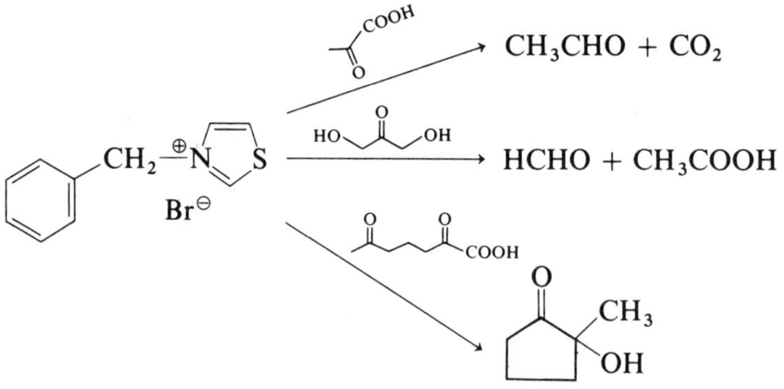

$CH_3CHO + CO_2$

$HCHO + CH_3COOH$

These reactions amplify the role of the active portion of thiamine as a condensing agent.

By studying other thiazolium compounds in the catalyzed acetoin condensation, it was found that blocking the 2-position either with a bulky neighboring substituent (isopropyl) or by substitution with a methyl group at this position destroys the catalytic abilities of the thiazolium salt (656).

Furthermore, the greater efficiency of catalysis of the *N*-benzyl analog to the corresponding *N*-methyl analog led Breslow to suggest that the benzyl group, through an inductive effect, increases the reactivity. A similar role is postulated for the pyrimidine ring of thiamine.

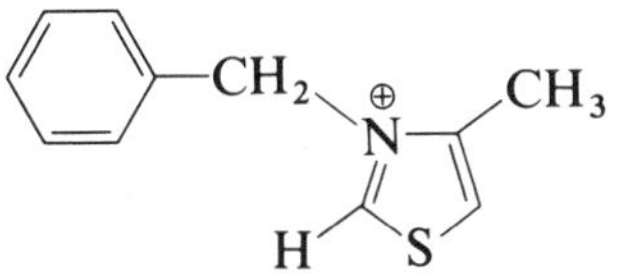

active 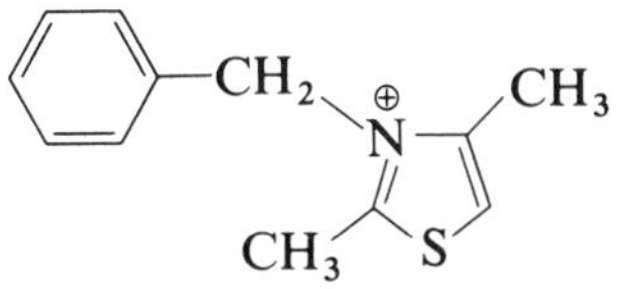 inactive

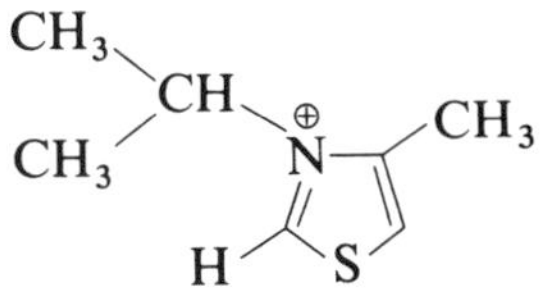

inactive 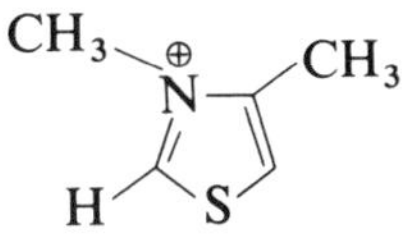 active

Another interesting model is the following:

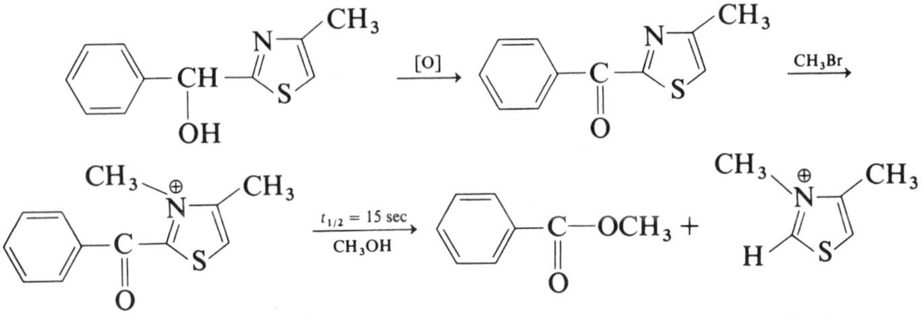

$\Delta G = -63$ kJ/mol (-15 kcal/mol)

Here, the thiamine analog, as a good leaving group, behaves as a "high-energy" acylating agent (655).

In a different approach, the reactivity of thiazolium-salts-derived acyl anion equivalents (biological "active aldehyde") toward sulfur electrophiles has been examined recently (657) and provides a model for the thioester-forming step catalyzed by the lipoic acid-containing enzymes. The results suggest that the biological generation of thioesters of coenzyme A from α-keto acids occur via the direct reductive acylation of enzyme-bound lipoic acid by the "acitive aldehyde," as already shown on page 567.

In order to define the chemical basis of the binding function of thiamine-dependent enzymes, Kluger and Pike (658) studied the reaction of thiamine-PP and methyl acetylphosphonate in aqueous sodium carbonate. This substrate analog of pyruvic acid binds to pyruvate dehydrogenase but cannot undergo all the enzyme multistep reaction processes.

Indeed, the phosphonate adduct resembles the reactive intermediate α-lactyl-TPP (see normal process, page 563) but the reaction cannot proceed further because the necessary leaving group species would be methyl metaphosphate, and that certainly involves a large energy barrier (658).

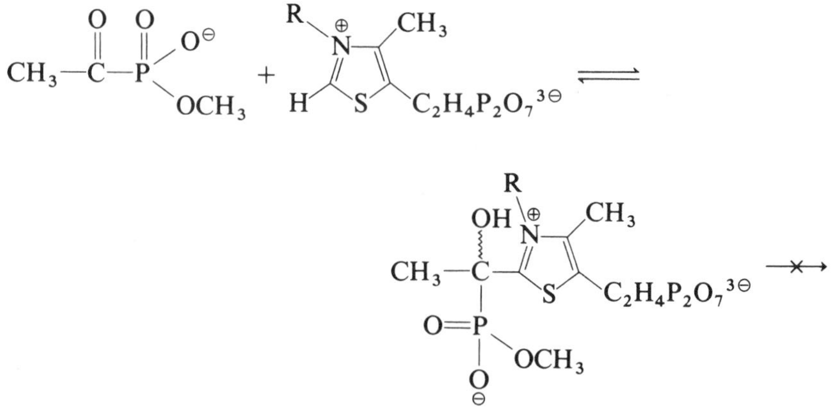

Furthermore, this reaction generates a chiral center where the thiamine-PP adduct is a racemic mixture. Kinetic data showed that both enantiomers bind to the apoenzyme but only one is being converted back to thiamine-PP. This provides information about the enzyme's active site. Of course, the absolute stereochemistry is unknown. This example illustrates a concept of growing importance in bioorganic chemistry: the formation of *enzyme-generated "reactive-intermediate analogs."*

Continuing brilliant efforts by Kluger's group to elucidate the enzymatic and nonenzymatic mechanism of substrate decarboxylation by thiamine-PP led to the isolation and resolution of racemic 2-(lact-2-yl)thiamin diphosphate as the 2,3-dibenzoyltartrate salt (659,660). The two isomeric

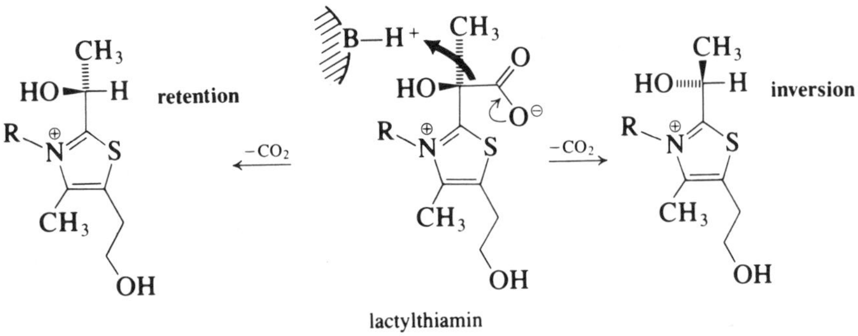

retention — $-CO_2$ — lactylthiamin — $-CO_2$ — inversion

acids were allowed to undergo spontaneous decarboxylation under a variety of conditions (H_2O solution, EtOH solution, DMF solution) and in all cases the product was racemic 2-(1-hydroxyethyl)thiamin. These results require that the reaction proceed via a preferential formation of a symmetrically solvated *achiral* intermediate. Since it is known that the corresponding enzymatic decarboxylation produces a single enantiomer of hydroxyethlthiamin, the stereospecificity of the enzmatic process is due to the chirality of the enzyme active site and not the intrinsic mechanism of the decarboxylation reaction.

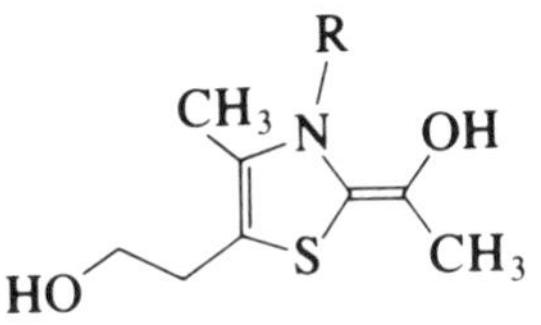

E-form

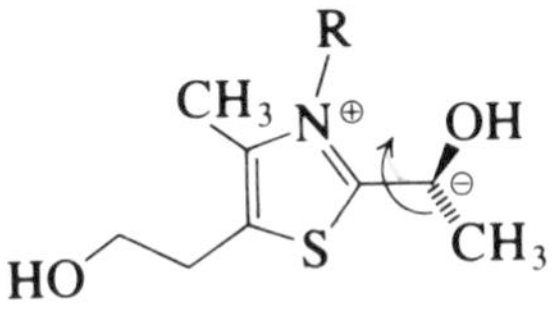

non-planar intermediate

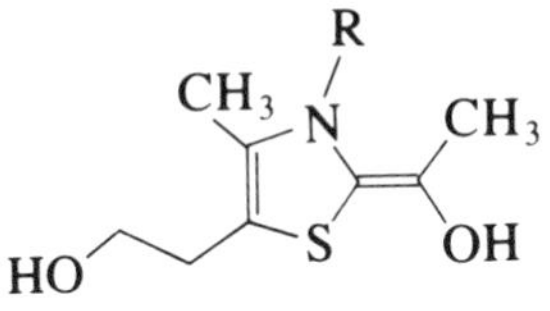

Z-form

Whether the *E*- or the *Z*-form of the enamine will predominate on the enzyme is still open to question. In all these transformations, the stereochemical course of the breakdown of intermediates must obey stereoelectronic control. Considering that the catalysis of decarboxylation mediated by the coenzyme thiamin-PP seems to involve the formation of a series of charged species that eventually evolve to a neutral enamine, Gutowski and Lienhard reasoned that the transition state of enzyme-bound lactyl-thiamin diphosphate should resemble the resultant neutral enamine intermediate. On that basis they designed a transition state analog with a neutral thiazole ring (661).

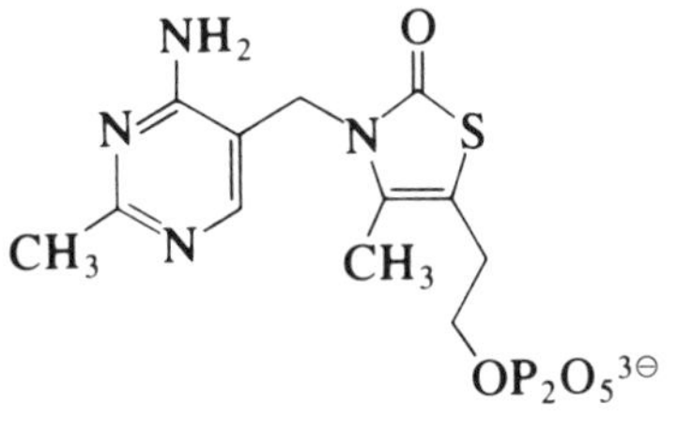

The fact that the material binds slowly and very tightly to pyruvate dehydrogenase is consistent with a hydrophobic nonpolar active site that accommodates uncharged species.

In order to mimic the hydrophobic environment of thiamin-dependent enzymes, Hilvert and Kool in Breslow's group prepared a series of β-γ-cyclodextrins having a covalently linked thiazolium group (662). Although these compounds were found to catalyze the benzoin condensation of *tert*-butylbenzaldehyde, they were generally less effective turnover catalysts than the analogous simple thiazolium salts lacking the artificial β-CD binding site. This is because the addition of a second molecule of substrate to the cavity might be hindered by the presence of the first molecule still covalently bound to the thiazolium moiety. However, they all speeded up the rate of tritium exchange from suitably labeled aromatic aldehydes and in some cases, rate enhancements of up to 40-fold were observed for the thiazolium-catalyzed oxidation of *tert*-butylbenzldehyde by ferricyanide ion. This represents cooperative binding of substrate by cyclodextrin in the thiazolium-catalyzed reaction.

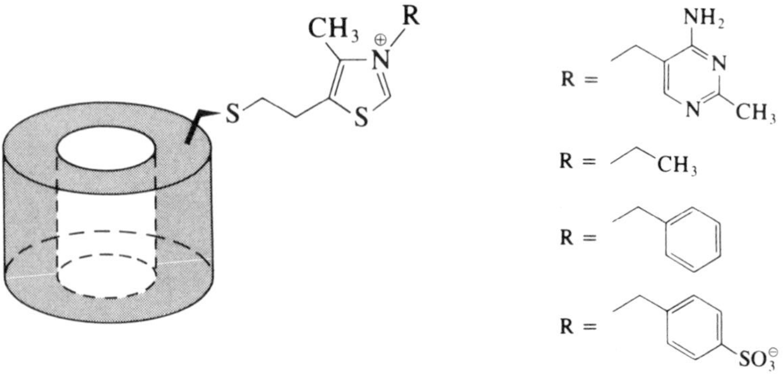

According to Lowe and Ingraham (614), it is very likely that thiamine-PP must have optimized, through chemical adaptation, its chemical structure for its biological role as a result of chemical evolution. For instance, an alkylated Schiff base is not basic enough to be a possible catalyst in the benzoin condensation.

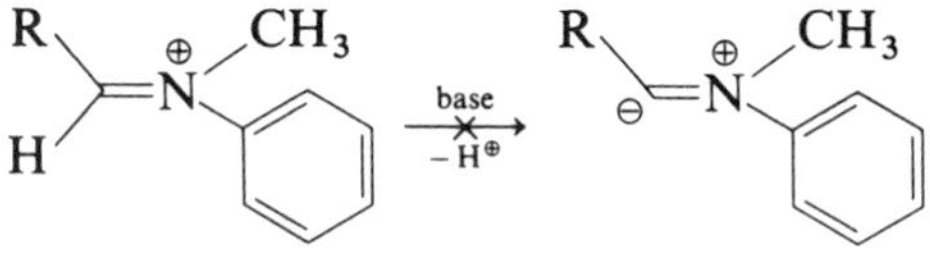

Schiff base

On the other hand, the cyclic form of a thioamide (enolic form) could be a possible candidate but was found to be extremely unstable in water, reacting like a Schiff base to give the open ring system.

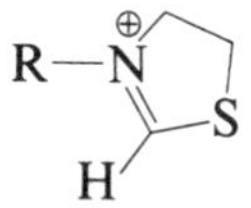

cyclic thioamide

Possibly, if the sulfur is replaced by a nitrogen, the positive charge will be stabilized by both nitrogens and the hydrogen will remain acidic enough for the compound to act as a catalyst.

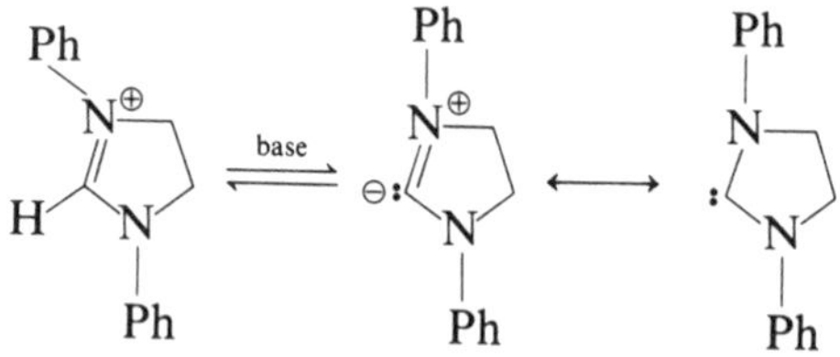

ylid carbene

Such a compound is found to be stable in water and the hydrogen acidic enough to be exchanged with ^2H_2O. The acidity is ascribed to the carbene resonance structure which helps stabilize the carbanion intermediate. However, the compound has no catalytic activity toward benzaldehyde. Instead, an addition product is formed that can be explained by the following mechanism (614):

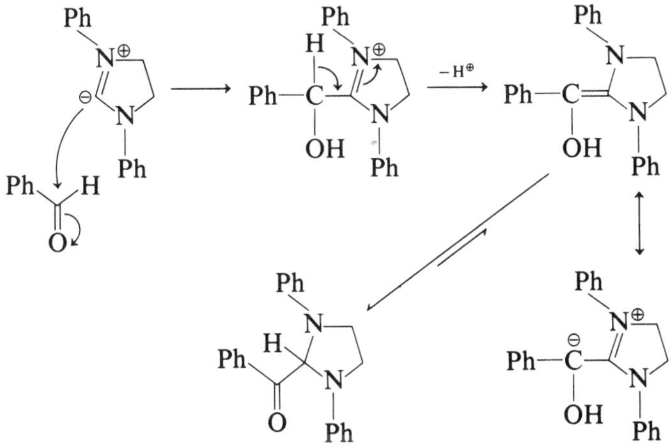

Therefore, the carbanionic intermediate is not nucleophilic enough to react with another molecule of benzaldehyde. It prefers to ketonize. What is needed is a compound that can do the same thing without so much carbanion stabilization. If too much resonance contribution is present, a good nucleophile does not form and internal ketonization takes place. Obviously, the solution of the problem is to have only partial aromatic character in the ring. A sulfur atom can help to do this because it does not have much π-character. So a thiazolium ion is the only remaining choice.

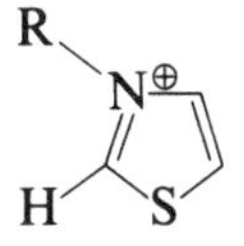

thiazolium ion

As mentioned earlier, the compound with R = benzyl, for instance, has been synthesized and found to act as an excellent catalyst in the benzoin condensation. Interestingly, oxazolium salts are also well known by organic chemists and they exchange protons faster than thiazolium salts. π-Overlap to stabilize the cation and σ-electron withdrawal (inductive effect) to stabilize the anion are both superior in oxygen than in sulfur chemistry.

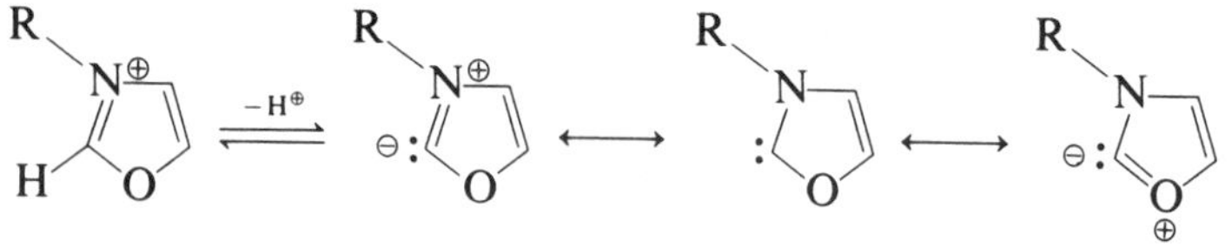

oxazolium ion

Consequently, one can wonder if nature made a mistake in choosing a thiazolium ring system rather than an oxazolium ion? Of course the answer is no because oxazolium salts simply *do not* catalyze the benzoin condensation. No products are formed because the system is too stable and thus unreactive toward benzaldehyde. In other words, the oxazolium system is not unstable enough to react with a carbonyl group. Consequently, thiamine-PP is a particularly well-designed catalyst by nature and its conception corresponds to a compromise among many alternatives (614). This analysis on how thiamine-PP has adapted to its role should help in understanding how to plan and design future bioorganic models of coenzymes in general.

7.5 Biotin

Biotin was first isolated and identified as a yeast growth factor in 1935. Contrary to the rapid progress made in the field of water-soluble vitamins in the 1940s, the function of biotin remained a mystery until 1959 when F. Lynen and his colleagues from Munich observed that bacterial β-methylcrotonyl-CoA carboxylase carries out the carboxylation of free (+)-biotin in the absence of its natural CoA-thioester substrate.

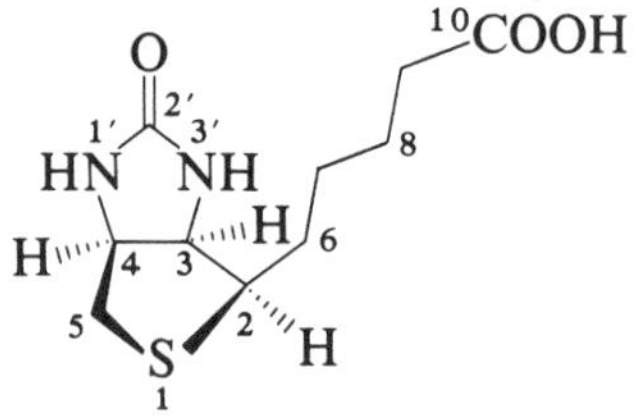

(+)-biotin

At the same period, X-ray work on the active (+)-form revealed that the two rings are *cis* fused, and the hydrogens of all three asymmetric carbons are in a *cis* relationship. The structure shows that the N-3′ of the ureido function is hindered from reaction by the five carbon side chain of valeric acid. The distance between N-3′ and C-6 is only 0.28 nm. Only then did the mechanism of action of biotin become clear; a CO_2-transfer reaction is achieved through the reversible formation of 1′-*N*-carboxybiotin.

The sulfur atom of biotin can be readily oxidized to the sulfoxide or the sulfone which both have biological activity. Desthiobiotin, a biosynthetic precursor of biotin, in which the sulfur is not yet present and is thus replaced by two hydrogen atoms, and oxybiotin, in which the sulfur has been replaced by an oxygen atom, are also both biologically active for many organisms. Consequently, the sulfur atom does not seem to be a stringent requirement for biological activity.

Biotin (referred to as vitamin H in humans) is an essential cofactor for a number of enzymes that have diverse metabolic functions. Almost a dozen different enzymes use biotin. Among the most well-known are acetyl-CoA carboxylase, pyruvate carboxylase, propionyl-CoA carboxylase, urea carboxylase, methylmalonyl-CoA decarboxylase, and oxaloacetate decarboxylase. Biotin serves as a covalent bound "CO_2 carrier" for reactions in which CO_2 is fixed into an acceptor by carboxylases. Then this carboxyl group in an independent reaction can be transferred from the acceptor substrate to a new acceptor substrate by transcarboxylases, or the carboxyl group can be removed as CO_2 by decarboxylases.

Later, in 1968, Lynen showed that ^{14}C-bicarbonate ion is a better substrate than free $^{14}CO_2$, being incorporated into the product much more rapidly. ATP and Mg(II) are also essential for the transformation to occur. Presently there is ample evidence that the overall reaction catalyzed by the biotin enzyme complex proceeds in two discrete steps (E refers to the enzyme):

$$\text{E-biotin} + \text{ATP} + \text{HCO}_3^{\ominus} \xrightleftharpoons{\text{Mg}^{2+}} \text{E-biotin}{-}\text{CO}_2^{\ominus} + \text{ADP} + \text{P}_i \qquad (7\text{-}8)$$

$$\text{E-biotin}{-}\text{CO}_2^{\ominus} + \text{acceptor} \rightleftharpoons \text{E-biotin} + \text{acceptor}{-}\text{CO}_2^{\ominus} \qquad (7\text{-}9)$$

$$\text{ATP} + \text{HCO}_3^{\ominus} + \text{acceptor} \xrightleftharpoons{\text{Mg}^{2+}} \text{ADP} + \text{P}_i + \text{acceptor}{-}\text{CO}_2^{\ominus} \qquad (7\text{-}10)$$

The initial step involves the formation of carboxybiotinyl enzyme. In the second step, carboxyl transfer from carboxybiotinyl enzyme to an appropriate acceptor substrate takes place, the nature of this acceptor being dependent on the specific enzyme involved. In brief, the function of biotin is to mediate the coupling of ATP cleavage to carboxylation. This is accomplished in two stages in which a carboxybiotin intermediate is formed. In transcarboxylation ATP is not needed because "activated carbonate," and not HCO_3^-, is the substrate.

These biochemical transformations occur on a multienzyme complex composed of at least three dissimilar proteins: biotin carrier protein (MW = 22,000), biotin carboxylase (MW = 100,000), and biotin transferase (MW = 90,000). Each partial reaction is specifically catalyzed at a separate subsite and the biotin is covalently attached to the carrier protein through an amide linkage to a lysyl ε-amino group of the carrier protein (663,664). In 1971, J. Moss and M. D. Lane, from Johns Hopkins University, proposed a model for acetyl-CoA carboxylase of *E. coli* where the essential role of biotin in catalysis is to transfer the fixed CO_2, or carboxyl, back and forth between two subsites. Consequently, reactions catalyzed by a biotin-dependent carboxylase proceed through a carboxylated enzyme complex intermediate in which the covalently bound biotinyl prosthetic group acts as a mobile carboxyl carrier between remote catalytic sites (Fig. 7.14).

Basically, biotin behaves as a CO_2-carrier between two sites. Schematically, the biotin carboxylase subsite catalyzes the carboxylation of the biotinyl prosthetic group on the carrier protein. Following the translocation of the carboxylated functional group from the carboxylase subsite to the carboxyl transferase subsite, carboxyl transfer from CO_2-biotin to acetyl-CoA occurs. Presumably free (+)-biotin can gain access to the carboxylation site because the prosthetic group is attached to the carrier protein through a 1.4 nm side chain and can oscillate in and out of the site. However, intensive biophysical studies on the transcarboxylase that catalyzes the transfer of a carboxyl group from methylmalonyl-CoA to pyruvate show that the carboxybiotin moves, at most, only 0.7 nm during the transfer of CO_2 (665). The role of the long arm (1.4 nm) thus appears to be to place the transferred carboxyl group at the end of a long probe, permitting it to traverse the gap that occurs at the interface of the three subunits and to be located between the coenzyme A and pyruvate sites.

What kind of chemical mechanism can explain the two partial reactions (carboxylation and transcarboxylation) mentioned previously? Biotin is a CO_2 activator where the CO_2 molecule is covalently bound to an enolizable position of the cofactor. Any mechanism must agree with the experimental observation that if oxygen-labeled bicarbonate is used ($HC^{18}O_3^-$), two of the labeled oxygens ($\frac{2}{3}$) are found in the end product and the third one ($\frac{1}{3}$) ends up in the inorganic phosphate that is split from ATP.

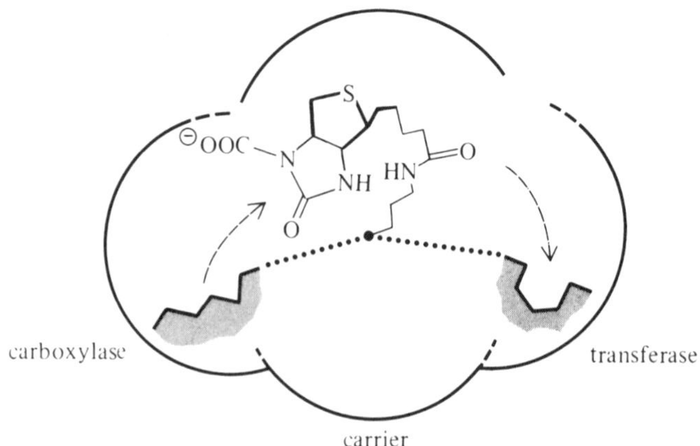

Fig. 7.14. A pictorial model of acetyl-CoA carboxylase of *E. coli* (663).

In fact, three schools of thought exist, all consistent with the above observations. In 1962, S. Ochoa first proposed a two-step mechanism that involves the activation of bicarbonate by ATP to produce carbonyl phosphate. This activated bicarbonate then reacts with biotin-bound enzyme to give 1′-*N*-carboxybiotin, another activated form of carbonate. These two processes could occur in a concerted way and have been generally accepted for some time. The 1′-*N*-carboxybiotin intermediate is stable under basic conditions but decarboxylates readily in acid medium. In the above example, carboxybiotin reacts with the enol form of acetyl-CoA to give malonyl-CoA and regenerated biotin. As expected, one oxygen from bicarbonate appears in the form of phosphate ion, the other two are in the carboxyl group of malonyl-CoA.

very fast
– ADP
high-energy anhydride intermediate
carbonyl phosphate
Mg^{2+}
$CoA—S—C(=O)—CH_2—C●●H$ +
malonyl-CoA
acetyl-CoA (enol form) via nucleophilic attack

In 1976, Kluger and Adawadkar (666), from the University of Toronto, proposed an interesting variation where the first step involves the formation of an *O*-phosphobiotin, a "masked" form of carbodiimide (Fig. 7.15). A "masked carbodiimide" was already described in Chapter 3 as an *O*-phosphoryl urea.

The isoimide intermediate (*O*-acyl urea) reacts with bicarbonate and then decomposes to give 1′-*N*-carboxybiotin and phosphate ion. Kluger

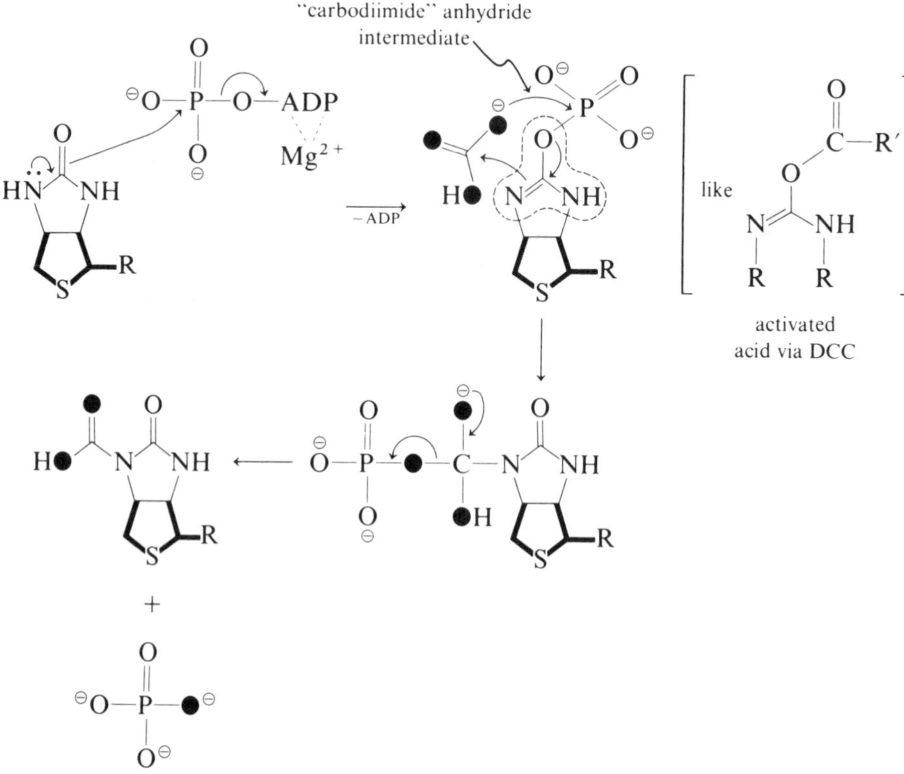

Fig. 7.15. Possible mode of action of biotin according to Kluger and Adawadkar (666).

synthesized an interesting model compound to prove the existence of his intermediate (see page 585).

Finally, in 1977, W.C. Stallings (667), from the Institute for Cancer Research in Philadelphia, proposed a third mechanism based on the analysis of the hydrogen bonding observed in the crystal structure of biotin (Fig. 7.16). He argues that biotin and bicarbonate make complementary hydrogen bonds to form a "stable" intermediate. Approach of ATP triggers carboxylation by bond polarization and a keto–enol tautomerism of the ureido moiety, which enhances the nucleophilicity of the N-1′ position. The involvement of an enol form of biotin is supported by the crystallographic studies on biotin. Notice that the previous mechanism also displays an enol intermediate.

Therefore, the ATP participation shifts the equilibrium toward the enol form. This last mechanism represents an example of activation by a substrate. The binding of bicarbonate, the substrate in question, increases the nucleophilicity of the biotin cofactor. However, the formation of an

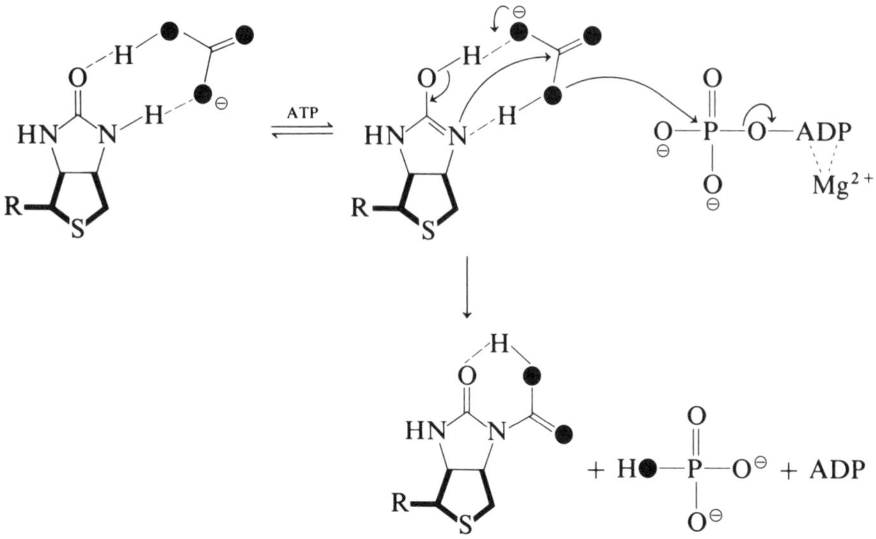

Fig. 7.16. Stallings' hypothetical mode of action of biotin (667).

activated bicarbonate, as proposed in the first mechanism, cannot be excluded. In all three cases examined, the role of Mg(II) ions could be one of binding to ATP and promoting the formation of ADP. It is also remarkable that in all three cases the action of biotin is mediated by activation by phosphorylation, reminiscent of some reactions presented in Chapter 3.

Now that we have examined objectively three mechanisms that agree with the experimental results, we should examine the real problem in biotin chemistry. Indeed, in recent years a controversy has arisen regarding the exact site on biotin to which the carboxyl group is attached. During isolation and characterization by Lynen of the relatively unstable free carboxybiotin, particularly at acidic pH, the product was converted to the more stable dimethyl ester with diazomethane (668,669). This derivative was subsequently identified as 1′-*N*-methoxycarbonyl-(+)-biotin methyl ester. The same product was also obtained by enzymatic degradation of enzyme-bound biotin. Figure 7.17 shows some of these transformations.

All these experiments with several different biotin enzymes concluded that the site of carboxylation on biotin is the 1′-ureido-N atom, in conformity with the original observations made by Lynen in 1959 (669). However, in 1970, T.C. Bruice and A.F. Hegarty (670), from the University of California at Santa Barbara, pointed out that this structural assignment of the 1′-*N*-carboxyl as the reactive group is equivocal. They argued that carboxylation occurred first at the 2′-ureido-O atom and that, during the methylation used to isolate carboxybiotin, the O-carboxylated

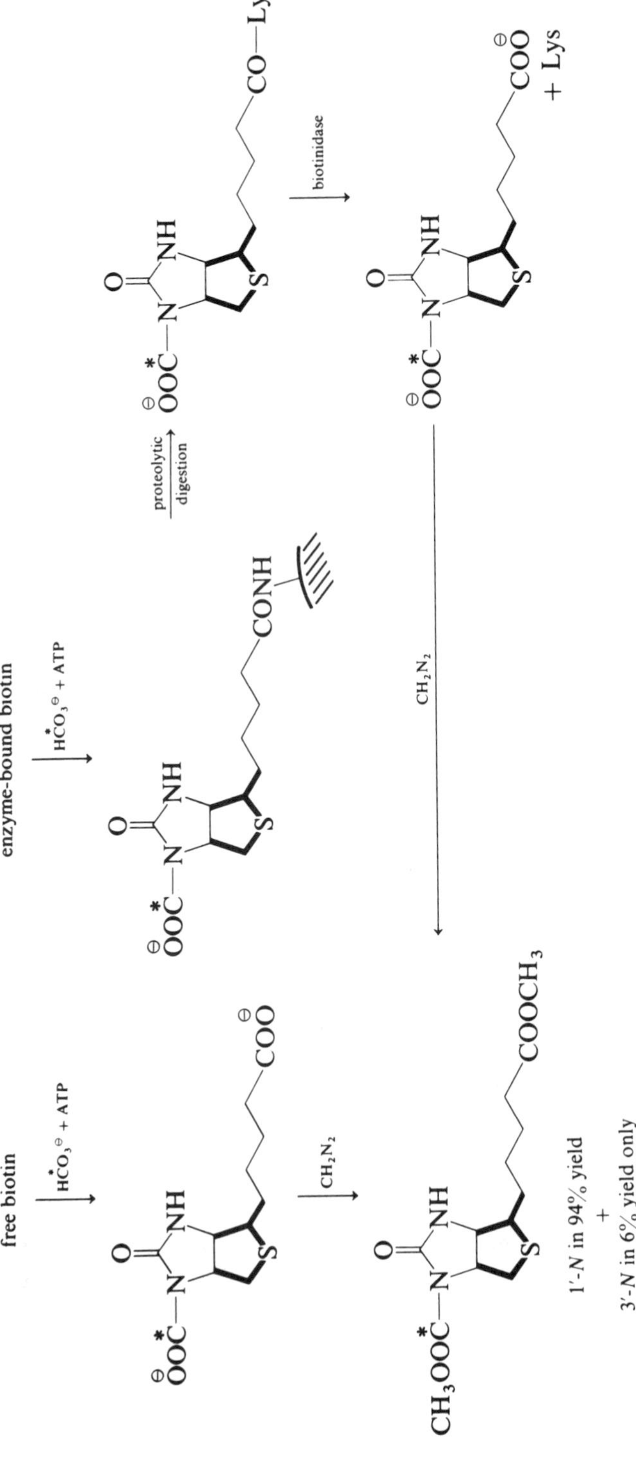

Fig. 7.17. Identification of the site of carboxylation of free and enzyme-bound biotin (668).

product could have rearranged by transferring the carboxyl group to the more thermodynamically stable 1′-ureido-N-substituted biotin derivative.

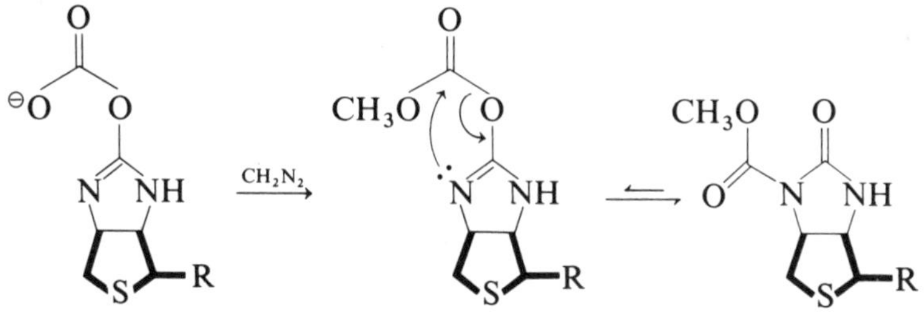

The question is: What is involved first, an N-carboxylation (Lynen) or an O-carboxylation (Bruice) on the most nucleophilic site, the urea oxygen, followed by migration to the nitrogen? Obviously, bioorganic models are needed to understand the pivotal role of the ureido function.

7.5.1 Model Studies

Since biotin appears to participate in carboxylation reactions as a nucleophilic catalyst, Caplow and Yager (671) used a 2-imidazolidone structure as a biotin analog.

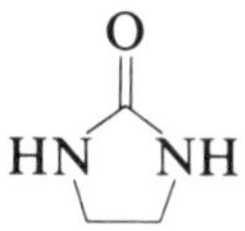

2-imidazolidone

Decarboxylation studies of *N*-carboxy-2-imidazolidone revealed the poor leaving ability of the imidazolidone anion. Metal ions such as Cu(II) or Mn(II) prevent decarboxylation. The sensitivity of carboxyimidazolidone to specific-acid catalysis is probably the most remarkable feature of this compound and could account for the fact that carboxybiotin–enzyme intermediates are unstable. The neutral form, however, can decarboxylate in a unimolecular pathway via a six-membered transition state.

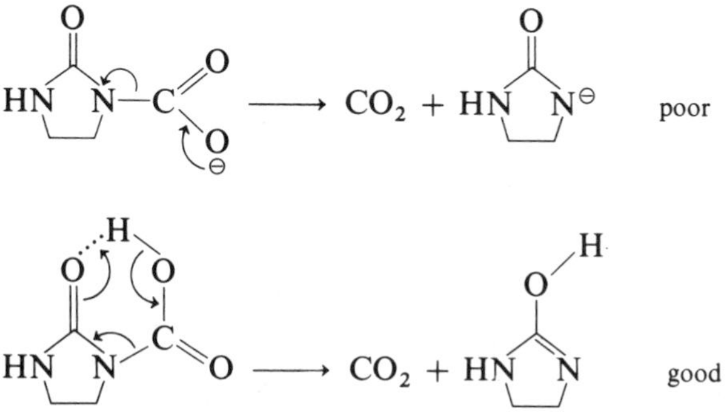

$\longrightarrow CO_2 +$ poor

$\longrightarrow CO_2 +$ good

The failure for the model compound to be acylated by *p*-nitrophenyl acetate, acetylimidazole, or acetyl-3-methylimidazolium chloride also indicated that the ureido-N atom of biotin is not highly nucleophilic. Nonetheless, an intramolecular model reaction in which the interacting functional groups are properly juxtaposed might more closely mimic the enzymatic process.

For instance, Bruice's group studied the intramolecular nucleophilic attack of a neutral or anionic ureido function on acyl substrates with leaving groups of varying basicity (670).

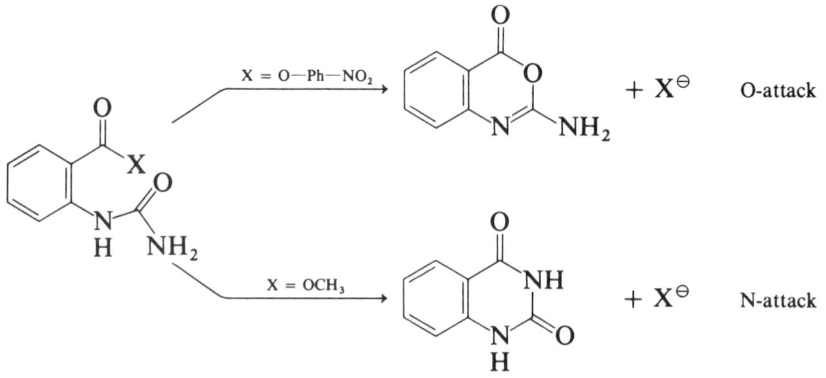

It was observed that O-attack was favored over N-attack if a good leaving group is present. These results support his proposal of O-carboxylation of biotin where a high-energy carbodiimide-type compound (isoimide or isourea) is first formed and rapidly rearranges to the more stable *N*-acyl form.

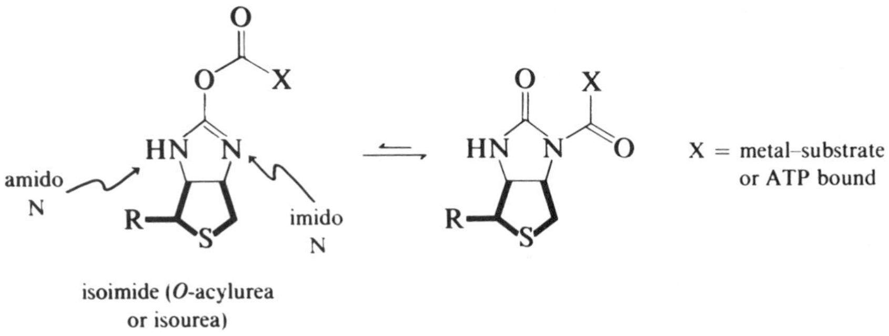

isoimide (*O*-acylurea or isourea)

X = metal–substrate or ATP bound

However, since it is not known to what extent the enzyme contributes to the stabilization of the leaving group anion X in the biotin-dependent carboxylation reactions, a conclusion cannot be drawn a priori as to whether O- or N-attack occurs. Clearly, simple model compounds are not always reliable indicators of reactivity in the environment of an enzyme (668). Maybe enzyme-bound biotin reacts in its high-energy isourea form to increase the nucleophilicity of the nitrogen atom. After all, an imido nitrogen is known to be a better nucleophile than an amido nitrogen.

R′ = phosphate or ATP
X = electrophile
($H^{\oplus}$ or metal ion)

Returning to the original question as to whether the postulate of *O*-carboxybiotin by Bruice is the true intermediate in the first half-reaction of biotin chemistry, the group of M.D. Lane from Johns Hopkins provided very convincing evidence in 1974 that, in fact, 1′-*N*-carboxybiotin must be a true biochemical intermediate (672). Using two subunits of acetyl-CoA carboxylase of *E. coli*, the biotin carboxylase and the carboxytransferase,* they showed quite conclusively that 1′-*N*-carboxybiotin is active as a carboxyl donor (Fig. 7.18). They synthesized 1′-*N*-methoxycarbonyl-(+)-biotinyl acetate and characterized it chemically and spectroscopically by mass spectroscopy and ^{1}H- and ^{13}C-NMR. Alkaline hydrolysis afforded 1′-*N*-carboxy-(+)-biotinol. It was observed that biotinol, the reduced form (alcohol) of the valeric acid group of biotin, is much more effective than biotin. They also showed that the reaction mediated by the carboxylase occurred from right to left with synthesis of ATP in the presence of authentic 1′-*N*-carboxy-(+)-biotin. Furthermore, authentic 1′-*N*-carboxy-(+)-biotin had stability properties indistinguishable from those of the carboxybiotin formed enzymatically.

These convincing data showed that the 1′-N-ureido position of biotin serves as the site for carboxyl transfer with biotin enzymes. Lane also correctly pointed out that N → O carboxyl migration might have preceded the participation of carboxybiotin in the enzymatic process. However, the well-established thermodynamic and kinetic stabilities of *N*-acyl and *N*-carboxy-2-imidazolidone derivatives render this possibility unlikely. Moreover, the urea carboxylase component of ATP-amidolyase, also a biotin-dependent enzyme, reversibly carboxylates urea to form *N*-carboxyurea, a known example of carboxylation at the N-ureido position (673).

$$H_2N{-}CO{-}NH_2 + HCO_3^{\ominus} \xrightarrow[\text{urea carboxylase}]{\text{ATP}} H_2N{-}CO{-}NH{-}COO^{\ominus} \xrightarrow[\text{allophanate amidolyase}]{H_2O} 2\,HCO_3^{\ominus} + 2\,NH_4^{\oplus}$$

N-carboxyurea
(allophanate)

* Each subunit of the multienzyme complex can be used separately in model studies.

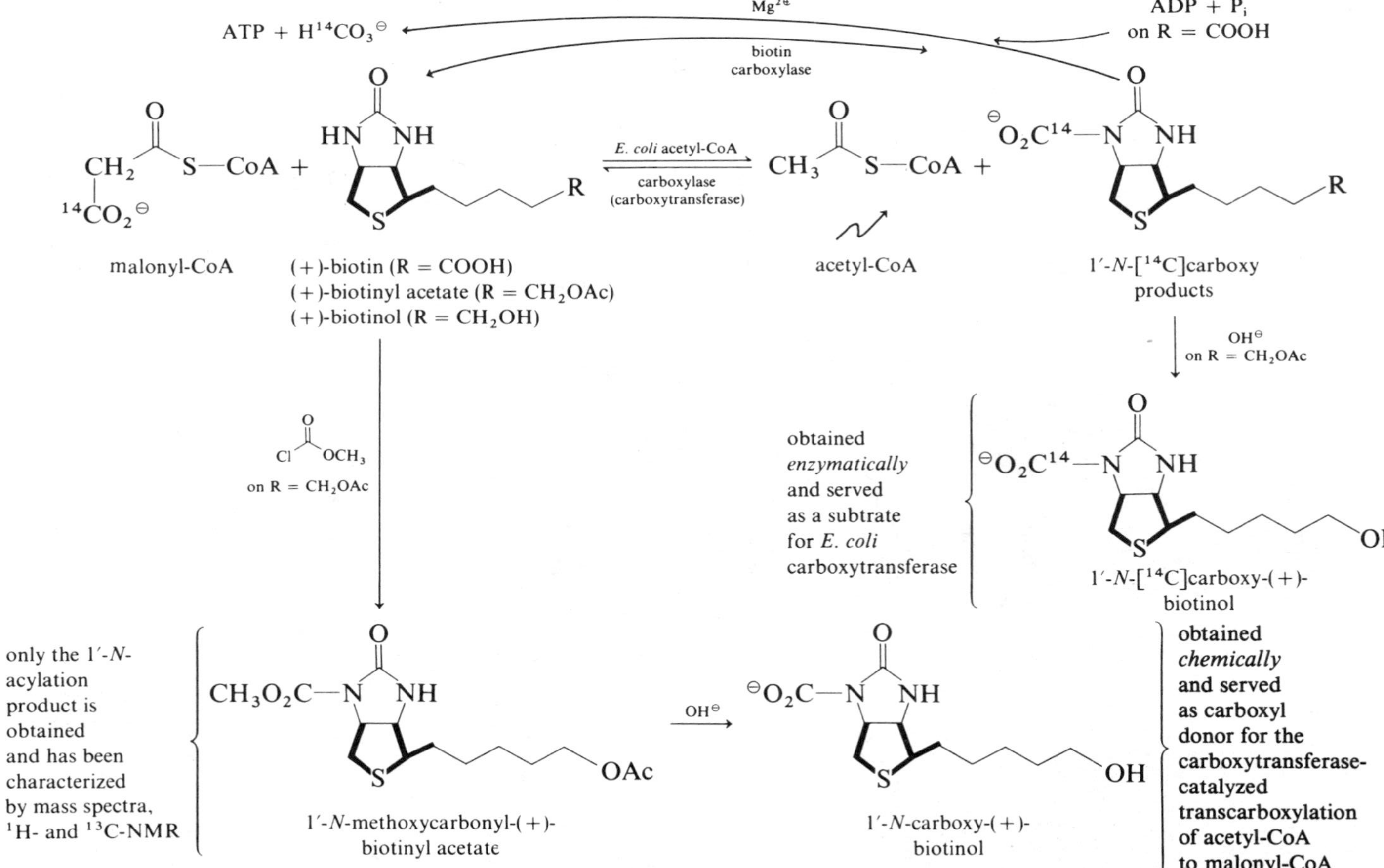

Fig. 7.18. The methodology used by Lane (672) to prove that biotin is carboxylated at position N-1′.

Lane and coworkers also speculated as to why the 1′-N, rather than the apparently more reactive 2′-O, carboxybiotinyl prosthetic group evolved as the carboxylating agent in these enzymatic processes. They suggest that possibly the consequent exposure of the mobile carboxyl carrier chain to the solvent would oblige the carboxylate group to possess low, rather than high, carboxyl transfer potential to prevent O → N migration during the transfer process.

We should now return to the unsolved original problem of finding the method by which biotin and carboxybiotin are activated. Earlier in this section we presented three possible mechanisms for carboxylation of a substrate by biotin. One of them, by Kluger and Adawadkar (666), suggests the presence of an *O*-phosphobiotin intermediate. Kluger has now developed an interesting model for such a mechanism, in which a phosphoryl group, via a methylene bridge, is attached to a ureido nitrogen. The starting material was synthesized from dimethylurea, and LiOH hydrolysis led to a model compound possessing the reactive portions of biotin and ATP bound in the same disposition:

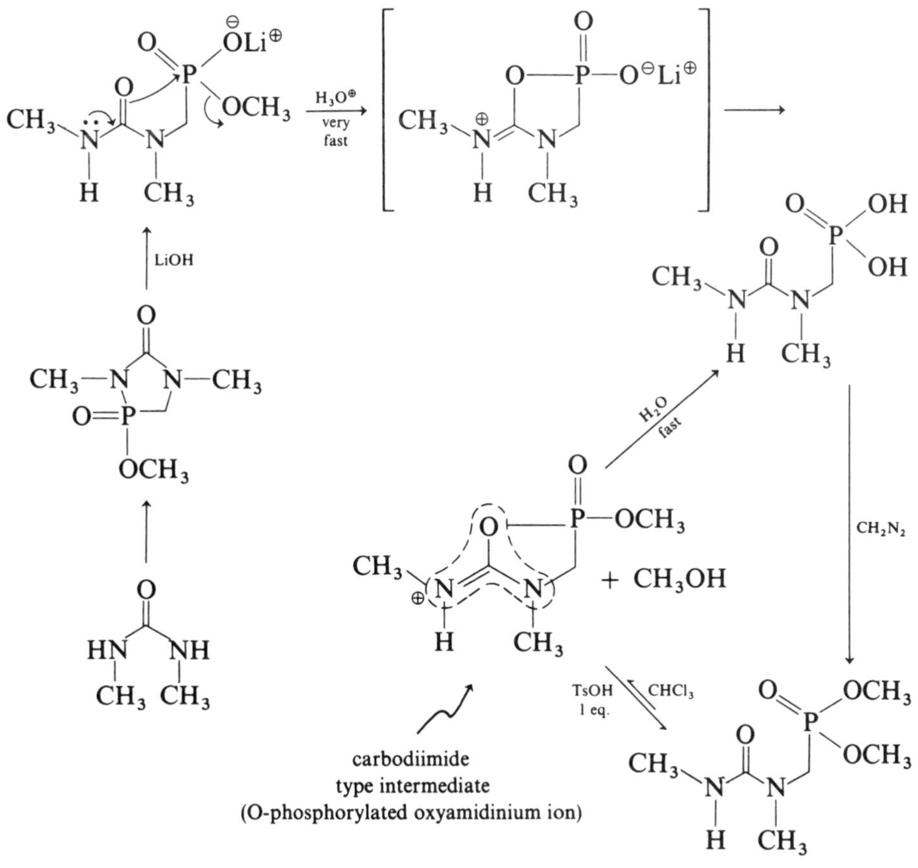

$H_3O^{\oplus}$ very fast

LiOH

H_2O fast

CH_2N_2

$+ CH_3OH$

TsOH 1 eq.

$CHCl_3$

carbodiimide type intermediate (O-phosphorylated oxyamidinium ion)

Aqueous acid hydrolysis of the product is very fast and proceeds through the formation of a cyclic phosphourea intermediate. On the basis of its hydrolysis and other properties of related compounds, he proposed that the urea moiety of biotin is nucleophilic toward phosphate derivatives. Thus, ATP, HCO_3^-, and biotin can bind in close proximity at the carboxylation subunit to give an *O*-phosphobiotin, a "masked" carbodiimide dehydrating agent, that can condense with HCO_3^- and rearrange to 1′-*N*-carboxybiotin (see Fig. 7.15). In fact, the possibility of an *O*-phosphobiotin mechanism was first proposed by Calvin in 1959. However, these new available data are appealing because they demonstrate that the 2′-O-ureido function of biotin is a nucleophile toward phosphorus.

In 1977, R.M. Kellogg (674), from the University of Groningen, synthesized an interesting model to mimic an essential aspect of the second half-reaction, namely, the transfer of the carboxyl group from biotin nitrogen to an acceptor molecule. He found that the lithio derivative of *N*-methylethyleneurea (or thiourea), a model for the isourea form of biotin, is capable of causing the rearrangement of 1-methyl-4-methylene-3,1-benzoxazin-2-one to 4-hydroxy-1-methyl-2-quinolinone. The *N*-methyl derivative improves the solubility in THF and allows the abstraction of only one hydrogen. This isomerization does not occur with organic bases such as $R_4N^+OH^-$, pyridine, or LiOH. But if the isourea is activated by BuLi, the reaction proceeds in 49% yield in refluxing THF for 15 min and the following adduct was isolated in 10% yield:

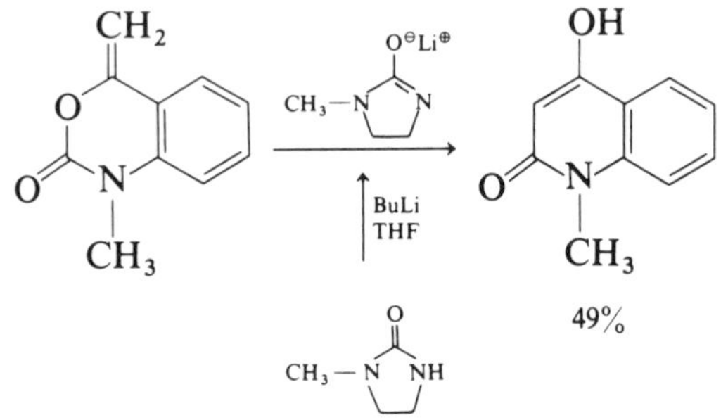

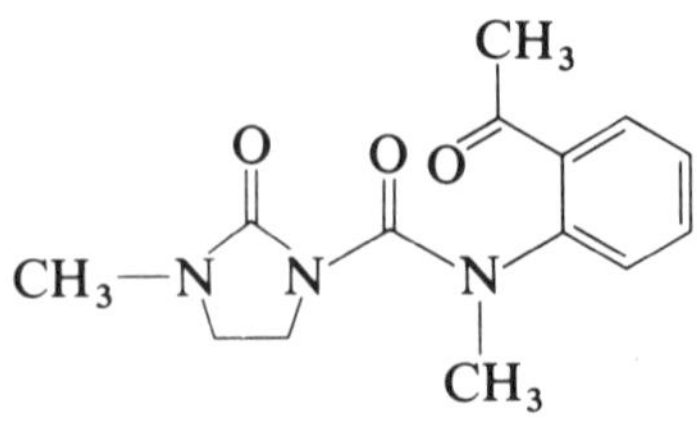

A reasonable mechanism for this cooperative process is:

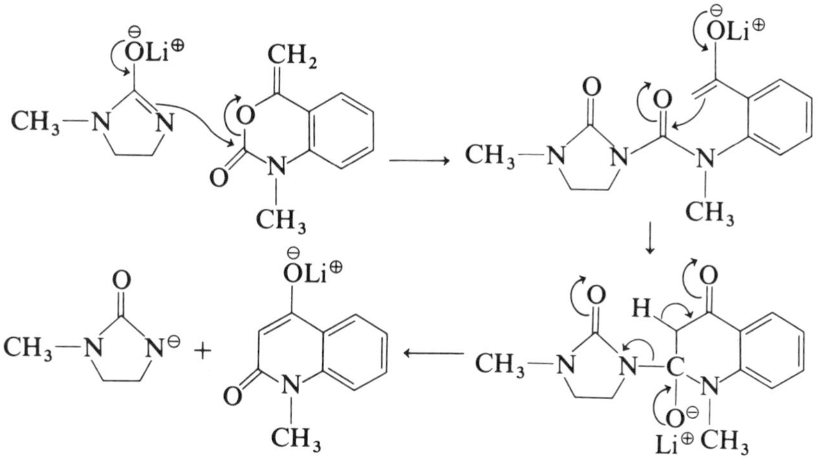

Therefore, in a number of respects this transformation mimics the carboxylation of biotin on nitrogen and the subsequent carboxyl transfer to a carbon atom bonded to a carbonyl group.

In 1979, Kluger's group examined in more detail his mechanism of urea participation in phosphate ester hydrolysis (675). In addition to contributing further evidence for the potential involvement of *O*-phosphobiotin in ATP-dependent carboxylations in enzymatic systems, they have also considered the stereochemical pathways of carboxylation of *O*-phosphobiotin in the enzymatic reactions, leading to the formation of *N*-carboxybiotin and phosphate ion. Since a pentacovalent intermediate must be involved, two mechanisms are possible (refer to Chapter 3 for notions of pseudorotation): an adjacent mechanism, not involving free carboxyphosphate, and an in-line mechanism, involving free carboxyphosphate.

Adjacent mechanism leads to retention of configuration at phosphorus:

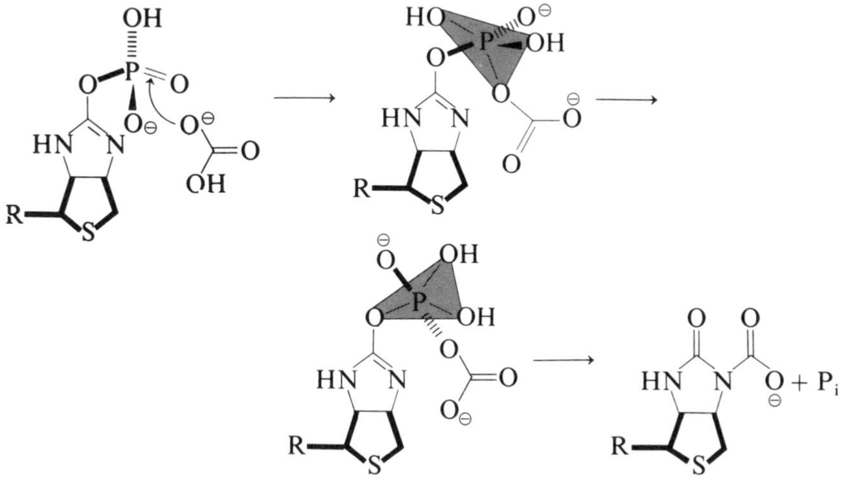

In-line mechanism causes an inversion of configuration at phosphorus:

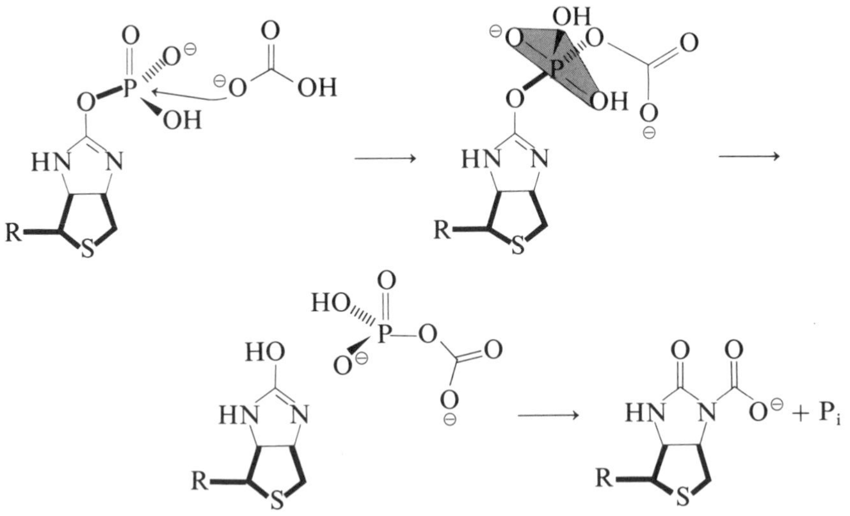

$$\longrightarrow \quad \longrightarrow \quad \longrightarrow \; + \; P_i$$

In the in-line mechanism, bicarbonate attacks *O*-phosphobiotin *trans* to the biotin ureido function. If this is the case, a concerted in-line attack followed by carboxylation is stereochemically impossible. Consequently, in the in-line mechanism carboxyphosphate must be formed. On the other hand, the adjacent mechanism allows the reaction to proceed without formation of carboxyphosphate because of a more favorable stereochemical disposition of bicarbonate. Although this mechanism corresponds to the one they had previously suggested (666), more work is needed to solve this remaining ambiguity related to the stereochemistry of carboxylation of *O*-phosphobiotin in enzymatic processes. Structural analysis of a biotin-dependent enzyme may resolve this problem.

In conclusion, the formation of *O*-phosphobiotin from ATP is likely to occur with inversion of configuration at the γ-phosphorus. Since the subsequent transfer to bicarbonate can occur with retention or inversion, the route via *O*-phosphobiotin can account for net inversion (inversion and retention) or net retention (inversion and inversion) at the phosphorus atom (675).

Furthermore, one could envisage a possible transannular sulfur coordination to the carbonyl to increase the basicity of the 1′-N nitrogen but no evidence for such a bridged intermediate could be found (676).

Finally, a word should be said about the stereochemistry of the transcarboxylation reaction. In 1966, Arigoni and co-workers (677) first showed by tritium labeling that carboxylation of propionic acid operates by *retention* of configuration.

Like other biotin-dependent carboxylations, this reaction involves the abstraction of the α-proton (tritium here) and its replacement by CO_2.

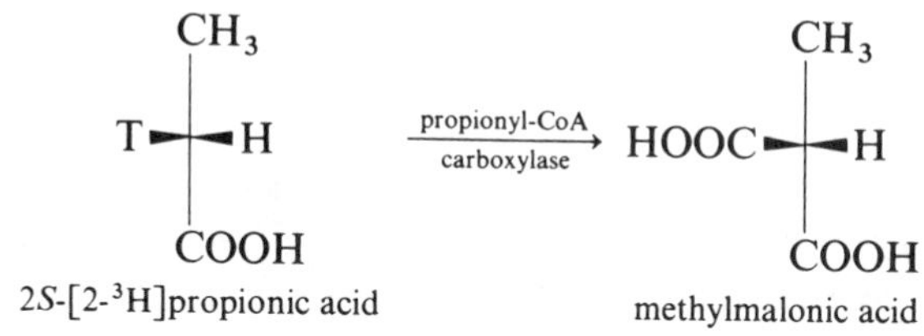

$$\text{T}\blacktriangleright\!\!\underset{\text{COOH}}{\overset{\text{CH}_3}{|}}\!\!\blacktriangleleft\text{H} \xrightarrow[\text{carboxylase}]{\text{propionyl-CoA}} \text{HOOC}\blacktriangleright\!\!\underset{\text{COOH}}{\overset{\text{CH}_3}{|}}\!\!\blacktriangleleft\text{H}$$

2*S*-[2-^{3}H]propionic acid — methylmalonic acid

This aspect was also analyzed by Rétey and Lynen (678) and by Rose and coworkers (679). They proposed that proton abstraction and carboxylation occur in a concerted process (Fig. 7.19).

The concerted mechanism, although it could account for the retention of configuration in the carboxyl transfer during the process, is stereochemically unreasonable.

Such transformations imply that the carbonyl group of biotin serves as a proton acceptor in one direction and a proton donor in the other. A better mechanism is to have an external base. Thus, an alternative to the concerted mechanism is a stepwise process involving α-proton abstraction followed by carboxylation. To shed light on this possibility, Stubbe and Abeles in 1979 (680) examined the action of propionyl-CoA carboxylase on β-fluoropropionyl-CoA. This substrate is particularly suitable to determine whether proton abstraction occurs in the absence of carboxylation because (a) fluorine is a small atom and will not cause steric problems and (b) when a carbanion will be generated β to the carbon bearing the fluorine atom, fluoride will be eliminated. Fluoride elimination will thus be indicative of carbanion formation.

When β-fluoropropionyl-CoA is incubated in the presence of HCO_3^-, ATP, and Mg(II) and added to the enzyme, ADP is formed and F^- is released. However, the rate of F^- release is 6 times that of ADP formation and no evidence for the formation of fluoromethylmalonyl-CoA was obtained. The release of a F^- ion is an indication of an abstraction of an α-proton of the substrate, and the formation of ADP corresponds to the formation of biotin-CO_2. Therefore, these results indicate that hydrogen abstraction can occur without concomitant CO_2 transfer from

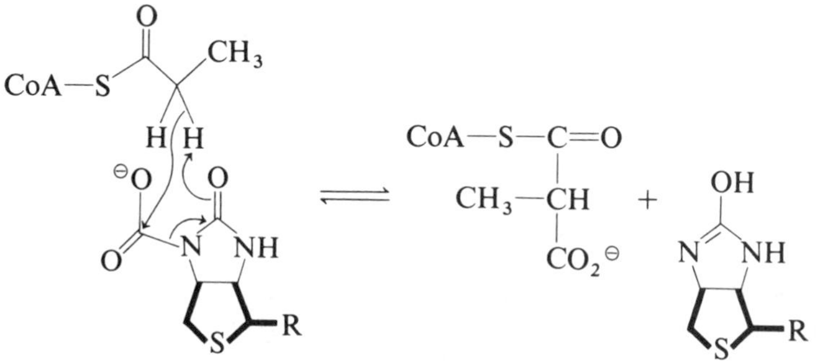

Fig. 7.19. Concerted mechanism for carboxylation of propionyl-CoA.

biotin-CO_2 to the substrate. The concerted mechanism proposed earlier is thus not applicable when propionyl-CoA carboxylase acts on β-fluoropropionyl-CoA. The authors also suggest that the *nonconcerted process should occur* with "normal" substrates and that there must be a group at the active site of the enzyme that functions as proton acceptor. Furthermore, it was shown that this site does not exchange protons with the solvent.

In summary, the mechanism (Fig. 7.20) of F^- elimination from β-fluoropropionyl-CoA can be proposed (680) where a base (B) on the enzyme abstracts the α-proton of the substrate to form an enzyme-bound carbanion.

The elimination of F^- is more rapid than the transfer of CO_2 from biotin to the carbanion. After the elimination, the complex can be decomposed through several possible routes such as decarboxylation and release of acrylyl-CoA.

A further study concluded that oxaloacetate is required and that the rate of pyruvate formation is equal to that of acrylyl-CoA formation (681).

Conclusive evidence for a stepwise mechanism in enzymatic biotin-mediate carboxylation came from O'Keefe and Knowles's investigation using doubly labeled (^2H and ^{13}C) pyruvic acid in pyruvate decarboxylation (682). The kinetic isotope effect and the isotope distribution obtained were in agreement with a mechanism by which proton removal and carboxy group addition occur in different steps in the transcarboxylase-catalyzed reaction.

pyruvate-d_3

$^{\ominus}OOC-C(=O)-CD_2-D$ + $^{\ominus}OOC-N$ (biotin, ^{13}C) $\xrightarrow{\text{first step}}$ $^{\ominus}OOC-C(=O)-CD_2^{\ominus}$ $\xrightarrow{\text{second step}}$ $^{\ominus}OOC-C(=O)-CD_2-COO^{\ominus}$

malonate
doubly-labeled

We would like to end this chapter by mentioning an interesting speculation made by Visser and Kellogg (674) about the origin of biotin and other coenzymes. First, model studies indicated that biotinlike molecules to not have catalytic power in carboxylation unless they have been

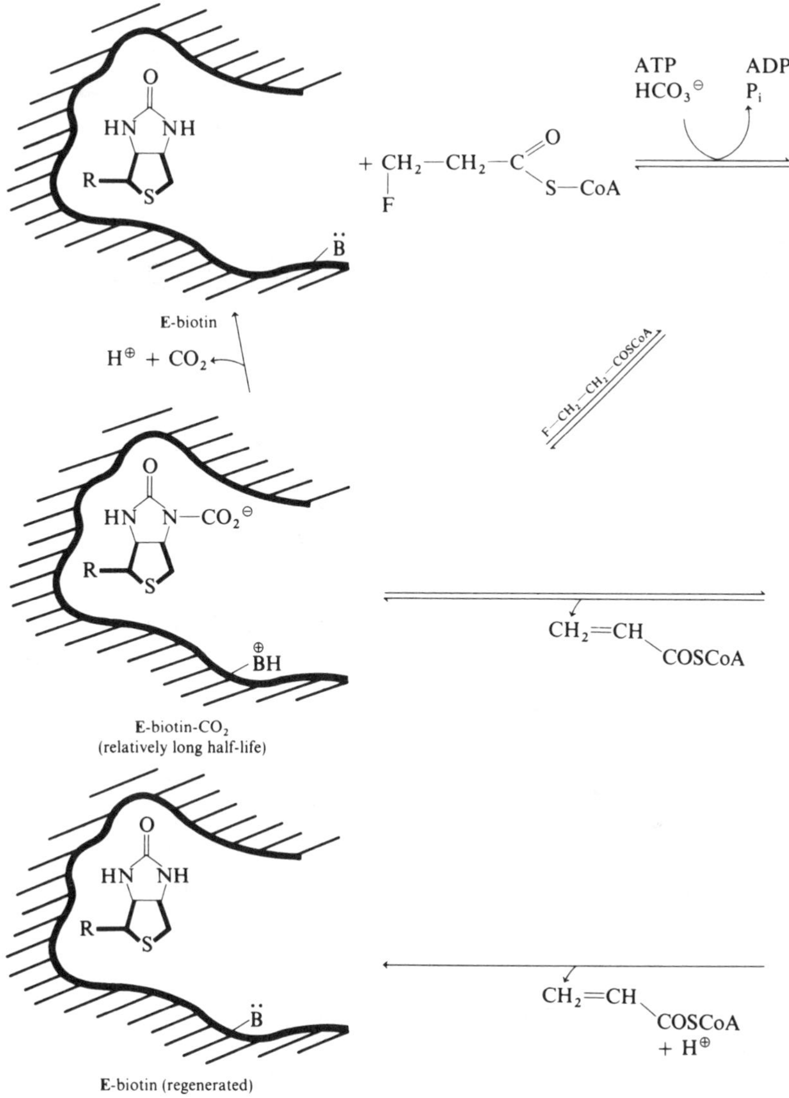

Fig. 7.20. Mechanism of fluoride elimination from fluoropropionyl-CoA (680). (See page 592 for the completion of this scheme.)

activated first to a high-energy tautomer, which is more nucleophilic. Consequently, biotin seems to act only as a CO_2 shuttle, linking two different active subunits together. Contrary to other transfering coenzymes, it does not contribute itself to the lowering of the activation energy of the reaction. Second, and more peculiar, biotin and the structurally related enzyme-bound lipoic acid bear no resemblance to a nucleotide. All other coenzymes contain at least a purine or a pyrimidine ring in their struc-

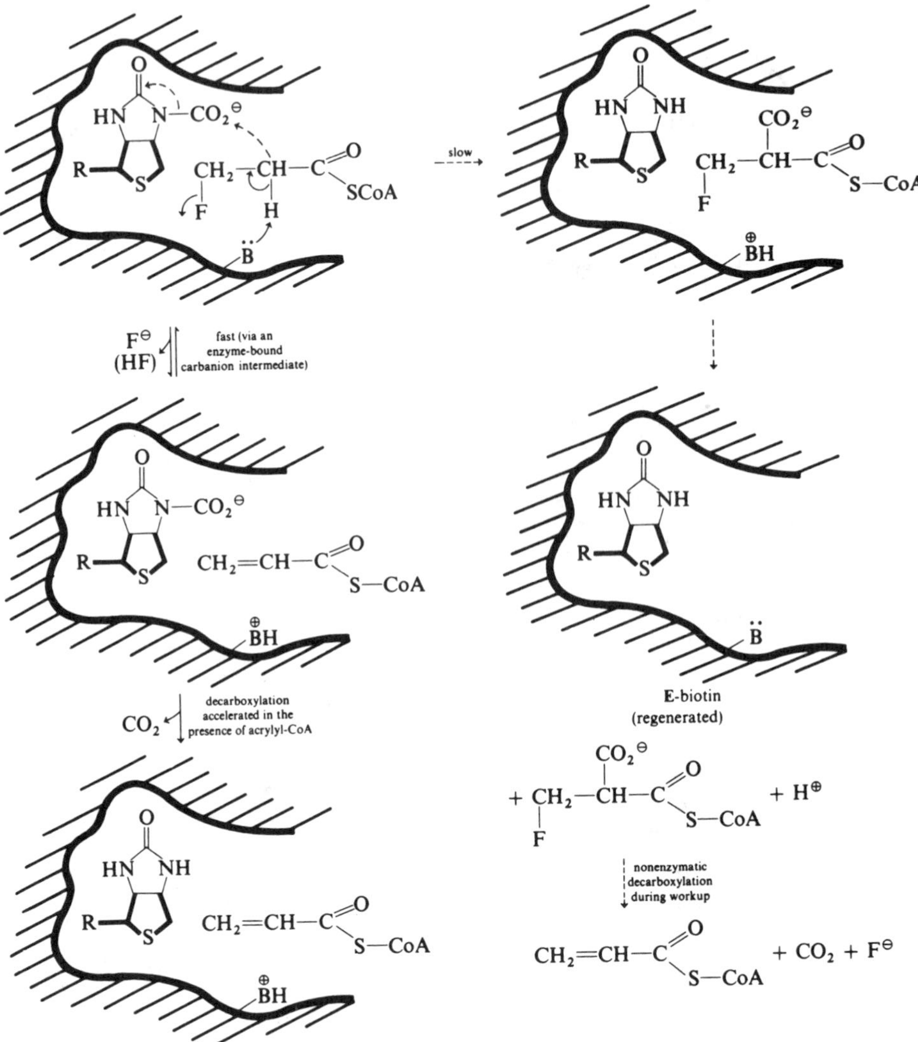

Fig. 7.20. (continued)

tures. In the speculation about the early stages of the origin of life it is believed that self-reproducing systems were made of nucleic acid-like molecules coding for nucleic acid enzymes, proteins being not yet important. According to Kellogg, present-day coenzymes may well be the vestiges from this polynucleotide enzyme period and thus still posses the catalytic capacity of group transfer reactions. However, biotin and lipoic acid have no tendency to act as a coenzyme unless they are covalently attached to a carrier protein, and thus they behave only as transferring agents in multienzyme complexes. Therefore, as prosthetic groups they may well have originated from essentially late evolutionary developments.

Chapter 8
Molecular Devices

"Biomimetic materials chemistry should be viewed as a malleable science capable of being shaped by inspirations emanating from both physical and biological sources."

S. Mann
Nature **365**, 499 (1993)

We have seen that *molecular recognition*, *transformation*, and *translocation*, represent the basic functions of supramolecular species. More complex functions may result from the interplay of several binding subunits in a polytopic coreceptor. In association with organized polymolecular assemblies and phases (films, layers, membranes, vesicles, liquid crystals, etc.), new functional supramolecules may lead to the development of *molecular devices* (683).

This chapter describes these aspects of supramolecular chemistry and sketches some lines for future development. Emphasis is on conceptual frameworks, classes of compounds, and types of processes, using a series of diagrams. At present, *supramolecular chemistry*, the orchestrated and designed chemistry of molecular associations through intermolecular bonds, is rapidly expanding at the frontiers of molecular science with physical and biological phenomena.

8.1 Introduction to Self-Organization and Self-Assembly

Beyond the preorganization used in the construction of molecular receptors lies *self-organization*. It involves the design of systems capable of spontaneously generating well-defined supramolecular entities by *self-assembly* from their components in a given set of conditions (684). The expression *self* refers here to a mutual recognition event. The information necessary for the process to take place and the program that it follows are based on molecular recognition events. Thus, these systems may be termed "*programmed supramolecular systems*" (685). In other words, the approach is molecular, but the operation is supramolecular and uses the concepts of self-assembly and self-organization.

Self-assembly and self-organization have recently been implemented in several types of organic and inorganic systems. By clever use of metal coordination, hydrogen bonding, and donor–acceptor interactions, researchers have achieved the spontaneous formation of a variety of novel and intriguing species, such as the inorganic double and triple helices termed helicates, catenanes, threaded entities such as rotaxanes, cage compounds, and so forth (684).

Furthermore, at this polymolecular level of complexity, chemists are now developing "*designer solids*," that is, molecular networks of indefinitely large sizes with tunable properties, such as the spacing of lattice and the chemistry of lattice members (686). Many scientists compare these molecular achitectures or devices to molecular meccano or tinkertoys at the nanometer scale using building block or "molecular units." Others, like J. Moore (University of Illinois), refer to these units as "three-dimensional nanoscaffolding" or molecular lattices that could serve as frameworks for catalysts and photosynthetic molecules (686).

In this context, chemical amplification in high-resolution *imaging systems*, with photolithographic processes using microlithographic materials and techniques, will play a central role in the development of more advanced devices containing even smaller structures (687). Progress is being made in this direction by J. Fréchet from Cornell University. Needless to say, the readily availability of inexpensive, high-density semiconductor devices have made the "computer age" possible.

For F. Stoddart (University of Birmigham), the ultimate goal in research on catenane-based devices (Section 8.3.2) is the development of *molecular computers* in which an array of nanoscale switches, perhaps like a *molecular shuttle*, might behave like the integrated circuitry of semiconductor chips. It is worth noting that back in 1985, an article entitled "Biochips: the ultimate computer" appeared in *Biotechnology* (688). This structural, but hypothetical, model was based on a regular array of highly specific antibody molecules. Their self-assemblies with different antigens into a network served as the structural basis for the construction of a molecular-scale electronic device. Future medical applications of molecular devices could include the development of *biosensors* in *gene probe technology* for genetic disease identification.

8.2 General Overview of the Approach

This section will provide an overview of the field of supramolecular chemistry in the form of diagrams adapted from slides used in a recent series of graduate-level lectures on bioorganic chemistry given at Peking University and at Nankai University in China in 1994.

As we recall (Section 1.4), supramolecular chemistry is divided into three levels of complexity (see Diagram 8.1). The basic concepts involved here use molecular and intermolecular interactions for the design of such new devices as *chemionics* (see Section 5.3.2 and Diagram 8.2).

Supramolecular Chemistry

The levels of complexity

- **Three levels of complexity**
  - Molecular level
    - Supramolecular level
      - Polymolecular level

The chemistry beyond the molecule

Diagram 8.1

Supramolecular Chemistry

Basic concepts

- **Requirements:**
  - molecular interactions
    - intermolecular interactions
      - systems based on chemical reactivities
- **Design:**
  - "devices" ☞ CHEMIONICS

Diagram 8.2

Because molecular recognition will play a major role, its basic principles and general pattern are presented in Diagram 8.3.

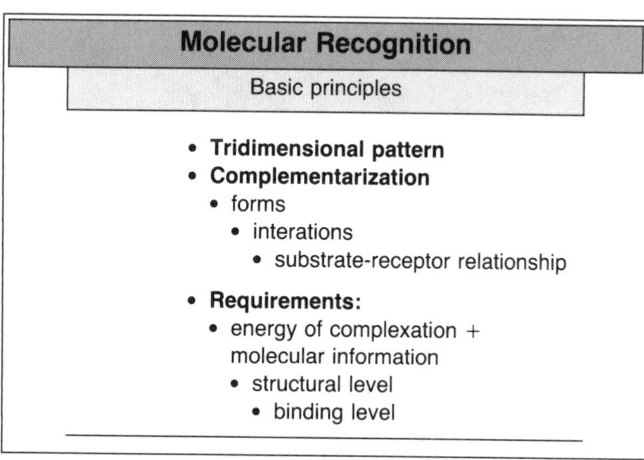

Diagram 8.3

Diagram 8.4 presents the events involved in molecular self-organization. This process is facilitated by among other things, hydrogen bonding and ion coordination to form helices. Molecular auto-association, on the other hand, is of two types: covalent and noncovalent. This is shown in Diagram 8.5.

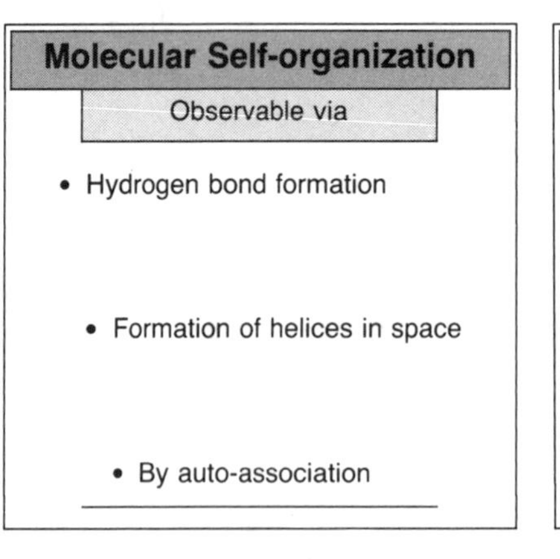

Molecular Self-organization
Observable via
• Hydrogen bond formation
• Formation of helices in space
• By auto-association

Diagram 8.4

Molecular Auto-association

Of two natures

- **Covalent associations**
  - macrocycles, cages
    - polycyclizations
      - metallic chelations
- **Non-covalent associations**
  - crystals
    - monolayers
      - charge transter complexes
        - substrate-receptor complexes

Diagram 8.5

Molecular assembly also has two complementary approaches and can operate either on small ionic species or on large supramolecules. Diagrams 8.6 and 8.7 summarize these possibilities.

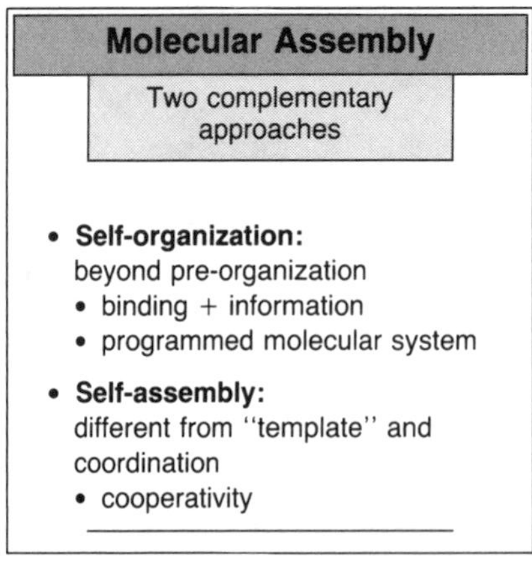

Molecular Assembly
Two complementary approaches
• Self-organization:
beyond pre-organization
• binding + information
• programmed molecular system
• Self-assembly:
different from "template" and coordination
• cooperativity

Diagram 8.6

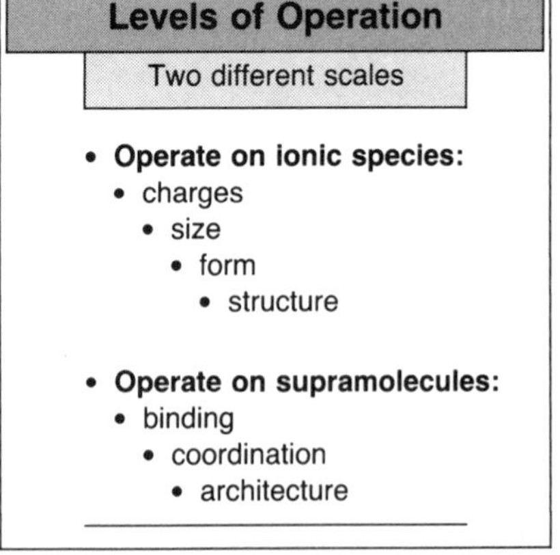

Levels of Operation
Two different scales
• Operate on ionic species:
• charges
• size
• form
• structure
• Operate on supramolecules:
• binding
• coordination
• architecture

Diagram 8.7

A more detailed examination of auto-association (or self-assembly) processes reveals that it can be subdivided into the four levels illustrated in Diagram 8.8.

Types of Auto-association

Four different levels

- **Irreversible covalent**
  biosynthesis of squalene (Corey)
- **Strict**
  double-helix helicates (Lehn)
- **With post-modifications**
  catenanes, knots (Sauvage)
  topochemical polymerizations
- **Directed**
  auto-replication (Rebek, Jr.)
  bi-dimensional assembly (Lehn)
  tri-dimensional assembly (Whitesides)

Diagram 8.8

The first type, the irreversible covalent self-assembly, is exemplified by the biosynthesis of steroids, a subject developed in Section 5.8. The so-called strict self-assembly will be presented in Section 8.3.1; while self-assembly with postmodification was already discussed in part in Section 6.5.1. Other aspects will also be treated in Section 8.3.

For the topochemical polymerization of the self-assembly of amphiphilic copolymers and the preparation of polymeric monolayers and multilayers, the work of H. Ringsdorf (University of Mainz) should be consulted (689).

At the level of directed self-assembly, J. Rebek, Jr., has been working on the development of a self-replicating system (690). The project was based on the molecular cleft approach presented in Section 4.10. It consisted of an adenine derivative covalently bonded to a monosubstituted Kemp imide. Such a molecule could then form an autodimer by self-complementarity via hydrogen bonding. Interestingly, the amide bond formation exhibited autocatalysis. Template-catalyzed replication is the source of autocatalysis (692), as outlined in Fig. 8.1.

As the authors point out, this can be regarded as a primitive sign of life, in which the system bridges the information of nucleic acids and the synthesis of amide bonds. The work was extended to a system in which one of the replicator units beared a photochemically active function (an *ortho*-nitrophenol) that was cleaved upon light irradiation (691). In this way, replication and mutation processes, two necessary elements of biological evolution, were explored with synthetic structures. The challenge in the exploration of new replicator molecules was reviewed in 1994 (692).

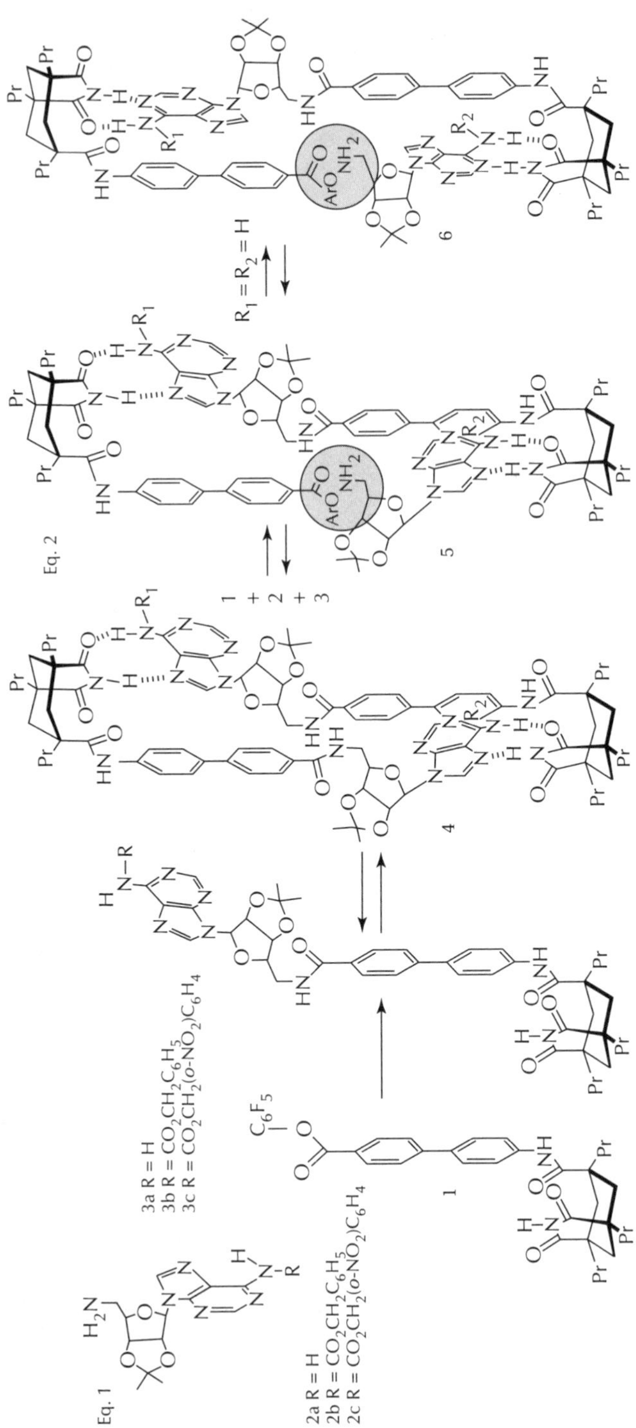

Fig. 8.1. A self-replicating system. Adapted from (691).

Two examples will be given of bi- and tridimensional assemblies. First, the work of J.M. Lehn (693), in which formation of polymeric supramolecular species are made by hydrogen-bonding association of two complementary ditopic components. This type of supramolecular process is depicted in Fig. 8.2. Similar assemblies can yield supramolecular rigid rods that present a lyotropic mesophase (694).

Examples of directed tridimensional assemblies come from the work of G.M. Whitesides (695,696), in which cyanuric acid and melamine units specifically hydrogen bond and assemble to form a stacking tubular module, such as the one presented in Fig. 8.3.

Such a complex is stable only because the enthalpy gained by forming so many hydrogen bonds (up to 36 in this example) in the complex is large enough to compensate for the entropic factor. This elegant work has been extended to other cyanuric acid-melamine lattices that are also held together by networks of noncovalent interactions (696). The strategy is closely modeled on the principles that determine secondary and tertiary structures in proteins and nucleic acids.

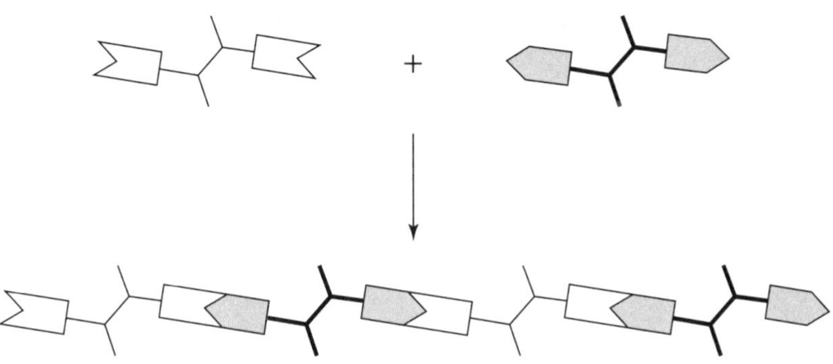

Formation of a polymeric supramolecular species by the association of two complementary ditopic components.

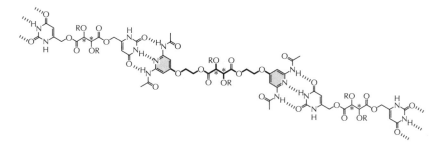

Fig. 8.2. Bidimensional self-assembly of polymeric supramolecular components. Adapted from (694).

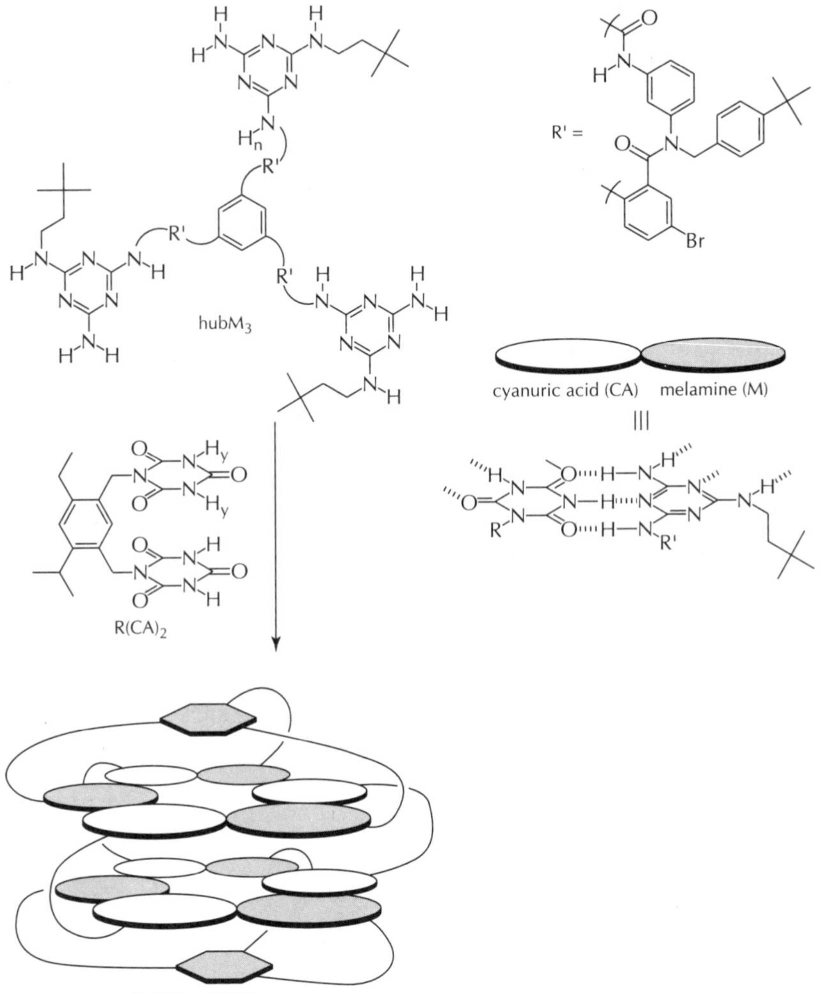

Fig. 8.3. Tridimensional self-assembly of cyanuric acid and melamine units (695). Reproduced with permission. Copyright © 1991 by the American Chemical Society.

As mentioned previously, the selective binding of a substrate by a molecular receptor to form a supramolecule involves molecular recognition, which depends on the molecular information (see Section 1.4) stored in the interacting partners. Chapters 5 through 7 showed that a great variety of receptor molecules have been designed in the last decade for effecting the recognition of numerous and very diverse types of substrates (spherical, tetrahedral, linear or branched, charged or neutral, organic, inorganic or biological, etc.). In the past, most investigations focused on *endo-receptors*, in which the binding (information) sites are oriented into a molecular cavity. However, less attention has been focused on *exo-receptors*, with outward-directed sites, or "informed" surfaces, on which recognition may occur (693). These two different systems have been referred to as convergent and divergent processes, respectively, and are summarized in Diagram 8.9.

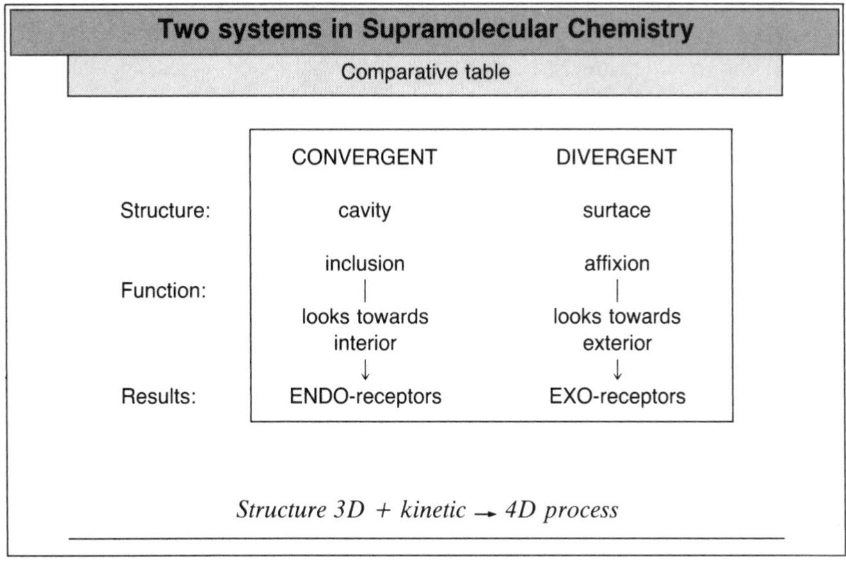

Two systems in Supramolecular Chemistry
Comparative table
CONVERGENT
DIVERGENT
Structure:
cavity
surtace
Function:
inclusion
looks towards interior
affixion
looks towards exterior
Results:
ENDO-receptors
EXO-receptors
Structure 3D + kinetic → 4D process

Diagram 8.9

By using the three-dimensional information storage/readout operating in molecular (endo- or exo-) recognition, in combination with substrate transformation and translocation, one might be able to design components for molecular devices that would be capable of processing information and signals at the molecular and supramolecular levels. This area of investigation may be termed *semiochemistry*: the chemistry of molecular signal generation, processing, transfer, conversion, and detection (693). This is illustrated in Diagram 8.10.

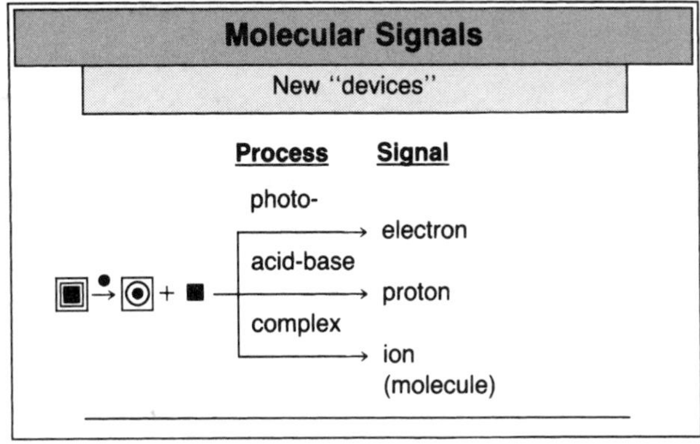

Diagram 8.10

An effort in this direction for the nondestructive detection of the neurotransmitter acetylcholine in protic media was recently published (697). The strategy consisted of building a polyphenolic (resorcinol) bowl-shaped cavity that could ionize in alkaline media and simultaneously bind to a pyrene-modified methylpyridinium substate (Fig. 8.4).

The presence of acetylcholine (no other neurotransmitters) displaces the pyridinium cationic species and induces a large fluorescence enhancement in protic media. This work represents the first artificial-signaling acetylcholine receptor that behaves as a molecular sensor by changing optical properties in response to the presence of a biologically important substance.

Diagrams 8.11 and 8.12 give a more complete picture of the types and functions of molecular devices that can be imagined and possibly developed. More elaborate molecular construction kits, based on networks such as crystals, are listed in Diagram 8.13, with the names of the main contributors and developers (686,698). These projects will be discused in the next section. Precise catenation in producing synthetic DNA knots relates to nanotechnology and biotechnology. As pointed out by N.C. Seeman (New York University), such molecular assemblies of novel biomaterials will eventually be used for analytical, industrial, and therapeutic purposes (699).

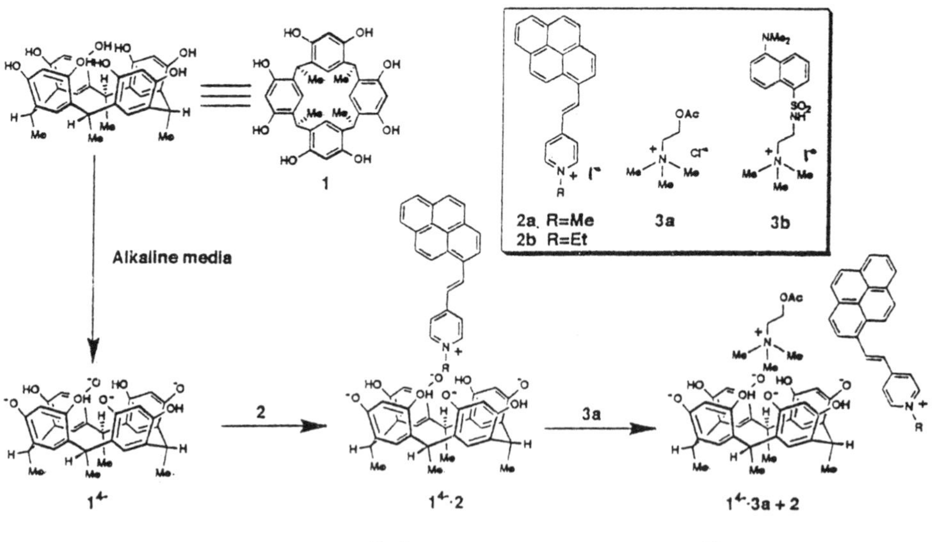

Fig. 8.4. Design of an artificial-signaling acetylcholing receptor (697). Reproduced with permission. Copyright © 1994 by the American Chemical Society.

Types of Molecular "Devices"

Structure-function relationship

Structures	Functions
electroactive	transformation
ionoactive	storage
photoactive	regulation
magnetoactive	(coupled to)
mechanoactive	informational
thermoactive	signal
chemioactive	

Diagram 8.11

Other Possibilities

More complex "devices"

- Molecular "switches" ("push/pull")
- Nonlinear optics
- Coupled transport process
  - pump
  - regulation
  - conversion
    - electron → ion
    - photon → ion
    - proton → ion
- Molecular chemionics
- Photoionic signal

Diagram 8.12

Designer Solids: "Haute Couture" in Chemistry

Molecular construction kits, Lego, Meccano

- "Programmed assembly" protocol, molecular units, design artificial crystals (Michl)
- Molecular nanoscaffolding (Moore)
- Molecular shuttles, nanoscale switches (Stoddart)
- Molecular tectonics (Mann, Wuest)
- Networking (3D beehives) with DNA (Seeman)

The goal is the deliberate synthesis of periodic matter, like crystals, on the nanometer scale based on molecular construction kits.

Diagram 8.13

Finally, the last Diagrams 8.14 amd 8.15 summarize the types of molecular devices that will be developed in the future.

Emphasis will be toward more efficient and selective recognition and binding of various anions and zwitterions. In this regard, a recent publication appeared on the synthesis of a difunctional receptor for the simultaneous complexation of anions and cations (700). In this case, the receptor serves as a phase transfer agent to recognize potassium phosphate (KH_2PO_4). Its structure is shown below. The two crown ether rings serve

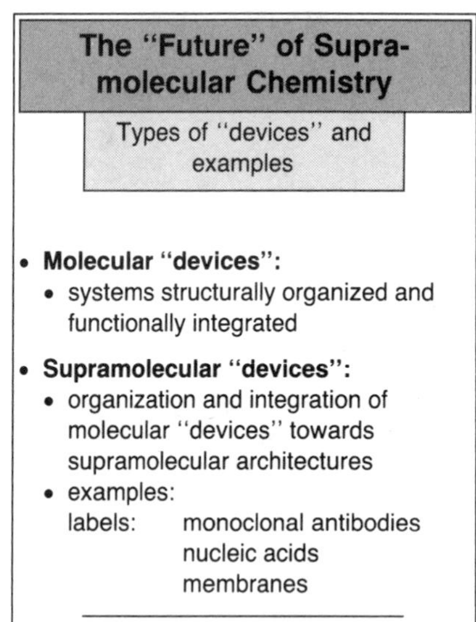

The "Future" of Supramolecular Chemistry
Types of "devices" and examples
• Molecular "devices":
• systems structurally organized and functionally integrated
• Supramolecular "devices":
• organization and integration of molecular "devices" towards supramolecular architectures
• examples:
labels: monoclonal antibodies
nucleic acids
membranes

Diagram 8.14

Fields to Develop . . .

For the future . . .

- Binding of anions
- Binding of zwitterions
- Add chirality to the binding process
- Kinetic → "proton wire"—membrane
- Molecular information → directional process

Diagram 8.15

to sandwich the K^+ ion, while the amide linkages and the redox-active UO_2 center contribute to the stability of the anionic phosphate group by hydrogen bonding and electrostatic interactions.

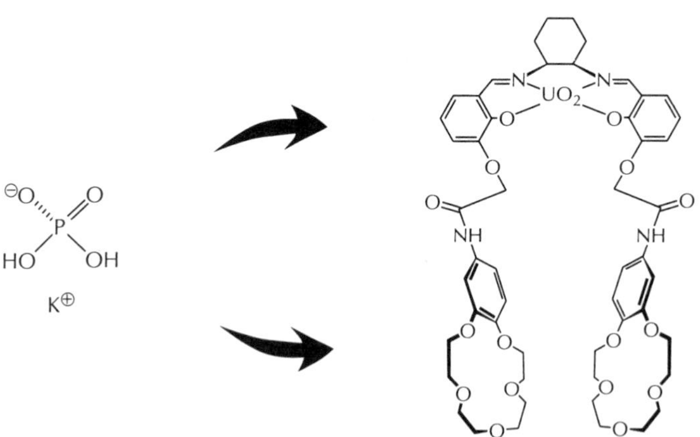

A crown ether-based ambiphillic receptor for KH_2PO_4 that incorporates a UO_2 center (700).

A second illustrative case is in the field of electrophotoswitching of luminescence and switched molecular wires. Indeed, the group of J.M. Lehn has described a dual-mode switching device based on both photochromic and electrochromic properties. This is illustrated by the following structures:

A dual-mode optical-electrical molecular switching device (701).

The device may be reversibly converted among three states (**A**, **B**, and **C**), which are all stable and exhibit distinct absorption spectra. The special features of the three species are, in principle, well suited as the basis for an optical memory system with nondestructive write–read–erase capacity. After photochemical writing with UV light (A → B), the information may be "safeguarded" by an electrical process (B → C). It may then be read many times, and finally, after electroreductive unlocking (C → B), the information may be erased with visible light (B → A). The safeguarding step also results in amplification, because the quinonoid form in **C** absorbs approximately twice as strongly as **B.** Furthermore, since the longest wavelength bands of **B** and **C** overlap, reading may be done with either the same visible light as that used to erase or, for higher sensitivity, at the absorption maximum of **B** (534 nm). This represents a EDRAW (erase direct read after write) process, as compared to the usual WORM (write once read memory) process (701).

At the macroscopic level, substances of this type are of interest as materials for reversible information storage. At the microscopic level, they represent, as shown here, optically and electrically addressable, dual-mode molecular devices.

As expressed recently by DeSilva and McCoy (740), in order to be useful in this field of information technology, these designed molecules must be capable of existing in at least two stable and distinguishable states.

More selective ion transport, molecular wire devices, and redox-active processes (as mentioned in Section 5.3.2) will also be developed by using the concept of molecular information and directional processes at the supramolecular level.

8.3 Specific Examples

We have just seen that self-assembly of well-defined supramolecular architectures rests on the design of molecular components that contain appropiate information and interact through the correct recognition algorithm (693). Metal coordination and/or hydrogen bonding are often used for the construction of such molecular arrays. The first example in this section is the self-assembly of a bisporphyrin supramolecular cage, induced by molecular recognition between complementary hydrogen bonding sites. This is illustrated in Fig. 8.5, where the synthesis of a porphyrin component was accomplished by the coupling of two dipyrroylmethanes. The addition of alkyl-substituted triaminopyrimidine (TAP) to a solution of the porphyrin units, bearing two rigidly attached complementary hydrogen bonding moities of the uracil type, leads to the self-assembly of a bisporphyrin cage-like structure. This bisporphyrin supramolecular cage entity and its Zn(II) derivative are both closed and functional. A marked change in fluorescence was observed on self-assembly. The accumulated evidence supports the conclusion that self-assembled bisporphyrin or its Zn(II) complex and TAP yields mainly a closed bisporphyrin species through hydrogen bonding recognition (702).

This interesting result is the first step in the controlled formation of organized multimeric and polymeric porphyrin arrays of a supramolecular nature. As pointed out by the authors, metalation of one of the two porphyrins would create a self-assembled donor–acceptor system. Alternatively, various acceptors, such as quinones (see Section 6.6), may be incorporated into the arrays. A variety of electron- and energy-transfer processes may result, giving rise to photoactive, electroactive, or ionoactive functional supramolecular devices.

8.3.1 Helicates

Nucleic acids, because of their key biological role in replication processes, are prime targets for the design of either analogs that may mimic some of their features or of complementary ligands that may selectively bind to and react with them for regulation or reaction. In this context, J.M. Lehn and his coworkers have designed a new class of artificial systems—the *deoxynucleohelicates* or DNH—a self-assembly of oligonucleosidic double-helical metal complexes (703). The approach was based on the discovery of simple double-helical complexes, the so-called *helicates*, that are formed

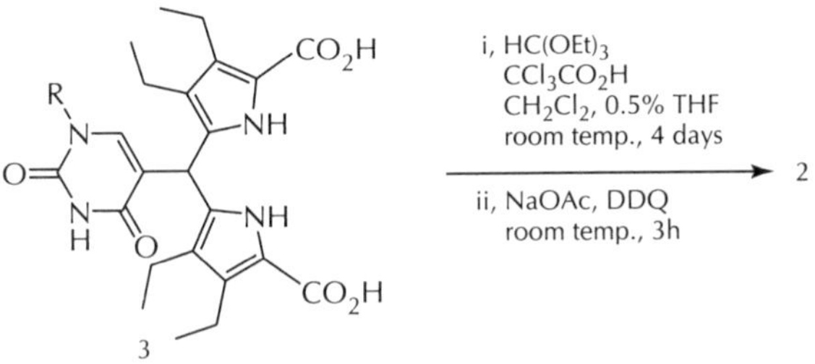

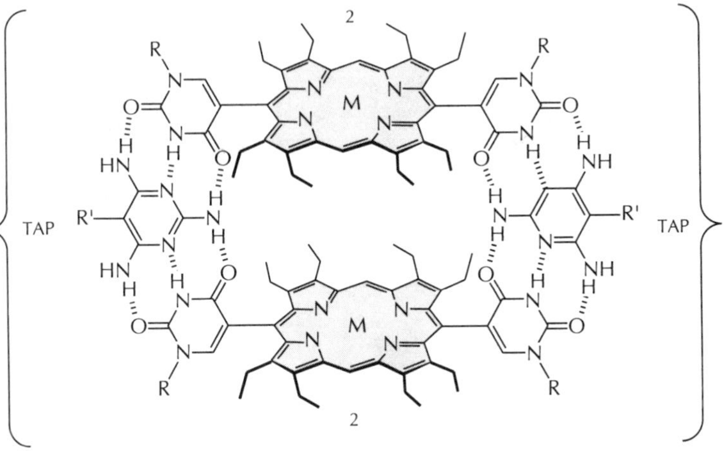

Fig. 8.5. Supramolecular cage-like structure (1), formed by the self-assembly of two porphyrins containing 5-alkyluracil recognition groups (2) and two alkyltriaminopyrimidine (TAP) units (702). R = decyl, R′ = octyl or decyl, M = $2H^+$, Fe(III)Cl or Zn(II). Reproduced by permission. Copyright © 1993 by the Royal Society of Chemistry.

spontaneously from tetra- or penta(bipyridine) ligands and Cu(I) ions in solution (704a) (see Fig. 8.6).

Whereas hydrogen bonds and stacking interactions are responsible for holding together the double helix of natural DNA, coordination interactions between tetrahedral Cu(I) and bipyridine units play the key role in determining the structure of these rudimentary double-stranded molecules. In normal DNA, the double helix is held together by hydrogen bonding between purine and pyrimidine units, and metal ions binding at the surface of the DNA ribbon. With helicates, it is the other way around. Indeed, what makes these molecules remarkable is that the ions are inside the double helix and the hydrogen bonding units are pointing toward the exterior. Each helicate can then extend to a three-dimensional network with neighbors via complementary hydrogen bonding.

Therefore, lengthening these ligands should lead to self-organizing nanostructures. The field is now wide open for manipulations and modi-

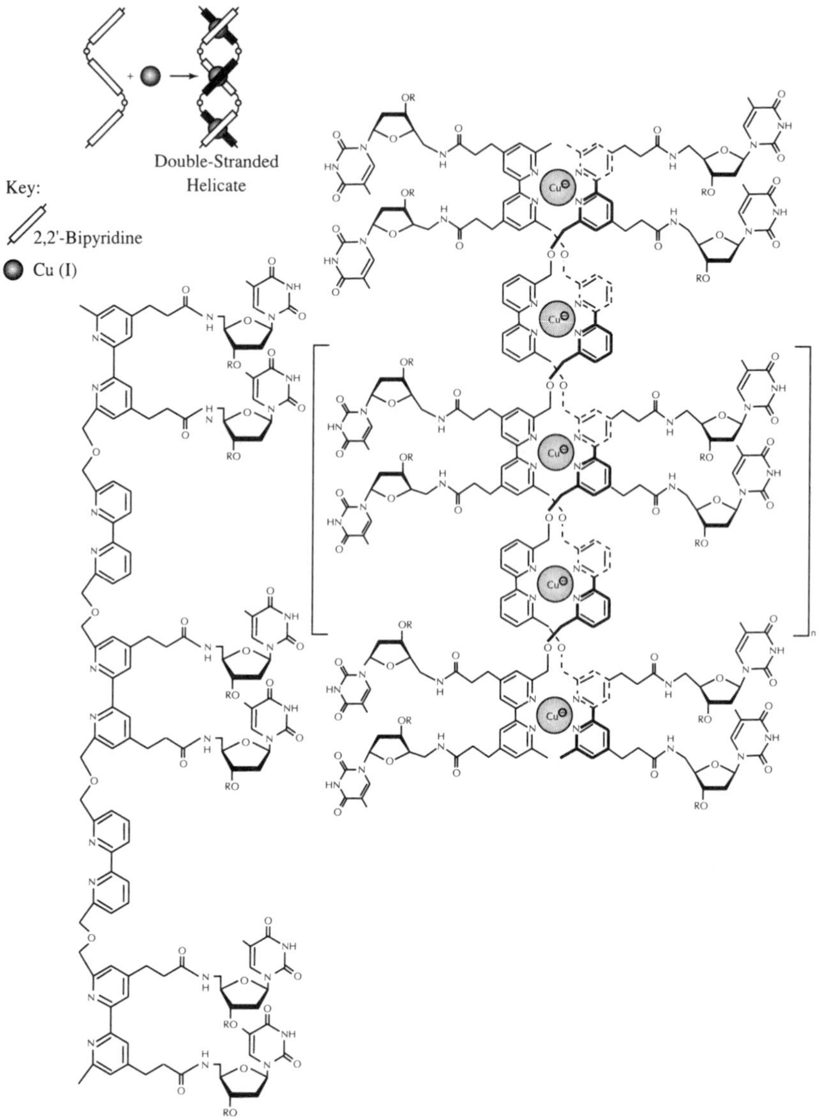

Fig. 8.6. The strategy for the construction of deoxyribonucleohelicates is based on molecular self-organization, n = 0, 1; R = $tBuMe_2Si$, H. Adapted from (693).

fications of these subunits, complexation of other ions, investigations of binding cooperativity, interaction with DNA, and the construction of self-organizing and self-amplifying devices (704a). Indeed, changing the coordinating metal ion opens new perspectives. If Cu(I) is replaced by Ni(II), a triple-stranded helical metal complex is formed by the spontaneous assembly, that time, of three linear polypyridine ligands and nickel ions, which display an octahedral coordination geometry (704b).

8.3.2 Catenanes and Rotaxanes

Catenanes and rotaxanes were mentioned briefly in Sections 5.5.1 and 6.5.1 respectively. They consist of either interlocking rings or rings on a molecular thread. Such molecular edifices, using chemically interlocked components, have gained considerable attention in recent years, and chemists are exploiting the properties of these supramolecules for making devices, such as molecular switches, diodes, and transistors, as well as molecular wires to join them together. One of the most interesting possibilities embodied in this work is the construction of computer units based on supramolecules interacting on thin, wet-film supports (705). Although these thoughts are impressive, F. Stoddart, a leader in this area, believes we are still a long way from developing commercially useful molecular switching devices.

This research began in the 1980s, when scientists were looking for molecular components that could be stimulated by chemical, electrochemical, or photochemical input. In this regard, viologen and paraquat units are well-documented molecules that respond to these stimuli because of their characteristic redox properties. It was partly for this reason that Stoddart's group and Italian coworkers identified them as potential building blocks for the self-assembly of controllable catenanes and rotaxanes (706).

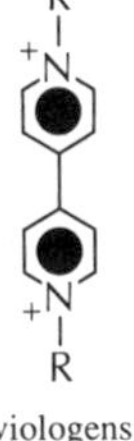

viologens

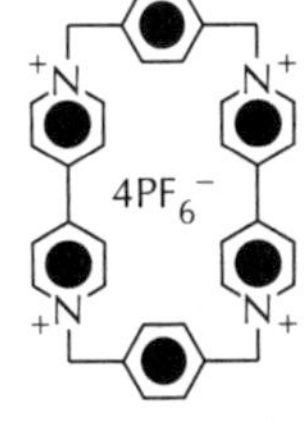

cyclobis(paraquat-*p*-phenylene)

By combining polyether chemistry (Sections 5.1 and 5.2) with the polycationic properties of paraquat, the British team designed a self-assembled molecular shuttle consisting of a ring-shaped molecule (a paraquat) that encircles a molecular string (a polyether) and switches back and forth between two points along the polyether chain. By better controlling this movement, researchers believe that more complicated versions of the system could be made. This would be the basis of "molecular machines" (707) or "molecular shuttles," an expression coined by Stoddart. Figure 8.7 gives an overview of the process.

The strategy for creating the string molecule begins with a polyether chain that has two phenol groups at each end. Then two more polyether

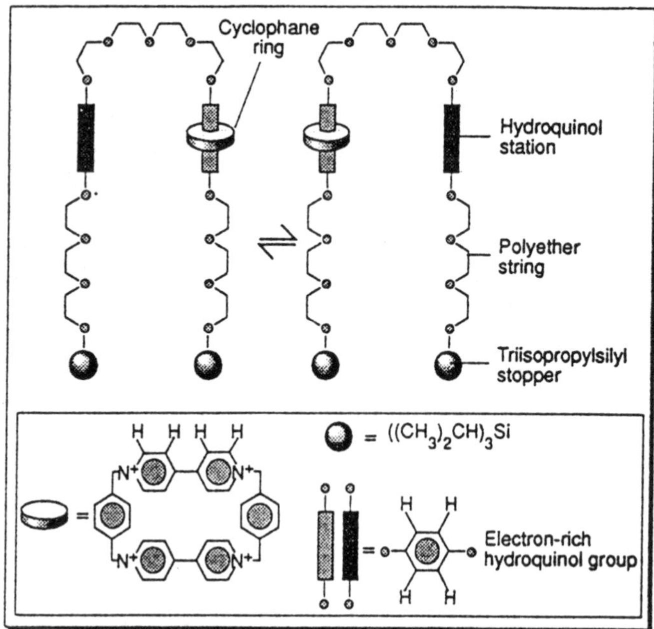

Fig. 8.7. Pictogram of a molecular shuttle (707). The charged cyclophane ring (paraquat) self-assembles on the polyether molecular string, then shuttles back and forth between two hydroquinol "stations" that are rich in electrons. Reproduced with permission. Copyright © 1991 by IPC Magazines Limited, England.

chains are then attached to these ends to complete the string, and bulky triisopropylsilyl groups, such as steric stoppers, are added at the ends. The cyclophane molecule is formed by bridging a U-shaped bis (pyridinium) ion with 1,4-dibromomethylbenzene, to form the deep orange paraquat compound. During the ring-forming reaction, the polyether string maneuvers the U-shaped ion into a good position for the dibromo compound to close the loop around the polyether and to make a pseudorotaxane.* The bulky triisopropylsilyl groups prevent the ring from falling off. This reaction takes more than a week in solution to complete, and the polyether string acts as a template for the formation of the ring. In this system, the hydroquinol groups form weak electrostatic bonds with the positively charged cyclophane ring. Because the ring is equally attracted to both, the ring shuttles between them several times a second.

* The name *rotaxane* derives from the Latin *rota* meaning wheel and *axis* meaning axle. The addition of the prefix *pseudo* indicates that the wheels are free to dissociate from the axle as a more conventional complex (708, 709).

More elaborate supramolecular arrays have been prepared since. One example is [2]pseudorotaxane, which is built up from the same macrocyclic cyclophane containing two aromatic π-acceptors, but with an acyclic polyether that contains three aromatic naphthalene π-donors. This array has the ability to undergo self-organization on crystallization to form a layered superstructure as presented below:

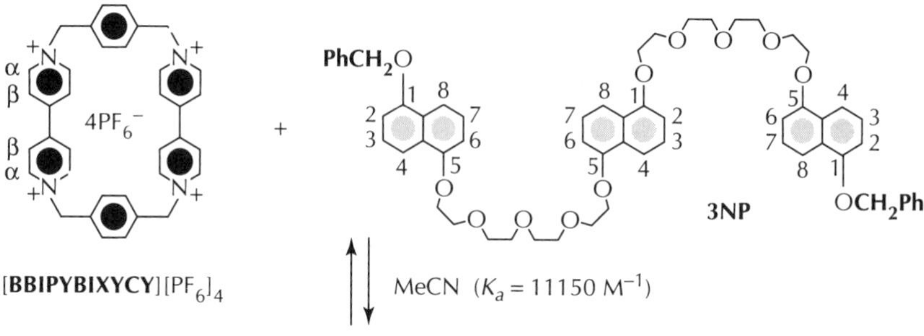

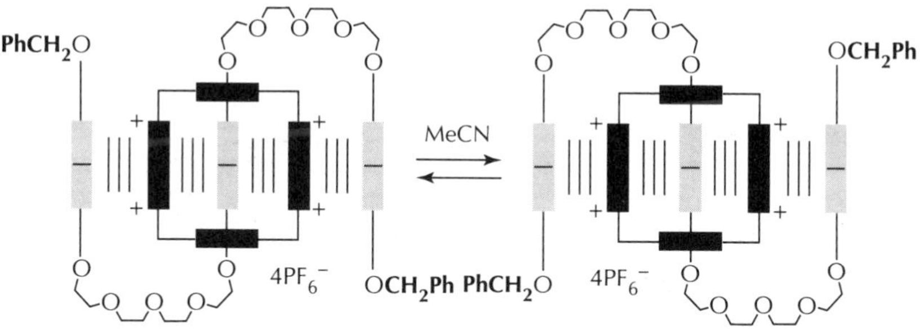

[BBIPYBIXYCY.3NP][PF$_6$]$_4$

Formation of a [2]pseudorotaxane (709)

These crystals, which are suitable for X-ray structural analysis, revealed that in the solid state, the polyether portion of the molecule is not only inserted through the center of the tetracation, so that the middle, 1,5-dioxynaphthalene unit is encircled, but its polyether chains also curl back on themselves around the cyclophane. This enables the other two π-donors to stack against the sides of the π-accepting bispyridinium residues.

The [2]pseudorotaxane has crystallographic C_i symmetry (709). Furthermore, Fig. 8.8 clearly shows that in the crystal lattice, the [2]pseudorotaxanes are arranged to form infinite two-dimensional sheets sustained by a combination of π-π stacking and aromatic (CH · · · π) interactions that extend to create a grid-like pattern. This "edge-to-face" interaction of

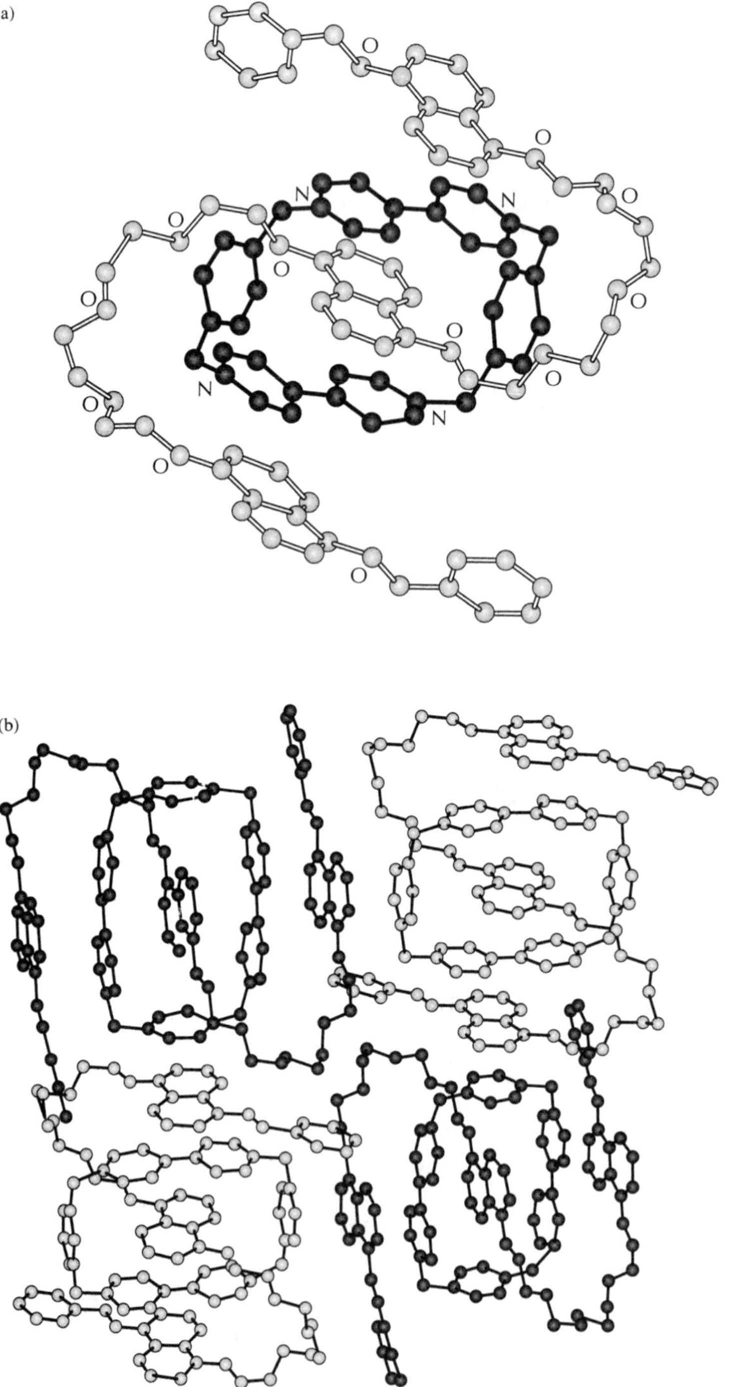

Fig. 8.8. The X-ray crystal structure of (a) [2]pseudorotaxane and (b) the arrangement of the tetracations in the crystalline state (709). Reproduced by permission. Copyright © 1994 by the Royal Society of Chemistry.

two phenyl rings with no π-stacking is an important observation in cyclophane chemistry for the design of aromatic guests.

This type of [2]pseudorotaxane molecule not only exhibits high stability and a remarkable degree of internal organization, but also provides the ingredients, in the form of appropriately sited π-acceptors and π-donors, to generate supramolecular order on the nanometer scale in the crystalline state. Two-dimensional arrays like this one, in which the molecules are ions of an aromatic nature that can form continuous networks, could ultimately play a role in the conception of more elaborate nanosystems containing molecular machinery.

This is particularly important in the development of *molecular-based electronic devices* at the nanometer scale. The creation of nanometer-scale devices has fascinated the scientific community for more than a quater of a century (710). However, current electronic technology does not allow silicon components in electric circuitry to be closer than 100 nanometers (1000 Å) apart. It would thus be advantageous to reduce this gap by a factor of 10. This is where new materials and molecular devices come into play. The requirements of molecular electronics demand that the chemist not only provide molecular scale bistable devices (switches), but also permit the communication infrastucture (input/output) necessary to exchange information with the outside world. Moreover, the switches must be completely controllable at the molecular level. It follows that the challenge is not only to provide the means for creating molecular devices, but also to construct the necessary molecular architectures and machineries to support such devices. To be able to meet these challenges, one must understand and apply the rules of molecular self-organization, self-synthesis, self-assembly, and self-replication.

Furthermore, the chemist is not alone in requiring nanometer-scale devices. Biological systems display a diverse array of functioning nanometer-scale structures in the range 1 to 10,000 nm. The specificity and precision displayed by biological systems is derived from the highly directed mutual recognition displayed by the constituent components of a structure. Clearly, the ultimate goal for the chemist seeking to match the achievements of biological systems in creating and maintaining nanometer-scale structures is to control the assembly and function of synthetic structures from components with the same precision as that displayed by nature (710).

Other applications of nanometer-scale devices are in *nonlinear optics*. Nonlinear optical phenomena are of major importance because they form the basis for the optical processing and storage of data and images. They are therefore crucial in the development of future generations of communication systems and computers (711). Nonlinearity in the optical response comes from a very strong interaction between the light waves (laser) and the material (polymeric surface) in which they propagate. The material needs to have delocalizable or polarizable π-electrons or else have a dipolar nature. The incident strong light polarizes the electrons, thereby modifying the refraction index of the material and the frequency and speed of the reemitted light. Supramolecular arrays having large π-systems are thus particularly good candidates for this growing technology.

Although polyrotaxanes have received considerable attention, a less convincing strategy has been proposed for the synthesis of polycatenanes. Polycatenanes embody one of the goals of polymer chemists, that is, the construction of polymer chains in which the components are a series of interlocked rings that are reminiscent of our familar mechanical chains. Recently, Stoddart's group made considerable progress in this area by developing efficient methodologies for the preparation, by self-assembly, of [2]-, [3]-, [4]- and [5]catenanes (712). This is outlined in Fig. 8.9 for the template-directed synthesis of a [2]catenane and a [3]catenane.

After many trials and errors, a one-step self-assembly of [2]- and [3]catenanes was made to occur via a cesium-ion promoted, phase-transfer-catalysis; the products were isolated as red PF_6 salts. These macrocyclic polyethers were obtained this way in 14% yield from readily available starting materials (diphenol and ditosylate). This work was also extended to the synthesis, in a two-step self-assembly, of [4]- and [5]catenanes at ambient temperature and atmospheric pressure (712). The [4]catenane molecule resembles the logo of Audi automobiles. These accomplishments demonstrate what Stoddart calls structure-directed synthesis. This means that the molecular information is "preprogrammed" into the reactants for efficient assembly of large molecules. Consequently, in these syntheses, the difficult steps are at the beginning. The whole approach is analogous to building a house by joining together prefabricated panels.

The persevering and ingenious efforts of Stoddart's group have demonstrated that by subtle and appropriate modification of "intelligent molecular components," efficient molecular recognition and assembly processes can be applied successfully to the construction of more highly catenated molecules. These efforts were rewarded recently by the synthesis and characterization of the first molecular compound consisting of a linear array of five interlocked rings (713). This [5]catenane supramolecule resembles the well-known symbol of the International Olympics, and Stoddart proposed to call it olympiadane.

Another molecular curiosity is the construction by M.J. Gunter (University of New England, Australia) of several porphyrin-catenanes (714). Since π-π electron interactions are of fundamental importance in catenane building for both paraquat complexation and subsequent catenane formation, Gunter reasoned that the replacement of one of the electron-rich hydroquinol or naphthoquinol units of the crown ether by a porphyrin ring, with its large π-electron-rich nucleus, should enhance paraquat complexation as well as encourage catenane formation in the usual manner. This stategy worked, and these compounds have much to offer

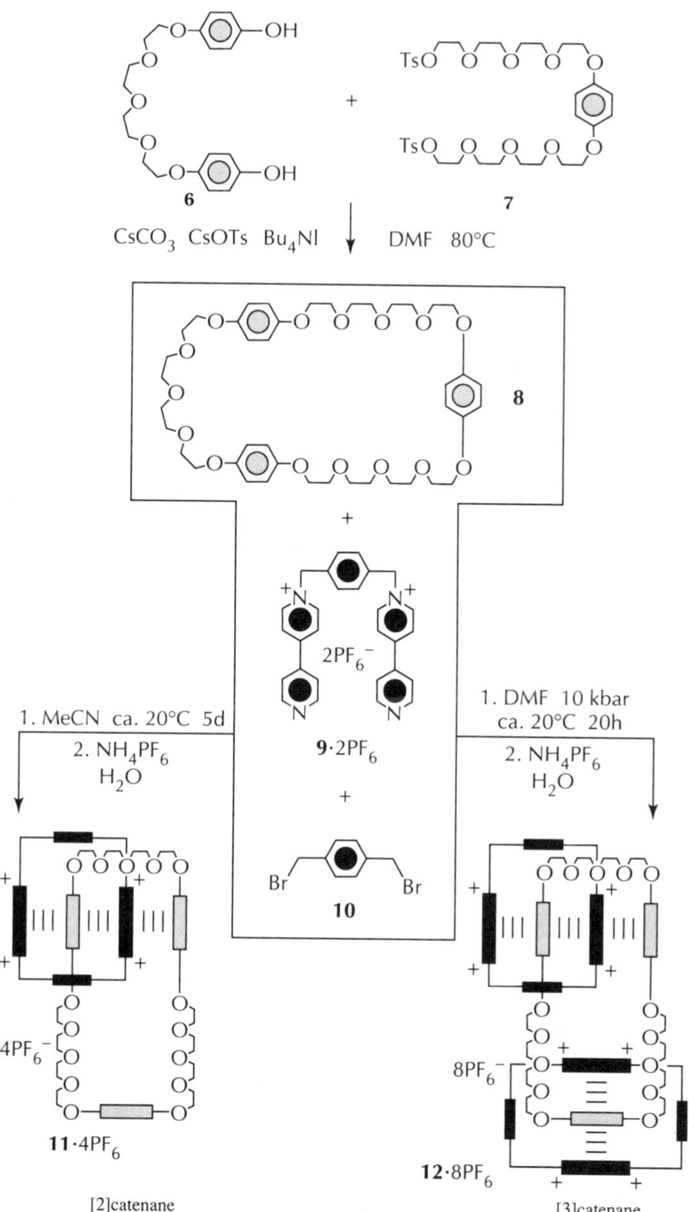

Fig. 8.9. The cesium template-directed synthesis of the macrocyclic polyether and the subsequent template-directed synthesis of [2]catenane and [3]catenane (712). Reproduced with permission. Copyright © 1994 by VCH, Weinheim.

as new materials for energy storage and transfer of information at the molecular level.

8.3.3 Dendrimers

With the development and construction of arborols and dendrimers,* new frontiers in multibranched macromolecules appeared on the horizon. These new compounds are synthesized with branching strategies that allow the construction of large molecules that resemble polymers but exhibit precisely controlled size, shape, and molecular weight (715). D.A. Tomalia (now at the Michigan Molecular Institute, Midland), G.R. Newkome (University of South Florida), and J.M.J. Fréchet (Cornell University) are the pioneers in this area. Tomalia developed and patented a family of compounds that he calls "starburst dendrimers." These macromolecular compounds have three distinguishable substructures: an initiator core; an interior or layers of repeating units radially attached to the core; and an outer surface of terminal functionality. The following diagram shows the potential and possible applications of these structurally controlled polymers.

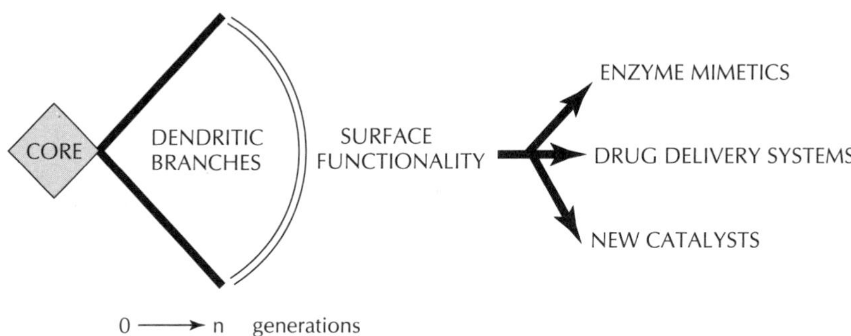

General structure and applications of dendrimers

Dendrimer synthesis is a stepwise process. The starburst (cascade) approach of Tomalia uses polyamidoamines made from ammonia and ethylenediamine (initiator core). Reaction with methyl acrylate, via a Michael addition, leads to a triester. In a second step, the triester is exhaustively amidated with a large excess of ethylenediamine to form a triamidoamine. This first generation compound is the starbranch oligomer. The stepwise procedure is then repeated to form succeeding generations

*The word dendrimer derives from *dendron*, the Greek word for tree, whereas arborol comes from *arbor*, the Latin word for tree.

of starburst dendrimers, each with twice as many terminal groups as it predecessor. However, it is essential that all the ethylenediamine be removed before the next acylation steps. Otherwise, the residual ethylenediamine molecules will act as initiator cores, degrading the monodispersity of the product by leading to the formation of molecules with lower-than-desired molecular weight. As the molecular size increases, it becomes harder and harder to remove all the ethylenediamine. At the same time, the number of terminal groups is growing at a rate faster than the surface area, and the surface-dense packing and departure from stoichiometry start to show up by the tenth generation (see Fig. 8.10). Although these higher-generation starburst dendrimers are fairly solid at the surface, they are porous and somewhat hollow in the interior. They thus tend to trap small molecules (or solvent) inside.

Following the approach of Tomalia, other chemists, such as G.R. Newkome, J. Fréchet, F. Vögtle (Institute für Organische Chemie, Bonn), and T.X. Neeman (AT&T Bell Laboratories, New Jersey) have developed more convenient synthetic strategies. Vögtle's "cascade molecules" strategy has been referred to as the *divergent method* (see Fig. 8.11),

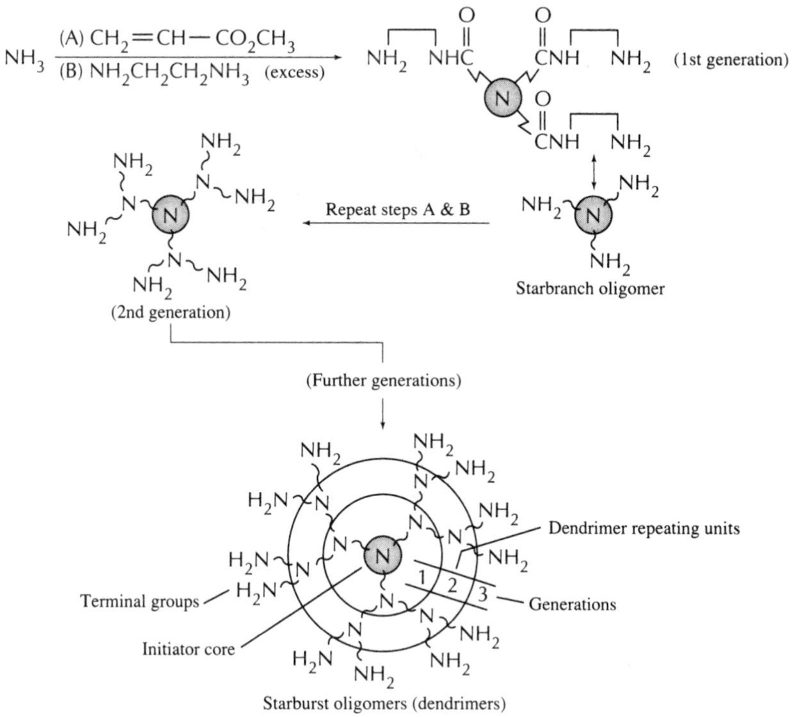

Fig. 8.10. General presentation of the starburst dendrimer synthesis (715). Reproduced with permission. Copyright © 1988 by the American Chemical Society.

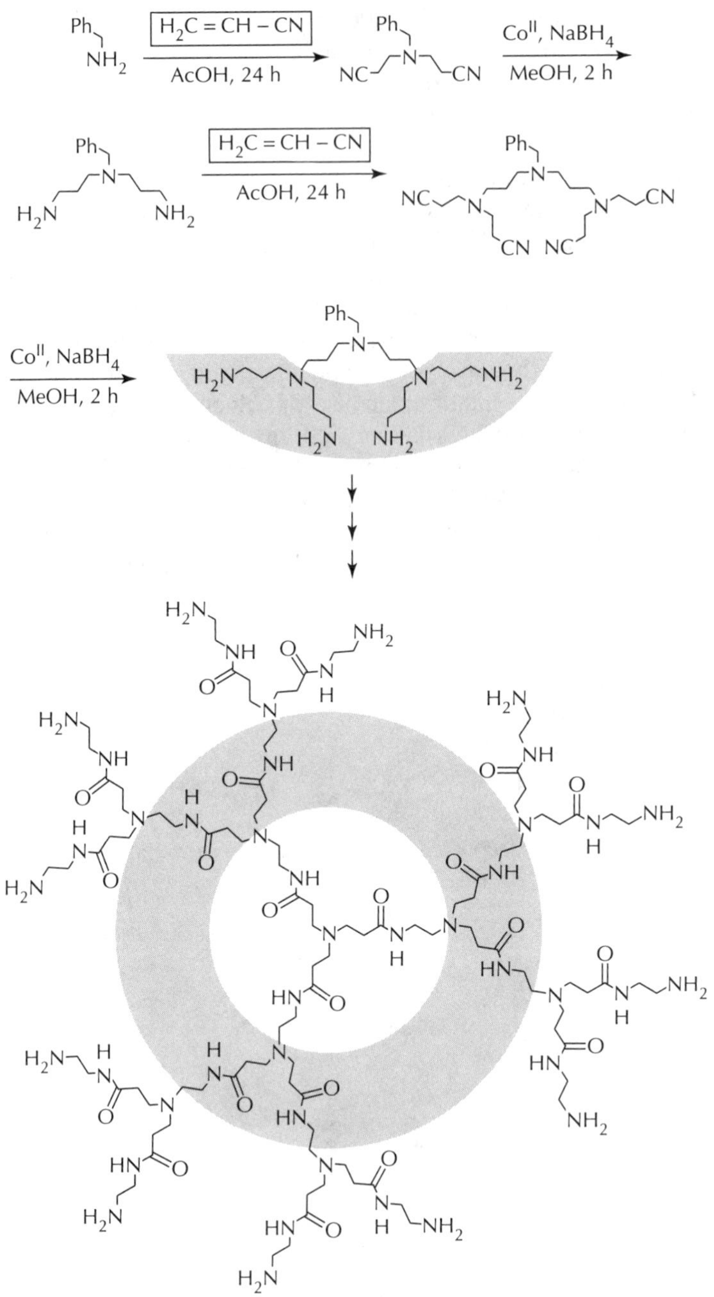

Fig. 8.11. The Vögtle divergent method of dendrimer synthesis (716). The figure shows the cascade-like synthesis of branched polyamines and the result after the second generation. Reproduced with permission. Copyright © 1992 by VCH, Weinheim.

while that of Fréchet and Neeman has been called the *convergent method* (see Fig. 8.12) of dendron and dendrimer synthesis.

A wide variety of dendrimer families have been synthesized by the divergent method and include poly(amidoamines), poly(ethers), poly(siloxanes), poly(thioethers), poly(amidoalcohols), poly(amines), and recently poly(organometallic) types (719,720).

The alternative convergent method of Fréchet and Neeman involves two stages: a reiterative coupling of protected/deprotected branch cells to produce a focal-point functionalized *dendron*, followed by divergent core

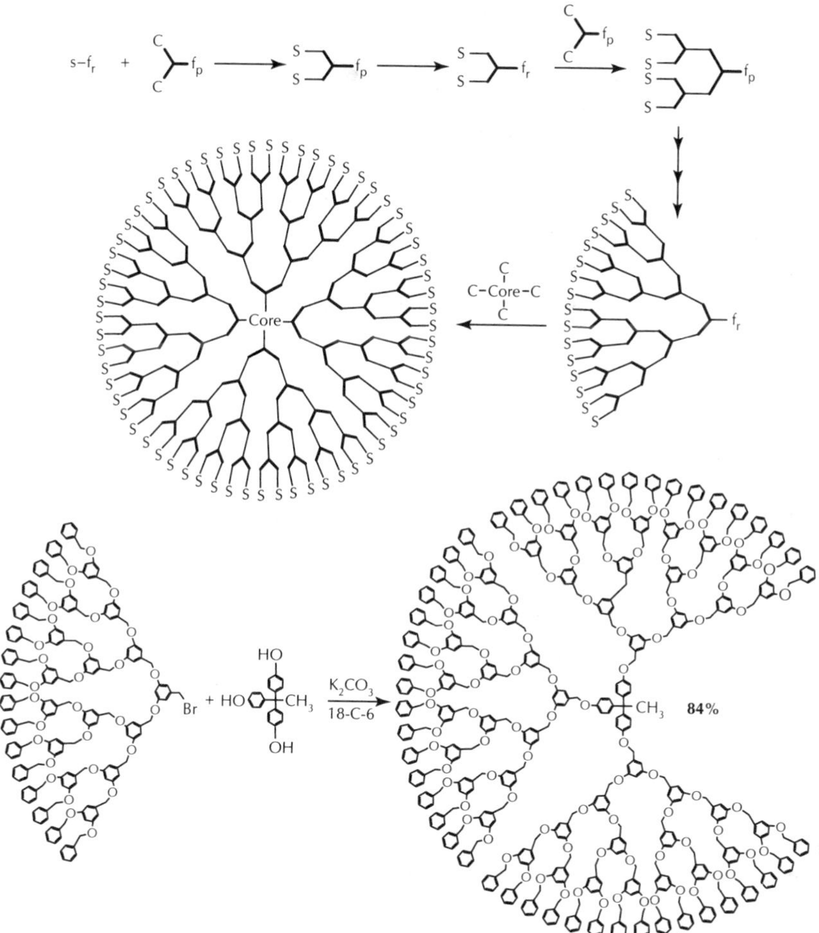

Fig. 8.12. The Fréchet (717) and Neeman (718) convergent strategy for the synthesis of dendrimers and a specific example using a triphenylmethane core. f_r = reactive groups, f_p = protected groups, S = surface and c = interior. Reproduced with permission. Copyright © 1990 by the American Chemical Society.

anchoring to produce various multidendron dendrimers. This convergent dendron/divergent core anchoring method has the unique advantage of allowing one to differentiate dendron sectors as a function of branch cell chemical composition as well as surface functionality. In all dendrimer syntheses, a reiterative branch cell assembly scheme is necessary for developing concentric (shells) generations and advancing the starburst (cascade) architecture to various levels. By this means, "directionality reactivity" can be designed into the surface features of the final dendrimer structure. This is an important feature for the construction of well-defined "nanoscopic" (10 to 1000 Å) or nanometer-scale dendrimeric building blocks. Furthermore, these supermolecular compounds, clusters, and assemblies are large enough to be observed directly by electron microscopy. In some instances, suitable dendrimer surfaces have led to entirely new *supramolecular assemblies* never before observed in nature (720).

G.R. Newkome has been independently pursuing parallel investigations along different chemical pathways that have led to the synthesis of *arborols*. Bis-homologues of tris(2)- and of tris(3-hydroxypropyl)aminomethane have been successful, but Newkome's arborols do not build up generationally in the same way as Tomalia's dendrimers (715).

Regardless of their differences, the starburst dendrimers and the arborols have much in common. Their respective macromolecules, with hydrophilic surfaces and hydrophobic interiors, are reminiscent of micelles (see Section 5.3), with interior crevices that are perfect loci for molecular inclusion. But in contrast to ordinary micelles, which are fragile agglomerations of separate molecules held together by van der Waals forces, the dendrimers and the arborols are sturdy unimolecular assemblies held together by covalent bonds. In fact, lower generations of dendrimers resemble micelles, whereas higher generations, with their densely packed surface groups, are more reminiscent of the bilayers of liposomes with the ability to enclose or exclude materials. Although the dendrimers tend to be highly uniform in shape, some deviate from the normal spherical shape. Specialists in this field are already looking at ways to build hemispherical or oval molecules. Arborols find use in areas as diverse as solubilizing agents, gelation agents, and carriers for catalytic sites like *zeolites*. For dendrimers, the applications are in the controlled and targeted delivery of drugs and agricultural chemicals. In addition they might be used as adhesives, as adsorbents, or for surface modification of plastics or structural materials. However, new issues, such as *nanoscopic steric effects*, *nanoscopic chirality*, new *nanoscopic architectures, nanoreceptors*, and *nanoscopic recognition* require further understanding. These studies will undoubtedly involve the evolution of new rules and concepts that may directly impact our understanding of the behavior and characteristics of many biological nanostructures that are so intimately involved in creating and sustaining life (720).

Chiral dendrimers, using optically active derivatives of tris-hydroxymethylmethane, have recently been developed by D. Seebach and coworkers (E.T.H., Zurich). This effort represents the first synthesis of a starburst dendrimer with chiral core units. The structure of one of these dendrimers is presented below. It was observed that its optical activity decreases as the size of the dendrimer increases.

MW 2717 Da
$[\alpha]_D$ +4°

An optically active third generation dendrimer with twelve terminal phenyl groups (721). The 3-hydroxybutanoic acid skeleton and the spacer are marked in dark gray. The dendron, introduced by etherification of the triol precursor with the corresponding benzylic bromide, is marked in light gray.

In contrast, in the following "fully chiral" small dendrimer, the optical rotation is approximately the same as that of the trihydroxy building block.

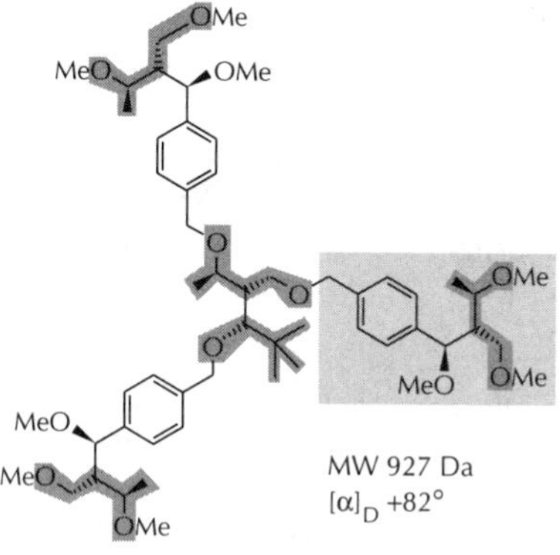

A "fully chiral" dendrimer (721)

D. Seebach hopes that this work will help answer the following three questions:

1. Will a chiral core in a dendrimer with otherwise achiral building blocks cause the dendrimer to have a chiral shape or to be optically active?
2. Will there be enantioselective clathration or host–guest interaction near the core of such a dendrimer?
3. Will a dendrimer containing chiral and unsymmetrical branching units really have the predicted surface with fractal dimensions and, thus, a pronounced capacity for chiral recognition?

Preliminary results suggest that the synthesis of higher generations of chiral dendrimers should provide answers to these questions and deepen our understanding of organized complex molecules and their interactions with small molecules (721).

Beside chiral precursors, other functionalized core units can be imagined. For instance, one can think of the presence of a porphyrin ring for developing new redox-active reactions or a cyclophane ring for a novel catalytic cavity within a dendrimer network. F. Diederich's group (E.T.H., Zurich) is currently making progress along these lines. His group has constructed a photoactive metal porphyrin-core dendrimer based on peptide branching of polyether chains in order to study the luminescence and fluorescence of such a porphyrin-dendrimer and the corresponding Zn(II)-bound derivative (722).

The growing dendrimer network uses a simple peptide bond coupling between the 12 ester termini and a new branching of a free amino-polyether chain to give in the third generation, 48 new ester termini (Fig. 8.13). At the Zn-complex stage, the supermolecule assembly has up to 832 carbon atoms, a molecular weight of about 19,000, and a spherical shape (40 Å in diamater) the size of the iron-dependent enzyme cytochrome C (see Fig. 8.14). Interestingly, the metal porphyrin center is sterically shielded from possible reduction from the outside and consequently this novel Zn-porphyrin-dendrimer shows differences as high as 500 mV in the first oxidation potential, as compared to tetra-alkylamide porphyrin-Zn(II) alone. Such a porphyrin-Zn(II)-dendrimer opens the way for the study of new devices for the tuning of redox potentials by a biomimetic approach and for the building of electrochemical and photochemical "molecular gadgets" such as "molecular energy collector" or other antenna-based molecular devices that can be designed.

Even more interesting is the prospect that chemists can use dendrimers to produce molecules that demonstrate heredity (723). The strategy is based on abiotic genealogically directed molecular growth from simple branched molecules to true dendrimer architecture. Genealogically directed processes can be viewed as an evolution in the molecular design from abiotic divergent dendrimer synthesis to biotic cell amplification.

Color Plates

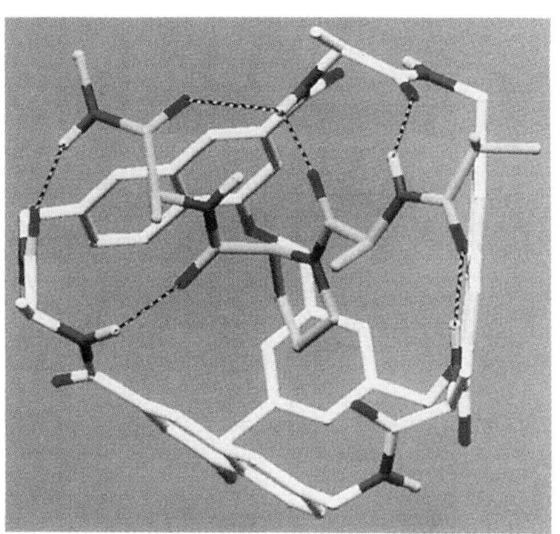

Fig. 5.14. A presentation of the global minimum of the complexation of iPrCO-(L)Ala-(L)Pro(L)Ala-NHMe (in blue-green) in the peptide receptor cavity (ref. 359). The dotted lines show the hydrogen bondings. Reproduced with permission. Copyright © 1994 by the American Chemical Society.

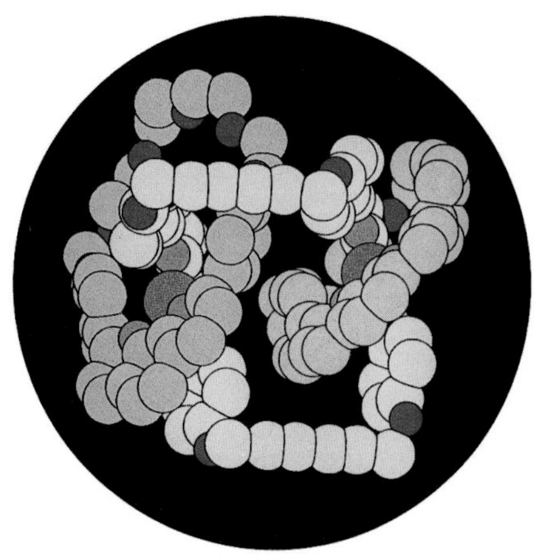

Fig. 6.9. Structure of di-Cu(I) [3]-catenate (ref. 530). Reproduced with permission. Copyright © 1987 by Verlag Chemie GmBH.

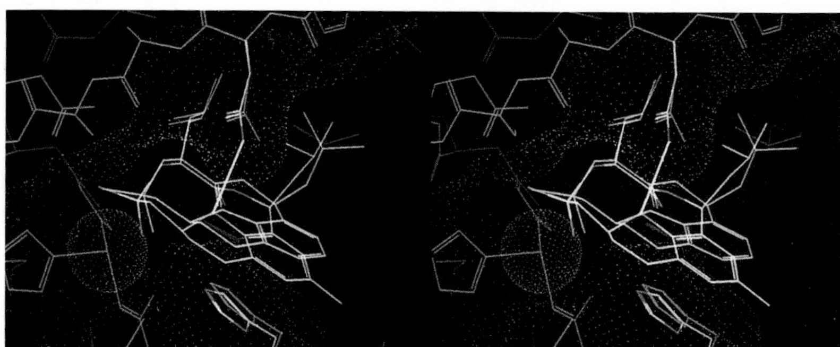

Fig. 7.12. A stereoview of a computer-assisted molecular modeling of a structure-based design of a semirigid peptidomimetic inhibitor of the zinc-dependent enzyme thermolysin. This figure compares modeled (orange) and observed (color-coded) conformations of the *S,S*-inhibitor in the active site of thermolysin (blue). The protein structure is taken from the complex with Cbz-Glu^{P}-Leu-Leu; color-coded representations of the side-chains of His-231 and Asn-112 (below and above the inhibitor, respectively) indicate the different orientations they adopt in the inhibitor complex (ref. 651). Reproduced with permission. Copyright © 1994 by the American Chemical Society.

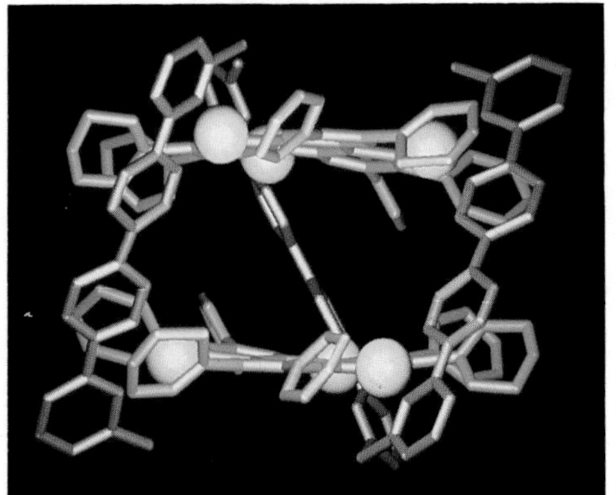

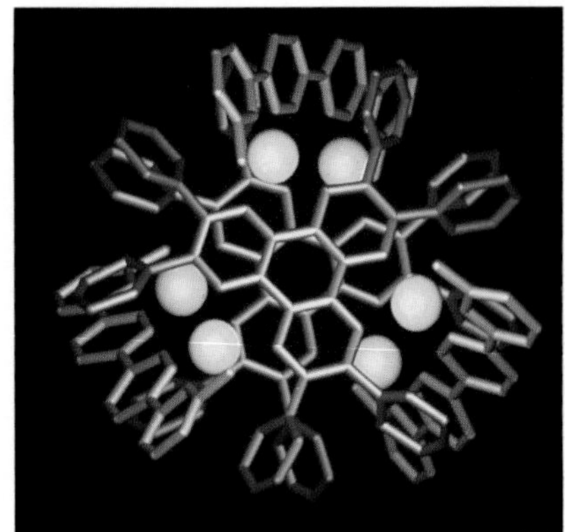

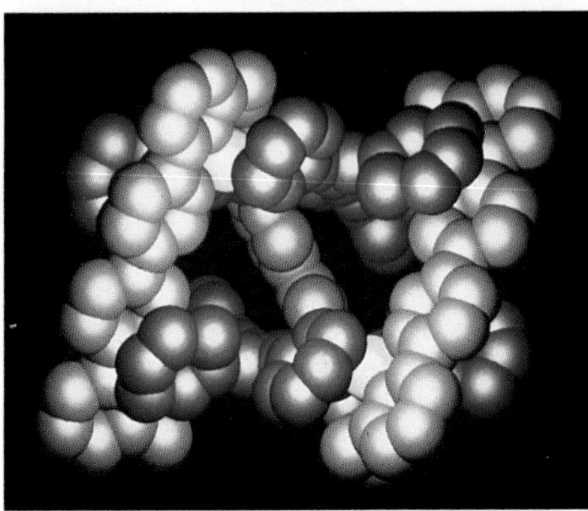

Fig. 8.18. Model of the crystal structures of a cylindrical complex made of five ligands and six metal ions. Top left: Side view. Top right: View along the vertical axis. Bottom: CPK representation of the side view (ref. 732). Reproduced with permission. Copyright © 1993 by VCH, Weinheim.

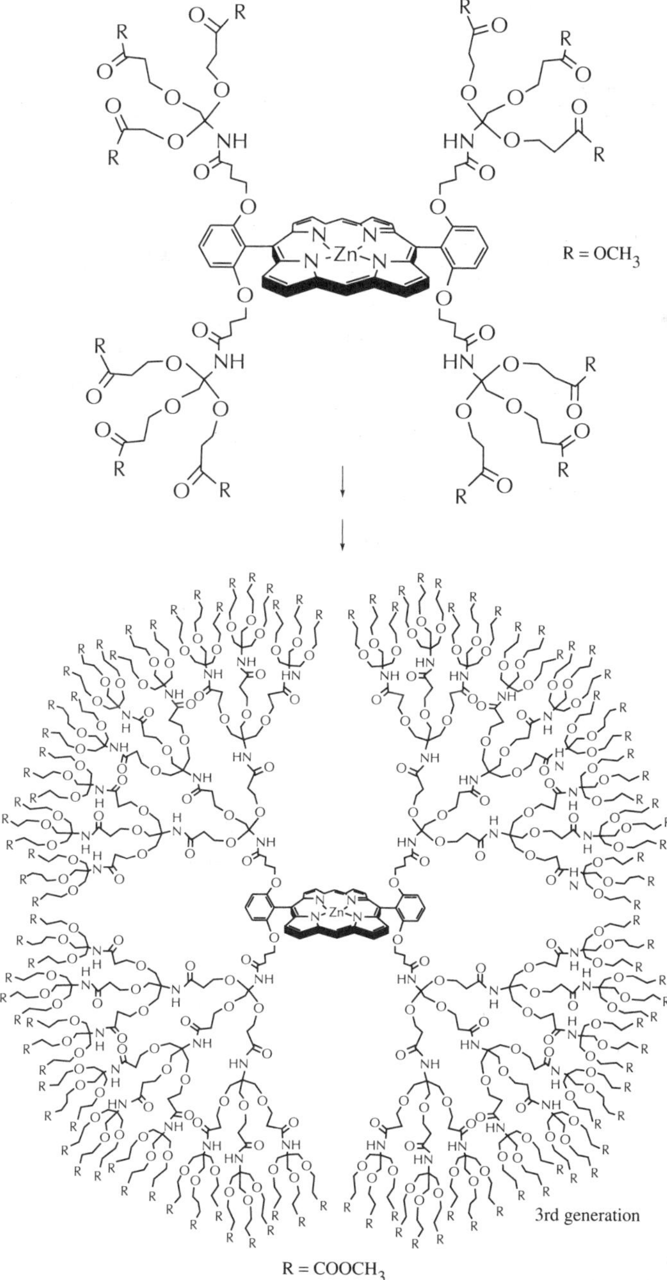

Fig. 8.13. Third generation in the synthesis of a porphyrin-branched dendrimer (722).

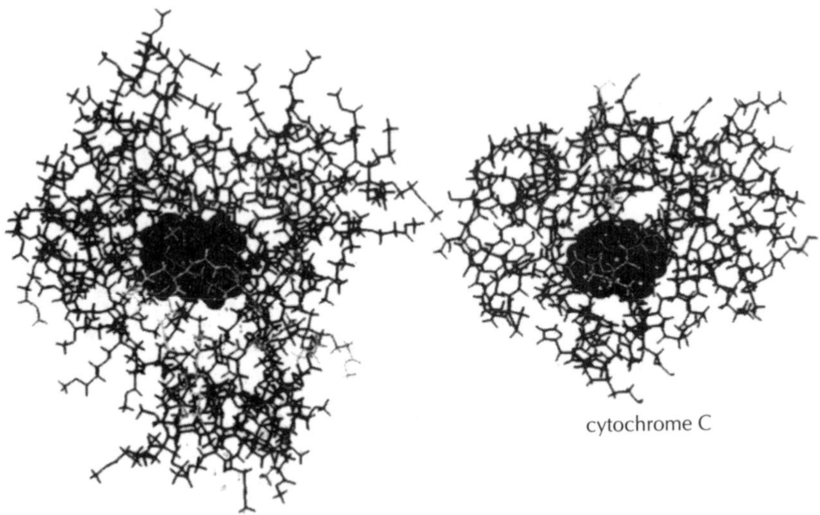

Fig. 8.14. Computer-generated model of the dendritic Zn-porphyrin and comparison with the X-ray crystal structure of cyctochrome C at 2.3 Å resolution (722). The porphyrin nucleus is represented by the dark CPK spheres. Reproduced with the permission of the authors.

8.3.4 Molecular Tectonics

Supramolecular chemistry explores the relatively weak noncovalent interactions that can exist between molecules (see Sections 1.4 and 8.1). When these intermolecular bonds act in a cooperative and convergent way, organized supramolecular systems can be constructed by self-assembly. In addition, there is a growing belief in inorganic materials chemistry that the traditional methods of constructing materials will not be able to address the requirements of future advanced materials. Consequently, a new approach to materials science, based on molecular design and organized assembly, is also needed. In other words, the traditional "soft organic" technology of chemo- and electroactive polymers is incompatible with the "hard-and-stiff" approach of advanced materials engineering an robotics (724). As a consequence, the new molecular domain overlaps the traditional frontiers of both bioorganic and bioinorganic chemistry. Thus, while synthetic chemists are beginning to build new molecular devices beyond the molecule, materials chemists are developing skills in molecular engineering. Both perspectives are striving toward the chemical construction of higher-order architectures. In a review article in 1994, the expression *molecular tectonics** was proposed

* The expression comes from the Greek word *tekton*, for builder.

to define this new development (724), although the term *tecton* has appeared in the literature since 1991 (731).

Chemists in many laboratories now espouse the principles of building large molecular assemblies by joining preorganized units with organic as well as inorganic composites, across the whole face of structural chemistry. We are in an age where *biomineralization* is applied to materials chemistry. Indeed, the study of biomineralization offers valuable insights into the scope and nature of materials science at the inorganic–organic interface. In other words, the interest is in how these organic architectures can be associated with inorganic solids to give unique and exquisite biomaterials (for example, coccoliths, seashells, bone, and so on) in which the structure, size, shape, orientation, texture, and assembly of the mineral constituents are precisely controlled (725).

As mentioned previously in this chapter, the ability to construct organized nanoscale, microscopic, and bulk-mixed inorganic–organic materials from molecular components is of prime importance in electronics, catalysis, magnetism, and sensory devices. In particular, the current thrust toward "intelligent materials" relies on the interplay among molecular structure, molecular organization, and molecular dynamics in determining novel functional responses (724). As mentioned by S. Mann, much of the activity in "intelligent materials" is centered around the biomimicking of, for example, such tissues as muscle fiber and skin.

Figure 8.15 summarizes the interconnections among molecular tectonics, biomineralization, and biomimetic materials chemistry (724). Biomimetic strategies for nanoscale synthesis are based on supramolecular preorganization and interfacial recognition. Crystal engineering is based on the integration of processing and recognition, and microstructural fabrication on the combination of preorganization and processing.

Diagram 8.13 in Section 8.2 summarizes clearly the goal of what has been called *designer solids*. It is the deliberate synthesis of periodic matter, like crystals, on the nanometer scale based on molecular construction kits. For this, we could take another example from nature, because the control of crystal formation has been developed to a remarkable degree by many organisms (725). More than sixty different minerals are known to be formed by living organisms. These include amorphous minerals, inorganic crystals, and organic crystals. Calcium minerals alone represent some 50% of all known biogenic minerals. Nature has always been rich in inspiration. Figure 5.11 illustrates this point. Many more examples and illustrations can be found (725). Look, for example, at the marine organism, clams, which make high-strength ceramic composites. These mollusks engineer a shiny and tough shell by using a series of proteins that assemble themselves at a nucleation site by adsorption on a scaffolding matrix. The scaffolding guides tiny ceramic plates, created by the mollusk, into precise shell layers. Naturally, scientists would like to emulate and duplicate this self-assembly process, because it offers tre-

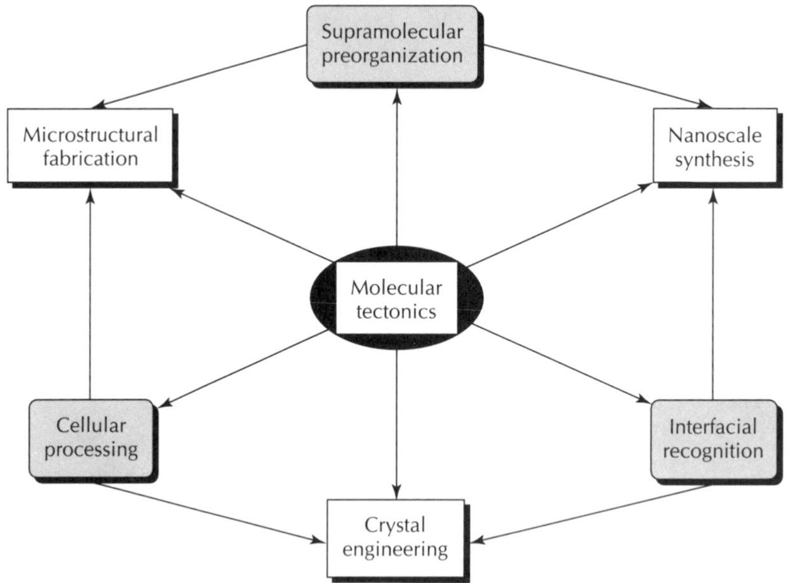

Fig. 8.15. The connections among molecular tectonics, biomineralization (white boxes), and biomimetic material chemistry (gray boxes). Adapted from (724).

mendous advantages in control and economy over conventional manufacturing processes (726).

The diversity in texture, composition, and state of mineralized biological materials is enormous. Furthermore, these materials have adapted, through evolution, to a wide variety of functions. Identifying the underlying mechanism common to the formation processes of different mineralized materials provides insight into key biological phenomena (725). Therefore, the results of such investigations may well be a rich source of inspiration for improving our ability to fabricate novel synthetic materials.

Although scientists are nowhere near duplicating nature's most elegant self-assemblies, they are already beginning to register their first practical successes. Indeed, several drug companies are in late-stage chemical trials with self-assembled microscopic vesicles that ferry potentially lifesaving drugs to cancer patients. For instance, Liposome Technologies of Menlo Park, California, and The Liposome Company of Princeton, New Jersey, have both shown that encasing the anticancer drug doxirubicin in *liposomes* and injecting them into cancer-stricken mice can increase survival rates of the mice over others that received the drug alone (726). Therefore, materials that assemble themselves are beginning to make their presence felt as novel *drug-delivery* vehicles.

This trend is also true in the development of new electronic components. Indeed, in the field of molecular electronic devices one can think of microtubules, which are made of modified phospholipids and/or synthetic diacetylenic lipids, as suitable microscopic templates. Coated with metals, they become stronger and have electronic properties that may allow development of new *microchips* (726).

Electronic applications are also driving researchers to experiment with self-assembly as a way of layering sheets of materials with various arrangements. This could lead to such devices as solar cells and light-emitting diodes. In 1992, H. Katz and his colleages at AT&T used the so-called "metal-phosphonate" precursors of T.E. Mallouk (University of Texas at Austin) to emulate part of a solar cell (727, 728). Layers of porphyrin and viologen molecules were used to capture sunlight, producing electrons and holes. The porphyrins donate electrons to the viologen molecules, allowing the charges to remain separated. The elaboration of the "metal-phosphonate" self-assembling layers is illustrated in Fig. 8.16.

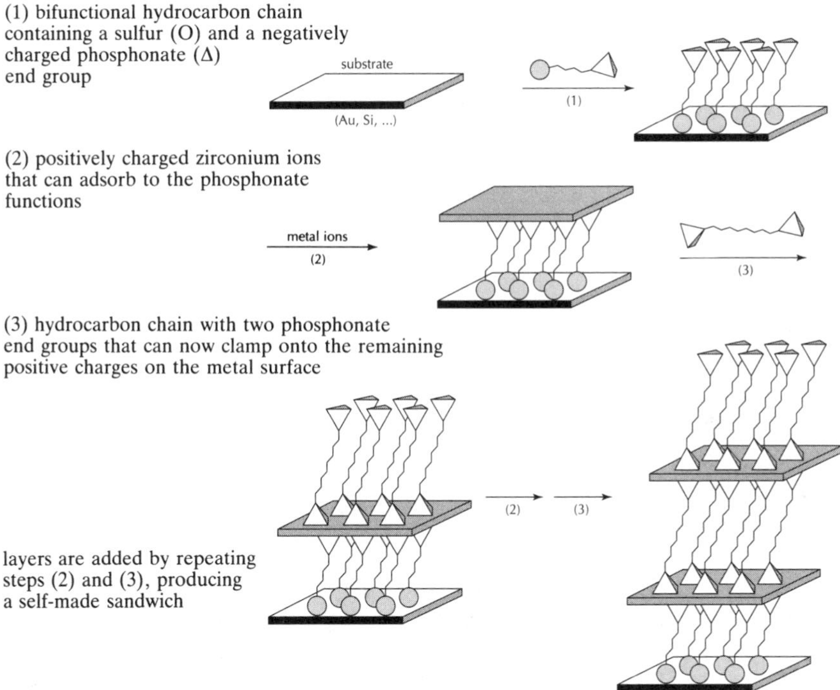

(1) bifunctional hydrocarbon chain containing a sulfur (O) and a negatively charged phosphonate (Δ) end group
substrate
(Au, Si, ...)
(1)
(2) positively charged zirconium ions that can adsorb to the phosphonate functions
metal ions
(2)
(3)
(3) hydrocarbon chain with two phosphonate end groups that can now clamp onto the remaining positive charges on the metal surface
(2)
(3)
layers are added by repeating steps (2) and (3), producing a self-made sandwich
This type of self-assembly system has been used to bring together electronically active materials (726).

Fig. 8.16. Construction of self-assembling multilayers. T. Mallouk, personal communication.

Bifunctional or ditopic hydrocarbon chains containing sulfur and phosphonate end groups are used as building blocks and are layered on a gold-plated silicon bed by dipping the metallic base into a solution containing the organic molecules. The sulfur atoms become tightly bonded to the gold surface, leaving the phosphonate groups sticking out at the surface. Since these phosphonates carry a net negative charge, the material is then placed in a bath containing positively charged zirconium ions that will bind to the phosphonate groups to form a single-atom-thick layer of sheet metal. By dipping the structures in a bath now containing hydrocarbon chains with phosphonate groups at both ends, layers are starting to fill up, leaving another series of unbound phosphonate groups sticking out at the top and ready to make a self-made sandwich by simply repeating the process (726). Such metal-phosphonate film has a strong analogy to the well-known Langmuir–Blodgett film in membrane chemistry. In general, layered structures are essential for semiconductor devices. By the same token, researchers are trying to get self-assembling molecules to aid in orienting liquid crystal polymers for *optical displays*, as well as in making nonlinear optical materials (see Section 8.3.2), which are essential for routing optically transmitted data (726). Indeed, challenges in the area of new *photonic* device technologies and self-assembled chromophoric superlattices typify important themes in contemporary chemistry (741).

A few more examples of recent and promising research projects aimed at developing new devices and technologies should prove instructive. F.M. Menger has used the synthetic amphiphilic molecule 5-hexadecyloxyisophthalic acid for the assembly of long organic fibers (729). The molecule is soluble in a hot solution of potassium hydroxide and forms the mono-potassium salt. As the solution is cooled, a tangled mass of fibers forms. Although wet fibers are easily bent without obvious structural damage, dry fibers are extremely fragile. The entangled fibers can be manually "stretched" to form a filament several centimeters long. Furthermore, the mono-potassium salt of this fatty diacid self-organizes into disks that then stack one upon the another. The component disks are stabilized by a combination of hydrogen bonding and hydrophobic forces. A model of this is presented in Fig. 8.17. The model is vaguely reminiscent of discotic mesophases, except that the hydrophobic chains point inward instead of radiating outward. The organic whiskers, with their remarkable length-to-width ratios, exemplify the synthesis of interesting states of matter via the rational positioning of molecules controlled by noncovalent bonding (729).

The group of J.D. Wuest (Université de Montréal) has made imaginative use of hydrogen bonds to control molecular aggregation of 2-pyridone units into three-dimensional networks with large chambers (731). Crystallization of the 2-pyridone tecton from butyric acid/methanol/hexane provides regular plates. The X-ray crystallographic study has confirmed that the amide function of each pyridone interacts by hydrogen bonding

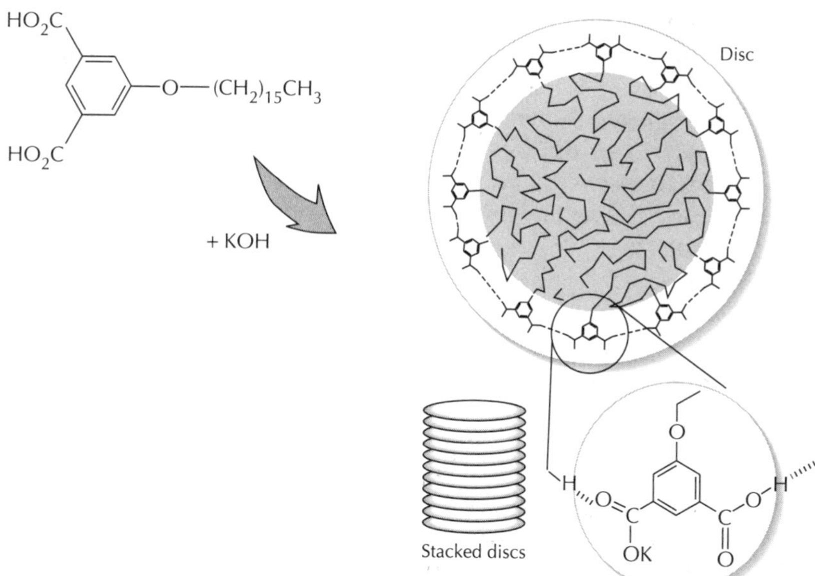

Fig. 8.17. As the hydrocarbon tails become entangled, the organic acids link to form disks, which assemble themselves into stacks (730). Reproduced with permission. Copyright © 1994 by IPC Magazines Limited, England.

in the expected way to induce self-assembly of a remarkable diamond-like network. The tetrahedral centers of adjoining tectons are separated by 19 to 20 Å, defining a network of enormous chambers and interconnecting windows. Each chamber enclathrates only two molecules of butyric acid, even though crystallization occurred in a mixed solvent. The interstitial guests are surprisingly well ordered and form two parallel columns in rectangular channels. Since the columns are retained within a porous host framework by van der Waals forces alone, loss of butyric acid occurs when the crystals are placed under vacuum.

tetrahedral
2-pyridone
tecton center

Self-assembly with this type of diamondoid network suggests that cleverly designed *tectons* can give chemists the elements of a powerful molecular-scale construction set. Presumably, that same strategy could be used to build predictably ordered materials with useful properties, including selective enclathration, microporosity with high ratios of strength to density, and cavity with catalytic activity. As pointed out by Wuest, the principal advantage of this approach is that complex structures with specific architectural or functional features are formed reversibly by spontaneous self-assembly, not by tedious bond-by-bond syntheses (731).

Further progress in understanding and controlling the self-assembly and self-organization of inorganic, as well as organic, superstructures requires systems capable of spontaneously generating well-defined mixed organic–inorganic complexes from a larger set of components that include, at least two types of ligands and/or metal ions. The design and choice of these components must fulfill criteria at all three levels of supramolecular programming and information input that determine and orient the output of the desired final species: that is at the recognition, orientation, and termination levels.

J.M. Lehn's group has reported one such multiligand, multimetal, self-assembly process involving two types of ligands and several Cu(I) ions (732). It was found that the assembly of two hexaphenylhexaazatriphenylenes (in gray) with three $[Cu(bpy)_2]^+$ species generated the molecular cylindrical complex presented below.

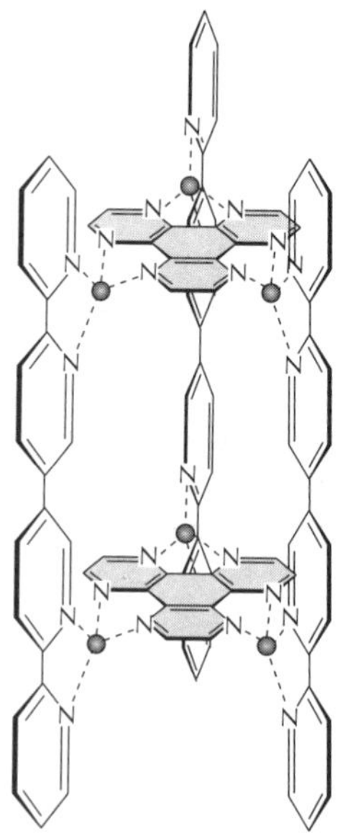

This picture demonstrates the remarkable self-organization of a closed organic–inorganic architecture by the spontaneous and correct assembly of altogether eleven particles, consisting of two types of ligands and one type of metal ion (732). Phenyl and methyl substituents are omitted for clarity. Cu ions are presented by gray dots.

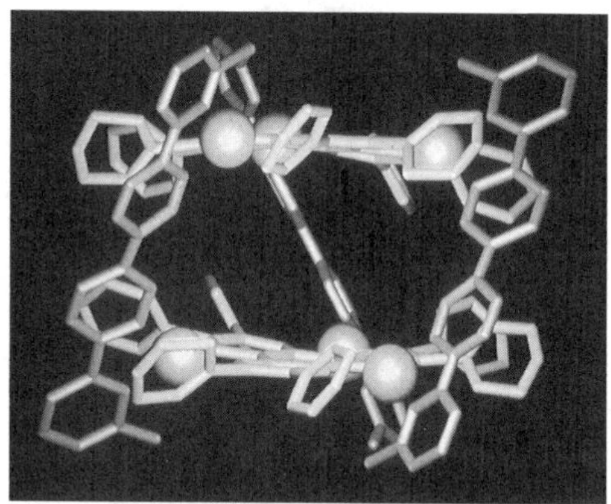

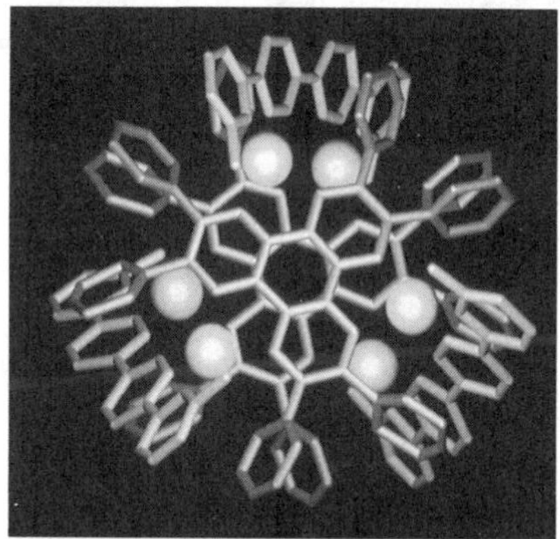

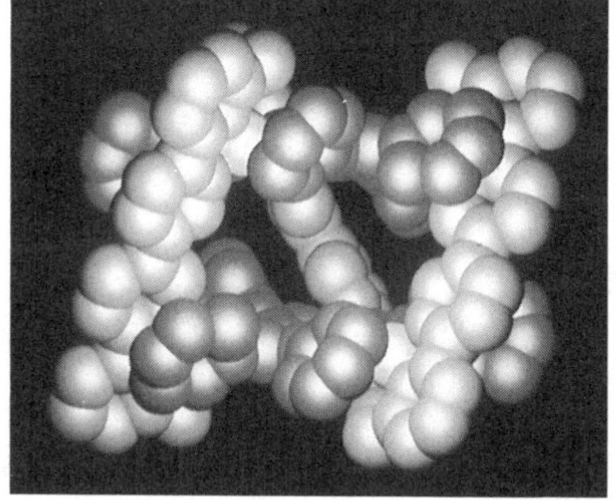

Fig. 8.18. Model of the crystal structure of a cylindrical complex made of five ligands and six metal ions. Top: side view, middle: view along the vertical axis, and bottom: CPK representation of the side view (732). Reproduced with permission. Copyright © 1993 by VCH, Weinheim. See color print.

A model of the crystal structure as well as a CPK representation of this self-assembled cylindrical inorganic–organic complex is shown in Fig. 8.18. In this complex, the cylindrical internal cavity has a height of 7.4 Å and a radius of about 5.5 Å. Taking the van der Waals radii into account, the cylindrical void has a geometry of about 4 Å by 4 Å, and one may envisage binding substrate molecules in this cavity, in particular, flat aromatic units of suitable dimensions. This complex could then behave as a self-assembled molecular receptor for appropriate substrate species.

Even more spectacular is the recent preparation, also by Lehn's group and again by self-assembly, of a superstructure built in one stroke by combining nine Ag(I) ions with six linear tritopic ligands made of 3,6-bis(2-pyridyl)pyridazine (733). The crystal structure confirmed the nature of the complex species $(Ag_9L_6)^{9+}$. The average distance between the mean planes of two neighboring ligand units is about 3.74 Å, that is, somewhat larger than that expected for van der Waals contact (~3.4 Å). This is portrayed in Fig. 8.19.

The generation of this (3 × 3) grid represents the spontaneous and correct assembly of a closed inorganic superstructure from a total of fifteen components: six ligands, and nine metal ions. It displays the expected three basic levels of operation of a programmed supramolecular system—namely recognition, orientation, and termination—and involves all-together geometrical, thermodynamical, and entropic factors.

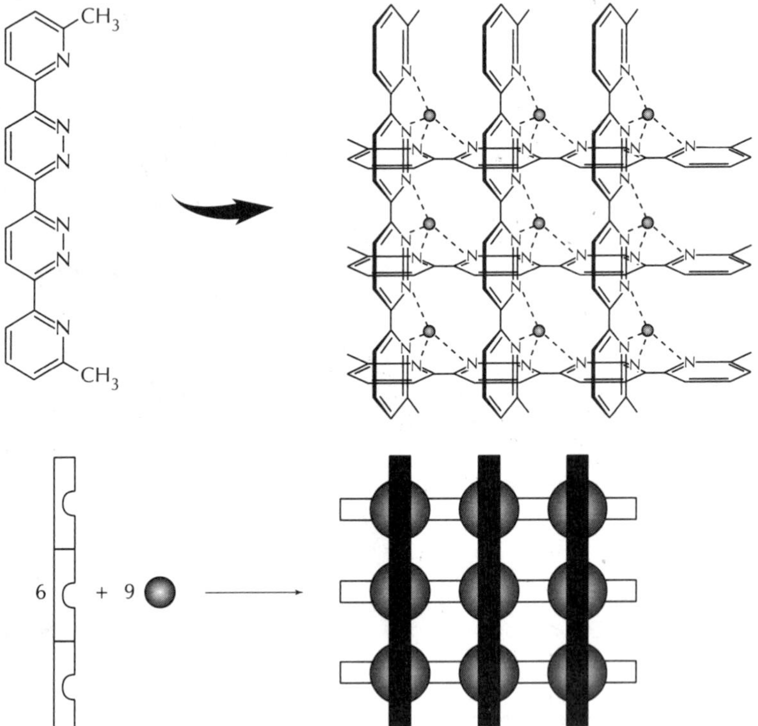

Fig. 8.19. Self-assembly and structure of a [3 × 3] inorganic grid made of nine Ag(I) ions and six ligand components. Underneath is a schematic representation of the grid. Silver ions are presented by gray dots. Adapted from (733).

This unique superstructure opens the way to a whole family of polynuclear inorganic supramolecular architectures that may present novel physicochemical properties as well as intriguing features as potential information *storage devices*. Finally, one may compare such grid-like structures with grids based on quantum dots that are of great interest in microelectronics. Here we have the equivalent of *ion dots* that do not necessitate microfabrication, but form spontaneously by self-assembly. Such architecture may represent multistate digital supramolecular chips for information storage. Extension into three dimensions involving stacks of grids may be envisaged, leading to layered arrays (733).

A last, but not least important, area of molecular design that is rapidly expanding is the study of polymeric carbon allotropes and the construction of carbon networks or carbon scaffoldings. Tremendous progress has been made in this area by F. Diederich (E.T.H., Zurich) and U.H.F. Bunz (Max-Planck Institute, Mainz) (734–736).

Allotropes are molecules with different forms and physical properties but of similar or identical chemical properties. Until recently, graphite and diamond were the only allotropic forms of carbon on earth that were available in macroscopic quantities and were structurally well characterized. With the synthesis of macroscopic quantities of buckminsterfullerene (C_{60}) and higher *fullerenes* and the exploration of the fascinating properties of these all-carbon spheres, this situation has completely changed. In the coming decade, the design, preparation, and study of novel molecular and polymeric allotropic forms of carbon will be a central topic in chemistry. Research in this area will dramatically advance the fundamental knowledge of carbon-based matter.

For example, diamond thin films have attracted great interest for use in electronic devices and in the propective coating of tools and materials (734). A whole variety of other, still unprepared, all-carbon network polymers have been predicted to exibit interesting material properties. One example will suffice to express this trend.

F. Diederich's group has recently made the following polyyne (737).

TIPS

H≡ ≡H

TIPS

TIPS

Ph— [ ] n —Ph

via Cu(I) and Pd(I) chemistry

via "end capping"

TIPS

n = 1 to 4
TIPS = triisopropylsilyl

Ph—≡H

At n = 5, the molecule is 49 Å, or about 5 nm long, and melts at 230°. The material is stable to air and is not oxidized. However, it could be reduced readily. In the solid state, the molecules tend to stack via the participation of the phenyl end groups to form rods of 5 to 15 nm in length. If doped with a metal ion, these rods have a metallic behavior and conduct electricity.

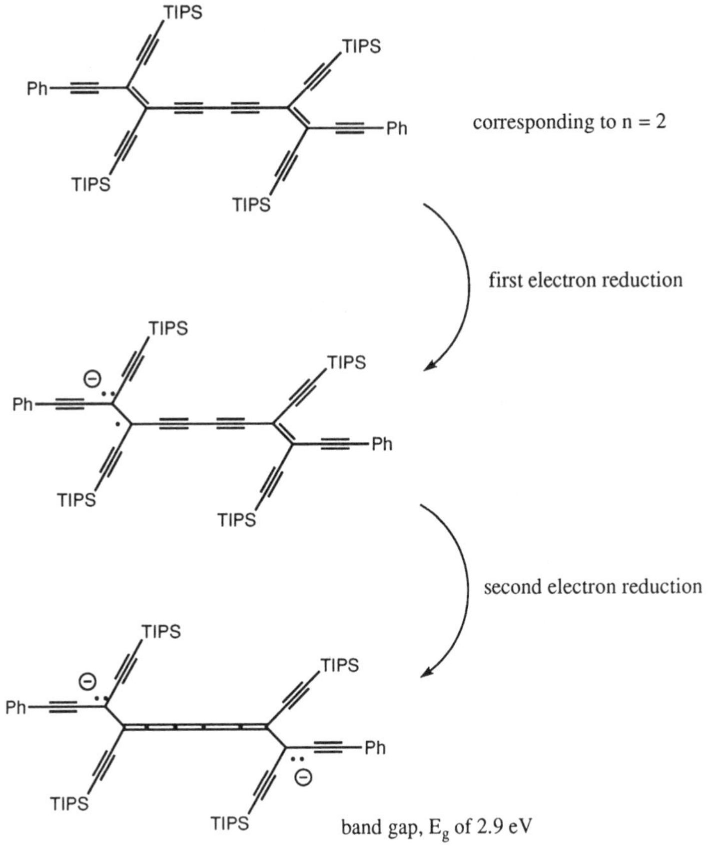

TIPS
TIPS
Ph
Ph
TIPS
TIPS
corresponding to n = 2

first electron reduction

TIPS
TIPS
Ph
Ph
TIPS
TIPS

second electron reduction

TIPS
TIPS
Ph
Ph
TIPS
TIPS
band gap, E_g of 2.9 eV

There is no doubt that this kind of approach, and others, will open new possibilities and strategies for the construction of new carbon networks that will generate materials with unpredicted properties and introduce new technological perspectives. Covalent hybrids of fullerene spheres and poly-acetylene chains are such examples (738). The demonstration that fullerenes, despite their thermodynamic instability, are stable carbon molecules because of high kinetic stabilization has fueled worldwide efforts to synthesize new molecular and polymeric allotropes.

It is undeniable that the preparation of acetylenic molecular and polymeric carbon allotropes and carbon-rich nanometer-sized structures opens new avenues of fundamental and technological research at the interface between chemistry and materials science. Unusual structures, high stability, and useful electrical and nonlinear optical properties are some of the desirable characteristics of these materials (735). The new carbon allotropes are predicted to have exciting material properties and, if synthesized in bulk quantities, could assume similar roles to diamond and graphite in technological applications. As mentioned in the preface, this and other research in supramolecular chemistry promise to be an active frontier in chemical sciences. These goals will only be reached by interdisciplinary approaches involving chemists and materials scientists. Just as the barriers between biology and chemistry are currently collapsing (739), those between materials science and chemistry (735) will become increasingly permeable.

References

General Reading

B. Alberts, D. Bray, J. Lewis, M. Raff, K. Roberts, and J.D. Watson (1983), *Molecular Biology of the Cell*. Garland Publishing Inc., New York.

R. Barker (1971), *Organic Chemistry of Biological Compounds*. Prentice-Hall, Englewood Cliffs, New Jersey.

M.L. Bender (1971), *Mechanisms of Homogeneous Catalysis from Protons to Proteins*. Wiley-Interscience, New York.

M.L. Bender and L.J. Brubacher (1973), *Catalysis and Enzyme Action*. McGraw-Hill, New York.

C. Branden and J. Tooze (1991), *Introduction to Protein Structure*. Garland Publishing, Inc., New York.

T.C. Bruice and S. Benkovic (1966), *Bioorganic Mechanisms*. Vols. 1 and 2. Benjamin, New York.

G.M. Coppola and H.F. Schuster (1987), *Asymmetric Synthesis—Construction of Chiral Molecules using Amino Acids*. J. Wiley & Sons, New York.

A. Fersht (1985), *Enzyme Structure and Mechanism*. 2nd Ed. Freeman, San Francisco.

R.P. Hanzlik (1976), *Inorganic Aspects of Biological and Organic Chemistry*. Academic Press, New York.

W.P. Jencks (1969), *Catalysis in Chemistry and Enzymology*. McGraw-Hill, New York.

J.B. Jones, C.J. Sih, and D. Perlman, Eds. (1976), *Application of Biochemical Systems in Organic Chemistry*, Vol. 10, Parts I and II. Techniques of Chemistry Series. Wiley-Interscience, New York.

K.D. Kopple (1971), *Petides and Amino Acids*. Benjamin, New York.

B. Lewin (1983), *Genes*. J. Wiley & Sons, New York.

J.N. Lowe and L.L. Ingraham (1974), *An Introduction to Biochemical Reaction Mechanisms*. Prentice-Hall, Englewood Cliffs, New Jersey.

D.E. Metzler (1977), *Biochemistry, The Chemical Reactions of Living Cells*. Academic Press, New York.
J.D. Morrison, Ed. (1985), *Asymmetric Synthesis*, Vol. 5, Chiral Catalysis. Academic Press Inc., New York.
E. Ochai (1977), *Bioinorganic Chemistry, an Introduction*. Allyn and Bacon, Boston.
J.D. Rawn (1989), *Biochemistry*. Carolina Biophysical Supply Co., Burlington, North Carolina.
E.E. van Tamelen, Ed. (1977), *Bioorganic Chemistry*, Vols. I–IV. Academic Press, New York.
C. Walsh (1979), *Enzymatic Reaction Mechanisms*. Freeman, San Francisco.
J.D. Watson (1976), *Molecular Biology of the Gene*, 3rd ed. Benjamin, New York.

Further Reading on Bioorganic Subjects

R.M. Izatt and J.J. Christensen, Eds. (1978), *Synthetic Multidentate Macrocyclic Compounds*. Academic Press, New York.
R.M. Izatt and S.J. Christensen, Eds. (1979), *Progress in Macrocyclic Chemistry*, Vol. 1. J. Wiley & Sons, New York.
K.G. Scringeour (1977), *Chemistry and Control of Enzyme Reactions*. Academic Press, New York.
D. Dolphin, C. McKenna, Y. Murakami, and J. Tabushi (1980), *Biomimetic Chemistry*, Adv. Chem. Series 191. Amer. Chem. Soc., Washington, DC.
J.H. Fendler (1982), *Membrane Mimetic Chemistry*. J. Wiley & Sons, New York.
P. Dunnill, A. Wiseman, and N. Blakebrough, Eds. (1980), *Enzymic and Non-Enzymic Catalysis*. E. Horwood, Ltd., Chichester, England.
M.L. Bender, R.J. Bergeron, and M. Komiyama (1984), *The Bioorganic Chemistry of Enzymatic Catalysis*. J. Wiley & Sons, New York.
L.B. Spector (1982), *Covalent Catalysis by Enzymes*. Springer-Verlag, New York.
C.J. Suckling, Ed. (1990), *Enzyme Chemistry, Impact and Applications*. 2nd ed. Chapman and Hall, London.
B.S. Green, Y. Ashani, and D. Chipman, Eds. (1952), *Chemical Approaches to Understanding Enzyme Catalysis*. Elsevier Science Publishing Co., Oxford.
R.D. Gandour and R.L. Schowen, Eds. (1978), *Transition States of Biochemical Processes*. Plenum Press, New York.
G.G. Hammes (1982), *Enzyme Catalysis and Regulation*. Academic Press, New York.
W. Saenger (1984), *Principles of Nucleic Acid Structure*. Springer-Verlag, New York.
J. Rétey and W.G. Robinson (1982), *Stereospecificity in Organic Chemistry and Enzymology*. Verlag-Chemie.
M.L. Bender and M. Komiyama (1978), *Cyclodextrin Chemistry*. Springer-Verlag, New York.
G. Vanbinst, Ed. (1986), *Design and Synthesis of Organic Molecules Based on Molecular Recognition*. Springer-Verlag, Berlin.
F.L. Boschke (1985), *Biomimetic and Bioorganic Chemistry*, no. 128 of Topics in Current Chemistry. Springer-Verlag, Berlin.

F. Vögtle and E. Weber (1986), *Biomimetic and Bioorganic Chemistry II*, no. 132 of Topics in Current Chemistry, Springer-Verlag, Berlin.

R. Breslow, Ed. (1986), *International Symposium on Bioorganic Chemistry*, in *Ann. New York Acad. Soc.* **471**.

P. Deslongchamps (1983), *Stereoelectronic Effects in Organic Chemistry*, Pergamon Press, Oxford.

W. Bartmann and K.B. Sharpless, Eds. (1987), *Stereochemistry of Organic and Bioorganic Transformations*, VCH, Weinheim.

C. Bleasdale and B.T. Golding, Eds. (1990), *Molecular Mechanisms in Bioorganic Processes*. Royal Society of Chemistry, Cambridge.

S. Yoshikawa and Y. Murakami, Eds. (1990). *Supramolecular Assemblies*. Mita Press, Tokyo.

Y. Inoue and G.W. Gokel (1990), *Cation Binding by Macromolecules*. Marcel Dekker, Inc., New York.

H.J. Schneider, H. Dürr, and J.M. Lehn (1991), *Frontiers in Supramolecular Organic Chemistry and Photochemistry*. VCH, Weinheim.

L.F. Lindoy (1992), *The Chemistry of Macrocyclic Ligand Complexes*. Cambridge University Press, England.

H. Dugas, Ed. (1990), *Bioorganic Chemistry Frontiers*. Vol. 1, Springer-Verlag, New York.

H. Dugas, Ed. (1991), *Bioorganic Chemistry Frontiers*. Vol. 2, Springer-Verlag, New York.

H. Dugas, Ed. (1993), *Bioorganic Chemistry Frontiers*. Vol. 3, Springer-Verlag, New York.

F. Vögtle (1993), *Supramolecular Chemistry: An Introduction*. J. Wiley & Sons, New York.

Y. Murakami, Ed. (1994), Second International Symposium on Bioorganic Chemistry. *Pure & Appl. Chem.* **66**, No. 4.

D.J. Cram and J.M. Cram (1994), *Container Molecules and Their Guests*. Monograph in Supramolecular Chemistry, The Royal Society of Chemistry, Cambridge.

THE TOP 100 REVIEW ARTICLES IN BIOORGANIC CHEMISTRY

1. J.-M. Lehn (1978), Cryptates: Inclusion complexes of macropolycyclic receptor molecules. *Pure & Appl. Chem.* **50**, 871–892.
2. F. Vögtle and E. Weber (1979), Multidentate acyclic neutral ligands and their complexation. *Angew. Chem. Int. Ed. Engl.* **18**, 753–776.
3. Y. Imanishi (1979), Intramolecular reactions on polymer chain. *J. Polym. Sc., Macromol. Rev.* **14**, 1–205.
4. C.W. Wharton (1979), Synthetic polymers as models for enzyme catalysis. *Int. J. Biol. Macromol.* **1**, 3–16.
5. J.-M. Lehn (1979), Macrocyclic receptor molecules. *Pure & Appl. Chem.* **51**, 979–997.
6. R. Breslow (1979), Biomimetic chemistry in oriented systems. *Israel J. Chem.* **18**, 187–191.
7. R. Breslow (1980), Biomimetic control of chemical selectivity. *Acc. Chem. Res.* **13**, 170–177.

8. J.-M. Lehn (1980), Binuclear cryptates: dimetallic macropolycyclic inclusion complexes—concepts, design, prospects. *Pure & Appl. Chem.* **52**, 2442–2459.
9. T.C. Bruice (1980), Mechanisms of flavin catalysis. *Acc. Chem. Res.* **13**, 256–262.
10. C. Walsh (1980), Flavin coenzymes: At the crossroads of biological redox chemistry. *Acc. Chem. Res.* **13**, 148–155.
11. J.C. Vederas and H.G. Floss (1980), Stereochemistry of pyridoxal phosphate catalyzed enzyme recations. *Acc. Chem. Res.* **13**, 455–463.
12. F.H. Westheimer (1980), Models of enzyme systems. *ChemTech.* 748–754.
13. J.-M. Lehn (1980), Dinuclear cryptates. *Pure & Appl. Chem.* **52**, 2441–2459.
14. L.N. Ferguson (1981), Bio-organic mechanism II: chemo-reception. *J. Chem. Educ.* **58**, 456–461.
15. J.-M. Lehn (1981), La chimie supramoléculaire. *La Recherche* **12**, 1213–1223.
16. T.G. Traylor (1981), Synthetic model compounds for hemoproteins. *Acc. Chem. Res.* **14**, 102–109.
17. P. Bey (1981), Design of enzyme-activated irreversible inhibitors of pyridoxalphosphate dependent enzymes. *Chem. & Ind.* 139–144.
18. I. Tabushi (1982), Cyclodextrin catalysis as a model for enzyme action. *Acc. Chem. Res.* **15**, 66–72.
19. C. Walsh (1982), Suicide substrates: Mechanism-based enzyme inactivators. *Tetrahedron* **38**, 871–909.
20. Y. Murakami (1982), Biomimetic chemistry—Implication in organometallic chemistry and future development (in Japanese). *J. Synth. Org. Chem. Jpn.* **40**, 1082–1089.
21. S.J. Jolley, J.S. Bradshaw, and R.M. Izatt (1982), Synthetic chiral macrocyclic crown ligands: A short review. *J. Heterocyclic Chem.* **19**, 3–18.
22. V.K. Majestic and G.R Newkome (1982), Pyridinophanes, pyridinocrowns, and pyridinocryptands. In: *Topics in Current Chemistry*, Vol. 106, pp. 79–120. Springer-Verlag, New York.
23. R. Breslow (1982), Artificial enzymes. *Science* **218**, 532–537.
24. R. Breslow (1983), Artificial enzymes. *Chem. Britain* 126–131.
25. D.J. Cram (1983), Cavitands: Organic host with enforced cavities. *Science* **219**, 1177–1183.
26. R.C. Hayward (1983), Abiotic receptors. *Chem. Soc. Rev.* **112**, 285–308.
27. I. Tabushi and Y. Kuroda (1983), Cyclodextrin and cyclophanes as enzymes models. *Adv. in Catalysis* **32**, 417–466.
28. A.P. Croft and R.A. Bartsch (1983), Synthesis of chemically modified cyclodextrins. *Tetrahedron* **39**, 1417–1474.
29. R.H. Abeles (1983), Suicide enzyme inactivations. *Chem. & Eng. News* Sept. 19, 48–56.
30. J.L. Pierre and P. Baret (1983), Complexes moléculaires d'anions. *Bull. Soc. Chim. France* **II**, 367–380.
31. C.D. Gutsche (1983), Calixarenes. *Acc. Chem. Rev.* **16**, 161–170.
32. I. Tabushi (1984), Design and synthesis of artificial enzymes. *Tetrahedron* **40**, 269–292.
33. C.J. Suckling (1984), Reactive intermediates in enzyme-catalyzed reactions. *Chem. Soc. Rev.* **113**, 97–129.

34. R.M. Kellogg (1984), Les enzymes artificielles. *La Recherche* **15**, 819–829.
35. C. Sirlin (1984), Catalyse supramoléculaire en série cyclodextrine et polyéther macrocyclique. *Bull. Soc. Chim. France* **II**, 5–40.
36. A.R. Battersby (1984), Stereospecific synthesis using enzymes. *Chem. Britain* 611–616.
37. R.M. Kellogg (1984), Chiral macromolecules as reagents and catalysts. *Angew. Chem. Int. Ed. Engl.* **23**, 782–794.
38. J. Rebek, Jr. (1984), Binding forces, equilibria, and rates: New models for enzymic catalysis. *Acc. Chem. Res.* **17**, 258–264.
39. C.F. Meares and T.G. Wensel (1984), Metal chelates as probes of biological systems. *Acc. Chem. Res.* **17**, 202–209.
40. J.H. Fendler (1984), Membrane mimetic chemistry. *Chem. & Eng. News*, Jan. 2, 25–38.
41. E.T. Kaiser and D.S. Lawrence (1984), Chemical mutation of enzyme active sites. *Science* **226**, 505–511.
42. C.R. Beddell (1984), Designing drugs to fit a macromolecular receptor. *Chem. Soc. Rev.* **13**, 279–319.
43. R.M. Kellogg (1984), Chiral macrocycles as reagents and catalysts. *Angew. Chem. Int. Ed. Engl.* **23**, 782–794.
44. J.H. Fuhrhop and J. Mathieu (1984), Route to functional vesicle membranes without proteins. *Angew. Chem. Int. Ed. Engl.* **23**, 110–113.
45. J.-M. Lehn (1985), Supramolecular chemistry: Receptors, catalysts, and carriers. *Science* **277**, 849–856.
46. B. Dietrich (1985), Coordination chemistry of alkali and alkaline-earth cations with macrocyclic ligands. *J. Chem. Educ.* **62**, 954–964.
47. F. Vögtle, M.G. Löhn, J. Franke, and D. Worsch (1985), Host/guest chemistry of organic onium compounds-clathrates, crystalline complexes, and molecular inclusion compounds in aqueous solution. *Angew. Chem. Int. Ed. Engl.* **24**, 727–742.
48. J.H. Fendler (1985), Membrane mimetic systems that contain catalysts, semiconductors, and magnets, *ChemTech*. 686–691.
49. P.L. Luisi (1985), Enzyme hosted in reverse micelles in hydrocarbon solution. *Angew. Chem. Int. Ed. Engl.* **24**, 439–450.
50. H.G. Löhn and F. Vögtle (1985), Chromo- and fluoroionophores. A new class of dye reagents. *Acc. Chem. Res.* **18**, 65–72.
51. G.M. Whitesides and C.H. Wong (1985), Enzymes as catalysts in synthetic organic chemistry. *Angew. Chem. Int. Ed. Engl.* **24**, 617–638.
52. F. Montanari, S. Quici, P.L. Anelli, H. Molinari, and T. Beringhelli (1986), New lipophilic multidentate ligands: Effective complexing agents for anion activation in non-polar media. *Gazz. Chim. Ital.* **116**, 275–280.
53. J.B. Jones (1986), Enzymes in organic synthesis. *Tetrahedron* **13**, 3317–3403.
54. H.M. Colquhoun, J.F. Stoddart, and D.J. Williams (1986), Second sphere coordination—a novel rôle for molecular receptors. *Angew. Chem. Int. Ed. Engl.* **25**, 487–507.
55. S.J. Lippard (1986), The bioinorganic chemistry of rust. *Chem. Britain* 222–229.
56. S. Yasui and A. Ohno (1986), Model studies with nicotinamide derivatives. *Bioorg. Chem.* **14**, 70–96.

57. J.H. Fuhrhop and D. Fritsh (1986), Bolaamphiphiles from ultrathin, porous, and unsymmetric monolayer lipid membranes. *Acc. Chem. Res.* **19**, 130–137.
58. C.O. Dietrich-Buchecker and J.-P. Sauvage (1987), Interlocking of molecular threads: From the statistical approach to the templated synthesis of catenands. *Chem. Rev.* **87**, 795–810.
59. J. Rebek, Jr. (1987), Model studies in molecular recognition. *Science* **235**, 1478–1484.
60. D.J. Cram (1987), Molecular cells, their guests, portals, and behavior. *ChemTech*. 120–125.
61. V.T. D'Souza and M.L. Bender (1987), Miniature organic models of enzymes. *Acc. Chem. Res.* **20**, 146–152.
62. S. Butt and S.M. Roberts (1987), Opportunities for using enzymes in organic synthesis. *Chem. Britain* 127–134.
63. A. Akiyama, M. Bednarski, M.J. Kim, E.S. Simon, H. Waldmann, and G.M. Whitesides (1987), Enzymes in organic synthesis. *Chem. Britain* 645–652.
64. D.G. Gorenstein (1987), Stereoelectronic effects in biomolecules. *Chem. Rev.* **87**, 1047–1077.
65. J.F. Stoddart (1987), Chiral crown ethers. In: *Topics in Stereochemistry* (E. Eliel and S.H. Wilen, Eds.) Vol. **17**, pp. 207–288. J. Wiley & Sons, New York.
66. J.S. Pagington (1987), β-Cyclodextrin: The success of molecular inclusion. *Chem. Britain* 455–458.
67. R. Kluger (1987), Thiamin diphosphate: A mechanistic update on enzymic and nonenzymic catalysis of decarboxylation. *Chem. Rev.* **87**, 863–876.
68. R.A. Lerner and A. Tramontano (1987), Antibodies as enzymes. *Trends Biochem. Sci.* **12**, 427–431.
69. S. Shinkai (1987), Switch-functionalized systems in biomimetic chemistry: *Pure & Appl. Chem.* **59**, 425–430.
70. A. Collet (1987), Cyclotriveratrylenes and cryptophanes. *Tetrahedron* **43**, 5725–5759.
71. P.A. Sutton and D.A. Buckingham (1987), Cobalt(III)-promoted hydrolysis of amino acid esters and peptides and the synthesis of small peptides. *Acc. Chem. Res.* **20**, 357–364.
72. H.E. Moser and P.B. Dervan (1987), Sequence-specific cleavage of double helical DNA by triple helix formation. *Science*, **238**, 645–650.
73. D.N. Silverman and S. Lindskog (1988), The catalytic mechanism of carbonic anhydrase: Implications of a rate-limiting proteolysis of water. *Acc. Chem. Res.* **21**, 30–36.
74. T.A. Dix and S.J. Benkovic (1988), Mechanism of oxygen activation by pteridine-dependent monooxygenases, *Acc. Chem. Res.* **21**, 101–107.
75. H. Ringsdorf, B. Schlarb, and J. Venzmer (1988), Molecular architecture and function of polymeric oriented systems: Models for the study of organization, surface recognition, and dynamic of biomembranes. *Angew. Chem. Int. Ed. Engl.* **27**, 113–158.
76. A. Eschenmoser (1988), Vitamin B_{12}: Experiments concerning the origin of its molecular structure. *Angew. Chem. Int. Ed. Engl.* **27**, 5–39.

77. F. Diederich (1988), Complexation of neutral molecules by cyclophane hosts. *Angew. Chem. Int. Ed. Engl.* **27**, 362–386.
78. J.-M. Lehn (1988), Supramolecular chemistry-scope and perspective, molecules, supermolecules, and molecular devices (Nobel Lecture). *Angew. Chem. Int. Ed. Engl.* **27**, 89–112.
79. D.J. Cram (1988), The design of molecular hosts, guests, and their complexes (Nobel Lecture). *Science*, **240**, 760–767.
80. C.J. Pedersen (1988), The discovery of crown ethers (Nobel Lecture). *Science*, **241**, 536–541.
81. J.M. Lehn (1988), Supramolecular chemistry—scope and perspectives: molecules–supermolecules–molecular devices. *J. Inclusion Phenomena* **6**, 351–396.
82. M.W. Hosseini (1989), La catalyse supramoleculaire. *La Recherche* **206**, 24–32.
83. J.M. Lehn (1990), Perspectives in supramolecular chemistry—from molecular recognition towards molecular information processing and self-organization. *Angew. Chem. Int. Ed. Engl.* **29**, 1304–1319.
84. F. Stoddart (1991), Making molecules to order. *Chem. Britain* **27**, 714–718.
85. H.J. Schneider (1991), Mechanisms of molecular recognition: investigations of organic host-guest complexes. *Angew. Chem. Int. Ed. Engl.* **30**, 1417–1436.
86. J.M. Lehn (1993), Supramolecular chemistry—molecular information and the design of supramolecular materials. *Makromol. Chem. Macromol. Symp.* **69**, 1–17.
87. P.G. Schultz and R.A. Lerner (1993), Antibody catalysis of difficult chemical transformations. *Acc. Chem. Res.* **26**, 391–395.
88. J.D. Stewart, L.J. Liotta, and S.J. Benkovic (1993), Reaction mechanisims displayed by catalytic antibodies. *Acc. Chem. Res.* **26**, 396–404.
89. N.C. Seeman, J. Chen, S.M. Du, J.E. Mueller, Y. Zhang, T.J. Fu, Y. Wang, H. Wang, and S. Zhang (1993), Synthetic DNA knots and catenanes. *New J. Chem.* **17**, 739–755.
90. E.C. Constable (1994), Metallosupramolecular chemistry. *Chem. & Ind.* 56–59.
91. H. An, J.S. Bradshaw, R.M. Izatt, and Z. Yan (1994), Bis- and oligo(benzocrownether)s. *Chem. Rev.* **94**, 931–939.
92. R. Hoss and F. Vögtle (1994), Template syntheses. *Angew. Chem. Ed. Engl.* **33**, 375–384.
93. A.M. Reichwein, W. Verboom, and D.N. Reinhoudt (1994), Enzyme models. *Recl. Trav. Chim. Pays-Bas* **113**, 343–349.
94. D.B. Amabilino, I.W. Parsons, and J.F. Stoddart (1994), Polyrotaxanes. *Trends Polym. Sci.*, **2**, 146–152.
95. J. Rebek, Jr. (1994), A template for life. *Chem. Britain* 286–290.
96. E.A. Wintner, M.M. Conn, and J. Rebek, Jr. (1994), Studies in molecular replication. *Acc. Chem. Res.* **26**, 198–203.
97. R. Breslow (1994), Bifunctional acid-base catalysis by imidazole groups in enzyme minics—review. *J. Mol. Catalysis* **91**, 161–174.
98. J.M. Lehn (1994), Supramolecular reactivity and catalysis. *Applied Catalysis A—General* **113**, 105–114.
99. A.J. Kirby (1994), Enzyme mimics. *Angew. Chem. Int. Ed. Engl.* **33**, 551–553.

100. F. Diederich (1994), Carbon scaffolding: Building acetylenic all-carbon and carbon-rich compounds. *Nature* **369**, 199–207.

Selected References

Chapter 1

1. V.B. Schatz (1960), Isosterism and bio-isosterism as guides to structural variations. In: *Medicinal Chemistry* (A. Burger, Ed.), 2nd ed., pp. 72–88. Interscience, New York.
2. A. Korolkovas (1970), *Essentials of Molecular Pharmacology*, pp. 55–59. Wiley-Interscience, New York.
3. G.I. Birnbaum, M. Cygler, and D. Shugar (1984), Conformational features of acyclonucleotides: structure of acyclovir, an antiherpes agent. *Can. J. Chem.* **62**, 2646–2652.
4. R. Engel (1979), Phosphonates as analogues of natural phosphates. *Chem. Rev.* **77**, 349–367.
5. K.C. Tang, B.E. Tropp, and R. Engel (1978), The synthesis of phosphonic acid and phosphate analogues of glycerol-3-phosphate and related metabolites. *Tetrahedron* **34**, 2873–2878.
6. E. Frieden (1981), Iodine and the thyroid hormones. *Trends Biochem. Sci.* **6**, 50–53.
7. T.P. Singer, A.J. Trevor, and N. Castagnoli, Jr. (1987), Biochemistry of the neurotoxic action of MPTP. *Trends Biochem. Sci.* **12**, 266–270.
8. J.-M. Lehn (1985), Supramolecular chemistry: Receptors, catalysts, and carriers. *Science* **227**, 849–850.
9. H. Colquhoun, F. Stoddart, and D. Williams (1986), Chemistry beyond the molecules. *New Scientist*, May 1, p. 44.
10. K.L. Wolf, M. Duken, and K. Merkel (1940), Über Übermolekülbildung. *Z. Phys. Chem. Abt. B* **46**, 287–312.
11. J.M. Lehn (1993), Supramolecular chemistry—molecular information and the design of supramolecular materials. *Makromol. Chem. Macromol. Symp.* **69**, 1–17.
12. C.J. Suckling, K.E. Suckling, and C.W. Suckling (1979), *Chemistry Through Models*. Cambridge University Press, Cambridge.
13. S. Sasaki, M. Shionoya, and K. Koga (1985), Functionalized crown ether as an approach to the enzyme model for the syntheses of peptides. *J. Am. Chem. Soc.* **107**, 3371–3372.
14. B.A. Boyce, A. Carroy, J.-M. Lehn, and D. Parker (1984), Heterotopic ligands: Synthesis and complexation properties of phosphine-functionalized dipodal macrocycles. *Chem. Commun.* 1546–1548.

15a. T.R. Cech (1987), The chemistry of self-splicing RNA and DNA enzymes. *Science* **236**, 1532–1539.

15b. T.R. Cech (1986), RNA as an enzyme. *Sci. Amer.* Dec., pp. 64–75.

Chapter 2

16. R.J.P. Williams (1987), Missing information in bio-inorganic chemistry. *Coord. Chem. Rev.* **79**, 175–193.

17. G.M. Bodner (1986), Metabolism, parts I, II and III. *J. Chem. Educ.* **63**, 566–570, 673–677, and 772–775.
18. F. Lipmann (1973), Nonribosomal polypeptide synthesis on polyenzyme templates. *Acc. Chem. Res.* **6**, 361–367.
19. J.A. Walder, R.Y. Walder, M.J. Heller, S.M. Freier, R.L. Letsinger, and I.M. Klotz (1979), Complementary carrier peptide synthesis: General strategy and implications for prebiotic origin of peptide synthesis. *Proc. Nat. Acad. Sci. USA*, **76**, 51–55.
20a. D.S. Kemp, D.J. Kerkman, S.L. Leung, and G. Hanson (1981), Intramolecular O,N-acyl transfer via cyclic intermediates of nine and twelve members. Models for extensions of the amine capture strategy for peptide synthesis. *J. Org. Chem.* **46**, 490–498.
20b. D.S. Kemp and N. Fotouhi (1987), Peptide synthesis by prior thiol capture V. The scope and control of disulfide interchange during the acyl transfer step. *Tetrahedron Lett.* **28**, 4637–4640.
21. K. Johnsson, R.K. Allemann, H. Widmer, and S.A. Benner (1994), Synthesis, structure and activity of artificial, rationally designed catalytic polypeptides. *Nature* **365**, 530–532.
22. J.P. Vigneron, H. Kagan, and A. Horeau (1968), Synthèse asmétrique de l'acide aspartique optiquement pur. *Tetrahedron Lett.* 5681–5683.
23. E.J. Corey, R.J. McCaully, and H.S. Sachdev (1970), Studies on the asymmetric synthesis of α-amino acids. I. A new approach. *J. Amer. Chem. Soc.* **92**, 2476–2488.
24. M.D. Fryzuk and B. Bosnich (1977), Asymmetric synthesis. Production of optically active amino acids by catalytic hydrogenation. *J. Amer. Chem. Soc.* **99**, 6262–6267.
25. M.D. Fryzuk and B. Bosnich (1979), Asymmetric synthesis. Preparation of chiral methyl chiral lactic acid by catalytic asymmetric hydrogenation. *J. Amer. Chem. Soc.* **101**, 3043–3049.
26. N. Takaishi, H. Imai, C.A. Bertelo, and J.K. Stille (1978), Transition metal catalyzed asymmetric organic Synthesis via polymer-attached optically active phosphine ligands. Synthesis of *R* amino acid and hydratopic acid by hydrogenation. *J. Amer. Chem. Soc.* **100**, 264–267.
27. M.E. Wilson and G.M. Whitesides (1978), Conversion of a protein to a homogeneous asymmetric hydrogenation catalyst by site-specific modification with a diphosphinerhodium(I) moiety. *J. Amer. Chem. Soc.* **100**, 306–307.
28. M.E. Wilson, R.G. Nuzzo, and G.M. Whitesides (1978), Bis(2-diphenylphosphinoethyl)amine. A flexible synthesis of functionalized chelating diphosphines. *J. Amer. Chem. Soc.* **100**, 2269–2270.
29. W. Marckwald (1904), Veber asymmetrische synthese. *Ber. Chem*, **37**, 1368–1370.
30. D.A. Evans (1984), Stereoselective alkylation reactions of chiral metal enolates. In: *Asymmetric Synthesis* (J.D. Morrison, Ed.), Vol. 3. Academic Press, New York.
31. A.I. Meyers, G. Knaus, and K. Kamata (1974), Synthesis via 2-oxazolines. IV. An asymmetric synthesis of 2-methylalkanoic acid from a chiral oxazoline. *J. Amer. Chem. Soc.* **96**, 268–270.

32. M. Kitamoto, K. Miroi, S. Terashima, and S. Yamada (1974), Stereochemical studies. XXIV. Asymmetric synthesis of 2-alkylcyclohexamines via optically active lithioenamines. *Chem. Pharm. Bull.* **22**, 459–464.
33. T. Katsuki and K.B. Sharpless (1980), The first practical method of asymmetric epoxidation. *J. Amer. Chem. Soc.* **102**, 5974–5976.
34. D. Seebach, R. Imwinkelried, and T. Weber (1986), in R. Scheffold, Ed., *Modern Synthetic Methods*. Springer-Verlag, Berlin.
35. D.A. Evans, T.C. Britton, R.L. Dorow, and J.F. Dellaria (1986), Stereoselective aminatin of chiral enolates. A new approach to the asymmetric synthesis of α-amino and derivatives. *J. Amer. Chem. Soc.* **108**, 6395–6397.
36. F. Schöllkopf and U. Groth (1981), Enantioselective synthesis of (*R*)-α-vinylamino acids. *Angew. Chem. Int. Ed. Engl.* **20**, 977–978.
37. M.B. Buergi, J.D. Dunitz, J.-M. Lehn, and G. Wipff (1974), Stereochemistry of reaction paths at carbonyl centres. *Tetrahedron* **30**, 1563–1572.
38. F.M. Menger (1983), Directionality of organic reactions in solution. *Tetrahedron* **39**, 1013–1040.
39. J.E. Baldwin (1976), Rules for ring closure. *Chem. Commun.* 734–736.
40. D.R. Storm and D.E. Koshland, Jr. (1970), A source for the special catalytic power of enzymes: Orbital steering. *Proc. Nat. Acad. Sci. USA* **66**, 445–452.
41. F.M. Menger and L.E. Glass (1980), Contribution of orbital alignment to organic and enzymatic reactivity. *J. Amer. Chem. Soc.* **102**, 5404–5406.
42. S. Scheiner, W.N. Lipscomb, and D.A. Kleier (1976), Molecular orbital studies of enzyme activity. 2. Nucleophilic attack and carbonyl systems with comments on orbital steering. *J. Amer. Chem. Soc.* **98**, 4770–4777.
43. D. Seebach and R. Naef (1981), Enantioselective generation and diastereoselective reactions of chiral enolates derived from α-hetero-substituted carboxylic acids. *Helv. Chim. Acta* **64**, 2704–2708.
44. D. Seebach, M. Boes, R. Naef, and W.B. Schweizer (1983), Alkylation of amino acids without loss of the optical activity: Preparation of α-substituted proline derivatives. A case of self-reproduction of chirality. *J. Amer. Chem. Soc.* **105**, 5390–5398.
45. R.M. Wenger (1985), Synthesis of cyclosporine analogues: Structural requirements for immunosuppressive activity. *Angew. Chem. Int. Ed. Engl.* **24**, 77–85.
46. D.A. Evans and A.E. Weber (1986), Asymmetric glycine enolate aldol reactions: Synthesis of cyclosporine's unusual amino acid, MeBmt. *J. Amer. Chem. Soc.* **108**, 6757–6761.
47. D.A. Evans and T.C. Britton (1987), Electrophilic azide transfer to chiral enolates. A general approach to the asymmetric synthesis of α-amino acids. *J. Amer. Chem. Soc.* **109**, 6881–6883.
48. G.M. Coppola and M.F. Schuster (1987), *Asymmetric Synthesis*. J. Wiley & Sons, New York.
49. M.G. Finn and K.B. Sharpless (1985), On the mechanism of asymmetric epoxidation with titanium-tartrate catalysts. In: (J.D. Morrison, Ed.) *Asymmetric Synthesis*, Vol. 5. Academic Press, Inc., New York.
50. E.N. Jacobsen, I. Markó, W.S. Mungall, G. Schröder, and K.B. Sharpless (1988), Asymmetric dihydroxylation via ligand-accelerated catalysis. *J. Amer. Chem. Soc.* **110**, 1968–1970.

51. B. Bosnish (1984), Asymmetric Catalysis. *Chem. Britain* 808–810.
52. H. Brunner (1983), Rhodium catalysts for enantioselective hydrosilylation. A new concept in the development of asymmetric catalyses. *Angew. Chem. Int. Ed. Engl.* **22**, 897–907.
53. P.A. MacNeil, K.N. Roberts, and B. Bosnish (1981), Asymmetric synthesis. Asymmetric catalytic hydrogenation using chiral chelating six-membered ring bisphosphines. *J. Amer. Chem. Soc.* **103**, 2273–2380.
54. G.W. Parshall and W.A. Nugent (1988), Making pharmaceuticals via homogeneous catalysis. *ChemTech.* 184–190.
55. J.W. ApSimon and T. Lee Colhier (1984), Recent Advances in Asymmetric Synthesis II. *Tetrahedron* **42**, 5127–5254.
56. M. Levin-Pinto, J.L. Morgat, P. Fromageot, D. Meyer, J.L. Poulin, and H.B. Kagan (1982), Asymmetric tritiation of N-acetyl dehydrophenylalanyl(*S*)-phenylalanine (methyl ester) catalyzed with a rhodium (+)-DIOP complex. *Tetrahedron* **38**, 119–123.
57. R. Noyori, T. Ohkuma, M. Kitamura, H. Takaya, N. Sayo, H. Kumobayashi, and S. Akutagawa (1987), Asymmetric hydrogenation of β-ketone carboxylic esters. A practical, purely chemical access to β-hydroxy esters in high enantiomeric purity. *J. Amer. Chem. Soc.* **109**, 5856–5857.
58. T.T. Ngo and G. Tunnicliff (1981), Inhibition of enzymic reactions by transition-state analogs. An approach for drug design. *Gen. Pharmac.* **12**, 129–138.
59. K.T. Douglas (1983), Transition-state analogues in drug design. *Chem. & Ind.* 311–315.
60. G.M. Blackburn (1981), Phosphonates as analogues of biological phosphates. *Chem. & Ind.* 134–138.
61. P.A. Bartlett and C.K. Marlowe (1983), Phosphonamidates as transition-state analogue inhibitors of thermolysin. *Biochemistry* **22**, 4618–4624.
62. N.E. Jacobsen and P.A. Bartlett (1981), A phosphonamidate dipeptide analogue as an inhibitor of carboxypeptidase A. *J. Amer. Chem. Soc.* **103**, 654–657.
63. P.P. Giannousis and P.A. Bartlett (1987), Phosphorous amino acid analogues as inhibitors of leucine aminopeptidase. *J. Med. Chem.* **30**, 1603–1609.
64. M.L. Sinnott (1979), Inimitable enzymes? *Chem. Britain* **15**, 293–297.
65. S.D. Copley and J.R. Knowles (1985), The uncatalyzed Claisen rearrangement of chorismate to prephenate prefers a transition-state of chain-like geometry. *J. Amer. Chem. Soc.* **109**, 5306–5308.
66. S.D. Copley and J.R. Knowles (1987), The conformational equilibrium of chorismate in solution. Implications for the mechanism of the non-enzymic and the enzyme-catalyzed rearrangement of chorismate to prephenate. *J. Amer. Chem. Soc.* **109**, 5008–5013.
67. P.A. Bartlett and C.R. Johnson (1985), An inhibitor of chorismate mutase resembling the transition-state conformation. *J. Amer. Chem. Soc.* **107**, 7792–7793.
68. J. Jacobs, P.G. Schultz, R. Sugasawara, and M. Powell (1987), Catalytic antibodies. *J. Amer. Chem. Soc.* **109**, 2174–2176.

69. A. Tramontano, K.D. Janda, and R.A. Lerner (1986), Chemical reactivity at an antibody binding site elicited by mechanistic design of a synthetic antigen. *Proc. Natl. Acad. Sci.* **83**, 6736–6740.
70. S.J. Pollack, J.W. Jacobs, and P.G. Schultz (1986), Selective chemical catalysis by an antibody. *Science* **234**, 1570–1573.
71. A.D. Napper, S.J. Benkovic, A. Tramontano, and R.A. Lerner (1987), A stereospecific cyclization catalyzed by an antibody. *Science* **273**, 1041–1043.
72. V. Morell (1993), Enzyme may blunt cocaine's action. *Science* **259**, 1828.
73. R.M. Baum (1993), Antibodies catalyze reactions otherwise difficult to achieve. *Chem. & Eng. News*, April 19, 33–35.
74. J.-L. Reymond, J.-L. Reber, and R.A. Lerner (1994), Enantioselective, multigram-scale synthesis with a catalytic antibody. *Angew. Chem. Int. Ed. Engl.* **33**, 475–476.
75. D.A. Campbell, B. Gong, L.M. Kochersperger, S. Yonkovich, M.A. Gallop, and P.G. Schultz (1994), Antibody-catalyzed prodrug activation. *J. Amer. Chem. Soc.* **116**, 2165–2166.
76. K.E. Neet and D.E. Koshland (1966), The conversion of serine at the active site of subtilisin to cysteine a "chemical mutation." *Proc. Natl. Acad. Sec. USA* **56**, 1606–1611.
77. L. Polgar and M.L. Bender (1966), A new enzyme containing a synthetically formed active site. Thiol-subtilisin. *J. Amer. Chem. Soc.* **88**, 3153–3154.
78. P.I. Clark and G. Lowe (1978), Conversion of the active-site cysteine residue of papain into a dihydro-serine, a serine and a glycine residue. *Eur. J. Biochem.* **84**, 293–299.
79. J.H. Parish and M.J. McPherson (1987), Chemical and biochemical manipulation of DNA and the expression of foreign genes in micro-organisms. *Natural Prod. Rep.* **4**, 139–156.
80. M.J. McPherson and J.H. Parish (1987), Applications of recombinant DNA in biotechnology. *Natural Prod. Rep.* **4**, 205–224.
81. R. Baum (1986), Enzyme research advances protein engineering. *Chem. Eng. News*, Oct. 13, 23–25.
82. D.S. Lawrence and E.T. Kaiser (1984), Chemical mutation of enzyme active sites. *Science* **226**, 505–511.
83. J.A. Gerlt (1987), Relationships between enzymatic catalysis and active site structure revealed by applications of site-directed mutagenesis. *Chem. Rev.* **87**, 1079–1105.
84. J.T. Slama, S.R. Oruganti, and E.T. Kaiser (1981), Semisynthetic enzymes: Synthesis of a new flavopapain with high catalytic efficiency. *J. Amer. Chem. Soc.* **103**, 6211–6213.
85. H. Zemel (1987), A new semisynthetic esterase. *J. Amer. Chem. Soc.* **109**, 1875–1876.
86. G. Henderson and I. McFedzean (1985), Opioids—a review of recent developments. *Chem. Britain* 1094–1097.
87. G.J. Hite (1981), In: *Principles of medicinal chemistry* (W.O. Foye, Ed.), Chap. 12. Lea & Febiger, USA.
88. B. Belleau (1982), Stereoelectronic regulation of the opiate receptor. Some conceptual problems. In: *Chemical Regulation of Biological Mechanisms*. Burlington House, London.

89. J. Hughes, T.W. Smith, H.N. Kosterlitz, L.A. Forthergill, and B.A. Morgan (1975), Identification of two related pentapeptides from the brain with potent opiate agonist activity. *Nature* **258**, 577–579.
90. B.P. Roques, C. Garbay-Jaureguiberry, R. Oberlin, M. Anteunis, and A.K. Lala (1976), Conformation of Met^5-enkephalin determined by high field PMR spectroscopy. *Nature* **262**, 778–779.
91. C. Garbay-Jaureguiberry, D. Marion, E. Fellion, and B.P. Roques (1982), Refinement of conformational frequencies of Leu enkephalin and Tyr-Gly-Phe by ^{15}N-NMR. *Int. J. Pet. Protein Res.* **20**, 443–450.
92. I.L. Karle, J. Karle, D. Mastropaolo, A. Camerman, and N. Camerman (1983), [Leu^5]enkephalin: Four cocrystallizing conformers with extended backbones that form an antiparallel β-sheet. *Acta Crystallogr. Sect. B* **39**, 625–637.
93. H. Dugas, P. Mayers, and P. Couillard (1994), A molecular modeling study of the direct effect of Leu-enkephalin on *Amoeba proteus* membrane in the presence of ions. *Endocrine* **2**, 651–658.
94. P.W. Schiller and J. Di Maio (1982), Opiate receptor subclasses differ in their conformational requirments. *Nature* **297**, 74–76.
95. P.W. Schiller (1986). In: R.S. Rapaka, G. Barnett, and R.L. Hawks, Eds., Opioid Peptides: Medicinal Chemistry, pp. 291–311, *UIDA Research Monograph 69*.
96. P.S. Portoghese, D.L. Larson, G. Ronsisvalle, P.W. Schiller, T.M.D. Nguyen, C. Lemieux, and A.E. Takemori (1987), Hybrid bivalent ligands with opiate and enkephalin pharmacophores. *J. Med. Chem.* **30**, 1991–1994.

Chapter 3

97. A. De Mesmaeker, A. Waldner, J. Lebreton, P. Hoffmann, V. Fritsch, R.M. Wolf, and S.M. Freier (1994), Amides as a new type of backbone modification in oligonucleotides. *Angew. Chem. Int. Ed. Engl.* **33**, 226–229.
98. R.L. Baughn, O. Adelsteinsson, and G.M. Whitesides (1978), Large-scale enzyme-catalyzed synthesis of ATP from adenosine and acetyl phosphate. Regeneration of ATP from AMP. *J. Amer. Chem. Soc.* **100**, 304–306.
99. W.A. Blättler and J.R. Knowles (1979), The stereochemical course of glycerol kinase-phosphoryl transfer from chiral [γ-(*S*)-^{16}O, ^{17}O, ^{18}O]ATP, *J. Amer. Chem. Soc.*, **101**, 510–511.
100. G. Lowe, G. Tansley, and P.M. Cullis (1982), Synthesis of adenosine 5′(*R*)-α-^{17}O]triphosphate. *J. Chem. Soc. Chem. Commun.* 595–598.
101. W.A. Blättler and J.R. Knowles (1979), Stereochemical course of phosphokinases. The use of adenosine [γ-(*S*)-^{16}O, ^{17}O, ^{18}O] triphosphate and the mechanistic consequences for the reactions catalyzed by glycerol kinase, hexokinase, pyruvate kinase and acetate kinase. *Biochemistry*, **18**, 3927–3933.
102. P. Greengard and J.A. Nathanson (1977), "Second messengers" in the brain. *Sci. Amer.* **237**, 108–119.
103. J.A. Gerlt, N.I. Gutterson, P. Dalta, B. Belleau, and C.L. Penney (1980). Thermochemical identification of the structural factors responsible for the thermodynamic instability of 3′,5′-cyclic nucleotides. *J. Amer. Chem. Soc.* **102**, 1655–1660.

104. R. Singleton, Jr. (1973), Bioorganic chemistry of phosphorous. *J. Chem. Educ.* **50**, 538–544.
105. F.H. Westheimer (1968), Pseudo-rotation in the hydrolysis of phosphate esters. *Acc. Chem. Res.* **1**, 70–78.
106. C.A. Deakyne and L.C. Allen (1979), Role of active-site residues in the catalytic mechanisms of ribonuclease A. *J. Amer. Chem. Soc.* **101**, 3951–3959.
107. D.A. Usher, E.S. Erenrich, and F. Eckstein (1972), Geometry of the first step in the action of ribonuclease A. *Proc. Nat. Acad. Sci. USA* **69**, 115–118.
108. D.A. Usher and D.I. Richardson, Jr. (1970), Absolute stereochemistry of the second step of ribonuclease action. *Nature* **228**, 663–665.
109. R. Breslow and M. Labelle (1986), Sequential general base-acid catalysis in the hydrolysis of RNA by imidazole. *J. Amer. Chem. Soc.* **108**, 2655–2659.
110. K. Taira (1987), Stereoelectronic control in the hydrolysis of RNA by imidazole. *Bull. Chem. Soc. Jpn.* **60**, 1903–1909.
111. F. Eckstein (1983), Phosphorothioate analogues of nucleotides. Tools for the investigation of biological processes. *Angew. Chem. Int. Engl. Ed.* **22**, 423–506.
112. P.A. Frey (1982), Stereochemistry of enzymatic reactions of phosphates. *Tetrahedron* **38**, 1547–1567.
113. D. Yee, V.W. Armstrong, and F. Eckstein (1979), Mechanistic studies on deoxyribonucleic acid dependent ribonucleic acid polymerase from *E. coli* using phosphorothioate analogues. I. Initiation and pyrophosphate exchange reactions. *Biochemistry* **18**, 4116–4120.
114. S. Mehdi and J.A. Gerlt (1982), Oxygen chiral phosphodiesters. 7. Stereochemical course of the reaction catalyzed by staphylococcal nuclease. *J. Amer. Chem. Soc.* **104**, 3223–3225.
115. S.L. Buchwald and J.R. Knowles (1980), Determination of the absolute conformation of [^{16}O, ^{17}O, ^{18}O] phosphate monoesters by using ^{31}P-NMR. *J. Amer. Chem. Soc.* **102**, 6601–6602.
116. W.G. Wadsworth and W.S. Wadsworth, Jr. (1983), A cyclic phosphate: base and metal catalyzed ring opening. *J. Amer. Chem. Soc.* **105**, 1631–1637.
117. K.W.Y. Abell and A.J. Kirby (1983), Intramolecular general acid catalysis of intramolecular nucleophilic catalysis of the hydrolysis of a phosphate diester. *J. Chem. Soc. Perkin Trans. II*, 1171–1174.
118. D.G. Gorenstein, R. Rowell, and J.B. Findlay (1980), Stereoelectronic effects in reactions of epimeric 2-aryloxy-2-oxy-1,3,2-dioxaphosphorinanes and oxozaphosphorinanes. *J. Amer. Chem. Soc.* **202**, 5077–5081.
119. M. Feughelman, R. Langridge, W.E. Seeds, A.R. Stokes, H.R. Wilson, C.W. Hooper, M.H.F. Wilkins, R.K. Barclay, and L.D. Hamilton (1955), Molecular structure of deoxyribose nucleic acid and nucleoprotein. *Nature* **175**, 834–838.
120. J.D. Watson and F.H.C. Crick (1973), The structure of DNA. *Cold Spring Harbor Symp. Quant. Biol.* **18**, 123–131.
121. K.T. Douglas (1984), Anticancer drugs, DNA as a target. *Chem. & Ind.* 738–742.
122. K.T. Douglas (1984), Anticancer drugs, DNA-intercalation and free radical attack. *Chem. & Ind.* 766–771.

123. B.E. Bowler, L.S. Hollis, and S.J. Lippard (1984), Synthesis of DNA binding and photomimicking properties of acridine orange linked by a polymethylene ether to (1,2-diaminoethane) dichloroplatinum(II). *J. Amer. Chem. Soc.* **106**, 6102–6104.
124. J.W. Lown and C.C. Hanstock (1982), Structure and function of the antitumor antibiotic carzinophilin A: The first natural intercalative bisalkylator. *J. Amer. Chem. Soc.* **104**, 3213–3214.
125. A.H.J. Wang, G. Ughetto, G.J. Quigley, T. Hakoshima, G.A. van der Marel, J.H. van Boom, and A. Rich (1984), The molecular structure of a DNA-triostin A complex. *Science* **225**, 1115–1121.
126. D. Léon, C. Garbay-Jaureguiberry, J.B. LePecq, and B.P. Roques (1985), A new and improved synthesis of a potential antitumor 7H-pyrido [4,3-c]carbazole analog. *Tetrahedron Lett.* **26**, 4929–4932.
127. G.J. Atwell, G.M. Stewart, W. Leupin, and W.A. Denny (1985), A diacridine derivative that binds by bisintercalation at two contiguous sites on DNA. *J. Amer. Chem. Soc.* **107**, 4335–4337.
128. G.J. Atwell, B.C. Baguley, D. Wilmanska, and W.A. Denny (1986), Potential antitumor agents. 45. Synthesis, DNA-binding intercalation and biological activity of triacridine derivatives. *J. Med. Chem.* **29**, 69–74.
129. J.W. Lown (1982), Newer approaches to the study of the mechanisms of action of antitumor antibiotics. *Acc. Chem. Res.* **15**, 381–387.
130. P.B. Dervan and M.M. Becker (1968), Molecular recognition of DNA by small molecules. Synthesis of bis(methidium) spermine. a DNA polyintercalating molecule. *J. Amer. Chem. Soc.* **100**, 1968–1970.
131. J. Stubbe and J.W. Kozarich (1987), Mechanisms of bleomycin-induced DNA degradation. *Chem. Rev.* **87**, 1107–1136.
132. R.P. Hertzberg and P.B. Dervan (1982), Cleavage of double helical DNA by (methidium-EDTA) iron(II). *J. Amer. Chem. Soc.* **104**, 313–315.
133. P.G. Schultz, J.S. Taylor, and P.B. Dervan (1982), Design and synthesis of a sequence-specific DNA cleaving molecule. (Distamycin-EDTA) iron(II). *J. Amer. Chem. Soc.* **100**, 6861–6863.
134. M.W. van Dykes, R.P. Hertzberg, and P.B. Dervan (1982), Map of distamycin, netropsin, and actinomycin binding sites on heterogeneous DNA: DNA cleavage-inhibition patterns with methidiumpropyl-EDTA.Fe(II). *Proc. Natl. Acad. Sci. USA*, **79**, 5470–5474.
135. P.B. Devan and J. Sluka (1980), In: *Internat. Kyoto Conf. on Organic Chem.* Processing, "New Synthetic Methodology and Functionally Interesting Compounds," pp. 307–322. Elsevier, Amsterdam.
136. J.H. Griffin and P.B. Dervan (1987), Metalloregulation in the sequence specific binding of synthetic molecules to DNA. *J. Amer. Chem. Soc.* **109**, 6840–6842.
137. R.S. Youngquist and P.B. Dervan (1987), A synthetic peptide binds 16 base pairs of A,T double helical DNA. *J. Amer. Chem. Soc.* **109**, 7564–7566.
138. M. Mrksich and P.B. Dervan (1994), Design of a covalent peptide heterodimer for the sequence recognition in the minor groove of double-helical DNA. *J. Amer. Chem. Soc.* **116**, 3663–3664.
139. M.G. Oakley, K.D. Turnbull, and P.B. Dervan (1994), Synthesis of a hybrid protein containing the iron-binding ligand of bleomycin and the DNA-binding domain of Hin. *Bioconjugate Chem.* **5**, 242–247.

140. R.M. Baum (1989), Transition-metal complexes probe DNA conformation. *Chem. & Eng. News*, June **12**, 22–25.
141. W.A. Saffran, J. Goldenberg, and C.R. Cantor (1982), Site-directed psoralen crosslinking of DNA. *Proc. Natl. Acad. Sci. USA* **79**, 4594–4598.
142. G.D. Cimino, H.B. Gamper, S.T. Isaacs, and J.E. Hearst (1985), Psoralens as photoactive probes of nucleic acid structure and function: Organic chemistry, photochemistry and biochemistry. *Ann. Rev. Biochem.* **54**, 1151–1193.
143. A. Basak and H. Dugas (1986), Design and synthesis of DNA intercalating crown ether molecules. *Tetrahedron Lett.* **27**, 3–6.
144. R.L. Letsinger and M.S. Schott (1981), Selectivity in binding a phenanthridinium-dinucleotide derivative to homopolynucleotides. *J. Amer. Chem. Soc.* **103**, 7394–7396.
145. A.D. Hamilton and D. Van Engen (1987), Induced fit in synthetic receptors: Nucleotide base recognition by a "molecular hinge." *J. Amer. Chem. Soc.* **109**, 5035–5036.
146. S.C. Hirst, P. Tecilla, S.J. Geib, E. Fan, and A.D. Hamilton (1992), Molecular recognition of phosphate esters: a balance of hydrogen bonding and proton transfer interactions. *Israel J. Chem.* **32**, 105–111.
147. T. R. Kelly and M.P. Maguire (1987), A receptor for the oriented binding of uric acid type molecules. *J. Amer. Chem. Soc.* **109**, 6549–6551.
148. U. Pindur, M. Haber, and K. Sattler (1993), Antitumor active drugs as intercalators of deoxyribonucleic acid. *J. Chem. Educ.* **70**, 263–272.

Chapter 4

149. J. Hine (1978), Bifunctional catalysis of α-hydrogen exchange of aldehydes and ketones. *Acc. Chem. Res.* **11**, 1–7.
150. T.C. Bruice and S. Benkovic (1966) *Bioorganic Mechanisms*, Vol. 1, p. 134. Benjamin, New York.
151. P. Cossee (1962), Stereoregularity in heterogeneous Ziegler-Natta catalysis. *Trans. Faraday Soc.* **58**, 1226–1232.
152. W.W. Cleland (1975), What limits the rate of an enzyme-catalyzed reaction? *Acc. Chem. Res.* **8**, 145–151.
153. K.R. Hanson and I.A. Rose (1975), Interpretations of enzyme reaction stereospecificity. *Acc. Chem. Res.* **8**, 1–10.
154. G.E. Schultz and R.H. Schimer (1979), *Principles of protein structure*, Springer-Verlag. New York.
155. R.J.D. Miller (1994), Energetics and dynamics of deterministic protein motion. *Acc. Chem. Res.* **27**, 145–150.
156. A. Ferscht (1977), *Enzyme Structure and Mechanism*, pp. 44–48. Freeman, San Francisco.
157. P.A. Srere (1984), Why are enzymes so big? *Trends Biochem. Sci.* **9**, 387–390.
158. J.R. Knowles and W.J. Albery (1977), Perfection in enzyme catalysis: The energetics of triosephosphate isomerase. *Acc. Chem. Res.* **10**, 105–111.
159. W.L. Alworth (1972), *Stereochemistry and Its Application in Biochemistry*, Chap. 3. Wiley-Interscience, New York.
160. A.G. Ogston (1948), Interpretation of experiments on metabolic processes, using isotopic tracer elements. *Nature* **162**, 963.

161. F.A. Loewus, F.H. Westheimer, and B. Vennesland (1953), Enzymatic synthesis of the enantiomorphs of ethanol-1-*d*. *J. Amer. Chem. Soc.* **75**, 5018–5023.
162. D.M. Blow (1976), Structure and mechanism of chymotrypsin. *Acc. Chem. Res.* **9**, 145–152.
163. M.I. Page (1981), Enzymes-binding energy, catalysis and inhibition. *Chem. & Ind.* 144–150.
164. A.R. Fersht and A.J. Kirby (1980), Intramolecular catalysis and the mechanism of enzyme action. *Chem. Br.* **16**, 136–142.
165. T.C. Bruice and V.K. Pandit (1960), The effect of general substitution ring size and rotamer distribution on the intramolecular nucleophilic catalysis of the hydrolysis of monophenyl esters of dibasic acids and the solvolysis of the the intermediate anhydrides. *J. Amer. Chem. Soc.* **85**, 5858–5865.
166. W.P. Jencks (1975), Binding energy, specificity, and enzymic catalysis: The circle effect. *Adv. Enzymol.* **43**, 219–410.
167. F.M. Menger (1993), Enzyme reactivity from an organic perspective. *Acc. Chem. Res.* **26**, 206–212.
168. C.G. Swain and J.F. Brown (1952), Concerted displacement reactions. VII. The mechanism of acid base catalysis in non-aqueous solvents. *J. Amer. Chem. Soc.* **74**, 2534–2537.
169. K.A. Engdahl, H. Bivehed, P. Ahberg, and W.H. Saunders, Jr. (1983), Ratecontrolling two-proton transfer coupled with heavy-atom motion in the 2-pyridinone-catalyzed mutarotation of tetramethylglucose. Experimental and calculated deuterium isotopes effects. *J. Amer. Chem. Soc.* **105**, 4767–4774.
170. T. Higuchi, H. Takechi, I.H. Pitman, and H.L. Fung (1971), Intramolecular bifunctional facilitation in complex molecules, combined nucleophilic and general acid participation in hydrolysis of hexachlorophene monosuccinate. *J. Amer. Chem. Soc.* **93**, 539–540.
171. B.A. Cunningham and G.L. Schmir (1966), Iminolactones. II. Catalytic Effects on the nature of the products of hydrolysis. *J. Amer. Chem. Soc.* **88**, 551–558.
172. Y.N. Lee and G.L. Schmir (1978), Concurrent general acid and general base catalysis in the hydrolysis of an amide ester. 1. Monofunctional catalysis. *J. Amer. Chem. Soc.* **100**, 6700–6707.
173. Y.N. Lee and G.L. Schmir (1979), Concurrent general acid and general base catalysis in the hydrolysis of an imidate ester. 2. Bifunctional catalysis. *J. Amer. Chem. Soc.* **101**, 3026–3035.
174. W.P. Jencks (1972), Requirements for general acid-base catalysis of complex reactions. *J. Amer. Chem. Soc.* **94**, 4731–4732.
175. D.M. Blow, J.J. Birktoft, and B.S. Hartley (1969), Role of a buried acid group in the mechanism of action of chymotrypsin. *Nature* **221**, 337–340.
176. P.B. Sigler, D.M. Blow, B.W. Matthews, and R. Henderson (1968), Structure of crystalline α-chymotrypsin II. A preliminary report including a hypothesis for the activation mechanism. *J. Mol. Biol.* **35**, 143–164.
177. R.M. Garavito, M.G. Rossmann, P. Argos, and W. Eventoff (1977), Convergence of active center geometries. *Biochemistry* **16**, 5065–5071.
178. H.T. Wright (1973), Activation of chymotrypsinogen-A. An hypothesis based upon comparison of the crystal structures of chymotrypsinogen-A and α-chymotrypsin, *J. Mol. Biol.* **79**, 13–23.

179. M.W. Hunkapiller, M.D. Forgac, and J.H. Richards (1976), Mechanism of action of serine proteases: Tetrahedral intermediate and concerted proton transfer. *Biochemistry* **15**, 5581–5588.
180. H. Kaplan, V.B. Symonds, H. Dugas, and D.R. Whitaker (1970), A comparison of properties of α-lytic protease of *Sorangium* sp. and porcine elastase. *Can. J. Chem.* **47**, 649–658.
181. M.W. Hunkapiller, S.H. Smallcombe, D.R. Whitaker, and J.H. Richards (1973), Carbon nuclear magnetic resonance studies of the histidine residue in α-lytic protease. Implication for the catalytic mechanism of serine proteases. *Biochemistry* **12**, 4732–4743.
182. J.L. Markley and I.B. Ibañez (1978), Zymogen activation in serine proteinases. Proton magnetic resonance pH titration studies of the two histidines of bovine chymotrypsinogen A and chymotrypsin Aα. *Biochemistry* **17**, 4627–4639.
183. G. Robillard and R.G. Shulman (1972), High resolution nuclear magnetic resonance study of the histidine–aspartate hydrogen bond in chymotrypsin and chymotrypsinogen. *J. Mol. Biol.* **71**, 507–511.
184. G.D. Brayer, L.T.J. Delbaere, and M.N.G. James (1979), Molecular structure of the α-lytic protease from *Myxobacter* 495 at 2.8 Å resolution. *J. Mol. Biol.* **131**, 743–775.
185. W.W. Bachovchin and J.D. Roberts (1978), Nitrogen-15 nuclear magnetic resonance spectroscopy. The state of histidine in the catalytic triad of α-lytic protease. Implications for the charge-relay mechanism of peptide bond cleavage by serine proteases. *J. Amer. Chem. Soc.* **100**, 8041–3047.
186. M. Komiyama and M.L. Bender (1979), Do cleavages of amides by serine proteases occur through a stepwise pathway involving tetrahedral intermediates? *Proc. Nat. Acad. Sci. USA* **76**, 557–560.
187. M.A. Porubcan, W.M. Westler, I.B. Ibañez, and J.L. Markley (1979), (Diisopropylphosphoryl) serine proteinases. Proton on phosphorus-31 nuclear magnetic resonance-pH titration studies. *Biochemistry* **18**, 4108–4115.
188. J. Kraut (1977), Serine proteases: Structure and mechanism of catalysis. *Annu. Rev. Biochem.* **46**, 331–358.
189. W.W. Bachovchin (1986), ^{15}N-NMR spectroscopy of hydrogen-bonding interactions in the active site of serine proteases: Evidence for a moving histidine mechanism. *Biochemistry* **25**, 7751–7759.
190. K. Kanamori and J.D. Roberts (1983), ^{15}N-NMR studies of biological systems. *Acc. Chem. Res.* **16**, 35–41.
191. W.P. Jencks (1980), When is an intermediate not an intermediate? Enforced mechanism and general acid-base catalyzed, carbocation, carbanion, and ligand exchange reactions. *Acc. Chem. Res.* **13**, 161–169.
192. R.A. McClelland and L.J. Sanrtry (1983), Reactivity of tetrahedral intermediates. *Acc. Chem. Res.* **16**, 394–399.
193. A.L. Fink and P. Meehan (1979), Detection and accumulation of tetrahedral intermediates in elastase catalysis. *Proc. Nat. Acad. Sci. USA* **76**, 1566–1569.
194. T.C. Bruice and J.M. Slurtevant (1959), Imidazole catalysis. V. The intramolecular participation of the imidazolyl group in the hydrolysis of some esters and the amide of γ-(4-imidazolyl)-butyric acid and 4-(2′-acetoxyethyl)-imidazole. *J. Amer. Chem. Soc.* **81**, 2860–2870.

195. G.A. Rogers and T.C. Bruice (1973), Isolation of a tetrahedral intermediate in an acetyl transfer reaction. *J. Amer. Chem. Soc.* **95**, 4452–4453.
196. G.A. Rogers and T.C. Bruice (1974), Control of modes of intramolecular imidazole catalysis of ester hydrolysis by steric and electronic effects. *J. Amer. Chem. Soc.* **96**, 2463–2472.
197. G.A. Rogers and T.C. Bruice (1974), Synthesis and evaluation of a model for the so-called "charge-relay" system of the serine esterases. *J. Amer. Chem. Soc.* **96**, 2473–2480.
198. G.A. Rogers and T.C. Bruice (1974), The mechanisms of acyl group transfer from a tetrahedral intermediate. *J. Amer. Chem. Soc.* **96**, 2481–2488.
199. M.L. Schiling, H.D. Roth, and W.C. Herndon (1980), Zwitterionic adducts between a strongly electrophilic ketone and tertiary amines. *J. Amer. Chem. Soc.* **102**, 4271–4272.
200. M. Komiyama and M.L. Bender (1977), General base-catalyzed ester hydrolysis as a model of the "charge-relay" system. *Bioorg. Chem.* **6**, 13–20.
201. L.D. Byers and D.E. Koshland, Jr. (1978), On the mechanism of action of methyl chymotrypsin. *Bioorg. Chem.* **7**, 15–33.
202. C.J. Belke, S.C.K. Su, and J.A. Shafter (1971), Imidazole catalyzed displacement of an amine from an amide by a neighboring hydroxyl group. A model for the acylation of chymotrypsin. *J. Amer. Chem. Soc.* **93**, 4552–4561.
203. M.P. Gamcsik, J.P.G. Mathouse, W.U. Primrose, N.E. Mackenzie, A.J.F. Boyd, R.A. Russell, and A.I. Scott (1983), Structure and stereochemistry of tetrahedral inhibitor complexes of papain by direct NMR observation. *J. Amer. Chem. Soc.* **105**, 6324–6325.
204. W.U. Primrose, N.E. MacKenzie, J.P.G. Mathouse, and A.I. Scott (1985), ^{13}C-Nuclear magnetic resonance observations on the interaction between *p*-amidinophenylpyruvic acid and trypsin. *Bioorganic Chemistry* **13**, 335–343.
205. G.E. Hein and C. Niemann (1962), Steric course and specificity of α-chymotrysin-catalyzed reactions. I. *J. Amer. Chem. Soc.* **84**, 4487–4494.
206. G.E. Hein and C. Niemann (1962), Steric course and specificity of α-chymotrypsin-catalyzed reactions. II. *J. Amer. Chem. Soc.* **84**, 4495–4503.
207. H. Dugas (1969), The stereospecificity of subtilisin BPN′ towards 1-keto-3-carbomethoxy-1,2,3,4-tetrahydroisoquinoline. *Can. J. Biochem.* **47**, 985–987.
208. M.S. Silver and T. Sone (1968), Stereospecificity in the hydrolysis of conformationally homogeneous substrates by α-chymotrypsin. *J. Amer. Chem. Soc.* **90**, 6193–6198.
209. S.G. Cohen and R.M. Schultz (1968), The active site of *a*-chymotrypsin. *J. Mol. Biol.* **243**, 2607–2617.
210. B. Belleau and R. Chevalier (1968), The absolute conformation of chymotrypsin-bound substrates. Specific recognition by the enzyme of biphenyl asymmetry in a constrained substrate. *J. amer. Chem. Soc.* **90**, 6864–6866.
211. P. Elie (1977), Ph.D. Dissertation, McGill University, Montreal, Canada.
212. D.C. Phillips (1967), The hen egg-white lysozyme molecule. *Proc. Nat. Acad. Sci. USA* **57**, 484–495.
213. R.E. Dickerson and I. Geis (1969), *The Structure and Action of Proteins*, p. 71. Harper and Row, New York.
214. M.R. Pincus and H.A. Scheraga (1979), Conformational energy calculation of enzyme-substrate and enzyme-inhibitor complexes of lysozyme. 2. Calcula-

tion of the structures of complexes with a flexible enzyme. *Macromolecules* **12**, 633–644.

215. B. Capon and M.C. Smith (1965), Intramolecular catalysis in acetal hydrolysis. *Chem. Comm.* 523–524.
216. R. Kluger and C.H. Lam (1978), Carboxylic acid participation in amide hydrolysis. External general base catalysis and general acid catalysis in reactions of norbornenylanilic acids. *J. Amer. Chem. Soc.* **100**, 2191–2197.
217. I. Hsu, T.J. Delbaere, M.M.G. James, and T. Hofmann (1977), Penicillopepsin from *Penicillium janthinellum*. Crystal structure at 2.8 Å and sequence. Homology with porcine pepsin. *Nature* **266**, 140–145.
218. J.P. Street and R.S. Brown (1985), Biomimetic models for cysteine proteases. 1. Intramolecular imidazole catalysis of thiol ester solvolysis: A model for the deacylation step. *J. Amer. Chem. Soc.* **107**, 6084–6089.
219. P. Deslongchamps (1975), Stereoelectronic control in the cleavage of tetrahedral intermediates in the hydrolysis of esters and amides. *Tetrahedron* **31**, 2463–2490.
220. P. Deslongchamps (1977), Stereoelectronic control in hydrolytic reactions. *26th IUPAC*, Tokyo, Japan.
221. P. Deslongchamps (1977), Stereoelectronic Control in Hydrolytic Reactions. *Heterocycles* **7**, 1271–1317.
222. P. Deslongchamps (1983), *Stereoelectronic Effects in Organic Chemistry*. Pergamon Press, Oxford.
223. C.L. Perrin and O. Nunez (1986), Absence of stereoelectronic control in the hydrolysis of cyclic amidines. *J. Amer. Chem. Soc.* **108**, 5997–6003.
224. P. Deslongchamps, U.O. Cheriyan, J.-P. Pradère, P. Soucy, and R.J. Taillefer (1979), Hydrolysis and isomerization of syn unsymmetrical *N, N*-dialkylated immidate salts. Experimental evidence for conformational changes and for stereoelectronically controlled cleaves in hemi-orthoamide tetrahedral intermediates. *Nouv. J. Chim.* **3**, 343–350.
225. S.A. Bizzozero and B.O. Zweifel (1975), The importance of the conformation of the tetrahedral intermediate for the α-chymotrypsin-catalyzed hydrolysis of peptide substrates. *FEBS Lett.* **59**, 105–108.
226. D. Petkov, E. Christova, and I. Stoineva (1978), Catalysis and leaving group binding in anilide hydrolysis by chymotrypsin. *Biochim. Biophys. Acta* **527**, 131–141.
227. D.G. Gorenstein, J.B. Findlay, B.A. Luxon, and D. Kar (1977), Stereoelectronic control in carbon-oxygen and phosphorus-oxygen bond breaking processes. *Ab initio* calculation and speculations on the mechanism of ribonuclease A, staphylococcal nuclease and lysozyme. *J. Amer. Chem. Soc.* **99**, 3473–3479.
228. W.L. Mock (1976), Torsional-strain considerations in enzymology. Some applications to proteases and ensuing mechanistic consequences. *Bioorg. Chem.* **5**, 403–414.
229. A.J. Kirby (1984), Stereoelectronic effects and acetal hydrolysis. *Acc. Chem. Res.* **17**, 305–311.
230. W.N. Lipscomb (1982), Acceleration of reactions by enzymes. *Acc. Chem. Res.* **15**, 232–238.
231. C.B. Post and M. Karplus (1986), Does lysozyme follow the lysozyme pathway? An alternative based on dynamic, structural, and stereoelectronic considerations. *J. Amer. Chem. Soc.* **108**, 1317–1319.

232. K.P. Nanibiar, D.M. Stauffer, P.A. Kolodzies, and S.A. Benner (1983), A mechanistic basis for the stereoselectivity of enzymatic transfer of hydrogen from nicotinamide cofactors. *J. Amer. Chem. Soc.* **105**, 5886–5890.
233. J.P. Praly and R.U. Lemieux (1987), Influence of solvent on the magnitude of the anomeric effect. *Can. J. Chem.* **65**, 213–223.
234. D.G. Gorenstein (1987), Stereolectronic effects in biomolecules. *Chem. Rev.* **87**, 1047–1077.
235. B.J.F. Hudson (1975), Immobilized enzymes. *Chem. Ind.* pp. 1059–1060.
236. K.J. Skimer (1975), Enzymes technology. *Chem. Eng. News.* August 18, pp. 23–42.
237. K. Mosbach (1976), Applications of biochemical systems in organic chemistry. In: *Techniques of Chemistry Series* (J.B. Jones, C.J. Sih, and D. Perlman, Eds.), Vol. 10, Part II, Chap. 9. Wiley-Interscience, New York.
238. C.J. Suckling (1977), Immobilized enzymes. *Chem. Soc. Rev.* **6**, 215–233.
239. A. Pollak, R.L. Baughn, O. Adalsteinsson, and G.M. Whitesides (1978), Immobilization of synthetically useful enzymes by condensation polymerization. *J. Amer. Chem. Soc.* **100**, 302–304.
240. G.G. Guilbault and M.H. Sadar (1979), Preparation and analytical uses of immobilized enzymes. *Acc. Chem. Res.* **12**, 344–359.
241. K. Mosbach and P.O. Larsson (1970), Preparation and application of polymer-entrapped enzymes and microorganisms in microbial transformation processes with special reference to steroid 11-β-hydroxylation and Δ^1-dehydrogenation. *Biotechnol. Bioeng.* **13**, 19–27.
242. P.A. Suf, S. Kay, and M.D. Lilly (1969), The conversion of benzyl penicillin to 6-aminopenicillanic acid using an insoluble derivative of penicillin amidase. *Biotechnol. Bioeng.* **12**, 337–348.
243. G.H. Whitesides and C.H. Wong (1985), Enzymes as catalysts in synthetic organic chemistry. *Angew. Chem. Int. Ed. Engl.* **24**, 617–638.
244. J.B. Jones (1986), Enzymes in organic synthesis. *Tetrahedron* **92**, 3351–3403.
245. A.M. Klibanov (1990), Asymmetric transformations catalyzed by enzymes in organic solvents. *Acc. Chem. Res.* **23**, 114–120.
246. C.H. Wong and G.M. Whitesides (1983), Synthesis of sugars by aldolase-catalyzed condensation reactions. *J. Org. Chem.* **48**, 3199–3205.
247. W.E. Ladner and G.M. Whitesides (1984), Lipase-catalyzed hydrolysis as a route to esters of chiral epoxy alcohols. *J. Amer. Chem. Soc.* **106**, 7250–7251.
248. Y.F. Wang, C.S. Chen, G. Girdaukas, and C.J. Sih (1984), Bifunctional chiral synthons via biochemical methods. 3. Optical purity enhancement in enzymatic asymmetric catalysis. *J. Amer. Chem. Soc.* **106**, 3695–3696.
249. G. Sabbioni and J.B. Jones (1987), Enzyme in organic synthesis. 39. Preparation of chiral cyclic acid-ester and bicyclic lactones via stereoselective pig liver esterase catalyzed hydrolyses of cyclic *meso* diesters. *J. Org. Chem.* **52**, 4565–4570.
250. G. Wulff and I. Schulze (1978), Enzyme-analogue built polymers. IX. Polymers with mercapto groups of definite cooperativity. *Israel J. Chem.* **17**, 291–297.
251. K.J. Shea, E.A. Thompson, S.D. Pandey, and P.S. Beauchamp (1980), Template synthesis of macromolecules, synthesis and ammisty of function-

alized macroporous polydivinylbenzene. *J. Amer. Chem. Soc.* **102**, 3149–3155.

252. Y. Fujii, K. Matsutami, and K. Kikuchi (1985), Formation of a specific coordination cavity for a chiral amino acid by template synthesis of a polymer Schiff base cobalt(III) complex. *Chem. Commun.* 415–417.
253. C.R. Beddell (1984), Designing drugs to fit a macromolecular receptor. *Chem. Soc. Rev.* **13**, 279–319.
254. G. Wulff (1982), Selective binding to polymers via covalent bonds. The construction of chiral cavities as specific receptor sites. *Pure & Appl. Chem.* **54**, 2093–2102.
255. G. Wulff (1989), Main-chain chirality and optical activity in polymers consisting of C-C chains. *Angew. Chem. Int. Ed. Engl.* **28**, 21–37.
256. J. Rebek, Jr. (1987), Model studies in molecular recognition. *Science* **235**, 1478–1484.
257. J. Rebek, Jr., R.B. Askew, M. Killoran, D. Nemeth, and F.T. Lin (1987), Convergent functional groups. 3. A molecular cleft recognizes substrates of complementary size, shape and functionality. *J. Amer. Chem. Soc.* **109**, 2426–2431.
258. J. Rebek, Jr. and D. Nemeth (1986), Molecular recognition: Ionic and aromatic stacking interactions bind complementary functional groups in a molecular cleft. *J. Amer. Chem. Soc.* **108**, 5637–5638.
259. J. Rebek, Jr., D. Nemeth, P. Ballester, and F.T. Lin (1987), Molecular recognition: Size and shape specficity in the binding of dicarboxylic acids. *J. Amer. Chem. Soc.* **109**, 3474–3475.
260. R.D. Gandour (1981), On the importance of orientation in general base catalysis by carboxylate. *Bioorg. Chem.* **10**, 169–176.
261. J. Rebek, Jr., R.J. Duff, W.E. Gordon, and K. Parris (1986), Convergent functional groups provide a measure of stereoelectronic effects at carboxyl oxygen. *J. Amer. Chem. Soc.* **108**, 6068–6069.
262. J. Wolfe, A. Costero, and J. Rebek, Jr. (1992), Convergent functional groups XII. Arrays for catalysis. *Israel. J. Chem.* **32**, 97–104.
263. J. Rebek, Jr., B. Askew, P. Ballester, C. Buhr, S. Jones, D. Nemeth, and K. Williams (1987), Molecular recognition: Hydrogen bonding and stacking interactions stabilize a model for nucleic acid structure. *J. Amer. Chem. Soc.* **109**, 5033–5035.
264. A. Galan, J. de Mendoza, C. Toiron, M. Bruix, G. Deslongchamps, and J. Rebek, Jr. (1991), A synthetic receptor for dinucleotides. *J. Amer. Chem. Soc.* **113**, 9424–9425.
265. J. de Mendoza (1993), Molecular recognition of biomolecules: from amino acids to nucleic acids. *An. Quim.* **89**, 57–62.
266. S.C. Zimmerman (1991), Molecular tweezers: synthetic receptors for π-sandwich complexation of aromatic substrates. In: (H. Dugas, Ed.), *Bioorganic Chemistry Frontiers*. Vol 2. Springer-Verlag, New York.

Chapter 5

267. R. Breslow (1972), Biomimetic chemistry. *Chem. Soc. Rev.* **1**, 553–580.
268. T.H. Fife (1977), Intramolecular nucleophilic attack on esters and amides. In: *Bioorganic Chemistry* (E.E. van Tamelen, Ed.), Vol. 1, Chap. 5, pp. 93–116. Academic Press, New York.

269. C.W. Wharton (1979), Synthetic polymers as models for enzyme catalysis—a review. *Int. J. Biol. Macromolecules* **1**, 3–16.
270. M.L. Bender and M. Komiyama (1977). In: *Bioorganic Chemistry* (E.E. van Tamelen, Ed.), Vol. I, Chap. 2, pp. 19–57. Academic Press, New York.
271. C.J. Pedersen (1967), Cyclic polyethers and their complexes with metal salts. *J. Amer. Chem. Soc.* **89**, 2495–2496.
272. C.J. Pedersen (1967), Cyclic polyethers and their complexes with metal salts. *J. Amer. Chem. Soc.* **89**, 7017–7036.
273. G.W. Gokel and H.D. Durst (1976), Principles and synthetic applications in crown ether chemistry. *Synthesis*, 168–184.
274. G.R. Newkome, J.D. Sauer, J.M. Roper, and D.C. Hager (1977), Construction of synthetic macrocyclic compounds possessing subheterocyclic rings, specifically pyridine, furan, and thiophene. *Chem. Rev.* **77**, 513–597.
275. J.S. Bradshaw and P.E. Stott (1980), Preparation of derivatives and analogs of the macrocyclic oligomers of ethylene oxide (crown compounds). *Tetrahedron* **36**, 461–510.
276. D.J. Cram and J.M. Cram (1974), Host-guest chemistry. *Science* **183**, 803–809.
277. D.J. Cram and J.M. Cram (1978), Design of complexes between synthetic hosts and organic guests. *Acc. Chem. Res.* **11**, 8–14.
278. D.J. Cram (1976), Applications of biochemical systems in organic chemistry. In: *Techniques of Chemistry Series* (J.B. Jones, C.S. Sih, and D. Perlman, Eds.), Vol. 10, Part II, Chap. 5, pp. 815–874. Wiley-Interscience, New York.
279. D.J. Cram *et al.* (1977), Host-guest complexation. 1. Concept and illustration. *J. Amer. Chem. Soc.* **99**, 2564–2571.
280. D.J. Cram *et al.* (1977), Host-guest complexation. 3. Organization of pyridyl binding sites. *J. Amer. Chem. Soc.* **99**, 6392–6398.
281. A. Warshawsky, R. Kalir, A. Deshe, H. Berkovitz, and A. Patchornik (1979), Polymeric pseudocrown ethers. 1. Synthesis and complexation with transition metal anions. *J. Amer. Chem. Soc.* **101**, 4249–4258.
282. W.D. Curtis, D.A. Laidler, and J.F. Stoddart (1977), To enzyme analogues by lock and key chemistry with crown compounds. I. Enantionmeric differentiation by configurationally chiral cryptands synthesized from L-tartaric acid and D-mannitol. *J. Chem. Soc. Perkin Trans. I*, 1756–1769.
283. J.F. Stoddart (1979), From carbohydrates to enzyme analogues. *Chem. Soc. Rev.* **18**, 85–142.
284. S. Hanessian (1979), Approaches to the total synthesis of natural products using "chiral templates" derived from carbohydrates. *Acc. Chem. Res.* **12**, 159–165.
285. R.G. Pearson (1968), Hard and soft acids and bases, HSAB, Part I. *J. Chem. Educ.* **48**, 581–587.
286. R.G. Pearson (1968), Hard and soft acids and bases, HSAB, Part II. *J. Chem. Educ.* **48**, 643–648.
287. J. Sunamoto, H. Kondo, H. Okamoto, and Y. Murakami (1977), Catalysis of the deacylation of *p*-nitrophenyl hexadecanoate by 11-amino [20] paracyclophane-10-ol in neutral and alkaline media. *Tetrahedron Lett.* 1329–1332.
288. Y. Murakami, Y. Aoyama, M. Kada, and J.I. Kikuchi (1978), Macrocyclic enzyme model system: Catalysis of ester degradation by a [20] paracyclo-

phane bearing nucleophilic and metal-binding sites. *Chem. Commun.* 494–496.

289. J.P. Kintzinger, J.-M. Lehn, E. Kauffmann, J.L. Dye, and A.I. Popov (1983), Anion coordination chemistry, ^{35}Cl-NMR studies of chloride anion cryptates. *J. Amer. Chem. Soc.* **105**, 7549–7553.
290. Y. Chao and D.J. Cram (1976), Catalysis and chiral recognition through designed complexation of transition states in transacylations of amino ester salts. *J. Amer. Chem. Soc.* **98**, 1015–1017.
291. J.-M. Lehn and C. Sirlin (1978), Molecular catalysis: Enhanced rate of thiolysis with high structural and chiral recognition in complexes of a reactive receptor molecule. *Chem. Commun.* 949–951.
292. J.-M. Lehn and C. Sirlin (1987), Catalyse supramoléculaire: Coupure des esters activés d'aminoacides liés à un récepteur macrocyclique portant des résidus cysteinyles. *Nouveau J. Chimie* **11**, 693–702.
293. J.-M. Lehn (1978), Cryptates: The chemistry of macropolycyclic inclusion complexes. *Acc. Chem. Res.* **11**, 49–57.
294. K. Kirch and J.-M. Lehn (1975), Selective transport of alkali metal cations through a liquid membrane by macrobicyclic carriers. *Angew. Chem. Int. Ed. Engl.* **14**, 555–556.
295. F. Vögtle and E. Weber (1979), Multidentate acyclic neutral ligands and their complexation. *Angew. Chem. Int. Ed. Engl.* **18**, 753–776.
296. C.M. Deber and E.R. Blout (1974), Amino acid-cyclic peptide complexes. *J. Amer. Chem. Soc.* **96**, 7566–7568.
297. J.F. Stoddart (1987), Chiral crown ethers, In: *Topics in Stereochemistry*, Vol. 17, pp. 207–287. J. Wiley & Sons, New York.
298. J.P. Behr, J.-M. Lehn, D. Moras, and J.C. Thierry (1981), Chiral and functionalized face-discriminated and side-discriminated macrocyclic polyethers. Syntheses and crystal structures. *J. Amer. Chem. Soc.* **103**, 701–703.
299. J.P. Behr, C.J. Burrows, R. Heng, and J.-M. Lehn (1985), Synthesis of novel macrobicyclic polyfunctional cryptands. *Tetrahedron Lett.* **26**, 215–218.
300. J.P. Behr, M. Bergdoll, B. Chevrier, P. Dumas, J.-M. Lehn, and D. Moras (1987), Macrotricyclic and macropentacyclic ditopic receptor molecules. Synthesis, crystal structure and substrate binding. *Tetrahedron Lett.* **28**, 1989–1992.
301. J.P. Kintzinger, F. Kotzyba-Hibert, J.-M. Lehn, A. Pagelot, and K. Saigo (1981), Dynamic properties of molecular complexes and receptor—substrate complementarity. Molecular dynamics of macrotricyclic diammonium cryptates. *Chem. Commun.* 833–836.
302. A.D. Hamilton, J.-M. Lehn, and J.L. Sessler (1986), Coreceptor molecules. Synthesis of metalloreceptors containing porphyrin subunits and formation of mixed substrate supermolecules by binding of organic substrates and metal ions. *J. Amer. Chem. Soc.* **108**, 5158–5167.
303. M. Dhaenens, L. Lacombe, J.-M. Lehn, and J.P. Vigneron (1984), Binding of acetylcholine and other molecular cations by a macrocyclic receptor molecule of speleand type. *Chem. Commun.* 1097–1099.
304. J. Canceill, A. Collet, J. Gabard, F. Kotzyba-Hibert, and J.-M. Lehn (1984), Speleands. Macropolycyclic receptor cage based on binding and shaping subunits. Synthesis and properties of macrocycle-cyclotriveratylene combinations. *Helv. Chim. Acta* **65**, 1894–1897.

305. F.P. Schmidtchen (1984), Synthesis of an abiotic ditopic receptor molecule. *Tetrahedron Lett.* **25**, 4361–4364.
306. F.P. Schmidtchen (1986), Probing the design of a novel ditopic anion receptor. *J. Amer. Chem. Soc.* **108**, 8249–8255.
307. M.W. Hosseini, J.-M. Lehn, and M.P. Mertes (1983), Efficient molecular catalysts of ATP-hydrolysis by protonated macrocyclic polyamines. *Helv. Chim. Acta* **66**, 2454–2466.
308. M.W. Hosseini and J.M. Lehn (1985), Cocatalysis: pyrophosphate synthesis from acetylphosphate catalyzed by a macrocyclic polyamine. *Chem. Commun.* 1155–1157.
309. P.G. Yohannes, M.P. Mertes, and K.B. Mertes (1985), Pyrophosphate formation via a phosphoramidate intermediate in polyammonium macrocycle/metal ion-catalyzed hydrolysis of ATP. *J. Amer. Chem. Soc.* **107**, 8288–8289.
310. M.W. Hosseini and J.-M. Lehn (1987), Supramolecular catalysis and phosphoryl transfer: Pyrophosphate synthesis from acetyl phosphate mediated by macrocyclic polyamines. *J. Amer. Chem. Soc.* **109**, 7047–7058.
311. M.P. Mertes and K.B. Mertes (1990), Polyammonium macrocycles as catalysts for phosphoryl transfer: The evolution of an enzyme mimic. *Acc. Chem. Res.* **23**, 413–418.
312. J.L. Kraus, A. DiPaola, and B. Belleau (1984), Cyclic tetrameric clusters of chemotactic peptides as superactive activators of lyzozyme release from human neutropyils. *Biochem. Biophys. Res. Comm.* **124**, 939–944.
313. B.D. White, K.A. Arnold, and G.W. Gokel (1987), One- and two-armed lariat ether peptide derivatives: Synthesis and cation binding properties. *Tetrahedron Lett.* **28**, 1749–1752.
314. B.D. White, F.R. Fronczek, R.D. Gandour, and G.W. Gokel (1987), Molecular structures of 4, 13-diaza-18-crown-6 derivatives having glycine-glycine sidearms: Two potassium iodide complexes. *Tetrahedron Lett.* **18**, 1753–1756.
315. R.A. Schultz, D.M. Dishong, and G.W. Gokel (1982), Lariat ethers. 4. Chain length and ring size effects in macrocyclic polyethers having neutral donor groups on flexible arms. *J. Amer. Chem. Soc.* **104**, 625–626.
316. R.A. Schultz, D.M. Dishong, and G.W. Gokel (1981), Lariat ethers. 3. Macrocyclic polyethers bearing donor groups on flexible arms attached at a nitrogen pivot point. *Tetrahedron Lett.* **22**, 2623–2626.
317. V.J. Galto and G.W. Gokel (1984), Synthesis of calcium-selective, substituted diaza-crown ethers: A novel, one step formation of bibracchial lariat ethers (BiBLEs). *J. Amer. Chem. Soc.* **106**, 8240–8244.
318. J. Desroches, H. Dugas, M. Bouchard, T.M. Fyles, and G.D. Robertson (1987), A new series of mono-acid and dipiperdine, oxazoline, and oxazolidine crown ethers. Stability constants, ion transport rates, and NMR studies. *Can. J. Chem.* **65**, 1513–1520.
319. T.W. Robinson and R.A. Bartsch (1985), Sidearm participation in crown phosphonate monoethylalkali metal cation complexes. *Chem. Commun.* 990–991.
320. M. Kim and G.W. Gokel (1987), A molecular box, based on bibracchial lariat ether having adenine and thymine sidearms, that self-assembles in water. *Chem. Commun.* 1656–1688.

321. O.F. Schall and G.W. Gokel (1994), Molecular boxes derived from crown ethers and nucleotide bases: probes for Hoogsteen vs Watson–Crick H-bonding and other base–base interactions in self-assembly processes. *J. Amer. Chem. Soc.* **116**, 6089–6100.
322. G.W. Gokel, D.M. Goli, C. Minganti, and L. Echegoyen (1983), Classification of the hole-size cation-diameter relationship in crown ethers and a new method for determining calcium cation homogeneous equilibrium binding constants. *J. Amer. Chem. Soc.* **105**, 6786–6788.
323. C.A. Chang, J. Twu, and R.A. Bartsch (1986), pH-Dependent metal ion selectivity by a crown ether carboxylic acid. *Inorg. Chem.* **25**, 396–398.
324. R.A. Bartsch, B.P. Czech, S.I. Kang, L.E. Stewart, W. Walkowiak, W.A. Charewicz, G.S. Heo, and B. Son (1985), High lithium selectivity in competitive alkali-metal solvent extraction by lipophilic crown carboxylic acids. *J. Amer. Chem. Soc.* **107**, 4997–4998.
325. K. Kimora, H. Sakamoto, S. Kitazawa, and T. Shono (1985), Novel lithium-selective ionophore, bearing an easily ionizable moiety. *Chem. Commun.* 669–670.
326. A. Hriciga and J.-M. Lehn (1983), pH Regulation of divalent/monovalent Ca/K cation transport selectivity by a macrocylic carrier molecule. *Proc. Natl. Acad. Sci. USA* **80**, 6426–6428.
327. Y. Nakatsuji, H. Kobaysashi, and M. Okahara (1983), Active transport of alkali metal cations: A new type of synthetic ionophore derived from crown ether. *Chem. Commun.* 800–801.
328. J.K. Beadle, D.M. Dishong, R.K. Khama, and G.W. Gokel (1984), Synthesis and properties of arenediazonium and anilinium cation lariat ether complexes: An "ostrich molecule" complex and evidence for intramolecular sidearm-macroring interaction. *Tetrahedron* **40**, 3935–3944.
329. M.J. Pugia, B.E. Knudsen, and R.A. Bartsch (1987), Lithium-selective, lipophilic, small-ring bis(crown ethers). *J. Org. Chem.* **52**, 2617–2619.
330. P.D. Beer (1986), The synthesis of a novel Schiff base bis(crown ether) ligand containing recognition sites for alkali and transition metal guest cations. *Chem. Commun.* 1678–1680.
331. N. Nakashima, I. Moriguchi, K. Nakano, and M. Takagi (1987), Design of a novel bilayer system responsible to chemical signals; selective discrimination of Na^+ by a spectroscopic method. *Chem. Commun.* 617–619.
332. H. Dugas, P. Brunet, and J. Desroches (1986), Design and synthesis of a novel bis-crown ether carrier molecule minic of (Na^+, K^+)-ATPase. *Tetrahedron Lett.* **27**, 7–10.
333. J.S. Bradshaw, R.M. Izatt, and Z. Yan (1994), Bis- and oligo(benzocrown ether)s. *Chem. Rev.* **94**, 939–991.
334. S. Shinkai (1987), Switch-functionalized systems in biomimetic chemistry. *Pure & Appl. Chem.* **59**, 425–430.
335. S. Shinkai, T. Nakaji, T. Ogawa, K.J. Shigematsu, and O. Manabe (1981), Photoresponsible crown ethers. 2. Photocontrol of ion extraction and ion transport by a bis(crown ether) with a butterfly-like motion. *J. Amer. Chem. Soc.* **103**, 111–115.
336. S. Shinkai, T. Minami, Y. Kusano, and O. Manabe (1982), Photoresponsible crown ethers. 5. Light-driven ion transport by crown ethers with a photoresponsible anionic cap. *J. Amer. Chem. Soc.* **104**, 1967–1972.

337. S. Shinkai, T. Yoshida, K. Miyazaki, and O. Manabe (1987), Photoresponsible crown ethers. 19. Photocontrol of reversible association-dissociation phenomena in "tail" (ammonium)-biting crown ethers. *Bull. Chem. Soc. Jpn.* **60**, 1819–1824.
338. S. Shinkai, K. Inuzuka, K. Hara, T. Sone, and O. Manabe (1984), Redox-switch and crown ethers. 1. Redox-coupled control of metal-ionophore interactions and their application to membrane transport. *Bull. Chem. Soc. Jpn.* **57**, 2150–2155.
339. M. Irie and M. Kato (1985), Photoresponsive molecular tweezers. Photoregulated ion capture and release using thioindigo derivatives having ethylenedioxy side groups. *J. Amer. Chem. Soc.* **107**, 1024–1028.
340. H.-G. Löhr and F. Vögtle (1985), Chromo- and fluoroionophores. A new class of dye reagents. *Acc. Chem. Res.* **18**, 65–72.
341. D.J. Cram (1983), Cavitands: Organic host with enforced cavities. *Science* **219**, 1177–1183.
342. D.J. Cram (1983), Preorganization from solvents to spherands. *Angew. Chem. Int. Ed. Engl.* **25**, 1039–1057.
343. C.D. Gutsche (1983), Calixarenes. *Acc. Chem. Res.* **16**, 161–170.
344. D.J. Cram and P.Y.S. Lam (1986), Host-guest complexation. 37. Syntheses and binding properties of transacylase partial mimic with imidazole and benzyl alcohol in place. *Tetrahedron* **42**, 1607–1615.
345. D.J. Cram and H.E. Katz (1983), An incremental approach to hosts that mimic serine proteases. *J. Amer. Chem. Soc.* **105**, 135–137.
346. S. Shinkai (1990), Functionalization of crown ethers and calixarenes: new applications as ligands, carriers, and host molecules, In: (H. Durgas, Ed.), *Bioorganic Chemistry Frontiers*, Vol. 1, Springer-Verlag, New York.
347. A. Ikeda and S. Shinkai (1994), On the origin of high ionophoricity of 1,3-alternate calix[4]arenes: π-donor participation in complexation of cations and evidence for metal-tunneling through the calix[4]arene cavity. *J. Amer. Chem. Soc.* **116**, 3102–3110.
348. Y. Murakami, Y. Aoyama, and M. Kida (1980), Macrocyclic enzyme models. A metallo[10.10]paracyclophane bearing two imidazolyl groups as an efficient, bifunctional catalyst for ester hydrolysis. *J. Chem. Soc. Perkin II*, 1665–1671.
349. I. Takahashi, K. Odashima, and K. Koga (1984), Diastereomeric host-guest complex formation by an optically active paracyclophane in water. *Tetrahedron Lett.* **25**, 973–976.
350. R. Dharanipragada and F. Diederich (1987), Diastereomeric complex formation between a novel optically active host and naproxen in aqueous solution. *Tetrahedron Lett.* **28**, 2443–2446.
351. F.M. Menger, M. Takeshita, and J.F. Chow (1981), Hexapus, a new complexing agent for organic molecules. *J. Amer. Chem. Soc.* **103**, 5938–5939.
352. F.H. Kohnke, A.M.Z. Slawin, J.F. Stoddart, and D.J. Williams (1987), Molecular belts and collars in the making: A hexaepoxyoctacosahydro[12] cyclocene derivative. *Angew. Chem. Ed. Engl.* **26**, 892–894.
353. J.P. Mathias and J.F. Stoddart (1992), Constructing a molecular LEGO set. *Chem. Soc. Rev*. 215–225.
354. S.N. Davey, D.A. Leigh, A.E. Moody, L.W. Tetler, and F.A. Wade (1994), C_{60}-Azacrown ethers: the first monoaminated fullerene derivatives. *Chem. Comunn.* 397–398.

355. J.I. Anzai, Y. Suzuki, A. Veno, and T. Osa (1985), Cation transport through liquid membranes mediated by photoreactive crown ethers. Effects of alkali metal cations on their photoreactivities and transporting properties. *Israel J. Chem.* **26**, 60–64.
356. R. Sinta, P.S. Rose, and J. Smid (1983), Formation constants of complexes between crown ethers and alkali picrates in apolar solvents. Application of crown ether network polymers. *J. Amer. Chem. Soc.* **105**, 4337–4343.
357. V. LeBerre, L. Angely, N. Simonet-Gueguen, and J. Simonet (1987), Anodic trimerization: A facile one-step synthesis of tris(15-crown-5) triphenylene. *Chem. Commun.* 984–986.
358. J.L. Dye (1987), Electrides, *Sc. Amer.* Sept., pp. 66–75.
359. A. Borchardt and W.C. Still (1994), Synthetic receptor for internal residues of a peptide chain. Highly selective binding of (L)X-(L)Pro-(L)X tripeptides. *J. Amer. Chem. Soc.* **116**, 7467–7468.
360. F.M. Menger (1979), On the structure of micelles. *Acc. Chem. Res.* **12**, 111–117.
361. C.A. Bunton (1976), Applications of Biochemical Systems in Organic Chemistry. In: *Techniques of Chemistry Series* (J.B. Jones, C.J. Sih, and D. Perlman, Eds.), Vol. 10, Part II, Chap. 4, pp. 731–814. Wiley-Interscience, New York.
362. E.H. Cordes and R.B. Dunlap (1969), Kinetics of organic reactions in micellar systems. *Acc. Chem. Res.* **2**, 329–337.
363. L.R. Fisher and D.G. Oakenfull (1977), Micelles in aqueous solution. *Chem. Soc. Rev.* **6**, 25–42.
364. F.M. Menger (1977). In: *Bioorganic Chemistry* (E.E. von Tamelen, Ed.), Vol. III, Chap. 7, pp. 137–152. Academic Press, New York.
365. J.H. Fendler and E.J. Fendler (1975), Catalysis in micelles and macromolecular systems. Academic Press, New York.
366. J. Baumucker, M. Calzadilla, M. Centeno, G. Lehrmann, M. Urdanela, P. Lindquist, D. Dunham, M. Price, B. Sears, and E.H. Cordes (1972), Secondary valence force catalysis. XII. Enhanced reactivity and affinity of cyanide ion toward N-substituted 3-carbamoylpyridinium ions elicited by ionic surfactants and biological lipids. *J. Amer. Chem. Soc.* **94**, 8164–8172.
367. F.M. Menger, J.A. Donohue, and R.F. Williams (1973), Catalysis in water pools. *J. Amer. Chem. Soc.* **95**, 286–288.
368. J.R. Escabi-Pérez and J.H. Fendler (1978), Ultrafast excited state proton transfer in reversed micelles. *J. Amer. Chem. Soc.* **100**, 2234–2236.
369. R.G. Shorenstein, L.S. Pratt, C.J. Hsu, and T.E. Wagner (1968), A model system for the study of equilibrium hydrophobic bond formation. *J. Amer. Chem. Soc.* **90**, 6199–6207.
370. R.A. Moss, R.G. Mahas, and T.L. Lukas (1978), A cysteine-functionalized micellar catalyst. *Tetrahedron Lett.* 507–510.
371. W. Tagaki and H. Hara (1973), Acyloin condensation of aldehydes catalyzed by *N*-laurylthiazolium bromide. *Chem. Comm.* 891–892.
372. C.A. Bunton, L. Robinson, and M.F. Stam (1971), Stereospecific micellar catalyzed ester hydrolysis. *Tetrahedron Lett.* 121–124.
373. J.M. Brown and C.A. Bunton (1974), Stereoselective micelle-promoted ester hydrolysis. *Chem. Comm.* 969–971.

374. R.A. Moss, Y.S. Lee, and T.J. Lukas (1979), Micellar stereoselectivity cleavage of diastereomeric substrates by functional surfactant micelles. *J. Amer. Chem. Soc.* **101**, 2499–2501.
375. P. Läuger (1985), Mechanisms of biological ion transport-carriers, channels, and pumps in artificial lipid membranes. *Angew. Chem. Int. Ed. Engl.* **24**, 905–923.
376. P. Tundo, K. Kurihara, D.J. Kippenberger, M. Politi, and J.H. Fendler (1982), Chemically dissymmetrical, polymerized surfactant vesicules: Synthesis and possible utilization in artificial photosynthesis. *Angew. Chem. Int. Ed. Engl.* **21**, 81–82.
377. J.H. Fendler (1984), Membrane mimic chemistry. *Chem. in Britain* 1098–1103.
378. P.N. Tyminski, I.S. Ponticello, and P.F. O'Brien (1987), Polymerizable dienoyl lipids as spectroscopic bilayer membrane probes. *J. Amer. Chem. Soc.* **109**, 6541–6542.
379. Y. Murakami, J.I. Kikuchi, T. Takaki, and K. Uchimura (1985), Novel globular aggregates composed of synthetic peptide lipids. *J. Amer. Chem. Soc.* **107**, 3373–3374.
380. S. Shinkai, S. Nakamura, O. Manabe, T. Yamada, N. Nakashima, and T. Kunitake (1986), Synthesis and properties of membranous surfactants bearing an anion-capped crown ring as a head group. *Chem. Lett.* 49–52.
381. J.H. Fuhrhop and J. Mathieu (1984), Route to functional vesicule membrane without proteins. *Angew. Chem. Int. Ed. Engl.* **23**, 100–113.
382. J.H. Fuhrhop, H.H. David, J. Mathieu, U. Liman, H.J. Winter, and E. Boekema (1986), Bolaamphiphiles and monolayer lipid membranes made from 1,6,19,24-tetraoxa-3,21-cyclophexatriacontadiene-2,5,20,23-tetrone. *J. Amer. Chem. Soc.* **108**, 1785–1791.
383. J.M. Dolfino, C.J. Stankovic, S.L. Schreider, and F.M. Richards (1987), Synthesis of a bipolar phosphatidylethanolamine: A model compound for a membrane-spanning probe. *Tetrahedron Lett.* **28**, 2323–2326.
384. M.D. Houslay and K.K. Standlay (1982), *Dynamics of Biological Membranes*. J. Wiley & Sons, New York.
385. T.M. Fyles, T.D. James, and K.C. Kay (1990), Biomimetic ion transport: On the mechanism of ion transport by an artificial ion channel mimic. *Can. J. Chem.* **68**, 976–978.
386. T.S. Arrhenius, M. Blanchard-Desce, M. Drolaitzky, J.-M. Lehn, and J. Malthete (1986), Molecular devices: Caroviologens as an approach to molecular wires—Synthesis and incorporation into vesicle membranes. *Proc. Natl. Acad. Sci. USA* **83**, 5355–5359.
387. P. Lüthi and P.L. Luisi (1984), Enzymatic synthesis of hydrocarbon-soluble peptides with reverse micelles. *J. Amer. Chem. Soc.* **106**, 7285–7286.
388. M.D. Bednarski, H.K. Chenault, E.S. Simon, and G.M. Whitesides (1987), Membrane-enclosed enzymatic catalysis (MEEC): A useful, practical new method for the manipulation of enzyme in organic synthesis. *J. Amer. Chem. Soc.* **109**, 1283–1285.
389. Y. Imanishi (1979) Intramolecular reactions on polymer chains. *J. Polym. Sci. Macromo. Rev.* **14**, 1–205.
390. G. Manecke and W. Storck (1978), Polymeric catalysis. *Angew. Chem., Int. Ed. Engl.* **17**, 657–670.

391. C.G. Overberger and J.C. Salamone (1969), Esterolytic action of synthetic macromolecules, *Acc. Chem. Res.* **2**, 217–224.
392. H.C. Kiefer, W.I. Congdon, I.S. Scarpa, and I.M. Klotz (1972), Catalytic accelerations of 10^{12}-fold by an enzyme-like synthetic polymer. *Proc. Nat. Acad. Sci. USA* **69**, 2155–2159.
393. H. Frank, G.C. Nicholson, and E. Bayer (1978), Chiral polysiloxanes for resolution of optical antipodes. *Angew. Chem. Int. Ed. Engl.* **17**, 363–365.
394. M.L. Bender and M. Komiyama (1978), *Cyclodextrin Chemistry*. Springer-Verlag, Berlin and New York.
395. R. Breslow and P. Campbell (1969), Selective aromatic substitution within a cyclodextrin mixed complex. *J. Amer. Chem. Soc.* **91**, 3085.
396. M. Komiyama, E.J. Breaux, and M.L. Bender (1977), The use of cycloamylose to probe the "charge-relay" system. *Bioorg. Chem.* **6**, 127–136.
397. J. Emert and R. Breslow (1975), Modification of the cavity of β-cyclodextrin by flexible capping. *J. Amer. Chem. Soc.* **97**, 670–672.
398. K. Fujita, A. Shinoda, and T. Imoto (1980), Hydrolysis of phenyl acetates with capped β-cyclodextrins: Reversion from *meta* to *para* selectivity. *J. Amer. Chem. Soc.* **102**, 1161–1163.
399. M. Komiyama and M.L. Bender (1978), Importance of apolar binding in complex formation of cyclodextrins with adamantanecarboxylate. *J. Amer. Chem. Soc.* **100**, 2259–2260.
400. R. Breslow, J.B. Doherty, G. Guillot, and C. Lipsey (1978), β-Cyclodextrinylbisimidazole, a model for ribonuclease, *J. Amer. Chem. Soc.* **100**, 3227–3229.
401. I. Tabushi, K. Shimokawa, N. Shimizu, H. Shirakata, and K. Fujita (1976), Capped cyclodextrin. *J. Amer. Chem. Soc.* **98**, 7855–7856.
402. I. Tabushi, K. Shimokawa, and K. Fujita (1977), Specific bifunctionalization on cyclodextrin. *Tetrahedron Lett.* 1527–1530.
403. I. Tabushi, Y. Kuroda, and A. Mochizuki (1980), The first successful carbonic anhydrase model prepared through a new route to regiospecifically bifunctionalized cyclodextrin. *J. Amer. Chem. Soc.* **102**, 1152–1153.
404. R. Breslow and L.E. Overman (1970), An "artificial enzyme" combining a metal catalytic group and a hydrophobic binding cavity. *J. Amer. Chem. Soc.* **92**. 1075–1077.
405. J. Boger, D.G. Brenner, and J.R. Knowles (1979), Symmetrical triamino-per-*O*-methyl-α-cyclodextrin: Preparation and characterization of primary trisubstituted α-cyclodextrins. *J. Amer. Chem. Soc.* **101**, 7630–7631.
406. J. Boger and J.R. Knowles (1979), Symmetrical triamino-per-*O*-methyl-α-cyclodextrin: A host for phosphate esters exploiting both hydrophobic and electrostatic interactions in aqueous solution. *J. Amer. Chem. Soc.* **101**, 7631–7633.
407. R. Breslow, M. Hammond, and M. Lauer (1980), Selective transamination and optical induction by a β-cyclodextrin-pyridoxamine artificial enzyme. *J. Amer. Chem. Soc.* **102**, 421–422.
408. N. Saenger, M. Noltemeyer, P.C. Manor, B. Hingerty, and E.B. Klar (1972), "Induced-fit"-type complex formation of the model enzyme α-cyclodextrin. *Bioorg. Chem.* **5**, 187–195.
409. I. Tabushi, Y. Kuroda, K. Fujita, and H. Kawakubo (1978), Cyclodextrin as a ligase-oxidase model Specific allylation-oxidation of hydroquinone derivatives included by β-cyclodextrin. *Tetrahedron Lett.* 2083–2086.

410. I. Tabushi, Y. Kuroda, and K. Shimokawa (1979), Cyclodextrin having an amino group as a rhodopsin model. *J. Amer. Chem. Soc.* **101**, 4759–4760.
411. I. Tabushi, K. Yamamura, K. Fujita, and H. Kawakubo (1979), Specific includsion catalysis by β-cyclodextrin in the one-step preparation of vitamin K_1 or K_2 analogues. *J. Amer. Chem. Soc.* **101**, 1019–1026.
412. R. Breslow (1992), Bifunctional binding and catalysis in host-chemistry. *Israel J. Chem.* **32**, 23–30.
413. J.F. Stoddart (1992), Cyclodextrins, off-the-shelf components for the construction of mechanically interlocked molecular systems. *Angew. Chem. Int. Ed. Engl.* **31**, 846–847.
414. D.D. Sternbach and D.M. Rossana (1982), Cyclodextrin catalysis in the intramolecular Diels-Alder reaction with the furan diene. *J. Amer. Chem. Soc.* **104**, 5853–5854.
415. H.J. Schneider and N.K. Sangwan (1987), Host-guest chemistry. 13. Changes of stereoselectivity in Diels-Alder reactions by hydrophobic solvent effects and by β-cyclodextrin. *Angew. Chem. Int. Ed. Engl.* **26**, 896–897.
416. F.M. Menger and M. Lakida (1987), Origin of rate accelerations in an enzyme model: The *p*-nitrophenyl ester syndrome. *J. Amer. Chem. Soc.* **109**, 3145–3146.
417. R. Breslow, P. Bovy, and C.L. Hersch (1980), Reversing the selectivity of cyclodextrin bisimidazole ribonuclease mimics by changing the catalyst geometry. *J. Amer. Chem. Soc.* **102**, 2115–2117.
418. A. Ueno, K. Takahashi, and T. Osa (1981), Photocontrol of catalytic activity of capped cyclodextrin. *Chem. Commun.* 94–96.
419. I. Tabushi, Y. Kuroda, K. Yokota, and L.C. Yuan (1981), Regiospecific A,C- and A,D-disulfonate capping of β-*cyclodextrin. J. Amer. Chem. Soc.* **103**, 711–712.
420. I. Tabushi, Y. Kuroda, and T. Mizutani (1986), Artificial receptors for amino acid in water. Local environmental effect on polar recognition by 6*A*-amino-6*B*-carboxy- and 6*B*-amino-6*A*-carboxy-β-cyclodextrins. *J. Amer. Chem. Soc.* **108**, 4514–4518.
421. I. Willner and Z. Goren (1983), Diaza-crown ether capped cyclodextrin. A receptor with two recognition sites. *Chem. Commun.* 1469–1470.
422. H. Ogino (1981), Relativity high-yield syntheses of rotaxanes. Syntheses and properties of compounds consisting of cyclodextrins threaded by α,ω-diaminoalkanes coordinated to cobalt(III) complexes. *J. Amer. Chem. Soc.* **103**, 1303–1304.
423. G. Wenz (1994), Cyclodextrins as building blocks for supramolecular structures and functional units. *Angew. Chem. Int. Ed. Engl.* **33**, 803–822.
424. J.P. Guthrie (1976), Application of Biochemical Systems in Organic Chemistry. In: *Techniques of Chemistry Series* (J.B. Jones, C.J. Sih, and D. Perlman, Eds.), Vol. 10, Part II, Chap. 3, pp. 627–730. Wiley-Interscience, New York.
425. J.P. Guthrie and S. O'Leary (1975), General base catalysis of β-elimination by a steroidal enzyme model. *Can. J. Chem.* **53**, 2150–2156.
426. J.P. Guthrie and Y. Ueda (1974), Electrostatic catalysis and inhibition in aqueous solution. Rate effects on the reactions of charged esters with a cationic steroid bearing an imidazolyl substituent. *Chem. Comm.* 111–112.

427. J.P. Guthrie, P.A. Cullimore, R.S. McDonald, and S. O'Leary (1981), Large hydrophobic interactions with clearly defined geometry. A dimeric steroid with catalytic properties. *Can. J. Chem.* **60**, 747–764.
428. J.P. Guthrie, J. Cossar, and B.A. Dawson (1986), A water soluble dimeric steroid with catalytic properties. Rate enhancements for hydrophobic binding. *Can. J. Chem.* **64**, 2456–2469.
429. R. Breslow and M.A. Winnick (1969), Remote oxidation of unactivated methylene groups. *J. Amer. Chem. Soc.* **91**, 3083–3084.
430. R. Breslow, J. Rothbard, F. Herman, and M.L. Rodriguez (1978), Remote functionalization reactions as conformational probes for flexible alkyl chains. *J. Amer. Chem. Soc.* **100**, 1213–1218.
431. M.F. Czarniecki and R. Breslow (1979), Photochemical probes for model membrane structues. *J. Amer. Chem. Soc.* **101**, 3675–3676.
432. R. Breslow, R.J. Corcoran, and B.B. Snider (1974), Remote functionalization of steroids by a radical relay mechanism. *J. Amer. Chem. Soc.* **96**, 6791–6792.
433. R. Breslow, B.B. Snider, and R.J. Corcoran (1974), A cortisone synthesis using remote oxidation. *J. Amer. Chem. Soc.* **96**, 6792–6794.
434. R. Breslow (1980), Biomimetic control of chemical selectivity. *Acc. Chem. Res.* **13**, 170–177.
435. R. Breslow, U. Maitra, and D. Heyer (1984), Remote functionalization of the steroid β-face. Attack on an angular methyl group, and into the side chain. *Tetrahedron Lett.* **25**, 1123–1126.
436. R. Breslow and D. Meyer (1983), Directed steroid chlorination catalyzed by an ion-paired template. *Tetrahedron Lett.* **24**, 5939–5042.
437. E.E. van Tamelen (1975), Bioorganic chemistry; Total synthesis of tetra- and pentacyclic triterpenoids. *Acc. Chem. Res.* **8**, 152–158.
438. W.S. Johnson (1976), Biomimetic polyene cyclizations. *Angew. Chem. Int. Ed. Engl.* **15**, 9–17.
439. W.S. Johnson (1976), Biomimetic Polyene Cyclizations. *Bioorg. Chem.* **5**, 51–98.
440. E.E. van Tamelen, J.D. Willett, R.B. Clayton, and R.E. Lord (1966), Enzymic conversion of squalene 2,3-oxide to lanosterol and cholesterol. *J. Amer. Chem. Soc.* **88**, 4752–4754.
441. E.E. van Tamelen and J.H. Freed (1970), Biochemical conversion of partially cyclized squalene 2,3-oxide types to the lanosterol system. Views on the normal enzymic cyclization process. *J. Amer. Chem. Soc.* **92**, 7206–7207.
442. E.J. Corey, W.E. Russey, and P.R. Ortiz de Montellano (1966), 2,3-Oxidosqualene, an intermediate in the biological synthesis of sterols from squalene. *J. Amer. Chem. Soc.* **88**, 4750–4751.
443. R.K. Mandogal, T.T. Tchen, and K. Bloch (1958), 1,2-Methyl shifts in the cyclization of squalene to lanosterol. *J. Amer. Chem. Soc.* **80**, 2589–2590.
444. M. Nishizawa, H. Takenaka, and Y. Hayashi (1985), Experimental evidence of the stepwise mechanism of a biomimetic olefin cyclization: Trapping of cationic intermediates. *J. Amer. Chem. Soc.* **107**, 522–523.
445. E.E. van Tamelen, E.J. Leopold, S.A. Marson, and H.R. Waespe (1982), Action of 2,3-oxidosqualene lanosterol cyclase on 15′-nor-18,19-dihydro-2,3-oxidosqualene. *J. Amer. Chem. Soc.* **104**, 6479–6480.

446. W.S. Johnson, S.J. Telfer, S. Cheng, and U. Schubert (1987), Cation-stabilizing auxiliaries: A new concept in biomimetic polyene cyclization. *J. Amer. Chem. Soc.* **109**, 2517–2518.
447. P.V. Fish and W.S. Johnson (1994), The first examples of nonenzymatic, biomimetic polyene pentacyclizations. Total synthesis of the pentacyclic triterpenoid sophoradiol. *J. Org. Chem.* **59**, 2323–2335.
448. K. Baettig, A. Marinier, R. Pittelond, and P. Deslongchamps (1987), Synthesis and transannular Diels-Alder reaction of a *trans-cis-cis*-13-membered macrocyclic trienone. *Tetrahedron Lett.* **28**, 5253–5254.

Chapter 6

449. B.L. Vallee and R.J.P. Williams (1968), Enzyme action: Views derived from metalloenzyme studies. *Chem. Br.* **4**, 397–402.
450. M.M. Jones and T.H. Pratt (1976), Therapeutic chelating agents. *J. Chem. Educ.* **53**, 342–347.
451. E.W. Ainscough and A.M. Brodie (1976), The role of metal ions in proteins and other biological molecules. *J. Chem. Educ.* **53**, 156–158.
452. Y. Pocker and D.W. Bjorkquist (1977), Comparative studies of bovine carbonic anhydrase in H_2O and D_2O. Stopped-flow studies of the kinetics of interconversion of CO_2 and HCO_3^-. *Biochemistry* **16**, 5698–5707.
453. C.C. Tang, D. Davalian, P. Huand, and R. Breslow (1978), Models of metal binding sites in zinc enzymes. Synthesis of tris [4(5)-imidazolyl] carbinol (4-TIC), tris (2-imidazolyl)-carbinol(2-TIC), and related ligands and studies on metal complex binding constants and spectra. *J. Amer. Chem. Soc.* **100**, 3918–3922.
454. P. Woolley (1980), Models for metal ion function and evolution of the catalytic step in carbonic anhydrase, In: *Biophysics and Physiology of Carbon Dioxide* (C. Bauer, G. Gros, and H. Bartels, Eds.), pp. 216–225. Springer-Verlag, Berlin.
455. F.A. Quiocho and W.N. Lispcomb (1971), Carboxypeptidase A: A protein and an enzyme. *Adv. Prot. Chem.* **25**, 1–78.
456. D.W. Christianson and W.N. Lipscomb (1989), Carboxypeptidase A. *Acc. Chem. Res.* **22**, 62–69.
457. S.J. Gardell, C.S. Craik, D. Hilvert, M.S. Urdea, and W.J. Rutter (1985), Site-directed mutagenesis shows that tyrosine 248 of carboxypeptidase A does not play a crucial role in catalysis. *Nature* **317**, 551–555.
458. R. Breslow, D.E. McClure, R.S. Brown, and J. Elsenach (1975), Very fast zinc catalyzed hydrolysis of an anhydride. A model for the rate and mechanism of carboxypeptidase A catalysis. *J. Amer. Chem. Soc.* **97**, 194–195.
459. R. Breslow and D.E. McClure (1976), Cooperative catalysis of the cleavage of an amide by carboxylate and phenolic groups in a carboxypeptidase A model. *J. Amer. Chem. Soc.* **98**, 258–259.
460. R. Breslow, J. Chin, D. Hilbert, and G. Trainor (1983), Evidence for the general base mechanism in carboxypetidase A–catalyzed reactions: Partitioning studies on nucleophiles and $H_2{}^{18}O$ kinetic isotope effects. *Proc. Natl. Acad. Sci. USA* **80**, 4585–4589.
461. R. Breslow and A. Schepartz (1987), On the mechanism of peptide cleavage by carboxypeptidase A and related enzymes. *Chem. Lett. (Japan)* 1–4.

462. R. Breslow and D. Wernick (1976), On the mechanism of catalysis by carboxypeptidase A. *J. Amer. Chem. Soc.* **98**, 259–261.
463. T.H. Fife and V.L. Squillacote (1978), Metal ion effects on intramolecular nucleophilic carboxyl group participation in amide and ester hydrolysis. Hydrolysis of *N*-(8-quinolyl)phthalamic acid and 8-quinolyl hydrogen glutarate. *J. Amer. Chem. Soc.* **100**, 4787–4793.
464. M.W. Makinen, L.C. Kuo, J.J. Dymouski, and S. Jaffer (1979), Catalytic role of the metal ion of carboxypeptidase A in ester hydrolysis. *J. Biol. Chem.* **254**, 356–366.
465. D.W. Christianson, P.R. David, and W.N. Lipscomb (1987), Mechanism of carboxypeptidase A: Hydration of a ketonic substrate analogue. *Proc. Natl. Acad. Sci. USA* **84**, 1512–1515.
466. D.W. Christianson and W.N. Lipscomb (1987), Carboxypeptidase A: Novel enzyme-substrate-product complex. *J. Amer. Chem. Soc.* **109**, 5536–5538.
467. R. Breslow and C. McAllister (1971), Intramolecular bifunctional catalysis of ester hydrolysis by metal ion and carboxylate in a carboxypeptidase model. *J. Amer. Chem. Soc.* **93**, 7096–7097.
468. G.J. Lloyd and B.S. Cooperman (1971), Nucleophilic attack by zinc(II)-pyridine-2-carbaldoxime anion on phosphorylimidazole. A model for enzymatic phosphate transfer. *J. Amer. Chem. Soc.* **93**, 4883–4889.
469. D.S. Sigman and C.T. Jorgenren (1972), Models for metalloenzymes. The zinc (II)-catalyzed transesterification of *N*-(β-hydroxyethyl)ethylenediamine by *p*-introphenyl picolinate. *J. Amer. Chem. Soc.* **94**, 1724–1730.
470. D.A. Buckingham and J.P. Collman (1967), Hydrolysis of N-terminal peptide bonds and amino acid derivatives by the β-hydroxoaquotriethylenetetramine cobalt(III)ion. *J. Amer. Chem. Soc.* **89**, 1082–1087.
471. D.A. Buckingham, F.R. Keen, and A.M. Sargeson (1974); Facile intramolecular hydrolysis of dipeptides and glycinamide. *J. Amer. Chem. Soc.* **96**, 4981–4983.
472. E. Kimura (1974), Sequential hydrolysis of peptides with β-hydroxoaquotriethylenetetraminecobalt(III) ion. *Inorg. Chem.* **13**, 951–954.
473. M.W. Göbel (1994), Binuclear metal complexes as efficient intermediated in biochemically relevant hydrolysis reactions. *Angew. Chem. Int. Ed. Engl.* **33**, 1141–1443.
474. F. Frieden (1975), Non-covalent interactions. *J. Chem. Educ.* **52**, 754–761.
475. W.A. Hendrickson (1977), The molecular architecture of oxygen carrying proteins. *Trends Biochem. Sci.* **2**, 108–111.
476. J.C. Kendrew (1961), The three-dimensional structure of a protein molecule. *Sci. Amer.* **205**, December, 96–110.
477. S.J. Lippard (1986), The bioinorganic chemistry of rust. *Chem. Britain* 222–229.
478. E.F. Epstein, I. Bernal, and A.L. Balch (1970), Activation of molecular oxygen by a metal complex. The formation and structure of the anion $[Ph_3POFe(S_2C_2\{CF_3\}_2)_2]^{\ominus}$. *Chem. Comm.* 136–138.
479. J. Almog, J.E. Baldwin, R.I. Dyer, and M. Peters (1975), Condensation of tetraaldehydes with pyrrole. Direct synthesis of “capped” porphyrins. *J. Amer. Chem. Soc.* **97**, 226–227.

480. J. Almog. J.E. Baldwin, and J. Huff (1975), Reversible oxygenation and autooxidation of a "capped" porphyrin iron(II) complex. *J. Amer. Chem. Soc.* **97**, 227–228.
481. C.K. Chang and T.G. Traylor (1973), Synthesis of the myoglobin active site. *Proc. Nat. Acad. Sci. USA* **78**, 2647–2650.
482. (a) C.K. Chang and T.G. Traylor (1973), Solution behavior of a synthetic myoglobin active site. *J. Amer. Chem. Soc.* **95**, 5810–5811. (b) C.K. Chang and T.G. Traylor (1973), Neighboring group effect in heme-carbon monoxide binding. *J. Amer. Chem. Soc.* **95**, 8475–8477. (c) C.K. Chang and T.G. Traylor (1973), Proximal base influence on the binding of oxygen and carbon monoxide to heme. *J. Amer. Chem. Soc.* **95**, 8477–8479.
483. T.G. Traylor, D. Campbell, S. Tsuchiya (1979), Cyclophane porphyrin. 2. Models for steric hindrance to CO ligation in hemoproteins. *J. Amer. Chem. Soc.* **101**, 4748–4749.
484. E. Bayer and G. Holzbach (1977), Synthetic homopolymers for reversible binding of molecular oxygen. *Angew. Chem. Int. Ed. Engl.* **16**, 117–118.
485. J.P. Collman (1977), Synthetic models for the oxygen-binding hemoproteins. *Acc. Chem. Res.* **10**, 265–272.
486. J.P. Collman, J.L. Brauman, E. Rose, and K.S. Suslick (1978), Cooperativity in O_2 binding to iron porphyrins. *Proc. Nat. Acad. Sci. USA* **75**, 1052–1055.
487. N. Farrell, D.H. Dolphin, and B.R. James (1978), Reversible binding of dioxygen to ruthenium(II) porphyrins. *J. Amer. Chem. Soc.* **100**, 324–326.
488. F.S. Molinaro, R.G. Little, and J.A. Ibers (1977), Oxygen binding to a model for the active site in cobalt-substituted hemoglobin. *J. Amer. Chem. Soc.* **99**, 5628–5632.
489. J.P. Collman, J.I. Rauman, K.M. Doxsee, T.R. Halbert, S.E. Hayes, and K.S. Suslick (1978), Oxygen binding to cobalt porphyrins. *J. Amer. Chem. Soc.* **100**, 2761–2766.
490. T.G. Traylor, N. Koga, and L.A. Deardurff (1985), Structural differentiation of CO and O_2 binding to iron porphyrins: Polar pocket effects. *J. Amer. Chem. Soc.* **107**, 6405–6510.
491. C.K. Chang (1977), Stacked double-macrocyclic ligands. 1. Synthesis of "crowned" porphyrin. *J. Amer. Chem. Soc.* **99**, 2819–2822.
492. S. Takagi, T.K. Miyamoto, and Y. Sasaki (1985), A new synthetic model for myoglobin: "Tulip garden" porphyrin. *Bull. Chem. Soc. Jpn.* **58**, 447–454.
493. J.E. Baldwin, M.J. Crossley, T. Klose, E.A. O'Rear III, and M.K. Peters (1982), Synthesis and oxygenation of iron(II) "strapped" porphyrin complexes. *Tetrahedron* **38**, 27–39.
494. A.R. Battersby, S.A.J. Bartholomew, and T. Nitta (1983), Models for hemoglobin–myoglobin: Studies with loosely and tightly strapped imidazole ligands. *Chem. Commun.* 1291–1293.
495. I. Tabushi, S.-I. Kugimiya, M.G. Kinnaird, and T. Sasaki (1985), Artificial allosteric system. 2. Cooperative 1-methylimidazole binding to an artificial allosteric system, zinc-gable porphyrin-dipyridylmethane complex. *J. Amer. Chem. Soc.* **107**, 4192–4199.
496. J.P. Collman, C.S. Bencosme, C.E. Barnes, and B.D. Miller (1983), Two new members of the dimeric β-linked face-to-face porphyrin family: FTF_4* and FTF_3. *J. Amer. Chem. Soc.* **105**, 2704–2710.

497. G.M. Dubowchik and A.D. Hamilton (1985), Controlled conformational changes in covalently-linked dimeric porphyrins. *Chem. Commun.* 904–906.

498. E. Tsuchida, H. Nishide, and M. Yuasa (1986), pH-Induced oxygen uptake and evolution by aqueous synthetic heme-lipid solution. *Chem. Commun.* 1107–1108.

499. I. Hamachi, K. Nakamura, A. Fujita, and T. Kumitake (1993), Anisotropic incorporation of lipid-anchored myoglobin into a phospholipid bilayer membrane. *J. Amer. Chem. Soc.* **115**, 4966–4970.

500. I. Hamachi, S. Tanaka, and S. Shinkai (1993), Light-driven activation of reconstituted myoglobin with a ruthenium tris(2,2′-bipyridine) pendant. *J. Amer. Chem. Soc.* **115**, 10458–10459.

501. R. Malkin (1973), *Iron-Sulfur Proteins*, Vol. II, Academic Press, New York.

502. (a) R.H. Holm (1977), Synthetic approaches to the active sites of iron-sulfur proteins. *Acc. Chem. Res.* **10**, 427–434.
(b) R.H. Holm (1987), Metal-centered oxygen atom transfer reactions. *Acc. Chem. Res.* **87**, 1401–1449.

503. J.A. Ibers and R.H. Holm (1980), Modeling coordination sites in metallobiomolecules, *Science* **209**, 223–235.

504. S.J. Lippard (1988), Oxo-bridged polyion centers in biological chemistry. *Angew. Chem. Int. Engl. Ed.* **27**, 344–361.

505. A.L. Feig and S.J. Lippard (1994), Reactions of non-heme iron(II) centers with dioxygen in biology and chemistry. *Chem. Rev.* **94**, 759–805.

506. W.H. Rastetter, T.J. Erickson, and M.C. Venuti (1981), Synthesis of iron chelators. Enterobactin, enantioenterobactin, and a chiral analogue. *J. Org. Chem.* **46**, 3579–3586.

507. V.L. Pecoraro, F.L. Weitl, and K.N. Raymond (1981), Ferric ion-specific sequestering agents. 7. Synthesis, iron-exchange kinetics, and stability constants of N-substituted, sulfonated catechoylamide analogues of enterobactin. *J. Amer. Chem. Soc.* **103**, 5133–5140.

508. P. Stutte, W. Kiggen, and F. Vögtle (1987), Large molecular cavities bearing siderophore type functions. *Tetrahedron* **43**, 2065–2074.

509. F.P. Guengerich and T.L. MacDonald (1984), Chemical mechanisms of catalysis by cytochromes P-450: A unified view. *Acc. Chem. Res.* **17**, 9–16.

510. J.P. Collman and S.E. Groh (1982), "Mercaptan-tail" porphyrins: Synthetic analogues for the active site of cytochrome P-450. *J. Amer. Chem. Soc.* **104**, 1391–1403.

511. A.R. Battersby, W. Howson, and A.D. Hamilton (1982), Model studies on the active site of cytochrome P-450: An Fe(II)-porphyrin carrying a strapped thiolate ligand. *J. Chem. Soc. Chem. Commun.* 1266–1268.

512. J.I. Stesune and D. Dolphin (1987), Organometallic aspects of cytochrome P-450 metabolism. *Can. J. Chem.* **65**, 459–467.

513. J.P. Collman, J.I. Brauman, B. Meunier, and S.A. Raybuck (1984), Epoxidation of olefins by cytochrome P-450 model compounds: Mechanism of oxygen atom transfer. *Proc. Natl. Acad. Sci. USA* **81**, 3245–3248.

514. I. Tabushi and K. Morimitsu (1984), Stereospecific, regioselective, and catalytic monoepoxidation of polyolefins by the use of a P-450 model, H_2-O_2-TPP-Mn-colloidal platinum. *J. Amer. Chem. Soc.* **106**, 6871–6872.

515. J.T. Groves, G.D. Fate, and J. Lahiri (1994), Directed multi-heme self-assembly and electron transfer in a model membrane. *J. Amer. Chem. Soc.* **116**, 5477–5478.
516. G.L. Eichhorn (1973), *Inorganic Biochemistry*, Vols. 1 and 2, Elsevier, New York.
517. K.D. Karlin and Y. Gultneh (1987), Binding and activation of molecular oxygen by copper complexes. *Prog. Inorg. Chem.* **35**, 219–327.
518. A.R. Amundsen, J. Whelan, and B. Bosnich (1977), Biological analogues. On the nature of the binding sites of copper-containing proteins. *J. Amer. Chem. Soc.* **99**, 6730–6739.
519. R.R. Gagné, J.L. Allison, R.S. Gall, and C.A. Koval (1977), Models for copper-containing proteins: Structure and properties of novel five-coordinate copper(I) complexes. *J. Amer. Chem. Soc.* **99**, 7170–7178.
520. D.A. Buckingham, M.J. Gunter, and L.N. Mander (1978), Synthetic models for bis-metallo active sites. A porphyrin capped by a tetrakis (pyridine) ligand system. *J. Amer. Chem. Soc.* **100**, 2899–2901.
521. Y. Agnus, R. Louis, and R. Weiss (1979), Bimetallic copper(I) and (II) macrocyclic complexes as mimics for type 3 copper pairs in copper enzymes. *J. Amer. Chem. Soc.* **101**, 3381–3384.
522. P.K. Coughlin, J.C. Dewan, S.J. Lippard, E. Watanabe, and J.-M. Lehn (1979), Synthesis and structure of the imidazolate bridged dicopper(II) ion incorporated into a circular cryptate macrocycle. *J. Amer. Chem. Soc.* **101**, 265–266.
523. J.-M. Lehn (1980), Dinuclear cryptates: Dimetallic macropolycyclic inclusion complexes. Concepts-design-prospects. *Pure & Appl. Chem.* **52**, 2441–2459.
524. M.J. Gunter and J.M. Mander (1981), Synthesis and atropisomer of porphyrin containing functionalization at the 5,15-meso positions: Application to the synthesis of binuclear ligand systems. *J. Org. Chem.* **46**, 4792–4795.
525. J.P. Collman, A.O. Chong, G.B. Jameson, R.T. Oakley, E. Rose, E.R. Schmitton, and J.A. Ibers (1981), Synthesis of "face-to-face" porphyrin dimers linked by 5,15-substituents: Potential binuclear multielectron redox catalysts. *J. Amer. Chem. Soc.* **103**, 516–533.
526. D. Sellmann, W. Soglowek, F. Knoch, and M. Moll (1989), Nitrogenase model compounds: [μ-N_2H_2\{Fe("N_HS_4")\}$_2$], the prototype for the coordination of diazene to iron sulfur centers and its stabilization through strong NH···S hydrogen bonds. *Angew. Chem. Int. Ed. Engl.* **28**, 1271–2272.
527. C.O. Dietrich-Buchecker and J.P. Sauvage (1987), Interlocking of molecular threads: From the statistical approach to the template synthesis of catenands. *Chem. Rev.* **87**, 795–810.
528. C.O. Dietrich-Buchecker, J.P. Sauvage, and J.P. Kintzinger (1983). Une nouvelle famille de molécules: Les métallo-catenanes. *Tetrahedron Lett.* **24**, 7095–5098.
529. C.O. Dietrich-Buchecker, A. Khemiss, and J.P. Sauvage (1986), High-yield synthesis of multiring copper(I) catenates by acetylenic oxidative coupling. *Commun.* 1376–1378.
530. C.O. Dietrich-Buchecker, J. Guilhem, A.K. Khemiss, J.P. Kintzinger, C. Pascard, and J.P. Sauvage (1987), Molecular structure of a [3]-catenate:

Curling up of the interlocked system by interaction between the two copper complex subunits. *Angew. Chem. Int. Ed. Engl.* **26**, 661–663.

531. H.M. Colquhoun, J.F. Stoddart, and D.J. Williams (1986), Second-sphere coordination—a novel rôle for molecular receptors. *Angew. Chem. Int. Ed. Engl.* **25**, 487–507.
532. C.O. Dietrich-Bruchecker and J.P. Sauvage (1989), A synthetic molecular trefoil knot. *Angew. Chem. Int. Ed. Engl.* **28**, 189–192.
533. J.C. Chambon, V. Heitz, and J.P. Sauvage (1992), A rotaxane with two rigidly held porphorins as stoppers. *Chem. Commun.* 1131–1133.
534. A.R. McIntosh, A. Siemiarczuk, J.R. Bolton, M.J. Shilman, T.F. Ho, and A.C. Weedon (1983), Intramolecular photochemical electron transfer. 1. EPR and optical absorption evidence for stabilized charge separation in linked porphyrin-quinone molecules. *J. Amer. Chem. Soc.* **105**, 7215–7223.
535. J.H. Fendler (1987), Atomic and molecular clusters in membrane mimetic chemistry. *Chem. Rev.* **87**, 877–899.
536. M.R. Wasielewski, W.A. Svec, and B.T. Cope (1978), Bis(chlorophyll) cyclophanes, new models of special pair chlorophyll. *J. Amer. Chem. Soc.* **100**, 1961–1962.
537. R.E. Overfield, A. Scherz, K.J. Kaufmann, and M.R. Wasielewski (1983), Photophysics of bis(chlorophyll) cyclophanes: Models of photosynthetic reaction centers. *J. Amer. Chem. Soc.* **105**, 4256–4260.
538. R.R. Bucks and S.G. Boxer (1982), Synthesis and spectroscopic properties of a novel cofacial chlorophyll-base dimer. *J. Amer. Chem. Soc.* **104**, 340–343.
539. J. Haggin (1986), New source of gaseous fuels remain goal of researchers. *Chem. & Eng. News.* Jan. 20, 49–51.
540. B. Morgan and D. Dolphin (1985), The synthesis of porphyrins doubly linked to quinones by hydrocarbon chains. *Angew. Chem. Int. Ed. Engl.* **24**, 1003–1005.
541. J.S. Lindsey and D.C. Mauzerall (1982), Synthesis of a cofacial porphyrin-quinone via entropically favored macropolycyclization. *J. Amer. Chem. Soc.* **104**, 4498–4500.
542. T.A. Moore, D. Gust, P. Mathis, J.C. Mialocq, C. Chachaty, R.V. Bensasson, E.J. Land, D. Doiri, P.A. Liddell, W.R. Lehmann, G.A. Nemeth, and A.L. Moore (1984), Photodriven charge separation in a caratenoporphyrin quinone triad. *Nature* **307**, 630–632.
543. J. Deisenhofer, O. Epp, K. Miki, R. Huber, and H. Michel (1984), X-Ray structure analysis of a membrane protein complex. Electron analysis map at 3 Å resolution and a model of the chromophores of the photosynthetic reaction center from *Rhodopseudomonas viridis*. *J. Mol. Biol.* **180**, 385–398.
544. G.M. Dubowchik and A.D. Hamilton (1986), Towards a synthetic model of the structure of the photosynthetic reaction centre. *Chem. Commun.* 1391–1394.
545. G.M. Dubowchik and A.D. Hamilton (1987), Synthesis of tetrameric and hexameric *cyclo*-porphyrins. *Chem. Commun.* 293–295.
546. I. Abdalmuhdi and C.K. Chang (1985), A novel synthesis of triple-deckered triporphyrin. *J. Org. Chem.* **50**, 411–413.
547. H.A. Staab, M. Tercel, R. Fischer, and C. Krieger (1994), Synthesis and properties of a vertically stacked porphyrin-quinone(1)-quinone(2) cyclophane. *Angew. Chem. Int. Ed. Engl.* **33**, 1463–1466.

548. H.A. Staab and T. Carell (1994), Synthesis and properties of a vertically stacked porphyrin(1)-porphyrin(2)-quinone cyclophane. *Angew. Chem. Int. Ed. Engl.* **33**, 1466–1468.
549. J.C. Rodriguez-Ubis, B. Alpha, D. Plancherel, and J.-M. Lehn (1984), Photoactive cryptands. Synthesis of the sodium cryptates of macrobicyclic ligands containing bipyridine and phenanthroline group. *Helv. Chim. Acta.* **67**, 2264–2269.
550. B. Alpha, J.-M. Lehn, and G. Mathis (1987), Energy transfer luminescence of europium(III) and terbium(III) cryptates of macrobicyclic polypyridine ligands. *Angew. Chem. Int. Ed. Engl.* **26**, 266–267.
551. R.H. Abeles and D. Dolphin (1974), The vitamin B_{12} coenzyme. *Acc. Chem. Res.* **9**, 114–120.
552. J. Rétey, A. Ulmani-Ronchi, J. Seibl, and D. Arigoni (1966), Zum mechanismus der Propandioldehydrase-Reaktion. *Experentia* **22**, 502–503.
553. G.N. Schrauzer (1968), Organocobalt chemistry of vitamin B_{12} model compounds (cobaloximes). *Acc. Chem. Res.* **1**, 97–103.
554. G.N. Schrauzer (1976), New developments in the field of vitamin B_{12}: Reactions of the cobalt atom in corrins and in vitamin B_{12} model compounds. *Angew. Chem. Int. Ed. Engl.* **15**, 417–426.
555. G.N. Schrauzer (1977), New developments in the field of vitamin B_{12}: Enzymatic reactions dependent upon corrins and coenzyme B_{12}. *Angew. Chem. Int. Ed. Engl.* **16**, 233–244.
556. P. Dowd, B.K. Trivedi, M. Shapiro, and L.K. Marwaha (1976), Vitamin B_{12} model studies. Migration of the acrylate fragment in the carbon-skeleton rearrangement leading to α-methyleneglutaric acid. *J. Amer. Chem. Soc.* **98**, 7875–7877.
557. T. Toraya, E. Krodel, A.S. Mildran, and R.H. Abeles (1979), Role of peripheral side chains of vitamin B_{12} coenzymes in the reaction catalyzed by dioldehydrase. *Biochemistry* **18**, 417–426.
558. J. Halpern, S.H. Kim, and T.W. Leung (1984), Cobalt-carbon bond dissociation energy of coenzyme B_{12}. *J. Amer. Chem. Soc.* **106**, 8317–8319.
559. R.B. Silverman and D. Dolphin (1973), A direct method for cobalt-carbon bond formation in cobalt(III)-containing cobalamins and cobaloximes. Further support for cobalt(III) π-complexes in coenzyme B_{12} dependent rearrangements. *J. Amer. Chem. Soc.* **95**, 1686–1688.
560. R.B. Silverman and D. Dolphin (1974), Reaction of vinyl ethers with cobalamins and cobaloximes. *J. Amer. Chem. Soc.* **96**, 7094–7096.
561. R.B. Silverman, D. Dolphin, T.J. Carty, E.K. Krodel, and R.H. Abeles (1974), Formylmethycobalamin. *J. Amer. Chem. Soc.* **96**, 7096–7097.
562. L. Salem, O. Eisentein, N.T. Anh, H.B. Burgi, A. Devaquet, G. Segal, and A. Veillard (1977), Enzymatic catalysis. A theoretically derived transition state for coenzyme B_{12}-catalyzed reaction. *Nouv. J. Chim.* **1**, 335–347.
563. P. Dowd and R. Hershline (1986), Carbon-13 labeling study of the methylitaconate $\rightleftharpoons$ α-methyleneglutarate model rearrangement reaction. *Chem. Commun.* 1409–1410.
564. H. Flohr, N. Paunhorst, and J. Rétey (1976), Synthesis, structure determination, and rearrangement of a model for the active site of methylmalonyl-CoA mutase with incorporated substrate. *Angew. Chem. Int. Ed. Engl.* **15**, 561–562.

565. R. Breslow and P.L. Khanna (1976), An intramolecular model for the enzymatic insertion of coenzyme B_{12} into unactivated carbon–hydrogen bonds. *J. Amer. Chem. Soc.* **98**, 1297–1299.
566. E.J. Corey, N.J. Cooper, and M.L.H. Green (1977), Biochemical catalysis involving coenzyme B_{12}: A rational stepwise mechanistic interpretation of vicinal interchange rearrangements. *Proc. Nat. Acad. Sci. USA* **74**, 811–815.
567. K. Sato, E. Hiei, S. Shimizu, and R. Abeles (1978), Affinity chromatography of N^5-methyltetrahydrofolate-homocysteine methyltransferase on a cobalamin-Sepharose. *FEBS Lett.* **85**, 73–76.
568. Y. Murakami, Y. Hisaeda, and T. Ohno (1991), Hydrophobic vitamin B_{12}. Part 9. An artificial holoenzyme composed of hydrophobic vitamin B_{12} and synthetic bilayer membrane for carbon-skeleton rearrangements. *J. Chem. Soc. Perkin Trans. II*, 405–416.
569. Y. Murakami, Y. Hisaeda, A. Ogawa, T. Miyajima, O. Hayashida, and T. Ohno (1993), Aggregation behavior and reactivity of hydrophpbic vitamin B_{12} covalently bound to lipid in aqueous media. *Tetrahedron Lett.* **34**, 863–866.
570. Y. Murakami, Y. Hisaeda, and T. Ohno (1990), Hydrophobic vitamin B_{12}. Part 8. Carbon-skeleton rearrangement reactions catalyzed by hydrophobic vitamin B_{12} in octopus azaparacyclophane. *Bioorg. Chem.* **18**, 49–72.
571. J.M. Wood (1974), Biological cycles for toxic elements in the environment. *Science* **183**, 1049–1052.
572. J.S. Thayer (1977), Teaching bio-organometal chemistry. II. The metals. *J. Chem. Educ.* **54**, 662–665.
573. S. Krishnamurthy (1992), Biomethylation and environmental transport of metals. *J. Chem. Educ.* **69**, 347–350.

Chapter 7

574. A.E. Metzler (1977), *Biochemistry, The Chemical Reactions of Living Cells*, Chap. 8. Academic Press, New York.
575. B.O. Söderberg, O. Tapia, and C.I. Brändén (1976), Three-dimensional structure of horse liver alcohol dehydrogenase at 2.4 Å resolution. *J. Mol. Biol.* **102**, 27–59.
576. G. DiSabato (1970), Adducts of diphosphopyridine nucleotide and carbonyl compounds. *Biochemistry* **9**, 4594–4600.
577. D.J. Creighton and D.S. Sigman (1971), A model for alcohol dehydrogenase. The zinc ion catalyzed reduction of 1,10-phenanthroline-2-carboxaldehyde by *N*-propyl-1, 4-dihydronicotinamide. *J. Amer. Chem. Soc.* **93**, 6314–6316.
578. U. Grau, H. Kapmeyer, and W.E. Trommer (1978), Combined coenzyme-substrate analogues of various dehydrogenases. Synthesis of (3*S*)- and (3*R*)-5-(3-carboxy-3-hydroxypropyl) nicotinamide adenine dinucleotide and their interaction with (*S*)- and (*R*)-lactate-specific dehydrogenases. *Biochemistry* **17**, 4621–4626.
579. J.T. van Bergen and R.M. Kellogg (1977), A crown ether NAD(P)H mimic. Complexation with cations and enhanced hydride donating ability toward sulfonium salts. *J. Amer. Chem. Soc.* **99**, 3882–3884.

580. R.M. Kellogg (1984), Chiral macrocycles as reagents and catalysts. *Angew. Chem. Int. Ed. Engl.* **23**, 782–794.
581. A.I. Meyers and J.D. Brown (1987), The first nonenzymatic stereospecific intramolecular reduction by an NADH mimic containing a covalently bound carbonyl moiety. *J. Amer. Chem. Soc.* **109**, 3155–5135.
582. J.P. Behr and J.-M. Lehn (1978), Enhanced rates of dihydropyridine to pyridinium hydrogen transfer in complexes of an active macrocyclic receptor molecule. *Chem. Comm.* 143–146.
583. M. Kojima, F. Toda, and K. Hattori (1981), β-Cyclodextrin-nicotinamide as a model for NADH dependent enzymes. *J. Chem. Soc. Perkin I*, 1647–1651.
584. G.A. Hamilton (1971). In: *Progress in Bioorganic Chemistry* (E.T. Kaiser and T.J. Kézdy, Eds.), Vol. 1, pp. 83–137. Wiley-Interscience, New York.
585. C.J. Suckling and H.C.S. Wood (1979), Should organic chemistry meddle in biochemistry? *Chem. Britain* **5**, 243–248.
586. A.J. Irwin and J.B. Jones (1976), Stereoselective horse liver alcohol dehydrogenase catalyzed oxidoreductions of some bicyclic[2.2.1] and [3.2.1] ketones and alcohols. *J. Amer. Chem. Soc.* **98**, 8476–8482.
587. A.J. Irwin and J.B. Jones (1977), Regiospecific and enantioselective horse liver alcohol dehydrogenase catalyzed oxidations of some hydroxycyclopentanes. *J. Amer. Chem. Soc.* **99**, 1625–1630.
588. J.B. Jones and J.F. Beck (1976), Application of Biochemical Systems in Organic Chemistry. In: *Techniques of Chemistry Series* (J.B. Jones, C.J. Sih, and D. Perlman, Eds.), Part I, pp. 107–401. Wiley, New York.
589. V. Prelog (1964), Specification of the stereochemistry of some oxidoreductase by diamond lattice sections. *Pure & Appl. Chem.* **9**, 119–130.
590. J.B. Jones and J.J. Jakovac (1982), A new cubic-space section model for predicting the specificity of horse liver alcohol dehydrogenase-catalyzed oxidoreductions. *Can. J. Chem.* **60**, 19–28.
591. J.B. Jones (1993), Probing the specificity of synthetically useful enzymes, *Aldrichimica Acta* **26**, No. 4, 105–112.
592. C. Walsh (1980), Flavin coenzymes; At the crossroads of biological redox chemistry. *Acc. Chem. Res.* **13**, 148–155.
593. D.J. Creighton, J. Hajdu, G. Mooser, and D.S. Sigman (1973), Model dehydrogenase reactions. Reduction of *N*-methylacridinium ion by reduced nicotinamide adenine dinucleotide and its derivatives. *J. Amer. Chem. Soc.* **95**, 6855–6867.
594. P. Hemmerich, V. Massey, and G. Weber (1967), Photo-induced benzyl substitution of flavins by phenylacetate: A possible model for flavoprotein catalysis. *Nature* **213**, 728–730.
595. M. Brüstlein and T.C. Bruice (1972), Demonstration of a direct hydrogen transfer between NADH and a deazaflavin. *J. Amer. Chem. Soc.* **94**, 6548–6549.
596. J. Fisher and C. Walsh (1974), Enzymatic reduction of 5-deazariboflavine from reduced nicotinamide adenine dinucleotide by direct hydrogen transfer. *J. Amer. Chem. Soc.* **96**, 4345–4346.
597. D. Eirich, G. Vogels, and R. Wolfe (1978), Proposed structure of coenzyme F_{420} from Methanobacterium. *Biochemistry* **17**, 4583–4593.
598. J.M. Sayer, P. Conlon, J. Hupp, J. Fancher, R. Bélanger, and E.J. White (1979), Reduction of 1,3-dimethyl-5-(*p*-nitrophenylimino) barbituric acid by

thiols. A high-velocity flavin model reaction with an isolable intermediate. *J. Amer. Chem. Soc.* **101**, 1890–1893.

599. S. Shinkai, K. Kameoka, K. Ueda, and O. Manabe (1987), "Remote control" of flavin reactivities by an intramolecular crown ether serving as a metal-binding site. *J. Amer. Chem. Soc.* **109**, 923–924.
600. S. Shinkai, K. Kameoka, K. Ueda, O. Manabe, and M. Onishi (1987), "Remote control" of flavin reactivities by an intramolecular crown ether serving as a metal-binding site: Relationship between spectral properties and dissociation of the 8-sulfonamide group. *Bioorg. Chem.* **15**, 269–282.
601. J. Takeda, S. Ohta, and M. Hirobe (1985), Rate-enhancing effect of intramolecular linkage of flavin-porphyrin on reduction by 1,4-dihydropyridine. *Tetrahedron Lett.* **26**, 4509–4512.
602. J. Takeda, S. Ohta, and M. Hirobe (1986), Synthesis and properties of novel flavin-linked porphyrin. *Heterocycles* **24**, 269.
603. W.H. Rastetter, T.R. Gadek, J.P. Tane, and J.W. Frost (1979), Oxidations and oxygen transfer effected by a flavin N(5)-oxide. A model for flavin-dependent monooxygenases. *J. Amer. Chem. Soc.* **101**, 2228–2231.
604. H.W. Orf and D. Dolphin (1974), Oxaziridines as possible intermediates in flavin monooxygenases. *Proc. Nat. Acad. Sci. USA* **71**, 2646–2650.
605. G. Eberlein and T.C. Bruice (1982), One- and two-electron reduction of oxygen by 1,5-dihydroflavins. *J. Amer. Chem. Soc.* **104**, 1449–1452.
606. D. Vargo and M.S. Jorns (1979), Synthesis of a 4a,5-epoxy-5-deazaflavin derivative. *J. Amer. Chem. Soc.* **101**, 7623–7626.
607. D.M. Jerina, J.W. Daly, B. Witkop, S. Zaltzman-Nirenberg, and S. Udenfriend (1969), 1,2-Naphthalene oxide as an intermediate in the microsomal hydroxylation of naphthalene. *Biochemistry* **9**, 147–156.
608. G. Guroff, J.W. Daly, D.M. Jerina, J. Rensen, B. Witkop, and S. Udenfriend (1967), Hydroxylation-induced migration: The NIH shift. *Science* **157**, 1524–1530.
609. J.L. Fox (1978), Chemists attack complex organic mechanisms. *Chem. Eng. News* May 22, pp. 28–30.
610. W. Adam, A. Alzérreca, J.E. Liu, and F. Yany (1977), α-Peroxylactones via dehydrative cyclization of α-hydroperoxy acids. *J. Amer. Chem. Soc.* **99**, 5768–5773.
611. C. Kemal and T.C. Bruice (1976), Simple synthesis of a 4a-hydroperoxy adduct of a 1,5-dihydroflavine: Preliminary studies of a model for bacterial luciferase. *Proc. Nat. Acad. Sci. USA* **73**, 995–999.
612. S.P. Schmidt and G.B. Schuster (1978), Dioxetanone chemiluminescence by the chemically initiated electron exchange pathway. Efficient generation of excited singlet states. *J. Amer. Chem. Soc.* **100**, 1966–1968.
613. B.P. Branchaud and C.T. Walsh (1985), Functional group diversity in enzymatic oxygenation reactions catalyzed by bacterial flavin-containing cyclohexanone oxygenase. *J. Amer. Chem. Soc.* **107**, 2153–2161.
614. J.N. Lowe and L.L. Ingraham (1974), *An Introduction to Biochemical Reactions Mechanisms*, Chap. 3. Foundation of Molecular Biology Series. Prentice-Hall, Englewood Cliffs, New Jersey.
615. O.A. Gansow and R.H. Holm (1969), A proton resonance investigation of equilibra, solute structures, and transamination in the aqueous systems

pyridoxaminepyruvate-zinc(II) and aluminium(III). *J. Amer. Chem. Soc.* **91**, 5984–5993.

616. M. Blum and J.W. Thanassi (1977), Metal ion induced reaction specific in vitamin B_6 model systems. *Bioorg. Chem.* **6**, 31–41.
617. C. Walsh (1978), Chemical approaches to the study of enzymes catalyzing redox transformations. *Annu. Rev. Biochem.* **47**, 881–931.
618. B. Belleau and J. Burba (1960), The stereochemistry of the enzymic decarboxylation of amino acids. *J. Amer. Chem. Soc.* **82**, 5751–5752.
619. H.C. Dunathan L. Davis, P.G. Kury, and M. Kaplan (1968), The stereochemistry of enzymatic transamination. *Biochemistry* **7**, 4532–4536.
620. H.C. Dunathan and J.G. Voet (1974), Stereochemical evidence for the evolution of pyridoxal-phosphate enzymes of various functions from a common ancestor. *Proc. Nat. Acad. Sci. USA* **71**, 3888–3891.
621. J.N. Roitenan and D.J. Cram (1971), Electrophilic substitution at saturated carbon. XLV. Dissection of mechanisms of base-catalyzed hydrogen-deuterium exchange of carbon acids into inversion, isoinversion, and racemization pathways. *J. Amer. Chem. Soc.* **90**, 2225–2230.
622. J.N. Roitenan and D.J. Cram (1971), Electrophilic substitution at saturated carbon. XLVI. Crown ethers' ability to alter role of metal cations in control of stereochemical fate of carbanions. *J. Amer. Chem. Soc.* **90**, 2231–2241.
623. D.J. Cram, W.T. Ford, and L. Gosser (1968), Electrophilic substitution and saturated carbon. XXXVIII. Survey of substituent effects on stereochemical fate of fluorenyl carbanions. *J. Amer. Chem. Soc.* **90**, 2598–2606.
624. M.D. Broadhurst and D.J. Cram (1974), A model for the proton transfer stages of the biological transaminations and isotopic exchange reactions of amino acids. *J. Amer. Chem. Soc.* **96**, 581–583.
625. D.A. Jaeger, M.D. Broadhurst, and D.J. Cram (1979), Electrophilic substitution at saturated carbon. 52. A model for the proton transfer steps of biological transamination and the effect of a 4-pyridyl group on the base-catalyzed racemization of a carbon acid. *J. Amer. Chem. Soc.* **101**, 717–732.
626. R. Breslow, A.W. Czarnik, M. Lauer, R. Leppkes, J. Winkler, and S. Zimmerman (1986), Mimics of transaminase enzymes. *J. Am. Chem. Soc.* **108**, 1969–1979 and references therein.
627. I. Tabushi, Y. Kuroda, M. Yamada, H. Higashima, and R. Breslow (1985), A-(modified B_6)-B-[ω-amino(ethylamino)]β-cyclodextrin as an artificial B_6 enzyme for chiral amino transfer reaction. *J. Amer. Chem. Soc.* **107**, 5545–5546.
628. S.E. Brown, J.H. Coates, C.J. Easton, S.J. van Eyk, S.F. Lincoln, B.L. May, M.A. Stile, C.B. Whalland, and M.L. Williams (1994), Tryptophan anion complexes of β-cyclodextrin (cyclomaltaheptaose), an aminopropylamino-β-cyclodextrin and its enantioselective nickel(II) complex. *Chem. Commun.* 47.
629. S.C. Zimmerman and R. Breslow (1986), Asymmetric synthesis of amino acids by pyridoxamine enzyme analogues utilizing base-acid catalysis. *J. Amer. Chem. Soc.* **106**, 1490–1491.
630. B.R. Baker (1967), *Design of Active-Site-Directed Irreversible Enzyme Inhibitors*. J. Wiley & Sons, New York.
631. R.H. Abeles and A.L. Maycock (1976), Suicide enzyme inactivators. *Acc. Chem. Res.* **9**, 313–319.

632. G. Schoellmann and E. Shaw (1963), Direct evidence for the presence of histidine in the active center of chymotrypsin. *Biochemistry* **2**, 252–255.
633. V. Chowdhry and F.H. Westheimer (1979), Photoaffinity labeling of biological systems. *Annu. Rev. Biochem.* **48**, 293–325.
634. M.P. Goeldner, C.G. Hirth, B. Kieffer, and G. Ourisson (1982), Photosuicide inhibition—a step towards specific photoaffinity labeling. *Trends in Biochem. Sci.* 310–313.
635. L. Stryer (1978), Fluorescence energy transfer as a spectroscopic ruler. *Annu. Rev. Biochem.* **47**, 819–846.
636. R.R. Rando (1975), Mechanisms of action of naturally occurring irreversible enzyme inhibitors. *Acc. Chem. Res.* **8**, 281–288.
637. R.R. Rando (1974), Chemistry and enzymology of k_{cat} inhibitors. *Science* **185**, 320–324.
638. P. Fasella and R. John (1969), Substrate analogues as specific inhibitors of pyridoxal-dependent enzymes. *Proc. 4th Int. Congr. Pharmacol.* **5**, 184–186.
639. Y. Morino and M. Okamoto (1973), Labeling of the active site of cytoplasmic aspartate amino transferase by β-chloro-L-alanine. *Biochem. Biophys. Res. Commun.* **50**, 1061–1067.
640. M.J. Jung and B.W. Metcalf (1975), Catalytic inhibition of γ-aminobutyric acid α-ketoglutarate transaminase of bacterial origin by 4-aminohex-5-ynoic acid, a substrate analog. *Biochem. Biophys. Res. Commun.* **67**, 301–306.
641. R.R. Rando and J. de Mairena (1974), Propargyl amine-induced irreversible inhibition of non-flavin-linked amine oxidases. *Biochem. Pharmacol* **23**, 463–466.
642. D. Kuo and R.R. Rando (1981), Irreversible inhibition of glutamate dicarboxylase by α-(fluoromethyl)glutamic acid. *Biochemistry* **20**, 506–511.
643. C.T. Walsh (1983), Suicide substrates: Mechanism-base enzyme inactivators with therapeutic potential. *Trends in Biochem. Sci.* 254–257.
644. P.K. Chakravarty, G.A. Krafft, and J.A. Katzenellenbogen (1982), Haloenol lactone: Enzyme-activated irreversible inactivators for serine proteases. *J. Biol. Chem.* **257**, 610–612.
645. R.H. Abeles (1983), Suicide enzyme inactivations. *Chem. & Eng. News*, Sept. **19**, 48–56.
646. J.L. Adams and B.W. Metcalf (1984), The synthesis of a 3-diazobycyclo-[2.2.1]heptan-2-one inhibitor of thromboxane A_2 synthetase. *Tetrahedron Lett.* **25**, 919–922.
647. A. Nagahisa, W.H. Orme-Johnson, and S.R. Wilson (1984), Silicon-mediated suicide inhibition: An efficient mechanism-base inhibitor of cytochrome P-450 oxidation of cholesterol. *J. Amer. Chem. Soc.* **106**, 1166–1167.
648. M.D. Varney, G.P. Marzoni, C.L. Palmer, J.G. Deal, S. Webber, K.M. Welsh, J. Bacquet, C.A. Bartlett, C.A. Morse, C.L.J. Booth, S.M. Hermann, E.F. Howland, R.W. Ward, and J. White (1992), Crystal-structure-based design and synthesis of benz[*cd*]indole-containing inhibitors of thymidylate synthase. *J. Med. Chem.* **35**, 663–676.
649. S.H. Reich, M.A.M. Fuhry, D. Nguyen, M.J. Pino, K.M. Welch, S. Webber, C.A. Janson, S.R. Jordan, D.A. Mathews, W.W. Smith, C.A. Bartlett, C.L.J. Booth, S.M. Herrmann, E.F. Howland, C.A. Morse, R.W. Ward, and J. White (1992), Design and synthesis of novel 6,7-imidazotetrahydroquinoline inhibitors of thymidylate synthese using iterative protein crystal structure analysis. *J. Med. Chem.* **35**, 847–858.

650. W.J. Thompson, P.M.D. Fitzgerald, M.K. Holloway, E.A. Emini, P.L. Darke, B.M. McKeever, W.A. Schleif, J.C. Quintero, J.A. Zugay, T.J. Tucker, J.E. Schwering, C.F. Homnick, J. Nunberg, J.P. Springer, and J.R. Huff (1992), Synthesis and antiviral activity of a series of HIV-1 protease inhibitors with functionality tethered to the P_1 or P_1' phenyl substituents: X-ray crystal structure assisted design. *J. Med. Chem.* **35**, 1685–1701.

651. B.P. Morgan, D.R. Holland, B.W. Mathews, and P.A. Bartlett (1994), Structure-based design of an inhibitor of the zinc peptidase thermolysin. *J. Amer. Chem. Soc.* **116**, 3251–3260.

652. R. Breslow (1958), On the mechanism of thiamine action. IV. Evidence from studies on model systems. *J. Amer. Chem. Soc.* **80**, 3719–3726.

653. A.A. Gallo and H.Z. Sable (1974), Coenzyme interactions. VIII. ^{13}C-NMR studies of thiamine and related compounds. *J. Biol. Chem.* **249**, 1382–1389.

654. F. Jordan and Y.H. Mariam (1978), N^1-Methylthiaminium diiodide. Model study on the effect of a coenzyme bound positive charge on reaction mechanism requiring thiamine pyrophosphate. *J. Amer. Chem. Soc.* **100**, 2534–2541.

655. F. White and L.L. Ingraham (1962), Mechanism of thiamine action: A model of 2-acylthiamine. *J. Amer. Chem. Soc.* **84**, 3109–3111.

656. T.C. Bruice and S. Benkovic (1966), *Bioorganic Mechanisms*. Vol. 2, Chap. 8, p. 217. Benjamin, New York.

657. W.H. Rastetter, J. Adams, J.W. Frost, L.J. Nummy, J.E. Frommer, and K.B. Roberts (1979), On the involvement of lipoic acid in α-keto acid dehydrogenase complexes. *J. Amer. Chem. Soc.* **101**, 2752–2753.

658. R. Kluger and D.C. Pike (1979), Chemical synthesis of a proposed enzyme-generated "reactive intermediate analogue" derived from thiamine diphosphate. Self-activation of pyruvate dehydrogenase by conversion of the analogue in its components. *J. Amer. Chem. Soc.* **101**, 6425–6428.

659. R. Kluger, K. Marimian, and K. Kitamura (1987), Chiral intermediates in thiamin catalysis. The stereochemical course of the dicarboxylation step in the conversion of pyruvate to acetaldehyde. *J. Amer. Chem. Soc.* **109**, 6368–6371.

660. R. Kluger (1987), Thiamine diphosphate: A mechanistic update on enzymic and nonenzymic catalysis of dicarboxylation. *Chem. Rev.* **87**, 863–876.

661. J.A. Gutowski and G.E. Lienhard (1976), Transition-state analogs for thiamin pyrophosphate-dependent enzyme. *J. Biol. Chem.* **251**, 2863–2866.

662. (a) D. Hilvert and R. Breslow (1984), Functionalized cyclodextrins as holoenzyme mimics of thiamine-dependent enzyme. *Bioorg. Chem.* **12**, 206–220. (b) R. Breslow and E. Kool (1988), A γ-cyclodextrin thiazolium salt holoenzyme mimic for the benzoin condensation. *Tetrahedron Lett.* **29**, 1635–1638.

663. J. Moss and M.D. Lane (1971), The biotin-dependent enzymes. *Adv. Enzymol.* **35**, 321–442.

664. H.G. Wood and R.E. Barden (1977), Biotin enzymes. *Annu. Rev. Biochem.* **46**, 385–413.

665. A.S. Mildvan (1977), Magnetic resonance studies of the conformations of enzyme-bound substrates. *Acc. Chem. Res.* **10**, 246–252.

666. R. Kluger and P.D. Adawadkar (1976), A reaction proceeding through intramolecular phosphorylation of a urea. A chemical mechanism for enzymic carboxylation of biotin involving cleavage of ATP. *J. Amer. Chem. Soc.* **98**, 3741–3742.
667. W.C. Stallings (1977), The carboxylation of biotin. Substrate recognition and activation by complementary hydrogen bonding. *Arch. Biochem. Biophys.* **183**, 179–199.
668. H.G. Wood (1976), The reactive group of biotin catalysis by biotin enzymes. *Trends. Biochem. Sci.* **1**, 4–6.
669. F. Lynen, J. Knappe, E. Lorch, G. Jütting, and E. Ringelmann (1959), Die biochemische Funktion des Biotins. *Angew. Chem.* **71**, 481–486.
670. T.C. Bruice and A.F. Hegarty (1970), Biotin-bound CO_2 and the mechanism of enzymatic carboxylation reactions. *Proc. Nat. Acad. Sci. USA* **65**, 805–809.
671. M. Caplow and M. Yager (1976), Studies on the mechanism of biotin catalysis. II. *J. Amer. Chem. Soc.* **89**, 4513–4521.
672. R.B. Guchhait, S.E. Polakis, D. Hollis, C. Fenselau, and M.D. Lane (1974), Acetyl coenzyme A carboxylase system of *E. coli*. Site of carboxylation of biotin and enzymatic reactivity of 1′-*N*-(ureido)-carboxybiotin derivatives. *J. Biol. Chem.* **249**, 6646–6656.
673. P.A. Whitney and T.G. Cooper (1972), Urea carboxylase and allophanate hydroxylase. Two components of ATP: Urea-lyase in *S. cerevisiae*. *J. Biol. Chem.* **247**, 1349–1353.
674. C.M. Visser and R.M. Kellogg (1977), Mimesis of the biotin mediated carboxyl transfer reactions. *Bioorg. Chem.* **6**, 79–88.
675. R. Kluger, P. Davis, and P.D. Adawadkar (1979), Mechanism of urea participation in phosphonate ester hydrolysis. Mechanistic and stereochemical criteria for enzymic formation and reaction of phosphorylated biotin. *J. Amer. Chem. Soc.* **101**, 5995–6000.
676. A. Berkessel and R. Breslow (1986), On the structures of some adducts of biotin with electrophiles: Does sulfur transannular interaction with the carbonyl group play a role in the chemistry of biochemistry of biotin? *Bioorg. Chem.* **14**, 299–262.
677. D. Arigoni, F. Lynen, and J. Rétey (1966), Stereochemie der enzymatischen Carboxylierung von (2*R*)-2-^{3}H-Propionyl-CoA. *Helv. Chim. Acta.* **49**, 311–316.
678. J. Rétey and F. Lynen (1965), Zur biochemischen Funktion des Biotins. IX. Der sterische Verlauf der Carboxylierung von Propionyl-CoA. *Biochem. Z.* **342**, 256–271.
679. I.A. Rose, E.L. O'Connell, and F. Solomon (1976), Intermolecular tritium transfer in the transcarboxylase reaction. *J. Biol. Chem.* **251**, 902–904.
680. J. Stubbe and R.H. Abeles (1979), Biotin carboxylations concerted or not concerted? That is the question! *J. Biol. Chem.* **252**, 8338–8340.
681. J. Stubbe, S. Fish, and R. Abeles (1980), Are carboxylations involving biotin concerted or nonconcerted? *J. Biol. Chem.* **255**, 236–242.
682. S.J. O'Keefe and J.R. Knowles (1986), Enzymatic biotin-mediated carboxylation is not a concerted process. *J. Amer. Chem. Soc.* **108**, 328–329.
683. J.M. Lehn (1988), Supramolecular chemistry—scope and perspectives: molecules–supermolecules–molecular devices. *J. Inclusion Phenomena* **6**, 351–396.

684. J.M. Lehn (1993), Supramolecular chemistry. *Science* **260**, 1762–1763.

685. J.M. Lehn (1993), Supramolecular chemistry—molecular information and the design of supramolecular materials. *Makromol. Chem. Macromol. Symp.* **69**, 1–17.

686. I. Amato (1993), Designer solids: haute couture in chemistry. *Science* **260**, 753–755.

687. S.A. McDonald, C.G. Willson, and J.M.J. Fréchet (1994), Chemical amplification in high-resolution imaging systems. *Acc. Chem. Res.* **27**, 151–165.

688. J. Van Brunt (1985), Biochips: the ultimate computer. *Biotechnology* **3**, No. 3, 209–215.

689. A. Laschewsky, H. Ringsdorf, G. Schmidt, and J. Schneider (1987), Self-organization of polymeric lipids with hydrophilic species in side groups and main chains: investigation in momolayers and multilayers. *J. Amer. Chem. Soc.* **109**, 788–796.

690. T. Tjivikua, P. Ballester, and J. Rebek, Jr. (1990), A self-replicating system. *J. Amer. Chem. Soc.* **112**, 1249–1250.

691. J.I. Hong, Q. Feng, V. Rotello, and J. Rebek, Jr. (1992), Competition, cooperation, and mutation: improving a synthetic replicator by light irradiation. *Science* **255**, 848–850.

692. E.A. Wintner, M.M. Conn, and J. Rebek, Jr. (1994), Studies in molecular replication. *Acc. Chem. Res.* **27**, 198–203.

693. J.M. Lehn (1990), Perspectives in supramolecular chemistry—from molecular recognition towards molecular information processing and self-organization. *Angew. Chem. Ed. Engl.* **29**, 1304–1319.

694. M. Kotera, J.M. Lehn, and J.P. Vigneron (1994), Self-assembled supramolecular rigid rods. *Chem. Commun.* 197–199.

695. C.T. Seto and G.M. Whitesides (1991), Self-assembly of hydrogen-bonded 2 + 3 supramolecular complex. *J. Amer. Chem. Soc.* **113**, 712–713.

696. C.T. Seto and G.M. Whitesides (1993), Molecular self-assembly through hydrogen bonding: supramolecular aggregates based on the cyanuric acid-melamine lattice. *J. Amer. Chem. Soc.* **115**, 905–916.

697. M. Inouye, K. Hashimoto, and K. Isagawa (1994), Nondestructive detection of acetylcholine in protic media: artificial-signaling acetylcholine receptors. *J. Amer. Chem. Soc.* **116**, 5517–5518.

698. J. Moore and S. Lee (1994), Crafting molecular based solids. *Chem. & Ind.* 556–560.

699. Y. Zhang and N.C. Seeman (1994), Construction of a DNA-truncated octahedron. *J. Amer. Chem. Soc.* **116**, 1661–1669.

700. D.M. Rudkevich, Z. Brzozka, M. Palys, H.C. Visser, W. Verboom, and D.N. Reinhoult (1994), A difunctional receptor for the simultaneous complexation of anions and cations; recognition of KH_2PO_4. *Angew. Chem. Int. Ed. Engl.* **33**, 467–468.

701. S.H. Kawai, S.L. Gilat, and J.M. Lehn (1994), A dual-mode optical-electrical molecular switching device. *Chem. Commun.* 1011–1013.

702. C.M. Drain, R. Fischer, E.G. Nolen, and J.M. Lehn (1993), Self-assembly of a bisporphyrin supramolecular cage induced by molecular recognition between complementary hydrogen bonding sites. *Chem. Commun.* 243–245.

703. U. Koert, M.M. Harding, and J.M. Lehn (1990), DNH deoxyribonucleohelicates: self-assembly of oligonucleosidic double-helical metal complexes. *Nature* **346**, 339–342.

704a. J.M. Lehn and A. Rigault (1988), Helicates: tetra- and pentanuclear double helix complexes of Cu(I) and poly(bipyridine) strands. *Angew. Chem. Int. Ed. Engl.* **27**, 1095–1097.
704b. J.M. Lehn (1994), Perspectives in supramolecular chemistry: From molecular recognition towards self-organization. *Pure & Appl. Chem.* **66**, 1961–1966.
705. D. Bradley (1993), Will future computers be all wet? *Science* **259**, 890–892.
706. P.R. Ashton, R. Battardini, W. Balzani, M.T. Gandolfi, D.J.F. Marquis, L. Pérez-Garcia, L. Prodi, J.F. Stoddart, and M. Venturi (1994), The self-assembly of controllable [2]catenanes. *Chem. Commun.* 177–180.
707. D. Bradley (1991), How to make a molecular shuttle. *New Scientist* July 27, 20.
708. G. Schill (1971), *Catenanes, Rotaxanes, and Knots*. Academic Press, New York.
709. P.R. Ashton, D. Philp, N. Spenser, J.F. Stoddart, and D.J. Williams (1994), A self-organized layered superstructure of arrayed [2]pseudorotaxane. *Chem. Commun.* 181–184.
710. J.F. Stoddart (1993), Molecular recognition and self-assembly. *An. Quim.* **89**, 51–56.
711. J.L. Brédas (1994), Molecular geometry and nonlinear optics. *Science* **263**, 487–489.
712. D.B. Amabilino, P.R. Asthon, A.S. Reder, N. Spencer, and J.F. Stoddart (1994), The two-step self-assembly of [4]- and [5]catenanes. *Angew. Chem. Int. Ed. Engl.* **33**, 433–436.
713. D.B. Amabilino, P.R. Asthon, A.S. Reder, and J.F. Stoddart (1994), Olympiadane. *Angew. Chem. Int. Ed. Engl.* **33**, 1286–1290.
714. M.J. Gunter, D.C.R. Hockless, M.R. Johnston, B.W. Skelton, and A.H. White (1994), Self-assembling porphyrin [2]-catenanes. *J. Amer. Chem. Soc.* **116**, 4810–4823.
715. W. Worthy (1988), New families of multibranched macromolecules synthesized. *Chem. & Eng. News* Feb. 22, 19–21.
716. H.B. Mekelburger, W. Jaworek, and F. Vögtle (1992), Dendrimers, arborols, amd cascade molecules: breakthrough into generations of new materials. *Angew. Chem. Int. Ed. Engl.* **31**, 1571–1576.
717. C.J. Hawker and J.M.J. Fréchet (1990), Preparation of polymers with controlled molecular architecture. A new convergent approach to dendritic macromolecules. *J. Amer. Chem. Soc.* **112**, 7638–7647.
718. T.M. Miller and T.X. Neeman (1990), Convergent synthesis of monodisperse dendrimers based upon 1,3,5-trisubstituted benzene. *Chem. Mater.* **2**, 346–349.
719. J. Issberner, R. Moors, and F. Vögtle (1994), Dendrimers: From generations and functional groups to functions. *Angew. Chem. Int. Ed. Engl.* **33**, 2413–2420.
720. D.A. Tomalia (1993), Starburst^{TM}/cascade dendrimers: fundamental building blocks for a new nanoscopic chemistry set. *Aldrichimica Acta* **26**, No. 4, 91–101.
721. D. Seebach, J.M. Lapierre, K. Skobridis, and G. Greiveldinger (1994), Chiral dendrimers from tris(hydroxymethyl)-methane derivatives. *Angew. Chem. Int. Ed. Engl.* **33**, 440–442.

722. P.J. Dandliker, F. Diederich, M. Gross, C.B. Knobler, A. Louati, and E.M. Sanford (1994), Dendritic porphyrins: modulating redox potentials of electroactive chromophores with pendant multifunctionality. *Angew. Chem. Int. Ed. Engl.* **33**, 1739–1742.
723. P.R. Dvornic and D.A. Tomalia (1994), A family tree for polymers. *Chem. Britain* 641–645.
724. S. Mann (1994), Molecular tectonics in biomineralization and biomimetic materials chemistry. *Nature* **365**, 499–505.
725. L. Addadi and S. Weiner (1992), Control and design principles in biological mineralization. *Angew. Chem. Int. Ed. Engl.* **31**, 153–169.
726. R.F. Service (1994), Self-assembly comes together. *Science* **265**, 316–318.
727. K. Aoki, L.C. Brousseau, and T.E. Mallouk (1993), Metal phosphonate-based quartz crystal microbalance sensors for amines and amonia. *Sensors & Actuators B* **B14**, 703–704.
728. H.C. Yang, K. Aoki, H.G. Hong, D.D. Sackett, M.F. Arendt, S.L. Yau, C.M. Bell, and T.E. Mallouk (1993), Growth and characterization of metal (II) alkanebisphosphonate multilayer thin film on gold surfaces. *J. Amer. Chem. Soc.* **115**, 11855–11862.
729. F.M. Menger and S.J. Lee (1994), Long organic fibers obtained by non-covalent synthesis. *J. Amer. Chem. Soc.* **116**, 5987–6988.
730. J. Emsley (1994), Tangled tale of a self-organizing fiber. *New Scientist* Aug. 6, 17.
731. M. Simard, D. Su, and J.D. Wuest (1991), Use of hydrogen bonds to control molecular aggregation. Self-assembly of three-dimensional networks with large chambers. *J. Amer. Chem. Soc.* **113**, 4696–4698.
732. P. Baxter, J.M. Lehn, A. DeCian, and J. Fischer (1993), Multicomponent self-assembly: spontaneous formation of a cylindrical complex from five ligands and six metal ions. *Angew. Chem. Int. Ed. Engl.* **32**, 69–71.
733. P.N.W. Baxter, J.M. Lehn, J. Fisher, and M.T. Youinou (1994), Self-assembly and structure of 3 × 3 inorganic grid from nine silver ions and six ligand components. *Angew. Chem. Int. Ed. Engl.* **33**, 2284–2287.
734. F. Diederich and Y. Rubin (1992), Synthetic approaches toward molecular and polymeric carbon allotropes. *Angew. Chem. Int. Ed. Engl.* **31**, 1101–1123.
735. F. Diederich (1994), Carbon scaffolding: building acetylenic all-carbon and carbon-rich compounds. *Nature* **369**, 199–207.
736. U.H.F. Bunz (1994), Polyynes—fascinating monomers for the construction of carbon networks. *Angew. Chem. Int. Ed. Engl.* **33**, 1073–1076.
737. J. Anthony, C. Boudon, F. Diederich, J.P. Gisselbrecht, V. Gramlich, M. Gross, M. Hobi, and P. Seiler (1994), Stable conjugated carbon rods with a persilylethynylated polytriacetylene backbone. *Angew. Chem. Int. Ed. Engl.* **33**, 763–766.
738. H.L. Anderson, R. Faust, Y. Rubin, and F. Diederich (1994), Fullerene-acetylene hybrids: on the way to synthetic molecular carbon allotropes. *Angew. Chem. Int. Ed. Engl.* **33**, 1366–1368.
739. R. Baum (1993), Subtle tensions in organic chemistry emerge at conference. *Chem. Eng. News*, Nov. 29, 50–51.

740. A.P. DeSilva and C.P. McCoy (1994), Switchable photonic molecules in information technology. *Chem. & Ind.* 992–996.

741. T.J. Marks and M.A. Ratner (1995), Design, synthesis, and properties of molecule-based assemblies with large second-order optical nonlinearities. *Angew. Chem. Int. Ed. Engl.* **34**, 155–173.

Index

A

absolute configuration of bound substrate, 199
abzymes, 86
acetate, 23
acetate kinase, 118
acetoin, 565
acetylcholine, 8, 98, 282
  receptor, 602
acetyl-CoA, 23, 31, 520, 589
acetyl-CoA carboxylase, 576, 589
acetyl imidazole, 35
acetyl lipoamide, 567
acetyl phosphate, 35
acridine orange, 143, 245
acrylyl-CoA, 590
action potential, 123
"activated acetate," 520
"activated carbonate," 575
"active aldehyde," 564
"active ester," 120
active site, 159, 169, 171
active-site-directed irreversible inhibitors, 542
active transport, 273, 388
Acyclovir, 10
N-acylaminocrylic acid, 57
acyl fission, 126
acyl-phosphate, 31
adementane carboxylic acid, 349
adenine, 24
adenosine, 80
adenosine arabinoside, 9, 143
adenosine kinase, 120
adenosine 5′-phosphosulfate (APS), 518
S-adenosylmethionine (SAM), 32, 40, 478
adenylate kinase, 120
adjacent mechanism, 133
ADP, 118
adriamycin, 144
affinity chromatography, 231, 261, 476
affinity labels, 542 ff.
$[Ag_9L_6]^{9+}$, 362
Ag^+ tunneling, 310
agonists, 97, 102
AIBN, 231
alamethicin, 275
D-alanine, 57
L-alanine, 234
alanine dehydrogenase, 234
Ala-tRNA^{Cys}, 37
alcohol dehydrogenase, 174, 175, 493

aldol condensation, 522, 564
aldolase, 28, 239, 336
alkyl fission, 126
allosteric system, 294, 425
allostery, 408
allotropes, 633
amidines, 216
amine capture, 48
amino acid,
  essential, 51
  ester hydrolysis of, 398
aminoacyl-oligonucleotides, 48
aminoacyl site, 38
aminoacyl-tRNA, 36ff., 113
α-amino fatty acid, 74
6-aminopenicillanic acid, 236
ammonio-alkyl lariat, 299
amoxicillin, 556
AMP, 119
analgesic, 97
anchimeric assistance, 127
anomeric effect, 229
antagonists, 9, 97, 101
antamanide, 275
antibodies, 85ff.
antibonding orbital, 229
antiinflammatory, 11, 311
antimalarial, 12
antimetabolite, 9
antisense, 114
antitumor drugs, 144, 158
antiviral activity, 9, 144
AOT, 335
apical position, 129
apoenzyme, 476, 482
arachidonic acid, 554
arborols, 620
archaebacteria, 331
arene oxide, 508
asenite, 566
"artificial enzyme," 269, 350, 354
aryl sulfatase, 343
A site, 39, 47
Aspartame, 78
L-aspartate monomethyl ester, 52
aspartic acid, 237
assisted hydrolysis, 5
asymmetric syntheses of α-amino acids, 50ff.
  with biotin Rh(I) catalyst, 62
  via Corey's method, 51
  via Kagan's synthesis, 52
  via polymer-supported Rh(I), 61
  via rhodium (I) catalyst, 55
atactic, 171
ATP, 2, 24, 26, 31, 39, 116ff., 283
  chiral γ-phosphate, 120
  high energy, 35, 117
  synthesis, 119, 120
ATPase, 296
atropine, 98
Augmentin, 555
auto-association, 596
auxiliary ligand, 63
8-aza-adenine, 24
azaallylic anion, 534
azaallylic intermediate, 537
azamethine rearrangement, 536, 537
azo-coupling, 293

B

Baeyer-Villiger reaction, 517
Baldwin's rules, 68
bastatin, 82
Belleau's biphenyl compound, 201
bell-shaped curves, 182, 339
5-benzamido-2-benzyl-4-oxopentanoic acid (BOP), 398
benzitramide, 99ff.
benzo-15-crown-5, 294
benzoin, 162, 324
N-benzoyl-glycine, 36
N-benzoyl-L-phenylalanine, 398
N-benzyl-NADH, 506
benzyl succinate, 80
benzyl-valine, 243, 244
benzylchloropyrones, 553
BiBLE, 289
BINAP, 78
biochips, 594
bioisosteric groups, 8ff.
biological "active aldehyde," 563
"biological cyanide," 567
biological methylation, 478
biological role of phosphate, 111, 112
biomimetic approach, 252, 373ff.
biomimetic cyclizations, 386
biomimetic material chemistry, 626

biomimetic polyene cyclizations, 373, 377, 387
biomimetic strategies, 625
biomineralization, 625
bioorganic models, 16
biosensors, 594
biotechnology, 114, 430, 602
biotin, 28, 31, 574ff.
  carboxylated biotin, 582, 590
  mode of action, 576
  model studies, 581
  pseudo-rotation in, 587
bis-benzocrown ethers, 297
bis-dimensional assembly, 597
bis-2-carboxyphenyl phosphate, 140
bis-(2-ethylhexyl) sodium sulfosuccinate, 335
bis-intercalants, 145, 146
bis-(methidium) spermine, 148
bisporphyrin cage, 607
Bleomycins, 148, 152
blood-brain barrier, 13
Bohr effect, 429
bolaamphiphiles, 331
Bolton's model, 453
bradykinine, 104
butadiene-2,3-dicarboxylic acid, 473
butanomorphinam, 102
butyl-thymine, 155

C

cadaverine, 281
Cahn-Ingold-Prelog priority rules, 173, 215
calcium selectivity, 292
calixarene, 302, 308
CAMPOS, 77
cancer chemotherapy, 143
carbachol, 8
carbamoyl phosphate, 577
carbanion chemistry, 535, 561, 573
carbodiimide intermediates, 577, 585
carbohydrates, 24
carbon monoxide poisoning, 408
carbonic anhydrase, 390, 391
  cyclodextrin model of, 349
carbon scaffolding, 633
N-carboxybiotin, 574ff.
carboxylation reactions, 575
carboxypeptidase A, 80, 81, 204, 236, 248, 255, 390ff.
  biomodels of, 395
O-carboxyphenl β-D-glucoside, 4
bis-2-carboxyphenylphosphate, 140
"carcerand," 305
β-carotene, 454
caroviologens, 334
carzinophilin A, 147
casette mutagenesis, 92
catalysis, 16, 159, 160
  concerted, 181
  covalent, 165
  electrophilic, 165
  general acid or base, 163
  heterogenous, 160, 168
  homogenous, 161, 168
  intramolecular, 167, 177
  multifunctional, 181
  multiple, 167
  nucleophilic, 161
catalytic antibody, 88
catechol sulfate, 343
catenanes, 445, 609
  metallo-, 445
cationic lipid, 295
cavitands, 302ff.
C_{60}-azacrown ether, 314
cellular protein synthesis, 33
channel mechanism, 334
charge relay system, 186ff.
  with cyclodextrins, 347
  with thiamine, 563
charge-transfer complex, 18
chemical adaptation, 8, 572
chemical mutation, 90, 93
chemiluminescence, 516
  models of, 517
chemionic devices, 595
chemionics, 335
chiral, chirality, 4, 173
chiral auxiliary, 64
chiral organometallic catalyst, 77
chiral recognition, 259, 267
chiral specificity, 204
  in host, 253
  in recognition, 259
chirality, 4
(*S,S*)-chiraphos, 55, 77

chirasil-Val, 344
chlorophyll, 413, 448
  dimer, 450
chorand, 265, 303
chorismic acid, 82
chromosomes, 113
α-chymotrypsin, 91, 184ff., 248, 255, 306, 335, 552
  bound substrate in, 199
  cyclodextrin model, 347
  polyimidazole model, 341
  stereoelectronic control in, 224
circular DNA, 445
cistron, 113
Claisen rearrangement, 82
clavulanic acid, 556
cobalamin on solid support, 477
cobaloxime, 467
cobalt in catalysis, 460ff.
  in vitamin B_{12}, 460ff.
cobalt polymer, 244
cobalt-carbon bond, 467
cocaine, 88
codeine, 97
coenzyme A, 23, 520
coenzymes, 7, 160, 482ff.
  bound to cyclodextrin, 351
  evolution, 591
  nonenzymatic recycling, 493ff.
coformycin, 80
complementary bifunctional catalysis, 5
complementary carrier method of peptide synthesis, 47ff.
π-complex, 169
conducted-tour mechanism, 536
convergent evolution, 190
convergent method, 619
cooperativity, 407
copper complex with oxygen, 437
copper ion, 437
  biomodels, 438ff.
coreceptors, 280
Corey's methodology, 64
corrin, 462
cortisol, 235
cortisone, 11, 372
coupling agent, 41
covalent catalysis, 165
creatine phosphate, 116
critical micelle concentration, 318
crown ethers, 265, 277ff., 303
  aza, 266, 280, 283, 331, 361
  bis, 293
  bis-benzocrown, 315
  carbohydrate precursors, 263
  in catalysis, 267
  chiral, 489
  lariat, 286
  peptide, 286
  photoresponsive, 297ff.
  polymeric, 262, 308
  psoralen, 153
cryoenzymology, 190, 398
crypta-spherand, 305
cryptand, 265ff., 272, 303
  methylcoumario-, 291
  photoactive, 459
cryptate, 265
  azide, 267
  effects, 265
$[Cu(bpy)_2]^+$, 630
cyclic AMP, 24, 116, 122ff.
cyclic GMP, 116, 122ff.
cyclic peptides, 106ff.
cyclic phosphates, 139ff.
cyclo bis(paraquat-*p*-phenylene), 609
cyclobutane dicarboxylic acid, 242
cyclodextrin, 345ff., 355
  amino-carboxy, 360
  capped, 347, 348, 359, 539
  carbonic anhydrase model, 351
  chiral induction, 354
  crown ether, 361
  meta-para-selectivity, 349
  NADH, 491
  photo-irradiation, 359
  pyridoxal, 539ff.
  pyridoxamine bound, 354, 539
  rhodopsin model, 354
  ribonuclease A model, 350
  rotaxane, 361
  symmetrically substituted, 352
  thiazolium, 572
cyclophane, 421, 422
  porphyrin-quinone, 458
  pyridoxamine, 540
cyclosporine, 71
cylindrical complex, 630

cystathionine, 544
cytochrome b_6-f, 449
cytochrome C, 624
cytochrome C oxidase, 444
cytochrome P-450, 434ff., 555
cytochromes, 407
cytosine, 8
cytosine arabinoside, 143

D
daunomycin, 144
DDQ, 382
deazaflavin analogs, 504, 510
decarboxylation of oxaloacetic acid, 50
Demerol, 13, 98
dendrimer, 616
  chiral, 621
  convergent method, 619
  divergent method, 617
  porphyrin bounded, 623
  starburst, 616
dendron, 619
deoxyadenosine, 9
deoxynucleohelicates (DNH), 606
deoxyribofuranosyl-5-iodo-uracil, 9
desaturase, 366
designer drug, 12
designer solids, 594, 603, 615
detoxification, 35
β-deuterated amino acids, 59
dextran-NAD$^+$, 334
DFP, 184, 188, 543
diacetylenic lipids, 627
diastereotopic, 51, 176
diazoketone inhibitor, 554
dibenzo-crown ether, 290
cis-dichlorodiamine platinum (III), 145
di-1,5-cyclooctadiene rhodium (I), 58
Diels–Alder reaction, 312, 355, 387, 414
dienoyl lipid sulfonate, 330
dienoylphosphatidycholine, 330
diethyl tartrate, 75
dihydroisoallozazine, 508
dihydroxyacetone, 237
3,4-dihydroxybutyl-1-phosphonic acid (DHBP), 10
dimethylmercury, 479
dimethyl phosphate, 114
dinactin, 275
dinuclear receptors, 443
dipeptide, 34
DIOP, 77
directed self-assembly, 597
directional effect, 402
distamycin-EDTA, 149
dithiane, 566
divergent method, 617
DNA, 112, 142ff.
  amide linkage analog, 114
  antitumor drugs, 158
  double helix, 142, 145, 150, 446
  drug intercalation, 142ff.
  knots, 602
  replication, 113
Dolphin's model, 453
L-DOPA, 78, 123, 241, 551
dopamine, 123, 247
double helix, 142
Dowd's proposal, 472
Droperidol, 99
drug-bearing peptide, 9
drug delivery, 616, 626
drug design, 95, 107
dynamical complementarity, 559

E
effective concentration, 178
electrides, 316
electro-active devices, 603
electrochromic properties, 605
electron channel, 334
"electron sink," 522, 529, 564
electron spin resonance, 450
electron-transfer assemblies, 429
electron tunnelling, 429, 455
electron wire, 429
electrophilic catalysis, 165
electro-photoswitching, 604
electrostatic interactions, 18, 172, 205, 392
enantiomeric differentiation, 270
enantiomeric discrimination, 254
enantiomeric excess, 64
enantiotopic, 176
endoperoxide intermediate, 554
endorphin, 104
endo-receptors, 601

enkephalins, 104ff.
enolate chemistry, 68, 70
enol ether hydrolysis, 88
enterobactin, 432
entropy, 178, 180
enzyme, 3, 159, 171
  active center, 171
  active site, 159, 168, 171
  binding site, 176
  coenzyme, 160
  electrode, 234
  hydrolytic, 204
  inhibitor, 160
  immobilized, 230
  models, 252ff.
  reactor, 234
  specificity, 160, 175
  substrate, 160
enzyme-activated irreversible inhibitors, 545
enzyme-analog polymer, 240
enzyme catalysis in organic solvents, 237
enzyme engineering, technology, 233, 493
enzyme-generated "reactive-intermediate analogs," 570
enzyme models, 252ff.
ephedrine, 120
epoxidation, 75
epoxy ester, 238
EPR, 437
  in Cu-enzyme, 441
  in flavin models, 513
ethidium bromide, 148
ethylene phosphate, 127
bis(2-ethylhexyl) sodium sulfosuccinate (AOT), 335
europium cryptate, 460
Evan's methodology, 65, 71
exciton, 450
exo-anomeric effect, 229
exo-receptors, 601

F

FAD, 26, 497ff.
  biomodels, 501ff.
  Goddard's proposal, 515
  Hamilton's mechanism, 501
  nitroxyl radical (formation of), 514
  oxaziridine, 510
  oxene reactions, 507ff.
Fentanyl, 99
ferrocenylacrylate, 356
ferrodoxin, 430, 447, 449
flavin N(5)-oxide, 513
flavin remote control, 505
flavins, 495ff.
flavopapain, 93
fluorescence energy transfer, 543
5-fluorocytosine, 9
fluoromethyl-3,4-dihydroxyphenylalanine, 551
α-fluoromethyl-glutamic acid, 551
fluoropropionyl-CoA, 591
5-fluoro-uracil, 89
foods, 22
formyl cobalamin, 471
free energy of hydrolysis, 34
fullerene 314, 633, 635

G

GABA, 283, 552
GABA γ-acetylenic analog, 549
GABA transaminase, 549
gabaculline, 552
β-galactosidase, 233
gene probe technology, 594
genes, 113
genetic code, 113
genetic information, 113
genome, 113
geranygeranyl acetate, 436
Glaser reaction, 445
glucokinase, 125
glucose, 23
glutamate mutase, 462
glutamic acid, 92
glutamine, 92
glutathione, 27
glyceraldehyde 3-phosphate, 561
glycerol kinase, 122
sn-glycerol-3-phosphate, 11
glycolysis, 22
Goddard's proposal in flavin chemistry, 515

Gouy–Chapman layer, 319
gramicidin, 334
gramicidin S, 42
Griffith's model, 418

H
haloenol, 553
Hamilton's mechanism in flavin chemistry, 501
hard and soft acid and bases(HSAB), 264
haptens, 85
helicates, 606ff.
  triple stranded, 608
heme, 407
hemerythrin, 407, 410
hemiketal intermediate, 199
hemispherands, 304
hemocyanin, 407, 437
hemoglobin, 407, 421
hemoprotein, 407
heroin, 97
herpes simplex, 10
heterogeneous polymerization, 168, 470
5-hexadecyloxyisophthalic acid, 628
hexaphenylhexaazatriphenylene, 630
hexapus, 312
high dilution technique, 278
high energy (ATP), 36, 38
  acylating agent with, 569
hippuric acid, 36
HIV-1 protease inhibitor, 588
HLADH, 493
holoenzyme, 476
homogeneous catalysis, 163
homolytic cleavage, 468
homoserine phosphate, 527–528
homotopic, 176
Hoogsteen hydrogen bonding, 290
host-guest complexation, 255
hydrazonolactone, 53
hydride ion (transfer), 174, 176
hydrogen bonds, 18, 247
hydrolase, 237
hydrolysis of amino acid esters, 398
hydrolytic pathways, 125
hydrophobic bonds, 18
hydrophobic forces, 172
β-hydroxydecanoyldehydrase, 545
(hydroxyethyl)thiamine, 570
hypochromism, 143

I
imaging systems, 594
imidazolidone, 581
4-(4′-imidazolyl)butanoic phenyl ester, 6
immobilized coenzymes, 234
immobilized enzymes, 119, 230ff., 336
immunoglobulins, 85-86
inclusion complex, 345, 443
indolepyruvic acid, 540
induced-fit effect, 392, 395
information,
  function, 16
  storage, 605
  structural, 16
  technology, 605
inhibitor, 160
  mechanism-based design, 557
  structure-based design, 557
  suicide, 557
inosine, 80
intelligent materials, 614, 625
intercalant molecules, 145
  peptide heterodimer, 152
intramolecular catalysis, 167, 177
intraannular chirality transfer, 66
intrinsic binding energy, 179
ion carrier, 273, 333
ion dots, 663
ion pores, 333
ion pump, 288
ion selectivity, 290, 291
ion transport, 333
ion transporting machine, 297
ionic channels, 101, 333
ionoactive devices, 603
ionophores, 274ff.
iron in proteins, 407ff.
  ferredoxin, 430
iron oxide, 409
iron-oxo bridge, 431
iron-sulfur proteins, 430
irreversible inhibitors, 79
irreversible self-assembly, 597

isoeuphenol, 378
isoinversion, 533, 534
iso-levorphanol, 100
isomerase, 237
isoniazid, 544

K
Kagan's methodology, 64
KCTI, 199
Kemp's triacid, 246
α-ketovaleric acid, 539
Kolbe reaction, 70
Koshland's theory, 68
Krebs' cycle, 23, 28

L
β-lactamase inhibitors, 556
lactate dehydrogenase, 234
lactic acid, 67, 234
lactylthiamin, 570
Langmuir-Blodgett film, 628
lanosterol biosynthesis, 373
lanosterol cyclase, 385
lariat, 286ff., 505
  ammonio-alkyl, 299
  bibracchial, 286
LDA, 69
lead compounds, 558
leu-enkephalin, 106,110
leucotriene TLC_4, 74
levorphanol, 100
Lewisite, 389
ligands, 398
ligase, 237
light conversion devices, 460
light-driven ion transport, 298
Lindsey's model, 454
lipase, 336
lipoic acid, 518ff.
liposomes, 426, 626
liquid-phase peptide synthesis, 414
lithium selectivity, 291, 293
living cells, 22
Lomotil, 99
LTD_4, 74
luciferin, 516
lyase, 237
lysozyme, 26, 91, 205ff., 226, 248, 255
α-lytic protease, 188

M
macrobicyclic effect, 265
macrocyclic effect, 265
macrocyclic polyamines, 280
malonyl-CoA, 580, 587
(+)-mandelic acid, 53, 67
madelic ester, 325
α-mannose, 240
MAO, 13
MeBmt, 71
mechanism-based inhibitor, 545, 557
MEEC, 335
membrane chemistry, 317
membrane-enclosed enzymatic catalysis, 335
membrane mimic, 328ff.
membrane-spanning electron channel, 334
meperidine, 13
mercury methylation, 479
Merrifield peptide synthesis, 33
metal ions, 388ff.
  promotor role of, 406
metallo-catenanes, 445
metallo-receptor, 281, 445
metal-phosphate precursors, 627
methadone, 99
methidiumpropyl-EDTA, 149
met-enkephalin, 104, 105
bis-(methidium) spermine, 148
6-methoxy-8-aminoquinoline, 12
methyl donors, 479
methylcobalamin, 461, 478
α-methyleneglutaric acid, 473
N-methylmaleimide (NMM), 526
N-methyltetrahydrofolate, 478
methylviologen, 282
Met-myoglobin, 93
fMet-*t*RNA^{fMet}, 38, 39
Meyer's methodology, 64
micelles, 317ff.
  chiral, 325
  critical micelle concentration, 318
  membrane model, 328
  reverse micelles, 321
  stereochemical recognition with, 325
  "water pools," 322
  zwitterionic, 319
Michael addition, 526, 550

Michaelis-Menten kinetics, 79, 80, 346
microchips, 627
microelectronics, 633
microlithographic materials, 594
microscopic reversibility, 176, 217, 223
mitochondria, 22
mitomycin C, 144
model building, 2
molecular adaptation, 8ff.
molecular assembly, 17, 594
molecular asymmetry, 173
molecular belts, 312
molecular box, 289
molecular bracelet, 446
molecular catalysts, 3
molecular chambers, 629
molecular clefts, 244
molecular computers, 594
molecular devices, 16, 593
molecular electronic devices, 613, 627
molecular electronics, 335
molecular gadgets, 622
molecular hinge, 156
molecular machines, 609
molecular memory, 241
molecular modeling, 107, 317, 559
molecular pharmacology, 204
molecular recognition, 14, 95, 593, 595
  anion and cation, 604
molecular scissors, 150
molecular shuttle, 594, 609
molecular signal, 601
molecular switching devices, 609, 614
molecular tectonics, 603, 624, 626
molecular topology, 444
molecular tweezers, 250, 301
molecular wires, 334, 604
monactin, 275
monoamine oxidase (MAO), 13, 546
monoclonal antibody, 86
Moore's model, 454
morphan, 100
morphinan, 101
morphine, 13, 97ff.
MPP^+, 14
MPPP, 13
MPTP, 14
MSH, 104
multiple catalysis, 167
muscarine, 8
mutant, 92
myoglobin, 408, 421
  Baldwin's model, 410
  Bayer's model, 414
  Collman's model, 416
  Traylor's model, 410, 413

N

NAD^+, 24, 26, 175, 176, 483ff.
  crown ether model, 488
  in micelles, 321
  models, 486ff.
  on solid support, 234
  structure, 484
NADH, 227
NADH mimic, 489
$NADP^+$, 383ff.
naloxone, 103
nanometer scale, 594, 613, 620
nanoreceptors, 620
nanoscaffolding, 594
nanoscale switches, 594, 620
nanoscopic,
  chirality, 620
  recognition, 620
  steric effects, 620
nanotechnology, 603
Naproxen, 311
nerol acetate, 436
netropsin-EDTA, 149
neuroleptics, 98
neuropeptide, 103, 107
neurotransmitters, 8, 97, 123
Niemann's compound, 199ff.
nigericin, 274
NIH shift, 29, 511
nitrogen fixation, 304
nitrogenase model, 444, 613
p-nitrophenyl ester syndrome, 355
NMR, 190
  ^{13}C, 190, 280, 473
  ^{35}Cl, 267
  host-guest complex, 259
  ^{15}N, 190
  ^{31}P, 139
  of serine protease, 190
nomenclature (E) and (Z), 215

nonactin, 275
nonenzymatic recycling, 493
nonlinear optics, 613, 628
nonribosomal, 48
nonribosomal polypeptide synthesis, 42ff.
non-Watson-Crick base pairing, 48
normorphinam, 101
norvaline, 539
nucleophilic attack, 67, 68
nucleophilic catalysis, 163
nucleotide, 112

O

Ogston effect, 174
oleic acid, 365
oligonucleotide, 48
olympiadane, 614
optical displays, 628
optical memory system, 605
optical purity, 41, 52
orbital steering, 68, 180, 210
organoboranes, 74
organometallic catalysts, 77
orientation effect, 177, 180
oripavine, 105
orthoesters, 212
ostrich molecule, 293
oxaloacetate, 50
oxanion hole, 198
oxaziridine, 510ff.
oxazoline, 65
oxazolium ion, 573
oxidative decarboxylation, 511–512, 560
oxidative phosphorylation, 500
oxidoreductase, 237, 483
oxygen transport, 407ff.
oxygenases, 483
oxene reactions, 507

P

pain-killer, 97
papain, 96, 255
paracyclophanes, 264, 310ff.
pargyline, 547
passive transport, 274
Pauling's model, 418
penicillanic sulfone, 556
penicillase, 556
penicillin G, 236, 556
pepsin, 209
peptide bond, 25ff.
peptide hydrolysis by metal ions, 398ff.
peptide synthesis, 39, 46
peptidomimetic, 107
  HIV-1 inhibitor, 558
peptidyl site, 38, 47
phenylalanine, 77, 539
phenylalanyl-glycinal, 198
phenylpyruvic acid, 539
phosphates, 114ff.
O-phosphobiotin, 577, 584, 588
phosphoenolpyruvate, 81
phosphoglycollate, 81
phosphonamidate, 81
phosphonamidate cyclic analog, 559
phosphonate, 87, 88
phosphorothioate, 121
phosphorylcholine, 87
phosphoryl fission, 126
phosphoryl transfer, 285
photoactive cryptand, 459
photoactive molecular devices, 460
photoaffinity labeling, 543
photochromic properties, 605
photo-driven charge separation, 300
photoinduced electron transfer, 458
photo-irradiation, 359
photolabels, 460
photolithographic processes, 594
photonic devices, 628
photoresponsive crown ether, 297ff.
photoresponsive electron channels, 335
photosynthesis, 304, 447, 459
photovoltaic devices, 451
pig liver esterase (PLE), 239ff.
plastocyanin, 437
plastoquinone, 447
podand, 265, 304
polar effect, 422
poly-A, 155
polymerized vesicules, 329, 426, 449
polymers, 337ff.
  chirasil-Val, 344
  isotactic, 171
  main chain chirality, 244

poly(ethylimine), 342
poly(vinylbenzimidazole), 338
poly(vinylimidazole), 338
supported Rh(I), 61
syndiotactic, 171
polynuclear inorganic grid, 632
poly-triphenylene crown ether, 315
POMC, 104
porcine pancreatic lipase, 239
pore, ion, 333
porphyrin,
adamentane, 421
"capped," 410, 414, 440
crowned, 424
cyclo-, 456, 458
cyclophane, 415
di-zinc, 452
doubly bridged, 424
face-to-face, 426, 445
flavin-linked, 506
gable, 425
manganese, 435
mercaptan-tail, 435
picket fence, 417, 418, 427, 440
quinone, 453
strapped, 421
thiolate, 435
triple-deckered, 455
tulip garden, 423
porphyrin crown ether, 281
potassium selectivity, 291, 292
PPC mechanism, 500
PPM mechanism, 500
prebiotic origin of organic molecules, 48, 592
prednisolone, 235
pregnenolone, 555
prephenic acid, 82
primaquine, 12
primary optical specificity, 201
prochirality, 173
pro-drug, 98
proflavin, 144
progesterone, 582
proline, 66, 69
proline amide, 222
propargylamine, 550
propionyl-CoA carboxylase, 575, 589, 590
prosthetic groups, 482
protein-based technology, 429
proton-driven cation transport, 290
proximity effect, 4, 176, 179, 180
pseudo-crown ether, 261
pseudorotation, 125, 130, 350
in biotin, 587
pseudorotaxane, 611
P site, 39, 47
psoralen, 153
pterin, 481
purines, 112, 246
putrescine, 280
2-pyridone units, 629
pyridoxal phosphate, 26, 32, 520ff.
azaallylic anion, 534
biological role, 525
biomodels, 531
conducted-tour mechanism, 536
mode of action, 524
pyridoxamine, 521, 532
pyrimidine, 112
pyruvate decarboxylase, 526
pyruvic acid, 23, 234
nonoxidative decarboxylation of, 560, 561
pyruvic acid labeled, 590

Q

QSARS, 90
8-quinolyl hydrogen glutarate, 397
quinone, 452, 453

R

racemization, 71
receptor, 16, 19, 97, 256, 273
acetylcholine, 602
dinucleotide d(AA), 249
endo-exo, 601
KH_2PO_4, 604
nucleotides, 157
phosphate ester, 156
tripeptides, 316
tryptophan, 541
receptor site, 12
recombinant DNA, 91

Reichstein's compound, 235
remote control in flavin, 505
remote oxidation (functionalization), 366
  desaturase model, 366
respiratory chain, 500
reversed micelles, 321
reversible inhibitors, 79
rhizobitoxine, 544
Rhodopseudomonas viridis, 455
ribonuclease A, 26, 121, 132ff., 230, 255, 357
  cyclodextrin model of, 350
ribosomal surface, 39
ribosome, 36
RNA, 112, 164ff.
RNase A, 132
rotaxanes, 361, 609
  with porphyrin, 446
ruthenium chemistry, 152
ruthenium complex, 450
Ru(II) protoheme, 429

S
SAM, 38, 40
sarcosine amide, 222
SDS, 320, 368
secondary structural specificity, 202
Seebach's methodology, 65
self-amplifying devices, 614
self-assembly, 593, 596, 613
  directed, 597, 598
  irreversible, 597
  multilayers, 629
  post-modification, 597
  strict, 597
self-organization, 593, 594
self replicating system, 597
self-reproduction of the stereogenic center, 69
semiochemistry, 601
semiquinone radical, 500
semisynthetic enzymes, 93
sensory devices, 625
sequence-specific DNA, 149
serine proteases, 184
  NMR of, 189
  stereoelectronic control in, 217
serine-*O*-sulfate, 548
Sharpless' epoxidation, 75
Sharpless' methodology, 65
siderophores, 432
signal peptides, 429
signal transduction, 429
silicon-mediated suicide inhibitor, 555
site-directed chemical mutation, 90
site-directed mutagenesis, 91
SKEWPHOS, 77
sodium pump, 274, 288
solar cells, 627
sonification, 329
sophoradiol, 387
specificity, 175
  primary optical, 201
  primary structural, 201
  regiospecificity, 175
  secondary structural, 202
  stereospecificity, 175
  tertiary structural, 202, 203
speleands, 282
speleates, 282
spherands, 302
squalene, 373ff.
stacked discs, 629
staphyococcal nuclease, 138
starburst dencrimer, 616
stearic acid, 366
stereochemical recognition, 325
stereoelectronic control, 210ff.
  in orthoester hydrolysis, 212
stereogenic center, 69
stereoselective oxidoreduction, 385ff.
stereoselective transport, 270ff.
steric compression, 180
steriod, 361, 370ff., 373ff.
  benzophenone, 367, 370
  bis(11)-ammonium, 366
    as enzyme model, 361
    remote functionalization of, 370–371
    template, 361ff.
  sulfates, 373
storage devices, 633
Stork-Eschnemoser hypothesis, 375, 383
structure-base design of inhibitor, 557

substance P, 104
substrate, 160
subtilisin, 91, 92
suicide inhibitors, 542ff.
  examples of, 29, 548ff.
super acid, 388, 399, 406
superoxide dismutase, 444
supramolecular assemblies, 620
supramolecular catalyst, 270
supramolecular chemistry, 15, 593, 595
supramolecular chips, 633
supramolecular devices, 606
supramolecular level, 14
supramolecules, 17
surfactants, 317, 325, 329
switched molecular wires, 605
synapse, 123
syndiotactic, 171
synthetic biochemistry, 234
synzyme, 343

T
tail-biting, 292
TCA cycle, 22
tectons, 625, 629
template effect, 35
  with metal ions, 399, 402
template synthesis method, 240
terpene synthesis, 29
tertiary structure, 320
tetra-acid 18-crown 6, 278
tetrahedral intermediate, 190ff., 215
tetrahydrofoliate, 481
tetrahydrofolic acid, 32, 40, 481
tetrapyrroles, 455
thermolysin, 81, 558
thiamine pyrophosphate, 31, 559ff.
  α-lactyl derivative, 563, 569
  hydroxyethyl derivative, 564
  model design, 568
  suicide inhibitor, 551
  transition state analog, 571
thiazolium ring, 562ff.
thiol capture, 49
thio-subtilisin, 91
threonine, 242, 521, 528, 538
thromboxane A_2, 554
thymidine, 9
thyroxine, 12
2,2′-tolancarboxylic acid, 5
topoisonmerases, 445
tosyl-L-phenylalanine chloromethyl-ketone (TPCK), 542
toxins, 544
transferase, 237
transition state, 159ff.
transition-state analogs, 79, 88
translational entropy, 179
transmethylation, 478
transport, 273
  active, 274
  passive, 274
transporter, 16
tricarboxylic acid (TCA), 22
tri-dimensional assembly, 597
triosephosphate isomerase, 81, 174
triostin A, 147
trisyl-N_3, 72
tRNA, 26, 35
Trochodiscus longispinus, 305
trypsin, 189, 255, 390
tryptophan, 360, 540
tyrosine, 123

U
ubiquinones, 500
uric acid, 158

V
valine, 66
valinomycin, 274, 286, 334
van der Waals forces, 18
α-vinylamino acid, 66
vinylboronic acid, 241
viologens, 329, 609
vitamin B_1, 560
vitamin B_6, 521, 532
vitamin B_{12}, 30, 40, 460ff.
  biomodels, 467, 471
  Breslow's model, 475
  Corey's model, 476
  Dowd's experiment, 472
  homolytic cleavage, 468ff.
  Retey's model, 474
  1,2 shift in, 463, 477
vitamin C, 27

vitamin H, 587
vitamins, 482

W
Watson-Crick base pairs, 50, 142, 250, 289, 290
water splitting, 449ff.
Wilkinson's catalyst, 55

X
xylulose 5-phosphate, 561

Y
Yamada's methodology, 64
Yang triplet reaction, 368

Z
zeolites, 620
Ziegler-Natta catalysis, 33, 76, 169
zinc in catalysis, 390ff.
  dual role, 395
zymogen, 185